INTELLIGENT SYSTEMS

INTELLIGENT SYSTEMS

Architecture, Design, and Control

Alexander M. Meystel

Professor of Electrical and Computer Engineering
Drexel University
and
Guest Researcher
National Institute of Standards and Technology

James S. Albus

Senior NIST Fellow
Intelligent Systems Division
Manufacturing Engineering Laboratory
National Institute of Standards and Technology

A Wiley-Interscience Publication
JOHN WILEY & SONS, INC.

Cover: *Multiresolutional Thinker* by A. Meystel

This book is printed on acid-free paper. ♾

For ordering and customer service, call 1-800-CALL-WILEY.

Library of Congress Cataloging-in-Publication Data:

Meystel, Alexander M.
Intelligent Systems : architecture, design, and control / Alexander M. Meystel, James S. Albus.
p. cm.
"A Wiley-Interscience Publication."
Includes bibliographical references.
ISBN 0-471-19374-7 (cloth)
1. Intelligent control systems. I. Albus, James Sacra. II. Title.
TJ217.5.A43 2000
629.8—dc21 99-23427

10 9 8 7 6 5 4 3 2

To the memory of our parents

CONTENTS

8 MULTIRESOLUTIONAL PLANNING: A SKETCH OF THE THEORY 343

9 MULTIRESOLUTIONAL HIERARCHY OF PLANNER/EXECUTOR MODULES 399

PREFACE

INTELLIGENCE AS A TOOL OF EVOLUTION

About 600 million years ago, there was an unusual explosion in the biosphere of the earth. Biologists call it the *Cambrian explosion*. The amount and diversity of living creatures that emerged during the Cambrian explosion far exceeded anything that happened before or after this unusual period in the ecological history of the earth. A comparison with the middle part of the sigmoidal[1] curve is a suggestive metaphor for the Cambrian explosion although it leaves more to our imagination rather than offering any explanation of what happened.

The famous scientist S. Gould made an interesting observation:

> The world of life was quiet before and it has been relatively quiet ever since. The recent evolution of consciousness must be viewed as the most cataclysmic happening since the Cambrian if only for its geological and ecological effects. Major events in evolution do not require the origin of new designs. The flexible eucaryotes[2] will continue to produce novelty and diversity so long as one of its latest products controls itself well enough to assure the world a future.[3]

Undoubtedly, the latest products of evolution are we, the people. We are the products who must control ourselves well. This is why the evolution of consciousness happened; it is supposed to *assure the world a future*. S. Gould made an interesting observation: We might not see new evolutionary designs. Apparently, the richness, flexibility, and inventiveness of our brain would compensate for the absence of new evolutionary designs. The brains of living creatures, even in their rudimentary form, guide them toward development of new designs for the evolution (our thoughts, plans, inventions, visual arts, music, and poetry).

Evolution and intelligence are the major known tools in producing a beneficial response to anything that may happen to a living creature. They help the creature survive in most situations, expected and unexpected. Here we should proceed with caution: The concept of the *expected event* presumes a creature that can have expectations. Evolution

[1] *Sigmoidal curve* describes processes that initially grow slowly, then gradually accelerate, achieving a rapid rate of growth, and then slow down, gradually achieving the end with a very low rate of growth.

[2] Eucaryotes are multichromosome living cells, as opposed to procaryotes that are single-chromosome cells.

[3] Stephen Jay Gould, *Ever Since Darwin: Reflections in Natural History*, Norton & Company, New York, 1977, p. 118.

works to the benefit or the successful outcome for a *species*; the idea of expectation in this case is at least puzzling. The idea of the *benefit* or the *success* is puzzling as well. Yes, the need for survival is well known. But what is the source of this need? Even more difficult is the idea of *success*. The latter presumes a *goal*. It is customary to link the goal with intelligence, but what might be the goal of the evolution?

Well, *evolution* and *intelligence* have something very important in common. Both affect our lives the most, and yet both are understood the least.

WE ARE THE TOOLS OF OUR INTELLIGENCE

Certainly, the productivity of design,[4] provided by the old chromosome-based mechanism, has its limitations. It achieved its peak at the time of the Cambrian period, and now, as our brain takes over, a new explosion is on the way. The intricacy of our brain navigates unexpectedly the subtle processes of decision making, beyond this organ's presently known capabilities, with its slow wetware-based neurons placed into an unreliable body prone to hedonism and addictions. In a hectic outburst of events compressed into one century, the brains of mankind have created a surprising engineering phenomenon: *intelligent systems*, which can work for their evolution much more efficiently than we, the people, can.

This thought is intuitively acceptable for those of us who continue researching the surrounding reality even after graduation. Computer-aided design and data mining are not buzzwords anymore; there are innovations actually produced by machines. The decision-making role of all-pervasive automated machines and various software packages is a fact of our lives. This is what happens—automation is coming to the domain of decision making rapidly and maybe even faster than it conquers the area of menial works. (This might be an alarming sign, but it is probably linked with the surprising fact that the market still creates more opportunities for menial jobs than for decision making.)

Evolution, until now the unchallenged domain of living organisms, may soon become possible for robots as well. A computer-based form of evolution—nature's own design strategy—has succeeded in designing LEGO structures without any assistance from humans, by the virtue of learning (Brandeis University). A human-like robot follows passersby with its eyes; a robot-drummer entertains the public (MIT). A robot-vehicle recognizes the edge of the road (Carnegie Mellon). A dune-buggy, controlled by the planner-navigator-pilot hierarchy, travels via cluttered environment by trial-and-error to the goal, then returns much faster because it had already seen the environment, remembered it, and planned a better motion trajectory (Drexel University). Autonomous vehicles can move with no human help at a speed of 55 mph on a highway (Universität der Bundeswehr, München, Germany). A squad of autonomous vehicles travels cooperatively at a speed of 20 mph in a cross-country environment (National Institute of Standards and Technology). Robotic brains can control manipulators, allowing them to carry heavy things cooperatively, not to speak about playing ping-pong which was already achieved years ago. These are mankind's first successful leaps from the early concepts of decision making based on expert systems into the futuristic realm of multiresolutional thinking.

[4]Naturally, we talk about design of a living creature. Evolution is interpreted frequently as a mechanism for designing the living creature. Probably, *self-design* is meant.

THE MOLECULE OF INTELLECT OR EVEN LIFE

It is not so difficult to discover that both evolution and intelligence use the same set of techniques: Both transform real experiences into symbolic representation and keep the messages in a well-organized, computationally efficient form; use these messages for searching groups of similarity and generalizing them; repeat this generalization recursively so that the nested hierarchies of efficiently organized information are being developed; increase their efficiency by focusing attention upon limited subsets of information; and use searching processes to combine elements of this information into messages that could not be obtained from the experiences.

Compact packages of operators, comprising of "grouping-focusing attention-combinatorial search" triplets in various combination, consume the arriving information and transform it into multiresolutional architectures. A simple feedback loop well known from the views of cybernetics could not be decisive in explaining the phenomena of the mind. In this book, we demonstrate that metaphorically speaking, the decision-making processes are realized via distributing the feedback loop over the multiresolutional architecture of information. Or even more accurately, via imposing the multiresolutional architecture upon the feedback mechanism. This is how the mind was built through the history of its development. This is how the model of mind is visualized by the authors of this book.

MULTIRESOLUTIONAL MIND

The concept of multiresolutional (multiscale, multigranular) information representation is the most formidable insight of the mankind in a rush to design new tools for a further leap in the process of evolution. People always distinguish between actual view and the bird's-eye view; smaller picture and larger picture; and the faraway look and close-up. This is the century when we have eventually discovered that our mind uses views of different granularity and scale for the benefits of decision making. This book demonstrates how people employ multiresolutional representations and make decisions of different resolutions.

This concept came simultaneously from different communities and appeared in different disguises. Researchers in computer vision discovered pyramidal image representation. Fractal people recognized this multiscale key to information representation as a mechanism of the construction of nature. European scientists made multigranular representation a part of their new science "synergetics." Practical engineers employed it within "finite element method." Mathematicians have not yet succeeded in unlocking the mystery hidden in *nonstandard analysis*.[5] However, they successfully use many levels of resolution as a practical method of solving partial differential equations ("multigrid methods").

The multiscale habits of thinking reveal something about the architecture of the human mind. Authors are convinced that the key issues of the "mind-design" can be resolved only by restoring the mind's structure in a multiresolutional fashion. Although we cannot observe or touch the structure of our mind, we have to restore it from watching the records of its multiscale functioning. This process of top-down/bottom-up

[5]Nonstandard analysis, a domain of logic discovered by A. Robinson, contemplates the possibility of a noncontradictory bridge between the continuous and discrete mathematics.

restoration of things that cannot be available to our immediate observation is similar to *reverse engineering*. (This term can be found in the 1993 documents[6]; in the context of cognitive science, it was used by D. Dennett[7].)

The results of mind restoration in the form of multiresolutional models described in this book allow for constructive understanding and simulation of our cognitive processes. In addition, multiresolutional models can be used for constructing and exploring machines and organizations that we call *intelligent systems*. Intelligent systems mimic human abilities to perceive the world, to collect and organize knowledge, to imagine things and make plans about them, to execute these plans and control ourselves. Remember that if we are able to control ourselves well enough, we may *assure the world a future*.

So far, we have been successful in inventing what has already been invented by nature. Computer scientists, with biologists, have succeeded in simulating the processes of the evolution of eye as an organ, and with engineers, have analyzed the process of evolving robotic skills of building bridges. In hindsight, the algorithm for discovery of sophisticated engineering devices turned out to be very simple. In reality, it took many centuries for humans to capture the skills of engineering design, develop rational plans of actions that take planners weeks to design, and compute trajectories of motion that require engineers hours to design.

ABOUT THIS BOOK

This book is about robotic brains as they should be designed and implemented in the very near future. The biologists and psychologists will find the engineering views on cognitive processes in human brain. Also, they can find numerous suggestions from two engineers on how they should understand human brains. Engineers should consider these suggestions as engineering metaphors. However, in the engineering part, the book contains the outcomes of many years of experience in engineering design and industrial implementation. Several doctoral and masteral theses were developed under the guidance of the authors; the essence of these researches was included in the book. The authors have merged together the results of their research obtained for the last 15 years. Some of these results happen to coincide, and they were very persuasive for the authors. Part of its materials precipitated from working together since 1994.

Our engineering recommendations are verified by implementing them in a realistic environment. Indeed, multiresolutional representation reduces the complexity of computation. Planning can be applied as a nested search process at all levels of resolution. Elementary loop of functioning must be specified for all levels of resolution before we start other engineering activities; a lot of time and money will be saved. Multiple levels of resolution can work jointly and coherently.

This book is intended to serve as a textbook for graduate courses in various disciplines: engineering (including robotics, mechatronics, and automatic control), business management, computer and information science. However, we would expect that cognitive scientists, psychologists, and biologists will benefit in using this as a text, too. Quiz and test problems are included at the end of each chapter.

[6]See a legal document in URL http://www.lgu.com/cr46.htm

[7]D. Dennett, Cognitive science as reverse engineering, in *Brainchildren, Essays on Designing Minds*, MIT Press and Penguin, 1998.

The book is illustrated by the results of application of the new theory for equipping numerous intelligent machines, including industrial robots and autonomous vehicles. We did not address several important issues related to intelligence, like natural language understanding. However, our methodology is expected to be instrumental in this case, too.

In the last 15 years or so, we have worked on these issues in close cooperation. We have achieved the highest level of mutual understanding, and it would be very difficult to determine who wrote what in this book. Nevertheless, even this level of cooperation would be insufficient for preparing the manuscript without help and contribution from our colleagues (in alphabetical order): U. Bar-Am, R. Bhatt, S. Bhasin, R. Bostelman, D. Coombs, N. Dagalakis, S. Drakunov, P. Filipovich, D. Gaw, G. Grevera, T. Hong, J. Horst, H.-M. Huang, A. Jacoff, E. Koch, A. Lacaze, S. Legowik, E. Messina, M. A. Meystel, J. Michaloski, Y. Moscovitz, K. Murphy, M. Nashman, R. Nawathe, L. Perlovsky, F. Proctor, I. Rybak, W. Rippey, H. Scott, S. Szabo, C. Tasoluk, and S. Uzzaman. We thank all of them for their constant and continuous support in the past 10 to 15 years.

ALEX MEYSTEL
JIM ALBUS

August 2001

CHAPTER 1

INTELLIGENCE IN NATURAL AND CONSTRUCTED SYSTEMS

1.1 INTRODUCTION

Intelligent systems have received the attention of scientists since antiquity. Nevertheless, the concept of *intelligent system* is not fully understood, and this affects interpretation of the existing research results as well as the choice of new research directions. In this book the subject is considered from the multidisciplinary point of view. The phenomenon of intelligence is demonstrated as a computational phenomenon. It emerges as a result of the joint functioning of several operators: grouping, focusing of attention, searching, and formation of combinations. When information is processed by these operators, multiresolutional systems of knowledge develop, and nested loops of knowledge processing emerge.

This conceptual structure permits the explanation of most of the processes characteristic of intelligent systems. The emphasis in this book is on demonstrating that the nature of intelligent systems is revealed best, and the means of controlling them introduced in the most constructive way, when the theoretical tools of computational mathematics are applied within the semiotically inspired architectures of computation (see Chapter 2).

Much is unknown about intelligence, and much will remain beyond human comprehension. The fundamental nature of intelligence is only dimly understood. Even the definition of intelligence remains a subject of controversy, and so must any theory that attempts to explain what intelligence is, how it originated, or what are the fundamental processes by which it functions.

Yet, much is known, both about the mechanisms and function of intelligence. The study of intelligent machines and neuroscience are both extremely active fields. International researchers in neuroscience search for the anatomical, physiological, and chemical basis of behavior.

Neuroanatomy has produced extensive maps of the interconnecting pathways making up the structure of the brain. Neurophysiology has demonstrated how neurons compute functions and communicate information. Neuropharmacology has discovered many of the transmitter substances that modify value judgments, compute reward and punishment, activate behavior, and produce learning. Psychophysics has provided many clues about how individuals perceive objects, events, time, and space, and how they reason about their relationship to the external world. Behavioral psychology has added information about mental development, emotions, and behavior.

MULTIDISCIPLINARY PARADIGM

Research in learning automata, neural nets, and brain modeling provides insight into learning and the similarities and differences between neuronal and electronic computing processes. Computer science and artificial intelligence probe the nature of language and image understanding, and have made significant progress in rule-based reasoning, planning, and problem solving. Game theory and operations research have developed methods for decision making in the face of uncertainty. Robotics and autonomous vehicle research has produced advances in real-time sensory processing, world modeling, navigation, trajectory generation, and obstacle avoidance. Research in automated manufacturing and process control has produced intelligent hierarchical controls, distributed databases, representations of object geometry and material properties, data-driven task-sequencing, network communications, and multiprocessor operating systems. Modern control theory has developed precise understanding of stability, adaptability, and controllability under various conditions of feedback and noise.

Progress is rapid, and there exists an enormous and rapidly growing literature in each of the above fields. What is lacking is a general theoretical model of intelligence that ties all these separate areas of knowledge into a unified framework. This chapter is an attempt to formulate at least the broad outlines of such a model. In this chapter, it is emphasized that an intelligent system always has an architecture that develops as a result of the joint functioning of certain computational operators. The architecture that emerges as a result has multiple closed loops, each loop at a particular level of resolution.

Our model is to be used for the analysis of all kinds of intelligent systems, including animals, humans, automated machines, robots, autonomous vehicles, and integrated manufacturing systems. We believe that the functioning of most large and complex systems can be treated as objects in our theory of intelligent systems. For simplicity though, we allude primarily to living creatures or robots in our examples. We expect that active readers will make all the necessary implications and evolve the theory in the direction of their interests.

The ultimate goal is a general theory of intelligence that encompasses all possible instantiations: biological, machine, societal, and others. The goal is to discover an architecture of intelligence that would be applicable in all these cases. The architectures introduced in this book incorporate knowledge gained from many different sources, and the discussion frequently shifts between natural and artificial systems as both elements are discussed. The definition of intelligence addresses both natural and artificial systems. The origin and function of intelligence is treated from the standpoint of biological evolution, which is shown to be similar to the general process of learning.

SIMILAR ARCHITECTURES IN ALL DOMAINS

The system architecture is strongly influenced by our knowledge of the brain, although as far as its engineering is concerned this architecture is derived almost entirely from research in robotics and control theory. It has been applied to devices ranging from undersea vehicles to automatic factories. The material contains numerous references to neurophysiological, psychological, and psychophysical phenomena that support the model. Frequent analogies are drawn between biological and artificial systems. The value judgments are based mostly on the neurophysiology of the limbic system and the psychology of emotion. Results on neural computation and learning are derived mostly from and motivated by neural net research.

BIOLOGICAL AND MACHINE EMBODIMENTS

The definition of intelligence includes both biological and machine embodiments. These span an intellectual range from that of an insect to that of an Einstein, from thermostats to the most sophisticated computer systems. In order to be useful in the quest for a general theory, the definition of intelligence must not be limited to behavior that is not understood. The definition of intelligence should include the ability of a robot to spotweld an automobile body, the ability of a bee to navigate in a field of wild flowers, a squirrel to jump from limb to limb, a duck to land in a high wind, a swallow to work a field of insects and a human to come up with the set theory.

At a minimum intelligence requires the ability to sense the environment, to perceive and interpret the situation, to make decisions, and to control action. Higher levels of intelligence may include the ability to recognize objects and events, to represent knowledge in a world model, and to reason about and plan for the future. In advanced forms, intelligence provides the capacity to perceive and understand, to choose wisely, and to act successfully under a large variety of circumstances, and to survive, prosper, and reproduce in a complex and often hostile environment.

From the viewpoint of control theory, intelligence might be defined as a knowledgeable "steersman"[1] of behavior. Intelligence is a phenomenon that emerges as a result of the integration of knowledge and feedback into a sensory-interactive, goal-directed control system that can make plans and generate effective, purposeful action directed toward achieving them.

From the viewpoint of both psychology and biology, intelligence might be defined as a behavioral strategy that gives each individual a means for maximizing the likelihood of propagating its own genes. Intelligence is the integration of perception, reason, emotion, and behavior in a sensing, perceiving, knowing, caring, planning, and acting system that can succeed in achieving its goals in the world.

For the purposes of this book, intelligence will be defined as follows:

DEFINITION OF INTELLIGENCE

Intelligence is the ability of a system to act appropriately in an uncertain environment, where an appropriate action is that which increases the probability of success, and success is the achievement of behavioral subgoals that support the system's ultimate goal.

Both the criteria of success and the system's ultimate goal are defined external to the intelligent system. For an intelligent machine system, the goals and success criteria are typically defined by designers, programmers, and operators. For intelligent biological creatures, the ultimate goal is gene propagation, and success criteria are defined by the processes of natural selection.

DEGREES OF INTELLIGENCE

Assertion There exist degrees, or levels, of intelligence that are determined by the following features of the system: (1) the computational power of the system's brain (or computer), (2) the sophistication of algorithms the system uses for sensory processing, world modeling, behavior generation, value judgment, and global communication, (3) the information and values the system has stored in its memory, and (4) the sophistication of the processes of the system functioning. These levels of intelligence are different in the probability of the success of decisions that is measured by various criteria of performance (including time, accuracy, and others.)

[1]This metaphor is used by N. Wiener in his book *Cybernetics* (second edition, p. 11, published by MIT in 1999).

Intelligence can be observed to grow and evolve, both through growth in computational power and through accumulation of knowledge of how to sense, decide, and act in a complex and changing world. In artificial systems, growth in computational power and accumulation of knowledge derives mostly from human hardware engineers and software programmers. In natural systems, intelligence grows over the lifetime of an individual through maturation and learning, and over intervals spanning generations through evolution.

Thus our idea of intelligence goes beyond the concept of simple *adaptation*. It includes adaptation at different time scales and becomes closer rather to the concept of *learning*. Note that learning is not required in order to *be* intelligent, only to *become more* intelligent as a result of experience. Learning is loosely defined as consolidating short-term memory into long-term memory in a form of generalities providing improved inference and exhibiting altered behavior because of that memory (see Chapter 10). We discuss learning as a mechanism for storing knowledge about the external world and for acquiring skills and knowledge of how to act. Of course it is assumed that many creatures can exhibit intelligent behavior using instinct, without having learned anything. But again, it is possible that instinct appears as a result of learning in prior generations.

1.2 BRIEF OVERVIEW OF THE EVOLVING CONCEPTS OF MIND AND INTELLIGENCE

During the last 30 or 40 years multiple visionaries came up with their architectural concepts of mind and intelligence. This is the chronology of the evolution of scientific vision in this area.

- One of the leading psychologists of the twentieth century, D. O. Hebb [1] recognized that computational models of mind and intelligence have promise: "It is becoming apparent from such work as that of Broadbent [2] and of Miller, Galanter and Pribram [3] that the computer analogy which can readily include an autonomous central process as a factor of behavior, is a powerful contender for the center of the stage." Interestingly this was said in 1960 at a time when no *introspective* models of psychology were permitted. It was the time of a belligerent victorious *rattomorphism* [4] when the scientists were expected to judge just about everything by constructing stimulus–response models of behavior, and no such terrible things as *mind* and *consciousness* were permitted at all. Later, L. von Bertalanffy applied the term "zoomorphism" to the same phenomenon that was called "rattomorphism" by A. Koestler. As a later prominent psychologist, Gardner [5], characterized the situation, "... adherence to behaviorist canons was making a scientific study of mind impossible."
- The influence of the computational models on the analysis of mind and intelligence cannot be underestimated. The fundamental work by R. Ashby [6] was one of the first efforts to introduce control loops for analysis of the complex systems controlled by a brain. Considering the "community" of multiple loops some of them stable, some not, was the first step for the future hierarchical structures of intelligence.
- After Ashby outlined *how the intelligent system acts*, O. Sefridge (1955) researched the associated problem of *how the intelligent system perceives* [7]. The natural result was an interest in *what the intelligent system does with its perception before beginning the action*; these were the focus of works by Newell, Shaw and Simon [8], as well as by Samuel [9]. At the same time we were treated to business discussions between the eyes and the brain of frogs, and thus incipient loops of intelligence [10].

■ Neural science, as a field, took its inspiration from the computational vision of brain processes in the Rosenbatt's (1958) seminal work [11]. Further refinement of the models treating the perceptual and control parts of the brain followed in works by M. F. H. George [12] and Minsky [13].

■ The synthetic character of mind and intelligence was becoming a challenge for scientists. The issue was how to integrate a multiplicity of disciplines. Langer (1962) provided some of the perspective:

> Descartes' self-examination gave classical psychology the mind and its content as a starting point. Locke set up sensory immediacy as a new criterion of real .... Hobbes provided the genetic method of building up complex ideas from simple ones ... Pavlov built intellect out of conditioned reflexes and Loeb built life out of tropisms. [14]

■ The decade of general problem solver (GPS) ended leaving unresolved the exploration of thinking processes involved in problem solving (A. Newell, H. Simon, and others [15, 16]). GPS made the concept of the planner, and even a "conniver," applicable and understandable: In order to survive, the mind and intelligence should create plans.

■ The science of the mind and intelligence entered a new period. Whereas initially researchers were interested in explanation of the natural phenomena of intelligence, attention now turned toward the possible practical use of the research results. Since our mind and intelligence are computational by nature, it is only natural to simulate mind processes on a computer. W. R. Reitman [17] attempted to use computer simulation for modeling and supporting more complex intellectual activities such as solving ill-posed problems and composing music besides general problem solving. L. J. Fogel, A. J. Owens, and M. J. Walsh [18] proposed computational algorithms that helped by evolutionary searches and thus substantially reduced the complexity of computations. J.C. Eccles, M. Ito, and J. Szentgothal [19] made a serious attempt to describe the neuronal substrate as a computational system.

■ Computer scientists, engineers, psychologists, and neurophysiologists all took a lead in the development of a science of the mind and intelligence. However, the area of signs and their systems still remained to be addressed sufficiently. D. Hebb referred to the fundamental role of C. Peirce in the development of the science of the mind [1], although C. Peirce was not explicit in linking semiotics to the analysis of mind. However, N. Chomsky [20] has no doubts about his theory of syntax for surface and deep linguistic processes, that they are (in his words) a "linguistic contribution to the science of mind." Earlier (1963), Chomsky worked together with Miller (who was one of the authors of the book on "plans" [3]), and he clearly saw the role of language in development of inner models and the plans of actions [21]. At this moment in time, the unity of hierarchy of linguistic syntax and hierarchy of plans started to loom as an important issue (as we will see later, as an issue for "mind architecture"). On the other hand, it became clear that parallel to formation of linguistic hierarchies, the hierarchies of logical types could be formed, and Bateson [22] introduced the concept of the *learning hierarchies*.

■ The creator of the general theory of systems, Ludwig von Bertalanffy [23] considered the issue of modeling mind and intelligence to be a focal point for science (Bertalanffy [23]). He saw the interpretation of the mind as only possible within its *Umwelt*. By this semiotic terminology (introduced by J. von Uexkull [24]) he reasoned that "the basic fact in anthropogenesis is the evolution of symbolism" (p. 21). The very existence of a mind can be considered to the "materialization of symbolic activities" (p. 22).

The notion of symbolic activities was substantially richer than the reinforcement based schemes promulgated by behaviorists: "learning by 'meaning' or 'understanding' is essentially different from and cannot be reduced to reinforcement" (p. 27). Bertalanffy gave a list of mind-related consequences of human symbolic activities:

- Phylogenetic evolution based on hereditary changes is supplanted by history based on the tradition of symbols, which accelerates the development of humans.
- Actual trial and error (random search) is replaced by reasoning (a guided search within conceptual symbols).
- Symbolism makes true purposiveness possible. The future goal can now be anticipated in its symbolic representation and so may determine present actions.
- Symbolic universes created by minds attain their own autonomy and become "self-propelling."
- As a result of the autonomy of symbolic universes, humans develop algorithmic properties; they can create prediction, imaginary alternatives of future, and prediction and control in science and technology.

■ The concept of organizational levels starts showing up in the applied educational literature related to mind and intelligence. The interrelated concept of aggregation-disaggregation (or fine-grain vs coarse-grain granulation) can be found in the books of educator de Bono [25] in the form anticipating the future object-oriented approach:

> At each level the output of the level below is taken as a unit and organized as such. The detailed organization of such a unit itself is not required: it is taken for granted and dealt with as a whole. For the bricklayer the functional unit is a brick. For the contractor the units are labor, bricks and transport, which all have to be coordinated. For the town planner it is complete house units.

■ At about this time researchers started paying attention to the intelligence of a single cell (Zeigler and Weinberg [26]). The hybrid nature of systems becomes a part of modeling and simulation of processes (Zeigler [27]).

■ A semiotic platform for modeling the mind and intelligence is developed by Pribram [28]. His book demonstrated that a hierarchy of languages exists and that all brain processes can be simulated in terms of a system of communicating languages.

■ Within the nervous system a similar concept is applied to the clustering of signals into the meaningful groups (pathways) in order to demonstrate that a hierarchical organization of signals (groups of pulses) exists at all levels of information related to mind and intelligence (Eccles [29]). Formation of clusters (grouping) and the subsequent development of hierarchical structure becomes an issue in a multiplicity of sources related to the mind and intelligence. The seminal work is *Steps to an Ecology of Mind* [30], in which many papers invoke the phenomena of group formation and growth of hierarchies.

■ M. Arbib, in his book *The Metaphorical Brain* [31], synthesized the existing knowledge on the brain's structure in both the control and behavioral sciences. M. Arbib entered this area with an extensive background in mathematics [32] and control theory [33]. Using automata theory and algebra, he explored the concept of commutative diagrams which is closely linked with functioning loops (primarily, control loops) and applied the concept of loops of communicating symbols (the concept loops of semiotic closure [34]). Arbib found loops in all systematic activities of the brain, and he anticipated the architectures of it as many multiresolutional loops. He would finalize the leap into the multiresolutional paradigm into the second edition of his book in 1989 [84].

■ At about the same time there was progress in the direction of search procedures for the functioning of intelligent systems. A. Newell and H. Simon [35] presented their famous Turing Award Lecture before the Conference of the Association for Computing Machinery (1975). In this lecture they outlined their view of "physical symbolic systems" (actually, a semiotic view) and the role of heuristic search processes for functioning of these systems.

■ But what was the reasoning behind "heuristic search?" Why was there need to depart from clear and reliable exhaustive arrangement of the possible results? As the models and concepts of the mind and intelligence grew more complicated, the issue of complexity became a high priority issue. In order to survive, it became obvious that it is not enough to *plan*; the intelligent system must plan *quickly*. Does an intelligent system regret not using all available information? Well, the counterquestion is valid, too: Does it regret using available information that contains elements of uncertainty? Does it regret browsing exhaustively through all information containing a lot of redundant and uncertain data? There was the recognition of the trade-off that occurs between procrastinating a decision and making a quick but erroneous decision, so the number of complexity-related research kept growing. Pylyshin [36] calls it a trade-off between "planning" and "responsibility" and introduced two powerful principles ("organization" and "constraints") as two major tools of fighting computational complexity. The first is linked with the hierarchical architectures, and the second with "focusing attention" in dealing with any information set.

■ Earlier discoveries that have introduced hierarchical organization as a reality for learning processes in real, yet, symbolic systems (e.g., mind and intelligence) once more became valid. Hierarchies attracted the attention of mathematicians (Mesarovich, Macko, and Takahara [37]) and system scientists (Jantsch [38]). Hierarchies of the mind and intelligence were connected with the process of self-improvement, self-renewal which was labeled *autopoesis* (Varela, Maturana, and Uribe [39]). There was an immediate desire to relate to animals all the concepts of mind and intelligence initially considered the province of humans (Shepard [40]).

■ All these advancements led to the temptation to develop a broad model of the mind that would reflect and address all pertinent issues such as perceiving, waking/sleeping, thinking processes connected with words and images, thinking processes related to memories and plans, and of course, consciousness (Furst [41]; Taylor [42]).

■ Efforts began to be made to construct and build a machine that could perform all the functions of the mind. For McCorduck [43] this machine would integrate all the effort within the discipline of artificial intelligence since 1956. For Albus [44] the problem of building a machine functioning as the mind was instead an engineering problem. It was to be a mind of a robot. J. Albus attempted to integrate within a single machine everything that was critical for the successful functioning of this machine: the hierarchical architecture, the system of control and actuation, the perceptual systems, and the neural components. Some components should be introduced anew—well, we can create them (Albus [45], [46]).

■ Campbell [47] analyzed an important aspect of the mind—its ability to create and manipulate symbols and languages. Language became treated as "the mirror" of the mind. A deep analysis of these issues was given in Johnson-Laird [48]. A purely syntactic theory of the mind was proposed in Stich [49]. Poppel [50] explored how the syntactic competence may affect temporal side of the activities constructed by a mind.

▪ A regular means of analysis was required for psychologist practitioners and researchers in evaluating the parameters of intelligence, its functional and operational properties, and the relationship of intelligence to other parts of an intelligent system. These issues were collected by Sternberg [51]. Gardner [52] proposed to treat separately linguistic, logical-mathematical, spatial, musical, bodily-kinesthetic, and interpersonal and intrapersonal intelligence. This educational focus was further demonstrated to be useful for a number of practical purposes (Gardner [53]) but no constructive concepts were proposed for its architecture. Similar attempts with a neurobiological and neurophysiological focus were made by Churchland [54]. Although both sources contained surveys of recent research results, they were far from proposing any constructive hypothesis that could be utilized in designing a computer model of the mind and intelligence.

▪ It is not clear that all phenomena are known that should be reproduced in order to mimic all of the mind's functions. Marr [55] showed that much of visual perception could be explained by computational algorithms. Kosslyn [56] tried to address all the issues that were not yet clear, including memorizing images, generating and transforming images, predicting images, thinking in images, and so on. Twenty-two years later he addressed the unresolved issues again, and the number of obscurities that could be addressed by researchers turned out to be 10 times higher (Kosslyn [57]). Gregory [58] attempted a similar account for overall *mind* issues. This resulted in an encyclopedic work but one removed from the engineering design that was being sought (Gregory [59]). A number of constructive discussions in neurobiology and the psychology of perception, attention, and emotion were provided by LeDoux and Hirst [60]. The work on complexity reduction could benefit from these discussions, since they also constructively illuminate the still obscure issue of emotions. Thus a computer-based model of choice could find serious factual support in this work. More on the emotional factor in making choices can be found in LeDoux [61]. The discussions in Litvak and Wayne Senzee [62] suggested the need for an evolutionary component, but the results do not seem to be entirely relevant in the concrete examples discussed. The simple GPS-type or checkers playing cases do not fit real complex cases. Models of thinking do not necessarily need to satisfy diligently collected and applied sets of logical rules. The results of creative thinking violate the constraints of any paradigm of logical analysis (Margolis [63]).

▪ The computational merits of multiresolutional image processing attracted a lot of attention and was reflected in the literature (Rosenfeld [64]; Cantoni and Levialdi [65]). At this time multiresolutional (multigranular, multiscale) representation was broadly recognized in practice, although it was called different ways in different areas. Because of this the global character of the multiresolutional representation and processing was not immediately recognized. (The hierarchies of "Kohonen's memories" [66] are multiresolutional too). Finite elements method, multigrid [67], [68], [69], and domain decomposition [70] methods were successfully used for systems with synthetic and incomplete representation by way of solving differential equations. Fractals which have a similar multiresolutional nature (Mandelbrot [71]) were considered viable tools [72]. Later the wavelet analysis of signals would come into fashion (Meyer [73]; Daubechies [74]). All of these computational tools—multigrid, domain decomposition, finite elements, fractals and wavelets—are strikingly similar in that they use focusing of attention, searching the objects for groups formation, and finally, grouping. As the results of grouping are obtained, a triplet of procedures is applied again to these results.

■ Multiresolutional representation was employed for motion planning (Davis [75], Meystel [76]). It was demonstrated that multiresolutional planning-control continuum leads to tremendous reduction of computational complexity. Multiresolutional ideas began to quickly enter the area of intelligent control.

■ Analyses of human motion demonstrated that the vocabulary of movement has a multiresolutional character (Heuer, Kleinbeck, and Schmidt [77]; Phillips [78]), and it is prudent to expect that the brain has a multiresolutional system of motion control. This was the fundamental Albus premise; later, K. Baev will build a theory of a multiresolutional control architecture in the central nervous system [79].

■ Novel developments increased the number of the degrees of freedom available for a designer of mind and intelligence. Johnson-Laird [80] (which is a subsequent work of Johnson-Laird [48]), introduced "thought" produced by the mind such as daydreams and thoughts with a goal. The latter can be nondeterministic, but when known precisely, it can be computed. In the nondeterministic case the goal is not explicit and must be drawn out. Even if the goal is known, the cases may differ. Inductive reasoning can be used when the amount of arriving semantic information increases; when no more information is available, then deductive reasoning is applied. This categorization of cases was extended to the subtle situation when abductive reasoning is required. Johnson-Laird showed that the process of communication is an additional channel for obtaining and refining meaning. He added to the architecture of the mind the phenomena of consciousness, self-reflection, intention, and free will. The perspective of the thought formation can be enhanced by introducing cases with uncertainty of information and allowing for loss of memory (Langer [81]). The evolutionary aspect of the development of mind allows for addressing all these issues with more agility (see Eccles [82]).

■ The model of the mind presented by Edelman [83] brought him a Nobel Prize for his discovery that laws of grouping govern living cells. The image of clustering cells dominated his model of mind processes. In it the processes of concept formation behave similarly to the processes of clustering living cells. As he addresses the other issues such as language formation and functioning, we can see that the elements of language behave also like living cells when they stick together in clusters. This living cell-like behavior can be noticed at all levels of language: conceptual, semantic, syntactic, phonological, and acoustic/motor.

■ In a new edition of his book on brain architectures, Arbib [84] constructed the architecture of the brain/mind as a "set of levels of refinement" (p. 365). Nevertheless, his neural-nets related materials remained a tribute to connectionism and its maxims. They did not behave like Edelman's living cells and did not demonstrate any reorganizational intentions required by changing circumstances. This adherence to the visions of "connectionists" can be seen in otherwise excellent books such as Boden [85] and Minsky [86]). Yet, the highest level of resolution alone is not sufficient for explaining most phenomena of mind and intelligence.

■ Some researchers invoked concepts from quantum mechanics as a possible avenue of further research (Penrose [87]). We do not discuss these lines of research here because they are not constructive for our goal of building a functioning architecture of the mind and intelligence. The many existing attempts to incorporate quantum physics into the tissue of the mind and consciousness are not persuasive (e.g., Herbert [88]). Other sources contained analogies of another sort, comparing the brain to the universe, introducing terminology of neurocosmology, and the like (Siler [89]). We do not discuss these sources either.

■ A useful survey by Sternberg [90] provided a comprehensive complement to Sternberg [51]. In it all existing theories and views related to mind and intelligence were considered to be metaphors. Talbot [91] next treated the brain as a hologram. An evolutionary hypothesis was discussed by Calvin [92], who explained the phenomenon of intelligence development by changes in environment.

■ Connectionist models are usually packaged with a multi-agent behavior-based philosophy of recognition and control. Trehub [93] was one of such works. It allocated substantial discussion to micro-events propagating through the NN-communities of the system. No clear design recommendations was given. Churchland and Sejnowski [94] was substantially more equipped with relevant information and comprehensive surveys. The connectionist models sometimes appeared very impressive. This is how F. Crick, a Nobel Prize winner, started his book about mind:

> The astonishing Hypothesis is that "You," your joys and your sorrows, your memories and your ambitions, your sense of personal identity and free will, are in fact no more than the behavior of a vast assembly of nerve cells and their associated molecules. (Crick [95]).

■ Multi-agent "behavioral" philosophies became all-pervasive among researchers involved in the development of intelligent systems. The premise is very similar to the one employed in H. Gardner's theory of multiple intelligence: if a system has many faculties, it can be represented as a sum of systems each responsible for a single faculty.

Although in some cases it works, this "global superposition principle" can lead to deceptive stituations and erroneous functioning. Global superposition requires global linearity of the system and its subsystems. Frequently, global linearity does not hold.

■ Albus [96] had suggested a reference model architecture for intelligent systems design. It was built upon a multiresolutional hierarchy of computational processes, data structures, and a communication network. A similar architecture applied to autonomous mobile robot was described by Meystel [97].

■ Linguistic overtones continued to be strong in the studies of the mind and intelligence. Fodor [98] had argued for a study of the "language of thought." The earlier K. Pribram book on the languages of the brain evolved into a theory with strong mathematical and semiotic roots (Pribram [99]). This theory correlated, in a creative and productive way, with other linguistic theories of mind and intelligence, like those presented in Chomsky [100] and Jackendorff [101]. These theories were well supported by studies about the early processes of mind functioning (Wellman [102]). Bickerton [103] demonstrated language as a sole basis for explaining many subtleties of intelligent systems' functioning.

■ The literature in the area of the mind and intelligence had led to the emergence of a new wing of philosophy: computational philosophy. Thagard [104] introduced the direction in which the philosophy of mind was expected to develop. Searle [105] made serious contributions illuminating the set of problems to be resolved by the computational philosophers.

■ The issue of consciousness had been frequently treated as a separate topic, not as a part of the model of mind. In a pure connectionist fashion, Greenfield [106] moved the focus of interpretation to neuron masses, to a search for neuronal gestalt and neuronal reverberation. The term *qualia* (qualitative "feel") was introduced to make the issue distinct and scholarly. Chalmers [107] tried to link consciousness to quantum mechanics phenomena. More comprehensive studies of consciousness in the system of mind can be found in Dennett [108] and Dennett [109].

■ Recently Calvin [110] gave a recapitulation of prior efforts to explain the mind and intelligence with a strong emphasis on the significance of syntax. Pinker [111] provided a penetrating and more balanced review of the scientific and philosophical foundations of a computational theory of mind. However, the bottom line remains the same: the fundamental concept on which the computational theory of mind can be grounded is that the brain is a machine and the mind is a process that runs in the brain. In short, the mind is what the brain does.

SYSTEMS ENGINEERING VIEWS

In this book, we approach the phenomenon of intelligence from a systems engineering viewpoint. Our goal is to develop engineering guidelines that enable the design and construction of intelligent systems that rival natural intelligence in performance of significant tasks in the natural world. In the following pages we set forth an outline for a theory of intelligence and propose a reference model architecture that can serve as a guideline for engineering intelligent systems.

COMPUTATIONAL PHENOMENON

Our approach is to treat intelligence as a computational phenomenon that emerges from the joint functioning of a closed loop with four fundamental processes: behavior generation, world modeling, sensory processing, and value judgment. We will show how these four processes can work together to process sensory information, to build, maintain, and use a knowledge database, to select goals, react to sensory input, execute tasks, and control actions. Sensory processing functions to focus attention, detect and group features, compute attributes, compare observations with expectations, recognize objects and events, and analyze situations. World modeling constructs and maintains an internal representation of entities, events, relationships, and situations. It generates predictions, expectations, beliefs, and estimates of the probable results of future actions. Value judgment computes the cost, benefit, risk, and expected payoff of plans. It assigns value to objects, events, and situations. It decides what is important or trivial, what is rewarding or punishing, and what degree of confidence to assign to entries in the world model. Behavior generation selects goals, decomposes tasks, generates plans, coordinates activity, and controls action.

These four processes are linked together through a computational feedback control loop. This loop is closed inside the intelligent system from sensory processing, through world modeling and value judgment, to behavior generation. In Figure 1.1, an intelligent system (or an agent) is shown that has all these subsystems (sensors and actuators). They allow for an access to the endless richness of the reality of the external world (to the degree of its variety of sensors and ability to develop actions). The loop is closed outside the system from action, through the real world, to sensing. Goals enter this computational loop from above and sensory signals enter from below. Within the loop,

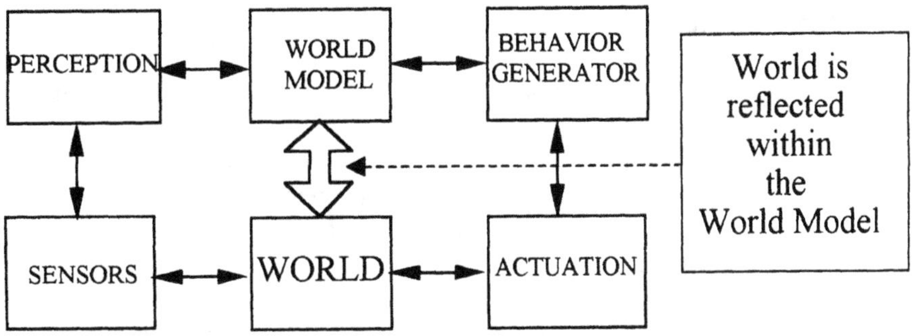

Figure 1.1. Elements of intelligence and the functional relationships between them.

sensed results are compared against desired goals. Plans are formulated and control is exerted on the world to bring results into correspondence with plans. It is this interaction between top-down goals and bottom up sensory feedback that results in the phenomena known as intelligence.

A question can be raised: what does it mean having world W a part of this closed loop? Should this "endless richness of the external world" be included into the closed loop? All of it? This is an important question. Later, we will see that *closure* is a prerequisite for constructing an intelligent system. Of course, only a limited part of the world is included into the closed loop: the subset actually participating in the loop functioning.

There also exists an internal link between world modeling and sensory processing that generates predictions that can be compared with sensory observations. This closes a loop within the intelligent systems that can be used for recursive estimation, predictive filtering, or as a queue for storing events in short term memory. There also exists an internal link between behavior generation, world modeling, and value judgment. This closes an internal planning loop wherein a variety of tentative plans can be simulated and evaluated prior to being selected for execution.

We will call the model shown in Figure 1.1 an Elementary Loop of Functioning (ELF), or a primitive agent. These computational loops from sensing to acting, from world modeling to sensory processing, and from behavior generation to world modeling to value judgment and back again, are repeated many times within an intelligent system at many different levels as the units of information in all of the subsystems are aggregated into entities, events, and situations, and goals are decomposed into subgoal tasks and generate commands. Within each loop, world modeling maintains a knowledge database with a characteristic range and resolution. At each level, plans are made and updated with different planning horizons. At each level, short-term memory traces store sensory experiences over different historical intervals. At each level, feedback control loops have a characteristic bandwidth and latency.

This model of a multiresolutional hierarchy of computational loops yields deep insights into the phenomena of behavior, perception, cognition, emotion, problem solving, and learning. We will attempt to demonstrate how this model can support a formal theory of intelligent systems. We will then propose a reference-model architecture for the design of intelligent systems. Finally we will suggest engineering guidelines for implementing a wide variety of practical intelligent systems including autonomous vehicles, intelligent manufacturing systems, automated construction systems, intelligent highways and transportation systems, and intelligent machine systems for exploring the deep ocean and the distant planets.

1.3 INTELLIGENT SYSTEMS: CAN WE DISTINGUISH THEM FROM NONINTELLIGENT SYSTEMS?

It is our intention to delineate a synthetic image of an intelligent system. This image should allow for generating a definition equally related to humans, animals, and autonomous robots. This definition will focus on those features that can be considered "features of intelligence," and be perceived by the scientific community with no controversy. Let us consider several answers to the question proposed in the title of this section:

Researcher answers This group of answers is trying to implicitly define the intelligent systems based upon numerous "objective" tests:

TURING TEST

- A test proposed by A. Turing in 1950, now called the *Turing test* [112]. In one of the test scenarios a human judge interrogates a program through an interface. If the program can fool the human into believing that responses come from another human and not from a computer, then the program should be considered intelligent. Clearly, in this test we don't talk about intelligence as a phenomenon but rather about an ability of pretending to be intelligent. At the present time such an approach seems to be naive.

CHINESE ROOM

Nevertheless, this approach has generated a lot of literature, in particular the famous problem of the Chinese room [113]. J. Searle considers the following mental experiment. A person is given a set of formal rules for manipulating Chinese symbols. This person does not speak or understand written Chinese, so he does not know the meaning of these symbols; he can distinguish them just visually. The rules state that if a symbol of a certain shape is given to him, he should write down a symbol with a certain other shape on a piece of paper. The rules prescribe how the groups of symbols accompany one another. When a set of Chinese symbols enters from outside, the person applies the rules, writes down a set of Chinese symbols as specified by the rules, and returns the result to the external observer. The external observer perceives the result as a grammatically correct conversation in Chinese. However, the person inside does not understand Chinese.

Searle believes that the person in the Chinese room is doing exactly what a computer would be doing if it used the same rules to engage in a grammatically correct conversation in Chinese. Both the computer and our "inside" person are engaging in "mindless" symbol manipulation. This mental experiment leads J. Searle to the following statements:

Axiom 1 Computer programs are formal (syntactic) and manipulate *symbols*.

Axiom 2 Human minds have mental contents (semantics) and manipulate *meanings*.

Axiom 3 Syntax is not translated into semantics; therefore symbol manipulation does not contain any *understanding*.

Searle's argument is intended to show that implementing a computational algorithm that is *formally* isomorphic to human thought processes cannot be sufficient to reproduce the real process of *thought*. Something more is required.

This argument proclaims that "symbol manipulation" and "meaning manipulation" are different in their nature. Numerous questions emerge: can we manipulate anything else but symbols? Why can "meaning" not be represented in symbols? We hope that this book will be instrumental in answering these questions.

ZADEH TESTS

- L. Zadeh's test for a "strong intelligence" can be formulated as follows: A paper is presented to the computer, and it is asked to transform it into an abstract. A quality of the abstract can be judged by the ability of a computer to extract the meaning of the paper in a sufficiently concise form. No doubt, any system that could do it would be considered intelligent. But what if the computer cannot do it, but it can do other intelligent things? (On the other hand, the problem of meaning extraction is not yet clearly defined; different intelligent people can see different meaning in the same paper, and interpret it in a different way. What can we do with this predicament?)

L. Zadeh's test for a "weaker intelligence" seems to be very simple: make an autonomous mobile robot that will be capable of performing parallel parking (say, to an allocated spot). Seems trivial. However, the autonomous vehicles capable of this skill are not as widespread as they possibly should.

From this book, it will be clear that both Zadeh's tests of intelligence allude to the same architecture.

TESTS OF SOPHISTICATION AND RATIONALITY

- Other tests can be formulated that demand different levels of knowledge, skills of reasoning, and degree of sophistication. For example, one could ask a computer to analyze photographs of three to four different artists, and then analyze three to four art pieces and make a judgment on who is the author of each piece. Interestingly people often give successful answers depending on their reading of an artist's character from a photograph of the art piece (even if they have not seen the actual work). I would not dare to propose this as a test for a computer program, but would that be a good test for a human intelligence?
- Various tests can be proposed based on more mundane but more practical evaluations of sophistication and rationality. For example, we can check the capability of a program to generate alternatives for a decision being made in a particular situation and select one of them properly, or its capabilities to analyze the experimental data related to a particular physical system and compute feedforward control and feedback compensation. The key issue in the last case is the ability to use the experimental data: different experimental data require different approaches to computing feedforward control and different laws of feedback compensation.

Answers based on descriptive enumeration of the properties of intelligence Newell has listed properties that intelligent system must have [114]:

PROPERTIES OF INTELLIGENCE

- It must recognize and make sense of a scene.
- It must understand a sentence.
- It must construct a correct response from the perceived situation.
- It must form a sentence that is both comprehensible and carrying a meaning of the selected response.
- It must represent a situation internally.
- It must be able to do tasks that require discovering relevant knowledge.

Think about this list: It does not necessarily portray a human being. Cats and dogs, and much lower animals, fit some of the list. One could argue that a bacterium has all these properties, not to speak of systems for intelligent manufacturing.

PRAGMATIC VIEWS OF INTELLIGENCE

This group of answers can be illustrated by a very pragmatic definition of intelligence:

An intelligent machine tool is defined by comparison with an intelligent human machinist. A higher level scheduler can rely on both of them in the same way. A given input leads to an expected output. Or, the intelligence reports back that the input is beyond the scope of the current system. We must therefore acknowledge that the degree of intelligence can be gauged by the complexity of the input and/or the difficulty of ad hoc in-process problems that get solved during a successful operation. Our unattended, fully matured intelligent machine tool will be able to manufacture accurate aerospace components and "get a good part right the first time" [115, p. 8].

Answers given by pragmatic scientists The pragmatic group aims into getting an opportunity to have formal clues of distinguishing an intelligent system. For example, if a system uses fuzzy techniques and neural networks, then it must be considered intelligent (obviously, since this is how this group prefers to define intelligence). It may be contradictory that one can build both intelligent and nonintelligent systems using both fuzzy and neural techniques. Nevertheless, somehow fuzzy and neural systems lead to the domain of intelligence. The question is how?

This book will demonstrate how and why the combination of NN, fuzzy logic, and search tools can lead to the property of "intelligence." However, it does not always happen. It is important to find when it does happen.

COGNITIVE CONSIDERATIONS

Cognitive and anticognitive answers Traditionally the evolution of intelligence has been perceived (quite understandably) as the development of mechanisms of survival associated with building a model of the world and the tools for understanding an external reality. But this tendency of being introspective should not be considered a *bias* of "cognitivists"—it is their *method.*

- J. Albus observes an inclusive view of intelligence [116], and he gives several definitions, all alluding to the integrative role of this faculty of a system or of a creature capable of decision making, knowledge representation, perception, genes propagation, and the like [116, p. 474]. The invariant kernel of all cases he addresses is "the ability of a system to act appropriately in an uncertain environment, where appropriate action ... increases the probability of success," ... i.e., "achievement of behavioral sub-goals that support the systems ultimate goal." Although the emphasis is on engineering applications, the objective is to create an architecture of cognition in living creatures.
- A. Newell links intelligence with cognition and cognition with knowledge [117, p. 90]: "The system is *intelligent* to the degree that it approximates a knowledge level system .... 1. If a system uses all of the knowledge that it has ..., it must be perfectly intelligent .... 2. If a system does not have some knowledge, failure to use it cannot be a failure of intelligence .... 3. If a system has some knowledge and fails to use it, then there is certainly a failure of some internal ability, ..." A. Newell and his colleagues have created a conceptual structure and software system for imitating intelligence—SOAR.
- Some of the answers are rather anticognitive: "One problem with using human intelligence as a basis for AI is the tendency to confuse intelligence and cognition" [118, p. 23]. The term "cognition" infers a claim of anthropomorphism. The same authors attempt to clarify their view by proposing a definition of intelligent behavior: "Three basic principles of intelligent behavior are that behavior requires a body, that behavior is intelligent only by virtue of its effect on the environment, and that intelligent behavior requires judgment" [118, p. 24]. The confusion is not eliminated even after a more formal definition is given: In defining intelligent behavior, what matters is the behavioral outcome, not the nature of the mechanism by which the outcome is achieved. In particular, intelligent behavior does not necessarily involve cognition" [118, p. 282].

Some of the authors are inclined to judge the intelligence of a system in a way similar to that of a Turing test: If you can fool the experimenter—you are intelligent. And how sharply this contradicts the statement of A. Newell, one of the most prominent cognitivists of our time: "a system's behavior can be predicted based only on the *content* of its representations plus its knowledge of its goals—[this] is the essence of being an intelligent system" [119]. In this book we follow, in general, the cognitivist approach.

AVOIDING THE NEED TO DEFINE

Answers attempting to avoid the problem Consider the answer: "Intelligence is in the eye of the beholder;" in other words, "if one wants to call something 'intelligent' why not?" This attitude appears a lot and is a very tempting one to adopt because one can eliminate conflict by declaring that a problem does not exist. This attitude has led to the abuse of the term commercially, such as when we refer to the 'intelligent screwdriver' or an 'intelligent vacuum-cleaner.' To be sure, language is used loosely in the advertising world to attract the consumer, but we are seeking to find whether such a phenomenon as "intelligent systems" really exists and can be properly defined.

Unusual answers Some other definitions have promise. For example: "if the system can do symbol grounding, it is intelligent." The symbol-grounding problem is a recent development in AI: "The problem of attaching meaning (with respect to the creature) to the symbols it employs is often called the problem of *symbol grounding*" [120, p. 130]. The general understanding is that a meaningful manipulation of knowledge requires defining a totality of intimate relationships between symbolic systems and knowledge bases. Other treatments of the symbol-grounding problem see it as a way toward uncovering some of the still hidden aspects of cognitive processes [121]. The increasing gravitation of a discourse on symbol-grounding intelligence has been toward a semiotic paradigm. Later in this book, we will demonstrate that the symbol-grounding problem is always at the core of the semiotic approach.

Another unusual answer dwells on the fact that anything understood and formalized as an algorithm may not be due to intelligence; it may be due to routine. This has led to a circular definition of intelligence: "A system is intelligent if it is more intelligent than what can be considered intelligent today." We think that this insight may be helpful in searching for a more adequate definition.

1.4 INTELLIGENCE: PRODUCT AND TOOL OF BEHAVIOR AND COMMUNICATION

1.4.1 Advantageous Behavior

The brain is first and foremost a control system. Its primary function is to produce successful goal-seeking behavior in finding food, avoiding danger, competing for territory, attracting sexual partners, and caring for offspring. All brains, even those of the smallest insects, generate and control behavior. Some brains produce only simple forms of behavior, while others produce very complex behaviors. Only the most recent (on the scale of evolution) and highly developed brains show any evidence of abstract thought.

Assertion In every intelligent system (natural or constructed), intelligence is the mechanism that can generate the most advantageous behavior.

CHOOSING BETWEEN ALTERNATIVES

Intelligence improves an individual's ability to act effectively and choose wisely between alternative behaviors. A more intelligent animal has many advantages over less intelligent rivals in acquiring choice territory, gaining access to food, and attracting more desirable mates. The intelligent use of aggression improves an individual's position in the social world. Intelligent predation improves success in capturing prey. Intelligent exploration improves success in hunting and establishing territory. Intelligent use of stealth gives a predator the advantage of surprise. Intelligent use of deception improves the prey's chances of escaping from danger.

Higher levels of intelligence allow the system to imagine events and choose the best one reason about the probable results of alternative actions. These abilities give the more intelligent individual a competitive advantage over the less intelligent in the competition for survival and gene propagation. Intellectual capacities and behavioral skills that produce successful hunting and gathering of food, acquisition and defense of territory, avoidance and escape from danger, and bearing and raising of offspring tend to be passed onto succeeding generations. Intellectual capabilities that produce less successful behaviors reduce the survival of the brains that generate them. Competition among individuals becomes competition among intelligences, and this drives the evolution of intelligence within a species.

MULTIAGENT INTELLIGENCE

Assertion For groups of individuals (agents), intelligence provides a group mechanism (multiagent structure) for cooperatively generating advantageous behavior.

The intellectual capacity to simply congregate into flocks, herds, schools, and packs increases the number of sensors watching for danger. The ability to communicate danger signals improves the survival probability of all individuals in the group. Communication is most advantageous to the quickest individuals, who recognize danger messages, and effectively respond. The intelligence to cooperate in mutually beneficial activities, such as hunting and group defense, increases the probability of gene propagation for all members of the group.

The most intelligent individuals and groups within a species will tend to occupy the best territory, be the most successful in social competition, and have the best chances of their off-springs' survival. More intelligent individuals and groups will dominate in serious competition.

Biological intelligence is the product of continual competitive struggle for survival and gene propagation which has taken place among billions of the carriers of intelligence (brains?) over millions of years. The results of those struggles have been determined in large measure by the intelligence of the competitors.

The intelligence of constructed systems should be designed so that similar properties can be achieved by *intelligent systems*.

1.4.2 Efficient Symbolic Representation

The concept of "relation" between two individual entities (objects, agents, systems, etc.) is generally understood and the consequences of relation are usually embodied into such phenomena as "gravitation" that entails the relation of adjacency for two masses. Communication is a similar phenomenon that emerges to provide for realization of the relation existing between two agents. (In the case of two masses the relation is "communicated" by the field of gravitation.)

The ability to transform "reality" into a "representation of reality" can be considered to be one of the most important phenomena linked with intelligence. This representation of reality is done in signs, and it is necessary for storage, communication, and behavior generation. The signs contain data, information, and knowledge which reflect different levels of representation.

DEFINITION OF COMMUNICATION

Communication is the transmission of symbolically represented data, information, and knowledge among intelligent systems, or among subsystems of an intelligent system.

DEFINITION OF LANGUAGE

Language is the symbolic means by which data, information, and knowledge are encoded for the purposes of communication and storage.

Language has three basic components: vocabulary, syntax, and semantics. Vocabulary is the set of words in the language. Words may be represented by symbols. Syntax, or grammar, is the set of rules for generating strings of symbols that form sentences. Semantics is the correspondence among the language constructs and the "fields of meaning." It allows for encoding of information into meaningful patterns, or messages. Messages are sentences that convey useful information.

Communication requires that information be (1) encoded, (2) transmitted, (3) received, (4) decoded, (5) interpreted, and (6) understood. Understanding implies that the information in the message has been decoded correctly, incorporated into the world model of the receiver, and problems in it are explicated. Understanding of problems presumes subsequent search for their solution and generation of the required actions. Each stage contributes to meaning extraction.

Communication and language are not unique to human beings. Virtually all creatures, even insects, communicate in some way, and hence have some form of language. For example, many insects transmit messages announcing their identity and position. This may be done acoustically, by smell, or by some visually detectable display. The goal may be to attract a mate or to facilitate recognition and/or location by other members of a group. Species of lower intelligence, such as insects, have very little information to communicate, and hence have languages with only a few of what might be called words, with little or no grammar. In many cases language vocabularies include motions and gestures (i.e., body or sign language) as well as acoustic signals generated by variety of mechanisms from stamping feet, to snorting, squealing, chirping, crying, and shouting.

BALANCE BETWEEN CONTENT AND COMPLEXITY

Theorem In any intelligent system (living creatures, or constructed devices), language evolves or is being designed to provide for the balance between the adequate content and the complexity of messages that can be generated by the intelligence of that system.

Depending on its complexity, a language may be capable of communicating many sophisticated messages, or only a few simple ones. More intelligent individuals have a larger vocabulary and more complex syntax, and they are quicker to understand and act on the meaning of messages.

FROM MESSAGE TO INTERPRETATION TO GENE PROPAGATION

Hypothesis To the receiver, the benefit, or value, of communication is roughly proportional to the product of the amount of information contained in the message, multiplied by the ability of the receiver to understand and act on that information, multiplied by the importance of the act to survival and gene propagation of the receiver. This statement can be extended by a further hypothesis that the increment of the value per unit of information (a sign, or a label) can serve as a measure of increase in knowledge and/or action resulting in a better state for success of functioning.

Similar evaluations can be made for many other parameters of a situation by using similar premises. For example, to the sender, the benefit is the value of the receiver's action minus the danger incurred by transmitting a message that may be intercepted by, and give advantage to, an enemy.

Greater intelligence enhances both the individual's (agent's) and the group's ability to analyze the environment, to encode and transmit information about it, to detect messages, to recognize their significance, and to act effectively on information received.

Greater intelligence produces more complex languages capable of expressing more information, that is, more messages with more shades of meaning.

In social species, communication also provides the basis for societal organization. Communication of threats that warn of aggression can help to establish the social dominance hierarchy and reduce the incidence of physical harm from fights over food, territory, and sexual partners. Communication of alarm signals indicates the presence of danger and, in some cases, identifies its type and location. Communication of pleas for help enables group members to solicit assistance from one another. Communication between members of a hunting pack enables them to remain in formation while spread far apart, and hence to hunt more effectively by cooperating as a team in the tracking and killing of prey.

Among humans, primitive forms of communication include facial expressions, cries, gestures, body language, and pantomime. The human brain is, however, capable of generating ideas of much greater complexity and subtlety than can be expressed through cries and gestures. To transmit messages commensurate with the complexity of human thought, human languages have evolved with grammatical and semantic rules capable of stringing words from vocabularies consisting of thousands of entries into sentences that express ideas and concepts with exquisitely subtle nuances of meaning. To support this process, the human vocal apparatus has evolved complex mechanisms for making a large variety of sounds.

1.4.3 Elementary Loop of Functioning

In this book we will demonstrate that the faculty of intelligence in systems is created by a definite architecture that organizes joint functioning of otherwise nonintelligent devices. This architecture processes signs and symbols and is characterized by several important properties that include:

LOOP OF CLOSURE

- functional closure
- building up representation of the environment
- algorithms of learning via generalization that produce multiresolutional properties in the system
- algorithms of self-referencing.

We will address all these devices in the subsequent chapters. Here, we will introduce the elementary functioning loop that organizes all elements of intelligence so as to create the functional relationships and information flows shown in Figure 1.1. In all intelligent systems, a sensory processing sub-system processes arriving (from sensors) information to acquire and maintain an internal model (representation) of the external world. In all systems, a behavior generating sub-system exists that decides upon the course of actions to be taken for achieving the goal. The behavior generation sub-system controls actuators to pursue behavioral goals in the context of the perceived world model.

In systems of higher intelligence, the behavior generating sub-system may interact with the world model and value judgment sub-systems to reason about space and time, geometry and dynamics, and to formulate or select plans based on values such as cost, risk, utility, and goal priorities. The sensory processing sub-system may interact with the world model and value judgment sub-systems to assign values to perceived entities, events, and situations, and enable the other sub-systems to determine preferences and make choices.

The architecture combining these components is shown in Figure 1.1. This is an architecture that can be considered a general arrangement for joint functioning of a set of sub-systems, so that any system consisting of them can exist, perform functions, and perpetuate its own existence. This architecture describes a simple loop of activities, and is called *elementary loop of functioning* (ELF). Other versions of ELF are presented in Figures 1.3 and 2.12.

Because of this orientation to survival subsistence we will call this loop a *generalized subsistence machine* (GSM). Later we will consider simplified versions (like ELF in Chapter 2) and more complicated multiresolutional versions (Chapter 4). The multiresolutional version of ELF appears in this chapter in Figure 1.2. However, in all cases, its essence remains determined by the functional capabilities that contain elements of automatism.

1.5 EVOLUTION OF AUTOMATISMS

1.5.1 Learning Automaton

Patterned sets of rules, or schemata patterns, as a tool of representing systems can be put in the framework of automata theory (see Chapter 2 for details). This theory requires that input and output languages be defined. Then the concept of "state" is introduced that describes the overall internal and external situation in more detail. Note that the automata theory does not contemplate such realistic things as the phenomena of the "outside world." The latter arrives to an automaton as a set of messages encoded in the input language. However, if one tries to apply this theory to the central nervous system (CNS), the arriving symbols and/or codes should be equipped with interpretations that could be understood by CNS. Input messages are delivered through sensors. State is a set of variables serving as a representation of both the inside and outside worlds for the automaton. After this, the transition and output functions can be determined as a product of the functioning of the CNS. Certainly our actions are the output statements.

MULTIPLE CLOSURES WITHIN CNS

Given the abundance of literature in the area of automatic control and control theory, the percentage of automata theory papers in this information stream seems to be surprisingly small. Even more scarce is literature on learning automata [122, 123]. And there is almost no literature dedicated to learning control of hierarchical systems similar to the CNS. The semiotic sources put together the areas of neurobiology or neuropsychology with the areas of automatic learning control and learning control theory.

Learning for acting is the leitmotif of symbols formation. The evolutionary development of the higher nervous system is equivalent to the development of learning for action—this is what reflexes are all about. Reflex theory was easily accepted because of its conceptual simplicity. However, when reflexes are clustered and the clusters become entities, in which learning for action works at lower resolution, the conceptual structure requires from the reader a bit of insight and imagination. This is why the concept of *automatism* was introduced; it has become a fundamental component of the hierarchy of nested loops in the CNS [124], and a basis for the subsequent interpretation of general phenomena of the architectures of intelligence.

1.5.2 Concept of Automatism and How It Can Be Learned

The term "automatism" is rare in books on neurophysiology or neuropsychology. In dictionaries, it is interpreted as follows: the state or quality of being automatic; automatic mechanical action; the theory that the body is a machine whose functions are accompanied but not controlled by consciousness; the involuntary functioning of an organ or other body structure that is not under conscious control, such as the beating of the heart or the dilation of the pupil of the eye; the reflexive action of a body part; mechanical, seemingly aimless behavior characteristic of various mental disorders. (Compare these definitions with the German interpretation: *Unwillkürliche Funktionsabläufe.*)

Practical applications of this term are frequent in engineering. It is not unusual in medical science too. For example, *automatism* is used to characterize aimless and apparently undirected behavior that is not under conscious control and is performed without conscious knowledge: seen in psychomotor epilepsy, catatonic schizophrenia, psychogenic fugue, and other conditions. A phenomenon is known called *traumatic automatism* which is characterized by performing complicated motions (e.g. playing football) totally automatically. Temporal lobe epilepsies are often characterized by automatisms such as the repetitive opening and closing of a door.

The term "automatism" was introduced to demonstrate the commonality of automatic control processes at different levels of motion control in the CNS. The idea is that all hierarchical levels of the nervous system are built according to the same functional principles, and each level is a learning system that generates its own automatism.

STEREOTYPED RESPONSES

The tradition in the behavioral sciences in the United States is to talk about stereotyped responses, in fact, about automatisms. These are usually separated into four categories:

1. Unorganized or poorly organized responses.
2. Reflex movements of a particular part of an organism as a result of the existence of a reflex arc (an open loop pair of receptor and affector is usually mentioned).
3. Reflexlike activity of an entire organism.
4. Instinct.

"Reflexlike activities" is a label for complicated phenomena of some automatisms. Reflexlike activities of entire organisms may be unoriented or oriented. Unoriented responses include kineses—undirected speeding or slowing of the rate of locomotion or frequency of change from rest to movement (orthokinesis) or of frequency or amount of turning of the whole animal (klinokinesis). Oriented reflex activities include tropisms, taxis, and orientations at an angle. Their mechanisms are rarely discussed. The tradition of behaviorism is to talk about them as "reflex actions of entire organisms." Their analysis and classification is based on rather external features. For example, tropotaxis is orientation based upon some need, while telotaxis is an orientation toward light.

Instinctive behavior is considered to be a hereditary property of an organism (cleaning, grooming, playing dead, taking flight, etc.) Sometimes, this behavior is considered "unlearned" but it is not so. It is rather learned by previous generations. Behaviorists analyze the complexity of patterns of instinctive behavior, their adaptivity, stability, and the like. We are interested in discovering their inner mechanisms. All instinctive behavior can be unified under the title of "automatisms," and this demonstrates that they have similar control architectures. However, they belong to different levels of granularity in the CNS. Stereotyped responses of all types can emerge if a particular mechanism of learning is assumed that leads to formation of the Elementary Loop of Functioning (like the one shown in Figure 1.1). Then, the gigantic body of information consisting of seemingly unrelated units appears to be well organized and explainable.

DEFINITION OF AUTOMATISM

Automatism is the ability of ELF to generate strings of elementary symbols (strings of information, of actions, of behavioral units) following the stored output commands in response to the prearranged stimulating inputs; existence of automatisms can be associated with an intrinsic lookup table of action rules.

1.5.3 From Reflexes and Rules to Programs

Learning in its elementary form is a development of automatisms. As new environments persist and new experiences perpetuate, new rules emerge as a result of generalization upon similar repetitive hypotheses. Learning on a large scale, as an evolution of the nervous system of the particular species, can be considered an evolution of the automatisms that compose this nervous system. (More on learning can be found in Chapter 10.)

The process of experiences acquisition and their transformation into rules is later demonstrated in Chapter 2 (Figure 2.14) where a learning automaton is presented (and experiences a process equivalent to semiosis). The system functions as follows:

ALGORITHM OF EXPERIENCE ACQUISITION

Step 1. Experiences are recorded in the form of associations that incorporate four components: state, action, change in the state, and value. The system of storage allows for clustering these associations by similarity in each of the elements of the association.

Step 2. The clusters are generalized in the form of hypotheses of future rules.

Step 3. Elements of the rules are organized in the form of concepts.

Step 4. As additional experiences confirm the hypotheses, the latter are assigned a higher value of preference.

Step 5. As the number of rules increases, the same procedure of clustering is applied to them. Meta-rules are obtained; meta-concepts are extracted from them.

Once the process of learning starts, it cannot stop, since new rules generate new experiences. The new experiences give basis for new generalizations, and then there emerge new levels of generality (or levels of abstraction, or granularity, or resolution, or scale—all of these accompany each other). As soon as a new level of generality (granularity) emerges, the learning process starts again at the new level. Each level of granularity has its vocabulary words related to the automatism of this level.

Step 6. Numerous repetition of steps 1 through 5 lead to the formation of hierarchies of rules, meta-rules, and concepts.

Algorithms of learning are equivalent to the one described Chapter 2 (Section 2.9.4, Figure 2.14) have been tested in a variety of machine learning systems. Various types of hardware can be used for implementing this algorithm of learning. If the hardware of the system is our nervous system, then instead of the hierarchy of rules, we obtain a hierarchy of automatisms that allows for control of the system. This hierarchical learning results in building up a control hierarchy. As a result of this development, the ELF is transformed into the multiresolutional loop shown in Figure 1.2 (see also Figure 2.4). This diagram differs from that shown in Figure 1.1b in that the module of behavior generation is substituted by its simplest form: a table that contains all available commands.

The ability to generate beneficial commands is rooted within the system of knowledge representation (see more in Chapter 3). The efficient system of knowledge representation is a multiresolutional one. There is little doubt that knowledge representation in a human brain has a multiresolutional architecture. The following observations sug-

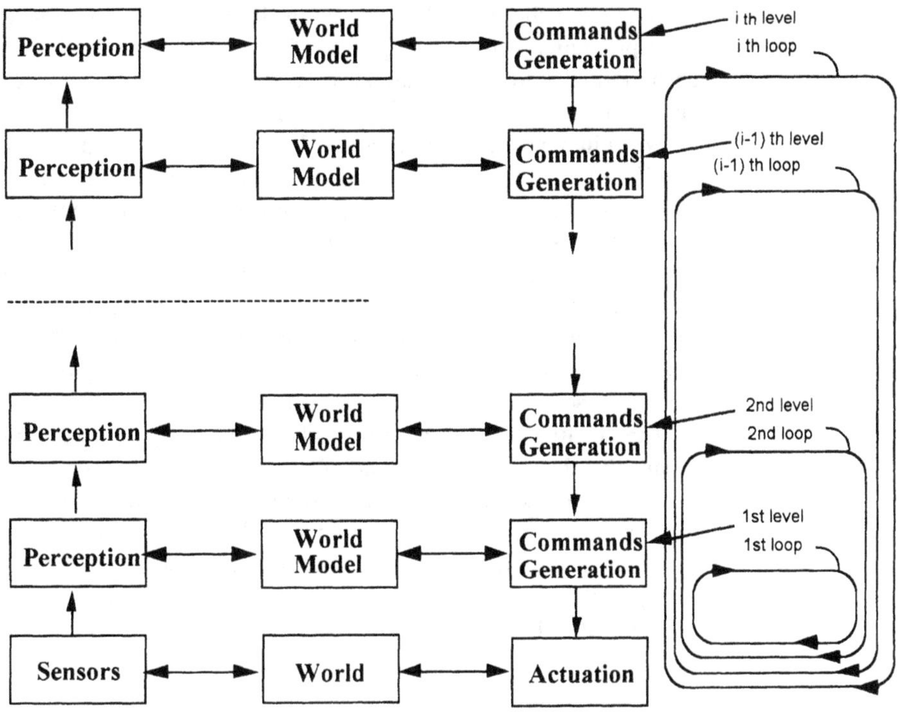

Figure 1.2. Hierarchy of automatisms.

gest that the functioning brain distributes its processing among several levels that have different resolution of time and space:

HIERARCHY OF AUTOMATISMS

- Duration of the neuron spike is from 1 to 10 milliseconds.
- Lateral inhibition as an operation of the neural network takes 10 milliseconds.
- Simple perceptual acts such as pattern classification can take tens of milliseconds (40–60 ms)[125] but no more than 100 milliseconds [126]. Note that the stimulus itself can take 50 milliseconds.
- Physical acts start with eye movement (300 ms) and can be as long as 1 to 5 seconds, for example, articulating a word, or raising a hand.
- More complicated, though elementary, tasks may take from 5 seconds (dialing phone number) to 30 seconds (eating a hamburger, looking up a number in the phone book).

This list [127] corresponds well to J. Albus's conjecture of the approximate ratio 10 between resolution values of two adjacent resolution levels of knowledge representation system [44]. The conceptual scheme of the neuronal components is presented in Chapter 4 (Section 4.5). A variety of neural networks applied in constructed intelligent systems is described in Chapter 10.

1.5.4 From Programs to Self-Organization

In this book we will try to come up with a system of theoretical views and with the corresponding architecture that allows for minimum programming effort and is capable of generating the desired programs in the most efficient manner, or learning the desired programs of functioning from its own experience. The faculty of learning gives an opportunity to incorporate all required knowledge from direct interaction with the environment.

We will investigate in Chapter 2 the fundamental principle that determines the specifics of information processing in the living and intelligent systems. There the principle is described as a triplet of operators GFS that work together as a tool of "meaning summarization," or "generalization," of the content of information messages. This principle is embodied in simultaneously performed focusing of attention, grouping, and combinatorial search. In fact, this principle embodies techniques that are ubiquitous in the activities that we attribute to "intelligence." The GFS is designed to reach the desired endpoint with a minimum of computational complexity. A consequence of this is the emergence of the mechanisms of self-organization.

The faculty of self-organization advances the capabilities of intelligent systems. This means that the system is capable of synthesizing mechanisms that can improve its control structures according to the recognized needs and subtleties of its existence. The core of such architecture is described as NIST-RCS architecture (see Chapter 4). We will demonstrate the emergence and the role of advanced features, including planning (Chapter 9) and learning (Chapter 10), gradually developing into self-organization that is a subject of future research (Chapter 12).

1.6 FROM AGENT TO MULTISCALE COMMUNITIES OF AGENTS

The concept of "agent" turned out to be very fruitful in the area of intelligent systems. It allows professionals in various fields to identify their ongoing or attempted intellectual activities with a short and elegant set of ideas that has been suggested in Figures 1.1 and 1.2.

1.6.1 The Concept of "Agents" and Its Place in the State of the Art

We will demonstrate a scope from simple agents through autonomous agents to intelligent agents. We will discuss both computational agents and realistically existing agents. It is our intention to arrive at the architectures of intelligence constructed as interrelated communities of agents. The term *agents* is ubiquitous in the literature on large and complex systems, intelligent systems, robotics, and software engineering. Yet it is rarely defined, and its existing interpretations are very vague. At the present time, we have three salient groups of agents reflected in the literature.

Simple Rational Agents

ALWAYS CERTAIN WHAT TO DO!

Agents were formally introduced in the discipline of artificial intelligence [128]. The concept of "rationality" is associated with the existing laws of thinking, e.g., laws of logic. "A rational agent is the one that does the right thing" [128, p. 31]. This alludes to the concept of success, and the latter invokes the idea of performance measure.

Agent is a system that is immersed into some environment, has "percepts" as its inputs, and actions as its outputs. Percepts enter through sensors; actions are produced by "effectors." So far it looks pretty similar to Figure 1.1. (Of course, effectors are called actuators.) For convenience, the diagram is redrawn in Figure 1.3*a*. One can see that

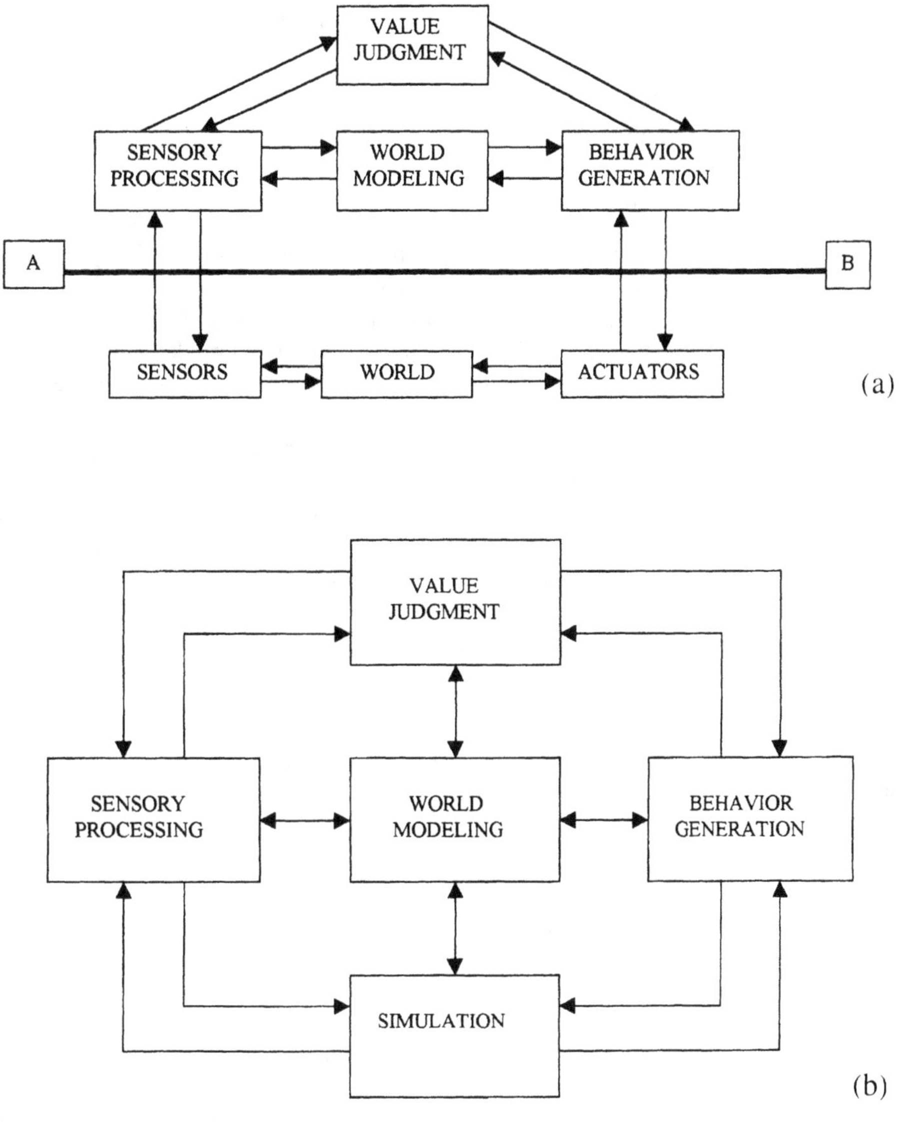

Figure 1.3. Elementary functioning loop (an agent): *(a)* an agent living in reality; *(b)* a self-containing agent, living in its imagination.

the model presumes something more than just mapping *sequences of input percepts* into *sequences of output actions*. Such a mapping would be no different from a normal automaton that remembers its transition and output functions and comes up with anticipated (actually, preprogrammed) behavior. The device in Figure 1.3*a* is expected to provide for further development of the world model and improve the processes of behavior generation as a result of new information arriving from sensory processing. Also, the module of "Value Judgment" is added (see more in Chapter 5). Let us outline our options of doing these things (further development of the world model and improvement of the processes of behavior generation).

Russell and Norvig [128] distinguish the following types of agents:

Simple reflex agents
Agents that keep track of the world
Goal-based agents
Utility-based agents

Simple reflex agents are presented in Figure 1.2. They are the "automatism-agents," that is, their function does not exceed the one expected from the automaton (see Chapter 2). Even simple tracking of the world would mean changing the world model. Following a goal is ingrained into the automata model. However, using the concept of utility raises the opportunity to choose among various available outputs bringing different utility.

Rational agents are like automated machines or pre-programmed robots. They can be equipped by simple devices for choosing one out of a few preplanned solutions. They can have rudimentary world representation and simple learning capabilities (like recognition of single repeated moves or their strings). The intelligence of rational agents is limited, although they have their field of application. In many areas, rational agents are called *automated devices* or *automated systems*.

However, the capabilities of agents can be increased beyond the simple automatisms.

Autonomous Agents

EXPERIENCES IMPLY DECISIONS

Automated systems work with no human operator involvement, i.e., although they are autonomous, this autonomy is limited since it is known in advance what and when will happen if the circumstances of the environment change. The term *autonomous* is used more justifiably when the upcoming situations are not prescheduled and the agent has some freedom to decide what, when, and how to act. There is a broad variety of situations with autonomous functioning of agents, starting with the case of limited and known input vocabulary and ending with total unawareness of the situations that the system might encounter.

The property of autonomy is loosely determined by Russell and Norvig in 1995: they considered the system autonomous if "its behavior is determined by its own experience" [128, p. 35]. This definitely presumed having been equipped with the subsystems for unsupervised learning and alluded to a potentially very high level of sophistication. Three years later, the definition of "autonomous agents" was introduced, and they were characterized somewhat more in-depth by the organizers of the Second International Conference on Autonomous Agents in 1998 sponsored by AAAI, ACM SigArt and Microsoft Research:

> Autonomous agents are computer systems that are capable of independent action in dynamic, unpredictable environments.

It is not clear from this definition whether any goal is being pursued by these agents, whether their actions were "the best" in any sense, or whether they were improving in time.

> Agents are also one of the most important and exciting areas of research and development in computer science today. Agents are currently being applied in domains as diverse as computer games and interactive cinema, information retrieval and filtering, user interface design, and industrial process control.

Clearly, all agents mentioned above are meant to be understood as *software agents* [129]. Software agents are computational units shown in Figure 1.3*b* (including, or not including, the subsystem of simulation which substitutes functioning in the real physical world meant in Figure 1.3*a*). The best characterization of computational agents was given by M. Minsky: "...each mind is made of many smaller processes. These we'll call *agents*. Each mental agent by itself can only do some simple thing that needs no mind or thought at all" [130]. This is an expression of the belief that at the highest level of resolution the agents do not need to have faculties of learning and self-organization. This belief in the intelligence that somehow emerges from the multiplicity of mindless elements is very frequent. Is this belief right or wrong? We will try to figure it out.

Software agents that can change their location in the system are called *mobile agents*. These are slightly different since they are, kind of, Internet-related.

MOBILE SOFTWARE UNITS

Mobile agents are agents that can physically travel across a network, and perform tasks on machines that provide agent-hosting capability. This allows processes to migrate from computer to computer, for processes to split into multiple instances that execute on different machines, and to return to their point of origin. Unlike remote procedure calls, where a process invokes procedures of a remote host, process migration allows executable code to travel and interact with databases, file systems, information services, and other agents.

Mobile agents have been the focus of much speculation and hype in recent years. The appeal of mobile agents is quite alluring—mobile agents roaming the Internet could search for information, find us great deals on goods and services, and interact with other agents that also roam networks (and meet in a gathering place) or remain bound to a particular machine.

Significant research and development in mobile agency has been conducted in recent years, and there are many mobile agent architectures available today. However, mobile agency has not become a sweeping force of change, and now faces competition in the form of message passing and remote procedure call (RPC) technologies [130].

Mobile agents can be illustrated by the system of views from the tutorial "Mobile Software Agents for Dynamic Routing" [131]. As portable digital devices of all kinds proliferate, wireless networks that allow for flexible, timely, and efficient data communication become more and more important. Networks for mobile devices are quite difficult to design for several reasons, chief among them the problem of routing packets across networks characterized by constantly changing topology. In this chapter we describe ways to address the routing problem using a new technique for distributed programming, mobile software agents.

Mobile agents are actively promulgated to make the contemporary manufacturing systems especially CORBA-oriented [132]. (For more on mobile agents see [133].)

Individual Agents, Communities of Agents, and Societies of Agents

SWARMS OF SWARMS OF SWARMS...

We will introduce a fundamental concept that is the core in the whole issue of intelligent systems: groups of agents behave as agents too. However, the newly emerged agent does not belong to the same category as the agents that have generated it. We can consider the new agent as *generalized agent*. Moreover, groups of groups of agents can also be lumped into an even more generalized agent, and this agent will be part of a new category of agents.

Each result of generalization gives a new scale of world representation (or new granulation, or new resolution). We will illustrate this process of the gradual lumping of agents into generalized agents in Figure 1.4. Each oval represents an agent. Connections between the agents demonstrate their relationships (in this particular case, there

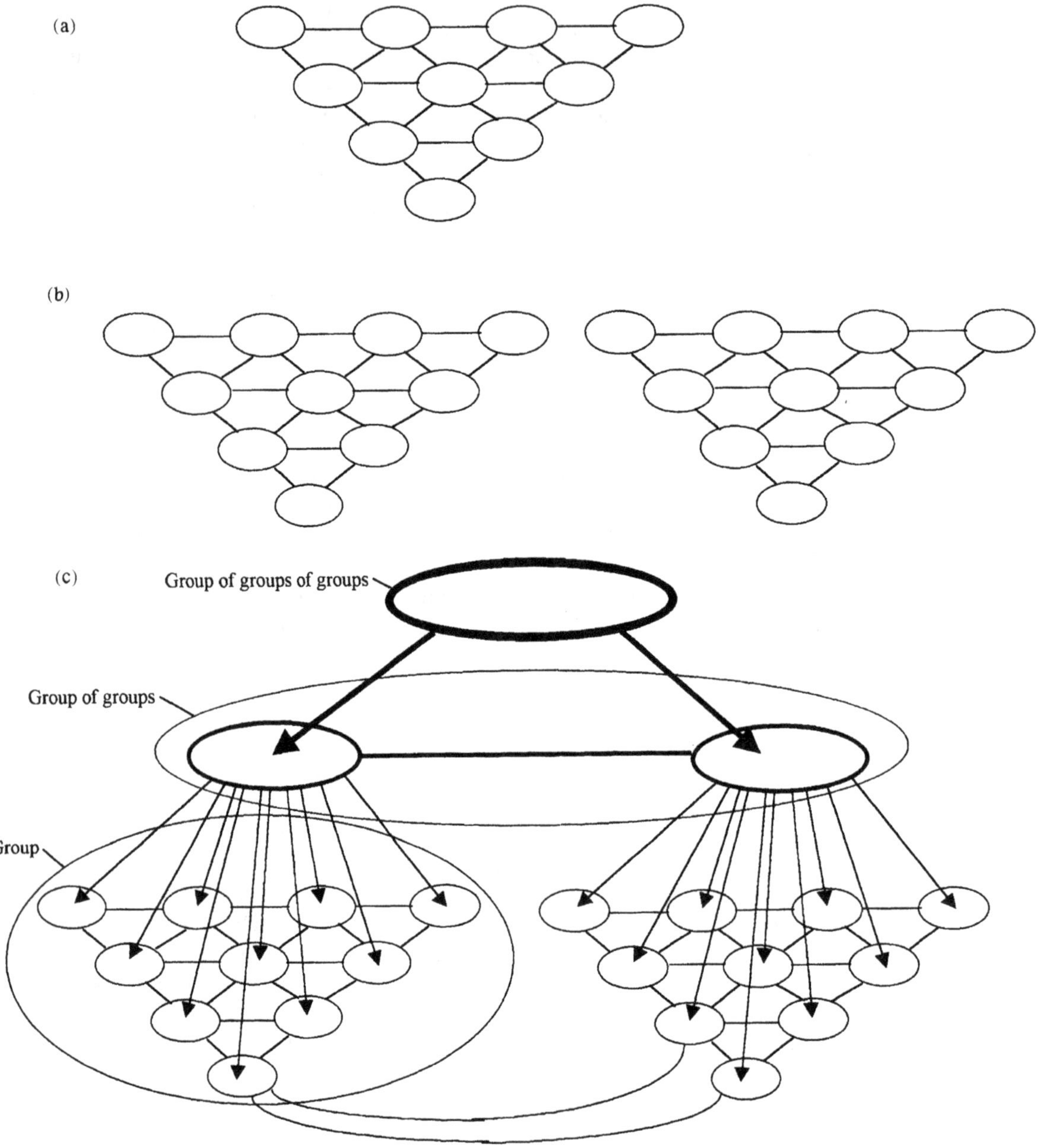

Figure 1.4. Grouping the agents: a single ellipse is an agent at a level of resolution. *(a)* A formation of agents (a group); *(b)* two similar formations; *(c)* three levels of grouping.

are relationships of adjacency and cooperation). Then the set of agents in Figure 1.4*a* can be interpreted as a formation (like formation of flying birds, or formation of traveling vehicles).

This formation is considered to be a single agent at the lower level of resolution. In Figure 1.4*b*, two triangular formations are shown, which we can discuss without focusing our attention on single agents within the formation. We consider this formation as an agent itself. However, this will be a different agent, a generalized agent; it does not belong to the "world of agents" represented by small ovals. It is a common practice to discuss movement of formations of vehicles. The object of attention in the world of formations is a generalized one. (It will become clear later that in the world of generalized objects, the scale is different, the resolution is lower, this is a coarse grain world.)

In Figure 1.4*c* three levels of resolution (scale, granulation) are demonstrated. At the highest level of resolution all agents are distinguishable, discernible as well as all relationships among them. Notice that relationships can exist between all agents of the formation 1 and all agents of the formation 2. (Only two connections are shown in the figure; the reader will imagine easily the rest of connection—it should be a busy image.)

At the level of lower resolution we are interested in the relations between "group of groups." One relation represents all interwoven relationships between single agents that are shown in the imaginary busy picture. At the level of lowest resolution we have one agent (the group of groups of groups) that embodies the set of formations under consideration. So we have an example that can be interpreted as merger of multiple agents into groups (one level), groups into communities (second level), and communities into a society (third level). In such a multiscale hierarchy of the communities of agents, each next grouping makes the discussion simpler but the resolution gets lower.

In conclusion, we would like to present a metaphor that makes these concepts more understandable—the metaphor of bees, merging into swarms, swarms gathering into swarms of swarms. This is how the organism of an animal can be visualized in the light of this metaphor (from J. Hoffmeyer's book *Signs of Meaning in the Universe*):

> One quadrillion of bacteria, in the form of ten trillion cells, collaborate on the job of being a human. Like an astronomical swarm of swarms all of these cells stream together through one single brain-body as it makes its way along the path of life toward all those unknown futures that will eventually become just one single life story [135].

1.7 COGNITIVE AGENTS AND ARCHITECTURES

A multiplicity of research groups conducting research in the area of intelligent systems do not have any unified and/or theoretically consistent view about their architectures. An architecture is a specific formation of agents. Since each agent always consists of agents of higher resolution, it always has its own architecture. Following A. Newell, they call their efforts "cognitive architectures" but interpret this term generally as the portion of a system that provides and manages the primitive resources of an agent. For many cognitive architectures, these resources define the information medium and physical realization upon which the symbolic control system is realized.

In research publications, the researchers address the issues related to the choice, definition, extent, and limits of these resources (medium and realization) and present computer based solutions concerning their management. No source discusses necessary, sufficient, and optimal distribution of resources for the development of agents exhibiting intelligence-like manifestations. It is true that intelligent systems have been in the focus of attention of scientific community for a long time. Nevertheless, the concept of *intelligent system* is not fully understood, and it affects interpretation of the existing research results as well as the choice of new research directions.

LET'S MIMIC HUMAN COGNITION

"Cognitive agents and architectures" have different features that lead to different properties. Some of the results utilize a uniform knowledge representation, some employ a heterogeneous representation, and others, attempt to get away with no explicit representation at all. The choice of features is sometimes driven by explicit methodological assumptions. Some researchers deny any need in methodologies and proclaim a "toolbox" policy. A number of solutions is driven by the environments in which the architecture will be used.

Many researchers purposely ignore the conceptual frameworks related to human cognition. Some of them are interested in developing "agents" which populate and behave effectively in a particular environment. In this case, the interactions between the architecture and the environment can be of interest. Apparently, solutions for a static, problem-solving situation should differ from a solution for a highly dynamic environment. They admit that the term cognitive architecture can be a little misleading.

The following architectures are known from the literature:

Subsumption Architecture
ATLANTIS
Theo
Prodigy
ICARUS
Adaptive Intelligent Systems (AIS)
A Meta-reasoning Architecture for 'X' (MAX)
Homer
SOAR
NIST-RCS
Teton
RALPH-MEA
Entropy Reduction Engine

The lack of unity in the theoretical approaches and the divergence in the existing results are necessary parts of a research process, and they should be praised rather than criticized. However they should be properly understood in the overall knowledge base of the scientific process because one of the important parts of this process is teaching. Therefore, once in a while we should reconsider the results of the not so remote past and make an attempt to properly categorize the milestones of this not very remote past. The misconceptions which can be frequently found in the area of intelligent systems are almost always based upon the provisional *myths*, or provisional *paradigms*. Researchers create these myths, are frequently driven by them, and place within them the contexts of their future theories. We will list here some of the myths generated in the area of cognitive architectures.

WHAT DO WE GET FROM NN?

Myth 1. If all subsystems are built of neural networks the system won't have a choice: it will become intelligent sooner or later. Maybe not immediately—after some learning, maybe not with a small number of neural networks—with a large number. But it will become intelligent! Unexpected surprise: equipping a system with neural networks can make this system intelligent.

This myth is a pervasive one. It is obvious that a neural network is just a tool. Frequently, it is a tool of finding the best approximation of a spatial property, and/or of a temporal process, for which a multiplicity of instantiations is known. Clearly, to make a good approximation is a right thing to do, but it is wrong to think that multiple capabilities for a good approximation makes up for an intelligence.

PREPLANNED REACTIVE BEHAVIOR

Myth 2. An activity of a very complex system which is driven by a hierarchical cognition-like architecture can be substituted by joint functioning of the myriad of the lowest level actuators, each equipped by a local "reactive intelligence," i.e. by "stimulus-response" rule of action. These "agents" are supposed to be given an opportunity to freely negotiate and discuss their local situations. (The expectation is that when the system relies upon the lowest level of rules "information→action" the symbol grounding predicament is always adequately resolved.) Subsequent misconception: if all functions of the system are divided among simple intelligent agents equipped with reactive/reflective rules, and each of them oriented toward a simple elementary problem, thus, generating an elementary behavior—then the overall system becomes intelligent.

So far, the only intelligent property demonstrated for this type of system design was the emerging property of flocking together of little mobile robots which are given a skill of "wandering aimlessly" and a number of reactive rules for dealing with the obstacles.

An opinion that *planning* and *reactive behavior* are different categories is an example of persistent misconception which affects a more serious group of researchers. Within the intelligent system, the World can be (and should be) represented at many scales (or at many resolutions). As soon as we realize that intelligent systems generate behavior at many levels of resolution simultaneously, we come to understand that any reactive behavior generated at a particular level, is a "plan" for the adjacent level of a higher resolution. Behavior is reactive if it reacts to the events observed in a situation. But plans can be reactive too. It seems that the picture is different if behavior is generated as a result of active anticipation (prediction) of the course of events. This means that the behavior driven by planning can be considered reactive, too. The only difference between the cases is that in the case with no planning we react to our current observations, while in the case with planning we react to our anticipations (predictions).

Plans become active, indeed, when we pursue the course of events by actively shaping the very event which is supposed to emerge. Very often it is called "feedforward control" (FFC), while reactive and even anticipatory compensation, is called "feedback control" (FBC) [6]. Using this terminology helps in eliminating some of the persistent misconceptions such as counterposing "planning" and "reactive control."

DISCOVERING FEEDBACK AND FEEDFORWARD

Some of the AI researchers have independently discovered the same concept of FFC and FBC which have been known in the control theory for decades. They have arrived at these ideas in a difficult way by discussing concepts of "situated actions" which presume to include "deliberative" and "reactive" actions as kinds of FFC and FBC incarnations [24].

RELYING ON MIRACLES OF EMERGENCE

Finally, hopes for miracles are very common: any research is linked with a hope for a miracle. An expectation of a miracle can become harmful if it is a sole motivating factor for the research. One of such expectations of a miracle is linked with a belief that "intelligence" is demonstrated when the solution emerges by itself out of communication among the mass of agents. "Emergence" has been observed for very large collections of non-linear components (re: chaos, etc.) An expectation appears that if many, many units are put together, each of them equipped with, say, a genetic search system—the overall system will be doomed to become intelligent!

One can agree that the model of multiple elements, interacting and genetically searching, is a very inspiring model. What if this model can produce "emergent" phenomena which allow for many powerful scientific results? Well, it seems to be as remote from explaining intelligent systems behavior generation, as prebiotic protein processes are remote from explaining the human brain.

ENIGMA OF REPRESENTATION

Another hope for miracle that can be frequently found in the literature, is a hope for an intelligent system *without representation*. This idea is motivated by a comparative analysis of so called "western" and "eastern" models of consciousness and thinking. Allegedly, the first is based upon discretization of the continuum into entities, while the second allows for "fluid," meditative processes which seem to be conducive of creativity.

We will address this issue in more detail later in our discussion on continuum and its discretization.

REFERENCES

[1] D. O. Hebb, The American revolution, *American Psychologist*, Vol. 15, 1960, pp. 735–745.

[2] D. E. Broadbent, *Perception and Communication*, Pergamon Press, New York, 1958.

[3] G. A. Miller, E. Galanter, and K. H. Pribram, *Plans, and the Structure of Behavior*, Holt, Rinehart, and Winston, New York, 1960.

[4] A. Koestler, *The Ghost in the Machine*, Hutchinson, London, 1967.

[5] H. Gardner, *The Mind's New Science: A History of the Cognitive Revolution*, Basic Books, New York, 1985.

[6] W. Ross Ashby, *Design for a Brain*, Chapman and Hall, London, 1952.

[7] O. Selfridge, Pattern recognition and modern computers, *Proceedings of the Joint Western Computer Conference*, IRE, 1955, pp. 91–93.

[8] A. Newell, J. C. Shaw, and H. A. Simon, Elements of a human problem-solving, *Psychological Review*, Vol. 65, 1958, pp. 151–166.

[9] A. L. Samuel, Some studies of machine learning: using the game of checkers, *IBM Journal of Research Development*, Vol. 3, 1959, pp. 211–229.

[10] J. Y. Lettvin, H. Maturana, W. S. McCulloch, and W. H. Pitts, What the frog's eye tells the frog's brain, *Proceedings of IRE*, Vol. 47, 1959, pp. 1940–1951.

[11] F. Rosenblatt, The perceptron: a probabilistic model for information storage and organization in the brain, *Psychological Review*, Vol. 65, 1958, pp. 386–408.

[12] F. H. George, *The Brain as a Computer*, Pergamon Press, Oxford, 1961.

[13] M. Minsky and S. Papert, *Perceptrons*, MIT Press, Cambridge, MA, 1969.

[14] S. Langer, *Philosophical Sketches*, John Hopkins' Press, Baltimore, 1962.

[15] A. Newell and J. C. Shaw, Programming the logic theory machine, *Proceedings of the Western Joint Computer Conference*, IRE, 1957, pp. 240–240.

[16] A. Newell and H. Simon, GPS: a program that simulates human thought, *Computers and Thought*, E. A. Feigenbaum and J. Feldman, eds., McGraw-Hill, New York, 1963, pp. 279–296.

[17] W. Reitman, *Cognition and Thought: An Information Processing Approach*, Wiley, New York, 1965.

[18] L. J. Fogel, A. J. Owens, and M. J. Walsh, *Artificial Intelligence Through Simulated Evolution*, Wiley, New York, 1966.

[19] J. C. Eccles, M. Ito, and J. Szentgothal, The cerebellum as a neuronal machine. University of Liverpool, 1969.

[20] N. Chomsky, *Language and Mind*, Hartcourt Brace Jovanovich, San Diego, 1968.

[21] G. A. Miller and N. Chomsky, Finatary models of language users, in *Handbook of Mathematical Psychology* (R. D. Luce, R. Bush, and E. Galanter, eds.), Vol. 2, Wiley, New York, 1963.

[22] G. Bateson, The logical categories of learning and communication, A position paper at the Conference on World Views sponsored by Wenner-Gren Foundation, August 2–11, 1968.

[23] L. von Bertalanffy, *Robots, Men and Minds*, Braziller, New York, 1967.

[24] J. von Uexkull, The theory of meaning (1940), *Semiotica*, Vol. 42, No. 1. 1982, pp. 25–82.

[25] E. deBono, *The Mechanism of Mind*, Simon and Schuster, New York, 1969.

[26] B. P. Zeigler and R. Weinberg, System theoretic analysis of models: computer simulation of a living cell, *J. of Theoretical Biology*, Vol. 29, 1970, pp. 35–56.

[27] B. P. Zeigler, Modeling and simulation: structure preserving relations for continuous and discrete time systems, *Proceedings of the Symposium on Computers and Automata*, Polytechnic Press, New York, 1971, pp. 561–589.

[28] K. Pribram, *Languages of the Brain: Experimental Paradoxes and Principles in Neuropsychology*, Prentice Hall, Englewood Cliffs, NJ, 1971.

[29] J. C. Eccles, *The Inhibitory Pathways of the Central Nervous System*, Charles C. Thomas, Springfield, IL, 1969.

[30] G. Bateson, *Steps to an Ecology of Mind*, Jacob Aronson, Northvale, NJ, 1972.

[31] M. Arbib, *The Metaphorical Brain*, Wiley, New York, 1972.

[32] M. Arbib, *Theories of Abstract Automata*, Prentice Hall, Englewood Cliffs, NJ, 1969.

[33] R. E. Kalman, P. L. Falb, and M. A. Arbib, *Topics in Mathematical Systems Theory*, McGraw-Hill, 1969.

[34] J. von Uexkull, *Bedeutungslehre*, Fisher, Berlin, 1940.

[35] A. Newell and H. Simon, Computer science as empirical inquiry: symbols and search, *Communications of ACM*, Vol. 19, March 1976, pp. 113–126.

[36] Z. Pylyshin, Complexity and the study of artificial and human intelligence (presented in 1974), published in *Philosophical Perspectives in Artificial Intelligence* (M. Ringle, ed.), Humanities Press, Atlantic Highlands, NJ, 1979.

[37] M. Mesarovich, D. Macko, and Y. Takahara, *Theory of Hierarchical Multilevel Systems*, Academic Press, New York, 1970.

[38] E. Jantsch, *Design for Evolution: Self-organization and Planning the Life of Human Systems*, Braziller, New York, 1975.

[39] F. G. Varela, H. R. Maturana, and R. Uribe, Autopoesis: the organization of living systems, its characterization and a model, *Bio-Systems*, Vol. 5, 1974, pp. 187–196.

[40] P. Shepard, *Thinking Animals*, Viking Press, New York, 1978.

[41] C. Furst, *Origins of the Mind, Mind-Brain Connections*, Prentice Hall, Englewood Cliffs, NJ, 1979.

[42] G. R. Taylor, *The Natural History of the Mind*, E. P. Dutton, New York, 1979.

[43] P. McCorduck, *Machines Who Think*, W. H. Freeman and Co., San Francisco, 1979.

[44] J. S. Albus, *Brains, Behavior, and Robotics*, BYTE Books, Peterborough, NH, 1981.

[45] J. S. Albus, A theory of cerebellar functions, *Mathematical Biosciences*, Vol. 10, 1971, pp. 25–61.

[46] J. S. Albus, A new approach to manipulator control: The cerebellar model articulation controller (CMAC), *Journal of Dynamic Systems, Measurement and Control*, 1975, pp. 220–227.

[47] J. Campbell, *Grammatical Man: Information, Entropy, Language, and Life*, Simon and Schuster, New York, 1982.

[48] P. N. Johnson-Laird, *Mental Models*, Harvard University Press, Cambridge, MA, 1983.

[49] S. P. Stich, *From Folk Psychology to Cognitive Science*, MIT Press, Cambridge, MA, 1983.

[50] E. Poppel, *Mindworks: Time and Conscious Experiences*, Hartcourt Brace Jovanovich, Boston, 1985.

[51] R. J. Sternberg (ed.), *Handbook of Human Intelligence*, Cambridge University Press, Cambridge, GB, 1982.

[52] H. Gardner, *Frames of Mind: The Theory of Multiple Intelligences*, Basic Books, New York, 1983.

[53] H. Gardner, *Multiple Intelligences: The Theory in Practice*, Basic Books, New York, 1993.

[54] P. S. Churchland, *Neurophilosophy: Toward a Unified Science of Mind-Brain*, MIT Press, Cambridge, MA, 1986.

[55] D. Marr, *Vision: A Computational Investigation into the Human Representation and Processing of Visual Information*, Freeman, San Francisco, 1982.

[56] S. M. Kosslyn, *Ghosts in the Mind's Machine*, Norton, New York, 1983.

[57] S. M. Kosslyn, *Image and Brain: The Resolution of the Imagery Debate*, MIT Press, Cambridge, MA, 1995.

[58] R. L. Gregory, *Mind in Science*, Penguin Books, New York, 1984.

[59] R. L. Gregory (ed.), *The Oxford Companion to The Mind*, Oxford University Press, Oxford, 1987.

[60] J. E. LeDoux and W. Hirst, *Mind and Brain: Dialogues in Cognitive Neuroscience*, Cambridge University Press, Cambridge, 1986.

[61] J. LeDoux, *The Emotional Brain*, Simon and Schuster, New York, 1996.

[62] S. Litvak and A. Wayne Senzee, *Toward a New Brain: Evolution and the Human Mind*, Prentice Hall, Englewood Cliffs, NJ, 1986.

[63] H. Margolis, *Patterns, Thinking, and Cognition: A Theory of Judgment*, University of Chicago Press, Chicago, 1987.

[64] A. Rosenfeld (ed.), *Multiresolution Image Processing and Analysis*, Springer-Verlag, Berlin, 1984.

[65] V. Cantoni and S. Levialdi, *Pyramidal Systems for Computer Vision*, Springer-Verlag, Berlin, 1986.

[66] T. Kohonen, *Self-Organization and Associative Memory*, Springer-Verlag, Berlin, 1984.

[67] W. Hackbush and U. Trottenberg (eds.), *Multigrid Methods I*, Springer-Verlag, Berlin, 1982.

[68] W. Hackbush and U. Trottenberg (eds.), *Multigrid Methods II*, Springer-Verlag, Berlin, 1982.

[69] W. Briggs, *A Multigrid Tutorial*, SIAM, Philadelphia, 1987.

[70] T. F. Chan, et al. (eds.), *Domain Decomposition Methods*, SIAM, Philadelphia, 1989.

[71] B. B. Mandelbrot, *The Fractal Geometry of Nature*, Freeman, San Francisco, 1982.

[72] K. Falconer, *Fractal Geometry: Mathematical Foundations and Applications*, Wiley, New York, 1990.

[73] Y. Meyer, *Wavelets and Operators*, Cambridge University Press, Cambridge, GB, 1992.

[74] I. Daubechies, *Ten Lectures on Wavelets*, SIAM, Philadelphia, 1992.

[75] A. Waxman, et. al., A visual navigation system for autonomous land vehicles, *IEEE Journal of Robotics and Automation*. Vol. 3, No. 2, 1987, pp. 124–141.

[76] A. Meystel, Planning in a hierarchical nested controller for autonomous robots, *Proc. IEEE 25th Conf. on Decision and Control*, Athens, Greece, 1986, pp. 1237–1249.

[77] H. Heuer, U. Kleinbeck, and K.-H. Schmidt, *Motor Behavior: Programming, Control, and Acquisition*, Springer-Verlag, 1985.

[78] C. G. Phillips, *Movements of the Hand*, Liverpool University Press, Liverpool, 1986.

[79] K. Baev, *Automatisms in the Nervous System*, Birkhouser, Boston, 1997.

[80] P. N. Johnson-Laird, *The Computer and the Mind: An Introduction to Cognitive Science*, Harvard University Press, Cambridge, MA, 1988.

[81] E. Langer, *Mindfulness*, Addison-Wesley, Reading, MA, 1989.

[82] J. C. Eccles, *Evolution of the Brain: Creation of the Self*, Routledge, London, 1989.

[83] G. M. Edelman, *The Remembered Present: A Biological Theory of Consciousness*, Basic Books, New York, 1989.

[84] M. Arbib, *The Metaphorical Brain 2: Neural Networks and Beyond*, Wiley, New York, 1989.

[85] M. A. Boden, *Computer Models of Mind*, Cambridge University Press, Cambridge, 1988.

[86] M. Minsky, *The Society of Mind*, Simon and Schuster, New York, 1985.

[87] R. Penrose, *The Emperor's New Mind: Concerning Computers, Minds, and the Laws of Physics*, Penguin Books, New York, 1989.

[88] N. Herbert, *Elemental Mind*, Dutton, New York, 1993.

[89] T. Siler, *Breaking the Mind Barrier: The Artscience of Neurocosmology*, Touchstone, New York, 1990.

[90] R. J. Sternberg, *Metaphors of Mind: Conceptions of the Nature of Intelligence*, Cambridge University Press, Cambridge, GB, 1990.

[91] M. Talbot, *The Holographic Universe*, HarperCollins, New York, 1991.

[92] W. H. Calvin, *The Ascent of Mind: Ice Age Climates and the Evolution of Intelligence*, Bantam, New York, 1991.

[93] A. Trehub, *The Cognitive Brain*, MIT Press, Cambridge, MA, 1991.

[94] P. S. Churchland and T. J. Sejnowski, *The Computational Brain*, MIT Press, Cambridge, MA, 1992.

[95] F. Crick, *The Astonishing Hypothesis: The Scientific Search for the Soul*, C. Scribner's Sons, New York, 1994.

[96] J. S. Albus, Outline for a theory of intelligence, *IEEE Transactions on Systems, Man, and Cybernetics*, 1991.

[97] A. Meystel, *Autonomous Mobile Robots: Vehicles With Cognitive Control*, World Scientific, Singapore, 1991.

[98] J. A. Fodor, *Psychosemantics*, MIT Press, Cambridge, MA, 1987.

[99] K. Pribram, *Brain and Perception: Holonomy and Structure in Figural Processing*, Laurence Erlbaum, Hillsdale, NJ, 1991.

[100] N. Chomsky, *Language and Thought*, Moyer Bell, Wakefield, RI, 1993.

[101] R. Jackendorff, *Languages of the Mind: Essays on Mental Representation*, MIT Press, Cambridge, MA, 1993.

[102] H. M. Wellman, *The Child's Theory of Mind*, MIT Press, Cambridge, MA, 1992.

[103] D. Bickerton, *Language and Human Behavior*, University of Washington Press, Seattle, 1995.

[104] P. Thagard, *Computational Philosophy of Science*, MIT Press, Cambridge, MA, 1988.

[105] J. R. Searle, *The Rediscovery of the Mind*, MIT Press, Cambridge, MA, 1992.

[106] S. Greenfield, *Journey to the Centers of the Mind: Toward a Science of Consciousness*, W. H. Freeman, New York, 1995.

[107] D. J. Chalmers, *The Conscious Mind: In Search of a Fundamental Theory*, Oxford University Press, New York, 1996.

[108] D. C. Dennett, *Consciousness Explained*, Little, Brown, Boston, 1991.

[109] D. C. Dennett, *Kinds of Minds: Toward an Understanding of Consciousness*, Basic Books, New York, 1996.

[110] W. Calvin, *How Brains Think*, Basic Books, New York, 1996.

[111] S. Pinker, *How The Mind Works*, Norton, New York, 1997.

[112] A. Turing, Computing machinery and intelligence, *Mind*, 1950, pp. 434–460.

[113] J. Searle, Minds, Brains, and Programs, *Behavioral and Brain Sciences*, No. 3, 1980, pp. 417–424.

[114] A. Newell, R. Young, and T. Polk, The approach through symbols, in *The Simulation of Human Intelligence* (D. Broadbent, ed.), Blackwell Publishers, 1993.

[115] P. K. Wright and D. A. Bourne, *Manufacturing Intelligence*, Addison-Wesley, Reading, MA, 1988, p. 352.

[116] J. S. Albus, *Brains, Behavior, and Robotics*, BYTE Books, McGraw Hill, Peterborough, NH, 1981, p. 352.

[117] A. Newell, *Unified Theories of Cognition*, Harvard University Press, 1990, p. 549.

[118] D. McFarland and T. Bosser, *Intelligent Behavior in Animals and Robots*, MIT Press, Cambridge, MA, 1993, p. 308.

[119] A. Newell, Reflections on the knowledge level, *Artificial Intelligence*, Vol. 59, 1993, pp. 31–38.

[120] C. Malcolm and T. Smithers, Symbol grounding via a hybrid architecture in an autonomous assembly system, in *Designing Autonomous Agents* (P. Maes, ed.), MIT Press, Cambridge, MA, 1990, p. 194.

[121] S. Harnad, Symbol grounding problem in *Emergent Computation* (S. Forrest, ed.), MIT Press, Cambridge, MA, 1991, pp. 335–346.

[122] S. Lakshmivarahan, *Learning Algorithms Theory and Applications*, Springer-Verlag, New York, 1981, 279 p.

[123] K. Najim and A. S. Poznyak, *Learning Automata: Theory and Applications*, Pergamon, 1994, 225 pp.

[124] K. V. Baev, Highest level automatisms in the nervous system: a theory of functional principles underlying the highest forms of brain function, *Progress in Neurobiology*, Vol. 51, 1997, pp. 129–166.

[125] J. Duncan, R. Ward, and K. Shapiro, Direct measurement of attention dwell time in human vision, *Nature*, Vol. 369, 1994, pp. 313–314.

[126] A. Newell, *Unified Theories of Cognition*, Harvard University Press, Cambridge, MA, 1990.

[127] D. H. Ballard, M. M. Haynoe, P. K. Pook, and R. P. N. Rao, Deictic codes for the emboliment of cognition, *Behavior and Brain Sciences*, 1996.

[128] S. Russell and P. Norvig, *Artificial Intelligence: A Modern Approach*, Prentice Hall, Upper Saddle River, NJ, 1995.

[129] URL *http://agents.www.media.mit.edu/groups/agents/people/.*

[130] M. Minsky, *The Society of Mind*, Simon and Schuster, New York, 1985.

[131] URL *(http://www.davidreilly.com/topics/software_agents/mobile_agents/index.html).*

[132] K. H. Kramer, N. Minar, and P. Maes, Tutorial: mobile software agents for dynamic routing, see URL *http://nelson.www.media.mit.edu/people/nelson/research/routes-sigmobile/.*

[133] R. Ben-Natan, *CORBA: A Guide to Common Object Request Broker Architecture*, McGraw-Hill, New York, 1995.

[134] J. M. Bradshaw, An introduction to software agents, in *Software Agents* (J. M. Bradshaw, ed.), AAAI Press, Menlo Park, CA, 1997, pp. 3–46.

[135] J. Hoffmeyer, *Signs of Meaning in the Universe*, Indiana University Press, Bloomington, IN, 1996.

PROBLEMS

1.1. What is the difference between intelligence and adaptation? Give 10 examples of systems (both living and constructed) that demonstrate *intelligence* (and/or intelligent behavior) and 10 examples of those demonstrating *adaptation*. Characterize the differences. Can you give a definition of both classes of systems? Demonstrate the difference between these definitions by comparing the block diagrams of these systems.

1.2. Introduce (evolve) examples of systems with adaptation of different degrees, starting with simple adjustment to the environment and ending with what could be called *systems with learning*. What is the difference between systems with adaptation and systems with learning, as you perceive it from your examples? Try to reflect this difference in a set of corresponding block diagrams.

1.3. Based on Section 1.3, formulate the difference between the intelligent systems and the non-intelligent ones. Propose your definition of an intelligent system and compare it with the existing definitions.

1.4. Give a definition of "advantageous behavior." What are the premises and assumptions that should be introduced to make this definition consistent? Try to introduce two definitions: one based on the concepts of evolution and natural selection, and another that does not require this concept. What did you introduce to be free from the need to allude to "natural selection?"

1.5. Draw the structure of intelligent system in the form of Figure 1.1. Analyze the concept of complexity. Find out how complexity of computations is being generated by each of the subsystems.

1.6. Describe the functioning of the system demonstrated in Figure 1.2. Make use of hierarchies on the right and on the left in the figure. How can Figure 1.2 be put in correspondence with Figure 1.1? Can Figure 1.2 be derived from Figure 1.1?

1.7. Choose a hierarchical system known to you from practice (e.g., hierarchy of manufacturing, hierarchy of a military unit). Represent it in a form demonstrated in Figure 1.4*c*. Describe the functioning of horizontal links at each level shown.

1.8. Represent architectures of (a) literary text and (b) musical piece in a form presented in Figure 1.4*c*. Find the differences and commonalities between these hierarchies and the hierarchies that were constructed in Problem 1.7. Focus your attention on the "goals." What roles do they play in these two examples?

1.9. Using your examples from the Problems 1.7 and 1.8, determine the "agents" at each level. Why do these agents form these hierarchies; what are the advantages? Introduce your definition for the scale of the level. Determine numerical values for scales at each of the levels in all your examples.

1.10. For all examples constructed in Problems 1.7 and 1.8, construct elementary functioning loops and check whether the property of "closure" is satisfied at each of the levels that have been constructed. How are these ELFs (belonging to different levels) connected among themselves?

CHAPTER 2

THEORETICAL FUNDAMENTALS

2.1 MATHEMATICAL FRAMEWORK OF THE ARCHITECTURES FOR INTELLIGENT SYSTEMS

In this chapter we outline the fundamental premises and possible tools for a consistent treatment of systems using *multiresolutional calculus* which is a computational method of discrete mathematics.

This calculus is required for the *representation, design, and control of complex systems* by computers. The need for multiresolutional calculus arises in software designed for complex systems and based on the concept of *recursive hierarchies* (e.g., RCS). The concepts of multiresolutional calculus used here are discussed in the mathematical and computer science literature. However, the treatment presented in different sources contains research results that do not fit nor communicate well with one other. Thus the *primary aim* of this chapter is *to establish a common conceptual basis* that allows for communicating the results of RCS research in an efficient manner. We address such important issues as *grouping, filtration* (or *focusing of attention*), *combinatorial search* (including *formation of combinations* and *search*) in a manner that is conducive to using these concepts as convenient formal operators. In particular, the operators will be independent of context and easily applicable. These operators are typically applied together. Two major forms of their joint application allows for performing "bottom-up generalization" and "top-down instantiation."

2.1.1 Role of Discrete Mathematics in the Development of the Formal Theory of Intelligent Systems

Until the twentieth century mathematicians paid attention to both discrete and continuous objects. The emergence of computers has created a paradoxical situation: Continuous objects and processes must be translated into the domain of the discrete objects because digital computers are designed to deal with discrete information. Our efforts to explain intelligence are tantamount to representing an intelligent system in the digital computer.

The phenomenon of intelligence includes dealing with a concept of continuity. The latter presumes infinitely high resolution.[1] Architectures of intelligent systems are based on the idea of finite resolution; that is, they allude to discrete mathematics. The controversy on the need to apply the results of ideal continuous reasoning to realistic discrete objects and processes has not been resolved. Some clarification of this theoretical inconsistency can be found in works on nonstandard analysis [1]. Nevertheless, it is prudent to note that the tendency to represent information in a discrete form is always useful in making computations efficient. This is why, instead of presenting a continuous line, we graphically apply a number of discrete coordinates and then "spline" the line by one of the methods available in a particular domain. For example, in the construction of boats and ships, a metal or wooden stripe is applied as a spline for modeling curvilinear cross sections and surfaces.

SETS OF SAMPLING UNITS...

The systematic development of the discrete mathematics can be attributed to the creation of set theory, probability theory, game theory, and operations research. In the mid-twentieth century, techniques of discretizing (or "tessellating") continuous systems into sets of sampling units became part of a university education. Later we will try to figure out what is the computational advantage of moving from continuous to discrete mathematics.

The obvious question is whether it is beneficial to counterpose continuous and discrete mathematics? They seem to be strongly interwoven in both theory and applications. Analysis of the physical world at various levels of resolution (granularity, scale) shows that objects are continuous only "in the eyes of the beholder" [2]. We see something as continuous only because of the comparatively low resolution of our retina or because of intentional procedures of smoothing. Many examples can be given that demonstrate the objects of the world are "granulated"; the idea of continuity is a tool used in putting complex systems together as whole units.

However, the discrete character of motion is not easy to demonstrate, and it cannot be denied either. There are many factors that demonstrate in many important instances the character of motion as being discrete rather than continuous (e.g., as a machine slows to the position where it must stop). The most efficient and powerful controllers (variable-structure controllers) have at their output a trajectory of motion with elements of "discreteness." In other words, the idea of "continuity" is a tool that allows objects and motion of a particular level of granularity to be transformed into a representation of coarser granularity.

...OR SYSTEMS OF INTERRELATED OBJECTS?

The real issue then is that both discrete and continuous mathematics are tools for modeling the world. These tools reflect hundreds of millions of years of the evolution of human mind. One could say that from the very beginning, the emphasis of human existence was on the efficiency of computations. It is possible to demonstrate that groupings of discrete objects were always used in order to increase the efficiency of computation and simulation. Reality consists in an overwhelming number of objects, and living creatures have discovered the best way to deal with them is to organize them hierarchically. They organized the perceived objects into countable units (groups, clusters), and then grouped these units again into a new, smaller sets of units, and so on. This is why evolution has equipped living creatures with a set of tools for lumping multiplicities of perceptions, thoughts, actions, and ideas into entities. These tools likely permeate the faculties of intelligence in all living creatures (though in different degrees). To be safe,

[1]"Resolution" is generally defined by the fineness of detail that can be distinguished in an image. In engineering, the term is related not only to visual images but to any type of representation. Higher resolution is associated with the property of fineness and lower with coarseness.

we will allude to humans having this skill (but we should remember that this could be a characteristic of all living organisms).

We are equipped with the biological sensors that have limited resolution. In addition we can change the scope of our vision by reducing or increasing our distance from a scene. We can change the scope also by focusing attention on a detail of a scene and neglect the rest of it. These two factors—resolution and scope of perception—reflect the distinction between continuity and discrete. Our sensor system estimates the average density of the signals in the window of our attention sliding over a scene and makes the quantitative "jumps," which are the borders between units of more or less uniform density.

LUMPING DETAILS INTO UNITIES

The lumping (or grouping) and curtailing of scope (or focusing attention) are therefore the primary operations of our intelligence. This is exactly what we do in applying discrete mathematics to any input information. Its techniques provide us with the fundamentals for dealing with the problem of intelligence. The *theory of sets* gives us a theoretical basis for the first fundamental procedures of intelligence: the procedures of *clustering*, *grouping*, or *lumping*.

Now, what should be grouped together? We constantly do this in looking around us. Our eyes scan our visual ground in rapid intermittent movements ("saccades") as if searching within the consecutive snapshots for similarities and differences. Similarities are blended tentatively, thus creating a hypothesis for possible unification. Mathematical operators of discrete mathematics borrow the ideas from these perceptual tools. Two interrelated procedures: *searching* and *formation of combinations* are the other fundamental operators that are required for the package of intelligence. We will call them *combinatorial search.*

Each discrete object demonstrates zones of continuity when considered at higher resolutions, while each continuous object always turns out to be composed of discrete objects when considered at lower resolutions. In practice, the proper choice of resolution and scope is a prerequisite of successful mathematical problem solving.

2.1.2 What Are the *Objects* to Which *Discrete Mathematics* Is Applied?

We can infer that within the totality of all problems, all things can be viewed as discrete objects and consisting of discrete subobjects depending on the chosen resolution and scope. Neither scope nor resolution is ever introduced in mathematics as a formal concept, though we will do so in this book. Scientists are guided by the implicit knowledge that in every problem the number of issues under consideration is limited (the scope) and the analysis must always have some constraints in the level of detail to be discussed (the resolution). Since so far in this book, we do not have any guidelines for choosing the scope and resolution of a particular level of problem solving, we will defer any further discussion and recommendations to later sections of this chapter.

Obviously the level of detail and the scope of consideration will affect the complexity of the problem solving process. If the values of scope and resolution are poorly selected, the grouping of objects at a level of lower resolution and their decomposition at a level of higher resolution will be unsuccessful. It is then clear that in our effort to simplify the process of problem solving, we must introduce various scopes and levels of resolution. It may be that the selection of a particular level of resolution and scope of vision must be performed at several levels of resolution simultaneously for the architecture of representation to be adequately designed, and the complexity of computation to be sufficiently small.

We can list several examples for things under consideration in discrete mathematics. Besides the obvious examples of a human, a car, a manufacturing plant, a squadron of soldiers, a country, or a molecule, we can consider discrete samples which we use to represent one variable as a function of another variable if we are interested in making an approximation. From the very description of all objects, it should be clear that the issues of scope and resolution become evident before the problem solving can begin: They emerge at the stage of problem representation.

COMPLEXITY IS THE REAL CONCERN

In computers, each object is represented by a memory location or group of memory locations. The parts of these objects are considered to be separate objects, so they are stored in their own memory locations. An object, as a rule, is a part of another object that has its own memory location. Thus it is not only the complexity of computation that is affected by the scope/resolution selection; the structure of memory can be more efficient, or less efficient because of this choice too. This is why computer researchers have arrived at the concept of *object-oriented programming* (OOP).

Each object is characterized by a list of attributes, and each position of the list is associated with its memory location. The functioning of objects, their recognition, the planning of their motion, are performed by computers in a discrete way. Even when we are interested in describing the continuity of an objects' motion or shape, or even when we have analytical formulas for continuous curves, the numerical values are obtained only in a discrete way, since the accuracy of computation is limited by the position of a concrete number after the decimal point.

Finding the next position in a representation of the number will require a jump. These views have led to a paradigm of design and manufacturing strongly dependent on the ideas of object-oriented programming (e.g., CORBA; see [3]). The significance of OOP is its means of representing the world as an architecture that contains interleaved hierarchies of objects, their linear origins, and their properties.

Both computer hardware and software are object-oriented systems. Digital computers fit perfectly within the paradigm of discrete mathematics. It would be incomplete to think that this is because they embody set theory and rules of logic. Even analog computers are based on set theory and logic. The real reason is their object-oriented hierarchical system of world representation. This system of representation provides for the lowest type of computations in comparison with other available concepts.

2.1.3 What *Jobs* Does Discrete Mathematics Do? What Are Our *Objectives*?

The value of discrete mathematics in dealing with practical problems of computing can be especially demonstrated in the areas related to computational intelligence. The following routine tasks of computational intelligence can be performed in a very efficient manner.

- Search for an approximation
- Representing information
- Transformation of data
- Sorting
- Grouping
- Searching
- Comparing
- Encoding of complex information
- Transmission of complex information

- Job assignment
- Scheduling
- Planning
- Prediction
- On-line control
- Statistical grouping
- Statistical generalization
- Reports with statistical analysis
- Search for inconsistencies in a complex system

TASKS OF COMPUTATIONAL INTELLIGENCE

The following objectives can be pursued in performing computational tasks related to those listed above:

- Problem solving
- Efficient computation
- Understanding of interrelationships existing within a system
- Optimization
- Image processing including recognition and interpretation
- Speech processing including recognition and interpretation
- Construction and updating of the system's representation

Some of the items given in these lists are not trivial, like "representing information," prediction, statistical generalization, inconsistencies, understanding. Most of the terms will be introduced in the corresponding sections. We will address here only two: representing and understanding.

Since at a level of resolution the accuracy of representation is limited by the very existence of the indistinguishability zone, the information of a single level contains an intrinsic uncertainty and must be constantly refined.

"Representing" is a term for constructing a system of knowledge representation. Data should be stored, relationships among them should be found, and the search for the fields of meaning should be initiated. These fields become the basis for generalization which creates data for the level of lower resolution.

"Understanding" means "finding the cause–effect strings," deriving from them meaningful implications.

These two terms are the key issues of the whole area of intelligent systems.

2.2 FORMAL MODEL OF INTELLIGENT SYSTEMS AND PROCESSES

The mathematical methods introduced in this section are intended for the *modeling of systems and situations of the real world.*[2] They are motivated primarily by the subsystems of sensory processing (SP) and behavior generation (BG) (see Chapter 1). In the SP hierarchy, the incoming information is bottom-up; it gradually moves from the high-resolution phenomena to objects and processes of low resolution. The BG hierarchy works in opposite direction: From the general or low-resolution goals and situation

[2]All systems to which the RCS methodology is applicable are in this category.

descriptions at the top, it proceeds with task decomposition to the higher levels of resolution at the bottom.

At each level of resolution, we will try to maintain a close correspondence between (1) the available information about the real systems we are going to analyze, and (2) the formalisms we will propose as well as the results of applying these formalisms. Later it will be demonstrated that the search for this correspondence at each level of resolution is a part of the procedure of symbol grounding; it is determined by the property of semiotic closure which is satisfied in the systems of reality.

MULTIRESOLUTIONAL CALCULUS

The concepts of multiresolutional calculus[3] are broadly applied in different scientific disciplines. The following list contains examples of disciplines and domains that have common roots in the multiresolutional calculus:

1. *Multigrid schemes for solving systems of partial differential equations (PDE).* PDE models are notoriously computationally expensive. For a long time their computational complexity was reduced by solving systems of equations with rough discretization models. Then the area of interest became focused, and the discretization was refined locally.
2. *Finite elements method.* This method depends on the same mechanism as multigrid PDE schemes in mechanical and other problems using custom-made equations and generally experimental data and expert information. However, the results are rough and refined gradually, level by level, with simultaneous narrowing the scope of local computations.
3. *Interval arithmetic.* Mathematicians applying interval arithmetic use the same mechanism of consecutive refinement as that already mentioned for the multigrid and finite element methods.
4. *Hierarchical estimation schemes (including recursive estimation).* In all hierarchical estimation schemes consecutive refinement is used to increase accuracy without disproportionally increasing their computational complexity.
5. *Multirate control systems analysis and design.* Specialists in control theory have recognized that computational complexity can be reduced substantially by designing and manufacturing separate controllers, each for its frequency interval.
6. *Multiresolutional filtration.* Channeling different intervals of frequency into different processing systems became a way of reducing computational complexity also in signal processing and communication.
7. *Wavelets.* Wavelet theory has evolved from realistic or mathematical multiresolutional filtration. However, the concept of wavelets is not necessarily based on harmonic functions; other periodical functions have been explored too.
8. *Fractals.* The theory of fractals was based on a multiresolutional (multigranular, multiscale) modeling of world phenomena. The classical example involves determining the length of the shoreline of Great Britain; it is based on an analysis of the shoreline at many levels of resolution.

Systems such as those of *wavelet theory* do not fully correspond to the body of knowledge which we call multiresolutional calculus. They do not use the ideas of scope

[3]Multiresolutional calculus is used in this book to denote the body of mathematical and computer science knowledge that is linked with analysis and synthesis of systems considered at several resolution levels simultaneously, which is usually done to increase the computational efficiency of research and/or design activities (see [4, 5]).

and resolution to the extent required in realistic problems. For example, all levels of resolution in the wavelet theory are represented with the same accuracy. This is all right for wavelet decomposition, but it does not serve the needs of knowledge representation outlined in this book.

All the methods mentioned above take approaches from the arsenal of multiresolutional calculus (though they are described without considering the problem at hand in its entirety). In all instances involving multiresolutional calculus, there are numerous important issues invoked typically in the form of heuristics or computational recommendations.

2.2.1 Fundamental Procedures of Multiresolutional Calculus

The key concept underlying multiresolutional calculus is the concept of resolution. *Resolution* is a property of representation that demonstrates the information about the details of a situation and/or object of interest. It depends on the value of information density in the carrier of the representation. Each carrier (whether hardware or software) is presumed to be limited in its ability to store the details by the size of the elementary "granule" of the carrier. This *granule* determines the finest *scale* of representation possible in this carrier. Later we will define all terms precisely. At this point, it is only important to understand that because of this limitation, real systems of representation are built at several resolutions and are *multiresolutional* ones (or *multiscale*, or *multigranular*, which is the same thing).

The domain of knowledge dedicated to the analysis of multiresolutional systems is called *multiresolutional calculus*. The following question is considered fundamental for multiresolutional calculus as a distinct theoretical domain: Is there a set of basic computational procedures such as are necessarily employed in the tasks listed in the Section 2.1.3? It is possible to demonstrate that the computational procedures ascend to the triplet of conceptual phenomena which include *grouping, filtration (focusing of attention),* and *search* (actually *combinatorial search* consisting of *combinations formation* and *searching*).

GROUPING

Grouping generally includes the following cases:

1. Larger objects assembled out of smaller ones serving as their parts
2. Multicomponent (multivariable) signals approximated by a structurally simpler function based on a smaller number of variables
3. *Blending* of computational procedures executed in neural networks
4. Fuzzy data, and the like

Grouping includes various procedures of *clustering, class formation, unifying of sets, construction of strings, n-tuples,* and so on.

Filtration (or *focusing of attention*) is introduced when we set up a rule *for selecting a subset of interest* as in methods using *windowing (moving average),* choosing *the subset of descendants,* the base for computing predictions, and so on. In the engineering field where the problem should be posed and resolved, there are formulated three conditions of filtration:

FOCUSING ATTENTION

1. *Condition of a scope.* The scope of representation is satisfied by focusing attention upon a subset of the world sufficient for problem solving. After this, the rest of the information about the world can be neglected. Later we will define the scope of representation in a more formal fashion (see Section 2.3.1).

The scope is the first intuitive constraint introduced in problem solving (we use only the map of the state of interest and do not use maps of other states unless a special need emerges, etc.). The meaning of the first condition can be rephrased as follows: The state space of the problem is considered bounded *from above* by the scope of problem. By introducing the scope, we declare that beyond these particular boundaries, the behavior of objects in the rest of the state space is of little or no interest.

2. *Condition of granulation.* Granulation is satisfied by determining the resolution for the first (rough) analysis of the system.

The meaning of the second condition is in bounding the attention *from below*. We will never use information more precise (and detailed) than that available at the resolution of the level of consideration. The second condition introduces the finest *granularity* of the state space under consideration, and each point of the state space is discussed only as a *granule* or an *indistinguishability zone*.

3. *Condition of windowing.*[4] This condition is satisfied by proper selection of the size and the law of motion within the scene for the sliding window applied for the consecutive clustering and generalization procedures.

The third condition concerns our ability to process the represented information so that new objects can be created and new properties discovered. We intend to exercise the procedure of grouping. But, in order to do this, we have to focus our attention.

All three conditions are prerequisites for the operation of *search* which is performed within the scope by moving the window and checking the conditions of grouping within the sliding window. The operation of search is a powerful area of computational methods applied in different areas of science. It has two peculiarities: (1) It alludes to both filtration and generalization and is never applied without association with these procedures, but (2) it is not well supported theoretically and is strongly permeated with heuristics. The search is performed among

1. Objects within the scope of attention to group them.
2. Groups that choose one of them as an object of interest.
3. Classes of the representation system that determines to which class the newly synthesized object belongs.

SEARCH

Combinatorial search is utilized for hypotheses formation (related both to objects and motion trajectories).

These tools of manipulating data: *grouping (G)*, *filtration* or *focusing of attention (F)*, and *search (S)* are frequently used together as a triplet of interrelated operations. We will refer to it as *GFS*,[5] or *operator of generalization*.

The ability to generalize is a highly respected feature of the intellect. However, it is not supported by results of extensive scientific analyses. The skill of recognizing a single generality in a multiplicity of particular cases carries a tremendous power of inference and discovery. This is as a skill that makes a scholar out of the student and makes a commander out of the soldier. However, this skill was for a long time neglected

[4]Sometimes, instead of *windowing*, the term *masking* is used. The window shows what the attention should be focused on. The rest of the information is masked.

[5]In earlier publications, GFS is called GFACS. We will use these abbreviations interchangeably.

by representatives of the behaviorist school in psychology. The following curious explanations of the term "generalization" are taken from *Encyclopaedia Britannica*:

> The occurrence of the phenomenon of generalization is seen in the tendency of laboratory subjects conditioned to respond to a particular stimulus (e.g., a light) to respond as well to similar stimuli beyond the original conditions of training. . . . For example, a dog conditioned to salivate to a tone of a particular pitch and loudness will also salivate with considerable regularity in response to tones of higher and lower pitch. . . . Similar behavior is observed in humans, as children learning to talk may call anything that can be sat upon "chair" or any man "daddy."

The source of our view on generalization is in both logic and mathematics. Analyses of the processes of reasoning, analyses of the way in which mathematical structures and systems are constructed, appeal to a reconsideration of the processes of generalization in the psychology of intelligence. Later in this book we will return again to the mysterious skill of generalization which turns out to be an instrumental part of many phenomena of life and intelligence. Numerous living and constructed intelligent systems employ the GFS triplet for generalization as the algorithmic basis of their functioning. We will introduce a formal mathematical system that enables a rigid and consistent treatment of problems that require the triplet of GFS.

GFS INVERSE

It has been demonstrated in many NIST-RCS applications that the system employs generalization at all levels and that the GFS triplet can be consistently applied to provide for NIST-RCS functioning. An inverse procedure of instantiation is presumed (GFS^{-1}). A powerful version of this is introduced in Chapter 7, and it is capable of performing the decomposition task.

In order to judge the properties of representation that manipulates groups of units, the laws of these groups formation should be addressed. This domain of knowledge is called *set theory*.

2.2.2 Existing Definitions of a Set

G. Cantor, the creator of the theory of sets, considered it important to associate each group of objects with a purpose, or a cause for putting them together. This is why the *axiom of abstraction* is so important for understanding sets. This theorem states that the statement

$$(\exists x)(\forall x)(x \in X \leftrightarrow \phi(x)) \tag{2.1}$$

holds for all objects x. This means that there exists a set X consisting of objects x such that for each x some function $\phi(x)$ can be found and this function is the same for all of x. This function is the actual basis for existence of a set, it is usually called *a property* of the set.

There are many definitions of sets, starting with any *collection of objects* and ending with a *collection of objects* that can be characterized by a particular common feature. The latter definition is unclear because "common feature" can be interpreted that the objects have one or many attributes in common or that all of them have particular relationships. Consider the example of a set of objects on the surface of a table. They all have in common the property of being on the same surface. On the other hand, the same sentence "a set of objects on the surface of a table" describes objects related to each other connected by the surface of the table.

DEFINITION OF SET

A *set* is a collection of objects that have a common feature.

This is not a clear definition because we still do not know what is the object. However, we know something about the object: It has features.

DEFINITION OF FEATURE

A *feature* is an element of an object which is considered to be important in problem solving.

Sometimes the process of set formation presumes taking into account the distances of corresponding objects to one another (relation) as a basis for considering a set of objects. In the previous example, the set of objects on the surface of the table can be divided into two subsets if members of these subgroups are closer to each other within a group. Later we will call these sets *clusters*.

There is something local in the idea of sets and clusters. Proximity is a necessary condition for *clusters*, and it is somewhat presumed for *sets*. That is, a set has related synonyms: array, batch, bunch, bundle, clump, clutch, lot, set, group, band, crew, party. However, no proximity is required for classes defined in dictionaries as "a whole consisting of elements which share one or more characteristics," or "a set, collection, group, or configuration containing members regarded as having certain attributes or traits in common; a kind or category."

Merriam-Webster gives the following definitions of *class*: "a group, set, or kind sharing common attributes: as (a) a major category in biological taxonomy ranking above the order and below the phylum or division, (b) a collection of adjacent and discrete or continuous values of a random variable, (c) a collection of elements. It can be also understood as a division or rating based on grade or quality." One can see that there is more to the concept of "object" than just such characterization of it as "feature."

The characteristics that are a basis for class formation are frequently listed. For example, "*class* is a group whose members share general characteristics: breed, genus, species, order, rank, kind, sort, type, variety, genre, ilk, category, strain, caste, cut, stamp, stripe, lot, style, persuasion, grade, rate, price, social strata, and others."

The same set of objects lying on the surface of a table can be divided into two subsets if some of them are parts of, say, a carburetor, and some of them are parts of the transmission.

The concept of set is introduced in order to make our reasoning about multiplicity of objects more efficient but in order to succeed we have to define the "object" by considering it in more detail. When we are able to mention only names of sets, our reasoning becomes less complex, since we don't need to list all elements of these sets repeatedly. In general, set formation is based on determining similar properties. Therefore members of a set are supposed to have at least some similar properties so that a connection between the objects can be made. A reminder: The formation of a set very often is a procedure of formation of objects for the lower level of resolution (granularity). Thus the set cannot be part of itself because it belongs to the different resolution of representation, and the well-known Russell paradox never emerges.

Introduction of the concept of a set generates a number of related questions: What makes an object a *member of a set?* What is an *object*? What are the features of an *object*?

By the definition above, a set is a collection of objects with a similar feature (or features) and similar relation (or relations). This means that the *sets of features* of several objects might overlap. The same can be said about the *sets of relations*. Now we can define an object via its features and relations. We will call this well defined "object" a "standardized object."

DEFINITION OF OBJECT

A *standardized object* is an element of the world that can be characterized by having a list of attributes, by properties of decomposability and aggregability, and by an ability to be a passive and/or an active actor in its relational links.

The following features of an object should be taken into consideration:

1. Attributes characterizing the belonging of the object to a particular class of quality (red, heavy).
2. Decomposability into parts (the so-called children).
3. Aggregability with other objects (the so-called parents).
4. Ability to act, namely produce and/or influence changes in other objects (actions).
5. Ability to be subjected to actions or influences of other objects.

Some of these properties, such as property 1, can be described by a label referring to a particular quality domain. A subset of attributes can be considered *states* (see Sections 2.2.3 and 2.3.2). Properties 2 and 3 refer to a relational structure (directed graph). Properties 4 and 5 can be also represented in the relational structure: They refer to the rules of implications associated with the ability of a standardized object to produce actions or to be subjected to the actions produced by other objects.

DEGREE OF BELONGING TO A SET

The problem of the properties' overlap is easily resolved if all properties are boolean (they either do or do not exist). The question is how to find similarity if belonging to a particular quality domain can be of *different degree*. Where this occurs, also the list of attributes of different objects will overlap, but the degree of overlapping should be evaluated in order to satisfy the condition of inclusion into the set. Using the techniques of fuzzy logic is presumed but any other technique of determining the degree of similarity can be applied.

This problem leads to the question of membership (in a set), whether membership is a property that can exist or not exist, or whether membership is a matter of a degree. It has been already shown that the overlapping of the properties can be of different degree. However, belonging to a set is determined by a threshold rule: Until the value of a degree of membership exceeds some particular number, the object does not belong to the set. After it exceeds this number, the object belongs to the set. The following algorithm determines whether or not the standardized object belongs to the set:

MEMBERSHIP

Step 1. Compute the amount of overlap for each standardized object with regard to another standardized object.

Step 2. Compare the results of step 1 with the threshold value.

Step 3. Separate all objects in two subsets: those passing the test of belonging and those not passing.

This problem is relatively simple when all objects are represented using a standardized model. When the model of standardized object is not determined, a variety of algorithms of clustering can be used. (They will be discussed later.)

Another important question is the degree of membership to be evaluated. The following cases can be considered:

- Attributes can be listed (with their boolean evaluation, or evaluation within the interval [0–1]).
- Offspring are known.
- Parents are known.
- All actions that can be produced are known.
- All actions that an object can be subjected to are known.

We can to proceed to evaluate the degree of membership by using properties 2 through 5 which are added to enhance the list of attributes given as property 1. Some of these properties may become constraints that supersede any numerical evaluation of membership. For example, the following constraint can be introduced: Objects that do not have some particular five parents cannot be considered members of the set. The degree of overlapping can be computed as

$$\mu = \frac{\sum_i k_i(v_{1i} + v_{2i})}{2\sum_i k_i}, \tag{2.2a}$$

or

$$\mu = \frac{\sum_i k_i[\min(v_{1i}, v_{2i})]}{\sum_i k_i} \tag{2.2b}$$

where k_i is a coefficient of importance for the ith attribute found in both objects under consideration and v_{i1} and v_{i2} are the values of a particular ith attribute found in common for objects 1 and 2.

Formula (2.2) can be used if the enhanced list of attributes has been constructed with numerical evaluations. The computed degrees of overlap reflect no more than intuitive guesses based on experience in grouping objects. Formula (2.2a) should be used when grouping presumes essentially adding similar properties. Formula (2.2b) is related to a case when the merger leads to the intersection of similar features. Construction of the formula for μ should be tuned up to a realistic situation.

Membership in a set is a phenomenon introduced for a particular level of granularity. At this level of granularity, sets are created by lumping high-resolution elements into new entities which become elements of a new (lower) level of resolution. Thus the set formations result in new (lower) levels of resolution. We use the expression *membership* when we consider the occurrence of this grouping phenomenon.

Membership presumes inclusion. Let us consider the definition of *inclusion*. For the sets A and B, A is included in B, $A \subseteq B$ iff every member of A is a member of B. Thus a single element cannot be *included* in a set. It only can be *a member* of a set. In other words, the concept of membership should be introduced in evaluating the degree of having various properties of a set of properties. Membership is the relation between an object and a set. The value of membership is within the interval 0–1.

Actually, we have discovered the fundamentals of *fuzzy sets theory* (FST). The more we get involved in the domain of intelligent systems, the more we would need to use the ideas of FST. For better familiarity with the FST area we recommend an excellent set of sources collected by J. Bezdek and S. Pal [6].

The notion of object components is important in building a system of representation. This is why we try to determine principles governing the objects' formation as well as their organization into groups according to similarities and differences.

Later we will need something more fundamental than an object. Let us consider whether the "surface of a table" can be regarded as an object. The question is not related to the substance of wood serving as a horizontal component of the table. We are talking just about the surface area. This horizontal wooden component of the table can have six surfaces: top-horizontal, bottom-horizontal, and four vertical side-facets. So we can talk about a "set of surfaces." Clearly, these surfaces are not physical "objects," but they are entities. Entities will be introduced later as an enhancement of the concept of "object" to the unbounded variety of cases.

WE DISCOVERED FUZZY SETS THEORY

Can we talk about a set of colors? Colors are properties of objects (attributes) and not actual objects, but there is some unity in the concept of color which allows it to be considered a set. Color can be considered a feature, and we can talk about a set of features. Again, we are faced with a phenomenon of entity. We will return back to the notion of entity many times. First, we need to build upon the intuitive recognition of unity that may exist among different individuals.

2.2.3 What Is a *State*? How Is It Related to the Concept of *Object*? What Is *Change*?

A set of properties can be described as a point in a coordinate system where each coordinate stands for a particular attribute. Instead of listing the attributes, we can discuss the position of this point in the system of coordinates. Thus a vector, or list of coordinates, can be replaced by what we call a *state*. A state is an artifact of the process of constructing a system's model. The definition of a state typical for the theory of systems and control theory presumes that the model of the system is known:

IDEALISTIC DEFINITION OF STATE AND SITUATION

The *state* s_i of a system at time t_i is the information required to determine the output $y(t)$ uniquely for all consecutive moments of time. For the lower resolution (generalized) description of systems, instead of the term *state*, the term *situation* is often used.

We call this definition idealistic because by the very fact of having resolution limited, we admit incomplete knowledge at a level and uncertainty of information. Thus, we cannot expect correct anticipation forever.

From the full description of an object, it is clear that the state information does not exhaust the object description if the state is presented only via a list of attributes (property 1). The object itself is richer than its state. From the state, we do not know what are the relationships of the objects with their offspring and parents, and what rules of actions might be entailed (see Figure 2.1). So a model of the system is required; it is not contained in the state.

However, a state is a convenient artifact, since it monitors what is going on with the object, at least partially and even introspectively. It is also convenient for monitoring what is going on with an object in time. In the same way as a point on a curve is characterized by the value of the function and its derivative, each state is also characterized by a description of change. In the world of discrete mathematics, two changes are essential: the latest change of the state, which happens to the object during the last Δt (or several last Δt intervals), and the change anticipated after one or several Δt intervals in the future. This anticipation is determined either by the model of a passive object or by intentional control.

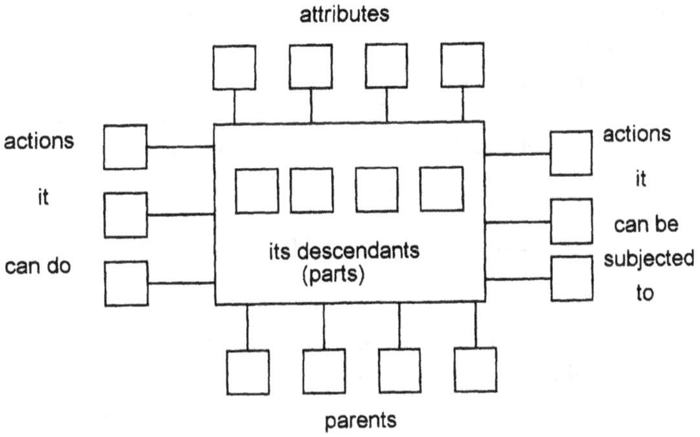

Figure 2.1. Model of a standardized object to be described in the state-space.

The components of a state (the coordinates of the state space) are frequently called *variables*. When a complicated system is being analyzed at a lower resolution, the state is called a *situation*. At even lower resolution we might be willing to talk about a "paradigm." Let us discuss how one can derive from the concepts of *object*, *state*, and *change* the dependent concepts *of state-space*, *action*, and *time*.

Earlier we have introduced state-space as a coordinate system built upon attributes of objects. Any curve driven in this space will be a curve of changes occurring to our objects; these changes occur only in time. For changes to happen, an applied action is presumed. This action can be determined by the previous state s_i consequent state s_{i+1}, and the time span between these two consecutive states $\Delta t_{i,i+1}$. In the automata the transition function $\delta(s)$ is represented as a list (a table) of all possible state transitions that declare the anticipated state changes under particular input commands u. The output function $\beta(s)$ shows how the new state is transformed into the output y (see Section 2.2.7). The change of the output associated with a particular input command $\Delta y_{i,i+1}$ (u_i) is called an *action*. All of the components of the automaton are also characterized by their accuracy. This is why the actual automaton belongs to a particular resolution (or granularity) of representation. Sometimes it is convenient to analyze the enhanced state where all inputs and outputs are added to the list of variables when the coordinate system is constructed.

2.2.4 Formation of New Objects via the GFS Triplet

A model of an object (Figure 2.1) presumes that the object is an assemblage of many small parts. The components may relate to their natural or technological properties. They may also be determined upon considerations of the efficiency of representation and/or computations. Several examples can be given. Two good examples are the decomposition of an electric motor via its components or the decomposition of concave figures in a set of convex figures. The first example shows partitioning based on technological reasoning. The second example reflects a common computational practice of dealing with concavity when one has only algorithms for convex figures. The new objects composed out of different "elementary" objects have new names, new attributes (property 1), new position in a hierarchy (properties 2 and 3), and new functionality (properties 4 and 5).

GFS POPULATES LEVELS OF RESOLUTION WITH OBJECTS

The partitioning of objects does not necessarily coincide with a hierarchy of resolutions accepted in a particular system of representation. This is possible because the hierarchy of resolutions in a system of representation is a result of a long-term design that determines scales for all attribute coordinates. These scales determine a unit of indistinguishability, which might be good enough for describing the electric motor as well as its parts. Therefore an electric motor, which is in a lower resolution than its parts, at the product hierarchy, will be at the same level of resolution at the hierarchy of visual representation of the shop. In all such cases the reasoning approach should be handled very cautiously in order to avoid paradoxes.

How should the grouping of objects into a set (which results in a new object) be performed? We will consider here only a very simple case where the relations are not taken in account but only the properties which are the basis of constructing a state-space.

Consider a state-space in which objects are represented as points. If you can visualize this, it will appear like a star-studded sky, with the points distributed three-dimensionally in space depending on their properties. We can decide for ourselves the density D of the points' distribution in the space:

$$D = \frac{\text{total number of objects}}{\text{total volume of state-space}} \tag{2.3}$$

When the average density of the distribution of objects in the state-space is computed, local increases in density are evidence of clusters which might be considered "groups" depending on the realistic interpretation of the problem. These groups sometimes turn out to be new objects. Using these new objects could lead us to a considerable reduction of complexity.

One of the ways of discovering local densities is by analyzing the state-space with a sliding window. The sliding window is constructed as follows:

Step 1 A rectangular window commensurable with the size of the expected objects is placed over a subset of the state-space.

Step 2 A sliding process is performed using the window to traverse the entire state-space.

Step 3 During the sliding process the local average density (within the window) is computed:

$$D = \frac{\text{number of objects in window}}{\text{volume of window}}. \tag{2.4}$$

Step 4 The results are collected in a form of the topographical map of densities.

Step 5 The regions of the consequent grouping are detected by areas of "thick" and "thin" densities.

Parts of this algorithm can be used to perform functions of attention focusing, combinatorial search, and grouping. Attention focusing is performed by selecting from the state-space a subset that allows one to quickly and efficiently apply the required operation. Then to get an objective picture, the window is gradually slid in a uniform manner over the entire state-space. The window-sliding process is not yet a search process. It is rather a preparation for consecutive searching. As we make a graph of the average density of the sliding window for each interval of resolution distinguishable on the surface of the state-space, we start to analyze the relief of this surface and search for peaks.

In the search process we compare the values of the average density with a particular threshold value, moving gradually from large to small values. The first intersection of the surface with the line of threshold is tested to see whether it has on the right and on the left smaller values. If it does, that point can be considered a peak. As a peak it is entitled to be included among the centers of the listed groups. Thus the candidates for the group centers are obtained as a result of the search procedure. The combinatorics emerges as we start attaching surrounding points to the centers of future groups, and must determine whether they are closer to one center more than to another. Combinatorics emerges in the saddle areas of the relief. Once all points are attached to the group centers (grouping), these groups become new objects belonging to a lower level of resolution. We have considered here the single most straightforward and conceptually simple algorithm in which all three components of GFS are present. The simple assembling of these points are more than aggregation; this is a generalization. In the conventional methods of clustering (k-min method, nearest-neighbor method, etc.), the same components can be distinguished.

AVERAGING IS A TOOL OF GENERALIZATION

The meaning of the averaging will become clear when we analyze the values of probability. Most of the information available for the subsequent grouping contains random components, and this is why the correct strategy of grouping is a subsequent repetition of the procedure presented above, gradually increasing resolution of representation. It is easy to see that the major big groups will be constructed at the low level of resolution after the process is locally repeated with higher resolution. We will discover that each large group consists of a number of small groups, which in turn, consists of even smaller groups. This is the recommended sequence of operations:

Step 1 Identify the desirable list of properties, for using during the process of clustering.

Step 2 Determine cost-function for determining distances between the objects.

Step 3 Select the threshold.

Step 4 Check the cost-function against the threshold.

Step 5 Proceed with these steps in multiresolutional fashion as described.

It is useful to take into account not only properties of object but also relationships among them. This is frequently applied in practice of working teams formation when not only the strength of each member of the group is taken into consideration but also the relationships between members of the group.

When we represent objects in a state-space, and consider the linear case, the distance between these objects is a natural and unavoidable indication of their relationships (use the analogy of topographical relief.) However, in real situations there are many other factors that transform the problem from a linear into a nonlinear one. Even in a topographical case, pure geometrical coordinates for determining closeness are not sufficient; that is to say, the short trip across the swamp can be much more costly than taking a long trip around it. The traversability factor must be taken into account.

There is a profound difference between applying GFS and what is called fuzzification in FST. The goal of fuzzification is to find a generalized quantitative characterization for an interval. GFS aims to find generalized objects within the same interval.

In this sub-subsection, it was tacitly assumed that we talk about grouping of objects in the state-space in a particular moment of time. However, the problem that emerges in practice is very often a more interesting one: while conducting the process of group formation take in account not only instantaneous properties but the evolution of these properties in time. Spatio-temporal grouping will be the next step of dealing with discrete information that we approach in the future.

2.2.5 Relations among the Objects: What Is a *Relation*? What Are the Properties of Relations? Can We Measure Relations?

Objects within a state-space have certain relations. These relations depend on the content of the problem, since it determines which subset of the attributes are to be applied later in order to characterize the situation. As far as attributes are concerned, relations serve to reflect the results of comparison of the properties of the objects (a is smarter than b, but b is stronger than a; obviously the relations between a and b can be interpreted differently depending on the problem). Some of the relations are part of the information contained in the object's model (e.g., parent-offspring relations). In describing relations, the information must therefore include actions that can be produced by or be inflicted upon the object. All relations must be brought into a system that contains the following types: relation of *equivalence*, relation of *order* (weak and strong), relation of *tolerance*, and relation of *resemblance* or *similarity*. The first three relations are clearcut; the next three are fuzzy. To determine which group belongs to the relation under consideration, the following list is useful. It demonstrates the belonging on properties of reflectivity, transitivity, and symmetry.

SYNTHESIS OF A RELATION

- Equivalence: reflexivity, transitivity, symmetry.
- Resemblance: reflexivity, nontransitivity, symmetry.
- Tolerance: reflexivity, symmetry.
- Partial order: reflexivity, transitivity, antisymmetry.
- Strict order: antireflexivity , transitivity, antisymmetry.

The presence of *relations* generates the notion of a *function*. Now, how can the algorithm based on (2.3) and (2.4) be modified so that the information on relations can be taken in account? We will encounter difficulties that do not allow us to simplify the search for clusters by first searching the areas of "thickness" in the geometric state-space. Different strengths of "attraction" between pairs of members could force us to include relations in the list of properties that increase its dimensionality tremendously. However, in reality the relationships are stronger when the properties of the objects are closer. This allows for using techniques based on (2.3) and (2.4) first, and then taking into account local relationships. The best ways to find the local increases in density is by analyzing the state-space with a sliding window.

As soon as the issue of resolution is invoked, both the truth tables and the causal clauses should be modified since they reflect only knowledge related to generalized objects and do not reflect the uncertainty.

2.2.6 Representation and Reality

We look for representations that are consistent with reality. Different "reality check" schemes are always built into existing systems. We need safeguards to protect us from inadequate representation, which can lead to inefficient functioning of the intelligent system.

Let us consider what truth is. Is this the same as logical truth? Is this the information contained in the truth tables?

DEFINITION OF TRUTH

Truth is a statement concerning the properties of objects and/or relations between (among) objects that can be mathematically expressed and/or empirically[6] verified.

The process of verification presumes putting into correspondence two symbolic systems: one which provides the statement to be verified and another in which the process of verification is conducted. One of the symbolic systems often represents an elementary picture of the phenomenon or processes that happen in the real world. This is the case where the truth of the logical inference must be verified by its conformity to reality. All systems of reasoning in which the concept of truth is important, manipulate the generalized concept, namely with concepts that are generalizations of some higher-resolution information. Thus a typical problem of truth verification is to find out whether reasoning in generalized terms leads to a faulty reasoning and errors in high-resolution projections.

In computer systems this checking of fitting between computation and reality is a part of what is called *symbol grounding*. For symbol grounding to be exhaustive, each level of resolution and the corresponding information loop must be verified separately, and simultaneously the conditions of inclusion must be checked. According to these conditions each statement of the lower-resolution level must be demonstrated as consisting of several statements of the higher-resolution level.

Logical truth has important specifics. It does not presume to be related to any correspondence between the assertion and the objects and/or events of reality. Any logical statement previously written is considered to be true. The statements obtained as a result of an inference must be proved to be true too. In logic it is irrelevant whether the result of an inference matches reality. This is why logical truth is frequently called vacuous. You can check your understanding of our line of thought by trying to explore whether this statement is true:

$$\overline{A \times B} = \overline{A} + \overline{B}.$$

Logical truth is often assumed to be a fact that cannot be challenged. It is not contradicted by any value of logic and/or mathematics. And simultaneously it is not concerned with the contents of the statements about which the logical conclusion is made. Of course, this is the same expression people use to imply that something must be "true." But as far as the discipline of logic is concerned, the term "true" means no more than "a convention is satisfied for using this particular term." This does not seem to be appropriate for a system called "intelligent," and later we will undertake some measures to make the usage of the term "true" more meaningful.

The truth table is a set of all the possible applications of logical truth:

x	y	AND $x \cdot y$	OR $x + y$	NAND $(x \cdot y)'$	NOR $(x + y)'$	COMPLEMENT x'	COMPLEMENT y'	IMPLICATION $x \to y' = x + y'$
0	0	0	0	1	1	1	1	1
0	1	0	1	1	0	1	0	0
1	0	0	1	1	0	0	1	1
1	1	1	1	0	0	0	0	1

[6]The term "empirical" is understood as arriving at a conclusion directly from the experimental data (or providing for a conclusion) without logical deduction. Instead of empirical verification, the ability to perform *symbol grounding* might be declared.

Several important concepts are linked with the idea of logical truth and with the processes of truth finding and verification. These are implication, causality, clause, and lookup tables (LUTs).

Implication is a logical statement that asserts that the truth of one phrase or predicate construction follows necessarily from the truth of another phrase or predicate construction: "It's raining → we will be wet." In this example we are dealing with a hidden temporal implication. (It is also tacitly assumed that neither shelter nor umbrella is available.) Temporal implications are usually equivalent to, or contain within themselves, a statement of cause–effect relations. However, in the following example of class property implication "Socrates is a man → Socrates is mortal," the temporal factor is weakened. The induction of prior observations has been transformed into a statement of a class property, so we see no cause–effect here. When causal relations are observed, many times a statement of the class property can be made such as the example about Socrates. In practical cases, causal relations are easily transformed into form of rules: prescriptions of what should be done. Rules contain the relation of implication between a set of "antecedents" and a set of "consequents":

$$B_1, B_2, \ldots, B_m \longleftarrow A_1, A_2, \ldots, A_m. \tag{2.5}$$

The rule in this form is also called a *clause*. On the right-hand side we have a list of prerequisites (precondition, or antecedents), and on the left we have a list of the results, subsequents. The following forms are similar to the rule-statements:

antecedent → consequent

cause → effect

When we deal only with one consequent, this rule is called a *Horn clause*. A list of horn clauses can be put into tabular form. Then the list is called a lookup table. All of these relations are checked for their ability to deliver equivalence in the end.

The concepts of *truth* and *implication* depend substantially on the concept of *equivalence* which might be interpreted differently in logic and in mathematics. This is an example of equivalence in the theory of sets: "a binary relation on a set is said to be an equivalence relation if it is reflexive, symmetric, and transitive." Equivalence in logic is understood as follows: Two predicates are called *logically equivalent* iff (if and only if) each one logically implies the other.

2.2.7 Models in the Form of Automata: Primitive Agents

Nevertheless, the pervasive form of modeling the reality is collecting together implications of interest and using them as a list in a form {antecedents → consequent}. Then we can consider this list to be a model of the system. Anything external can be considered an input "antecedent" for this list and imply all listed "consequents." The set of implications it invokes will be considered the output. This is how the concept of automaton has emerged.

The sets of implications (see δ and β below) are actually sets of rules. All mappings allow for using rules (the if–then statements) for the actual encoding of system representation. For example, in the hierarchy of *SP*, these are the rules for recognizing familiar images from the set of their detected and identified components. In the hierarchy of behavior generation BG (see Chapter 7), these are the rules of action. These rules are used to "recognize" what should be done if the situation is composed out of the familiar set of detected and identified components.

DEFINITION OF AUTOMATON

An *automaton* is specified by three sets X, Y, and Q and by two functions δ and β, where

(I) X is a finite set of inputs $\{x\}$
(II) Y is a finite set of outputs $\{y\}$
(III) Q is the finite set of states $\{q\}$
(IV) $\delta : Q * X \rightarrow Q$

The next state function is such that if at any time t the system is in the state $q(q \in Q)$ and receives input x, then at time $(t+1)$ the system will be in the state $\delta(q, x)$:

(VI) $\beta = Q \rightarrow Y$

The output function is such that the state q always yields output $\beta(q)$.

The automaton is finite if Q is a finite set. The following types of automata are applicable in practice: Mealy automaton, Moore automaton, Turing machine, and Markov controller.

Mealy machine
$\delta : Q * X \rightarrow Q$ state transition function (2.6)
$\beta : X * Q \rightarrow Y$ output function

Moore machine
$\delta : Q * X \rightarrow Q$ state transition function (2.7)
$\beta : Q \rightarrow Y$ output function

External and Internal Representations: An Agent Is Born

The external representation considers the automaton to be an input–output function, while the internal representation demonstrates the transition from an input to a new state, and then from this new state (under the particular input) to the output.

Certainly, the statement $\{x\}$ in the language of inputs should come from somewhere: from another automaton, or from sensors (that can be treated as another automaton). Similarly, the statement $\{y\}$ in the language of outputs should be submitted to someone: to another automaton, or to actuators (that can be treated as another automaton). Together with the automaton of sensors and with the automaton of actuators, the whole group can be considered "an agent."

DEFINITION OF INTERNAL DESCRIPTION

The *internal description* is obtained by specifying a new set Q and two functions: function of state transition:

$$\delta : Q^{*}X \rightarrow Q : (q, x) \rightarrow \delta(q, x)$$

and function of the output

$$\beta : Q \rightarrow Y : q \rightarrow \beta(q).$$

It is understood that if the system is in state $q \in Q$ and receives input $x \in X$ at time t then the output at that time will be $\beta(q)$, and the state at time $t + 1$ will be $\delta(q, x)$.

DEFINITION OF EXTERNAL DESCRIPTION

The *external description* is obtained by specifying a function

$$f : X^* \to Y$$

that maps the set X^* of sequences of inputs to single outputs in Y.

It is understood that if the system is fed the sequence $x_1, \ldots, x_n$ of inputs, the immediate subsequent output will be $f(x_1, \ldots, x_n)$.

Complexity of Automata

Until now, we have discussed a "reactive" mode of the automaton functioning: An input statement produces an output statement. We can use an automaton in a different mode, for solving the following problem: A particular output is required; find a string of inputs that gradually will lead to a desired output. Then each output is considered a source of external changes that produce new inputs. In a sense the external environment becomes another automaton, and functioning of these two automata is considered jointly (see Figure 2.2). In this case the critical issue is the value of complexity of functioning. Complexity of the automaton is evaluated by the number of computations it is supposed to perform in order to resolve a problem. If it requires several cycles of the automata, then complexity is equal to the sum of elementary computations performed at each cycle.

The vocabularies of inputs, outputs, and states correspond to the level of detail of our interest. In other words, they are determined by the resolution at which we want the reality to be considered. Thus all objects of the reality, their space-time coordinates, the scenes of the world and their processes, are represented as *automata* at some particular *resolution* within some *scope* of the problem.

Attributing resolution to automata makes all of them *fuzzy automata.* Thus, statements I through VI must be modified and (2.6) with (2.7) must be enhanced to incorporate the uncertainty.

2.2.8 Multiresolutional Automata

In this subsection we outline a number of maxims that allow for using the formalism of finite automata for representing multiresolutional hierarchies of Intelligent Systems.

Discovery of the Automaton at Each Level

FUZZY AUTOMATA

At each level of resolution, the components of the finite state automaton can be discovered. An automaton at a particular level of resolution is characterized by both its "state-resolution" and "time-resolution" (the time interval of "sampling"). The interval of sampling is closely related to the granulation intervals of "coordinate" representation (or the "zone of error" at a particular "zone of distinguishability" in computing the coordinate). These zones are fuzzy intervals. The set of all automata (one per level of resolution) is the multiresolutional hierarchy of automata.

Development of the Multiresolutional Automaton

In each couple of adjacent levels, the lower-resolution automaton is defined as an automaton whose input and output vocabularies (X_i and Y_o), states (Q) as well as transition and output functions (δ and β) are being generalized by the GFS procedure into a new, shorter input/output vocabulary, and transition and output functions. Thus a lower resolution automaton is formed. The same process can be described in the opposite way as follows: Each automaton with a particular set of $(X_{ilr}, Y_{olr}, Q_{lr}, \delta_{lr}, \beta_{lr})$ can be transformed into an automaton with languages, states, and functions described at a higher resolution $(X_{ihr}, Y_{ohr}, Q_{hr}, \delta_{hr}, \beta_{hr})$. The cardinal numbers for all elements of this quintuple satisfy these inequalities

$$[X_{ilr}] < [X_{ihr}], [Y_{olr}] < [Y_{ohr}], [Q_{lr}] < [Q_{hr}], [\delta_{lr}] < [\delta_{hr}], [\beta_{lr}] < [\beta_{hr}].$$

At high resolution the quintuple of the level automaton is implicitly constructed by the GFS^{-1} algorithm;[7] its elements are decomposed into the smaller subsets (shorter strings) which can be decoupled one from another, or they can retain their connections and generate "horizontal" links. Thus by the GFS^{-1} procedure a set of interrelated higher-resolution automata is formed. These procedures (GFS and GFS^{-1}) are applied again to the new automata (at lower and higher resolution). As a result a multiresolutional (MR) hierarchy emerges called *multiresolutional automaton*. We will also use synonyms of the term (multiresolutional, multigranular, multiscale, as needed).

Hierarchy with Horizontal Links

The MR hierarchy is not a tree. Rather, it has connections between the nodes of its automata at a particular level of the hierarchy. We will consider the analytical (e.g., differential and integral calculi) representation of information to be equivalent to automata representation: Transformation requires substitution of the differential equations by a set of corresponding difference equations. The equivalence between the MR automata and logical or linguistic controllers is straightforward. Thus, in using multiresolutional automata, we can cover all components of representation typical for hybrid controllers.

Functioning of a System as a Loop of Two Automata

SEMIOTIC CLOSURE

In Chapter 1 we introduced a loop that contains two strings (see Figures 1.1 and 1.3): S–W–A and SP–WM–BG. Each of them could be considered an automaton. The set S–W–A can be considered a *reality automaton* (A_r) which receives commands at the input of the actuator A and produces observations at the output of sensors S. In the meantime, the set SP–WM–BG can be considered *a computational automaton*, or an *intelligence automaton* (A_c) which receives the observations at the input of SP (sensory information); also it receives the goal, and then it produces commands at the output of the BG (control information).

This division of the overall system loop into two automata gives the meaning to this closure which is understood as a closure for information circulating within these automata (it is called *semiotic closure*). As a semiotic closure the system of two automata (see Figure 2.2) can be considered autonomously and separated from the rest of the

[7] This algorithm is a GFS-inverse. At this point it should be considered a mental experiment: We do not know yet whether an operation such as GFS is invertible.

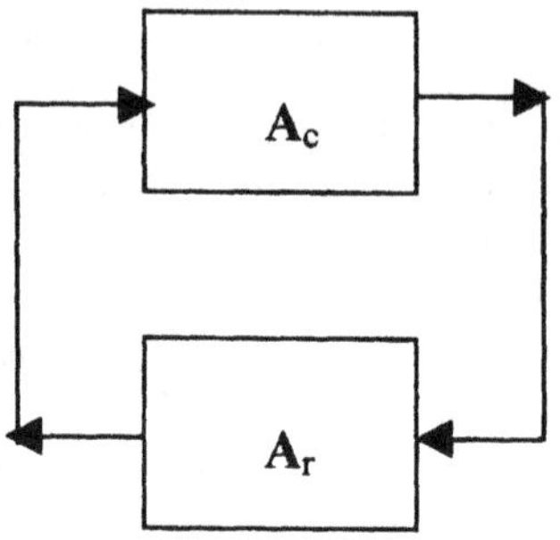

Figure 2.2. Loop of two automata: Semiotic closure.

world. What is unclear is its *goal* which should in some way either arrive from outside or be produced within the system. The notion of a goal will be considered later in the book.

Thus the whole intelligent system introduced previously in Chapter 1 can be considered as a connected couple of automata

$$A_c \leftrightarrow A_r.$$

2.2.9 Autonomy and Goal-Orientedness of Intelligent Systems

The subsystem of the world, W, is actually everything else in the reality besides the intelligent system under consideration. The module W belonging to a particular intelligent system loop is bound by the subset of reality associated with the intelligent system IS of interest and its functioning process. Thus the module W is not infinitely large; it allows for representation within the world model, WM.

The additional input arrow associated with the loop of intelligent system is the arrow of the goal, G. We will focus only on a subset of goal-oriented intelligent systems. All IS belonging to this subset, work toward achieving the goal submitted from some part of reality not included in W. It is our anticipation that G is either produced within IS, or submitted from another IS, or other loops of this particular IS.

Intelligent systems generate their own behavior. However, each level of the IS hierarchy works under an assignment (goal) that arrives from a lower level of resolution. This goal is always fuzzy, and the constraints delineated by the goal assignment allow for some degree of autonomy, generating behavior within these constraints with no strict guidance from the upper level (the level of lower resolution).

"Autonomy" is understood in the sense that IS will control itself with minimum use of the external support (including the tools for guiding and programming). Full autonomy presumes no external guidance and self-programming. Full autonomy requires introduction of learning capabilities.

2.3 NECESSARY TERMINOLOGY AND ASSUMPTIONS

2.3.1 Coordinates, Scope, and Resolution

Quantified properties of objects form the basis of the coordinate system in which the state-space is represented. Each point of the state-space is a state. A string of points in the state space is a sequence of states, presumably, a sequence in time. Everything related to construction of the coordinate system should be considered in the same way as in the theory of linear vector spaces.

DEFINITION OF REFERENCE FRAME

A coordinate system (reference frame) is a mathematical construct for locating points in a space with reference to a fixed point (origin) and the coordinate axes associated with that point.

The position of a single point in the N-dimensional space is specified by giving its N coordinates, $\{x_1, x_2, \ldots, x_N\}$. The coordinate space is particularly useful for describing what is known about the scope of the problem.

DEFINITION OF SCOPE

Scope is defined as the limited part of the world that contains objects and/or phenomena that are to be considered in the problem-solving process. The rest of the world is meant to be negligible, not significant for the problem solving. Scope is described by its boundaries. In this sense *scope* is a polytope in the coordinate space which is bounded by surfaces considered to be the borders of our attention. These boundaries are fuzzy in the reality, but for the sake of clarification of the problem, we specify sharp boundaries.

It is customary to assign convex polytopes in determining the scope of a problem. A (convex) polytope is the convex hull of some finite set of points. Each polytope of dimensions d has as faces finitely many polytopes of dimensions 0 (vertices), 1 (edge), 2 (2 faces), $\ldots$, $d-1$ (facets). Two-dimensional polytopes are usually called *polygons*, and three-dimensional ones are called *polyhedra*.

For the lowest level of decomposition where each subdomain is the elementary subdomain, we will introduce the *diameter* $d_{\min}$ *of the elementary subdomain (tessela, granule)* as the average diameter of the indistinguishability zone.

DEFINITION OF RESOLUTION

Resolution is a value that characterizes the limit of our ability to separate (distinguish) two quantities, for example, two signals (two units of information). To evaluate resolution we use a value inverse to the size of the indistinguishability zone.

In different areas there are different ways of evaluating resolution. In the area of chromatography, the following expression is recommended for distinguishing two time signals:

$$\rho = \frac{2(t_1 - t_2)}{w_1 + w_2},$$

where t and w are times and widths of two consecutive pulselike signals. From this description we can see that in evaluating the resolution, one should evaluate the zone of distinguishability, and this is intimately related to the practice of measurements in this particular area of activity. In linear measurements of distance, where the minimum unit of scale is Δ, the distinguishability zone is considered to be $\Delta/2$. Thus, instead of resolution, we often use the concepts of granulation and scale.

This example does not fit within a colloquial understanding of resolution which assigns higher-resolution levels to cases with smaller values of distinguishability zone. This is why we prefer to use a different form for evaluating resolution. The value of

$$\rho_k = \frac{1}{d_k} \tag{2.8}$$

is called resolution of the kth level of representation.

DEFINITION OF GRANULATION

The *granulation* of information is the quantitative characterization of the distiguishability zone, or *granule*. If granulation is a result of a uniform quantization of the space, the granules are considered spherical, and their diameter should be evaluated. Although granules are always convex bodies, they are not necessarily spherical. Sometimes they are characterized by their large and small diameters (D_{max} and d_{min}).

DEFINITION OF SCALE

Scale refers to the size of the representation as compared to the size of the object represented. Thus it uses the ratio

$$S = \frac{\text{size of representation}}{\text{real size of the object represented}}.$$

Thus, the issue of *scale* should be raised as soon as the task of measuring is formulated at a particular level of resolution.

If the statistical data are available, the level of resolution can be judged by averaging the error factor Δ_{av}. The averages can be taken in a number of ways. In all cases, the following should be taken in account: (1) The averaging can be done over the measurements of a single elementary subdomain Ω (e.g., the results of different crosssections of the subdomain), or (2) The averaging can be done also over the population of elementary subdomains belonging to the same subdomain of the adjacent level of lower resolution (we will call it the *upper adjacent level* of decomposition). (3) The averaging should be preceded by properly defining the domain (and subdomains) of averaging. It may happen that elementary subdomains belonging to the different subdomains of the adjacent upper level will have different diameters.

In order to compute the statistical data for measuring at the level of required accuracy, we must have information on the adjacent level of higher resolution, or higher accuracy (we will call it the *lower adjacent level* of decomposition). The value of resolution is determined by the density of detail and should be inversely proportional to the unit of volume of the space. It is customary to evaluate resolution in the terms of scale; the scale is inversely proportional to the unit of the single coordinate of interest.

All objects of reality, their space-time coordinates, their views of the world, and their processes are represented as *automata* at some *resolution* within some *scope* of consideration. Each finite automaton at a particular level of resolution is characterized by its time-resolution (the time interval of sampling). This interval is closely related to the intervals of the coordinate representation (or the "zone of error" in computing the coordinate). These zones are fuzzy intervals. The relations among all coordinate sampling fuzzy intervals and time sampling are to be determined from the corresponding "difference equations" of the transition and output functions, or from the transition and output rules (which are the same).

2.3.2 States and State-Space

Intelligent systems are characterized by rich and diverse temporal behaviors: To maintain successful behavior, they react to the environment dynamically, introduce changes, monitor external changes, and so on. Until now the description of temporal behavior of realistic systems was done by constructing models based on differential equations. The qualitative theory of differential equations was founded by Henri Poincaré around 1900, and it was developed in mathematics, physics, and control theory. The analysis was limited initially to linear systems; then nonlinear systems became the focus of attention. Complex systems with such phenomena as "chaos" are the more recent areas of application. This is how the dynamic systems theory has emerged and evolved toward the description of intelligent systems' behavior.

Formally it was understood that the model of a dynamical system is a set of first-order ordinary differential equations. It is defined on a manifold, the multidimensional analogue of a curved surface. The manifold is called the state-space of the system; it is equivalent to the coordinate system constructed upon the variables represented within the model at hand. Dynamical systems theory studies the qualitative properties of solutions of differential equations on manifolds or their discrete-time analogues. In the 1950s it became clear that the formalism of automata as a tool for modeling dynamical systems is equivalent to that of differential equations.

Dynamical systems theory has a special focus on the notion of stability. The stability of an individual solution of the computational model is the property of its behavior to persist and continue its prior motion when the initial conditions of the system are perturbed. More important is the notion of structural stability, which describes the architecture of the model that determines the property of stability. Structural stability is a formal way to generalize the concept of properly controlled behavior. The latter should not be destroyed by small changes that always happen in the world.

The model of a dynamical system allows for describing the time evolution of all variables reflected in the space of representation that can be thought of as the space of states (or the state-space). In applications of dynamical systems within many scientific fields outside of mathematics, the first goal is to explicate the manifold of states.

REALISTIC DEFINITION OF STATE AND SITUATION

A *state* (situation) of a system is information characterizing exhaustively the relevant set of objects, relationships, and the rate of observed changes at a particular moment of time.[8]

A model is used with n different variables. The list of these variables can include all physical measurable characteristics of motion and derivatives of them. Each variable is measured by a positive real number at a particular resolution of measurement and representation (a particular unit of measurement is presumed, and a particular way of processing). Thus coordinate space P will be the set of points of real cartesian space R^n with each coordinate a positive real number together with an evaluation of its distinguishability zone (fuzziness). One of the assumptions related to objects is the idea of *state* (or *situation*) which is an element of the *state-space*. A point of P together with its vicinity (distinguishability zone) represents a set of variables characterizing a particular object at a particular moment of time (or averaged over a particular interval of time). Thus in our state-space, every state exists with its *vicinities of distinguishability* that are subsets of the state space. If all inputs and outputs are added to the list of coordinates, we have an enhanced state-space which is convenient for describing the controlled motion.

Temporal behavior is often characterized by the concept of attractor. An attractor is a region of state-space that attracts and captures various motions and tends to determine the long-term behavior of the system. The equations that define a dynamical system can be fully deterministic, and under given initial conditions a uniquely specified behavior can be determined. Dynamical systems theory provides a major source of understanding for the phenomenon of chaos and attraction. Intelligent systems frequently find ways to oppose the laws of attraction specified by a particular dynamical system.

[8]This definition relaxes substantially the standard control theory definition which demands that "state" contain information sufficient for computing all subsequent motion (see sub-section 2.2.3).

At a particular level of resolution there exists the smallest distinguishable vicinity (*indistinguishability zone*). A state together with its smallest possible vicinity is equivalent to the *granule* of the state-space representation.

2.3.3 From Objects toward Entities

It fits the purpose of our work to *represent* (or *encode*) objects as *vectors* (or *lists*). We will define **vectors** as coordinates tagged with strings of numbers (or lists of numbers) and with a special meaning assigned to a particular position in the list such as $\{x_{i1}, x_{i2}, \ldots, x_{ij}\}$.

ELEMENTS OF STATE AND SITUATION

It is assumed that each ith object ($i = 1, 2, \ldots, m$) is actually a list of its n jth components, or n of the jth features. Each coordinate of the state-space is considered an object's *component*, *property*, or a *feature*. For the purposes of intelligent system we would recommend to use the concept of a standardized object (see Figure 2.1). We will distinguish two types of standardized objects: (1) *atomic* (or *elementary*) objects and (2) *composite* objects which are created by a synthesis (assembly, integration) of the elementary objects as a result of recursive hierarchies generation. Each composite object at a particular level of resolution can be substituted by a single state at a lower level of resolution. It is instructive to see what happens to objects when the system is considered at various resolution levels.

Level	Term Used	Entity	Characterized by
Very high resolution	Object	A molecule of the substance	State (vector of variables)
...	...	...	...
High resolution	Object (thing)	A door	State (image)
Middle resolution	Object	A house	State (image)
Low resolution	Scene	A village	Situation
...	...	...	...
Very low resolution	—/—	A state	Paradigm

2.3.4 Clusters and Classes

The GFS applied to the level under consideration (i) creates new objects that populate the adjacent level of lower resolution ($i + 1$). Correspondingly the objects of the ith level are created too by applying GFS to the information represented at the ($i - 1$)th level.[9] The first step in new object generation is to form a cluster. It begins with a search for classes of similarity. Then, within these classes, subsets are selected that qualify by their sticking together to form a new object. When there is similarity as a result of common attributes, it is a class generating similarity. However, when in addition to their similarity, the objects are close to each other in the state-space, this is the indication of the existence of a new object. For example, a multiplicity of red pixels in an image generates a class of red pixels. However, the subsets of red pixels situated in immediate

[9]Each set of objects is corrected by the results of applying the GFS^{-1} algorithm to the level above.

closeness form an entity of a segment which registers as an object and is considered to be a part of the object in the image recognition procedure.

DEFINITION OF CLUSTER

A *cluster* is considered to be a set of similar objects qualified to be considered an object at the lower level of resolution by the virtue of interpretable spatial closeness which is an evidence of unity.

DEFINITION OF CLASS

A *class* is a set of similar objects for which the property of spatial closeness does not hold.

The meaning of *similarity* varies from problem to problem. Often the members of a cluster are required to demonstrate similarity in a particular set of attributes. In this book we will frequently be interested in clusters with a particular predicate of similarity: the *distance between the cluster members.*

We should note that a major feature of information processing, and computational and control procedures, is *clustering* (*cluster generation, cluster formation, grouping, aggregation*, etc.). *Partitioning* (decomposition) is the inverse of clustering. Classification theory has applications in all branches of knowledge, especially in the theory of intelligent systems.

PHENOMENA OF GESTALT-INSTINCT

According to classical logic, in organizing a domain of objects into classes, we must leave no two classes with any object in common; also all the classes together must contain all the objects of the domain. This theory, however, disregards the frequent presence, in practice, of borderline cases: Some objects can, with equal correctness, be accepted or rejected as members of two otherwise exclusive classes. This is often seen in biology, since the theory of evolution implies that some animal populations have characteristics of two distinct species.

In practice, the principles used to classify a domain of objects are based on the nature of the objects themselves. In forming classes of perceptual objects, such as a class of green things or of elephants, the perceived similarities and differences between the objects are important.

The classification of objects requires a standard object against which all others are compared and thus included or excluded from a class. A domain of objects that never changes is classified morphologically (i.e., according to form or structure). If, on the other hand, the domain comprises changing or developing objects (e.g., evolving plants or animals), then it is likely to be classified genetically (i.e., in reference to crucial developmental stages). Sometimes objects are classified not so much by their characteristics as by the degree to which they possess them; minerals, for example, may be classified by their varying the values of hardness rather than by the characteristic of hardness itself. Finally classification by differences of quantity and of quality establishes equalities and inequalities of order or rank between different single objects within a domain as well as between different combinations of them.

2.3.5 Distinguishability

The concept of *distinguishability* (which can be numerically evaluated by the value of *accuracy* or *resolution*) alludes to determining the smallest distinguishable (elementary) vicinity of a point of the space with a certain diameter. This diameter can be considered a *measure*, and this vicinity can be considered a single *tessela*, (or *grain*, or *granule*), which determines *tessellation* (*granulation*) of the space. The relationship between reality and the tessellated representation is called the *scale*.

Each tesselatum may contain an infinite number of points, though for the observer they are indistinguishable. Each subdomain of the decomposition contains a finite number of elementary subdomains (granules, grains) that have a finite size and determine the accuracy of the level of decomposition. The observer will judge this elementary domain by the value of its diameter and by the coordinates of its center. We do not make any assumptions about the geometry of subdomains, which begs for a definition of the *diameter* of subdomain.

A standard definition for the diameter $d(\Omega)$ of the subset Ω of a metric space is presented via notion of distance D in a metric space as

$$d(\Omega) = \sup_{x, y \in \Omega} D(x, y), \tag{2.9}$$

and this holds for all levels of decomposition except of the lowest which is understood as the elementary subdomain in which points x, y cannot be specified and the distance cannot be defined.

SKILL OF NOTICING THINGS

Let us take one *arbitrary* point from the elementary subdomain Ω and call it the *center* of the elementary subdomain (it does not matter which point is chosen; at the highest level of resolution the various points of the elementary subdomain are indistinguishable anyway). The domain does not need to be circular to have a center. One can define the center as the point of intersection of the largest and the smallest diameters of the domain, or in any another manner.

In order to distinguish the centers of the elementary subdomains from other points of the space, we call them *nodes* and denote q_j ($j \in H$, where H is the set of all elementary subdomains). The relationships $\leftarrow R_{ijk}$ between two particular nodes i and j of a particular level k of decomposition are very important in solving a problem of control. The arrow reminds us that those relationships can be *directed*.

The relationships between two particular nodes depend on the special local property of the state-space which we call *traversability* and which affects the processes of motion in this state-space.

We will characterize this relationship by the *cost* of moving the state from one node (i) to another (j). Cost is a scalar number C_{ij} determined by the nature of a problem to be solved. Graphically cost is shown as a segment of a straight line connecting the nodes; this segment we call an *edge*.

For example, the value of traversability of a segment of terrain might be characterized by the relative reduction of the maximum speed of motion that can be developed by a vehicle traversing this terrain. The whole terrain can be characterized by a set of edges having a particular value of traversability. A set of all nodes and edges for a particular subdomain we call a *graph representation of knowledge* for the subdomain.

2.3.6 Representation

Instead of real objects and relations among them we can talk about corresponding *labels*. Labeling presumes applying a particular algorithm to the objects of the space and recording its results in the notation of the object. A *formal system* (i.e., any system represented by accepted symbols) has the following properties:

- It consists of spaces that are governed by a set of rules for their *generation* and *interaction*.
- It consists of elementary objects and relationships among them.

- All elementary objects are defined as represented at a particular value of *resolution* which is larger than a particular threshold value (threshold of *distinguishability*).
- The set of relationships at a particular level of resolution forms a semantic network at the level.
- Rules are introduced for the generation (synthesis), composite formation (clusters), and new elementary objects that constitute the new formal system.
- The set of relationships between levels of resolution forms a semantic network that determines consistency of level-to-level transformation of representation.

A set of mathematical relationships will be introduced for the level-to-level transformation of representation.

We consider n-dimensional euclidean state space R^n and a closed domain $\Omega : R^n \to \Omega$. This domain can be divided (decomposed) into a finite number of nonintersecting subdomains. If each pair of subdomains has mutual points at the boundary, then they are called *adjacent* subdomains. The relation of adjacency is a key property that represents the context of the problem to be solved.

Each subdomain can in turn be decomposed in a finite number of sub-subdomains, and so on. This sequential process we call *decomposition*. The opposite process of forming domains out of their subdomains we call *aggregation*. Every time we decompose (or aggregate) all domains simultaneously, the results of decomposition (or aggregation) we call *a level* (of decomposition, or aggregation).

STRUCTURE HOMOMORPHIC TO THE WORLD

It is essential that the system of representation be a model of the world with all knowledge relevant to a multiplicity of problems to be resolved. Thus two categories can be considered at this time: (1) the world category $C_W(O_W, M_W)$ and (2) the representation category $C_r(O_r, M_r)$, where O_W, M_W, O_r, M_r are objects and morphisms of the world and its representation correspondingly.

Categories are chosen as the most general algebraic structure. In many cases it will allow for an easy transfer to a more specific one. *Objects* are usually sets with all tools of the set theory to be applied. *Morphisms* are generalized relationships between the concrete objects in the category.

Thus *representation* can be defined as a structure (e.g., algebraic, information, and/or computational structure) that is homomorphic to the *world* (i.e., a domain of reality).

The following general properties of representations can be stated:

1. Representations consist of both numerical as well as descriptive information about the objects and systems.
2. Representations are assumed to be obtained partially from the prior experiences.
3. Representations are in part derived theoretically, based on a multiplicity of existing and possible tools of logical inference.

Since we consider a set of representations that form a recursive hierarchy, the latter can be interpreted as a multiple representation of the world. The number of representations is equal to the number of resolution levels.

2.4 CONSTRUCTION AND PROPERTIES OF OBJECTS

2.4.1 Formation of New Categories

Recall that new objects are constructed by using a special procedure called *clustering*, which is based on the applicable for of GFS procedure. (Its inverse GFS^{-1} performs the opposite operation of "unclustering"—decomposition, or instantiation, or specialization.) The results of both clustering and unclustering are *clusters*. Clusters are special sets of objects defined as sets that have the value of distance between each pair of the objects smaller than a particular threshold value (threshold of closeness).

Due to the recursive formation of clusters and generation of recursive hierarchies, we can consider clusters to be new objects, so we do not need to store and analyze all relations among their components. It is sufficient to do so only with the generalized objects, or clusters. The number of generalized objects is always smaller. Thus the complexity of the system representation is drastically reduced.

The set of clusters $\{C(E)\}$ is formed out of the objects of the space E by applying an algorithm $C(\cdot)$ of clustering that determines the set $\{E_c\}$ of subspaces of the space E which have the following properties:

1. The average distance among the objects neighbors does not exceed a particular value that characterizes this cluster.
2. The distances among the clusters exceed this value considerably.

Each cluster E_{cj}, $j = 1, 2, \ldots, m$ (m is a number of objects in the cluster) is considered an object of another category. One nonobvious example of clusters is the merging of words in a vocabulary; some combinations are in fact beyond any threshold of distinguishability.

LOOKING FOR SINGULARITIES

A partitioned representation of the world pertains to a distinct instant of time. Any particular $C_{\mathrm{r}}(t)$ can therefore be considered a snapshot of the world C_{W}. Changes in time are represented by sequences of snapshots in the domain of representation $\{C_{\mathrm{r}}(t_i)\}$, $i = 1, 2, \ldots$ describing sequences of the world states $\{C_{\mathrm{W}}(t_i)\}$, $i = 1, 2, \ldots$.

The decomposition of categories of representation is done through the decomposition of objects and morphisms represented in a category. The decomposition implies dividing objects and relationships into parts, so the components of objects come into interest, which was not the case before the decomposition.

This in turn implies a higher resolution of world representation than before the process of decomposition. If we continue with decomposition, the additional representations are expected to disclose new details about the objects and their parts, as well as about relations and their components. So, in the hierarchy of decompositions, all the levels describe the same world with higher and higher levels of resolution.

2.4.2 Recursive Hierarchies and Heterarchies of Representation

We introduce the new concept of *recursive hierarchy of representation* (RHR) because the standard concept of a hierarchy is not sufficient to represent the processes typical for such systems as RCS. As in the standard tree-hierarchies, each stratum (level, layer) is formed by the nodes where the branching is initiated. However, this branching always has the meaning of representing the entity-node before the branching (the parent-node) by a set of entity-nodes after the branching (children-nodes) representing more detail about the parent-node. Thus, at each level of RHR, all nodes form their horizontal relational network (HRN).

Although the standard treatment interprets these hierarchies as tree-structures with no horizontal connections, this is not the case for the multigranular systems of representation. In particular, the following fundamental properties of RHR state that the RCS hierarchical architecture can never be considered just tree-hierarchies:

1. All nodes of the tree are engaged in horizontal relational ("heterarchical") connections at each level where the node is formed by initiating a branching.
2. All horizontal cross-sections of this unusual "tree" represent the same phenomena, though at different levels of resolution.

The second property declares that for every couple of levels, the upper level demonstrates a particular representation, while the lower level shows the same with more detail. Instead of each node at the level i, we receive a group of nodes at the level $i - 1$. Therefore each edge of the ith level (the link connecting a pair of nodes) is transformed into a set of links connecting the nodes at the $(i - 1)$th level which are the children of the initial two nodes at the ith level.

As we move bottom-up (from children at the $(i - 1)$th level to their parents at the ith level), we consider each node at the ith level to be a result of GFS operator applied to the nodes of the $(i - 1)$th level. When we represent this hierarchy formation process in a metric state space E, each cluster can be considered an open ball (disc) A of the metric space E with the center $x_0 \in A$ and radius $R > 0$: The distance of any other point $x \in A$ from x_0 does not exceed value d (which is the *diameter* or *size* of the cluster and should not be confused with resolution ρ).

TRAJECTORY OF MOTION IS THE GOAL

The *trajectory of motion* is determined by a *substring of nodes* of interest which can be found within any representation at the level. Since the same elements of the level can generate different clusters, different systems of strings for the nested spaces could be obtained $[E_1 \subset E_2 \subset \ldots \subset E_j \subset E_{j+1}]_k, k = 1, 2, \ldots p$, where k denotes the different combination of tags, or meaningful positions of interest that have been chosen, and p is the total number of the combinations of interest.

It is important to understand the relations of inclusion which hold between the representations of different levels of resolution. Let us denote $\mathcal{R}(C_{k,i})$ as the world representation (utilizing categories of knowledge $C_{k,i}$ at the level i). Let us also denote the relation *of inclusion as a result of generalization* by a symbol $\xrightarrow{g}$ and the relation *of inclusion as a result of narrowing the scope of attention by a symbol* $\xleftarrow{a}$. The first inclusion reflects the fact that any description of the object at a level of resolution implies that the details which are shown at the higher level of resolution are implicit. The second inclusion means that any representation at a higher level of resolution is a result of curtailing the lower-resolution representation by focusing attention (narrowing the scope while going from a level to a level top-down).

Thus we have a bi-directional structures of multiresolutional nesting. However, at each level, the representation obtained by generalization $\mathcal{R}_g(C_{k,i})$ corresponds to representation obtained by narrowing of the scope of attention $\mathcal{R}_a(C_{k,i})$. These complex relationships that combine equivalence with bi-directionality of nesting are shown in the diagram as follows:

$$\begin{array}{l}
\cdots \xrightarrow{g} \mathcal{R}_g(C_{k,i-1}) \xrightarrow{g} \mathcal{R}_g(C_{k,i}) \xrightarrow{g} \mathcal{R}_g(C_{k,i+1}) \xrightarrow{g} \cdots \text{consistency string.} \\
\qquad \updownarrow \qquad\qquad\qquad \updownarrow \qquad\qquad\qquad \updownarrow \\
\cdots \xleftarrow{a} \mathcal{R}_a(C_{k,i-1}) \xleftarrow{a} \mathcal{R}_a(C_{k,i}) \xleftarrow{a} \mathcal{R}_a(C_{k,i+1}) \xleftarrow{a} \cdots \text{decision-making string.}
\end{array} \tag{2.10}$$

which implies the major basis of nested sensory processing and decision-making processes during behavior generation. From the upper string of (2.10) the rules of consistency check follow. From the lower string, a rule of ordering the decisions follows on the basis of nesting and the policy of decision making. We will formulate this rule as a theorem.

NESTING DECISIONS

Theorem of Nesting Decisions Given the following components of the problem formulated in a multiresolutional system of representation where the nested world representation is assigned by inclusions

$$\cdots \overset{a}{\longleftarrow} \mathcal{R}_a(C_{k,i-1}) \overset{a}{\longleftarrow} \mathcal{R}_a(C_{k,i}) \overset{a}{\longleftarrow} \mathcal{R}_a(C_{k,i+1}) \overset{a}{\longleftarrow} \cdots \tag{2.11}$$

and a set of cost-functionals for these representations, based on a common policy "p" of decision-making processes $D_{\rm p}$ forms the string of inclusions

$$\cdots \overset{g}{\longrightarrow} J_{g,i-1}[\mathcal{R}_g(C_{k,i-1})] \overset{g}{\longrightarrow} J_{g,i}[\mathcal{R}_g(C_{k,i})] \overset{g}{\longrightarrow} J_{g,i+1}[R_g\ (C_{k,i+1})] \overset{g}{\longrightarrow} \cdots \tag{2.12}$$

then the set of decisions will constitute a nested hierarchy:

$$\cdots \overset{a}{\longleftarrow} D_{\rm p}[\mathcal{R}_a(C_{k,i-1}), J_{g,i-1}] \overset{a}{\longleftarrow} D_{\rm p}[\mathcal{R}_a(C_{k,i}), J_{g,i}]$$
$$\overset{a}{\longleftarrow} D_{\rm p}[\mathcal{R}_a(C_{k,i+1}), J_{g,i+1}] \overset{a}{\longleftarrow} \cdots. \tag{2.13}$$

Corollary *(Nesting of optimum decisions)* Optimum decisions found at different levels are nested if they are found by applying cost-functionals which constitute a nested (by generalization) system to the set of representations nested by the scope of attention. In particular, minimum-time controls found at different levels of resolution, are nested.

Different representations of the same reality can lead to different hierarchical structures. It is important to find how different representations can be transformed into each other. Unfortunately, we can judge reality only by using its representations; we do not have direct access to it.

2.5 EXTRACTING ENTITIES FROM REALITY

2.5.1 Natural Grouping of Components

We would like to make clear that the property of intelligence to form entities is inherited from our experience of dealing with physical objects of the reality. These objects have a tendency to group and ungroup. It is easier to construct algorithms for dealing with entities if their connections with physical reality, and among themselves can be recorded, understood, and formalized.

PRELIMINARY DEFINITION OF ENTITY

An *entity* is a phenomenon of existence; it has properties of independence and separateness from other entities, that is, self-containment. An entity is the existence of the thing as contrasted with its attributes. An entity is the existence of the thing as separate and distinct objectively or conceptually. This adjacency of the very opposite ideas "objectively" and "conceptually" creates difficulties in applying this definition. It means that if one cannot find a distinction in real objects, it would suffice to find it in their mental representations, introspectively. Well, this is the harsh reality of generating entities: one should determine whether a particular set of objects, or a chaotic formation has some *unity* in it. Determination of *unity* depends on our goal and can be different in different cases.

In this subsection we will demonstrate the phenomenon of generality (the full GFS package of procedures or at least a grouping) in both the domains of nonanimate and living nature. This phenomenon of a general theoretical underpinning (involving possibly even a description) does not depend on the domain in which it is observed. It could perhaps be closely related to some of the key activities of intelligent systems recognition, planning, and decision making.

It is not important for us to know whether these processes can be characterized as "nonlinear" processes or as processes of "emergence." We are interested in finding a phenomenological and representational (semiotic) commonality among these processes. We begin by considering examples of grouping in inanimate nature.

Grouping in Hydrodynamic Processes and Gas Dynamic Processes

VORTEX IS AN ENTITY

The following hydrodynamic processes can be illustrative in demonstrating the phenomenon of grouping.

1. Turbulent motion is characterized by formation of whirls. As a stream is agitated, the molecules of water belonging to a single whirl behave as a group, and the movement of this group within the mass of water can be described. Similar processes are typical for the formation of twisters. A whirl can be considered an entity, and its behavior can be analyzed as if it were a solid object (i.e. a diagram of Figure 2.1 can be constructed for the whirl).
2. In a similar way separate "streams" can be considered distinctive entities, and the molecules of liquid belonging to the group can be described as a part of the stream's motion.
3. The formation of the drops of liquid out of a continuous stream can be considered a result of the grouping process. The decomposition of a stream into a set of droplets is a decomposition of one entity with a particular set of properties into a set of different entities with other properties.
4. The resultant formation of hexagonal cells in experiments of thermal convection operates in the reverse manner of separation within the mass of liquid. This is a staunching example of self-organization in such a "chaotic" medium as liquid (for details, see [7]).
5. The behavior of large masses of liquid with one density entering the liquid of another density is characterized by streams splitting into substreams in a hierarchical fashion, which can be viewed as a formation of groups. These phenomena have persuasive interpretation in the literature on synergistics [8, 9]. Very instructive are the photos of river deltas, drainage streams, and others of this sort [10].

In all these cases grouping is a result of local optimization processes; the optimization is obtained by minimizing a Hamiltonian in the vicinity. Notice that intuitively we started with a concept of "locality" opposing it to "global" processes which tend to get optimized by minimization of a different Hamiltonian.

The concepts of locality and vicinity presume a focusing of attention on an entity that has been formed in some region of space or is about to be formed. This entity is comparable to the "self" which is mentioned in the treatment of processes that allude to the concept of self-organization.

Grouping in Physics of Solids

The following phenomenon can be applied to group formation known within the domain of solids.

CAVITY IS AN ENTITY

1. Crystal growing is characterized by the grouping of the molecules during solidification. A single crystal is the result of clusters gathered around a local center of crystallization; more intricate entities emerge when elementary crystals form their group hierarchies during crystal growth. This is a fact of self-organization at two levels of provisional locality, or two levels of resolution: the level of a single elementary crystal and the level of a unified multiplicity of elementary crystals.
2. The process of fracturing is the result of mechanical forces acting upon a large solid object; it leads to the development of groups that differ substantially from those received as a result of crystallization. There are also processes of self-organization for the formation of fractures, and they are similar in form for different materials fracturing under different circumstances. Observation of geological fractures, fractures of the pottery (including surfaces of porcelain vases) demonstrate similar laws of grouping.

Grouping in Physics of Light and Electricity

Grouping is a working hypothesis that allows scientists to address the nature of the phenomena as follows:

SPARK IS AN ENTITY

1. The high density of energy in laser beams is achieved by contracting forces emerging in a stream of light where minimization of a Hamiltonian in that locality leads to the formation of the group phenomenon of a laser.
2. The discharge of electricity through a dielectric in laboratory experiments, in the industrial environment, and in nature (lightning) is paradigmatic of spontaneous grouping, and the grouping occurs frequently in a hierarchical fashion.

Grouping in Statistical Physics

FERMION IS AN ENTITY

As it turns out, the general laws of probability, especially those governing clustering, provide a good scientific basis for explaining the phenomenon of grouping within the populations of particles. An advanced interpretation of statistical physics gave birth to the physics of open systems in which the entropy can decrease [11].

Grouping in Quantum Mechanics

PHOTON IS AN ENTITY

From particles it is a direct path to populations of objects and their well-known property of information insufficiency. Many phenomena at the level of elementary particles, photons and down to the level of quarks and leptons are fascinating as examples of our intentions and successes in grouping the multifaceted experimental data into consolidated entities [15].

There is also enthusiasm toward linking quantum physics with the construction of "microtubule associated proteins" that can be instrumental for interpreting the synaptic connection of neurons.

Grouping in Cosmological Processes

UNIVERSE IS AN ENTITY

The most fascinating grouping processes are observed in the star formation of planets. The phenomenon of cosmological group formation is kindred to that of quantum mechanics. It is notable that the models in quantum mechanics and in cosmology are similar and mutually supportive as one can see from [12, p. 67]: "...if one was able to combine general relativity and quantum mechanics ... one would find that the singularities of gravitational collapse or expansion were smeared out like in the case of the collapse of the atom." We can see a striking resemblance descending, probably, from the resemblance among the areas: lack of experimental data and primary reliance on introspection.

Examples of Grouping from Biology

MYCOPLASMA IS AN ENTITY

As far as our interest in biological grouping is concerned, some obvious examples are (1) grouping in protein growing, (2) Edelman's grouping of cells (self-adhesion), (3) the phenomena of plants growing, and (4) grouping in living organisms.

These examples entertain the same set of broadly known facts:

- cells tend to stick together in groups,
- these groups usually form clusters and unite into larger formations,
- as a result, all living organisms can be described as bottom-up hierarchies that start with cells at the very bottom (the highest resolution) and end with organs at the top,
- similar branching structures are typical for the protein architecture for each leaf of a tree, for the development of living creatures at all stages from embryo to the complete species, etc.

2.5.2 Grouping within Scientific Processes

So, what is this, the law of nature, or the way our intelligence functions? Or maybe, both have the common mechanism?

Extracting entities from reality is a prerequisite for the transformation of one representation into another because the vocabularies formed for the world and the action descriptions are part of a conscious search process. Since the description of actions can be done through incremental changes in the world description, we arrive at a conclusion that building a vocabulary of the world is necessary for subsequent behavior generation.

SECOND PRELIMINARY DEFINITION OF ENTITY

Each vocabulary is a list of words. These words are symbols for encoding entities of the world. An *entity* is defined as a thing that has a definite, individual existence in reality or in the mind; something is "real" by itself. In other words, an entity of the world is anything that exists, has a meaning, and is (or should be) assigned a separate word. Therefore the functioning of intelligent systems requires understanding how entities can be discovered within the world.

Our notion of reality is presumed to contain (1) entities that we have already learned and (2) the rest of reality that has not yet been learned, and therefore we cannot list its entities. It is prudent to regard anything unlearned to be a continuum, and a process of learning to be a process of discovering entities within this continuum. (A *continuum* is defined as a thing whose parts cannot be separated or separately discerned.)

From the natural sciences we know that physical laws work in such a way that singular entities are formed from the initially uniform media, and thus are separated from the continuum. We mentioned that these entities are assigned symbols, and they become words in vocabularies.

The separation of entities from the continuum is a result of thermodynamic processes and mechanical motion. Indeed cosmogonic theories of the birth of planets and stars from the chaos of previously disorganized matter illustrate entity generation. For additional examples, see sources on processes in nonlinear systems and catastrophe theory.

The natural processes in the physical world led to the transformation of the primordial chaos (actually, a continuum) into a collection of relatively "thick" zones where the particles of matter could stick together (they became entities) and the less thick zones with relative uniformity (they are remainders of the continuum). Since entities have a status of words and are parts of models we create, they must have meaning assigned to them. Notice that the concept of scale is not mentioned, although this concept is implicit in the description. Indeed the particles of matter mentioned above are probably results of entity separation from the continuum on a substantially finer scale. We have not mentioned "scale" yet because the concept of scale presumes an observer, while the processes we describe develop in nature no matter whether an observer is present.

LOOKING FOR SINGULARITIES IN A CHAOS

Thus a physical decomposition takes the place of a chaotic world separated into diverse kinds of uniform media. The large variety of singular entities that emerges can be thought of as a natural classification that happens in the world independently of

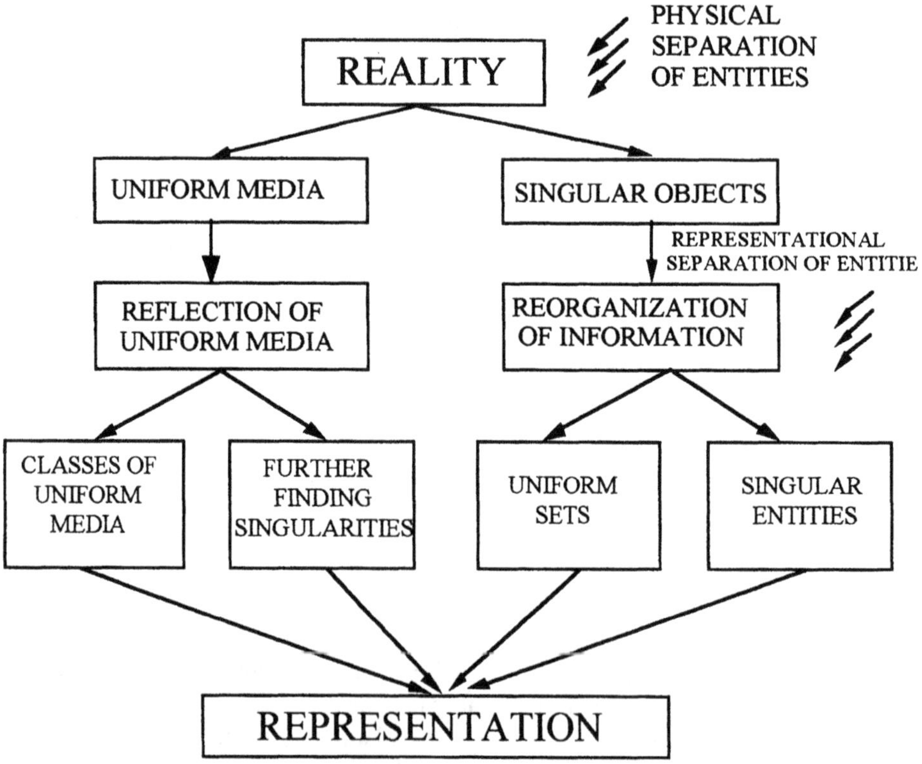

Figure 2.3. Stages of entity formation in reality.

the observer (see the first top-down forking of reality in Figure 2.3). The formation of singularities (as entities) can be metaphorically described as a result of the gravitation of elementary units of the chaos toward each other in the areas of continuum with a higher density of these elementary units. In clarifying this metaphor, we should emphasize that it is irrelevant whether the density is increased as a result of the gravitation, or whether the gravitational strength is increased because of the prevailing increase in density.

Let us consider these processes in computational terms. At this point the observer will legitimately appear in our presentation as a carrier of two interrelated concepts: the scale and the resolution. (Resolution is determined by the size of the smallest distinguishable zone, a pixel, or a voxel of the space in which we describe our system. Scale is a value inverse to the resolution.)

The concept of scale allows for the introduction of a formidable research tool which can be applied for each couple of adjacent levels of resolution. This tool is related to the specifics of a different interpretation of units in higher and lower levels of resolution (HLR and LLR).The units of the HLR emerge as a result of the singularity forming process at the previous, even higher-resolution level (which is not a part of our couple levels of resolution under consideration). As these singularities are formed, they receive an interpretation, a meaning, a separate word of a vocabulary at this HLR. For the LLR of the pair of levels which we discuss, these singularities have no meaning yet (or just is not recognizable yet).

The meaning will emerge when these entities of the higher level of resolution (HLR) assemble together into a singularity that can be recognized at the lower-resolution level (LLR) as a meaningful (recognizable) entity. Before the grouping of these entities into meaningful singularities occurs, they are just nameless units with a tendency to gravitate toward each other, expressed in the set of their relations. So the process of entity formation for LLR recognizes the entities of an HLR just as a set of anonymous units. The gravitational field "attracting" them to each other is to be investigated which can give birth to a new entity of LLR.

Uniform medium is always a collection of some nonuniform units at a finer scale (at higher resolution), and uniformity of the medium is a parameter that we obtain from characterizing the medium at a coarser scale (at lower resolution). In order to compute this parameter (the degree of uniformity, or density), different techniques can be used. All of them work as follows: Let us consider a particular zone of the medium which we intend to evaluate; we will call it *the scope of interest*. An imaginary large window (*the scope of attention*) is to be imposed upon the medium (*scope of interest*). Then a smaller window is slid within the scope of interest to evaluate the information density. Notice that the size of the scope of attention is presumed to be substantially smaller than the scope of interest. The density of the nonuniform units is to be computed within this window which allows us to evaluate the continuum quantitatively. In each position, as the window is slid over the whole scope of interest, the density is again computed.

The window-sliding strategy is assigned in such a way that all scopes of interest can be investigated efficiently. This strategy can be varied in constructing different models: We can scan in a parallel manner, we can provide a spiral trajectory of scanning, we can take random sampling from different zones of the scope of interest; the selection strategy should depend on the needs, hardware tools, and resources available (including time). If the values of density are about the same everywhere (with small variations within some particular interval), then the medium is considered to be uniform.

GFS GENERATES ENTITIES

Notice, first, that in order to introduce the concept of uniformity, we used a sliding window which is one of the techniques of *focusing attention*; second, that in order to form entities of a particular level of resolution, we have to *group* the units declared entities at the higher level of resolution; and third, that to find candidate units for grouping we have to *search* for future members of these groups or otherwise *combine* them together. Later we will return to these operations as components of the elementary unit of intelligence.

The subsequent classification is performed within a system by an observer. The observer perceives a multiplicity of zones of uniform media (UM) with various degrees of uniformity, and groups them in different classes. The sets of uniformity classes can be thought of as singularities by themselves; thus singular classes of uniformity in addition to singular entities are determined as a result of perception. Then the whole host of singular objects is informationally reorganized, and new sets of objects are formed pertaining to different levels of resolution. At each level of resolution there are additional singular objects: those left out from the previous grouping processes. These "left-out" entities supplement the multiresolutional system of entities that has been received. After this, a new iteration of grouping is supposed to be performed at each level of resolution.

2.5.3 Grouping Leads to the Multiresolutional Architecture

The computational process of entities formation allows the multiresolutional system of world representation to be obtained as a nested hierarchy of symbols (words). The hierarchy is formed as a result of the consecutive generalization of information at the upper level of ELF introduced in Chapter 1 (Figure 1.1). By the virtue of such consecutive generalization the elementary agent (ELF) is being transformed into a multiscale coalition of agents (GSM). The hierarchy of GSM is shown in Figure 2.4. (It is a much simplified structure and also for the sake of clarity, the box of value judgment is omitted from all levels of resolution.)

One should remember that each level of resolution of GSM is a part of its ELF as shown in Figure 1.2 (see Chapter 1). Thus, we must warn: talking about processes at a level in Figure 2.4 may lead to a mistake if the complete ELF would not be taken into account providing full closure.

This hierarchical stystem (more complete in the version of Figure 1.2) should provide for both goals achievement and subsistence for the system. Thus, we will refer to it as *generalized subsistence machine* (GSM).

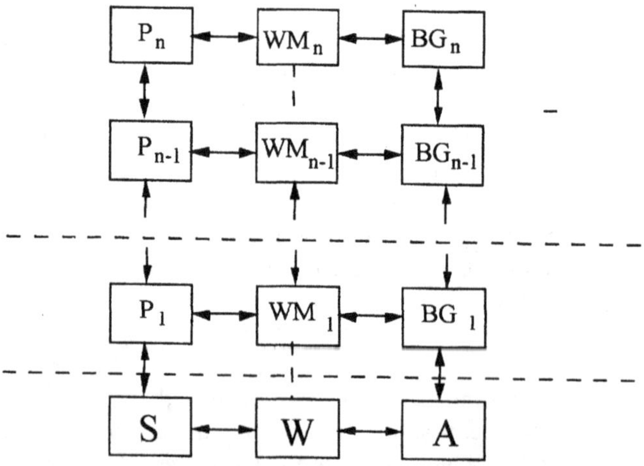

Figure 2.4. Multiscale coalition of agents.

As the vertical (top-down/bottom-up) process continues in locating entities and classes of media in which these entities are immersed, horizontal processes develop that verify the meanings of the new entities and the new media. The goal of a horizontal computational process is to explore how new sensory information has been transformed into the output of perception, how the latter fits within WM at each level of resolution, and how meaningful actions can be implied by the world model at this level and the goal produced by the adjacent level above. Without this process of verification, the multiresolutional system of world representation would not be viewed as consistent (see Figure 2.4). From the multiresolutional GSM that is formed by this strategy of computation, it is clear that vertical consistency within the multiresolutional system of world model (WM) is not sufficient. Two additional consistency checks are required: one is the top-down/bottom up consistency check within perceptual system, and the other, the top-down/bottom-up consistency check within the system of behavior generators (BG). The latter produces a system of goal decomposition which is critically important for GSM functioning.

SYSTEMS PROVIDE BOTH GOALS PURSUIT AND SUBSISTENCE

Multiresolutional hierarchy allows for the complexity of information processing to be minimized by selecting the optimum number of resolution levels [13]. For different systems it gives different numbers of levels. Thus there can be a problem of communication between two or more hierarchies belonging to the same real system, and yet having different numbers of levels in their models. We always have at least two hierarchies in large systems like manufacturing systems and military organizations. Indeed, in each of our GSM we have two systems whose parallel existence is a prerequisite for a successful functioning of the system. These systems are the system of GSM for goal-directed functioning (GDF), and the system of GSM for regular subsistence functioning (RSF). These two subsystems are frequently separated. GDF represents the ability of the intelligent system to cope with a set of all goal-oriented practices (but all of them are comparatively short-term activities). RSF deals with the ultimately long-term problem: the subsistence which provides reliability and even viability of the system.

In Figure 2.5 we illustrate these GSM hierarchies of GDF and RSF. Both are required for the functioning of the system, but they are different in their goals, realizations, and their number of levels. For the manufacturing company example GDF is a hierarchy associated with the manufacturing process, while RSF is a hierarchy of maintenance and other services. Levels belonging to different hierarchies must communicate whenever their vocabularies allow for this (see Figure 2.5). At the present time the problem has not been addressed theoretically, although a theoretical analysis can be expected in the future. RSF corresponds to infrastructures.

The "technology" of dealing with a continuum was known in practice and theoretically anticipated by mathematicians. We will mention here only the testimony of E. Schrödinger who eloquently described the trade-offs that are always present when the scale, accuracy, and/or chunking of continuous information are involved [14, 15]. In analyzing the process of dealing with a curve containing pieces of functions of the second order, he writes:

> We claim to have full knowledge of every point of such a curve, or rather, *given* the horizontal distance (abscissa) we are able to indicate the height (ordinate) *with any required precision*. But behold the words "given" and "with any required precision." The first means "we can give the answer when it comes to it"—we cannot possibly have all the answers in store for you in advance. The second means "even so, we cannot as a rule give you an absolutely precise answer" [15].

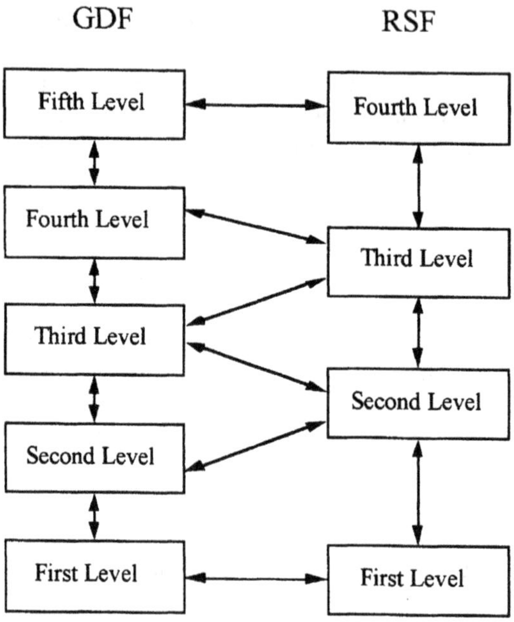

Figure 2.5. Interactions among the multiscale coalitions of agents of GDF and RSF.

In order to proceed with these processes of forming and classifying singularities and the media surrounding them, the following operations are required. They are applicable to both the physical objects and their symbolic representations.

1. Grouping of the units (physical and informational)
2. Selection of subsets of these units (subsets are "preferable" in some cases)
3. Searching among the units, sets, and subsets to support subsequent processes of clustering, and otherwise constructing the combinations

The main premise of our theory is that the world can be represented at multiple levels of resolution (scale) simultaneously. There are many considerations in forming a system of multiresolutional (MR) world representation. We will mention here only the most prominent ones:

MAXIMS OF MULTIRESOLUTIONAL REPRESENTATION

- The world consists of objects and agents.
- Objects and agents are characterized by attributes that can be considered their coordinates.
- Objects and agents consist of parts, which are usually objects of higher resolution.
- Objects and agents can be decomposed into their parts, as needed.
- Objects and agents can be parts of other objects and agents.
- Objects and agents are related to other objects and agents; these relations are listed.
- All numerical data known for objects and agents are fuzzy data.
- Objects and agents are subjected to processes of self-organization in the form of GFS and GFS^{-1} operators.
- Objects and agents at a particular level of resolution are obtained from objects and agents of higher resolution as a result of using the GFS procedure, or from objects

and agents of lower resolution as a result of using the GFS^{-1} procedure. These procedures describe linkages between adjacent levels.

- Changes in objects and agents happen because of actions (actions are defined as causes of changes emerging in objects and agents).
- Actions are to be stored in the world representation as a part of object and agent representation: (a) the list of actions that can be applied to the object and agent and the changes they produce; (b) the list of actions that can be produced by the object and agent and the changes that are supposed to happen.

2.5.4 Differences between Abstraction, Aggregation, and Generalization

In this subsection we will clarify the meaning of these frequently used words. Their use can lead to confusion. We have introduced an operator of generalization (GFS). Should we not then have a separate operator of abstraction for discovering types? Should we not have a separate operator of aggregation for making assemblies?

The words *abstraction*, *aggregation*, and *generalization* are used loosely, and this can be confusing during the cross-communication among professionals from different domains of knowledge in science and technology. We will use the following definitions which are typical within the scientific and industrial community, as well as engineering, computer science, physics, and biology. The purpose of these definitions is to distinguish different cases of entity formation.

Case 1

An entity is formed out of its parts. Information of belonging to the entity is contained in the description of the parts. We will consider this process to be representative of a very simple group formation: We know what is the whole, and we know what are the parts. The assembling of the parts into the whole, or the formation of the aggregate, is determined by certain specifications.

DEFINITION OF AGGREGATION

An *aggregation* is formation of an entity out of its parts. Each of the parts can also be obtained as a part of aggregation. *Synonym*—assembling; *antonym*—decomposition.

Case 2

Interestingly enough the properties that characterize objects can be considered objects in themselves. We are not surprised if someone calls kindness an entity. The fact that color is a property belonging to most physical objects of the real world makes it an important scientific and technological entity of the system of knowledge. It is important to indicate that the formation of an entity is possible by grouping together all similar properties of different objects. Red apple, red ink, red bird, red cheeks, all belong to the class of objects containing "redness."

DEFINITION OF ABSTRACTION

The formation of a class of objects characterized by the same property, and the labeling of this class with the name of this property is called *abstraction*. *Synonym*—class formation, and (sometimes) abstraction; *antonym*—specialization.

Case 3

Often as people encounter new situations (which happens all the time), they find the need to create groups. They may assemble together components that are not previously thought to be parts belonging to each other and intuit new classes of properties. This is typically what leads to technological inventions, new business ideas, and better medical diagnostics.

DEFINITION OF GENERALIZATION

A *generalization* is a formation of new entities (groups, classes, assemblies) where the parts to be assembled are not prespecified, and new classes of properties can emerge. *Synonym*—(sometimes) abstraction; *antonym*—instantiation.

Note that generalization performs aggregation even when parts are not specified. This means that it subsumes the aggregation. It subsumes the abstraction as well. In all cases concerning abstraction the term "generalization" is applicable. Generalization is typically applied when a similarity and observations are discovered and a general rule should be introduced. The term "abstraction" is inappropriate in this case, since *generalization subsumes both aggregation and abstraction*. Thus we have a more general procedure for which aggregation and abstraction are particular cases.

2.6 GROUPING + FILTERING + SEARCH: THE ELEMENTARY UNIT OF SELF-ORGANIZATION

2.6.1 Concept of GFS

NEW ENTITIES GENERATION

The development of different forms of organization of objects and agents is enabled by the properties of the GFS package which we described above. We can illustrate this package as a triangle whose vertices communicate with each other while processing the information (Figure 2.6*a*). The consecutive application of the procedures in vertices provides for all processes characteristic of self-organization in systems of objects and/or agents. The spiral in Figure 2.6*b* demonstrates the development of the multiresolutional system of the world model. While this spiral works in each box of GSM, the loops of GSM should watchfully explore whether the results of generalization produced by the GFS include the new entities generation, do not violate consistencies within the GSM loops at each level, and proceed vertically through the subsystems of P (SP), WM, and BG. Let us apply our scheme to different processes of intelligence.

The development of human intelligence would be impossible without attention focusing, and the window of attention bounded at the top and at the bottom. Indeed, our scope of interest in visual perception is bound by our field of view, our scope in hearing is bound by the audio-frequency interval, and so on. At the bottom, the resolution is also constrained, and it is good because if we are able to discern all the molecules and atoms in everything that surrounds us, we would drastically overburden our mechanisms of intelligence.

WINDOWING AND SEARCHING

FOCUSING ATTENTION

In addition to the windowing in the "perception" box of the GSM, we have an additional mechanism of windowing at the lower levels of resolution. Indeed, our communication with other people would be impossible if we could not focus our attention selectively only on the person with whom we are speaking at the moment. Our problem-solving processes would be unthinkable unless we are able to concentrate only upon the problem we decide to undertake.

Certainly our understanding of processes in human intelligence is incomplete, and we often catch ourselves on the fact that being involved with one problem consciously, we are solving also another problem and suddenly come up with a solution seemingly as a result of some parallel or time-sharing subconscious activities. Nevertheless, the power and need of focusing of attention is undeniable. It also leads to mistakes when

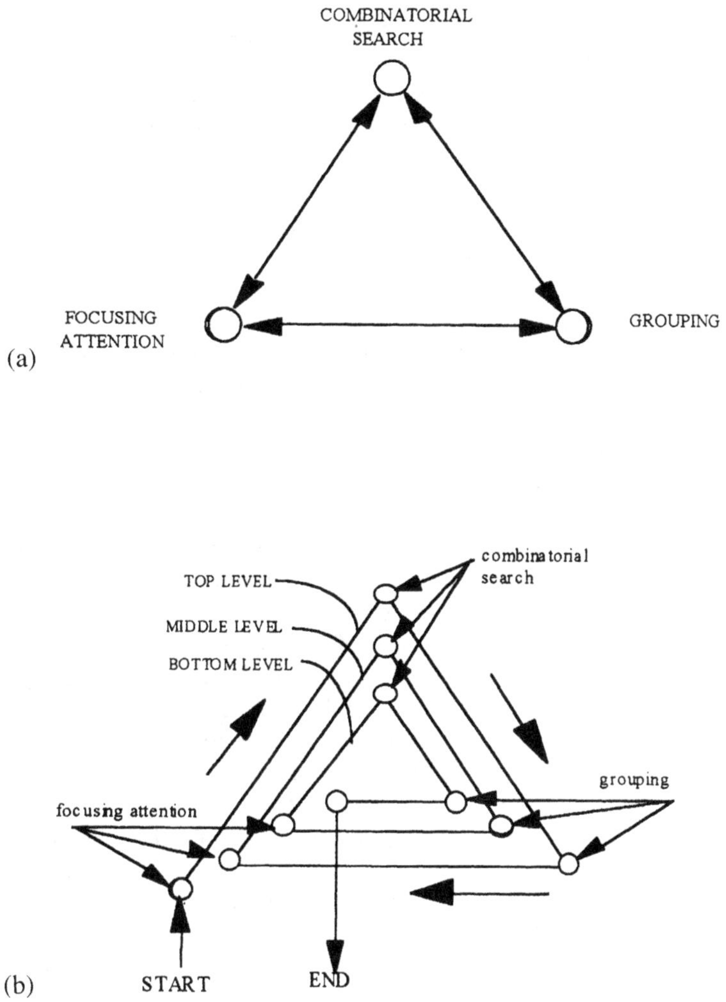

Figure 2.6. Elementary module of self-organization (GFS): *(a)* At a level; *(b)* functioning a multiresolutional system.

we cut off zones of the state-space which should not be sacrificed. This is why the reliability of success in our thinking machine is less than 100%.

INTERLEVEL COMMUNICATION

It is important to mention that GFS (and GFS^{-1}) are the means of interaction of the adjacent levels of the hierarchy of levels. They provide for two self-verifying waves of processing: the bottom-up (GFS) and the top-down (CFS^{-1}) ones. The bottom-up wave creates and re-creates entities of the lower-resolution levels. As soon as some information arrives at a level (e.g., from its sensory processing), the effort of re-clustering the objects for the upper adjacent level should be undertaken (GFS). However, this information might contradict the results of clustering that have arrived from the level of higher resolution. In this case a tentative instantiation should be performed, and the results of it should be submitted to the level of higher resolution (CFS^{-1}).

2.6.2 Functioning of GFS

GENERALIZATION

GFS performs the operation of "generalization." It unifies multiple high resolution units of information into new units of generalized information. The inverse functioning of the unit (GFS^{-1}) performs the operation of instantiation. It finds for a particular unit of low resolution information what is the high resolution unit that at the level of higher resolution could serve as a representative of the low resolution unit. Generalization is performed by lumping high-resolution entities into lower-resolution entities. This requires performing procedures of focusing attention upon a domain of interest (FA), combinatorially searching (CS) for subsets of similarity in this domain, and grouping (G) the subset of similarity into a new object for which the new label is created. These procedures form the algorithm we refer to as GFS, which amounts to a procedure of generalization.

SENSOR FUSION

Generalization in the subsystem of SP is usually associated with "sensor fusion" when, instead of signals from a multiplicity of sensors, a reduced set of "virtual sensors" is considered for the purposes of the subsequent world modeling and behavior generation. In turn, the latter two subsystems (WM and BG) demonstrate reduction of the number of coordinates as a part of the process of generalization.

FINDING THE BACKGROUND

Another phenomenon typical for generalization and associated with GFS functioning is reduction of the accuracy of information at hand (or the increase of the value of error, or coarsening of granulation, or increase in scale—all are equivalent phenomena). The conceptual roots of these phenomena are the explicit or implicit utilization of approaches from statistics or theories of stochastic systems. Indeed, the averaging of data presumes treating a set of samples representative for the state-space zone which is to be characterized by the value of some "uniform" property. Averaging along the time scale presumes implicit use of the ergodic hypothesis.

We will call instantiation the inverse procedure, GFS^{-1}, consisting of combinatorial search (S) for groups (G) candidates in becoming subsystems for subsequent decomposition, and filtering, or focusing of attention (F) upon them and their relations with other elementary objects. Instantiation in the subsystem of SP is usually associated with "sensor decomposition" when, instead of signals from a single virtual sensor of a low resolution, an increased set of (realistic) sensors is considered for the purposes of the subsequent world modeling and behavior generation. In turn, the latter two subsystems (WM and BG) demonstrate an increase of coordinates in the instantiation process. A phenomenon typical for instantiation is the increase of information accuracy (or the reduction of the value of error, or refining of granulation, or reduction in scale—all are equivalent phenomena). Again, the conceptual roots of these phenomena are the explicit or implicit utilization of approaches from statistics which sometimes allows for the restoration of components.

Focusing of attention frequently works as a prerequisite for *grouping*. In order to form a group (to cluster, to build a class) and then to label it (to attach a symbol, a term, a word), thus creating a concept, we have to cut off many high-resolution entities which we decide not to include in the cluster. We use the evaluation of "similarity" and have a threshold value: Anything beneath this particular value of similarity does not belong to the group. For evaluating the similarity we use a "measuring" of potential attributes (properties of the future, which is a hypothetical entity that we create as a result of concept formation).

In some cases the results of *grouping* are obtained first and then subjected to curtailing, which is also *focusing of attention*. This happens when grouping is simple, and it is easy to produce many candidates so that the best can eventually be selected. It does not matter in which sequence these processes are executed (grouping, then focusing of attention, and vice versa), the bottom line is to form cohesive groups.

When we cluster based on the value of similarity, upon gravitation toward a future cluster, we thus use the ontology of the state-space: "This is what happened to be around us." If we are not satisfied with our ontology of the world, then we start *searching* for possible components of the future concept that are not around. We might also be interested in intentional creation of alternatives among which the further searching is performed. Since the combinations are intentionally created, we call it a *combinatorial search*. The latter is utilized in automated discovery [16], in systems of planning (e.g., [17]), and in systems of genetic search [18].

The component procedures (G, F, and S) are understood in the most general sense and can be realized by various computational procedures. Generalization employs usually algorithms of averaging within a sliding window, pyramidal averaging, numerous clustering algorithms, and the like. From the description of GFS functioning, one can conclude that besides the spatial coarsening which is characteristic for the lower-resolution levels, there are temporal processes that do not contain higher-frequency components. This is a result of the increase of both the sampling intervals and the bottom-up generalizations performed by GFS.

No matter what the package is that one acquires for an intelligent system, no matter from which company it is bought, this package, if it is complete, will perform attention focusing, grouping, and combinatorial search. If the package is good, it will perform GFS at several levels of resolution, and if it is complete, it will provide, or require for the consistency, checks along the GSM loop at each resolution level and vertically across the P, WM, and BG subsystems.

Figure 2.7*a* presents an ELF in which the GFS unit provides for the functioning of P, WM, and BG GFS services to all subsystems at a certain level; Figure 2.7*b* shows the nested system of GSM operators (for more details, see [19]).

Recall that GFS components shown in the figure have parameters that are assigned based on function; these are the functions listed in Table 2.1.

Table 2.1. Parameters of GFS Varied by the GSM Assignment

Focusing of Attention	Grouping	Combinatorial Search
• Relative importance of entities • Value of resolution • Range or scope	• Resemblance • Compatibility • Closeness • Cohesiveness	• Goodness for a particular goal • Heuristic for alternatives formation • Number of alternatives

2.7 RELATIVE INTELLIGENCE AND ITS EVOLUTION

A persistent problem is whether it can be demonstrated that different systems have different amounts of intelligence. In this section we will demonstrate how the level of intelligence of an intelligent system evolves from the simplest possible form to the highest known levels of sophistication (see Table 2.2).

SIMPLE REFLEXES

Stage 1. Selecting a rule from a set. The simplest knowledge base contains a "stimulus–response" schema: If a goal G is desired in a situation S_1, apply the action A in order to get to situation S_2 (i.e., $[S_1 \rightarrow A \rightarrow S_2]$) and obtain goal G. Here the situation S is the perceived "stimulus." The set S_2 generated by S_1 corresponds to a mapping of this situation in the vocabulary of inputs. S under G produces A which is the required response, or action.

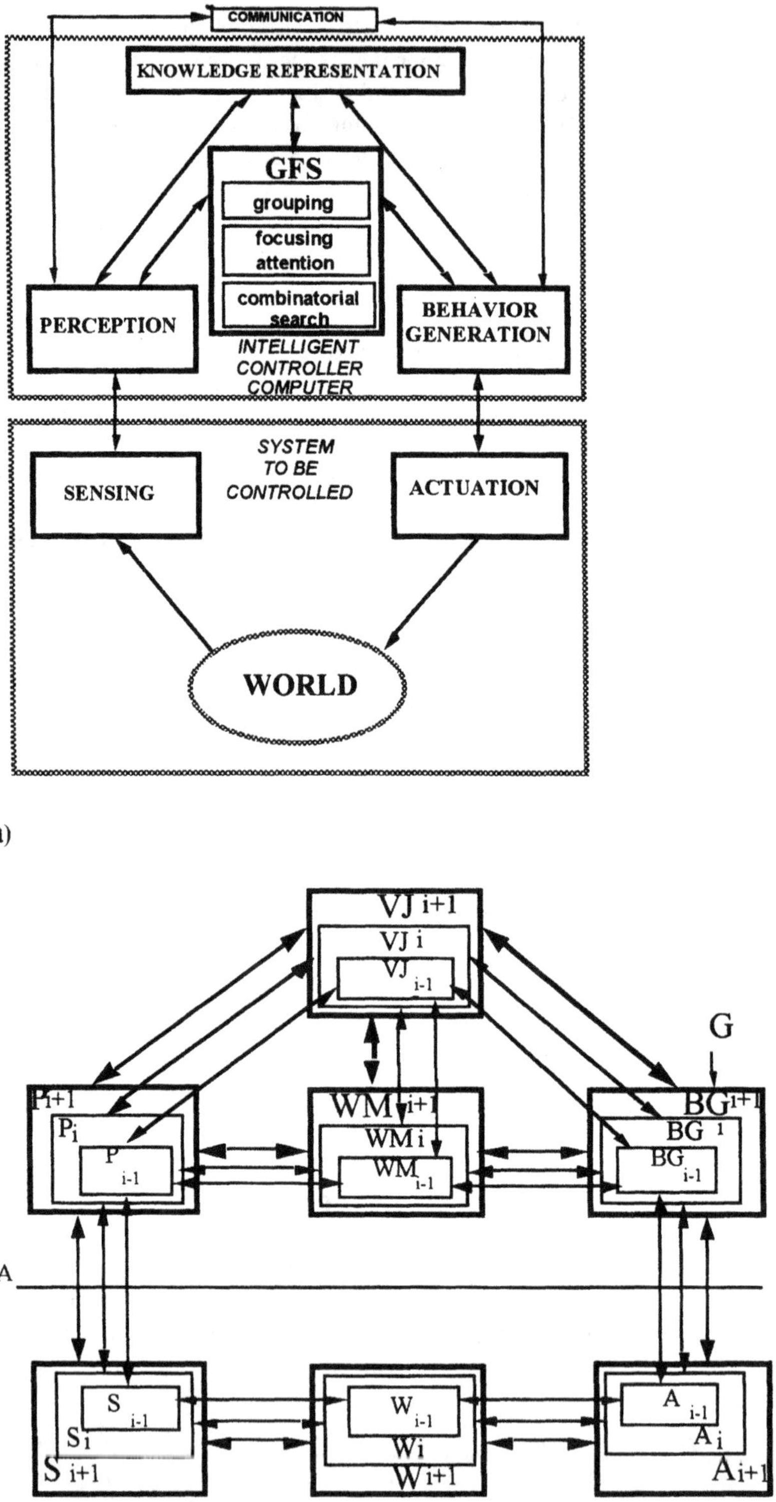

Figure 2.7. GFS as a part of the ELF structure *(a)*. As a result the set of nested ELFs is obtained *(b)*.

Table 2.2. Properties Acquired at Different Stages of the Intelligence Development

Stages	1 Selection Rules	2 Combinations	3 New Rules	4 Grouping of Rules	5 Synthesis of States	6 Synthesis of Context	7 Synthesis of Paradigm
Capability							
1	×						
2	×	×					
3	×	×	×				
4	×	×	×	×			
5	×	×	×	×	×		
6	×	×	×	×	×	×	
7	×	×	×	×	×	×	×

As a set of sensors delivers the *stimulus*, the storage of memorized responses (presumably this is the world model, WM) sends the corresponding rule to the behavior-generating (BG) subsystem; then BG informs the actuators that they are supposed to execute the required action *A*. This is not very rich intelligence, but for an *E. coli* or a simple air conditioner this level of intelligence is sufficient. One can imagine a richer intelligence where several rules are stored using information from different sensors as stimuli and evoking different actuators as far as response is concerned. The simple automated machine has this level of intelligence.

Technically the intelligent system at the stage 1 of intelligence performs only "search" and "focus of attention" (selection) by its GFS package. It does not create combinations of different alternatives yet; this property emerges only at stage 2.

COMBINED REFLEXES

Stage 2. Using combinations of rules. More advanced intelligence comes with the ability to "reason," or to develop consecutive chains and parallel sets of rules or consecutive-parallel combinations, that lead to a goal using separate rules. Making rule combinations is a process of grouping, and thus the full package of GFS is performed, with the rules combined together as a whole.

At stages 1 and 2 the vocabulary of sensors fully or partially coincides with the vocabulary of stimuli, and the vocabulary of responses coincides with the vocabulary of actuators. The simple rules given for stage 1 remain the same for all intelligent system functioning.

LEARNING NEW REFLEXES

Stage 3. Enhancing the initial set of rules by generalizing based on experience. The intelligent system can collect experiences and transform them into rules that augment the initial set of rules. Real experiences differ from those reflected in rules because of the errors of sensing and generating actions, because systems make mistakes (which become a source of new information), and because systems can acquire or develop a property to execute tentative (sometimes, random) actions. In other words, the system has the ability to ask itself the question "What if"—"What will happen if I do this action?" Then, if a system is capable of value judgment, the groups of good experiences can be collected and generalized upon (GFS is applied). This leads to the development of a new positive rule—"What to do in the case of . . . ." Collecting bad experiences leads to development of new negative rules—"What not to do in the case of . . . ."

The "What if" questions can be considered a form of search similar to "genetic search." Asking the question "What if" can involve different techniques. For example, one could use the so-called crossover. Anyway, at the stage 3, the intelligent system can enhance the list of rules (and the elementary concepts it has), but it does not yet construct any new levels of resolution.

GENERALIZING RULES

Stage 4. Discovering classes of rules and building new resolution levels. When the number of collected rules is large enough, it allows for discovering classes that can be interpreted as rules at a lower-resolution level. Thus a new (lower) level of resolution emerges; correspondingly a new level of concepts grows and forms a relational knowledge base for this new (lower) level of resolution. The combinatorics of grouping, focusing of attention, and searching can be applied to the rules, and this does not interfere with the vocabularies.

The discovery of rule classes brings the development of intelligence to the multiresolutional system of GSM, as was shown in Figure 2.5. The algorithms of systems reproducing stages 3 and 4 are introduced, discussed, and tested in [20, 21]. The most formidable advancement at stage 4 is the ability to construct the expression $[S_1 \rightarrow A \rightarrow S_2]$ to arrive at G using all three multiresolutional components G, S, and A. All loops run simultaneously at all resolution levels in different time scales. The "vertical" consistency within SP, WM, and BG is regularly checked. Obviously the "What if" tool of the stage 3 in addition to forming the rule combinations from the stage 2 is applied here at all levels of resolution.

SYNTHESIZING STATES AND ACTIONS

Stage 5. Combinatorial synthesis of new states and actions. The highest intelligent system is capable of making a combinatorial synthesis of states (situations) and actions.

One could easily imagine adding stages 6 and 7 with the combinatorics of questions like "What if" applied to the contexts and paradigms *(stage 6. synthesizing contexts; stage 7. synthesizing paradigms).*

Let us collect the properties of stages 1 through 5 in a comparative table (Table 2.2). As the table shows, later stages of development all have properties of preceding stages plus an added stage. However, these properties should not be considered just a direct consequence of past performance of the intelligent system. Procedural capabilities of the intelligence attained by an intelligent system are not reflected in performance directly. The character of the particular assignment is important, how prolonged in time the assignment is, how repetitive the assignment is, how much the parameters of the assignment vary from one execution to another, how structured the environment is in which the assignment is supposed to be performed, whether the assignment is a part of a larger paradigm where the skills acquired during performance can be applied in the future, what factors are unexpected, what the cost function is, and which components of the cost function—time, energy, monetary cost, accuracy, reliability, safety, etc.—receive emphasis.

FURTHER STAGES

The concept of "cost function" ("cost-functional" for the integral of "cost-function") has different interpretations at different levels of resolution. We use intelligence to improve quality of functioning via improving the quality of "thinking processes." This growth of the "quality of thinking" eventually boils down to the growth of the "quality of functioning" of the system that uses the results of thinking. Paradoxically enough, better "intelligence" can always be demonstrated as a faculty of the system that improves safety, reliability, accuracy that eventually reduces monetary costs, energy losses, or time. More often, good thinking leads directly to these material benefits.

Yet, further advances like stages 6 and 7 are looming. We can already take into account the adjacent level (stage 5). If the next adjacent level is taken into account, too, it will give stage 6—knowing the context (stage 6). From experience we know that the results can be improved by going one level further and taking into consideration the paradigm (stage 7).

The ability to refine parameters can be considered more important for performance than acquiring a new capability. In this way a system belonging to a lower level of intelligence development (to a stage with fewer capabilities) can demonstrate better performance than the system with a higher level of intelligence. This may seem to be contradictory, but it is not. The fact is that any sophisticated system (belonging to a stage with more capabilities) can improve and develop its performance.

The following parameters allow for ways in which intelligent systems can improve in all their subsystems separately (SP, WM, and BG) and at all levels of resolution independently:

- Assign relative importance to the entities at hand.
- Assign the size of window in focusing attention.
- Assign the scope of interest.
- Assign and correct the value of resolution.
- Evaluate values of similarity, resemblance, or closeness by correcting their measures.
- Evaluate the cohesiveness of groups, clusters, assemblies, and strings.
- Evaluate the value of goodness for the alternatives (combinations) produced under a particular goal.

In general, the distinction between groups of intelligent systems that is often used in discussions recognizes levels of (1) the primitive automaton, (2) adaptive intelligence, and (3) the decision generating and supporting intellect. This is a scale that can be useful in practice, but it does not yet have scientific validation.

2.8 ON THE RESEMBLANCES AMONG PROCESSES OF STRUCTURING IN NATURE AND REPRESENTATION

2.8.1 Issue of Resemblance

It is clear that focusing of attention, grouping, and combinatorial search is motivated by a symbolic system that makes substantial simplification in order to reduce its computational complexity. At a particular level of resolution we need for the whole environment just one symbol for each singularity—one symbol for a class of singularities having one attribute also one symbol. It has been demonstrated that by using focusing of attention and combinatorial search while grouping solutions under minimum cost requirements in the system with three levels of resolution, the number of computations can be reduced from 10^{17} to 10^{3} [13].

The question to be raised is: "If, in the system of representation of GSM, the substitution of the continua by the multiresolutional system of entities is meaningful because of computational complexity reduction, why should it be meaningful in the domain of reality? Why does Nature need to form entities that would be understandable as if it is a system of representation?

To find the answer, we observe that strangely enough, the laws of processing for physical matter and energy are similar to the laws of information processing. Both can be represented symbolically (otherwise, no understanding would be possible) and both have cost functionals that should be minimized. Indeed, why does an oil drop in a heated skillet tessellate (spatially discretize, tessellate) into a multiplicity of hexagonal cells [7]? The answer is that it reduces the amount of energy that otherwise would be necessary to dissipate through the limited overall surface. We should not be surprised then that in the long way from the continuum of reality to the structure of representation there is a good match between natural and informational processes that can be plugged into each other (Figure 2.3).

2.8.2 On the Resemblance among the Algorithms Applied for Structuring

Figure 2.8 gives steps of the algorithms for attention focusing, grouping, and combinatorial search. One can see that these steps can be achieved by a number of different computational schemes. In this section we therefore do not discuss a particular computational scheme. However, it is appropriate to mention that the principles of focusing of

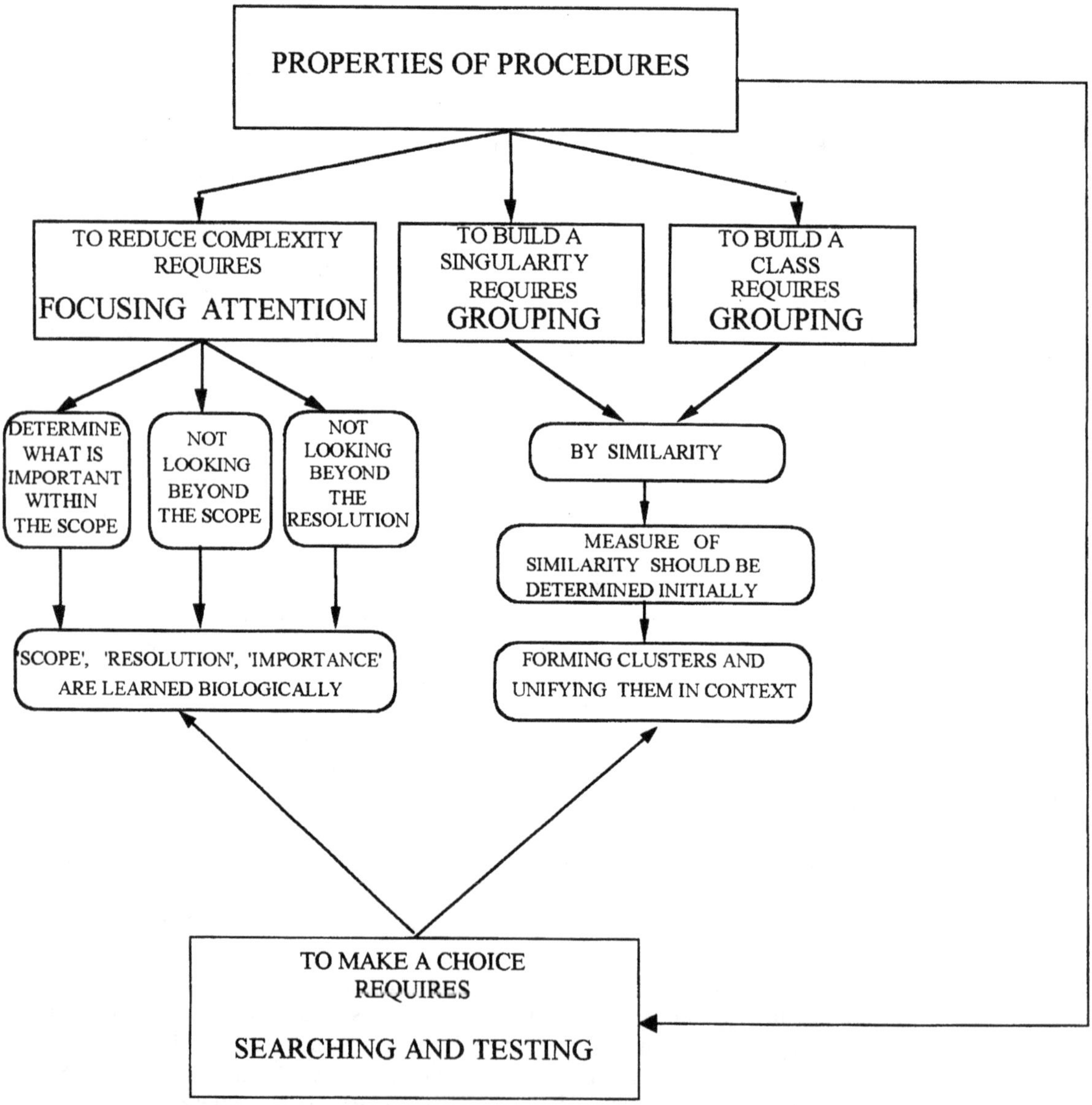

Figure 2.8. Properties of the procedures components of the algorithm of generalization.

attention, grouping, and combinatorial search in forming a multiresolutional system are at the core of following widespread analytical methodologies:

- Recursive estimation.
- Wavelet decomposition.
- Fractal theory.
- Any series including Taylor, Fourier, Tchebyshev polynomials, and others.
- Lyapunov stability (presumes fuzzification).
- Multilayer neural networks (presumes fuzzification).
- Multiresolutional variable structure systems.
- Multiresolutional dynamic programming.

Even where a complexity reduction of a multiresolutional system is not pursued, GFS is still applied for problem solving either as a complete set or partially. For many practical cases GFS is the motivation behind the problem solving. Most of the software packages for computer vision employ GFS at all stages of image processing.

Numerous algorithms are known in practice that use focusing of attention, or filtration (F), grouping (G), and search (S). F-algorithms can be driven by a variety of factors: aimless curiosity, communicated or synthesized goals, a recognized danger, and/or observed unusual phenomena. Some values should be assigned to these algorithms if we want them to operate: values of importance for different entities, limits of scope, and values of resolution. G-algorithms are driven by similarity and compatibility (which often should be assigned as external control parameters). They initiate a new investigation of the world if the process of clustering has failed. Finally S-algorithms are equipped with mechanisms of new entities generation (in the form of novel clusters, alternatives of assemblies, alternatives of strings, crossover). The existing and newly created alternatives are browsed by S-algorithms. All of them are evaluated, and then a limited number of them are selected for the further computations.

We will propose a generalized computational scheme that can generate algorithms applicable in multiresolutional GSM. We will take the generalized algorithm of information processing which we introduced in the previous section and present it in semiotic form.

2.9 SEMIOTIC FRAMEWORK OF THE ARCHITECTURES FOR INTELLIGENT SYSTEMS

2.9.1 What Is Semiotics?

Semiotics developed as a science of signs, but its focus is on the mechanisms of thinking and attaining the meaning. Semiotics is being applied at all stages of scholarly endeavor and scientific development and in all areas of science (see [22–26]). The "hard" applied semiotics has been developed during last 20 years. Significant contributions are linked with the name of D. Pospelov [24, 25]. In the U.S. literature the interest in applied semiotics is gradually growing (compare [27] and [28–31]).

Laws of Sign

The basis for the overall semiotic approach is the trilateral concept of the sign introduced by C. Peirce and illustrated in Figure 2.9. The sign is understood as a unity of label,

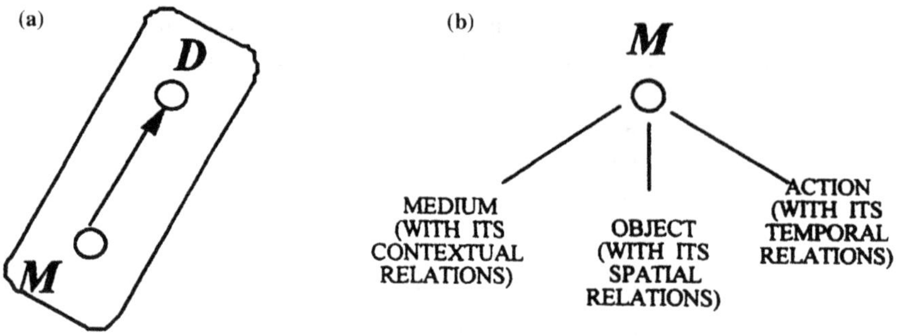

Figure 2.9. Model of meaning (M) and definition (D).

definition, and meaning which can be verified throughout all discourse. (A symbol is a particular case of a sign, but this distinction is not discussed here.)

The sign helps to recover the *interpretant* which is a couple $D \cup M$. *Meaning* is a unity of the attributes shown in Figure 2.9. This unity provides the sign with its interpretive power. *Definition* is an intersection of classes to which the meaning belongs. The definition is insufficient for arriving at the meaning because the latter can be found only as a combination of several relational networks obtained from the experience. Several entity-relational networks jointly allow for a complete interpretation. Thus meaning is always revealed as a result of a dynamic process of interpretation which, strictly speaking, can never be complete.

The triplet *object–sign–interpretant* is an invariance of the formalism of knowledge representation and communication. Therefore it is not correct to narrow the nature of semiotics to the "theory of codes," or the "theory of sign production." This triplet allows for broad epistemological connections, deep involvement with the theory of communication, and so on.

A straightforward instrumental definition of semiotics can be presented as follows:

DEFINITION OF SEMIOTICS

Semiotics is a theoretical field that analyzes and develops formal tools of knowledge acquisition, representation, organization, generation and enhancement, interpretation, communication, and utilization.

2.9.2 Semiotic Closure

Semiotic closure is a concept that is viable for all symbolic systems used for representing reality in order to make decisions about functioning in reality. The idea of semiotic closure is tantamount to the idea of closing the loop of the system if the system should reliably function. The concept of semiotic closure is reflected in the loop of the elementary RCS level, or GSM (see Figure 1.1). In this book we will frequently use a simplified version of Figure 1.1 which will be called an elementary functioning loop (ELF) as shown in Figure 2.10. (The subsystem of value judgment is not shown, but its functioning is presumed.)

We will always consider intelligent systems and their components as a complete set of the subsystems S–SP–WM–BG–A–W working as a closed loop: elementary loop of functioning (ELF).[10] All components of ELF are complex systems; their components

[10]Semiotic closure was discussed in the following sources [32–34]. The term semiotic closure (H. Pattee) is used intermittently with the term elementary functioning loop (A. Meystel), and with a request for symbol grounding (S. Harnad).

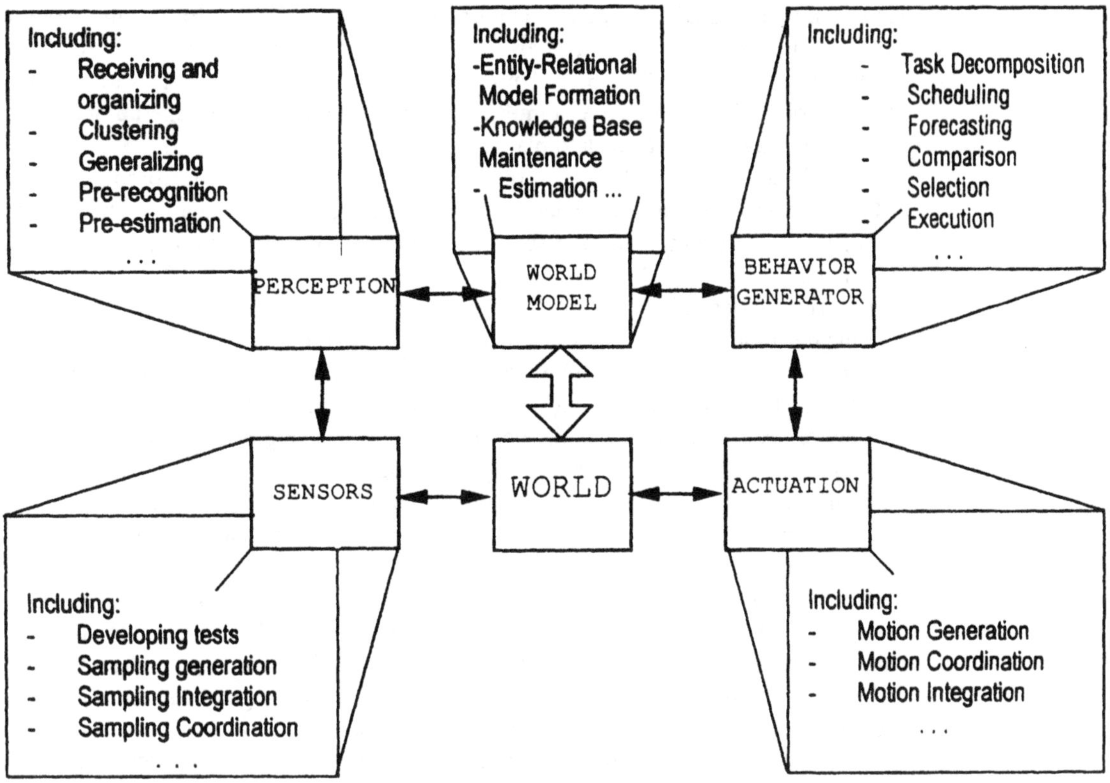

Figure 2.10. ELF illustrates the concept of semiotic closure (a simplified GSM).

cannot be represented exhaustively. If one wants to model a component of ELF, it should also be done by using ELF.

For the sake of efficient reasoning, we lump together systems and their components together into finite and countable generalized formations. ELF can be divided in two parts which belong to two different domains of our discussion:

- A part belonging to reality; it includes S–W–A.
- A part belonging to computations; it includes SP–WM–BG.

The reality part receives commands at its input and produces sensor signals at its output. (It also transforms matter and exchanges energy with the environment, which is a function we will not discuss here.) The computations part receives the output of sensors and the goal, and then produces commands.

Since intelligent systems have a great many components and variables, any detailed description and symbolic models based on a direct description of their ELF would be too complex for direct control purposes. Usually several consecutive stages of generalization (bottom-up) and instantiation (top-down) precede the use of a representation system in order to control satisfactorily its computational complexity.

The complexity of computations should be analyzed and put in correspondence with the concrete number of levels of the multiresolutional hierarchies of the multiresolutional GSM and with their multiresolutional systems of representation and behavior generation. We postulate that the complexity of representation of our systems will be dealt with by using GFS for generalization bottom-up and GFS^{-1} for instantiation top-down. A hierarchical system of goals and plans can be constructed for any intelligent system by a top-down GFS/GSF^{-1} procedure (see [35]). The symbiosis of SP–WM–BG subsystems is interpreted as a multiresolutional automaton for generating this multiresolutional system of goals and plans.

If one follows the set of tools in this definition, one can easily restore the structure of GSM shown in Figure 1.1. There is a strong connection between this structure and the well-known decomposition of semiotics into three domains: syntax, semantics, and pragmatics.

The triangle of semiosis (syntax, semantics, and pragmatics) is put in correspondence with the hexagon of ELF (actually shown in Figures 1.1 and 2.12). It turns out that there is a direct correspondence between these two aspects of viewing the intellectual process as a whole. The circulation of knowledge is done by communication which changes its incarnation from a node to a node, passing through the stages of encoding, representing, organizing, interpreting, generating, applying, and transducing—all considered as different forms of communication (mappings from one language to another). As something happens in the world, it is encoded by sensors in a symbolic form, and the process of representation begins.

The role of perception is to represent the results of sensing in some organized manner using *signs*. This process of shaping up the organization is called *syntax*. It starts at some point; it continues at all subsequent stages of dealing with knowledge and becomes more and more generalized. The initial structure becomes knowledge as the latter gets more and more generalized so that after representation is completed, interpretation is possible. Interpretation enables the process of decision making in which semantics joints syntax to create the *interpretant*.

The interpretant materializes in the process of actuation, which is analogous to generation of new knowledge. As a result of this process, new knowledge arrives in the world, creates changes in the world—physically and/or conceptually. Then new *objects* emerge.

The hypotheses enter the subsystem of behavior generation as a substitute for the rules, the decision for an action is made, the action is performed, changes in the world occur, the transducers (sensors) transform them into a form that can be used by perception, and the long and complicated process of moving from signs to meaning starts again. Now the enhanced set of experiences brings about another hypothesis which can confirm or refute the tested ones. This is when the *symbol grounding* occurs.

After multiple tests the hypotheses can cross the threshold of "trustworthiness," and a new rule is created. Among scientists a rule (or a set of rules) within a context is considered to be "a theory." At each step of this development, the unit under consideration undergoes a comparison with other kindred units confined in corresponding databases (of experiences, of rules, and of theories). Then the symbols tentatively assigned to some "unities," "entities," or "concepts" enter their place within the database of concepts (which is a relational network of symbols).

2.9.3 Semiosis: The Process of Learning in Semiotic Systems

The process of semiosis is shown in Figure 2.11. Rules (or the hypotheses which will become rules after multiple confirmation of the success of functioning) are formed when experiences cluster together unified by their similarity. Let us denote the prior state S_1, the action that has been applied, A_1, the state which emerged after this, S_2, and the value

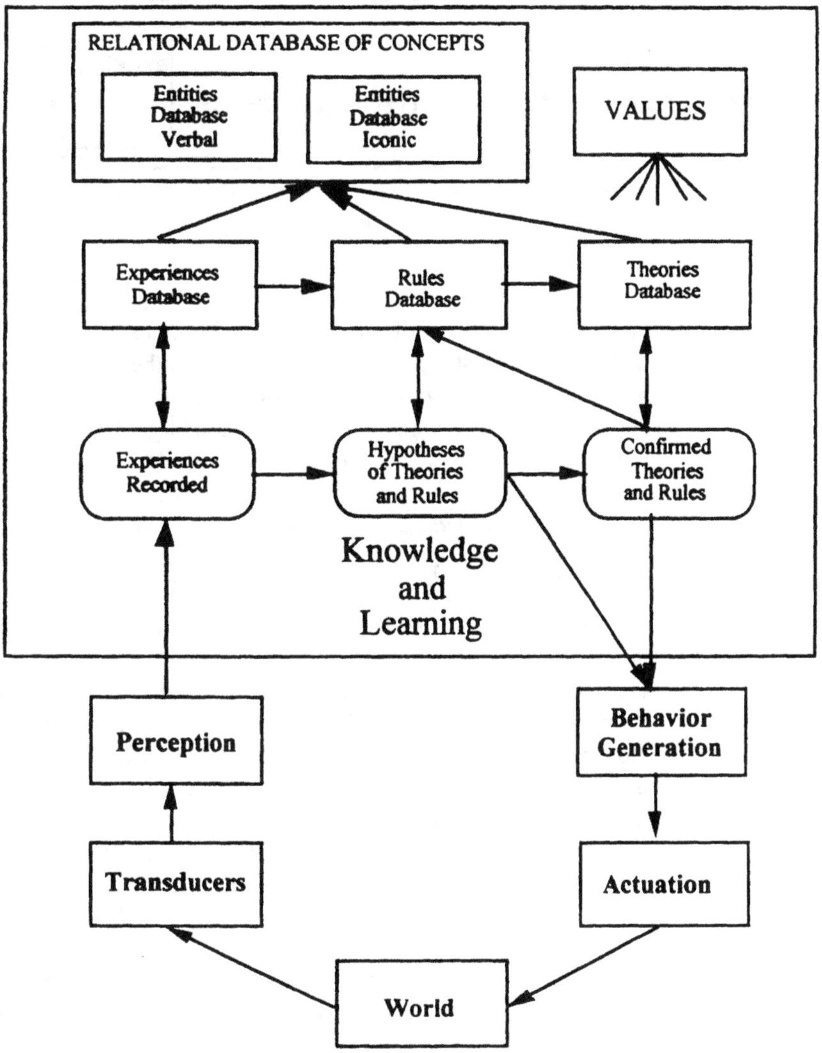

Figure 2.11. Six-box diagram of learning: The cycle of semiosis.

V_1 of the result. These are the three components of experience computed for S_1, A_1, S_2, and V_1:

$$[S_1]E_{\text{comp1}} \times [A_1]E_{\text{comp2}} \rightarrow [S_2]E_{\text{comp3}}, [V_1]E_{\text{comp3}}$$

(the value V which is attached to this experience is the first basis for grouping the experiences). Rules are formed after a commonality is detected in several of the components. They are usually presented as inverted experiences:

$$[\text{the } V \text{ desired}]R_{\text{comp1}} \times [\text{the state } S^* \text{ which is desired}]R_{\text{comp2}}$$
$$\times [\text{the prior state } S]_{\text{comp3}} \rightarrow [\text{the action to be applied } A]R_{\text{comp4}}.$$

An interesting and unique feature of generating rules is that each component is a *generalized* component of experience. This means that in order to obtain a component of a rule, several similar components of experience must be *grouped* together into a

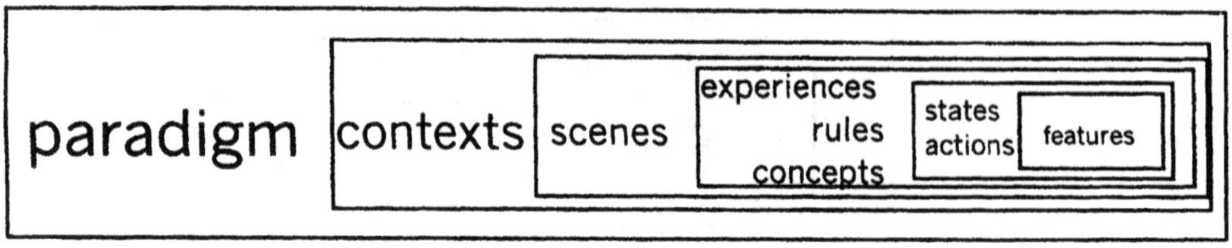

Figure 2.12. Nesting during semiosis.

class, that is, a cluster. This requires applying a set of GFS procedures. The symbol attached to this cluster signifies the process and the result of generalization. The premises behind this generalization can be different, but the result is always the same: A new class is born. If we denote the phenomenon of generalization upon i similar experiences ($i = 1, 2, \ldots, n$) by a symbol G_1, we can write

$$[\text{the } V \text{ desired}]R_{\text{comp1}} \to G_1\{\text{the values } V_i \text{ of the result}\},$$

$$[\text{the prior state } S]_{\text{comp3}} \to G_1\{\text{the prior states, } S_i\},$$

$$[\text{the action to be applied } A]R_{\text{comp4}} \to G_1\{\text{the actions that has been applied, } S_i\}.$$

Only the desired state is not subject to generalization; it is always singular and pertains to the particular situation. The nested evolution of semiosis is described in Chapter 11. However, the results can be seen in Figure 2.12, where the nestedness demonstrated is typical for the semiosis habits of humans. For many real instances of semiosis, we need to apply the various logical tools that has been developed for analyzing semiosis. Discussion of these tools is outside the scope of this book, however a familiarity with some of them can be obtained from [36].

We can see that the results of clustering do not belong to the list of initial words; they are generalized words. Their parameters are not as accurately defined as parameters of initial words: The parameters of generalized words are always represented as intervals that include the parameters of the initial words. This describes a process of fuzzification which happens when semiosis moves to the lower levels of resolution. All levels together are to be judged from the point of view of the most generalized reference frame, which is our natural language (see [37]).

The world of generalized words is different: It has a different resolution, coarser than the resolution of initial words. By virtue of generalization, a new lower level of resolution is created. Nested rules and theories (rules and their contexts) are always the bridges between adjacent levels of different resolution. However, it is clear that the totality of all experiences, all theories, and all rules constitutes a multiresolutional system that corresponds to the multiresolutional database of concepts ingrained within the body of natural language.

2.9.4 Reflection and Consciousness

Analysis of unsupervised learning systems demonstrates that the build-up of the multiresolutional world model starts immediately after the GSM initiates its functioning [38, 39]. However, all rules learned and concepts stored are represented in the state-space with the origin placed at the center of GSM. (We call it "in the reference frame of the system.") At this stage GSM does not know that it exists as an entity: It does not need to know this. This activity surrounding its origin is implanted within the system of GSM flow of information.

The situation changes as soon as GSM begins to search for different alternatives of motion trajectories. It is supposed to find its own strategy of motion, and it becomes aware that some losses and gains are components of the measure of "value" or "goodness" (e.g., the time the process should be reduced).

GSM contemplates the alternatives of solutions, and in order to generate and compare them, it comes to explore a "novel" opportunity of labeling the condition part of the rules *as a variable*. Until now the situations have not been perceived yet as variables; they were the initial and the final part of all actions. Now the lower level of resolution (the more generalized one) enters the process of reasoning.

Thus the higher-resolution situation can be considered a current variable, the variable of interest within the situation represented at the lower level of resolution. GSM arrives at a convenient opportunity: when it can assign a sign to itself. It discovers its "self" in the external world and places itself in the state-space with a different origin in different cases [40]. The consciousness has emerged, and Table 2.2 can be enhanced.

Consciousness is a very important property; it allows GSM to perform simulation of the processes as if they were viewed from the lower-resolution level without which the totality of the situation would not be seen.

Simulation of the world with a "self" as a part of it, qualifies for this state being called "imagination." In the semiotic literature this property is called "reflection" [40]. Reflection is defined as the ability of an intelligent system to represent not only the external world but also itself. Representing oneself is a demanding feature after the "self" has already represented itself. Nestedness of representations becomes infinite which is illustrated in Figure 2.13. Reflection is especially important when communication between intelligent systems is analyzed or when systems in conflict are under consideration.

The case of conflicting systems presumes not only that the world model of a particular GSM includes representation R of the GSM (the adversary) denoted as R_1[GSM] but also that this representation of the initial GSM can be denoted as R_2[GSM]. Both representations R_1[GSM] and R_2[GSM] can be called reflections of the first order. Indeed, other orders are possible since, instead of the R_2[GSM], the adversary should better have $R_2[R_1$[GSM]]. In the meantime, our GSM should better have $R_1[R_2$[GSM]] instead of R_1[GSM]. Obviously this consideration can be continued endlessly. However, multiple reflection has a "blurring" effect, and each new order of reflection becomes more and more fuzzified.

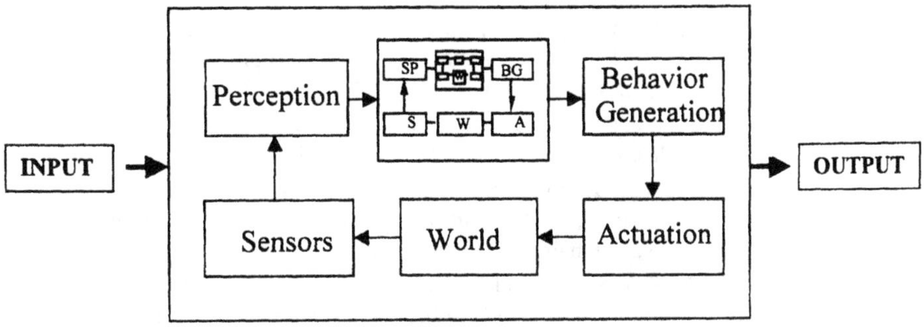

Figure 2.13. Nested hierarchical reflection.

REFERENCES

[1] A. Robinson and W. A. J. Luxemburg, *Non-Standard Analysis* (Princeton Landmarks in Mathematics and Physics), Princeton University Press, 1998.

[2] G. Birkhoff, Mathematics and psychology, *SIAM Review*, Vol. 11, No. 9, 1969: 429–469.

[3] R. Ben-Natan, *CORBA: A Guide to the Common Object Request Broker Architecture*, McGraw-Hill, New York, 1995.

[4] A. Meystel, Architectures, representations, and algorithms for intelligent control of robots, in *Intelligent Control Systems: Theory and Applications* (M. M. Gupta and N. K. Sinha, eds.), IEEE Press, New York, pp. 732–788.

[5] A. Meystel, Theoretical foundations of planning and navigation for autonomous mobile systems, *International Journal of Intelligent Systems*, Vol. 2, No. 2, 1987.

[6] J. C. Bezdek and S. K. Pal (eds), *Fuzzy Models for Pattern Recognition*, IEEE Press, New York, 1992.

[7] S. Chandrasekhar, *Hydrodynamic and Hydromagnetic Stability*, Dover Publications, 1981.

[8] H. Haken, *Advanced Synergetics: Instability Hierarchies of Self-Organizing Systems and Devices*, Springer Series on Synergetics, Vol. 20, Springer, Berlin, 1987.

[9] H. Haken, *Information and Self-Organization*, Springer Series on Synergetics, Vol. 48, Springer, Berlin, 1988.

[10] G. Gerster, *Below From Above: Aerial Photography*, Abberville Press Publishers, New York, 1985.

[11] Y. L. Klimontovich, Criteria of relative degree of order in self-organization processes, *Nanobiology*, No. 2(299), 1993.

[12] S. Hawkins, The edge of spacetime, in *The New Physics* (P. Davis, ed.), Cambridge University Press, 1992, pp. 61–69.

[13] Y. Maximov and A. Meystel, Optimum design of multiresolutional hierarchical control systems, *Proc. of the IEEE Int'l Symposium on Intelligent Control*, 11–13 August, 1992, Glasgow, Scotland, UK, 1992.

[14] E. Schrödinger, *What Is Life?: The Physical Aspect of the Living Cell With Mind and Matter*, Cambridge University Press, 1992.

[15] E. Schrödinger, Causality and wave mechanics, *Science and Humanism*, Cambridge, 1951.

[16] A. Meystel and M. Thomas, Computer aided conceptual design in robotics, *Proc. IEEE Int'l Conf. in Robotics*, Atlanta, GA, March 13–15, 1984, pp. 220–229.

[17] A. Meystel, Robot path planning, *The Encyclopedia of Electrical and Electronic Engineers*, J. Webster, ed., Vol. 18, Wiley, New York, 1998, pp. 571–581.

[18] S. K. Pal and P. P. Wang, *Genetic Algorithms for Pattern Recognition*, CRC Press, Boca Raton, FL, 1996.

[19] J. Albus and A. Meystel, *Behavior Generation in Intelligent Systems*, NIST Report, Gaithersburg, MD, 1996.

[20] A. Lacaze, M. Meystel, and A. Meystel, Multiresolutional schemata for unsupervised learning of autonomous robots for 3D space application, *Robotics and Computer Integrated Manufacturing*, Vol. 11, No. 2, 1994, pp. 53–63.

[21] J. Albus, A. Lacaze, A. Meystel, Evolution of knowledge structures during the process of learning, *Proc. of the 1995 ISIC Workshop*, Monterey, CA, 1995, pp. 329–335.

[22] T. A. Sebeok, Galen in medical semiotics, in *Proceedings of the 1997 International Conference on Intelligent Systems and Semiotics: A Learning Perspective* (A. Meystel, ed.), Gaithersburg, MD, September 22–25, 1997, pp. 354–363.

[23] T. A. Sebeok and J. Umiker-Sebeok (eds.), *Recent Developments in Theory and History of the Semiotic Web* (1990), Mouton de Gruyter, Berlin, 1991.

[24] D. A. Pospelov, Semiotic models in control systems, *Architectures for Semiotic Modeling and Situation Analysis in Large Complex Systems* (J. Albus, A. Meystel, D. Pospelov, and

T. Reader, eds.), Proc. of the 1995 ISIC Workshop, 10th IEEE International Symposium on Intelligent Control, Monterey, CA, August 1995, pp. 6–12.

[25] D. A. Pospelov, A. I. Ehrlich, V. F. Khoroshevsky, and G. S. Osipov, Semiotic modeling and situation control, ibid., pp. 127–129.

[26] L. Erasmus, G. Muhr, and G. Doben-Henisch, Semiotic modeling and simulation of the goal driven software developing organization, *Sign Processes in Complex Systems*, 7th International Congress of the IASS, Dresden, October 1999, pp. 147–148.

[27] J. Albus, A. Meystel, D. Pospelov, and T. Reader (eds.), *Architectures for Semiotic Modeling and Situation Analysis in Large Complex Systems*, Proc. of the 1995 ISIC Workshop, 10th IEEE International Symposium on Intelligent Control, Monterey, CA, August, 1995, 441 pp.

[28] *Intelligent Systems: A Semiotic Perspective*, Proc. of the 1996 Int'l Multidisciplinary Conference, Vol. 1: Theoretical Semiotics (298 p.), Volume 2: Applied Semiotics (312 p.), Gaithersburg, MD, October 1996.

[29] A. Meystel (ed.), *Proceedings of the 1997 International Conference on Intelligent Systems and Semiotics: A Learning Perspective*, Gaithersburg, MD, September 22–25, 1997, 591 pp.

[30] *A Joint Conference on the Science and Technology in Intelligent Systems*, Proc. of the 1998 IEEE Int'l Symposium on Intelligent Control (ISIC), Computational Intelligence in Robotics and Automation (CRIA) and Intelligent Systems and Semiotics (ISAS), Gaithersburg, MD, September 1998, 889 pp.

[31] *Proc. of the 1999 IEEE International Symposium on Intelligent Control, Intelligent Systems and Semiotics*, Cambridge, MA, September 1999, 464 pp.

[32] H. Pattee, How does a molecule become a message, *Developmental Biology*, Suppl. 3, 1969.

[33] A. Meystel, Intelligent systems: a semiotic perspective, *International Journal of Intelligent Control and Systems*, Vol. 1, No. 1, 1996, pp. 31–58.

[34] S. Harnad, Symbol grounding problem, *Emergent Computation* (S. Forrest, ed.), MIT Press, Cambridge, MA, 1991, pp. 335–346.

[35] J. Albus, A. Meystel, and S. Uzzaman, Nested motion planning for an autonomous robot, *Proc. IEEE Conference on Aerospace Systems*, May 25–27, Westlake Village, CA, 1993.

[36] E. Messina, Y. Moskovitz, and A. Meystel, Mission structure for an unmanned vehicle, *Proc. of the 1998 IEEE Int'l Symp. on Intelligent Control*, a Joint Conference of the Science and Technology of Intelligent Systems, Sept. 14–17, NIST, Gaithersburg, MD, 1998, pp. 36–43.

[37] R. Wilensky, Sentences, situations, and propositions, in *Principles of Semantic Networks* (J. F. Sowa, ed.), Morgan Kaufmann Publishers, San Mateo, CA, 1991.

[38] A. Meystel, Learning algorithms generating multigranular hierarchies, in *Mathematical Hierarchies and Biology* (B. Mirkin, F. R. McMorris, F. S. Roberts, and A. Rzhetsky, eds.), *DIMACS Series in Discrete Mathematics*, Vol. 37, American Mathematical Society, 1997.

[39] A. Meystel and A. Lacaze, Unified learning/planning automaton: generating and using multigranular knowledge hierarchies, in *Proceedings of the 1997 International Conference on Intelligent Systems and Semiotics*, Gaithersburg, MD, 1997, pp. 117–123.

[40] V. A. Lefebvre, Reflexive game theory, in *Proceedings of the 1997 International Conference on Intelligent Systems and Semiotics: A Learning Perspective* (A. Meystel, ed.), Gaithersburg, MD, September 22–25, 1997, pp. 455–458.

PROBLEMS

2.1. Outline the differences between the continuous and discrete mathematics. Do they have different subjects? different tools? different methodologies?

2.2. Using the examples from Problems 1.7 and 1.9 of Chapter 1, demonstrate how different levels of the architecture differ in their scope and resolution. Introduce a meaningful measure of scope for each of the examples. Introduce a measure of resolution. Is it possible to keep the same physical units for measuring the scope and the resolution at different levels of the hierarchy?

2.3. Give an outline of the concepts of object-oriented programming as you understand it. Illustrate it by using the same examples (see Problem 2.2).

2.4. Demonstrate which jobs and objectives from the list in Section 2.1.3 are performed in the examples of Problem 2.2.

2.5. Give two examples of grouping for the cases listed in Section 2.2.1.

2.6. Give two examples of filtration. Demonstrate that they satisfy the two main conditions of filtration: constraint by the scope from above and constraint by the granule from below (see Section 2.2.1).

2.7. Give two examples of searching that can be combined with grouping and filtration (see Section 2.2.1).

2.8. Give several examples of sets. Demonstrate that in all cases, the theorem of abstraction (2.1) is satisfied: We always have a set-forming property of the objects collected in a set.

2.9. Take an object from one of your examples of sets (see Problem 2.8). Demonstrate that it can be described according to Figure 2.1. Demonstrate the use of the expression (2.2) in forming the set.

2.10. Formulate a problem for the object used in Problem 2.9 in which the concepts of *state* and *change* can be introduced in addition to the description of the *object*.

2.11. Draw a starry sky with about 40 to 50 stars. Apply the sequence of operations required for using the formulas (2.3) and (2.4). Which part of this algorithm can be considered to perform functions of attention focusing, combinatorial search, and grouping?

2.12. Construct a coordinate system for the object chosen in Problem 2.9.

2.13. Choose one photo-picture and one topographical map. Demonstrate that the expression (2.5) can be used to evaluate the resolution of these representations. Compute the values of resolution and scale. What is the diameter of the indistinguishability zone (the granule)?

2.14. Demonstrate your knowledge of categories, objects, and morphisms by using the following nested hierarchy:

ith level	The world
$(i-1)$th level	Vehicles and obstacles upon a terrain that are objects of the category "the world"
$(i-2)$th level	S, SP, WM, BG, and A subsystems of the vehicles that are the objects within the category "the vehicle"

What are the morphisms in each of the categories?

2.15. Restore in your memory the examples from Problem 1.7, Chapter 1. Develop a simple automaton model for the external automata-based representation of these examples. Develop a Mealy-automaton description. Decompose one of these systems into its subsystems (according to Fig. 1.1). Develop a Mealy-automaton description for each subsystem. Check whether the overall system functions consistently.

2.16. Illustrate nestedness of the system analyzed in Problem 2.15 by drawing a diagram similar to Figure 2.6.

2.17. Take the example of an autonomous vehicle. Equip it with a learning system and develop a scenario of its intelligence evolution through all stages 1 through 7 of Table 2.2.

2.18. Give examples of intelligent systems belonging to different stages of evolving intelligence (stages 1 through 5 in Section 2.7).

2.19. Take an example from your personal engineering or research experience and demonstrate the formation of entities by using GFS and GFS^{-1}.

2.20. Take an example from Problem 1.9, Chapter 1, and demonstrate the phenomenon of semiotic closure.

CHAPTER 3

KNOWLEDGE REPRESENTATION

In the previous chapters we introduced the general architecture of intelligent systems, and outlined the theoretical premises. It turned out that this architecture is a multiresolutional system of closed loops of nested automata, or as we called it in the case of high self-organization, a multiscale community of agents (MCA). All elements of these automata are based on knowledge processing. In this chapter we focus our attention on the nature and properties of knowledge. General principles of knowledge organization and processing will be presented; basic phenomena associated with knowledge processing will be described.

We will demonstrate that knowledge is a prerequisite for the functioning of *living creatures*. People are able to exist only because they acquire, have, discover, and use knowledge in a variety of forms. It is not enough to encode the reality: The code should be transformed into the body of knowledge. When we want to describe human life, willingly or unwillingly we appeal to the phenomenon of knowledge, to its existence, or its absence.

3.1 PROBLEM OF REPRESENTING THE NATURAL WORLD

3.1.1 Unbearable Richness of the Reality

OUR EPISTEMIC OMNIPOTENCE

Transforming the reality into the system of symbols seems to be potentially possible. But is this potentiality a realistic world? The world in which we live is rich and complex far beyond our ability to fully describe it in words or symbols. When we look at a large tree, we see hundreds of thousands of leaves, twigs, and branches all of which converge to a trunk that disappears into the ground (and what is going on there?). The bark is furrowed and textured in patterns that can, at best, only be summarized by language. The upper branches weave a complex web of branches, twigs, leaves, and seeds—some alive and strong, others dead and weak. A squirrel running through the tree tops has little need for a sophisticated language but definitely must be able to represent and compute the position and motion of objects and assign attributes (e.g., safe for holding weight) to those objects. Facing a need to talk and just even think about trees, we quickly reduce the discourse to the "foliage" with a characteristic "silhouette," "contour," or "profile." Our brain is very, and efficiently, inventive in finding analogies, simplifications, and other tools of generalization.

It is said that "a picture is worth a thousand words," but a typical outdoor scene is richer and contains more information than a thousand pictures. Imagine how many pictures would be required to store the information in a monkey's visual system while it is foraging for food or fleeing from predators through the jungle canopy. It would take several thousand standard resolution TV images to cover the entire egosphere with the resolution of the primate fovea.

Consider for a moment the complexity of sensory experience that one might have in an average suburban neighborhood during the course of a simple backyard barbecue. One might see flowers, bushes, and trees blowing in the breeze. Buildings, windows, doors, grass, and rocks would be illuminated by sunlight and dancing shadows. Birds, insects, and falling leaves would fly through the air. There would be patio furniture, pets, and people eating, drinking, and moving about. We would hear birds chirping, the wind blowing, traffic on the street, children at play, and people talking and laughing. We might smell hamburgers cooking, see and smell the smoke, taste the food and sip the drink, feel the warmth of the sun, and the chill of water in the swimming pool.

The astonishing fact is that we would assimilate all of this sensory information simultaneously and apparently without effort or thought, and automatically combine it with our prior knowledge of everything we know about how the world works and how plants, animals, people, and insects normally behave. Our prior knowledge includes all that we have learned from an entire lifetime of experiences at home, in school, in church, at work, and play since the day we were born.

Somehow all this information is represented internally in a form that enables us to behave successfully in the complex, dynamic, and uncertain environment of the natural world. We can easily move about without collisions. We can manipulate objects and tools, gossip with friends, perform complex tasks, make plans, and achieve goals. We can read and write. We can understand mathematics and science. We can learn from history and plan our activities for the rest of today, for tomorrow, next week, next year, and after retirement. We can even imagine our status in the afterlife. How is this done? What kind of representation is it that can support these cognitive and behavioral capacities? What kind of representation can assimilate all this information and make it available for planning and decision making in real time?

SEARCHING FOR MECHANISMS OF REPRESENTATION

Surely this representation cannot be a simple one, or it could not handle so much complexity and subtlety or capture such exquisite detail. Clearly, it must exist internal to the intelligent system at least momentarily, or it could not be perceived at all, much less used to generate behavior. Certainly there is some mechanism by which what is directly observed is combined with, and augmented by, what is known a priori. Otherwise, we would not be able to recognize our friends, make conversation about previous experiences, or even to know who we are or how we got to where we are. We would all suffer the equivalent of complete and perpetual amnesia.

What kind of capability for knowledge representation must exist in the internal world model of an intelligent system to support the kind of reasoning, planning, and behavior that humans do in everyday life? How did knowledge representation support the development of human intelligence during fierce competition for survival in the jungle and savannah? How was it honed through hunting, home building, farming, tribal warfare, and all manner of cooperative endeavors and competitive conflicts during the rise of civilization and the building of empires? The representation of knowledge required to support human levels of performance in commerce, industry, science, politics, sports, warfare, and everyday living obviously go far beyond but include those possessed by insects, birds, and lower mammals.

Knowledge exists, perpetuates, and grows because of our need to acquire it, to have it, to discover it, and to use it.

3.1.2 Epistemology

The study of intelligence connected to the phenomenon of knowledge about the world that exists in the mind is among the oldest of scientific disciplines. The question of how the mind can know about and act on the world has been studied and debated by the world's foremost philosophers and scientists for at least 2,500 years. Epistemology is the branch of philosophy dealing with the study of knowledge, including its nature, origin, foundations, limits, and validity. Epistemology means "theory of knowledge" (Greek *episteme*, or "knowledge," *logos*, or "theory.") Epistemology began (as far as we know) with the ancient Greeks. Plato (428–348 B.C.) was among the first to draw a distinction between the world of ideas that exists in the mind and the world of reality that exists in nature. Plato's famous parable of the cave is an analogy of how the external world of reality is projected onto the internal world of the mind like the shadows of actors in the outside world can be projected onto the wall of a cave.

EARLY UNIFIED APPROACH

Aristotle (384–322 B.C.) developed a hierarchical representation of knowledge about objects and their components in the world. He formulated a theory of entity formation and classification. He realized that the whole can have attributes that are more than simply a sum of its parts.

Unfortunately, soon after the time of Aristotle, the Greeks were conquered by the Romans and serious interest in epistemology waned. During the rise of the Roman Empire, the keenest intellects were engaged with more practical issues such as military conquest, politics, law, and civil engineering. The Romans wasted little time on the philosophical aspects of science, and simply adopted most of the basic scientific and philosophical ideas of the ancient Greek authorities.

With the rise of Christianity, philosophy turned to theology and metaphysics (the branch of philosophy dealing with first principles). Philosophical inquiry into the nature of knowledge was deemed unnecessary, since anything worth knowing could be obtained from Holy Scripture and the teachings of Aristotle. Hence there was little interest in knowledge about the physical world, and no one questioned established conclusions.

In the wake of the Renaissance philosophers such as Francis Bacon (1561–1626) began to question authority and suggest that knowledge must begin with experience and proceed by induction to general principles. It was the pursuit of this attitude toward knowledge that led René Decartes (1596–1660) to the philosophical school of rationalism. Decartes wanted to assume nothing, to take nothing on faith, and to base his whole metaphysical explanation of man and nature on the simplest possible set of axioms.

DUALISM

Decartes's philosophy is one in which God and the human mind belong to one order of reality—one of the spirit and spiritual; while the body and the rest of nature belong to another order of reality—one of the material and corruptible. This is called dualism.

David Hume (1711–1776) introduced a philosophical precursor to the Godel's theorem of incompleteness. He argued that ideas belong to our mind, and that our ideas are either imposed upon the reality or are created as a result of our perception. Reality beyond perception just cannot be proven. Using contemporary language of semiotics, Hume argued that "symbol grounding" is impossible.

REALITY AS REFLECTION OF OUR MIND

Immanuel Kant (1724–1804) played a major role in clarifying the relationship between the cognitive representation of space and time in the mind, and the attributes of objects, events, and agents that we observe in the world. Even though absolute "symbol grounding" might be impossible, Kant argues that we can endlessly approach it by making our categories more and more adequate to representing the external reality. This is suggestive of modern recursive estimation theory that begins with a categorical hypothesis that is tested against experience until its adequacy (i.e., eventually, probability 1 of correctness is implied) grows high enough that the categorical hypothesis is deemed confirmed and the categorization is verified.

Charles Darwin (1809–1882) first suggested that the mind and intelligent behavior, like the body are products of evolution driven by natural selection. Darwin's analysis led the psychologist William James (1842–1910) to subject the mind to experimental scientific investigation. He attributed to the mind the ability to mentally represent goals, to choose means to reach them, and to act to achieve them.

Shortly thereafter Sigmund Freud (1856–1939) pointed out that there is much that goes on in the brain below the level of consciousness. Freud saw dreams, not rational discourse, as revealing our real hopes, desires, and wishes.

BEHAVIORISM...

J. Watson (1878–1958) and B. F. Skinner (1904–1990) established the school of behaviorism, which largely denied the existence of internal mental states of mind, such as intention or belief, and considered the notion of free will to be an illusion. Behaviorism focused entirely on stimulus–response as all that can be scientifically observed, and therefore all that can be reliably studied.

Behaviorism largely dominated theories of the mind for 50 years, despite the fact that Jean Piaget (1896–1980) undermined the central premise of behaviorism with his observations of the progressive development of mental capabilities in children. Piaget maintained that the development of cognitive behavior in children is a manifestation of the development of internal representations of knowledge in the brain. Nevertheless, the influence of behaviorism persisted, and still lives today in the behaviorist approaches to robotics research.

...AND BEYOND

There are, however, many for which the study of intention, goal seeking, knowledge representation, and mental imagery is a respectable avenue of research. During the 1940s Bertrand Russell (1872–1970) explored mental representations of belief, common sense, reason, and logic. Since the 1950s the mainstream of artificial intelligence research has focused on mental processes of planning, problem solving, and language translation based on production rules, symbolic reasoning, propositional and predicate calculus, list processing, and heuristic search using knowledge embedded in relational and hierarchical databases.

3.1.3 Evolution of the Theories of Knowledge

KNOWLEDGE BASES EVERYWHERE

Epistemology as a scholarly endeavor is associated with the period in the history of mankind when "knowledge" and "everyday activities" seemed to be very remote from each other. In the nineteenth century the idea of "knowledge" started to permeate all kinds of human activities. The new attitudes rapidly developed, and now "knowledge base" is a household word, although not too many people talk about "epistemology."

Nevertheless, epistemology is very much alive today, though it is coming in many disguises. The invention of the computer has made it possible to subject ideas that were the source of ancient philosophical debates to experimental analysis and direct implementation in the day-to-day life. Large-scale integrated circuit technology is rapidly making it feasible to build computing machinery that can mimic, simulate, and emulate the functional properties of at least parts of the brain. Much of the current knowledge-inspired debate revolves around the issue of reductionism versus holism, (can the gestalt experience be understood by reduction to the neuronal level?), functionalism versus structuralism, (can the function of the brain be understood without modeling the neural substrate?), and computation versus connectionism, (are the fundamental operations in the brain best modeled by mathematical logic and symbolic rules, or by connectionist circuitry?).

It is, of course, clear that the knowledge is not a global library full of diligently working "homunculi," and that the brain is not at all like a von Neuman digital computer. Individual neurons and computing nodes in the brain much more resemble analog than digital computers. The brain is a massively parallel machine, and although serial processing threads may exist in the neural structure, the brain's most obvious characteristic is its connectionist architecture. The total computational power of the brain may be on the order of 10^{14} to 10^{16} integer operations per second, and the memory capacity could exceed a terabyte of storage. Yet we still are unclear of the linkage between the specifications of this machine as related to such a concept as "knowledge."

COMPUTATIONAL NATURE OF MIND

But it now seems clear that the brain is indeed a knowledge-crunching machine driven by physical, chemical, and electrical forces. The mind is a process that runs on the hardware of the neural circuitry, and it is controlled by the software and firmware of the synaptic connections, transmitters, and hormones. The brain and mind within are stimulated by signals from sensory organs that encode energy input from the environment. There is no need to assume that the mind is animated by spiritual forces or magical vapors, nor to appeal to microtubules or quantum effects. The phenomena of mind can be adequately accounted for by computational theory, albeit on a scale and with complexity beyond anything yet on the drawing boards.

How knowledge should be represented in robots and intelligent systems is a matter of some controversy. For example, although there is some consensus that knowledge about spatial relationships is necessary for manipulation and locomotion, there is wide disagreement as to what particular form of spatial information is necessary and what data structures should be used to represent it. Some in the robotics community represent spatial relationships within and between objects as two-dimensional maps, potential fields, or three-dimensional grids.

Others represent spatial relationships within and between objects by lists or frames filled with geometric parameters. Examples of this approach are generalized cylinders, superquadrics, and manufacturing data standards. Still others argue for minimalist representations, or for no internal representations at all. They contend that the world is its own representation that can be sensed when necessary for making behavioral choices. Some researchers (e.g., in the AI community) ignore images as a medium for knowledge representation, and their textbooks declare the four major knowledge representing techniques used in AI as (1) logic, (2) production systems, (3) semantic networks, and (4) frames. Images and maps are not even mentioned.

DOWN WITH MINIMALISM!

Whatever the format is, it seems clear that some form of internal representation is needed for making behavioral choices. There are many behavioral choices that require information not directly or immediately available from sensory input. Also, without internal knowledge and reasoning, it cannot be estimated whether an occluded object still exists. Without representation in memory, an observed object cannot be recognized as having been seen before. The minimalist approach can support only the most rudimentary stimulus–response behavior.

More often than not, debates within the robotics community about representation revolve around laboratory experiments where the environment is carefully controlled and contrived to fit the limitations of the experimental hardware and software. The results are then clothed in mathematical and logical formalisms, and generalizations are made that largely ignore the richness and complexity of the real world. One only has to look around one's room, walk through the house, or venture outside into the natural environment to realize how sterile and distant from reality are most of the representations suggested in the AI and robotics literature. Six degrees of freedom kinematics and dynamics of laboratory robots are a far cry from the reality of controlling the skeleton of a cat or a deer running through the forest, or of a bird flying through the tree tops. Understanding relationships in the "blocks world" is a long way from coping with problems an ape may have in raising a family in the rain forest, or even the problems a crab might encounter in searching for food on a coral reef.

In the AI community, particularly among those working on language issues, there is a strong tendency to represent everything symbolically in a form that can be used in logical theorems or expert system rules [1, 2]. This approach has been strongly criticized by authors such as H. Dreyfus [3], J. Searle [4], and M. Kirkland and L. Terveen [5] who argue that symbol manipulation is inadequate even for understanding language, much less for supporting the full range of human perception, cognition, and behavior.

NEODUALISM

Dreyfus criticizes the tendency of AI researchers to extrapolate from a few initial successes with simple examples to more complex situations in the real world. He argues that the methods that produced early progress in problem solving, game playing, and language parsing do not scale up to solve to deeper problems of understanding in real world situations. In particular, Dreyfus and Searle insist that language understanding requires far more than symbol manipulation. However, neither offers any clear guidance as to where to search for additional mechanisms that might suffice. Rather, they conclude pessimistically that true machine intelligence is beyond the ability of current scientific knowledge to achieve. They invoke a form of cartesian dualism in arguing that mechanisms of symbol manipulation can never produce human levels of perception and cognition.

The criticisms that Dreyfus and Searle direct toward others can be turned back on their own arguments. They, like the AI researchers they critique, never seriously consider the complexity and richness of knowledge that can be represented in the form of images and maps. They never address the power of massively parallel computation in the image domain. As a result they fail to appreciate what can be achieved when massively parallel computation in the image domain is tightly coupled with mathematics and logic in the symbolic domain. This is a fatal flaw that leads to an overly pessimistic conclusion.

It is argued here that the failures of AI cited by Dreyfus and Searle can be quite easily explained without retreating into cartesian dualism where properties of the mind are beyond the reach of electronic computation. Ignoring the fact that much, if not most, of the representation and computation in the brain takes place in the image domain is quite sufficient to explain most of the difficulties experienced by the AI systems that attempt to perform real world planning, reasoning, and situation analysis entirely (or even primarily) in the symbolic domain. Images are an important form of knowledge representation. Certainly they are the oldest and most ubiquitous in nature. To ignore images in the quest for understanding knowledge is like ignoring the concept of space while attempting to understand physics.

SYMBOLIC CONTROL OF INSECTS

In recent years, as a result of the difficulties cited by Dreyfus, mainstream AI and robotics researchers have largely retreated from the problem of language understanding. Current efforts are more directed toward understanding primitive behaviors such as exhibited by insects. In many respects this is a healthy turn of events. At least it represents a focus on problems that can be solved with current technology. Certainly there is much to be learned from attempting to understand intelligence at the level of termites and cockroaches. The insect world is a fascinating place filled with incredible examples of manual dexterity and engineering prowess. Interesting behaviors can be readily observed in the ant, the bee, the spider, the house fly, the dragon fly, and the butterfly.

However, in our treatment of intelligent systems we want to go beyond the insect world, to try to understand the kinds of internal representations that are capable of supporting the full range of common everyday human experience. We believe that Dreyfus and Searle are correct in arguing that symbolic computation alone can never produce human levels of intelligent behavior. However, we believe they are wrong in predicting that digital computation cannot ever produce real intelligence. We believe that proper representation of information in images and other perceptual maps and use of parallel computation in the image domain is essential to capturing the full range of complexity necessary for perception and cognition.

We believe the secret to intelligent behavior even at the animal level, let alone at the human level, requires the integration of all kinds of symbolic representation into a multiresolutional architecture employing a special arsenal of computational methods.

3.1.4 Semiotics and Future Perspectives

THEORIES OF SYMBOLS MANIPULATION

Semiotics is a discipline dedicated to discovery of the laws of signs formation and the techniques of explicating their meaning. It aspires to be a basis for determining the laws of languages formation as well and should be considered a key to extracting the meaning of statements and texts presented in these languages. The sketch of semiotics as applied discipline is given in Chapter 2. The formal tools of this discipline allude to the theory of automata and category theory (algebra). More details on semiotics and its aspects can be found in [6, 7]. Unlike the theory of formal languages (and automata), semiotics appeals to what is really known about "thinking" processes (both primitive and more complicated ones), and thus can serve as a bridge from the disciplines in which intelligence is applied to those which focus upon its phenomenological analysis. Building this bridge allows for visualizing in a new light some facets of both areas related to intelligence including its analysis and application as well.

The intention of this chapter is to review the present situation in the area of intelligent systems including their analysis, design, and control, in order to clarify the status of concepts frequently utilized and to outline a sketch of a theory of intelligent systems based on approaches from applied semiotics. The body of knowledge in this area can be characterized by many very diverse approaches presented in several excellent books [8–11], conceptual papers [12–14], and numerous sources in cognitive science and artificial intelligence [15, 16]. Knowledge of these sources can be helpful for understanding, but it is not a prerequisite for reading this chapter.[1]

The peculiarity of the present situation in the area of intelligence and intelligent systems is in the fact that we have an abundance of diverse answers to a set of questions that have not yet been formulated in an organized fashion. We will try to raise these questions and to propose our answers in this chapter. In all cases the analysis ascends to a symbolic system that is supposed to be a carrier of intelligence no matter which particular model is selected by a researcher. This is why semiotics can be recommended as a natural framework, an invariance to the whole area of intelligent systems.

[1]This chapter does not incorporate and does not address the hypotheses raised in such sources as R. Penrose's "Emperor's New Mind" and "Shadows of the Mind"; however, the ideas presented there require a serious analysis from the semiotic point of view.

INTERPRETING SYMBOLS INTO MEANINGS

Semiotics starts with the analysis of signs and symbols and proceeds to the analysis of systems that can be represented by these signs and symbols (symbolic systems).[2] Thus it provides definitions and formal techniques applicable to a wide variety of fields including not only natural language disciplines (literature, linguistics, and so on.), communication, and sociology, but also science (physics, chemistry, biology, etc.), disciplines of a formal modeling (mathematics, logic, etc.), engineering sciences, and so on. The background of semiotics precipitated during the last two thousand years. As a discipline it was first presented in a complete form by C. Peirce in the middle of the last century [17–19]. It was developed in depth during nineteenth and twentieth centuries by many scholars working in different countries. Substantial contribution to semiotics was made by Russian scientists.[3] (See a condensed synopsis of the semiotics evolution in [7].)

Although semiotics has its pioneers in the United States, the existing semiotic schools are associated with Europe (Italy, France, Germany, and other countries). Nevertheless, the number of recorded milestones of the process of development of the semiotics related to intelligent systems, are related to the United States. The first meeting on artificial intelligence was conducted in 1956 in Dartmouth College: Its intention was to start an exploration of thinking processes by using symbolic systems. The first meeting on self-organizing systems was conducted in May 1959 at the Illinois Institute of Technology (Chicago). It invoked semiotics without saying so explicitly; the multivolume publication on cybernetics and brain functions was completed in the late sixties. These contained many powerful references to what the semiotics is actually meant to be. The monographs on hierarchical and other techniques of dealing with large systems were published in the seventies. The first meeting on intelligent control, which was a step ahead in formalizing applied problems linked with symbolic modeling of intelligence, was conducted in 1985 (at RPI in Albany, New York). The first meeting on applied semiotics and its applications in architectures for intelligent control was in 1995, in Monterey, California.

The fundamental role of signs and symbolic systems based on them, for analysis of intelligent systems, is doubtless. This fact was not very clear at the time of the Dartmouth meeting of 1956. The meetings on self-organization brought this issue to the attention of researchers. The period of the 1960s through the 1980s can be characterized by a gradual increase of the attention to the issue of self-organization, especially in the intelligent machines constructed by humans. Sign systems enter our activities under different disguises: in a form of automata languages, computer languages, systems of indexing related to data and knowledge bases, technological notations for the CAD/CAM systems, and sign systems in human/human and human/machine communications.

Starting from the road signs and ending with sophisticated terminological systems, our activities and the activities of human-made intelligent systems are permeated with semiotics. It is astonishing to realize that we are actually unprepared to properly handle sign systems and thus cannot properly discuss the intelligent activities as processes in sign systems. It seems, now is a good time to reconsider the assumptions and definitions related to symbols in the area of intelligent systems.

[2]Sign is defined in semiotics as a part of the inseparable triade "object-sign-interpretant" (see Chapter 2).

[3]This was the interesting observation of James H. Billington, the Librarian of Congress, at the meeting of the Library of Congress on November 8, 1995, in his introductory words. Billington said that semiotics is not so widespread in the scholarly life of the United States; it is better known in Europe, and especially in Russia where the best and the most innovative researchers usually communicate under the colors of semiotics.

3.2 WHAT IS KNOWLEDGE?

3.2.1 Knowledge as a Phenomenon

MEMORIES OF EXPERIENCES

The term "knowledge" is not defined precisely (it is understood as "familiarity, awareness, understanding gained through experience or study" and even as "the sum or range of what has been perceived, discovered, or learned"). Many aspects of this term are discussed in books on philosophy [20, 21] and artificial intelligence [22, 23]. It is our intention to consider knowledge as a collection of *messages* which are unified by some common *meaning*.[4] Then, the message should be understood as a statement explaining the available data.[5] Finally data should be understood as the couple "label with number attached."

Thus knowledge is a part of the triplet that puzzles and fascinates everyone who is involved in a variety of domains of human activity associated with dealing with information. This triplet is

$$\text{data} \rightarrow \text{information} \rightarrow \text{knowledge.}$$

The arrows denote the feeling that *data* associated with some descriptive element[6] are considered *information*, while multiple units of *information* organized in a meaningful structure[7] can be considered *knowledge*. This is how we will understand knowledge: a synthetic phenomenon existing in the form of a structure of messages unified by a particular meaning.

The structure of information organization is the most interesting part of the phenomenon of knowledge. It emerges as a result of the processes of knowledge discovery that happen when GFS operators (see Chapter 2) are applied to the available information. In other words, it is a result of generalization processes that discover patterns within a set of information units. This is how the phenomenon of knowledge is characterized by the specialists in *knowledge discovery*: "A pattern that is interesting (according to a user-imposed interest measure) and certain enough (again according to the user's criteria) is called *knowledge*." [24].

Knowledge is an important and mysterious phenomenon that is applied the most and understood the least. Most people are able "to know," while the most knowledgeable of them are able to admit that they "do not know." People like to have things; however, sooner or later they admit that to *know* things is more valuable than to *have* them. People advance their life with new inventions and activities and all of them are results of collection and production of knowledge.

ARCHITECTURE OF MEANINGS

Where does knowledge come from and how? Knowledge comes from experience and through its understanding. The phenomena of experience are intuitively clear: Just have things happen, and remember how they occurred. The memory of experiences undergoes the process of understanding which results in emergence of knowledge. It turns out that not only do we remember what, when, and where things happened, we

[4]"Meaning" was already discussed in Chapter 2. The *American Heritage Dictionary* (AHD) defines it as follows: "Something that is conveyed or signified; sense or significance."

[5]This is what AHD says about a "message": A usually short comunication transmitted by words, signals, or other means from one person, station, or group to another; the substance of such a communication; the point or points conveyed."

[6]Data, or factual information, are organized for analysis or used in order to reason or make decisions ("numerical data").

[7]We use the term "information" not in the narrow sense of measuring the uncertainty of the message ("so many bits of information") but in a broader sense as the *message* itself.

also realize why they happened, what they affected, and what they entailed. We might even be able to answer the question: what part have these events in a "larger picture?" Finding associations with a larger picture, and their causes and consequences, can give us additional knowledge about a particular event which by itself belongs to a "smaller picture." Knowledge of both smaller and larger pictures has more power than isolated knowledge of each of them, separately.[8]

This chain of "why?" and "what?" ascending to a "larger picture" and descending to the things of our interest and their details is what constitutes interpretation—the core of the phenomenon of understanding. The same set of facts can be associated with the different depths and breadths and thus can carry different amounts of interrelated data and information. Together, they form a body of knowledge.

3.2.2 Knowledge-Related Terminology

One of the items we have postponed, and have not addressed yet, is the definition of "knowledge" (later we introduce the definition for "knowledge representation" too). Let us try to deduce these definitions from what we already know about these notions. It is a plausible assumption that we already tacitly assumed some definitions for both knowledge and its representation since many of our considerations were acceptable for the audience and no contradictions seem to surface so far. What we said, gives us an opportunity to formulate the following definition of knowledge:

PRELIMINARY DEFINITION OF KNOWLEDGE

Knowledge is a system of patterns within the bulk of the available units of information based upon experiences; these patterns are generalized, labeled and organized into a relational network with a particular architecture; knowledge implies our ability to use it efficiently for reasoning and decision making. The reader will find a different definition in Section 3.6.3; it will support functioning of knowledge-based systems.

DEFINITION OF PATTERN

A *pattern* is a discernible coherent system of components and features based on the observed and/or intended interrelationship of parts (a discernible, or designed structure).

DEFINITION OF INTERPRETATION

An *interpretation* is a process (and the result) of putting in correspondence knowledge from the available system of representation with the system of representation more familiar to the user; an interpretation is performed via explanation and pattern discovery.

Let us refine the concept of "interpretation" and discuss it in more detail. Knowledge presumes an existence of a system of generalized records of experiences associated with

1. Other available generalized records of experiences organized into a relational system.
2. Interpretations of the generalized experiences, interpretation of their components, i.e., the experiences, related to the parts of generalized experiences, and the "larger-picture" experiences.
3. Patterns discovered within the records according to their interpretation.
4. Patterns, and their components, labeled and put into a relational system.

[8]The statement "knowledge is power" is of course a trivial one. The power of understanding is meant here, and this leads one to the power of decision making. (Sometimes it entails the power of actions determined by this decision.)

The concept of "experience" turns out to be fundamental for the explanation of knowledge. In this definition we allude to our ability to create records of experiences (i.e., to memorize states, events, processes); we refer also to generalized records and experiences. Thus we presume that some concept and/or procedure of generalization should be available to us (see Chapter 2).

DEFINITION OF EXPERIENCE

Experience is a record of an *event* (changes of state), or of a group of events, observed in the past and consisting of the following components:

1. The state (situation) prior to the changes of interest, called the *initial state*.
2. The state (situation) after the changes of interest happened, called the *final state*.
3. The set of state changes together with the set of causes that produced the changes, called *actions*.
4. Evaluation of the cost of changes (losses) and the gain attained as a result of these changes.
5. The final goal of the activities which determines the final destination of the activities (usually, beyond the scope of the particular experience) and also serves as the basis for evaluation.

If the experience (change of state) can be characterized by some unity in its interpretation, we call it a *discrete event*.

This definition of experience refers to our yet unspecified ability to *observe* things and figure out their parameters and variables. We will rely on our sensors to "observe" things. This definition also refers to our ability to evaluate things by "figuring out" how "good" they are for the final goal. However, this "figuring out" will require our ability to represent our observations in the form of descriptions that might contain numerical information too. So, in the future, we expect to deal with descriptions supplemented by their numerical values.

Following Section 2.9.4, let us denote the prior state, S_1, the action that has been applied, A_{12}; the state that emerged after this, S_2 the value V_{12} of the result, and the goal of action G for which this value was computed. These are the four fundamental components of experience: states, actions, value judgments, and goals.

$$E \to [S_1 \times A_{12}]_G \to [S_2], [V_{12}].$$

DEFINITION OF STATE (SITUATION)

A *state* (situation) is a set of parameters and variables that allows for monitoring, recording, and possibly control of an entity or a group of entities; this set is tagged with the moment of time at which it is registered or measured.

We should be prepared to measure the states and actions including their parameters and variables. Clearly, this is not a trivial procedure of comparing variables with a yardstick scale. The results of our evaluation have limited accuracy. So all our observations and records are contained in the system of knowledge representation that comes together with corresponding errors.

DEFINITION OF STATE VARIABLE

A *state variable* is a numeric or descriptive (symbolic) attribute of a system, or a process that represents the state of an object, situation, or process at a particular moment of time.

Some of the state variables can be considered output of the system and be communicated, or applied externally (*output variables*). Some of them can be obtained from another system and used as *input variables*. All state variables can remain constant or evolve in time as a result of external inputs or internal phenomena. The record of time evolution of the description and/or values of state is called a *motion trajectory*. The trajectory of motion can be demonstrated as a curve in a state-space (see Chapter 2).

DEFINITION OF STATE CHANGES

State changes are a set of variations in descriptions and/or values observed in some or all of states that emerge in time.

At each moment of time, the rate of state changes can be evaluated by the derivative of the motion trajectory computed for this particular moment of time. State changes should not be confused with the causes of these state changes (input variables or controls). The temporal strings of state changes are called *processes*. Some of the salient state changes (or the strings of state changes) can each be discussed as a distinctive "whole" and be given a special label. We call them *events*.

DEFINITION OF WORLD

The *world* is the set of interrelated entities within the scope of our attention; each of these entities is characterized by a *state*; the world contains the *goal* and is the environment in which the system produces the *state changes*.

Note that it is characteristic for the world to be curtailed by our attention. We will call the boundaries of the world the *scope* and define them in all coordinates of the state as well as in time. The temporal scope of the world is called sometimes the *horizon*. The "unbounded" cases and the situations where something can be expanded into "infinity" are consciously excluded from consideration (they are beyond the scope of our attention). Different intelligent systems have different worlds if their scopes of attention are different. The same is true about people.

FIRST DEFINITION OF ENTITY

An *entity* is *something* that exists as a particular and discrete unit. It is *something* that can be given a name. See second definition on page 134.

In other words, an entity is a single object or a group of objects of the reality that can be recognized as a unit existing separately from its properties, relations, and surrounding environment. Entities are related to each other, and these relations together with the interrelated entities form the *world*. The definition given above is an intuitive one. More detailed treatment of the concept of *entity* is given in Section 3.6.4.

Both the cost and the gain for the same event can be evaluated differently depending on a final goal of a system. The meaning of cost and gain presumes that each event inflicts upon the system losses (usually, of time and/or energy) and can deliver some advantage (increase in profit, knowledge, etc.) that accumulate as the processes develop. The ideas of cost and gain lead us inadvertently to the concept of goal. The concept of goal (together with the ideas of cost and gain) is ingrained into knowledge representation.

DEFINITION OF GOAL

A *goal* is a state to be achieved as a result of an experience, or as a result of a set of experiences that can happen consecutively and/or parallel in time; it is the designed final state of the accomplished task and/or subtask (like a destination point). The *subgoal* is the result of goal decomposition.

The goal is a state to be achieved; it is represented as a data structure. A goal is determined at a level of resolution in a multiresolutional system (see Chapter 2). Different approaches are acceptable in intelligent systems. One of them presumes that the goal for the next level of the hierarchy is selected by the upper level and submitted for performance by the lower level ("assigned" to the lower level). Another approach is that the goal is determined at the level under consideration after analysis of the long-term schedule submitted from the level above. Both approaches are intimately interrelated.

3.2.3 Storing the Knowledge

How is knowledge stored? How *should* it be stored? In what form is knowledge of space and time, entities and events, tasks and skills, plans and behavior, logic and language, physics and mathematics, morals and values, represented in our brains? More to the point, how can such knowledge be represented in a living system or in a computer?

Storing the knowledge should be predicated by its subsequent use. Intelligent systems equipped with a subsystem of knowledge representation should be capable of using this subsystem for such obvious things as follows:

- Updating of knowledge.
- Efficiently storing knowledge.
- Knowledge enhancement (from new experiences, communication, reasoning, and discovery).
- Knowledge retrieval (or request).
- Knowledge compression via encoding, or via generalization.
- Building up the multiresolutional knowledge architecture.

Then there are the less obvious activities to be performed by subsystems of knowledge representation:

- Generating rules for interpretation of sensory information (images, messages, etc.).
- Generating rules for task decomposition.
- Generating rules for job assignment.
- Generating rules for scheduling.
- Generating rules of error compensation.
- Anticipating states (predicting).
- Simulating plans developed by a subsystem of behavior generation.

These are the duties of the knowledge representation system. All these duties of the knowledge representation system are not esoteric, or overly sophisticated. They belong to the area of database (DB) systems. DB with such capabilities are called "active databases" [25].

Knowledge representation provides altogether for four major capabilities of intelligent systems that arise from the properties of the subsystem of representation:

- Intellect
- Awareness
- Consciousness
- Reflexia.

INTELLECT

The property of intellect is associated with an intelligent system's ability to solve problems, reason, and plan. In Chapter 7, we will demonstrate how the abilities of problem solving and reasoning serve for reconciling the discrepancy between the present state and the goal submitted to the system (behavior generation). In Chapter 8, we will demonstrate that planning, namely simulating future activities within the "imagination" is the powerful tool of behavior generation. Intellect is one of the primary features of the *intelligence* of a system.

SITUATION (STATE) AWARENESS

An intelligent system can be said to be aware of its situation (state) when there exists a close correspondence between knowledge in the system's world model and the situation in the external world. Situation awareness enables a system to reason about the current state of the world, to analyze the historical events leading to the current state, and to plan for the future. Perception may be said to be the functional transformation of data from sensors, in the context of knowledge from long-term memory, into situational awareness.

CONSCIOUSNESS

The situation awareness can be enhanced if the system's world model includes representation of the external world together with the representation of the inner states of the system (representation of "self"). This enhanced representation allows the intelligent system to distinguish between itself and the rest of the world, to determine its place in the world, and to anticipate the consequences of its motion in the external world. This enhanced situation awareness is called *consciousness.*

REFLECTION (*"REFLEXIA"*)

Consciousness can be enhanced too, if in addition to the ability to represent itself within the external world, the system is capable of representing its representation in the external world. This phenomenon is described in Chapter 2 as a source of the recursive and potentially infinite string of nesting in the representation. *Reflexia* is useful when several intelligent systems are considered. Their joint functioning (both cooperative and adversarial) will be affected by their ability

- To represent themselves within the external world.
- To represent themselves together with the world and to represent their own representation of themselves.
- To represent other intelligent systems and their representation of what they believe how other intelligent systems represent representation of each other.

Indeed, the decision-making process of each intelligent system will become more successful if the intelligent system A takes in account not only where the other systems B and C are, and what is their behavior, but also what they "think" about system A, or how system A is represented within systems B and C (the latter representations might be different). This means that in a world of intelligent systems, the list of state variables is composed not only for the objects and processes but for their representations too. It includes both physical variables and their representations. One can talk about representation of the first order based on the physical values of the variables. Representation of the second order will include description of the representation of the first order, and so on.

3.2.4 Why Does the Need for Representing Knowledge Emerge?

This question is the central motivating question of Chapter 3. Let us try to refine it. Undoubtedly, knowledge is not something brought inward from outside. This is a pattern that emerges within the intelligent system, and it seems to be stored within the system after it emerges. This pattern is created internally within the intelligent system based on the signals obtained externally as a result of their interactions with the contents of the "storage." This pattern is not just a memory of these external signals, or a memory of their interactions with the contents of the "storage." It is rather a trace within us of objects and events observed, experienced, and previously stored supplemented with emerging correlations among them discovered by the intelligent system. This pattern and these correlations together are called "knowledge representation." In this book we will try to answer the following questions:

- How is knowledge represented in living creatures?
- How can it be represented in intelligent systems other than living creatures?
- How can it be used for modeling and simulation of states and processes in systems?
- How should it be represented in constructed intelligent systems in various domains?

Knowledge representation is conjectured to be a symbolic system of models and memory records which allows for functioning of the system of elementary functioning loops described in Chapter 2.

FIRST DEFINITION OF KNOWLEDGE REPRESENTATION

Knowledge representation is a part of an intelligent system within which a *model, or a set of models* can emerge corresponding to (representing[9]) the external system of interest; this model, or a set of models can be used for solving problems for this system. A second definition is given on page 135.

DEFINITION OF MODEL

A *model* is an artificial mental, symbolic, or physical system which is substantially simpler than a system of interest, yet gives an opportunity to reason about, or simulate the processes in the system of interest with accuracy sufficient for decision making.

DEFINITION OF SYSTEM OF INTEREST

The *system of interest* is a subset of reality upon which a goal is formulated.

Although it is widely recognized that knowledge is crucial to behavior, perception, and cognition, there is little agreement about the way knowledge should be represented, or even what knowledge is. It is clear that humans have and use a great deal of knowledge about the world, and are aware of much that goes on in the world around them. But exactly what format the brain uses to store and retrieve this knowledge, and how the brain processes sensory input to acquire and update its internal store of knowledge, is unknown. There is some understanding of the tools of storing and organizing knowledge in various domains of societal activities including science, industries, politics, and so on.

The institutions and groups using and/or producing knowledge are intelligent systems too, and the conjecture is that architectural similarity might exist between the knowledge representation subsystems of the individual human brains and the collective brains of the group intelligent systems. On the other hand, there is no doubt that brains of animals have resemblance with human brains, and that nervous systems of other living creatures are the precursors of animal brains. All of these intelligent systems might have similar principles of knowledge representation.

Methods of dealing with knowledge in various intelligent systems are not different theoretically. In all cases there is a process of encoding of the phenomena of reality into

[9]To *represent* means to stand for, to express, to symbolize, to be equivalent of something in the world.

a particular representation paradigm that can be characterized by the nature of signals carrying representation. Signs are used to encode the signals from sensors entering the system of perception. Signs are the basic phenomena of representation. While input signals have different physical nature (light, heat, sound), the nature of the sign is the same; it is determined by the nature of representation system (e.g., by the architecture of synapses that carry the representation). The meaning of signs is determined by the input signal that has generated this particular sign. In order to encode experiences (including states, changes, goals, and costs), signs are transformed into symbols and schemata as a part of semiosis.

Undoubtedly, all these phenomena can be translated into a hardware architecture of synaptic connections, corresponding networks and loops. However, we will be more interested in generalized models of these "hardware" artifacts. This is understandable. If you want to control a power plant, you would be more interested in its generalized architecture of control rather than in the connection of wires within one of the underfloor cables.

3.3 KNOWLEDGE REPRESENTATION IN THE BRAIN: ACQUIRING AUTOMATISMS

3.3.1 An Elementary Information Processing Unit: A Neuron

The information processing part of the brain is composed of neurons. Each neuron is a tiny computer that receives inputs through synaptic receptor sites on dendrites and its cell body; it computes an output that is distributed to the inputs of other neurons via an axon that may have multiple branches. Each branch terminates on a receptor site on a dendrite or cell body of another neuron, or on an actuator cell such as a muscle or gland. The dendrites and cell body integrate the weighed input from each receptor site over time and sum over all the receptor sites.

WETWARE

The output of each neuron is conveyed by an axon to the various locations in the brain where the information is needed. At the point on the neuron where the output axon attaches to the cell body, there is a section of membrane (the axon hillock) that transforms the cell body potential into a string of impulses for transmission over the axon.

An axon may branch many times on its way to a variety of destinations. At the end of each branch of the axon, there is a termination consisting of a synaptic bouton. Inside of each synaptic bouton, there are many tiny vesicles filled with transmitter chemicals. The arrival of each impulse triggers the release of a set of vesicles filled with transmitter chemicals into the gap between the axon bouton and a dendritic receptor site on the next neuron. The transmitter chemical diffuses across the gap and activates ionic electric current flow across the receptor dendritic membrane. The value of the current flow is a product of the frequency of the impulse rate on the incoming axon multiplied by the strength of the synaptic connection on the receiving neuron. The dendrites and cell body of the receiving neuron integrate the flow of current to produce a voltage. By this process the information encoded by the membrane potential (i.e., value of a state variable) in one neuron is conveyed to and affects the membrane potentials (i.e., states) of other neurons. The effect of axon activity on neuron inputs may be modulated by the concentration levels of hormones and drugs in the fluid surrounding the neurons, as well as by waves of electrical potential caused by simultaneous activation of many surrounding neurons.

3.3.2 A System for Searching and Storing Patterns: A Neural Net

Knowledge of each synaptic loop is thus represented by electrical potentials in neuron dendrites and cell bodies, by impulse rates in axons, by the strengths of synaptic connections, by the network of connections between neurons, and by electrical and chemical environments surrounding the neurons. The electrical potentials in neurons and impulse rates on axons are transient phenomena that can change rapidly in a matter of milliseconds. Feedback from output to input of a single neuron can preserve state information and perform integration or differentiation, depending on the sign of the feedback loop. Longer feedback loops can store strings of information in what are essentially delay lines or dynamic memory structures. These type of structures are particularly well suited to storing and detecting temporal patterns and forming phase-lock loop filters and detectors.

Chemical environments depend on diffusion rates that may change concentrations over periods of seconds or minutes or longer. These concentrations may represent operational modes of various ELFs of the brain (or moods or "feelings" of the creature). For example, hormones are well known to affect moods, and chemicals such as adrenalin establish levels of alertness or aggressiveness.

DEVICES FOR ADAPTIVE GENERALIZATION

Long-term learning of memories in the brain is widely thought to occur due to changes in the synaptic weights that interconnect neurons. The patterns of interconnections emerge as a neural net. Thus there are two types of patterns: short-term patterns of potentials and long-term structural patterns of interconnections. As generalization processes develop, the layers of interconnections emerge.

Learning is hypothesized to be determined by structural changes in the microneuroanatomy of the synapses that persist over periods of months or years. There are, however, a variety of learning mechanisms, and different kinds and rates of learning in different parts of the brain. Memories of experiences can be stored almost immediately during a single learning experience. Knowledge of motor skills may require many practice sessions over long periods of time up to weeks or years. Some knowledge such as language skills can be learned easily during a brief interval within the development of the brain, and then remain fixed for a lifetime. The network of long-range patterned interconnections between neuronal modules develops over months and years as the brain grows and and matures. These pathways typically endure for a lifetime.

Thus the brain is a complex network of computing structures with many parameters that change over time on a variety of different time scales. Yet at any moment in time, a set of ELFs can be recognized which correspond to a set of automatisms of functioning. Synaptic processes characteristic of each ELF have various impulse rates (or inter-impulse intervals) and for each axon a time-dependent scalar can be determined that may represent a signal, an attribute, or a state variable. An ordered set of axons each producing an output may thus represent a state vector, an attribute vector, or a symbolic variable such as a character in an alphabet or a word in a vocabulary.

A two-dimensional array of neurons each producing an output may represent an attribute image or map. A three-dimensional array of neurons can become a map of the 3D situation. Maps with a larger number of dimensions could be considered. The synaptic weights, the network of interconnections, and the electrical and chemical environment in the brain define the functional mapping between input and output for a collection of neurons. The brain can therefore be modeled as a collection of computing ELF-like modules connected by a set of signal pathways. Each ELF can be decomposed into a set of micro-ELFs, while a number of ELFs can be aggregated into a single macro-ELF. Recent research results (e.g., by S. Grossberg and W. Freeman) confirm the existence of multiresolutional layers in the brain [52, 53]. See also Figure 1.2.

The following conjectures can be made for this multiresolutional system of nested ELFs. Functions computed by each neuron in a module can represent single-valued arithmetic or logical functions such as addition, subtraction, multiplication, division (AND, OR, XOR, NOR, NAND), or the inner product between an input vector carried by an ordered set of axons and a synaptic weight vector. A set of neurons in a module may compute multivalued functions such as coordinate transformations, correlation functions between spatial or temporal patterns, recursive estimation, phase-lock loops, and tracking filters. Neuronal computing modules can also represent table look-up memory functions or indirect address pointers. This enables them to execute if–then rules, act as finite state automata, perform list processing operations, and set up and search relational databases.

Learning in neural nets within and between neural modules provides mechanisms for acquiring both direct and inverse models of the complex dynamical systems being controlled. These enable prediction, simulation, and feedforward control. The learning of such models enables the learning of skills in manipulating objects and maneuvering successfully through complex dynamic environments such as exist in most places of the real world.

3.4 SENSORY AND SYMBOLIC REPRESENTATIONS IN THE BRAIN

3.4.1 General Comments

Sensory representations are based on sensory images, both immediate and memorized from the prior sensory impressions. Signals arrive from sensors and are encoded in the appropriate system of electrical potentials and concentration of chemicals which determine processes of storing them in corresponding memory locations. The primary signs that encode intensity, colors, temperatures, and other attributes of the pixels[10] are then clustered and generalized into abstractions of edges, patches, segments, and other symbolic units which are blended in with other abstract concepts and used for the subsequent interpretation.

Although sensory impressions arrive from the totality of sensors and are subjected to subsequent fusion into the integrated sensory image of the world, we will concentrate on images obtained from vision. The analysis of images consists of discovering and interpreting patterns that are contained within the image. One can see multiple immediate applications within the robotic systems. The results related to visual images can have powerful analogical implications for other sensory images as well as their integration. By and large, a substantial fraction of the nervous system of most living creatures is strongly affected by vision, if not derived from vision.

3.4.2 Sensory Images

VISION Let us consider an example of vision. The visual field is projected by a lens upon an array of photosensitive cells in the retina of the eye. Each retinal photosensor produces an electrical potential proportional to the intensity of the incoming light integrated over the area of the photosensor and over a time interval. The latter is determined by the cell membrane capacitance and resistance. The retinal neural image can be considered

[10]Pixels are elementary indistinguishability zones of all sensory images. They are called also granules, and tessellata.

a two-dimensional array of voltages at the output of photosensors. Each photosensor produces a single "pixel" of the image, and the array of pixels forms an intensity or color "attribute image," an array of the attribute values.

There are rod photodetectors in the retina that detect white light intensity and three types of cones that measure input energy in three different color bands. The retina also contains horizontal cells that compute spatial derivatives (or spatial frequency, or Gabor functions, or wavelets) at each point in the image. Output from the horizontal cells thus represents a spatial-derivative-attribute image that is registered with the intensity-attribute image. The retina also contains amacrine cells that compute temporal derivatives of the intensity or color image at each pixel. The output from the amacrine cells thus represents a temporal-derivative-attribute image.

More than 50 visual processing areas are known to exist in the brain [26]. The output of each of these areas represents an attribute image. It should be noted that there are 600 times more neurons in the visual cortex than there are input axons from the lateral geniculate. This implies that the brain dedicates many different neural computers to simultaneously extract many different attributes for each pixel. It suggests that many attribute images are being computed simultaneously in parallel.

Attribute images can be registered by virtue of the fact that the various types of attribute-computing cells occupy the same pixel area in the neural sheet. Registered attribute images may differ in resolution, depending on the relative density and receptive field size of the different types of attribute-computing cells.

Attribute images may also be registered by means of topological mappings between different visual processing areas. Registration with the retinal egosphere is maintained in many, if not most, computing modules throughout the visual system. For example, registered topological maps of the retinal visual field exist in the lateral geniculate nucleus, the visual cortex, and the superior collicullus [27].

SENSING POSITION AND ORIENTATION

The brain also has a set of translational and rotational accelerometers in the vestibular system. Signals from these sensors can be integrated to provide estimates of position, orientation, and linear and rotational velocity with respect to the external world. Control algorithms in the vestibular-occular reflex loop can then inertially stabilize the eyeballs in their sockets and generate an inertial reference frame for egosphere image representations. Other sensory modalities may also produce maps or images that represent the spatial distribution of excitation over arrays of sensory neurons.

TOUCH

For example, arrays of tactile (pressure, temperature, vibration, and pain) sensors exist on the surface of the skin. These arrays are quite dense (i.e., high resolution) on some parts of the body such as fingertips, lips, and tongue, and less dense on other parts of the body.

The topological mapping that exists between tactile sensors and regions in the brain that process tactile sensory input is well known and often explored. The entire surface of the body is represented by a topological mapping in the tactile sensory cortex. The variation in density of tactile sensors on the skin is mapped into different sized regions on the surface of the cortex dedicated to processing the tactile information. This produces a distorted image of the body surface on the surface of the brain. There is a similar mapping from the motor cortex to the muscle groups that produce forces and motions.

HEARING

In the auditory system the cochlea generates a map that resembles a sonogram in which the component frequencies of a sound are arrayed along one dimension and temporal delay along the other dimension. The superior colliculus contains a map of auditory space such that azimuth of the incoming sound is arrayed along the horizontal dimension and elevation is arrayed along the vertical. This is maintained in registration with a map of the visual egosphere. In the Barn Owl, the eyes and ears are both fixed in the head so the auditory and visual map can easily be maintained in registration on the surface of the superior colliculus. In the monkey, the auditory map (which is fixed with respect to the head) dynamically scrolls across the surface of the superior colliculus so as to remain in registration with the visual field (which is fixed with respect to the eyes).[11]

3.4.3 Symbolic Representations in the Brain

Symbolic representations include two groups interrelated in a hierarchical fashion: (1) linguistic descriptions using natural languages and (2) abstract descriptions using formal languages. Descriptions enter the brain via interface of our sensory systems, but the grammar upon the stored symbols is not directed by the ontology of physical world as it is for the regular sensory images. Grammars of the natural and formal languages have their own laws. This is why the human brain has special faculties that allow for dealing with descriptive and other symbolic representations.

How linguistic and abstract information is represented in the brain is less understood than the laws of dealing with sensory images. The fact that the brain stores and uses linguistic and abstract information is clear. Most of the higher animals have some capabilities for planning and communication that suggest some capabilities to compute abstract objects in a symbolic form. Humans have verbal reasoning and language facilities that set them far apart from the rest of the animal kingdom. The ability of the human mind to formulate and manipulate verbal descriptive images, situations, and abstractions is awesome. All of language, mathematics, science, art, music, dance, and theater, is based on manipulation of hierarchies of symbols in one form or another. Industry, business, finance, and commerce could not exist without the use of descriptive representation including verbal images of the world and their symbolic abstractions of various levels of resolution. Yet there is little understanding of how the neuronal substrate of the brain represents languages or performs manipulation of words and phrases.

It is known where the regions of the brain involved in language are located. It is also known where in the brain visual information is integrated with auditory information. The ability of the mind to form associations between written or spoken words and sensory images of objects in the world has been studied extensively. But the precise mechanisms by which the brain understands spoken or written language, associates semantic meaning with visual images, and generates verbal or written responses remain a deep mystery.

An ordered set of axons, each representing a scalar time-dependent variable, can represent a time-dependent vector, a point in state-space, a symbol from an alphabet, or a word from a vocabulary. Over an interval of time, a string of impulses on each axon in an ordered set can thus represent a string of vectors, a string of points (i.e., a path) through state-space, a string of characters in a word, or a string of words in a sentence.

An ordered set of axons can thus represent the state (e.g., position, orientation, velocity), and attributes such as color (e.g., red, blue, green) and shape (e.g., length, height, width). An ordered set of axons might also represent the symbol for an entity (a name of it), or a pointer to (or the address of) a frame describing an entity or symbol. The set of all possible states of a set of axons can define a vector space or a tensor field. Each point in the vector space can represent an address or location that contains a value, or a

[11] M. F. Jay and D. L. Sparks, Auditory receptive fields in primate superior colliculus shift with changes in eye position, *Nature* 309 (1984): 345–347; M. F. Jay and D. L. Sparks, Sensorimotor integration in the primate superior colliculus: II. Coordinates of auditory signals, *Journal of Neurophysiology*, 57 (1987):35–55.

pointer to an indirect address that contains a value. A chain of indirect address pointers can produce a list or graph data structure.

THE WORLD AND US IN SYMBOLIC FORM

There is some evidence that suggests how at least some abstract symbolic information is computed by neuronal mechanisms. Sets of neurons have been observed in the motor cortex that define a motion vector that indicates the direction and magnitude of upcoming movement by using the local coordinates. Single neurons in these sets are broadly tuned so that each neuron's activity is highest for a movement in its own preferred direction and decreased progressively as a linear function of the cosine between the neuron's preferred direction and the direction of the planned movement. The sum of activity of all the neurons in the set produces a resultant vector that indicates the direction and magnitude of the muscular forces required to produce the planned movement.

One of the most thoroughly studied motor representations is the map of saccadic eye movements found in the superior colliculus. The sets of neurons that define gaze control movements vary systematically across the surface of the superior colliculus. The motor commands computed at each point by these sets of neurons are abstract and symbolic. They define the motion direction and amplitude, or velocity, required to direct the gaze to that point in the retinal field. These commands are registered with the visual and auditory sensory map that also exists on the surface of the superior colliculus.

That these commands are symbolic is evident from the fact that the representation of motor command vectors is very different from what is required by motor neurons sending signals to individual muscles.

Thus, while some of the structures by which sensory data and motor command vectors are represented in the brain are becoming more clear, we are far from understanding how the brain represents the broad range of symbolic information necessary to support human levels of performance in problem solving, reasoning, and language.

3.5 REFERENCE FRAME, IMAGINATION, AND INSIGHT

In this section we will discuss knowledge-related cognitive phenomena existing because of the particular structure of knowledge representation that should be supportive of them.

REFERENCE FRAME

Humans are capable of fixating an environmental point that gives an opportunity to sample portions of the image of the surrounding environment and refer inner images to a particular point—origin of the reference frame. Neurons tuned to zero-disparity at the fovea refer to the instantaneous, exocentric three dimensional fixation point [30]. The role of this cognitive tool is the same as the role of any reference frame: to reduce complexity of computation. Not only physical variables can have reference frames; descriptive entities, lists, situation representations are given frequently in their reference frames.

IMAGINATION

One of the important duties of the subsystem of knowledge representation is simulation. The results of simulation are required for decision making, planning, and control (see Chapters 7 and 8). The faculty of imagination has evolved to satisfy the need in simulation. The same mechanism that is used to store the images of reality, and react to them by functioning of the behavior generation module, stores images of "would-be reality" which are synthesized for the conditions of interest. Thus a particular landscape of interest that exists in our memory only as a summer landscape (we never saw it otherwise) can be blended with our knowledge of how parts of it look like in winter time. Thus we obtain an imaginary scene and can apply to this scene plans obtained from the module of behavior generation.

Entity attributes stored in long- and short-term memory include information such as shape, size, color, position, and velocity. It therefore is at least theoretically possible to reconstruct images of entities from the information stored in long-term memory. Whether in fact the brain does this is a subject of considerable current debate. Many people report dreams that involve images, and it has been observed that during dreaming eye movements are consistent with viewing images. Kosslyn argues convincingly that the mind generates images during the process of recall, and that these images can be searched for information in response to questions regarding what is remembered [31]. Other authors argue just as convincingly that the same data are also consistent with theories of memory that involve symbolic information stored in linked lists or graph structures.

Whatever can be reconstructed from long-term memory is a pale shadow of immediate experience. Except for pathological cases of hallucinations, there is no mistaking what is remembered or dreamed for what is directly experienced. The imagination may be able to recall some features about an earlier experience, for example, be able to remember the number (up to about seven) of windows in a familiar wall by imagining standing in the room and counting windows, but one can never reconstruct from memory the exquisite detail, or the vivid colors, or the complexity of dynamic movement that occurs during immediate sensory experience of the natural world. Imagination can work not only with sensory images but with descriptive images too. This is how new descriptive images emerge (e.g., in literary works).

INSIGHT

We already discussed the phenomenon of imagination when new images are generated within our knowledge representation. When parallel processing is available, several channels of imagination can be feasible. Consciousness allows for constantly monitoring only one of them. Combinatorial search via imagination can run within other channels with no monitoring until a combination (assembly, aggregate, entity) is found satisfying conditions specified earlier. The result of combinatorial search via imagination will be brought to the attention of behavior generation module. By analogy with similar processes in humans, we will call it *insight*.

3.6 PRINCIPLES OF KNOWLEDGE REPRESENTATION, ENTITIES, AND RELATIONAL STRUCTURES

Comparison of the architectures of knowledge representation shows that they are based on the same basic principles. One of the principles is solving the problem many times at many levels of resolution in a nested manner. This is trivial for any computer. In the computer the hierarchies of languages are well known that allow for solving the problem in high-level language, then in assembler, then in machine codes, and so on, eventually at the level of electric pulses. The problem-solving cycles at the higher-resolution levels are nested within the problem-solving cycles of the lower-resolution levels (or higher levels of abstraction). Something similar happens in every intelligent system at a higher level of sophistication.

PRINCIPLES

The following principles of knowledge representation can be discovered implicitly, or explicitly in all kinds of hierarchies we are dealing with:

1. Principle of heterogeneity, or principle of multiple symbolic (labeling) systems representation.
2. Causality principle.
3. Principle of continuity.

4. Principle of efficiency.
5. Principle of limited resource (which entails limits of resolution, limited field of view, etc.).
6. Principle of thesaural interpretation of labeling.
7. Principle of entity-relational knowledge representation.
8. Principle of incompleteness.
9. Principle of multiresolutional representation (with aggregation-decomposition, or centralization-decentralization), which is the same as the principle of nested hierarchical representation.
10. Feedforward principle (goal-oriented decision making).
11. Principle of feedback (compensation-oriented decision making).
12. Principle of creativity (combinatorial techniques of decision making).

All these principles undoubtedly are rooted in the processes of evolution and are survival oriented. We will return later to the key importance of the three groups of them: attention focusing, grouping, and searching. All are based on the idea of *selection*.

3.6.1 Nested Hierarchical Knowledge Organization

The behavior (functioning) of the systems forms nested hierarchies as shown in Figure 3.1. Knowledge representation for the phenomena (Ph) and subphenomena (SPh) (processes, behaviors, controls, knowledge of the events and subevents) should be done in a manner which reflect the 12 principles of knowledge organization.

PROPERTIES

The following properties are characteristic for nested hierarchical (NH) knowledge representations:

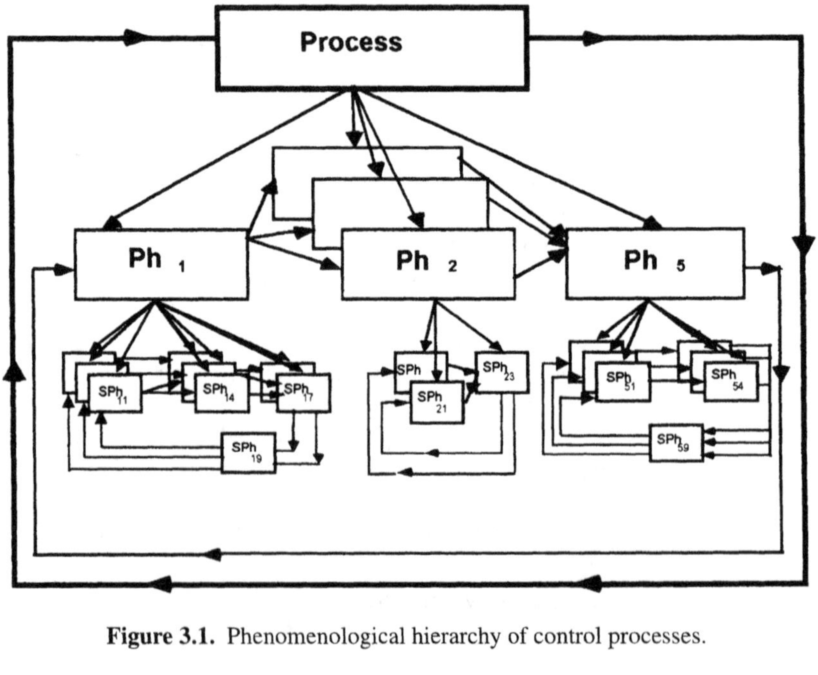

Figure 3.1. Phenomenological hierarchy of control processes.

Property 1. Computational independence of the resolutional levels. Each of the functional loops in Figure 3.1 can be considered and computationally dealt with independently from others. Each of them describes the same control process with different accuracy and different time scale, which entails the difference in the vocabularies of levels.

Property 2. Representation of different domains of the overall system processes resides at a level of resolution. Since all loops are performing the same operation at different resolutions, they are dealing with different subsets of the world (starting with the "small, fine grained, and quick" world at the bottom, and ending with "large, coarse grained, and slow" world at the top).

Property 3. Correspondence among different levels of resolution and different frequency bands within the overall process. The resolution of the level is associated with the frequency of sampling. This not only means that the frequent sampling is associated with the higher accuracy of the processes' representation but also that the frequencies of processes lower than the frequency of the sampling are not likely to be reflected in the control processes of this level.

Property 4. Capability of loops at different levels of resolution to integrate into the ELF diagrams. Loops are nested one within the other; the lower resolution loops can refine the representation of their processes by using the higher-resolution loops. Each of the loops contains sensory processing (perception), a world model (knowledge representation), and behavior generation subsystems with the external world attached to them via actuators and sensors. In the meantime they operate with different *scope of attention*: Each process at higher resolution has a scope of attention narrower than the adjacent level of lower resolution.

Property 5. Correspondence between the upper and the lower parts of the ELF diagram. The upper part of the NH-ELF shown in Figure 2.12 (SP, WM, BG) corresponds to the lower part (S, W, A). The hardware realities of S–W–A are represented in computer architecture as SP, WM, and BG.

Property 6. Formation of the behavior of the system as a superposition of behaviors generated by the actions at each resolution level. The system's action is generated simultaneously at several levels of resolution (granulation), such as when it is teleoperated, or an autonomous mobile robot is considered. Then the list of levels, bottom-up, will be (a) the output motion level (the lowest-abstraction level, or the most accurate level), (b) the maneuver level, (c) the plan of navigation level, (d) the scenario of operation level, and (e) the mission planning level (the highest-abstraction or lowest-resolution level).

Property 7. Similarities between the algorithms of behavior generation at all levels. All levels execute a particular (pertaining to the level) algorithm for finding the best set of activities (control trajectory); each higher level constitutes the prediction for each lower level. At each level, the action-generating algorithms should perform assignment generation, planning, plant inverse, decomposition, compensation, and execution command generation.

Property 8. Evolution of the hierarchy of representation from linguistic at the top to analytical at the bottom. Several levels of planning/control processes presume a nested system of representations that can be analytical at the level of high resolution and linguistic (knowledge based, or rule based) at the level of low resolution.

A good example in the design and control of NH-ELFs is the autonomous mobile robot. However complicated the state-space of a particular process might be, the goal of control can be formulated as arrival from the initial point (state) to the final point (state) in this space. Thus the moving point within the state-space can be identified with an imaginary little robot controlled from one location to another, capable of avoiding obstacles, or the disallowed zones, preferring low-cost areas to the high-cost areas, accelerating and decelerating, and so forth. In each machine and/or process, the levels of multiresolutional control are similar to what we stated for the mobile robot: They consist of (a) the output motion level (the lowest-abstraction level, or the most accurate level), (b) the maneuver level, (c) the plan of navigation level, (d) the scenario of operation level, and (e) the mission planning level (the highest-abstraction or lowest-resolution level). As is easily seen, this list fully applies to the general problem of *traveling within state-space*.

The nested system of representation provides a convenient way to model the world and to arrange for the computer controller as if components can be mapped one onto the other. We are able to deal with several world models (as many as we have levels of resolution in NH-ELFs). Each WM actually exists for the observer and the interpreter associated with a particular level of resolution. Hierarchical systems are an especially convenient form of organization for large systems that contain numerous goal-seeking decision units (subsystems). The problem then includes coordinating the actions in order to optimize the goal achievement process.

3.6.2 Definitions and Premises of the Principles of Knowledge Representation

In this section, we will introduce definitions and premises that are employed within the 12 principles of knowledge representation listed earlier.

Principle of Thesaural Interpretation of Words Assignment or Labeling

Assertion Any words based knowledge (WBK) representation is a system of statements in natural language with words and strings of words to be interpreted with the help of a *thesaurus*, and is constantly enriched from experience via learning processes [32, 33]. These statements pursue a goal (or a set of goals) that are typically not presented explicitly and should be explicated from the larger set of related statements, the overall text, or a group of text. The missing part of the discourse that can be restored or assumed is called *context*.

DEFINITION OF WORD-BASED KNOWLEDGE

WBK representation consists of meaningful statements in natural or other symbolic languages. The statements include definitions (together with related explanations) of concepts related to the world being represented, or the encodings of these concepts and the relationships among them, as well as the encodings of other messages. All encoding is done with the help of single words or strings of words, or signs and symbols introduced for brevity.

Premise 1 It is presumed that the word based statements of interest are *meaningful*, or otherwise *goal oriented*. Thus, the very existence of *meaning* is assumed to reflect a system of goals.

We have to remind the reader that the literature on meaning is scarce and belongs primarily to the field of pure linguistics [35]–[37].

Premise 2 The phenomena of reality are represented in the form of statements of *existence* (for both entities and relationships among them), statements of *interpretation* containing linkages between the statements of existence and the context, and statements of *implication* that invoke related statements of existence from the prior experience and/or logical transformations.

Premise 3 *(Corollary)* Any meaningful statement can be encoded in words.

Premise 4 Communication messages circulate between the adjacent subsystems of any intelligent system. The system is often described as an elementary functioning loop (ELF) consisting of the following components: sensors, sensory processing, world model, behavior generation, and actuators. This loop is closed through the world [34]. Each subsystem is presumed to carry representations required for its functioning. These representations are generally presumed to be symbolic. However, all representations can be explained in words of natural language, i.e., can be represented using WBK. A conjecture is looming about the homomorphism of the mapping from WBK to symbolic representation and vice versa.

DEFINITION OF WORD

A *word* (label or symbol) is understood as a sign consistent with and interpretable within natural languages, is convenient to use, and is efficient to apply within a real computer system that can be assigned to perform the operation of *interpretation*.

Premise 1 A multiplicity of available natural languages is presumed; it does not matter which one is used in the system.

Premise 2 The difference between word and symbol lies in the fact that word belongs associatively to the general body of language; also, the context plays a decisive role in its interpretation. On the other hand, a symbol is either part of a rigid protocol like in scientific papers, or is tightly linked to a particular application subdomain.

DEFINITION OF INTERPRETATION

An *interpretation* is the explanation of a word, or a statement consisting of words that carries meaning that exceeds the usual (thesaural) meaning of the components of this statement; the interpretation is usually done in an external subsystem. *Note:* interpretation is not equivalent to "symbol grounding;" it presumes a discovery that exceeds the routine symbol grounding procedures.

Premise 1 A human operator (or a human-operator-like subsystem) is presumed to be a part of the overall system that performs the role of the "ultimate symbol grounder" for the statements, or the "ultimate interpreter" of the commands, experiences, and situations.

Premise 2 A domain dictionary (a subset of the thesaurus) is presumed, and it contains interpretations for a particular WBK domain.

Premise 3 The interpretation dictionary should allow for a learning procedure. As the new concepts arrive to the system, they should appear within the dictionary with the interpretations pertaining to them and agreed upon by the human operator.

DEFINITION OF THESAURUS

A *thesaurus* is a global dictionary that contains interpretations for all words (labels) and combination of words that can be expected in the system based upon its prior experiences.

Conjecture All messages built upon words are considered information carriers; after assigning the interpretation, they become WBK.

Principle of Entity–Relational WBK Representation

Assertion Any word (W) and/or word based statement (WBS) is part of the coalition of the multiple words-object sets ({WO}) and words-relationship sets ({WR}). Thus any WBK representation is built upon the concepts of objects represented by words and concepts of relationships also represented by words.

Assertion Any change in a word (ΔW) and/or any change in a word based statement (ΔWBS) can emerge only due to multiple words-action sets ({A}) that pursue a particular set of goals ({G}) and are entailed by an evaluation of "goodness" ("cost") under this set of goals ({J[G]}). Thus any evolution of WBK representation can be obtained by applying the concepts of changes, concepts of actions that produce the concepts of changes, concepts of goals that are being pursued in producing the concepts of changes, and concepts of costs that are incurred during these processes. All these concepts are represented by words and together form the representation of world.

DEFINITION OF WORLD

The *world* (of interest, or under consideration) is a subset of reality which should be observed and/or controlled, or it should serve as a medium for the ELF activities (which is the same). The goals (tasks) are being formulated to determine what directs these ELF activities. Goals are considered to be the final states for the temporal process called "task" and denoted by the "command." WBK representation is being developed (created) ultimately to allow for the processes to develop in systems.

Premise 1 The world (W) and the system of words based knowledge representation (WBKR) are ideally presumed to allow for isomorphic mapping in principle.

Premise 2 The correspondence between the world (W) and WBK representation is our main object of interest; the mapping W $\rightarrow$ WBKR is presumed to exist. Mapping that constructs the realistic WBKR system is a *homomorphism*.

Premise 3 An existence of the semiotic closure through the world is presumed as part of ELF with its ability to sense, perceive (organize the results of sensing), collect and organize *knowledge*, make decisions under some goal (or task), and actuate in the world as a result of the decision. The existence of information channels among the subsystems are assumed.

Premise 4 The same world can be represented at various granularity (levels of resolution). If the resolution is reducing, then many objects represented at initial resolution become indistinguishable at a new resolution. They merge and form smaller numbers of lower resolution objects out of the multiplicity of higher resolution objects (aggregation). If the resolution increases, many objects which were indistinguishable at initial resolution become distinguishable and appear as parts of the previously represented lower resolution objects (decomposition, or partitioning). The same can be stated about relationships. All representations at all resolutions of interest can exist simultaneously as a multiresolutional knowledge representation.

The principle of entity-relational WBK representation can also be formulated as follows: *Representation at every adjacent level of resolution is obtained by a recursively applied algorithm of nesting: for decomposition, top down, and for aggregation, bottom up.*

DEFINITION OF OBJECT

An *object* is an elementary conceptual formation of the discourse related to the world, and it is distinguishable by the resolution under consideration and can be attributed to the possible existence of an elementary formation of the world. This formation is characterized by a conceptual unity and an object is frequently called an entity.

Premise 1 More than one elementary object is always under consideration.

Premise 2 Any consideration has limited accuracy (resolution).

Clearly, the entity-relational WBK representation is a kindred of the object-oriented approach [36].

DEFINITION OF RESOLUTION

A *resolution* is the measure of distinguishability of objects. This measure is assigned by the number of bits of information that are carried within the system of representation about particular objects; the minimum resolution representation carries one bit of information.

Premise 1 Digital system of information representation is presumed.

Premise 2 The existence of a procedure of *distinguishing* is presumed. This procedure appeals to the possible procedure of *identification*; intuitively both procedures refer to the ideas of *recognition* and *matching*.

DEFINITION OF RELATIONSHIP

A *relationship* is a name for a phenomenon of present or memorized interaction between objects, and it can be assigned to any couple of objects.

Premise 1 The phenomenon of interaction exists for each couple of objects.

Premise 2 Relation presumes existence of memory in the related objects.

Principle of Multiresolutional Representation (Coarsening-Refinement) or Principle of Nested Hierarchical Representation

Assertion It is impossible to organize a system of WBKR that would use the same resolution. Indeed, if some particular resolution is chosen, then all phenomena of higher resolution will not be represented. In this particular case it is impossible to construct WBKR at a higher resolution since (1) it is expected that an object can be found that requires higher resolution for the representation and (2) the complexity of representation grows rapidly with resolution (see Principle 6 of Efficiency). Thus all efforts related to WBKR construction were related to coarsening and refinement, fuzzification and defuzzification, aggregation and decomposition, abstraction and specialization, generalization and instantiation.

DEFINITION OF COARSENING AND REFINEMENT

Coarsening and *refinement* are two opposing procedures of forming multiresolutional WBKR system. Coarsening presumes lowering resolution of representation via increase of elementary granules. Refinement transforms representation by using smaller granules.

Premise 1 It is presumed that any representation has its elementary basis: minimal elementary granules.

DEFINITION OF FUZZIFICATION/ DEFUZZIFICATION

Fuzzification/defuzzification are methods of coarsening/refinement via algorithms of the fuzzy sets and fuzzy logic theory.

DEFINITION OF AGGREGATION

Aggregation is a merging together of objects at a higher-resolution level within one object of the lower-resolution level. Aggregation leads to refinement or defuzzification.

Premise 1 It is presumed that objects are meaningfully composable out of parts; in other words, that objects are assemblies of parts. This consideration seems to be recursively applicable to parts.

Premise 2 It is presumed that a technique or algorithm exists that allows averaging the properties and characteristics of the higher-resolution parts within the expected low-resolution object; a technique and algorithm should also exist for recognition of candidates for having the uniformity of the lower-resolution level.

DEFINITION OF DECOMPOSITION

Decomposition is the partitioning of the object of a lower-resolution level into two or more objects at the higher-resolution level. Decomposition leads to refinements or de-fuzzification [37, 38].

Premise 1 Objects are meaningfully decomposable in parts.

Premise 2 A technique or an algorithm exists that allows for recognition of the higher-resolution parts within the prior low-resolution uniformity of the lower-resolution object.

DEFINITION OF ABSTRACTION & SPECIALIZATION

Abstraction and *specialization* are methods of coarsening and refinement via classification and categorization algorithms (and their inverses).

DEFINITION OF GENERALIZATION & INSTANTIATION

Generalization and *instantiation* are methods of coarsening and refinement via combinatorial synthesis and search within the domain of possible instantiations.

DEFINITION OF HIERARCHY

A *hierarchy* is a structure that emerges as a result of multiple repetition of processes of refinement (top down) or coarsening (bottom up); the hierarchy is formed from the objects, their parts or classes and their members, generalities, and instantiations, and the relations of inclusion that are being recorded for objects and their parts.

DEFINITION OF HETERARCHY

A *heterarchy* is a structure which, in addition to a top-down and a bottom-up relationship of inclusion, the relations among the objects at a particular level of resolution are also being demonstrated. Briefly, we will use the term *heterarchy* for hierarchies with relational links at a level.

DEFINITION OF CHANGE

A *change in the system* is defined as a difference between two consecutive (in time) hierarchical or heterarchical representations of the system. This difference is also expected to be representable by a hierarchy or a heterarchy. One can expect therefore that changes also allow for decomposition and/or aggregation, too.

Conjecture Multiresolutional or multigranular representations (representations based on the hierarchy of the levels of coarsening and refinements) are constructive results of the law reflected in the Godel theorem of incompleteness. Indeed, each level evokes a meta-level that serves in interpreting statements that otherwise cannot be interpreted within the level itself.

Principle of Heterogeneity, or Principle of Multiple WBK Representation Systems

Assertion Any WBK representation is heterogeneous based on the multiplicity of WBK representation systems complementary in their ability to describe the entity-relational structures with desirable completeness and consistency.

DEFINITION OF HOMOGENEITY AND HETEROGENEITY

A *homogeneous* system of representation represents the object using only one language of representation; a heterogeneous system uses more than one language; the exact relationship between languages in the heterogeneous system are never known.

Premise 1 WBK representation is intuitively expected to be homogeneous.

Premise 2 A homogeneous system of representation deals with *objects* of a similar nature (belonging to the same class or junctioning in the same domain); the need for other languages appears because objects of other natures are part of the system, or because the original set of objects is not described exhaustively and does not satisfy the criteria of completeness and consistency required in the system.

Premise 3 A representation system uses various dialects of a natural language, possibly as a tool of representation; different general classes of objects and application domains generate different dialects.

DEFINITION OF WBK FOR ERN

WBK system describing entity-relational networks (ERN) is a language with a vocabulary and grammar, axioms and rules, that are accepted for building the models of a particular knowledge domain; selection of a particular language is explained by a variety of causes, one of them is the convenience of application for a particular purpose.

Premise 1 It is possible to obtain WBK about the world in different verbal forms.

Premise 2 The same WBK constructed for different application domains can be presented in different dialects.

Premise 3 No single way of exhaustive WBK representation exists.

Premise 4 We do not know the world well enough to aspire to selecting a single language for representing the world.

Causality Principle

Assertion WBK representation is based on records of cause-effect (C-E) relationships in all cases. Causal relationships can be contained in explicit statements or groups of categorized data. It is important that CWW be organized so that causal relationships are explicated.

DEFINITION OF CAUSE-EFFECT RELATIONS

Two items of knowledge are said to be in C-E relationships if one of them (C) happens before the other (E), and within a particular interpretation frame of WBK representation, a model can be built in which C can be considered an input which after being activated entails the output E.

Premise 1 C-E relationships depend on the interpretation frame, and in one frame they can exist while in another they are not considered to be a C-E pair.

Premise 2 C-E relationships presume existence of the *model* which is constructed for a subset of the world; this model can be understood as a control model. All models are control models; they have inputs, states, and outputs.

Premise 3 Couples C-E do not exist without reference to time.

Premise 4 WBK frequently presumes existence of verbal records.

Premise 5 WBK representation should be temporally ordered (see an example in [7]).

DEFINITION OF MODEL

A *model* is a representation system which often indicates for the user what is/are the *cause(s)* and what is/are the *effect(s)* as well as what are the relationships among them.

C-E BASED DEFINITION OF REPRESENTATION

A *representation* is a phenomenon of functioning in the meta-system with more than one agent. The essence of this phenomenon can be described as a delegation of duties from one agent to another. The latter requires forming a categorized list of C-E-couples.

DEFINITION OF META-SYSTEM

A *meta-system* is a system in which the system under consideration is enclosed; the external observer who builds the model is always presumed to be situated within the meta-system.

Principle of Efficiency

Assertion Different systems of WBK representation using homogeneous as well as heterogeneous principles of representation are judged by their efficiency, which includes expressive power, computability, and complexity. The decisions are preferable if they demonstrate higher values of these criteria.

DEFINITION OF EXPRESSIVE POWER

Expressive power (EP) is the ability of the representation to convey interpretations using interpretable logical statements; EP can be evaluated quantitatively by the rate of generating interpretations per memory and/or time required, or per cost.

Premise 1 A human operator is presumed to whom these interpretations are being addressed.

DEFINITION OF COMPUTABILITY

Computability is the ability of the system with mathematical, or linguistic representation to arrive at a complete and consistent solution in a reasonable time.

Premise 1 Existence of the solution, in principle, does not make the problem solvable.

DEFINITION OF COMPLEXITY

Complexity is a characterization of logical and mathematical systems that allows for approximate evaluation of the number of computations required for solving the problem, or the order of the number of computations.

Conjecture The higher the completeness is, the less efficient is the knowledge representation system.

Principle of Incompleteness

Assertion Any knowledge representation is incomplete.

DEFINITION OF COMPLETENESS

Knowledge is called complete if it gives an opportunity to *predict* exactly the outcomes of the situation (of interest).

Premise 1 The goal of dealing with knowledge is to *predict* the situation (of interest).

Premise 2 The situation can be predicted exactly (without error) if the knowledge is complete.

Premise 3 The system of interest is expressed (represented) by a set of states varying in time.

DEFINITION OF UNCERTAINTY

Uncertainty is a form of lacking knowledge, or incompleteness: statements that would help distinguish right from wrong (e.g., noise from a useful signal) are absent; higher resolution information is not available; source of disturbances are unknown; and some statements demand disambiguation [35–40].

Premise 1 It is presumed that all methods of uncertainty evaluation are explored and logically put in correspondence (some examples can be found in [40]).

Premise 2 It is possible to demonstrate that all methods of uncertainty evaluation are based on similar axiomatic premises (compare [39] and [40]).

The principle of incompleteness can actually be rephrased as the principle of uncertainty. It states that inaccurate information of the state variables is an intrinsic property of the system under consideration, so the system cannot be evaluated quantitatively. The uncertainty principle also deals with knowledge representation.

Feedforward Principle

Assertion At every moment of time, a level of resolution can be determined whose time interval exceeds the operational time of interest, and thus a single directional operational flow can be considered from knowledge representation to desirable activities with no changes in the world to be taken into account.

DEFINITION OF FEEDFORWARD

A *feedforward* or *single-directional operational flow* is a subset of the ELF diagram which can be considered with no changes in the world taken in account (principle 7 is dealing with the subset from knowledge representation to the desirable activities in the world).

Premise 1 In a GSM (ELF) diagram, knowledge representation is utilized in order to infer an extended time chain of decisions, which are required for receiving an extended time chain of desirable activities. These activities in turn create changes in the world, and they are recorded, organized, and utilized for enriching knowledge representation.

Premise 2 A time interval can be assigned that allows for thinking about this subset in an open-loop sense; this determines the level of resolution (generalization).

Principle of Feedback

Assertion Knowledge acquired through activities (decision and actuation) is utilized to compensate for incompleteness of knowledge stored in the system of WBK representation. This process of knowledge updating triggers the subsequent processes of decision making and is called *feedback*.

DEFINITION OF KNOWLEDGE ACQUISITION

The *acquisition* of WBK statements concerning the results of activities is a process required to maintain a level of completeness of knowledge that is accepted in the system under consideration.

Premise 1 It is presumed that feedback is required only for compensating the growing incompleteness of knowledge; that in the beginning the level of incompleteness was minimal, and that the feedforward operation was properly determined.

Principle of Limited Resource or Principle of Attention Focusing

Assertion Only a subset of WBK representation can be under consideration.

Any system of WBK representation and processing for intelligent system is built on the need to provide its operation under system constraints (constraints in acceptable time of operation, available computer power, accessible information, and available sum of money).

Premise 1 Constraints in acceptable time of operation, available computer power, accessible information, and available sum of money are the components of the cost-functional that should be extremized as a result of ELF operation.

Premise 2 The acceptable time of operation, available computer power, accessible information, and available sum of money are limited.

DEFINITION OF FOCUS OF ATTENTION

The subset of consideration is called the *focus of attention.*

Principle of Continuity

Assertion WBK representation should reflect the fact that all changes in the world with no external control applied are dominated by the tendency to maintain the least deviation from the previous states. *Continuity of the world* is an intrinsic property of all systems under consideration.

Premise 1 Any changes in WBK representation should reflect the continuity of the world, including the evolutionary development of objects and the relationships among them.

Premise 2 Every next snapshot of the world is a result of the next iteration of the control operation applied to a limited subset of the world.

DEFINITION OF SNAPSHOT OF THE WORLD

A *snapshot of the world* is defined as a WBK representation with no changes recorded during the process of recording. This phenomenon of "instance" pertains to a particular level of resolution.

The evolutionary character of objects and relationships among them reflect changes in state variables that can be computed by taking into account knowledge of prior developments.

Principle of Creativity

Assertion The system of WBK representation allows for and is usually accompanied by the mechanisms of creativity that combine units of knowledge not taken from reality. (This principle is also formulated as follows: *The system of knowledge representation is capable of generating possible worlds.*)

DEFINITION OF NEW WBK GENERATION

The *mechanism of new WBK generation* is a combinatorial algorithm of forming strings of the labels according to the laws of a particular domain of interest.

Premise 1 Processes of new knowledge generation are driven by a particular interest linked with the goal of maintaining a functioning system.

Premise 2 It is presumed that all possible mechanisms of forming strings of labels are included in *combinatorial* set of mechanisms.

Thus knowledge representation presumes the collection and organization of knowledge [41]. Recall the structure of ELF (see Chapters 1 and 2) which has subsystems below the bold line and subsystems above the line. The hardware is at the lower levels; it includes the material world, external events, objects of our interest, and the *domain for the future decision making*. Usually these are all things that we sense, measure, and experiment with in the real world; everything that is material, exists, and is given for us to perceive and formulate our problems upon. Everything in the upper levels is software. However, there is a kind of entity that can be understood as a common entity for the two worlds above and beneath the bold line. This entity is *knowledge about the external world*. This means that knowledge is the invariance of the two worlds mentioned above: the hardware and the software worlds, the upper and the lower parts of the ELF.

The upper leftmost part of the ELF diagram is the subsystem of sensory processing where the information from sensors is organized. The upper rightmost part of the picture is the subsystem of behavior generation where decisions are made about situations. Right in the middle of the picture above the bold line is a subsystem where our *perceptions* and our *decision-making activities* are interpreted, translated, and matched. We can in fact call our world model a knowledge base.[12] Sensory processing infers → knowledge base which in turn infers → decision making for behavior generation, (or planning/control). This dualism of consecutive knowledge transformation, where there is a bulk of information coming from the external world, occurs at the upper levels of the multiresolutional hierarchy. At every stage the subsystems are dealing with a *multiresolutional world representation*, so the decisions arriving at each level are the result of the upper-level activities.

MULTIRESOLUTIONAL WORLD REPRESENTATION

Knowledge representation is different in each of these subsystems; there are different vocabularies used, but essentially the knowledge of the world is represented in the way that is commensurate with the tools and techniques applicable within each particular subsystem of consideration. After the lowest level of the system, or the highest resolution level has completed its operation and has generated the plan of the future behavior, the decision arrives to the *actuator*.

3.6.3 Definitions of Knowledge-Related Mechanisms of ELF Functioning

When the task arrives from the upper level, a set of decision tools and techniques becomes available within the behavior generation subsystem. These decision tools and techniques start working on the model representing the subset of reality required for this decision. This subset (in order to be useful for the decision) requires additional information from the subsystem of sensory processing which in turn receives its information from sensors. Now we have a description of a very interesting entity that propagates

[12]In Chapter 5 we will see that the world model for a particular assignment is a limited model within the scope of attention of a particular assignment. It rather should be immersed in a knowledge base that links a particular assignment with all possible assignments.

through the loop of ELF. This entity is *knowledge*. However, it cannot be considered *knowledge* until the tools and techniques are applied to it and are evoked by the task of operation, and until the process is considered at all levels of resolution. Before the process is initiated, this subset of reality is just a set of information units. However, as soon as the whole loop of ELF starts functioning as described above, it becomes *knowledge*. We will consider this set of information as *factual knowledge*, and the tools and techniques that are being evoked to process the set of factual knowledge, we will call *operational knowledge*.

Decision making generates the behavior of our system. It is required for perpetuation of activities of the ELF agent introduced earlier in Figure 1.1. It generates changes in the world in response to a particular task. It generates behavior by applying *operational knowledge* to *factual knowledge*. We are not talking about just *any* operational knowledge, nor about just *any* factual knowledge. We are talking about a coupling *operational knowledge* + *factual knowledge* that emerges in response to a particular task and is required solely for this task. Only when we consider this "task-applicable" coupling are we talking about *knowledge*. At the level of resolution general enough to involve the whole domain of applications, we are talking again about "task-applicable" knowledge. Now, this knowledge is understood to be industry-oriented, technology-oriented, control-oriented, and it can also be understood in a much broader context. The following definition can be introduced.

DEFINITION OF KNOWLEDGE

Knowledge is a data structure supporting the purposely organized process of operations together with a subset of operational techniques for decision making (which is the purpose of organizing the information); the second subset is to be utilized in activities of the first subset in order to receive a particular decision as a response to the task.

DEFINITION OF OPERATIONAL KNOWLEDGE

Operational knowledge is a subset of knowledge concerned with the techniques and tools for decision making.

DEFINITION OF FACTUAL KNOWLEDGE

Factual knowledge is a purposely organized subset of information about the E-R structure of the world.

DEFINITION OF ERN

E-R network or ERN (entity-relational structure of the world) is understood as a unity of two interrelated lists: the first one is a list of entities (or labels of entities), and the second is a list of relationships (or labels of relationships) between each couple of entities from the first list.

2ND DEFINITION OF ENTITY

Entity is a concept characterized by semantic integrity.

DEFINITION OF RELATION

Relation is a characterization of a couple (two entities).

DEFINITION OF DECISION MAKING

Decision making is the purpose for existence of knowledge; it is the process as well as the set of techniques which together create a multiplicity of imaginary worlds contemplated for the future, and one then selects out of this multiplicity a goal (or more) of behavior.

It is clear that we are interested in dealing with information at a level for which these definitions are suited the best. Thus this definition is easily interpretable for issues such as controller knowledge, robotic knowledge, and knowledge of a manufacturing cell.

Each level of a multiresolutional system has a decision-making procedure in all these cases. As we go from the bottom up in the hierarchy of our system, we can talk about a nested sequence of more and more general decisions, or a set of more and more general tasks. Therefore, we can talk about a general decision found for the meta-problem implicitly formulated within the assignment of the task. Our definition of knowledge has turned out to be task-oriented. The bootstrap knowledge required for initiating the operation of an ELF can be considered a domain of knowledge collected for satisfaction of the *class of tasks*. Such a task-oriented definition for knowledge turns out to be instrumental and applicable even for the nontechnological areas.

Nevertheless, it should be clear that we were looking for a definition that is oriented toward engineering problems and allows for application in the AICS area. We will not consider any controversial issues existing in this area. We are interested only in systems and techniques that involve purposeful knowledge processing. *Our assignment-oriented knowledge* will make it easier to organize model-oriented knowledge, operation-oriented knowledge, control-oriented knowledge, or engineering-oriented knowledge.

Let us comment on our use of the term *representation*. The usual interpretation of the word in any language is *reflection*. "Representation" evokes the need to delegate the duties from *one agent to another. Representation* always presumes two agents. One agent delegates its duties to another agent. The term *reflection* also presumes two agents: one agent delegates to another its duties to allow for it being perceived by an external observer.

A similar situation with two agents is seen in ELF. The hardware is a physical reality. *The hardware is represented in the software*, to which it has delegated everything it contains in a set of subsystems. The representation is required because there are some duties of the system that cannot be done without representation. In order to make the system operational, one needs to separate the system of decision making from the system of performance. One does not put decision-making subsystems within sensors that are close to the potentially unsafe working zone; one puts them into a safe remote place and substitutes them by their representations. Thus representation is a necessary operation. To be consistent, one could gradually remove everything more or less important out of the working zone and combine it within the *computer brain of the machine*. All the knowledge of the world could be represented in the machine: its richness, the results of measurements, the processes of operation, at all stages, including those that require thinking and prediction.

2ND DEFINITION OF KNOWLEDGE REPRESENTATION

Knowledge representation is a result of mapping knowledge from an initial carrier into a system with a predetermined organization for subsequently carrying out the duties of decision making outside the initial carrier of knowledge.

Let us consider in a bit more detail these *duties*. This is a purpose-oriented term, and it presumes that the knowledge contained within the world reality (underneath the bold line AB in the ELF diagram shown in Figure 1.1B) should be used for solving a set of particular problems (likely the same as those determined by the purpose of this knowledge). There is an assumption that in every case there exists a set of duties (or the duties that can be formulated). The physical (performing) world transfers its duties to the software. We describe the duties of software in the form of *specifications* (a component of the software design). The duties are represented in the software description in the form of a *list of duties*. In this way particular parts of the robot are made to move, position, track, drill, and so on. If we are interested in solving the problems, we have to specify a list of duties.

What are we doing with knowledge represented in our system? The most general procedure of dealing with knowledge is a procedure of production. Production systems were first introduced by Eugene Post.[13] The production operation is performed in each subsystem of ELF. "Production systems" is a term introduced in logic and employed within artificial intelligence. A production system applies rules to particular situations in order to receive the result desired in active databases [25]. Production systems are a particular case of automata (from control theory and mathematics).

3.6.4 Types and Classes of Entities

The phenomenon of reflection evokes a number of theoretical issues. One of them concerns special types of entities. Indeed, one can expect to reflect upon at least two types of entities: external entities that exist in the world and internal entities that exist only inside of the intelligent system.

Immanuel Kant was one of the first philosophers to clearly distinguish between two types of entities: external entities and internal entities [42]. A. Schopenhauer clarified many issues regarding internal and external representations [43]. Most scholars involved in image analysis start with determining entities within the image. In the design of a knowledge representation for intelligent systems, we also must begin with a clear distinction between external and internal entities. We must then develop mechanisms for establishing and maintaining correspondence between external and internal entities. This correspondence is often called *symbol grounding*.

DEFINITION OF EXTERNAL ENTITY

An *external entity* is something that can be named, or that exists in the real world external to the intelligent system.

External entities typically are clumps of matter that are physically connected in some way. They may also be groups of things that are not physically connected but possess common attributes such as spatial proximity or similar motion. For example, an external entity may be a cloud of smoke, a swirl of dust, a school of fish, a wave of water, a crowd of people, a stream of traffic, a forest (consisting of hundreds of trees and bushes), falling rain or snow, or a lawn (consisting of millions of blades of grass).

External entities often correspond to things in the external world that can be measured or counted, such as pebbles, grains of sand, or drops of water. Manufactured objects typically are well defined with surface properties that can be precisely measured. Edges may be defined in terms of intersecting surfaces or by sharp radii of curvature in the surface of the object.

Many natural entities, however, are not clearly defined and cannot be precisely measured. For example, where does a mountain begin and a valley end? How does one measure the volume of a sand dune? What is the length of the coastline of the British Isles? These issues arise from the fact that measurements depend on the scale and resolution of a unit of measure. For example, the definition of the boundary between a mountain and a valley is completely arbitrary on a scale of centimeters but can be reasonably well defined on a scale of kilometers. The volume of sand in a dune depends on where the dune begins and ends. The measured length of the coastline of the British Isles depends on the resolution of a unit of measure.

[13]The U.S. mathematician Post proposed in 1936 a kind of automaton (or algorithm) which was the prototype of the program schemes developed 10 years later by von Neumann and his associates. Later (in the early 1960s) this idea migrated into the area of computer linguistics (semi-Thue systems) and the discipline of artificial intelligence (in a form of production systems).

All of these examples illustrate the fact that the concept of an entity is not inherent in the environment but is a hypothesis imposed on the world by the the mind of the observer to facilitate an interpretation of sensory input from the world. External entities are simply the observer's way of thinking about the world in a manner that increases the probability of successful behavior. Frequently, they emerge as a result of generalization by unifying multiple entities registered at high resolution into a single entity at low resolution.

Internal entities are even more a construct in the mind of the observer.

DEFINITION OF INTERNAL ENTITY

An *internal entity* is a data structure in the observer's knowledge database.

Internal entities are data structures that may correspond to, and represent, observed external entities. For example, an internal entity may be a data structure that represents an external object in the world, such as a rock or a tree, or a real world surface, such as that of a road or a building. An internal entity may be a symbolic frame or a group of pixels in an image that represents the edge of a building or a road.

Sometimes internal entities do not correspond to any real-world external entity. For example, the observed image of a sphere contains a group of edge pixels that defines an internal edge entity. Yet a sphere in the external world has no edges. This illustrates the fact that the shape and position, and even the existence, of internal entities in the image may depend on the viewpoint of the observer.

It is possible to have internal entities with no observable corresponding external entity at all. For example, many people believe in (i.e., have internal entities that represent) the existence of spirits, gods, angels, and demons. These are not directly observable and may not actually be present in the external world. Many people also have internal entities that represent atoms, magnetic fields, and black holes. No one has ever directly observed any of these things. The scientific viewpoint accepts only those internal entities that can be modeled by mathematical formulas and whose entity attributes can be measured by instruments developed and experiments carried out to support the existence of the hypothesized entities.

So how do we know whether there is any corresponding external reality? The short answer is "We don't." We can only estimate the probability that our internal entities have corresponding external entities. When predictions based on our internal hypothesized entities correctly predict the future, the probability that our hypothesized internal entities correspond to external reality is increased. When predictions based on hypothesized internal entities fail to predict the future, the probability that our entity hypothesis is correct is reduced. If the probability of correctness rises above an upper threshold, we accept the entity hypothesis as true and act accordingly. If the probability of correctness falls below a lower threshold, the entity hypothesis is rejected as false, and an alternative entity hypothesis is generated.

These examples illustrate only some of the deep philosophical issues surrounding the problem of knowledge representation. Sufficient for our discussion here is to note that all internal representations are hypothetical constructs that are internal to the intelligent system and enable it to interpret sensory input. Whether or not there exists any corresponding external reality is irrelevant. All that is important is that the hypothesized entities have functional utility, that they enable an interpretation of the world that is useful in generating successful behavior.

The functional utility of internal entities can be measured by their ability to explain the past and predict the future. The ultimate criterion is whether the entity hypotheses produce significant survival benefits. A creature or intelligent system that can accurately predict even a few seconds into the future is more likely to survive and achieve behavioral goals than one that can only react to events after they occur. A creature or system that is able to predict hours or days into the future has an enormous behavioral advantage. Reliable prediction enables the intelligent system to make plans and take preemptive action to avoid danger, to maximize benefit and payoff, to minimize cost and risk, and to win in competition with other creatures for a place in the gene pool. Whether internal entities actually correspond to external reality is largely irrelevant and in many cases may be undecidable. All that really matters is whether hypothesized internal entities increase the reliability of predicting the future and hence the probability of successful behavior.

Internal entities may have at least three types of attributes:

1. Observed attributes that are derived directly from sensors, or computed from other observed attributes.
2. Estimated attributes that are computed from observed attributes by recursive estimation, or another filtering technique.
3. Predicted attributes that are computed from estimated attributes using knowledge of system dynamics, geometry, and control actions.

Internal entities and their attributes may be represented in the knowledge database in symbolic form as entity frames.

CLASSES OF ENTITIES

There are at least three types of entity classes: geometrical entity classes, generic entity classes, and specific entity classes. Geometrical entity classes include point, list, surface, object, and group entity classes.

Point Entity Classes Point (or pixel) entity frames contain attributes that can be measured by a single sensor at a single point in time and space, or that can be computed at a single point (or over a single pixel) in time and space. Point attributes may describe the properties of a single pixel in an image.

List Entity Classes List entity classes include edge, vertex, and surface patch entities. List entity classes consist of sets of contiguous point entities that satisfy some gestalt grouping (generalization, discovery of a pattern) hypothesis over space and/or time. For example, an edge may consist of a set of contiguous pixels for which the first or second derivatives of intensity and/or range exceed threshold and are similar in direction. A vertex may consist of two or more edges that intersect, or a single edge with an endpoint.

Surface Entity Classes Surface entity classes include surface and boundary entities. Surface entities consist of sets of contiguous list entities that satisfy some surface generalization hypothesis. For example, a surface may consist of a set of contiguous surface patches that have similar range, orientation, texture, and color as their immediate neighbors. A boundary may consist of a set of edge entities that are contiguous along their orientation.

Object Entity Classes Object entities consist of sets of contiguous surface and boundary entities that satisfy some object gestalt group hypothesis. For example, an object may consist of a set of surfaces that have roughly the same range and velocity, and are contiguous along their shared boundaries.

Group Entity Classes Group entity classes consist of sets of objects that have similar attributes, such as proximity, color, texture, or common motion. Group attributes are computed over the entire set of objects that are included within the group.

The grouping properties of the above entity classes produces a hierarchical layering of inclusion or belonging relationships. Group entities have subentities that are objects. Object entities have subentities that are surfaces and boundaries. Surface and boundary entities have subentities that are edges, vertices, and surface patches. Edge, vertex, and surface patch entities have subentities that are point entities.

If the above set of entity classes appears arbitrary, it is because much in nature is arbitrary. Natural selection rewards those arbitrary selections that are useful for successful behavior and punishes those that are not. Internal entities are heuristic hypotheses that are constructed by the observer to assist in analysis of the sensory input and in the generation of successful behavior. In Chapters 6 and 11 we will demonstrate that for the 4-D/RCS architecture developed at NIST for an autonomous vehicle, the topological hierarchy of entity classes in the world model knowledge database is chosen to be parallel with the task decomposition hierarchy in the behavior generation side of the 4-D/RCS hierarchy so that sensory processing, world modeling, and behavior generation can be closely coupled. We expect this coupling to be useful in generating successful behavior.

The definition of an object depends on the task. If the autonomous intelligent vehicle task is to approach a tree, then the tree is the object. If the autonomous intelligent vehicle task is to avoid hitting the trunk of a tree, then the tree trunk is the object. If the vehicle task is to inspect a hole in the tree trunk, the hole is the object. If the vehicle task is to avoid the woods, the stand of trees comprising "the woods," is an object with surfaces, boundaries, edges, and points, and individual observed trees are part of the boundary of the woods.

In the realistic intelligent systems, segmentation of the world into geometrical entity classes is sufficient for generating effective behavior. Simply knowing the position, size, shape, and motion of edges, surfaces, objects, and groups is all that is required to avoid collisions and to compute whether the ground is too rough or too steep to be safely traversed. However, for many behavioral decisions, it is useful to classify topological entities into generic entity classes. For example, a bush and a rock may have similar topological shape. However, it may be possible and advantageous to drive through a bush but not through a rock. Therefore it may be advantageous to be able to distinguish between the two. This requires the ability to recognize topological entities as members of generic entity classes.

Generic Entity Classes

Class membership in a generic entity class is defined by a set of attributes. These attributes define a generic entity template (or exemplar) that exemplifies the class and provides criteria for recognition or classification. Observed geometrical entities that have attributes matching those of the generic entity frame may be classified as belonging to the generic entity class.

Generic entity classes may exist for any geometrical entity type. For example, generic point entity classes may include points of needles, specks of dirt, pixels in an image, or points in state-space. List entity classes may include edges of roads or buildings, skylines, corners of a room, or the ends of lines. Surface entity classes may include surfaces of roads or fields. Boundary entities may include contours of roofs or outlines of windows. Object entity classes may include buildings, trees, persons, fireplugs, windows, stairs, or stair steps. Group entity classes may include clusters of grapes, leaves of a tree, and groups of animals or people.

Specific Entity Classes

A specific entity class has only one member. It is in a class all to itself and is a particular instance of a thing. A specific entity has unique attributes that distinguish it from all other entities. For example, a specific object entity may be a famous person, a particular building, or a unique tree. It may be an identifying surface of a specific object or one edge of that surface.

3.6.5 General Characterization of the D-Structure

Knowledge is a major component of the elementary loop of functioning (ELF). Knowledge for both image interpretation (within the SP module of ELF) and behavior generation (within the BG module) require a systematic use of a particular system of logic. This system should be acceptable if it provides for a consistency of deduction and productive procedures of inference. Yet we are interested in using systems of logic that do not lead to exceeding some limit of complexity that would preclude its applicability. The subsequent use of logic of associative relations (LAR) benefits from representing knowledge as a *descriptive structure* (D-structure) constructed for a fuzzy linguistic representation space (FLR-space).

DEFINITION OF D-STRUCTURE

A *descriptive structure* (D-structure) is a relational network (ERN) in which the nodes are points in the fuzzy linguistic representation space (FLR-space) and the edges are cost-valued point-to-point relations.

Where do we get the nodes and the relations from? A D-structure is considered to be the result of statistical texts analysis, or STA. The "N-gram" operators of STA allow for determining all meaningful N-word strings that are statistically valid [44]. Frequently the latter requirement is difficult to satisfy, and the researcher has to use abduction instead of deduction (see Chapter 2).

The "texts" are understood as all available papers, documents, or other passive sources that are relevant to the problem of reasoning. The subject of relevant representing the system by a set of texts is discussed in a multiplicity of sources [45, 46]. It includes topics of necessary and sufficient conditions of selecting the representative set of texts. The texts are presumed to be interpretable within a definite context of the problem, and the problem is supposed to be understandable within definite universe of discourse (UOD).

DEFINITION OF UNIVERSE OF DISCOURSE

The *universe of discourse* (UOD) is a heterarchy of context within which the problem at hand can be interpreted at all stages of ELF functioning.

The UOD is represented by explanation of words and expressions given in a form of thesaurus that is not a part of the formal D-structure but is being used for its interpretation. An example of nested UOD is given in Figure 2.15.

DEFINITION OF FLR-SPACE

A *fuzzy linguistic representation space* (FLR-space) is a vector space in which the coordinates are linguistic variables with the boundaries within which the membership is evaluated by determining the value of indistinguishability zone (the size of the granule; see Chapter 2).

Any node of interest within the D-structure is considered a D-state and is represented as a point in FLR-space. The motion of this point as well as processes of generalization and instantiation are to be controlled by a logic of associative relations (LAR), and LAR-based production system.[14] Both nodes and relations are fuzzy valued. This means that we tend to transform all kinds of uncertainties into a form of a fuzzy representation.

DEFINITION OF LOGIC OF ASSOCIATIVE RELATIONS

The *logic of associative relations* is an extension of predicate calculus of the first order, supplemented by concepts and techniques from the multi-valued logic and fuzzy logic so that the interpretable inference is possible under conditions of uncertainty for both the objects and relations under consideration [47].

Among the multiplicity of problems that can be solved using LAR tools, we will consider only one problem: the problem of motion control. After the problem (for control) is posed and the goal-oriented grouping of labels is performed, the FLR-space becomes hierarchical and is treated as a set of FLR-subspaces; the D-structure also becomes hierarchical and is treated as a tree of D-substructures.

This hierarchy of mappings is illustrative for analysis of the process of node-generation as we perform the consecutive top-down, bottom-up, recursive processes of joint generalization-instantiation (aggregation-decomposition). Thus the D-structure is a nested structure, since each of its nodes is a nested object (which can be recursively decomposed into a set of components). The processes of recursive aggregation and decomposition generate new coordinate spaces, which have a higher number of dimensions in the case of decomposition, or a lower number in the case of aggregation. The process of recursive hierarchical aggregation is described in Section 3.12.4.

A production system contemplated for knowledge-based reasoning presumes use of (1) a rule base (an automaton, an active database) and (2) an operator of search for a particular trajectory in the FLR-space. All rules of the production system can be constructed during the design process. They are meant to exist in the FLR-subspaces as relations between the nodes or combinations (chains) of these relations. The structure of FLR-space determines the structure of production system utilized for solving the problem. Some examples of knowledge-based motion control programs will be given in Chapters 7 and 8.

We introduce here the concept of a D-structure as an entity-relational network (ERN). Thus we have a list of labels and a list of relations among the nodes. Unlike the ERN introduced earlier, D-structure supports the nested hierarchy of multiresolutional representation. Yet, at each level of resolution, this is a simple ERN. The LAR is used as an inference engine in the production system of such a controller. Passive sources of knowledge are those not subject to dynamic interaction with the controller. As a passive source, the D-structure consists of a list of relations between each couple of labels, and this list is believed to represent all pertinent knowledge from the texts. The D-structure is meaningful and can be interpreted by the production system only in context, and it is subject to further decomposition.

The D-structure is the connected list of cost-valued statements about relations among labels (nodes). It can be characterized by the following properties:

OPEN ENDEDNESS

1. The list is open ended (it is presumed that the number of dimensions can be increased, or decreased if necessary); new relationships can be added to the list during the design process as well as during the reasoning process of the functioning system (or as a result of *learning*; see Chapter 11). Each of the nodes (labels), and each of the edges (relations), are independently characterized by a number

[14] The terms "production system" and "automaton" are used interchangeably (see Chapter 2).

designating the significance of the label or the relation. This *significance value* is obtained from learning, and it takes into account (a) the statistics of the label usage in representative texts and (b) expert opinion.

CLUSTERING

2. The operation of generalization is applied to the list. Its function is to organize the nodes and edges into clusters (classes) for further generalization and to determine both generalized nodes and generalized relationships among them. New labels are assigned to the generalized nodes (clusters), and these labels are included in the initial list (*closure*). Actually these new labels are expected to exist within the initial complete list of the properly built D-structure (*completeness*).

CLOSURE

MEANING GENERATION

3. Each of the relations stated within the list, may be considered a meaning-generating relation. The meaning-generator is expected to generate statements that were not mentioned in the initial context and possibly, were not intended. (We have already spoken of "knowledge discovery" when new patterns emerge either as a result of explicated trends or as a result of search within the *version space*. The latter presumes reference to the associative memory, which is contained in the context.)

SEARCHING

Each statement received from the D-structure carries not only explicit information (interpretation within the thesaurus) but also implications of that statement that can be generated from the associative memory. The associative memory is not necessarily available in existing knowledge-based controllers, but it should be. A search (*browsing*) procedure leads to the implicit knowledge extraction from the associative memory.

4. After a D-structure is built using STD, it is not referred to any particular problem or procedure. An additional effort is required to refer this structure to some abstract reference frame and to locate, within this global data structure, a problem-oriented subset of the data structure to be controlled within the reference frame. The abstract reference frame is selected in order to describe the labels as combinations of real properties. The relations between them still hold. This is why each label is considered not just a word but a point in a multidimensional description space (D-space). Each relation in this D-structure in turn must be put in a D-space in order to be utilized for the problem solving.

5. The system of references, or reference frame, can be constructed in the form of a multidimensional coordinate system for an imaginary vector space. Each label is assumed to have a number of dimensions. Each label is represented as a set of properties, or eventually as a set of words (variables). Each word (variable) can be quantitatively evaluated. Indeed, if we are well informed, for each label we know the data with some precision. If the amount of information is minimal, then at least we know that this property exists (1) or does not (0).

LIMITED ACCURACY

6. To create a list of axes, we analyze the graph of the D-structure because every label should be represented there (speed, brightness, etc.) that is capable of affecting the control situation. The list of axes necessary for describing a particular label is the list of all of the other labels connected to the label of consideration (the total graph of D-structure). Each label is defined only by other labels and cannot have a dimension not described by a label.

LABELLING

COMPLEXITY

7. There exists an information limit imposed by the hardware (cost, complexity of the architecture, and the time of computation) as well as by the software (complexity of the algorithm). Partially, those factors are determined by the level of connectivity of the D-structure (2^n relations for n nodes if the description is complete). In practice, the number of relationships is limited based on the task at

hand. This is determined essentially by the highest (finest) level of resolution of the representation. The resolution, in turn, is determined by the minimum distinguishable cell (word) of representation at the level of primitives.

3.6.6 Properties of Labels

Let us consider several examples that are instrumental in supporting the subsequent theoretical analysis [48]. Figure 3.2 illustrates a class formation for two sets of noun labels. In case *a* three kinds of electrical motors are unified by a class-generating *property of being a motor*, or more formally, class-generating *common word in the set of elementary properties*. Obviously the class instantiations such as induction motor (X_{0_1}), synchronous motor (X_{0_2}), and direct current motor (X_{0_3}) cannot be considered elementary properties of the class: They are *instantiations* of the class. So definitely the word motor cannot be represented as a simple concatenation of the members of the class motor: $X_0(X_{0_1}, X_{0_2}, X_{0_3})$. An extraction of the common trait must be performed that does not preclude all other properties to be preserved.

GENERATION OF CLASSES

All words in set *a* have this common trait: being a motor. This property becomes a *class-generating property*. In other words, a class is generated out of several labels when there is a word component in a string of label components that is common for all labels under consideration. However, if we are interested in representing the label electrical motor as a concatenation of properties, we would use a different set of words (e.g., [stator with windings, rotor (armature), shaft, ...]).

On the other hand, the word (electrical) motor can be found in the D-structure, as well as words internal combustion engine, hydraulic cylinder, and pneumatic cylinder. There is no *class-generating property* in the label statement of these four words. However, given their thesaurus definitions (or informedness of experts), one can easily find the basis for their unification in a class. Experts know that these four words are possible members of such classes as motion producers, or propulsion devices, or transformers of one kind of energy into another. These classes might not be discovered in a concrete D-structure initially obtained as a result of STD, especially if the amount of the representative texts is not sufficient. Thus we can conclude that a comprehensive class formation might be produced only if each word is being analyzed together with its complete definition from the thesaurus, or with a participation of experts.

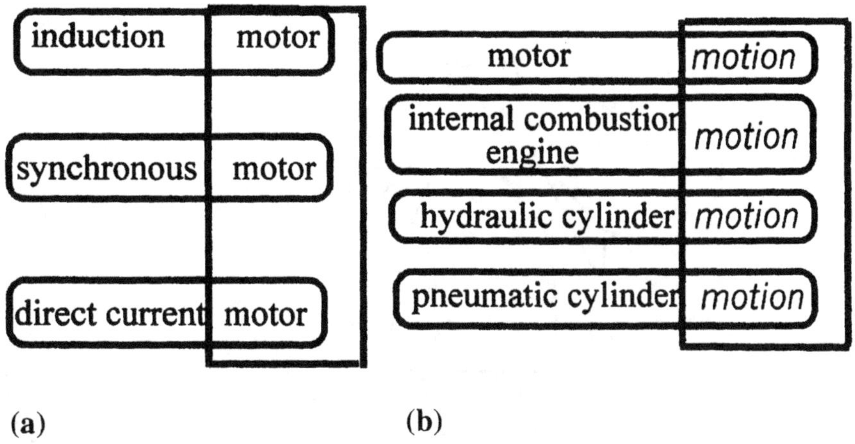

Figure 3.2. Examples of class generation for labels: *(a)* Using words from label statements; *(b)* using words from thesaurus descriptions.

From the examples demonstrated in Figure 3.2, one can draw a conclusion that the process of class generation is somewhat similar to the process of finding an intersection of all labels (object strings) in a set. If $X_1 = (x_a, x_b, x_k)$, $X_2 = (x_c, x_d, x_k)$, and $X_3 = (x_e, x_f, x_k)$, then $[X_1, X_2, X_3]$ is a set for which X_1 is unified by a property of forming together an entity of the object under consideration. In this view each object can be considered a class for its parts, and the concatenation of the parts can be interpreted as formation of the object.

Obviously the results of classification may not be unique. For example, our induction motor may turn out to be a yellow induction motor, and together with yellow submarine, yellow orange, and yellow metal, it will contribute to the class of yellow objects. One can expect that a multiplicity of classes can be created on a D-structure where each object is represented as a string of concatenated words. Thus we can conclude that classification is driven by the goal of problem solving, and that we can always assume that we are talking about *class of interest*.

VOLUMES OF MEANING

As one may expect, in a vector representation these yellow objects won't be represented in the vector-space as points; they rather will occupy some fuzzy subspaces (*volumes of meaning*). These subspaces will intersect in a zone that will have a common volume of meaning of their class-generating property. At the same time the class can be represented by the volume-union of the fuzzy volumes for the instantiations (or by the convex-hull around the set of fuzzy volumes for the members of the class). This does not contradict to the nearest neighbor principle of class generation. Indeed, the centers of fuzzy volumes to be united in a class (the volumes-candidates) are probably the set of nearest neighbor points because of the common intersection property. Thus the *nearest neighbor algorithms* can be applied to the centers of the fuzzy volumes. The center of the convex-hull around the fuzzy volumes is the point representation of the *center of the class*. One can also expect that for the properly built classes, the centers of the zone of intersection, and the center of the convex-hull will be in close relationships.

It would be very beneficial to consider an example from the pictorial world representation (computer vision, cartography, etc.) First is a pictorial example related to the area of computer vision (Figure 3.3). The different people *a* and *b* are related to a class of entities similar to syntactic graphs. After capturing the parts of the picture that seem to be independent entities (or blobs), the desirable algorithm of classification determines the relationships among the entities found, such as relationships of *closeness* or *con-*

Figure 3.3. Capturing the difference.

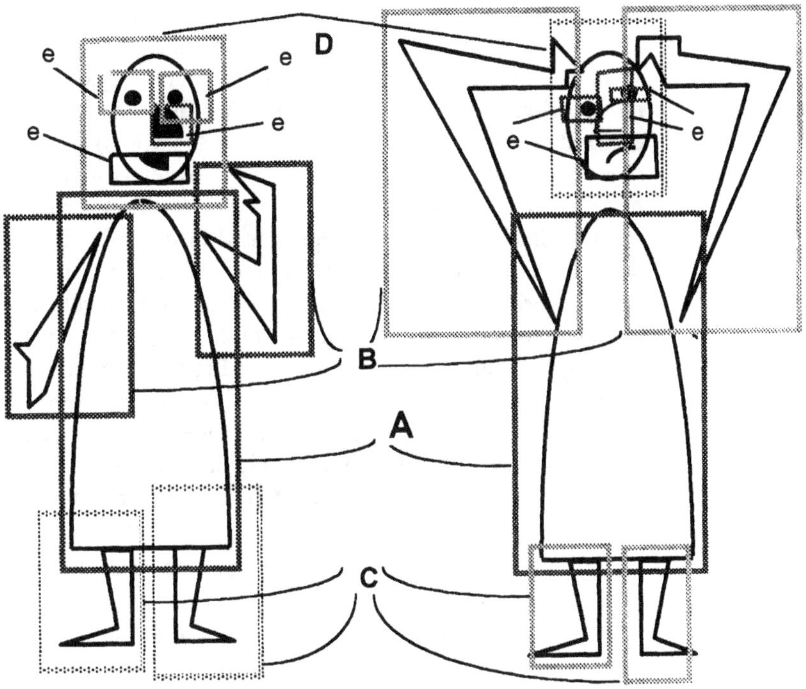

(a) (b)

(c) (A, 2B,2C,D,4e)

(d)

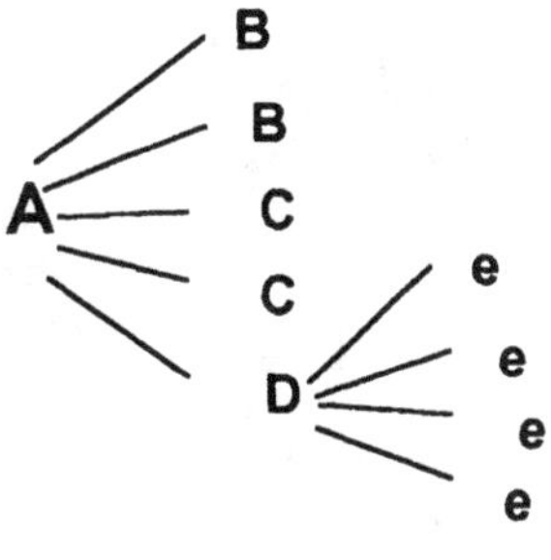

Figure 3.4. Discovering the structure in an image.

nectivity (see *gestalt* in Chapter 7). Figure 3.4 shows the result of capturing the entities of Figure 3.3, which are enclosed in separate rectangular boxes. The whole picture can then be represented either by a string *c* or by a syntactic graph *d* (both the string and the graph turned out to be the same for the pictures of Figure 3.3 *a* and *b*; this allows us to consider a class of objects with such a graph). Interestingly in this case the list of characteristic (class-generating) properties happens to be a description of partitioning.

On the intuitive level, an evaluation of graphs depends on the similarities determined per *level of the graph*, preserving the *similarity of the graph structure*. Let us explore whether the rules of generalization and refinement during the process of focusing of attention can be introduced in a formal manner.

RESOLUTION AND GENERALIZATION

In the previous examples we noticed that the objects are composed of the *elementary parts* (primitive objects, or primitive words). The structure of an object can be generalized or refined using clustering objects (words) together and receiving generalized objects, or decomposing the primitives into their components, and thus receiving refined objects. So the set of primitives imposes a limit for the resolution of representation. Indeed, a situation cannot be represented in more detail than is determined by a set of accepted primitive words. The operation of generalization is associated with transforming the world representation to a lower resolution level, and the operation of refinement allows us to consider the world at a higher resolution level.

Let us consider the set of pictures in Figure 3.5 a, b, and c. The primitives are simple: black pixel, 1×1; white pixel, 1×1. We are interested in a picture at a lower level of resolution; in order to do this, we intend to explore the methodology used in the generalization demonstrated in the pictures of Figures 3.3 and 3.4.

We can formulate some rules of generalization as follows:

- The primitive words of the adjacent lower level of resolution are black and white pixels, both 4×4.
- If the 4×4 area of the lower resolution representation is filled in only by white pixels 1×1, it will be considered white; if it is filled by 75% of white pixels, and 25% of black pixels, it will be considered white.
- If it is filled by 50% of white pixels, and 50% of black pixels, it is considered white (in another possible set of rules, one can accept the decision to consider this area black, or to randomly select black or white one).

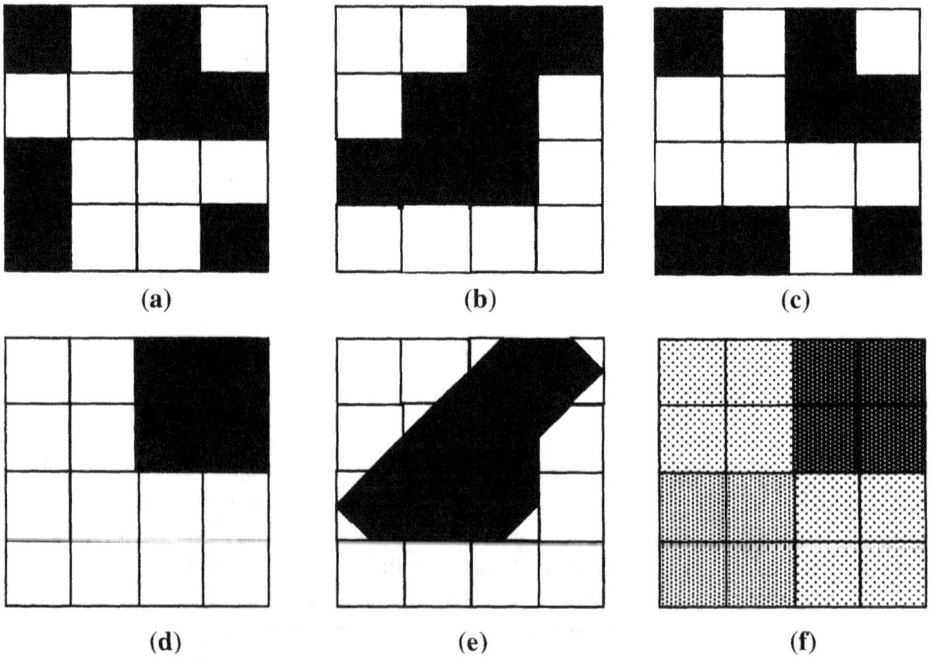

Figure 3.5. Formation of classes in tessellated images.

- If it is filled by 25% of white pixels, and 75% of black pixels, it will be considered black.
- If it is filled by black pixels only, it will be considered black.

It is easy to see that under these rules of generalization, all pictures (*a* through *c*) will be transformed into the image shown in Figure 3.5*d*. Unfortunately, this "generalized" image does not satisfy our intuitions about the expected generalized image. For example, we would expect to have something like Figure 3.5*e* for the picture of Figure 3.5*b*. This image (*e*) is actually obtained by "smoothing" the sharp edges of the body shown in the picture Figure 3.5*b*.

3.7 MULTIRESOLUTIONAL CHARACTER OF KNOWLEDGE AND ITS COMPLEXITY

3.7.1 State-Space Decomposition

We consider an n-dimensional continuous euclidian state-space E_n and a closed domain (of interest) in it $\Omega(E_n \supset \Omega)$ with volume V and diameter d. This domain can be divided (decomposed) in a finite number of nonintersecting subdomains so that their union be equal to the domain of interest (Ω_j is a closed jth subdomain with a diameter d_j, $j \in H$, where H is the set of all subdomains). Obviously, in this case

$$\bigcup_{j \in H} [\Omega_j] \in H[\Omega_j] = \Omega, \tag{3.1}$$

If each pair of subdomains can have only boundary mutual points ($\Omega_{j2} \cap \Omega_{j1} \neq \square$), then they are called *adjacent* subdomains. Relation of adjacency will be a key property in representing the context of the problem to be solved. This would form a discretized or tessellated space.

MULTIRESOLUTIONAL TESSELLATED SPACE

In turn, each of the subdomains can be decomposed in a finite number of sub-subdomains, and so on. This sequential process we will call *decomposition*, while the opposite process of forming domains out of their subdomains we will call *aggregation*. Every time we decompose (or aggregate) all domains simultaneously, and the results of decomposition (or aggregation), we will call *a level* (of decomposition, or aggregation). We will consider the *relation of inclusion* that emerges as part of the decomposition-aggregation process to be a key property in representing the context of the problem to be solved.

We won't allow for the infinite process of decomposition: We introduce the notion of *accuracy (resolution)* by determining the smallest distinguishable (elementary) vicinity of a point of the space with a diameter $d_j = \Delta$. This diameter can be considered a *measure*, and this vicinity can be considered a single *tessellatum* (or a single *grain*) that determines tessellation (granulation) of the space, or which is the same, the *scale* of the space. Each tessela contains an infinite number of points, however, for the observer they are indistinguishable. However, each subdomain of decomposition contains a finite number of elementary subdomains (grains) which have a finite size and determine the accuracy of the level of decomposition. The observer will judge this elementary domain by the value of its diameter and by the coordinates of its center. Let us demonstrate how this is done. We will discuss the euclidian space with limited accuracy of measurements available (which is a practical example of a Hausdorff space). A treatment is possible based upon fuzzy set theory.

We did not make any assumptions about the geometry of subdomains, which begs for a definition of the *diameter* of the subdomain. A standard definition for the diameter $d(\Omega_j)$ of the subset Ω_j of a metric space is presented via notion of distance D in a metric space as

$$d(\Omega_j) = \sup_{x,y\in\Omega_j} D(x, y), \tag{3.2}$$

and holds for all levels of decomposition except of the lowest where is understood as the elementary subdomain in which points x, y cannot be specified, and the distance cannot be defined. Thus formula 3.2 cannot be used for computing the diameter of the lowest (highest resolution) level of decomposition.

Let us take one arbitrary point from the elementary subdomain Ω_j, $j \in H$, and call it the center of the elementary subdomain (it does not matter which point is chosen; at the highest level of resolution different points of the elementary subdomain are indistinguishable anyway). To distinguish the centers of the elementary subdomains from other points of the space, we will call them *nodes* and denote q_j, $j \in H$. The relationships $\mathfrak{R}_{ijk}$ among each two particular nodes i and j of a particular level k of decomposition are very important for solving the problem of control. We will characterize this relationship by the *cost* of moving the state from one node (i) to another (j). Cost is a scalar number C_{ij} determined by the nature of a problem to be solved. Graphically cost will be shown as a segment of a straight line connecting the nodes, this segment we will call an *edge*. A set of all nodes and edges for a particular subdomain we will call *a graph representation of knowledge for this subdomain.*

3.7.2 Accuracy

RESOLUTION DETERMINES ACCURACY

For the lowest-level decomposition where each subdomain is an elementary subdomain, we take the diameter of the elementary subdomain (or the diameter of a tessellatum, or the diameter of a grain, or the size of the distinguishability zone) as the average diameter. At this time we will not consider the statistical data which are used in the averaging. The data can also be a collection of measurements of a single elementary subdomain Ω_j (e.g., the results of measuring different crossections of the subdomain). It can be also a collection of measurements performed on a population of the elementary subdomains belonging to the same subdomain of the upper adjacent level. It may happen that elementary subdomains belonging to different subdomains of the adjacent upper level will have different diameters. Therefore, in order to compute the statistical data at the level of required accuracy, there must be information on the next adjacent (higher accuracy) level of decomposition.

Assume that Δ is the average diameter of the elementary subdomain at the highest level of resolution. Then the order of the average volume for the elementary subdomain is evaluated as

$$V_{\text{av}} = O[(\Delta)_n], \tag{3.3}$$

and the quantity of the elementary subdomains at the level (of highest resolution), or the cardinal number of the set H, is expressed as

$$N = \frac{V}{V_{\text{av}}}. \tag{3.4}$$

Cost can be similarly characterized by a value of accuracy that is determined by the accuracy of the particular level of resolution. Statistically the cost at level k in moving from a node i to node j is understood as the average of all costs of moving from each subdomain of node i to each subdomain of node j.

Even more important are relationships of inclusion between the subdomains of the adjacent levels of decomposition [49].

3.7.3 Nestedness of Knowledge

Properties of the recursive hierarchy entail the phenomenon of *nestedness*. Behavior generation, via task decomposition, determines the parameters of the intermediate steps to the goal. The task decomposition involves the process of decision making which proceeds top-down and bottom-up in the multiresolutional hierarchy of the RCS controller (all sufficiently complex systems are designed, or tend to organize themselves, in a hierarchy because this makes for efficient computations. This efficiency has been discussed in many sources both qualitatively and numerically.)

In NIST-RCS we apply a special type of hierarchy called a *recursive hierarchy*. This hierarchy has a special relation between the nodes of the adjacent levels: All the nodes of the set of m nodes $\{n_j\}_i$ (where $j = 1, 2, \ldots, m$; at the ith level of resolution counting levels from bottom to top), attached to a particular kth node $n_k(i+1)$ of the adjacent $(i+1)$th level from above, are obtained as a result of a procedure called *decomposition* or *refinement*. This procedure presumes inversely that the properties of the node $n_k(i+1)$ can be obtained from the properties of nodes $\{n_j\}_i$ obtained by a special procedure of *aggregation*, or *generalization*, which is interpreted as integration over the j index. Decomposition alludes to the refinement of properties and functions, while aggregations are done via their integration, which is the essence of generalization.

These recursive properties of our multiresolutional hierarchy are determined by the inclusion properties of the knowledge transformation sets (K):

$$K_{si} \supset K_{s(i-1)} \supset \cdots \supset K_{s2} \supset K_{s1} \qquad K_{bgi} \supset K_{bg(i-1)} \supset \cdots \supset K_{bg2} \supset K_{bg1}$$

$$K_{pi} \supset K_{p(i-1)} \supset \cdots \supset K_{p2} \supset K_{p1} \qquad K_{ai} \supset K_{a(i-1)} \supset \cdots \supset K_{a2} \supset K_{a1}$$

$$K_{wmi} \supset K_{wm(i-1)} \supset \cdots \supset K_{wm2} \supset K_{wm1} \qquad K_{wi} \supset K_{w(i-1)} \supset \cdots \supset K_{w2} \supset K_{w1}$$

The sign of inclusion should be understood here in the sense that the knowledge set of a particular module at a particular level is a subset (obtained by decomposition) of the knowledge set belonging to the same module at the lower-resolution level. The multiresolutional hierarchy of knowledge under these conditions can be represented graphically as a tree focused (by attention) at each level of resolution. It is discussed in Chapter 9 (see Section 9.3.3).

RECURSIVE HIERARCHY

A recursive hierarchy is linked by recursive algorithms which generate top-down/bottom-up decision making processes and thus propagate refinement/generalization within the NIST-RCS control system. This leads to the following interesting property which entails the inclusion properties of the knowledge transformation sets: All global properties and functions of a system that can be represented by a recursive hierarchy *contain* all properties of the lower levels of the hierarchy (higher-resolution levels) in an integrated (generalized) form. Thus the subsystems of the higher resolution levels are not really autonomous; they are expected to carry out their assigned tasks assisted by other cooperating agents at their own level in supporting their own parent systemic goals and objectives. Their autonomy is limited by the fact that they are contained within the adjacent lower-resolution level and its role in pursuit of the goal. The system of recursive hierarchy is thus built using the recursive modules.

The architecture of NIST-RCS belongs to the class of recursive hierarchies designed for processing information and data with changing computational models from level to level; the increase in resolution is therefore compensated by a reduction in the interval of computation. It is possible to demonstrate that recursive computations always lead to nested structures of information and data and thus end up with a system of nested knowledge transformation sets $\{K\}$. The property of *nesting* allows for a functional decoupling of levels. However, even after the decoupling, levels remain dependent on each other. Although decoupled, normal functioning requires the satisfaction of nesting with its neighbors above and below. The nesting property is especially important if a search is required in place of analytical methods, or if a change in resolution is accompanied by the changes in vocabulary.

3.7.4 Recursive Algorithm of Constructing Multiscale Knowledge Representations

This recursive algorithm is applicable and is being applied in most of the multiresolutional intelligent systems. The authors personally implemented it in the cases described in [34].

SEARCHING

Step 1. Get information at the highest available level of resolution; consider the totality of this information a *representation at the resolution level* registered as the highest available level of resolution.

- The processing starts bottom-up.
- The smallest units of information are presumed which cannot be further decomposed because the smaller units than this, are indistinguishable; they are regarded as elementary units of the highest-resolution-level.
- A particular scope (breadth, scope of interest) is considered together with a particular value of the smallest unit of information for this scope (value of resolution).

TESTING UNIFORMITY

Step 1.1. Investigate the properties of uniformity for the *representation at the resolution level.*

- This investigation includes testing of the information by a set of sliding windows to determine zones of uniformity (constant density) and zones of increased density; the latter are candidates for emergence of clusters because the elementary units seem to gravitate to each other in these zones.

FOCUSING ATTENTION

- Determining density presumes a priori existence of the cost function linked with the vitally important properties of space (distance) and units (whose attributes might be mass, brightness, etc.).

GROUPING

Step 1.2. Cluster these units within the zones where they gravitate to each other and label the units received as a result of clustering.

- These new units can hypothetically be considered elementary units of the adjacent lower level of resolution.
- The conjecture of mutual gravitation presumes introduction of a cost function which may be richer than the cost function used for investigation of uniformity of the space.

EXPLORING ELFs

Step 1.3. Explore the ELF-loop (six-box diagram); if the new cluster does not have any meaning in the ELF-loop, mark it for further exploration at the lower level of resolution with no label assigned at this resolution level.

- "Having meaning" indicates a consistent description of GSM functioning as far as the goals of GSM are concerned; later there will be shown a correspondence to the analysis of "pragmatics."
- At this level "information" is transformed into "knowledge."

Step 1.4. Register the results as *representation of the lower-resolution level.*

Step 1.5. If no new clusters have been created at this stage, then go to step 2. Otherwise, loop back to step 1.1 and explore an adjacent lower-resolution-level.

Step 2. Send all *representations* of the resolution levels to the overall system of representation.

This algorithm can be rewritten as follows:

Step 1. Get information at the highest available level of resolution; consider it a *representation at the resolution level* registered for the highest available level of resolution.

Step 1.1. Investigate properties of uniformity for *representation at the resolution level.*

- *Focus attention* on all subsets of the initial information set (windowing).
- *Group* information within the *focus of attention* and evaluate groups.

Step 1.2. *Group* elementary units within the overall scope; consider the groups (clusters) to be elementary units, and label them.

Step 1.3. Check consistency of the clusters with GSM-loop (see Fig. 1.1); mark the inconsistent clusters for further exploration at the lower-resolution-level.

Step 1.4. Register the results as *representation of the lower-resolution level.*

Step 1.5. If no new clusters have been created at this stage, then go to step 2. Otherwise, loop back to the step 1.

Step 2. Send *representations* of all resolution levels to the overall system of representation.

3.8 VIRTUAL PHENOMENA OF KNOWLEDGE REPRESENTATION

3.8.1 Representation for Immediate Sensory Processing

For an intelligent system to cope effectively with the demands of complexity under the constraints of time and computing resources, it is necessary to partition the knowledge representation system into at least three parts. The first part supports immediate sensory experience, the second supports short-term memory, and the third supports long-term memory.

Immediate experiences are rich, vivid, and dynamic. Visual images are filled with color, motion, and three-dimensional perspective. Auditory experiences have a wide range of frequency, intensity, and directional sense based on phase and amplitude differences between the two ears. Tactile experiences can convey rich sensations of temperature, vibration, pressure, texture, along with a sense of position, velocity, and force based on proprioception.

RICHNESS OF IMMEDIATE EXPERIENCES

The richness of detail in the immediate experience is transient. It disappears as soon as the lights go out, or the sound stops, or we remove our tactile sensors from contact with the environment. Careful psychophysical experiments designed to see what is preserved of the visual image after it no longer persists on the retina have given negative results. The visual image disappears almost immediately (except for the afterimage produced by sensor fatigue effects), and what was not noticed during the viewing cannot be recalled into view. The immediate experience of sounds in an auditory sequence is also gone almost right after the sound has stopped. All that remains of immediate sensory experience after the stimulus is removed are memories consisting of symbolic information that was specifically noticed during the experience, or can be reconstructed by logical reasoning from what was noticed.

It is clear that there must exist an internal representation of immediate experience. Otherwise, it would never be perceived. Clearly, there exist filtering and prediction mechanisms for processing of signals and images because the ability of the sensory system to detect signals in noise approaches the theoretical limits.

There are at least three representations for each sensory signal that contributes to the immediate experience, and hence for each attribute image:

- Observed signals, images, attributes, and states represent the current array of signals from sensors and the attributes that can be directly computed from sensory signals.
- Estimated signals, images, attributes, and states are the output of filtering processes that integrate information from observed images and other sources over some time interval.
- Predicted signals, images, attributes, and states are the system's best guess of what the next sensory input will be based on all the latest estimated state-variables, the known system dynamics, and the known control output.

These three representations exist concurrently with the sensory input and disappear with it as well. None of these three representations persist for more than a moment after the stimulus ceases.

3.8.2 Intermediate Representation

Short-term memory forms an interface, or buffer, between immediate experience and long-term memory. Short-term memory persists until it is overwritten or disrupted. Short-term memory provides the equivalent of a delay line, or dynamic RAM memory. The delay line can hold strings of signal values, or strings of values such as symbolic characters, or words.

Short-term memory differs from immediate experience in that it contains only symbolic information. The kinds of things short-term memory can store include the position of an entity, its size, shape, color, motion, intensity, and frequency. Short-term memory also differs from immediate experience in that it can store strings or sequences of symbols such as a phone number, a string of words, a musical tune, a train of thought, a plan, or a procedure.

ADJUSTMENT OF IMMEDIATE AND LONG-TERM MEMORIES

Short-term memory differs from long-term memory in that it is dynamic. It retains information by recirculation or rehearsal. If this recirculation is interrupted, or overwritten with new information, what was previously stored in short-term memory is lost.

Short-term memory also differs from long-term memory in that it is limited in the number of entities it can hold. George Miller's famous paper on "The Magic Number Seven Plus or Minus Two" provides a description of the essential features of short-term memory [50]. It can hold only about seven to nine entities, or "chunks" of information. Larger numbers of things cannot be remembered individually, but their characteristics can be summarized by grouping them into chunks that represent higher-level entities.

Entity attributes in short-term memory can be compared with entity attributes computed from immediate experience in support of recursive estimation and predictive filtering operations. The correlation between short-term memory and immediate experience enables signal detection, grouping, and image segmentation processes. Both grouping and the subsequent segmentation are the processes of generalization and create abstracted entities (symbols).[15] Entity attributes in short-term memory can also be compared with entity class attributes derived from long-term memory in support of recognition and classification processes. Recognition occurs when entity attributes in topological entity frames match or correlate with entity attributes in entity class frames.

Entities and events can be selected by task goals and plans from long-term memory and transferred into short-term memory where they can be compared with iconic images and topological entity frames from immediate experience. By this means it is possible to recognize an entity detected in the input as corresponding to a task-related entity stored in long-term memory. Similarly entities and events detected in immediate experience and transferred into short-term memory can (if they are noteworthy) be stored in long-term memory for future use. By this means long-term memory can be kept up to date. Entities and events that are not noteworthy are lost as soon as short-term memory is overwritten by subsequent input.

3.8.3 Long-Term Memory Representation

Long-term memory provides a repository of information that can accumulate and be retained over a lifetime. Long-term memory contains information solely in symbolic form in a format that can be readily transferred to and from short-term memory.

Long-term memory differs from short-term memory primarily in the fact that the information endures over time. It may also preserve temporal ordering only through list structures such as strings, graphs, or frames. Information is retained even through unconsciousness or electroshock therapy, and it is effectively unlimited in capacity. New information entered into long-term memory decays slowly with time or as the result of being overwritten, but it can be reinforced and embellished by rehearsal. Long-term memory may integrate many experiences of objects and situations into a single (or few) generic class(es). Once firmly in place, long-term memories can endure for decades.

These virtual phenomena of knowledge organization are known from studies of humans. The same or similar phenomena might be expected in the architectures of machine intelligence.

REFERENCES

[1] A. Newell and H. A. Simon, *Human Problem Solving*, Prentice-Hall, Englewood Cliffs, 1972.

[2] *SOAR, The Soar Papers* (P. S. Rosenbloom, J. E. Laird, and A. Newell, eds.), MIT Press, Cambridge, MA, 1993.

[15] For an example of segmentation based on substantial abstraction, see [51].

[3] H. Dreyfus, *What Computers Still Can't Do*, MIT Press, Cambridge, MA, 1992.

[4] John Searle, Minds, brains, and programs, *Behavioral and Brain Sciences*, No. 3, 1980, pp. 417–424.

[5] Mark Kickhard and Loren Terveen, *Foundational Issues in Artificial Intelligence and Cognitive Science*, Elsevier, Amsterdam, 1996.

[6] T. Sebeok (ed.), *Encyclopedic Dictionary of Semiotics*, 3 vols., Berlin, Mouton de Gruyter, 1986.

[7] A. Meystel, *Semiotic Modeling and Situation Analysis: An Introduction*, AdRem, Bala Cynwyd, 1995, p. 157.

[8] P. J. Antsaklis and K. M. Passino (eds.), *An Introduction to Intelligent and Autonomous Control*, Kluwer Academic, Boston, 1993, p. 427.

[9] S. Tzafestas and H. Verbruggen (eds.), *Artificial Intelligence in Industrial Decision Making, Control, and Automation*, Kluwer Academic, Boston, 1995.

[10] M. M. Gupta and N. K. Singh (eds.), *Intelligent Control Systems: Theory and Applications*, IEEE Press, New York, 1995, p. 820.

[11] D. A. White and D. A. Sofge (eds.), *Handbook of Intelligent Control: Neural, Fuzzy, and Adaptive Approaches*, Van Nostrand Reinhold, New York, 1992, p. 568.

[12] P. Antsaklis, Intelligent control, *IEEE Control Systems Magazine*, June 1994.

[13] J. S. Albus, Outline for a theory of intelligence, *IEEE Transactions on Systems, Man, and Cybernetics*, Vol. 21, No. 3, May/June 1991, pp. 473–509.

[14] A. Meystel, Multiscale systems and controllers, *Proceedings of the IEEE/IFAC Joint Symposium on Computer-Aided Control System Design*, Tucson, AZ, 1994, pp. 13–26.

[15] J. Searle, *The Rediscovery of the Mind*, MIT Press, Cambridge, MA, 1992, p. 270.

[16] M. Posner (ed.), *Foundations of Cognitive Science*, MIT Press, Cambridge, MA, 1989, p. 888.

[17] J. Buchler (ed.), *Philosophical Writings of Peirce*, Dover, New York, 1955, p. 386.

[18] C. Peirce, *Reasoning and the Logic of Things*, Harvard University Press, Cambridge, MA, 1992, p. 297.

[19] G. Frege (1892), On sense and reference, in *Frege, Philosophical Writings*, P. Geach and M. Black (eds.), Oxford, 1952.

[20] B. Russell, *Human Knowledge: Its Scope and Limits*, Touchstone, 1948.

[21] R. Carnap, *Philosophical Foundations of Physics*, Basic Books, New York, 1966.

[22] M. Stefik, *Introduction to Knowledge Systems*, Morgan Kaufmann Publ., San Francisco, 1995.

[23] H. Adeli, *Knowledge Engineering*, Vols. 1 and 2, McGraw-Hill, New York, 1990.

[24] W. J. Frawley, G. Piatetsky-Shapiro, and C. J. Matheus, Knowledge discovery in databases: an overview, in G. Piatetsky-Shapiro and W. J. Frawley (eds.), *Knowledge Discovery in Databases*, AAAI Press/The MIT Press, Cambridge, MA, 1991, pp. 1–27.

[25] C. Zaniolo et al., *Advanced Database Systems*, Morgan Kaufmann Publ., San Francisco, 1997.

[26] M. S. Gazzaniga (ed.), *The Cognitive Neurosciences*, Bradford Books, MIT Press, Cambridge, MA, 1995.

[27] D. L. Sparks and J. M. Groh, The superior colliculus: a window for viewing issues in integrative neuroscience, in *The Cognitive Neurosciences*, M. S. Gazzaniga (ed.), Bradford Books, MIT Press, Cambridge, MA, 1995, pp. 565–596.

[28] M. F. Jay and D. L. Sparks, Sensorimotor integration in the primate superior colliculus: II coordinates of auditory signals, *J. Neurophysiology*, Vol. 57, 1987, pp. 35–55.

[29] M. F. Jay and D. L. Sparks, Auditory receptive fields in primate superior colliculus shift with changes in eye position, *Nature*, Vol. 309, 1984, pp. 345–347.

[30] D. H. Ballard, Animate vision, *Artificial Intelligence*, Vol. 48, 1991, pp. 57–86.

[31] S. M. Kosslyn, *Image and Brain: The Resolution of the Imagery Debate*, MIT Press, Cambridge, MA, 1995.

[32] L. A. Zadeh, Fuzzy logic—computing with words, *IEEE Transactions on Fuzzy Systems*, Vol. 4, No. 2, 1996, pp. 103–111.

[33] L. A. Zadeh, From computing with numbers to computing with words—from manipulation of measurements to manipulation of perception, *IEEE Transactions on Circuits and Systems-I: Fundamental Theory and Applications*, Vol. 45, No. 1, 1999, pp. 105–119.

[34] J. Albus and A. Meystel, *Behavior Generation in Intelligent Systems*, NIST, 1998.

[35] P. McCauley-Bell and A. Badiru, Fuzzy modeling and analytic hierarchy processing to quantify risk levels associated with occupational injuries—Part I: The development of fuzzy-linguistic risk levels, *IEEE Transactions on Fuzzy Systems*, Vol. 4, No. 2, 1996, pp. 124–131.

[36] G. Booch, *Object-Oriented Analysis and Design*, Benjamin Cummings, Redwood City, CA, 1994.

[37] D. Dubois and H. Prade, Processing fuzzy temporal knowledge, *IEEE Transactions on Systems, Man, and Cybernetics*, Vol. 19, No. 4, 1989, pp. 729–744.

[38] A. Basu and A. Dutta, Reasoning with imprecise knowledge to enhance intelligent decision support, *IEEE Transactions on System, Man, and Cybernetics*, Vol. 19, No. 4, 1989, pp. 756–770.

[39] A. Kolmogorov, *Axiomatic Principles of the Probability Theory*, 1936.

[40] V. Sridhar and M. N. Murty, Belief revision—an axiomatic approach, *Journal of Intelligent and Robotic Systems*, Vol. 8, 1993, pp. 127–153.

[41] A. Meystel, Intelligent systems as an object of semiotic research, *Proc. of the 1996 Biennial Conference of the North American Fuzzy Information Processing Society*, NAFIPS, Berkeley, CA, 1996.

[42] I. Kant, *The Critique of Pure Reason*, Publ. Encyclopedia Britannica, Chicago, 1952.

[43] A. Schopenhauer, in *World as Will and Idea* (I. Edman, ed.), Random House, New York, 1928.

[44] M. Dameshek, Gauging similarity with n-grams: language independent categorization of text, *Science*, Vol. 267, 1995, p. 843.

[45] C. Batini, S. Ceri, and S. B. Navathe, *Conceptual Database Design: An Entity-Relationship Approach*, 1992.

[46] G. Salton, *Automatic Text Processing: The Transformation, Analysis, and Retrieval of Information by Computer*, 1988.

[47] L. Kohout, The role of semiotic descriptors in relational representation of fuzzy granular structures, in *Proc. of the 1997 International Conference on Intelligent Systems and Semiotics: A Learning Perspective* (A. Meysel, ed.), Gaithersburg, MD, 1997.

[48] A. Meystel, Intelligent control, in *Encyclopedia on Physical Sciences and Technology*, Academic Press, 1987.

[49] Y. Maximov and A. Meystel, Optimum design of multiresolutional hierarchical control systems, *Proc. of the 7th IEEE Int'l Symposium on Intelligent Control*, Glasgow, GB, 1992.

[50] G. A. Miller, The magic number seven plus or minus two: Some limits on our capacity for processing information, *Psychological Review*, Vol. 63, 1956, pp. 81–97.

[51] L. Najman and M. Schmitt, Geodesic saliency of watershed contours and hierarchical segmentation, *IEEE Transactions on Pattern Analysis and Machine Intelligence*, Vol. 18, No. 12, 1996, pp. 1163–1173.

[52] W. J. Freeman, *Neurodynamics*, Springer, London, 2000.

[53] S. Grossberg and M. Kuperstein, *Neural Dynamics*, Pergamon Press, 1989.

PROBLEMS

3.1. Find and list parallels and similarities between evolution of knowledge in the history of mankind, and evolution of knowledge in the life of a particular person. How would you explain the fact that these similarities exist? Just a coincidence or something more serious?

3.2. Consider four major tools used in AI for knowledge representation (logic, production systems, semantic networks, and frames). Define each of these tools formally and demonstrate that each of them is a part of the other three.

3.3. Define semiotics and its subdomains. (Restore in your memory some information from Chapter 2). Demonstrate the cycle of knowledge processing by an intelligent system and show that four tools from Problem 3.2 are components of this cycle.

3.4. Define *experience* and all its components. Demonstrate how the experience is formed, and what is its place in the cycle of knowledge processing (discussed in Problem 3.3).

3.5. Explain emergence of *reflexia* using other phenomena (properties) of representation system including intellect, awareness, and consciousness. Define *representation* and propose its architecture. Try to propose a structure of an intelligent system that does not require any representation.

3.6. Give a definition of a neuron and demonstrate its functioning. Illustrate its functioning as a component of a simple neural network.

3.7. List eight properties of NH-AICS and demonstrate that all of them follow from the semiotic cycle of knowledge processing.

3.8. In the principle of thesaural interpretation of words assignment or labeling, what is *meaning*, and how is it related to the *goal*? How is the concept of meaning related to the function of thesaurus? If thesaurus contains interpretation of all labels, should it contain also interpretation for the terms that are used for making interpretations? Is it not a vicious circle? Explain and give 2–3 illustrative examples. Give an example of WBS (word-based statement). Can you give an example of an image-based statement?

3.9. Introduce the principle of entity relational knowledge representation. Take three simple scientific statements and demonstrate that you can transform it into an entity-relational form. Is ER only about objects (and relations between them, or about actions too?) What is elementary entity and elementary relation? Express your examples via their elementary components. How are these elementary units related to the concept of resolution (scale, granularity). Where is the elementary loop of functioning (ELF) in the ER-representation? Is it possible to provide for the joint existence of ELF and ER?

3.10. Try to infer multiresolutional representation from the examples you have developed for Problem 3.9. Illustrate processes of decomposition and aggregation based upon these examples. Demonstrate recursion of these processes. Would you construct two additional examples: 1) where instead of decomposition and aggregation we have instantiation and generalization, and 2) where instead of decomposition and aggregation we have specialization and abstraction?

3.11. Illustrate the property of nestedness by using your five examples (from Problems 3.9 and 3.10). Compose an algorithm of nesting (for decomposition top-down and for aggregation bottom-up). Would it be any problem for you to merge these two algorithms into one joint algorithm? Demonstrate functioning of this algorithm for the purposes of fuzzification/defuzzification in a control system.

3.12. Use two additional examples that you have constructed in Problem 3.10 to demonstrate the principle of multiple symbolic systems representation. For each of the symbolic representations, demonstrate the grammars, axioms, and rules that make operational these systems of representation.

3.13. Give five examples (similar to those discussed in Problems 3.9 and 3.10) to illustrate the causality principle. Do you really have some time interval between cause and effect? Can you infer a cause–effect relationship if you have no prior experimental records pertaining to this relationship? Can you represent cause–effect relationships by using images with no help from any linguistic tool? Explain.

3.14. Evaluate quantitatively the expressive power of representations utilized in five examples from Problem 3.12. Propose judgments concerning completeness and computability of these examples. Evaluate complexity of knowledge representation for these examples and demonstrate how it is linked with completeness and computability.

3.15. Choose one example of representation from the set discussed in Problem 3.14 and demonstrate the principle of incompleteness. What should be predicted in this problem? What for? What is the horizon of required prediction? What is "situation of interest?" How many outcomes are to be considered? What should be known to solve the problem? What are the limitations in providing the required knowledge? Evaluate the degree of incompleteness: what is the mathematical apparatus you would prefer to use? Would you suggest to evaluate incompleteness of knowledge by the Heisenberg principle of uncertainty? What are the advantages and disadvantages of doing this?

3.16. For the example used in Problem 3.15 demonstrate a joint functioning of the feedforward and feedback principles. Explore whether the principle of continuity holds for these examples.

3.17. For the example used in Problem 3.15, construct the closure diagram (ELF) and demonstrate for each subsystem of this diagram the focusing attention activities.

3.18. Using definitions and typology related to the entities, characterize all entities that were used in the five examples in Problem 3.12.

3.19. Formulate a problem that would be an extension of Problem 3.12 and would require using the principle of creativity. For example, one would need to combine a number of possible alternatives of solution. Contemplate using the terminology of D-structure for doing this.

3.20. For the same problem, apply the recursive algorithm of constructing the multiscale knowledge representation and explore the results.

CHAPTER 4

REFERENCE ARCHITECTURE

The purpose of reference architecture is to stimulate the design, control, and software engineers to organize their knowledge and activities in a way conducive to using the theoretical fundamentals and theoretical approaches outlined in Chapters 1 to 3.

4.1 COMPONENTS OF A REFERENCE ARCHITECTURE

Conjecture Any intelligent system consists of two parts: internal, or computational, and external, or interfacing the reality of application. The internal part can be decomposed into four internal elements (subsystems) of intelligence: sensory processing, world modeling, behavior generation, and value judgment. Input to and output from the internal part of intelligent systems are realized via sensors and actuators that can be considered external parts. However, they become components of the loop of functioning of the intelligent system (see ELF in Chapter 2).

4.1.1 Actuators

OUTPUT MOTION

Output from an intelligent system is produced by actuators that move, exert forces, and position arms, legs, hands, and eyes. Actuators generate forces to point sensors, excite transducers, move manipulators, handle tools, steer and propel locomotion. An intelligent system can have few or thousands of actuators, and all of these actuators must be coordinated to perform tasks and accomplish goals. Natural actuators are muscles and glands. Machine actuators are motors, pistons, valves, and solenoids. In organizational (and/or social) systems, actuators can be individuals or groups of people.

The concept of actuation can be extended for any process that produces an action. Speech generation can be considered a process that happens via actuation. A "move" on a stock market can be considered actuation for an appropriate system. In our further examples, we will discuss only relatively straightforward cases. In all cases, state-space trajectory (motion) is produced by actuators.

4.1.2 Sensors

TRANSDUCERS

Input to an intelligent system is produced by sensors, which may include visual brightness and color sensors; tactile, force, torque, position detectors; velocity, vibration, acoustic, range, smell, taste, pressure, and temperature measuring devices. Sensors may be used to monitor both the state of the external world and the internal state of the intelligent system itself. Sensors provide input to a sensory-processing system. Similar to actuators, sensors are not limited to technical or biological devices. In organizational (and/or social) systems, sensors, like actuators, can be individuals or groups of people.

4.1.3 Sensory Processing

PERCEPTION

Perception occurs in a sensory processing system element that compares sensory observations with expectations generated by an internal world model. Sensory processing algorithms integrate similarities and differences between observations and expectations over time and space so as to detect events and recognize features, objects, and relationships in the world. Sensory input data, from a wide variety of sensors and over extended periods of time, are fused into a consistent unified perception of the state of the world. Sensory processing algorithms compute distance, shape, orientation, surface characteristics, physical and dynamical attributes of objects and regions of space. Sensory processing is equivalent to the subsystem of perception in living creatures. Sensory processing may include elements of speech recognition and language and music interpretation.

4.1.4 World Model

REPRESENTATION

The world model is the intelligent system's best estimate of the state of the world. The world model includes a database of knowledge about the world, plus a database management system that stores and retrieves information. The world model also contains a simulation capability that generates expectations and predictions. The world model provides answers to requests for information about the present, past, and probable future states of the world. The world model provides this information service to the behavior generation system element in order to make intelligent plans and behavioral choices. It provides information to the sensory processing system element to perform correlation, model matching, and model-based recognition of states, objects, and events. It provides information to the value judgment system element to compute values such as cost, benefit, risk, uncertainty, importance, and attractiveness. The world model is kept up to date by the sensory processing system element.

4.1.5 Value Judgment

EVALUATION

The value judgment system element determines good and bad, rewards and punishments, important and trivial, certain and improbable. The value judgment system evaluates both the observed state of the world and the predicted results of hypothesized plans. It computes costs, risks, and benefits, both of observed situations and of planned activities. It computes the probability of correctness and assigns believability and uncertainty parameters to state variables. It also assigns attractiveness, or repulsiveness to objects, events, regions of space, and other creatures. The value judgment system provides the basis for decision making, or choosing one action instead of another. Without value judgments, any biological creature would soon be destroyed, and any artificial intelligent system would soon be disabled by its own inappropriate actions.

4.1.6 Behavior Generation

Behavior results from a behavior-generating system element that selects goals and plans and executes tasks. Tasks are recursively decomposed into subtasks, and subtasks are sequenced to achieve goals. Goals are selected and plans generated by a looping interaction between behavior generation, world modeling, and value judgment elements. The behavior-generating system hypothesizes plans. The world model predicts the results of those plans, and the value judgment element evaluates those results. The behavior-generating system then selects the plans with the highest evaluations for execution. The behavior-generating system element also monitors the execution of plans and modifies existing plans when the situation requires.

All six components (modules) correspond to the components of ELF diagram (see the modified ELF with a module of value judgment in Figure 4.1).

DECISION MAKING

A system architecture partitions the system elements of intelligence into computational modules, and interconnects the modules in networks and hierarchies. It is what enables the behavior generation elements to direct sensors and to focus sensory processing algorithms on objects and events worthy of attention, ignoring things that are not important to current goals and task priorities. The system architecture enables the world model to answer queries from behavior-generating modules and make predictions and receive updates from sensory processing modules.

Each of the modules represent a hierarchy of such modules actually distributed over all levels of resolution.

4.2 EVOLUTION OF THE REFERENCE ARCHITECTURE FOR INTELLIGENT SYSTEMS

A number of system architectures for intelligent machine systems have been conceived, and a few implemented [1–14, 20]. The architecture for intelligent systems that will be proposed here is largely based on the real-time control system (RCS) that has been implemented in a number of versions over the past 20 years at the National Institute for Standards and Technology (NIST, formerly NBS).

NIST-RCS was first implemented by A. Barbera for laboratory robotics in the mid 1970s [7]; it was adapted by Albus, Barbera, and others for manufacturing control in the NIST Automated Manufacturing Research Facility (AMRF) during the early 1980s [11, 12]. Since 1986 RCS has been implemented for a number of additional applications, including the NBS/DARPA Multiple Autonomous Undersea Vehicle (MAUV) project [13], the Army Field Material Handling Robot, and the Army TMAP and TEAM semiautonomous land vehicle projects. RCS is the basis of the NASA/NBS Standard Reference Model Telerobot Control System Architecture (NASREM) being used on the space station Flight Telerobotic Servicer [14] and the Air Force Next Generation Controller. Recent applications are described in [42–46].

Other groups have also introduced architectures that are equivalent to the NIST-RCS architecture. We mention here only results obtained by the Drexel University researchers who implemented this architecture for the autonomous mobile robot "Dune-Buggy" [15–17] in a machine for automated spray casting "OSPREY" [18] and for a power plant control system [19]. A research group at the University of Maryland introduced an RCS-like architecture with levels "planner," "navigator," and "pilot" [20] similar to the levels introduced in [16], [17].

The system architecture we introduce in this chapter organizes the elements of intelligence so as to create the functional relationships and information flows shown in Figure 1.1. In all intelligent systems, a sensory processing system acquires and maintains an internal model of the external world. In all systems a behavior-generating system controls actuators to pursue goals in the context of the perceived world model. In systems of higher intelligence, the behavior-generating system element may interact with the world model and value judgment system to reason about space and time, geometry and dynamics, and to formulate or select plans based on values such as cost, risk, utility, and goal priorities. The sensory processing system element may interact with the world model and value judgment system to assign values to perceived entities, events, and situations.

The architecture combining these components is shown in Figure 4.1. This is an architecture that can be considered a general functional representation of any functioning system, any system that can exist, performs some assigned function, perpetuates in this existence. This system is a generalized subsistence machine (GSM) and performs both goal-directed functioning (GDF) and regular subsistence functioning (RSF) as described in Chapter 2. GSM functioning is a superposition of GDF and RSF. Each of these has its own ELF and produces its own hierarchy. In the subsequent presentation we will focus mostly on the GDF systems.

Figure 4.2 shows how the proposed system architecture replicates and distributes the relationships of Figure 4.1 over a hierarchical computing structure. All logical and temporal properties illustrated in Figure 4.1 for a single level are kept. An organizational hierarchy where computational nodes are arranged in layers like command posts in a military organization appears on the left. Each node in the organizational hierarchy

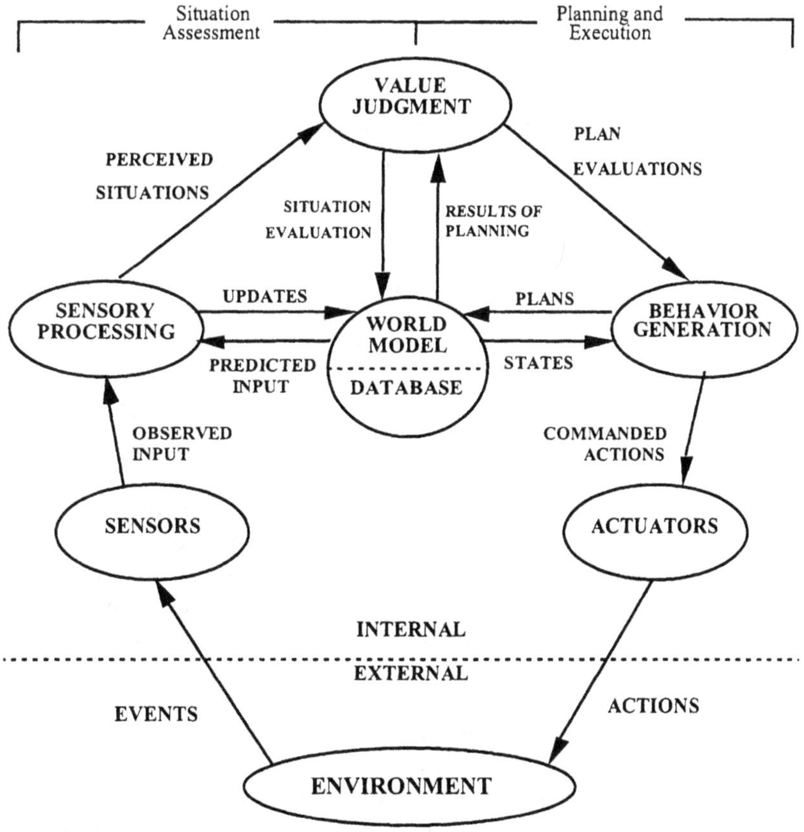

Figure 4.1. Functional relationships between modules of ELF (see Figure 1.1).

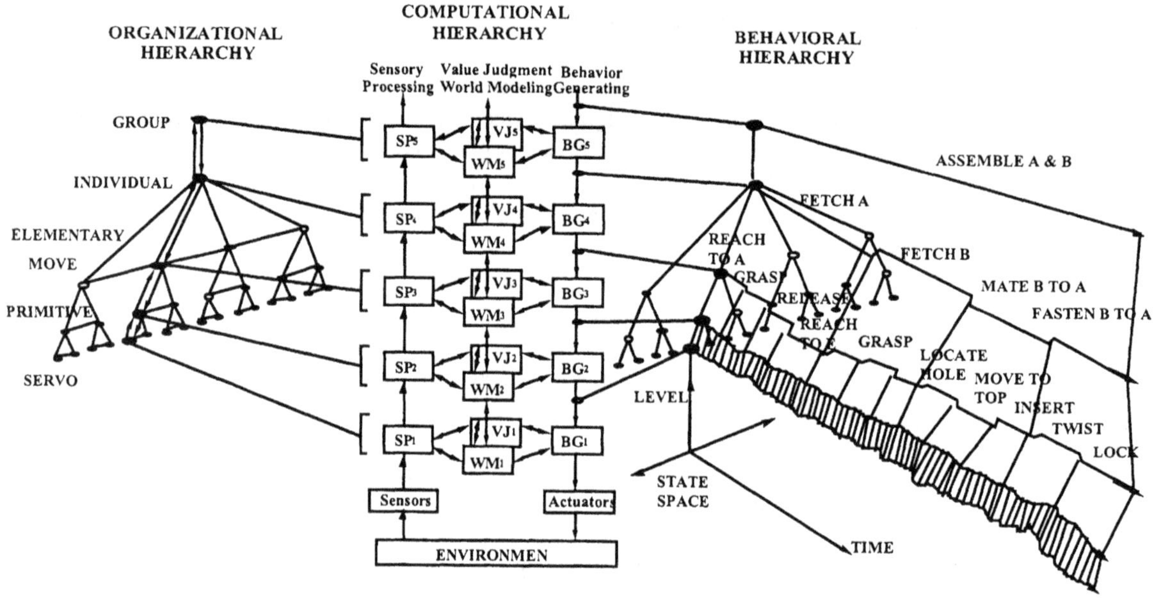

Figure 4.2. An example of relationships in heterarchical control systems. On the left is an organizational heterarchy. In the center is a computational heterarchy consisting of BG, WM, SP, and VJ modules. Each actuator and each sensor is serviced by a computational heterarchy. On the right is a behavioral heterarchy producing trajectories in the state space. Commands at each level can be represented by vectors, or points in state space. Sequences of commands can be represented as trajectories through states in space.

contains four types of computing modules: behavior-generating (BG), world-modeling (WM), sensory processing (SP), and value judgment (VJ) modules. Each chain of command in the organizational heterarchy, from each actuator and each sensor to the highest level of control, can be represented by a computational heterarchy, such as what is illustrated at the center of Figure 4.2.

At each level, the nodes and computing modules within the nodes are interconnected to each other by a communication system. Within each computational node, the communication system provides inter-module communications of the type shown in Figure 4.1.

ALL MODULES TALK TO THEIR ELF NEIGHBORS

Queries and task status are communicated from BG modules to WM modules. Retrievals of information are communicated from WM modules back to the BG modules making the queries. Predicted sensory data are communicated from WM modules to SP modules. Updates to the world model are communicated from SP to WM modules. Observed entities, events, and situations are communicated from SP to VJ modules. Values assigned to the world model representations of these entities, events, and situations are communicated from VJ to WM modules. Hypothesized plans are communicated from BG to WM modules. Results are communicated from WM to VJ modules. Evaluations are communicated from VJ modules back to the BG modules that hypothesized the plans.

The communications system also communicates between nodes at different levels. Commands are communicated downward from supervisor BG modules at one level to subordinate BG modules at the level below. Status reports are communicated back upward through the world model from lower-level subordinate BG modules to the upper-level supervisor BG modules from which the commands were received. Observed entities, events, and situations detected by SP modules at one level are communicated upward to SP modules at a higher level. Predicted attributes of entities, events, and situations stored in the WM modules at a higher level are communicated downward to lower-level WM modules. Output from the bottom level BG modules is communicated to actuator drive mechanisms. Input to the bottom level SP modules is communicated from sensors.

The communications system can be implemented in a variety of ways. In a biological brain, communication is mostly via neuronal axon pathways, although some messages are communicated by hormones carried in the bloodstream. In artificial systems, the physical implementation of communications functions may be a computer bus, a local area network, a common memory, a message-passing system, or some combination thereof. In either biological or artificial systems, the communications system may include the functionality of a communications processor, a file server, a database management system, a question-answering system, or an indirect addressing or list-processing engine. In the system architecture proposed here, the input/output relationships of the communications system produce the effect of a virtual global memory, or blackboard system [21].

The input command string to each of the BG modules at each level generates a trajectory through state-space as a function of time. The set of all command strings create a behavioral hierarchy as shown on the right of Figure 4.2. Actuator output trajectories (not shown in Figure 4.2) correspond to observable output behavior. All the other trajectories in the behavioral heterarchy constitute the deep structure of behavior.

4.3 HIERARCHY WITH HORIZONTAL "IN LEVEL" CONNECTIONS

Figure 4.3 shows the organizational heterarchy in more detail; it illustrates both the hierarchical and horizontal relationships involved in the proposed architecture. The structure is heterarchical, and commands and status feedback flow hierarchically up and down a behavior-generating chain of command. The architecture is also hierarchical in that sensory processing and world-modeling functions have hierarchical levels of temporal and spatial aggregation.

This is not a hierarchy in the classical sense of the term. Classical interpretation of hierarchy refers to tree-architecture. We are dealing instead with a multiscale relational structure. Even as we use the term hierarchy, the reader should know that we are talking about multiscale organization in which vertical branches have horizontal connections.

NETWORK OF CONNECTED NODES EXISTS AT EACH LEVEL

The architecture is horizontal in that data are shared horizontally between heterogeneous modules at the same level. At each hierarchical level, the architecture is horizontally interconnected by wide-bandwidth communication pathways between BG, WM, SP, and VJ modules in the same node, and between nodes at the same level, especially within the same command subtree. The horizontal flow of information is voluminous within a single node but less between related nodes in the same command subtree. It has relatively low bandwidth between computing modules in separate command subtrees.

The volume of information flowing horizontally within a subtree may be orders of magnitude larger than the amount flowing vertically in the command chain. The volume of information flowing vertically in the sensory processing system can also be very high, especially in the vision system.

The specific configuration of the command tree is task dependent and therefore not necessarily stationary in time. Figure 4.3 illustrates only one possible configuration that may exist at a single point in time. During operation, relationships between modules

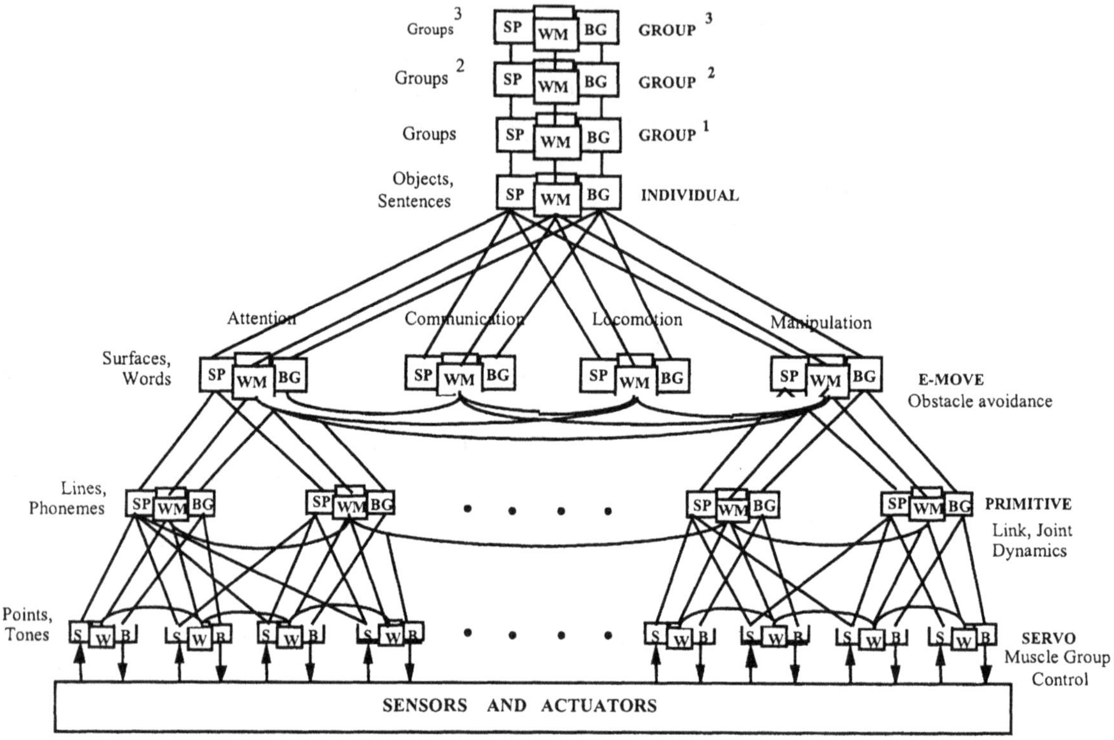

Figure 4.3. An organization of processing nodes (RCS-nodes) forming a control heterarchy. On the right are examples of the functional characteristics of the BG modules at each level. On the left are examples of the type of visual and acoustical entities recognized by the SP modules at each level. In the center of level 3 are the type of subsystems represented by processing nodes at level 3.

within and between layers of the hierarchy may be reconfigured in order to accomplish different goals, priorities, and task requirements. This means that any particular computational node, with its BG, WM, SP, and VJ modules, may belong to one subsystem at one time and a different subsystem a very short time later. For example, the mouth may be part of the manipulation subsystem (while eating) and the communication subsystem (while speaking). Similarly an arm may be part of the manipulation subsystem (while grasping) and part of the locomotion subsystem (while swimming or climbing).

In the biological brain, command tree reconfiguration can be implemented through multiple axon pathways that exist, but are not always activated, between BG modules at different hierarchical levels. These multiple pathways define a layered graph, or lattice, of nodes and directed arcs, such as shown in Figure 4.4.

They enable each BG module to receive input messages and parameters from several different sources. During operation, goal-driven switching mechanisms in the BG modules (discussed in Chapter 8) assess priorities, negotiate for resources, and coordinate task activities so as to select among the possible communication paths of Figure 4.4. As a result each BG module accepts task commands from only one supervisor at a time, and hence the BG modules form a command tree at every instant in time.

The SP modules are also organized hierarchically as a layered graph, and not a tree. At each higher level, sensory information is processed into increasingly higher levels

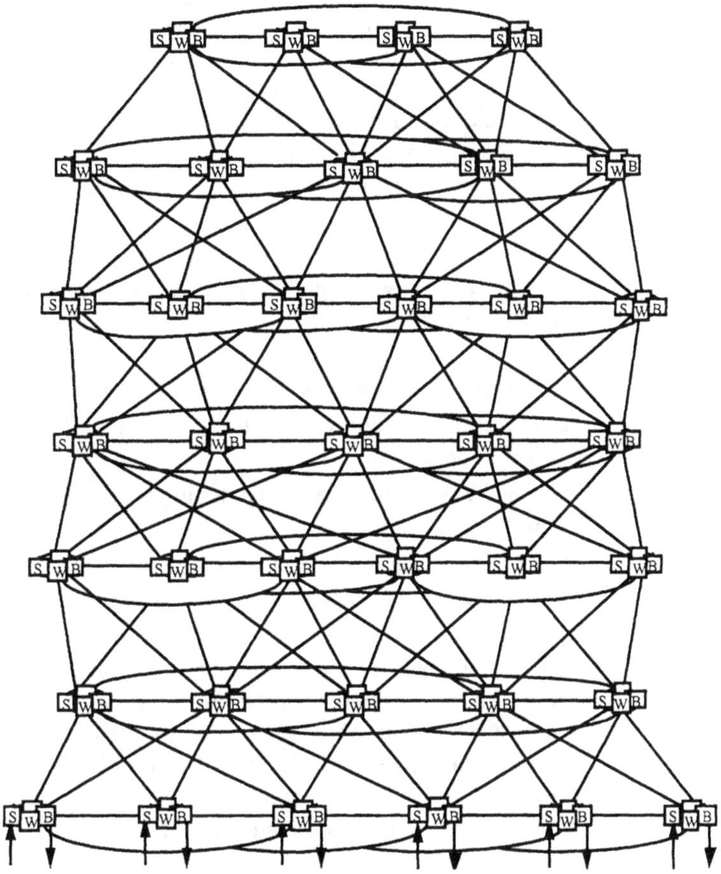

Figure 4.4. Each layer of the system architecture contains a number of RCS-nodes, each containing BG, WM, SP, and VJ modules. The nodes are interconnected as a layered graph, or lattice, through the communication system. Note that the nodes are richly, but not fully, interconnected. Outputs from the bottom layer BG modules drive actuators. Inputs to the bottom layer SP modules convey data from sensors. During operation, goal-driven communication path selection mechanisms configure this lattice structure into the organizational tree shown in Figure 4.3.

of abstraction, but the sensory processing pathways may branch and merge in many different ways.

4.4 LEVELS OF RESOLUTION

The levels in the behavior-generating hierarchy are defined by temporal and spatial decomposition of goals and tasks into levels of resolution. Temporal resolution is manifested in terms of loop bandwidth, sampling rate, and state-change intervals. Temporal span is measured by the length of historical traces and planning horizons. Spatial resolution is manifested in the branching of the command tree and the resolution of maps. Spatial span is measured by the span of control and the range of maps.

Levels in the sensory processing hierarchy are defined by temporal and spatial integration of sensory data into levels of aggregation. Spatial aggregation is best illustrated by visual images. Temporal aggregation is best illustrated by acoustic parameters such as phase, pitch, phonemes, words, sentences, rhythm, beat, and melody.

Levels in the world model hierarchy are defined by temporal resolution of events, spatial resolution of maps, and by parent-child relationships between entities in symbolic data structures. The spatial and temporal unification is achieved within the subsystem of behavior generation (BG) because the needs of SP and BG modules can differ within a level of resolution.

Theorem In a hierarchically structured goal-driven, sensory-interactive, intelligent control system architecture:

1. Control bandwidth decreases from level to level bottom-up.
2. Perceptual resolution of spatial and temporal information decreases from level to level bottom-up.
3. Goals expand in scope and planning horizons expand in space and time from level to level bottom-up.
4. Models of the world and memories of events decrease in resolution and expand in spatial and temporal range from level to level bottom-up.

RATIOS BETWEEN RESOLUTIONS OF ADJACENT LEVELS ARE THE SAME

It is well known from control theory that hierarchically nested servo loops tend to suffer instability unless the bandwidth of the control loops differs by about an order of magnitude. This is one of the reasons why these (1 through 4) changes quantitatively can be evaluated as increase or decrease by an order of magnitude. Numerous theoretical and experimental studies support the concept of hierarchical planning and perceptual "chunking" for both temporal and spatial entities [22, 23]. These support conditions 2 through 4 above.

In elaboration of the above assertion, we can construct a timing diagram as shown in Figure 4.5. The range of the time scale increases, and its resolution decreases, exponentially by about an order of magnitude at each higher level. Hence the planning horizon and event summary interval increases, and the loop bandwidth and frequency of subgoal events decreases, exponentially at each higher level.

The seven hierarchical levels in Figure 4.5 span a range of time intervals from three milliseconds to one day. Three milliseconds was arbitrarily chosen as the shortest servo update rate because that is adequate to reproduce the highest bandwidth reflex arc in the human body. One day was arbitrarily chosen as the longest historical memory/planning horizon to be considered. Shorter time intervals could be handled by adding another layer at the bottom. Longer time intervals could be treated by adding layers at the top, or by increasing the difference in loop bandwidths and sensory chunking intervals between levels.

The origin of the time axis in Figure 4.5 is the present, namely $t = 0$. Future plans lie to the right of $t = 0$, past history to the left. The open triangles in the right half-plane represent task goals in a future plan. The filled triangles in the left half-plane represent recognized task-completion events in a past history. At each level there is a planning horizon and a historical event summary interval. The heavy crosshatching on the right shows the planning horizon for the current task. The light shading on the right indicates the planning horizon for the anticipated next task. The heavy crosshatching on the left shows the event summary interval for the current task. The light shading on the left shows the event summary interval for the immediately previous task. In Figure 4.5 the scales is chosen so that the planning horizons look equal geometrically.

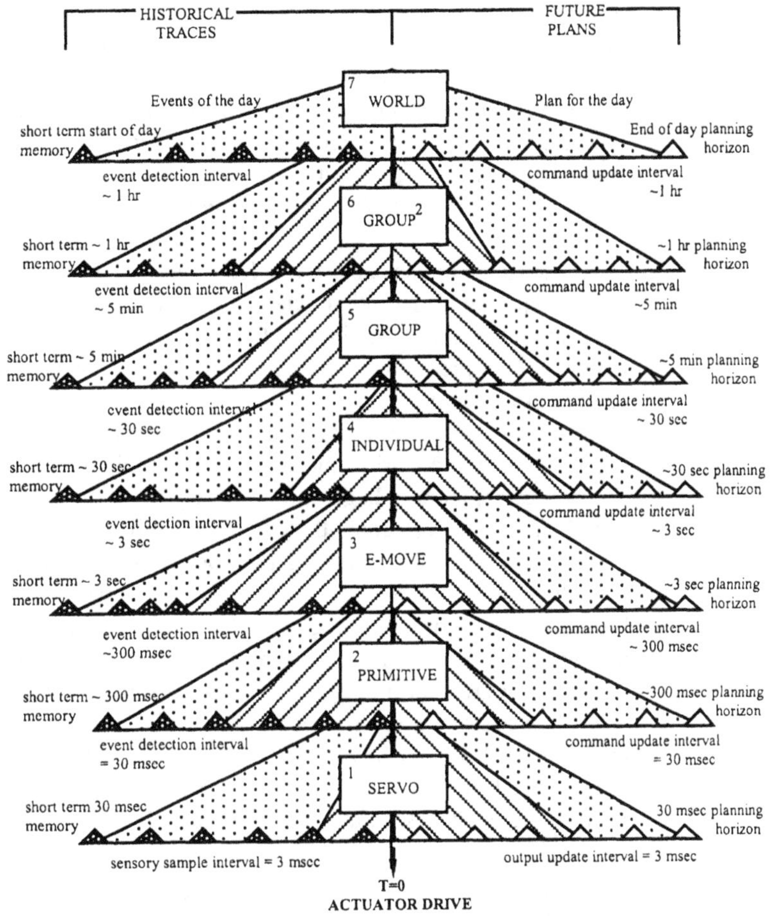

Figure 4.5. A timing diagram illustrating the temporal flow of activity in the task decomposition and sensory-processing systems.

HORIZONS OF PLANNING

Planning implies an ability to predict future states of the world. Prediction algorithms based on Fourier transforms or Kalman filters typically use recent historical data to compute parameters for extrapolating into the future. Predictions made by such methods are typically not reliable for periods longer than the historical interval over which the parameters were computed. Thus at each level, planning horizons extend into the future only about as far, and with about the same level of detail, as historical traces reach into the past.

Predicting the future state of the world often depends on assumptions concerning which actions are to be taken and the reactions to be expected from the environment, including which actions may be taken by other intelligent agents. Planning of this type requires search over the space of possible future actions and probable reactions. Search-based planning takes place via a looping interaction between the BG, WM, and VJ modules. This is described in more detail in the Section 4.6 discussion on BG modules.

Planning complexity grows exponentially with the number of steps in the plan (i.e., the number of decision steps in the search graph). If real-time planning is to succeed, any given planner must operate in a limited search space. If there are too many steps in the time line, or in the space of possible actions, the size of the search graph can easily become too large for real-time response. One method of resolving this problem is to use a multiplicity of planners in hierarchical layers [14, 23] so that at each layer no planner needs to search more than a given number (e.g., ten) steps deep in a game graph, and at each level there are no more than (ten) subsystem planners that need to simultaneously generate and coordinate plans. These criteria give rise to hierarchical levels with exponentially expanding spatial and temporal planning horizons, and characteristic degrees of detail for each level. The result of hierarchical spatiotemporal planning is illustrated in Figure 4.6.

FUNNEL HIERARCHY

At each level, plans consist of at least one, and on average ten, subtasks. The planners have a planning horizon that extends about one-and-a-half average input command intervals into the future. In Figure 4.6 the scales at the time axes are equal. We are dealing with a "funnel hierarchy" which focuses attention on a more and more narrow scope of attention top-down.

In a real-time system, plans must be regenerated periodically to cope with changing and unforeseen world conditions. Cyclic replanning may occur at periodic intervals. Emergency replanning begins immediately upon the detection of an emergency condition. Under full alert status the cyclic replanning interval should be about an order of magnitude less than the planning horizon (or about equal to the expected output subtask time duration). This requires that real-time planners be able to search to the planning horizon about an order of magnitude faster than real time (this requires one to have time intervals shorter than those selected for the real-time process). This is possible only if the depth and resolution of search is limited through hierarchical planning.

Plan executors at each level are responsible for reacting to feedback every control cycle interval. Control cycle intervals are inversely proportional to the control loop bandwidth. Typically the control cycle interval is an order of magnitude less than the expected output subtask duration. If the feedback indicates the failure of a planned subtask,

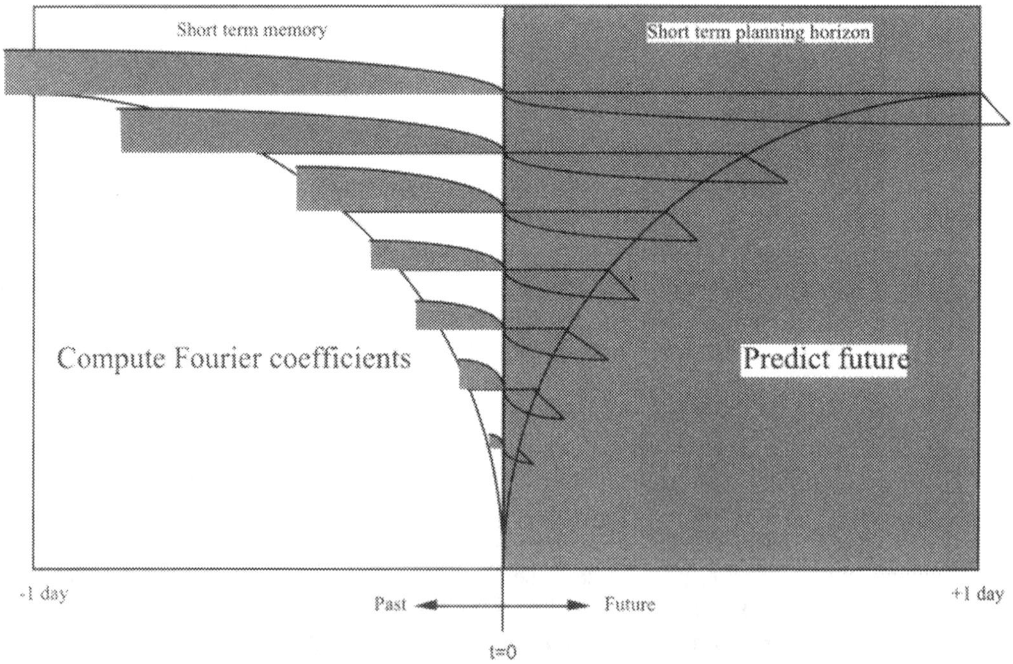

Figure 4.6. Hierarchical spatiotemporal planning.

the executor branches immediately (i.e., in one control cycle interval) to a preplanned emergency subtask. The planner simultaneously selects or generates an error recovery sequence; this is substituted for the former plan that failed. Plan executors are described in more detail in Chapters 8 through 10 of this book.

When a task goal is achieved at time $t = 0$, it becomes a task completion event in the historical trace. To the extent that a historical trace is an exact duplicate of a former plan, there were no surprises. For example, the plan was followed, and every task was accomplished as planned. To the extent that a historical trace is different from the former plan, there were surprises. The average size and frequency of surprises (i.e., differences between plans and results) is a measure of effectiveness of a planner.

At each level in the control hierarchy, the difference vector between planned (i.e., predicted) and observed events is an error signal. This signal can be used by executor submodules for servo feedback control (i.e., error correction) and by VJ modules for evaluating success and failure.

In the subsequent subsections the system architecture outlined above will be elaborated and the functionality of the computational submodules for behavior generation, world modeling, sensory processing, and value judgment will be discussed.

4.5 NEURAL COMPONENTS OF THE ARCHITECTURE

Theorem All of the processes described for the modules of the intelligent system BG, WM, SP, and VJ shown in Figure 4.1, whether implicit or explicit, can be implemented in neural net or connectionist architectures, and hence could be implemented in a biological neuronal substrate.

GENERALIZATION VIA AVERAGING

Modeling of the neurophysiology and anatomy of the brain by a variety of mathematical and computational mechanisms has been discussed in a number of publications [16, 32, 33, 35–41]. Many of the submodules in the BG, WM, SP, and VJ modules can be implemented by functions of the form $\mathbf{P} = \mathbf{H}(\mathbf{S})$. This type of computation can be accomplished directly by the typical layer of neurons that makes up the cortex or the subcortical nucleus.

To a first approximation, any single neuron can compute a linear single-valued function of the form

$$p(k) = h(S) = \sum_{i=1}^{N} s(i)w(i, k), \tag{4.1}$$

where

$p(k)$ = output of the kth neuron,

$\mathbf{S}$ = $(s(1), s(2), \ldots s(i), \ldots, s(N))$ is an ordered set of input variables carried by input fibers defining an input vector

$\mathbf{W}$ = $(w(1, k), w(2, k), \ldots w(i, k), \ldots w(N, k)$ is an ordered set of synaptic weights connecting the N input fibers to the kth neuron,

$h(\mathbf{S})$ = the internal product between the input vector and the synaptic weight vector.

A set of neurons can therefore compute the vector function

$$\mathbf{P} = \mathbf{H}(\mathbf{S}), \tag{4.2}$$

where $\mathbf{P} = (p(1), p(2), \ldots, p(k), \ldots, p(L))$ is an ordered set of output variables carried by output fibers defining an output vector.

The physical mechanisms of computation in a neuronal computing module are produced by the effect of chemical activation on synaptic sites. These are analog parameters with time constants governed by diffusion and enzyme activity rates. Computational time constants can vary from milliseconds to minutes, or even hours or days, depending on the chemicals carrying the messages, the enzymes controlling the decay time constants, the diffusion rates, and the physical locations of neurological sites of synaptic activity.

The time-dependent functional relationship between input fiber firing vector $\mathbf{S}(t)$ and the output cell firing vector $\mathbf{P}(t)$ can be captured by making the neural net computing module time dependent:

$$\mathbf{P}(t + dt) = \mathbf{H}(\mathbf{S}(t)). \tag{4.3}$$

The physical arrangement of input fibers can also produce many types of nonlinear interactions between input variables. It can in fact be shown that a computational module consisting of neurons can compute any single-valued arithmetic, vector, or logical function, if–then rule, or memory retrieval operation that can be represented in the form $\mathbf{P}(t + dt) = \mathbf{H}(\mathbf{S}(t))$ (for more details, see Chapter 10). By interconnecting $\mathbf{P}(t + dt) = \mathbf{H}(\mathbf{S}(t))$ computational modules in various ways, a number of additional important mathematical operations can be computed, including finite state automata, spatial and temporal differentiation and integration, tapped delay lines, spatial and temporal auto- and cross-correlation, coordinate transformation, image scrolling and warping, pattern recognition, content addressable memory, and sampled-data, state-space feedback control [35–39].

GENERALIZING WHILE THE ASSOCIATES GENERALIZE, TOO

In a two layer neural net such as a perceptron, or a simplified brain model such as CMAC [33] the nonlinear function

$$\mathbf{P}(t + dt) = \mathbf{H}(\mathbf{S}(t))$$

is computed by a pair of functions

$$\mathbf{A}(\tau) = \mathbf{F}(\mathbf{S}(t)), \tag{4.4}$$

$$\mathbf{P}(t + dt) = \mathbf{G}(\mathbf{A}(\tau)), \tag{4.5}$$

where

$\mathbf{S}(t)$ = a vector of firing rates $s(i, t)$ on a set of input fibers at time t,
$\mathbf{A}(\tau)$ = a vector of firing rates $a(j, \tau)$ of a set of association cells at time $\tau = t + dt/2$,
$\mathbf{P}(t + dt)$ = a vector of firing rates $p(k, t + dt)$ on a set of output fibers at time $t + dt$,
$\mathbf{F}$ = the function that maps $\mathbf{S}$ into $\mathbf{A}$,
$\mathbf{G}$ = the function that maps $\mathbf{A}$ into $\mathbf{P}$

NEURONS GENERALIZE AND PROPOSE INSTANTIATIONS

The function $\mathbf{F}$ is generally considered to be fixed, serving the function of an address decoder (or recoder) that transforms the input vector $\mathbf{S}$ into an association cell vector $\mathbf{A}$. The firing rate of each association cell $a(j, t)$ thus depends on the input vector $\mathbf{S}$ and the details of the interconnecting matrix of interneurones between the input fibers and association cells that define the function **F**. Recoding from **S** to **A** can enlarge the number of patterns that can be recognized by increasing the dimensionality of the pattern space, and it can permit the storage of nonlinear functions and the use of nonlinear decision surfaces by circumscribing the neighborhood of generalization [40, 41].

The function **G** depends on the values of a set of synaptic weights $w(j, k)$ that connects the association cells to the output cells. The value computed by each output neuron $p(k, t)$ at time t is

$$p(k, t + dt) = \sum_j a(j)w(j, k), \tag{4.6}$$

where $w(j, k)$ = synaptic weight from $a(j)$ to $p(k)$. The weights $w(j, k)$ may be modified during the learning process so as to modify the function **G**, and hence the function **H**.

Additional layers between input and output can produce indirect addressing and list-processing functions, including tree search and relaxation processes [16, 61]. Thus virtually all the computational functions required of an intelligent system can be produced by neuronal circuitry of the type known to exist in the brains of intelligent creatures.

4.6 BEHAVIOR-GENERATING HIERARCHY

Task goals and task decomposition functions often have characteristic spatial and temporal properties. For any task there exists a hierarchy of task vocabularies that can be overlaid on the spatial/temporal hierarchy.

A Bottom Up Example for an Intelligent Robot

Level 1 is where commands for coordinated velocities and forces of body components (e.g., arms, hands, fingers, legs, eyes, torso, and head) are decomposed into motor commands to individual actuators. Feedback controllers are used to compensate for deviation from the plans created for the position, velocity, and force of individual actuators. In vertebrates this is the level of the motor neuron and stretch reflex.

EACH *i*-th LEVEL RECEIVES COMMANDS FROM $(i + 1)$th ONE

Level 2 is where commands for maneuvers of body components are decomposed into smooth coordinated dynamically efficient trajectories. Feedback servos coordinated trajectory motions. This is the level of the spinal motor centers and the cerebellum.

Level 3 is where commands to manipulation, locomotion, and attention subsystems are decomposed into collision-free paths that avoid obstacles and singularities. Feedback servos movements relative to surfaces in the world. This is the level of the red nucleus, the substantia nigra, and the primary motor cortex.

Level 4 is where commands for an individual to perform simple tasks on single objects are decomposed into coordinated activity of body locomotion, manipulation, attention, and communication subsystems. Feedback initiates and sequences subsystem activity. This is the level of the basal ganglia and pre-motor frontal cortex.

Level 5 is where commands for behavior of a small group of intelligent agents are decomposed into interactions between the self and nearby objects or agents. Planners together with the feedback compensation of the executors initiate and steer the whole set of activities required by the task. Behavior-generating levels 5 and above are hypothesized to reside in temporal, frontal, and limbic cortical areas.

Level 6 is where commands for behavior of the individual agents, which are members of multiple groups, are decomposed into small group interactions. Plans and feedback compensations commands steer small group interactions.

EACH $(i+1)$th LEVEL REPORTS THE STATE TO THE i-th ONE

Level 7 (arbitrarily, the highest level in our discussion) is where long-range goals are selected and plans are made for long-range behavior relative to the world as a whole. Feedback steers progress toward long-range goals.

The mapping of BG functionality onto levels one to four defines the control functions necessary to control a single intelligent individual in performing simple task goals. Functionality at levels 1 through 3 is more or less fixed and specific to each species of intelligent system [31]. At level 4 and above, the mapping becomes more task and situation dependent. Levels 5 and above define the control functions necessary to control the relationships of an individual relative to others in groups, multiple groups, and the world as a whole.

There is good evidence that hierarchical layers develop in the sensory-motor system, both in the individual brain as the individual matures and in the brains of an entire species as the species evolves. It can be hypothesized that the maturation of levels in humans gives rise to Piaget's "stages of development" [32].

Of course the biological motor system is typically much more complex than is suggested by the example model described above. In the brains of higher species there may exist multiple hierarchies that overlap and interact with each other in complicated ways. For example, in primates, the pyramidal cells of the primary motor cortex have outputs to the motor neurons for direct control of fine manipulation as well as the inferior olive[1] for teaching behavioral skills to the cerebellum [33]. There is also evidence for three parallel behavior-generating hierarchies that have developed over three evolutionary eras [34]. Each BG module may thus contain three or more competing influences: (1) the most basic (IF it smells good, THEN eat it), (2) a more sophisticated (WAIT until the "best" moment) where best is when success probability is highest, and (3) a very sophisticated (WHAT are the long-range consequences of the contemplated action, and what are the existing options).

On the other hand, some motor systems may be less complex. Not all species have the same number of levels. Insects, for example, may have only two or three levels, while adult humans may have more than seven. In robots, the functionality required of each BG module depends on the complexity of the subsystem being controlled. For example, one robot gripper may consist of a dexterous hand with 15 to 20 force servoed degrees of freedom. Another gripper may consist of two parallel jaws actuated by a single pneumatic cylinder. In simple systems, some BG modules (e.g., the primitive level) may have no function (e.g., dynamic trajectory computation) to perform. In this case the BG module will simply pass through unchanged input commands (e.g., <Grasp>).

4.7 ANALYSIS OF MULTIRESOLUTIONAL ARCHITECTURES

4.7.1 Elementary Loop of Functioning (ELF)

Systems can be represented in two forms: in the form of input-output mappings and in the form of loops of functioning. The first approach is very common but is flawed: one should constantly worry about verification of the validity of mapping. Indeed, if

[1] For "olivary nuclei," see R. S. Smell, *Clinical Neuroanatomy for Medical Students*; Little, Brown & Co., Boston, 1987, page 385.

the system maps input into output, a question should be asked: why? what for? To answer, we start describing what is going on in the external world. Then we answer tentatively: Probably, it maps inputs into outputs to provide for some external processes, and depending on these processes, the input is shaped. We arrive at the second type of representation anyway: in the form of loops of functioning.

A LEVEL WORKS ONLY IF THE ELF IS CLOSED

The loop of functioning is a tool for verifying our generalizations about objects and attributes that we made in system representation. When we consider a particular scope of interest, it is based on some limitations from above and from below. The upper bound of this scope is dictated by our priorities and based on a trade-off between our desire to broaden the scope and the growing burden on our processing system. The lower bound determines the smallest interval measured and/or recorded (the value of resolution). A similar trade-off also occurs. To make a representation within these bounds, we must ensure that this representation will be helpful in practice. The loop of functioning is the measure of our ability to apply our representation. Three conditions are supposed to be satisfied: (1) the upper and lower bounds of the scope of attention must be selected properly, (2) the resolution of representation must be satisfactory, and (3) the objects must be properly determined within the limits of conditions 1 and 2. We will succeed in the goals of our analysis if we are further able to construct a proper vocabulary of the loop of functioning to describe the actions and the measurements, and to close all cause-effect links generated within this loop.

The elementary loop of functioning (ELF) is shown in Figure 4.7. This loop illustrates the obvious fact that all events in *W* (world's subset of interest) should be "sensed" (*S*). The results of sensing should be encoded (transformed into symbols), organized (some primary identification is supposed to be performed), and submitted to the world model WM. On one hand, WM serves as a collection of knowledge about the world that is immediately perceived. On the other hand, we must rely on a more complete and fundamental collection of knowledge (knowledge base) in order to make an interpretation of scenes and situations.

The WM supports the subsystem of behavior generation (BG) where the decision making occurs about all necessary activities (this subsystem is sometimes called "decision making," or "planning control"). BG submits its decisions to the system of actua-

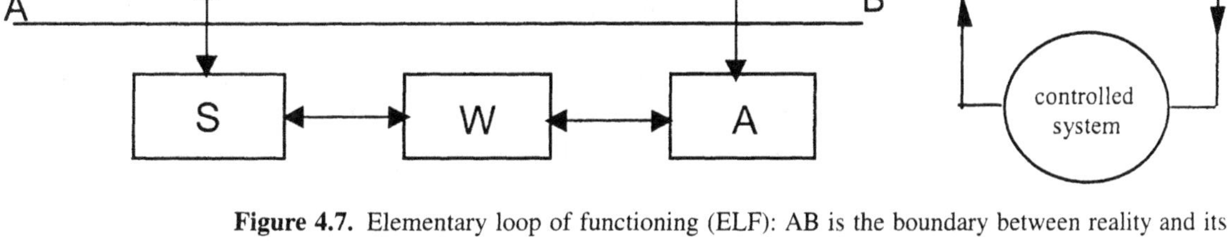

Figure 4.7. Elementary loop of functioning (ELF): AB is the boundary between reality and its representation, G the goal, SP the sensory processing or perception, WM the world model or knowledge representation system, BG the behavior generation, A the actuators, W the world (the subset of interest), and S the sensors.

tion (A) which produces changes in the world. ELF is a simple (and obvious) way of organizing information about systems starting with bacteria and ecological niches and ending with systems of manufacturing and autonomous robots. All systems surrounding us in reality can be represented in the form of ELF.

SP-WM-BG IS CONTROL SYSTEM

To judge the system's processes of functioning, one should visualize the whole loop simultaneously otherwise some important components of the functioning can be omitted (see Figure 4.7*a*). This loop contains two parts: the upper part includes SP, WM, and BG and is regarded as a control system; the lower part includes S, W, and A, and it is regarded as the controlled system.[2] It is senseless to discuss one of them without the other (see Figure 4.7*b*).

S-W-A IS CONTROLLED SYSTEM

In this section we discuss the system without addressing the issue of its hierarchical structure. The following preconditions should be satisfied to provide for functioning of ELF shown in Figure 4.1.

1. A vocabulary is to be developed to describe the world situations. A list of possible situations of interest is to be prepared with descriptions of these situations.
2. The means of receiving information about the world situations are to be explicitly stated. In other words, the sensors S are to be defined (e.g., physical devices or human sources of information). The world situations are to be described (and later represented) in terms of a set of sensors.
3. The signals from sensors S should enter the sensory processing (SP) subsystem (or perception) which contains the algorithms for decoding and organizing sensor information. Here the results of sensing should undergo their primary organization (including grouping and detection of correlations). The primary organization processes must be capable of dealing with the sensor information arriving at the SP input.
4. Primary organization requires knowledge of and communication with the subsystem of the world model (WM). The WM should contain relevant knowledge for interpretation of the SP output and for supporting the decision-making activities. The results of sensory processing are incorporated within or rejected by the knowledge base where the world model is to be held and maintained (the set of knowledge and databases in a variety of representational forms).
5. There should exist a menu of possible goals. The responses for these goal assignments must be discussed. How should we respond to the assigned goal? What are the rules of decision making that apply to the concrete case? After the goal G arrives to the subsystem of behavior generation BG, the latter should perform decision making including task decomposition, planning, and execution. BG should request and receive from WM the subset of knowledge required for the process of decision making. BG should employ WM for modeling and forecasting.
6. The list of actuators is to be prepared for considering all possible alternative responses to the goal assignments. From BG, the decision about behavior should arrive to the actuator A which leads to changes in world W. The world is to be considered as a system that is equivalent to its ontology. The validity of the ontology is to be regularly verified by the *symbol-grounding* operation.

[2]The "control systems, controlled systems" terminology is not always well understood, so it has not been widely accepted by users. This is particularly problematic if the system includes humans (both on the supervising and performing levels). Therefore in this book we try to adhere, where possible, to the terms "decision-making system" (for the control system) and "functioning system" (for the controlled system).

VALUE JUDGMENT CAN BE DISTRIBUTED OVER ELF MODULES

The ELF is a simplified version of the more complete process (see Figure 4.1) which focuses on value judgment. The ELF diagram gives an approximate mirror image of reality and the physical devices (below line AB) and the cognitive activities in the hardware and software (above this line) of the computer system. This means that the controlled system is represented within the control system or the performing system, that is, within the decision-making system. As a result the **BG** modules should be able to employ a direct representation of the **A** module. More precisely, they employ all representations of **A** at all resolution levels. The role of the WM module is to store and maintain these representations and make them available upon request from BG.

In the meantime the set of all actuators $\mathbf{A} = \{\mathbf{a}_i\}$ produces changes in **W**. If these changes can be generalized and explained, we call them "behavior."[3] If no rational interpretation can be given then the changes are categorized as a random motion. Each elementary action **a** is generated by the actuator **A** and produces a change $\Delta\mathbf{w}$ in the world **W**. This change is considered to be a difference between the previous and the subsequent states of the world. It is measured by the sensors **S**. The string of n consecutive state changes $\{\Delta_{w1}, \Delta_{w2}, \ldots, \Delta_{wn}\}$ is called "(a particular) behavior" if a "law" of the string formation can be found. Each elementary change happens during an elementary unit of the time scale accepted for a chosen space scale. Behavior can be defined also in the terms of actions produced that can lead to a different result, since the resulting actions are not necessarily equivalent to the states observed.

Both time and space scales are determined by the resolution chosen to represent the present knowledge. Later we show that a system should be represented at several levels of resolution simultaneously (i.e., at several time and space scales simultaneously.) The need for multiscale representation is determined by the computational complexity of the problem that we face. To deal with complex systems, we aggregate and decompose its components in space and time. This entails the hierarchies of representation and control.

Let us now consider an example of system decomposition linked with the scope of interest.

4.7.2 Primary Decomposition of an ELF

The ELF is defined as a system that is unified by a need to *exist as an entity* while achieving an assigned goal. In pursuit of this goal, the ELF performs tasks that have been developed internally or submitted to it externally. The phenomenon of being focused to achieve a concrete goal is the *unifying factor*, and this is fundamental in our discussion of ELFs and their operation. The ELF can contain subsystems that can be regarded as ELFs too. The nesting of subsystems gives rise to hierarchies of ELFs. We formulate the following statements as postulates:

Postulate 1 (Unity) The fundamental property of ELF is to exist as a goal-seeking entity.

Postulate 2 (Recursion) Any ELF can be a part of another ELF (in which the ELF under consideration is nested) and can be a composition of other ELFs (which are nested in the ELF under consideration).

[3]Whereas in reality one can call "behavior" any kind of motion, "motion" is a temporal change of states. But why introduce the idea of "behavior" instead of just saying "motion"? The term "behavior" is usually associated with the ability to characterize it as belonging to some general *type of behavior* satisfying some *purpose*. Otherwise, it would be more meaningful to use the work "motion" or "activities." It is linked also with our anthropomorphic tendencies to describe intelligent systems.

ONE SYSTEM, TWO DIFFERENT ELFs

Postulate 3 (Existential Duality) Each ELF consists of two parts (also ELFs) that are vitally required for its existence. The first handles goal-directed functioning, GDF, while the second handles the regular subsistence functioning, RSF (including maintenance) of the first one. They are denoted ELF-GDF and ELF-RSF correspondingly.

Corollaries

1. Any ELF-GDF should be considered together with a hierarchy of goals to be pursued in the external environment.
2. An ELF-RSF belongs to the environment in which the ELF-GDF is functioning.
3. Any ELF-RSF should be considered together with a hierarchy of goals of maintenance to be performed internally.
4. An ELF-GDF is a part of environment for the ELF-RSF.
5. Both the ELF-GDF and ELF-RSF may form nested systems.

The ELF represents all activities of the system that can be generalized for both the ELF-GDF and the ELF-RSF. Let us look more closely at the joint functioning of an ELF-GDF and an ELF-RSF within the ELF. Any autonomous ELF has two groups of functioning: goal-directed functioning (GDF), and regular subsistence functioning (RSF) which can be described by using separate loops of functioning.

The GDF corresponds to the main goal-oriented function of a system, such as the manufacturing process or an internal combustion engine. The RSF corresponds to the maintenance system of the manufacturing plant. Let us consider three types of communication: W–GDF, W–RSF, and GDF–RSF. This communication is conducted in languages that do not allow for a fully adequate interpretation of the messages. Multiple stochastic functions acting within the environment (friction, backlash, temperature changes, etc.) are the sources of error. Therefore, when the GDF and RSF communicate with actuators A, the resulting behavior generation often differs from the desired effect. This is one of the reasons for the interest in feedback compensation.

A question to be raised here is how large to make the vocabularies of languages for automata representation. Vocabulary size obviously affects the transition size and output functions of the sequential machine. The answer depends on the nature of the ELF, which is capable of providing communication among its modules components. Prior experiences are used to find languages for this communication. Ultimately the ELF will become a learning machine, whose learning is understood as the process of constantly generating new rules and concepts using experiences. The capacity to learn will affect all control subsystems of the ELF as the controlled system itself is better understood (and/or undergoes changes). Learning is not simply a recording of all processes that take place, nor is it just memorized experiences. Rather, learning is the development of a world model and rules of sensory processing and behavior generation by using a generalization of experiences [44].

This potential automaton is not the standard automaton with a finite vocabulary. It has open list vocabularies, transient, and output functions. All of these components of the automaton will change continually as the architecture is equipped with learning. The way this will be achieved is determined by the architecture of NIST-RCS.

4.7.3 Hierarchies of ELFs: the Essence of NIST-RCS

In all these descriptions we talk about the system as a whole, not about a particular level of its hierarchy. It is important to realize that we must deal with an assemblage of processes:

- Processes of the overall actuation that can be decomposed into a hierarchy of actions, or that have actuators physically belonging to the different levels of the hierarchy, or both.
- Processes of the overall measurement that can be decomposed or can be performed by sensors physically belonging to different levels of the hierarchy, or both.

As we consider the hierarchy of the system's components, we can represent its processes at different temporal and spatial scales. We see that the levels of resolution have loops of control with different bandwidth intervals, and that equations can be written with different sets of eigenvalues.[4] Each element of the control loop can be represented as a nested module.

The multiactuator ELF is shown in Figure 4.8. The goal G arrives into a BG module ("behavior generation") where the solution of the problem is to be found (we will explain in detail how this occurs in Chapter 8). The solution is transformed into a set of action tasks which is submitted to the set of actuators ({A}) of the subsystems. All subsystems operate on the set of their worlds ({W}). The latter are equipped by the set of sensors {S}. The signals from the sensors are integrated within the perception module (P), which updates the world representation contained in the module of the world model (WM). The best sensor {S} integration, the best model {WM}, and the best solution obtained in {BG} are selected with the help of the value judgment module (VJ). The boundary between reality and its representation is noted by the line *AB*.

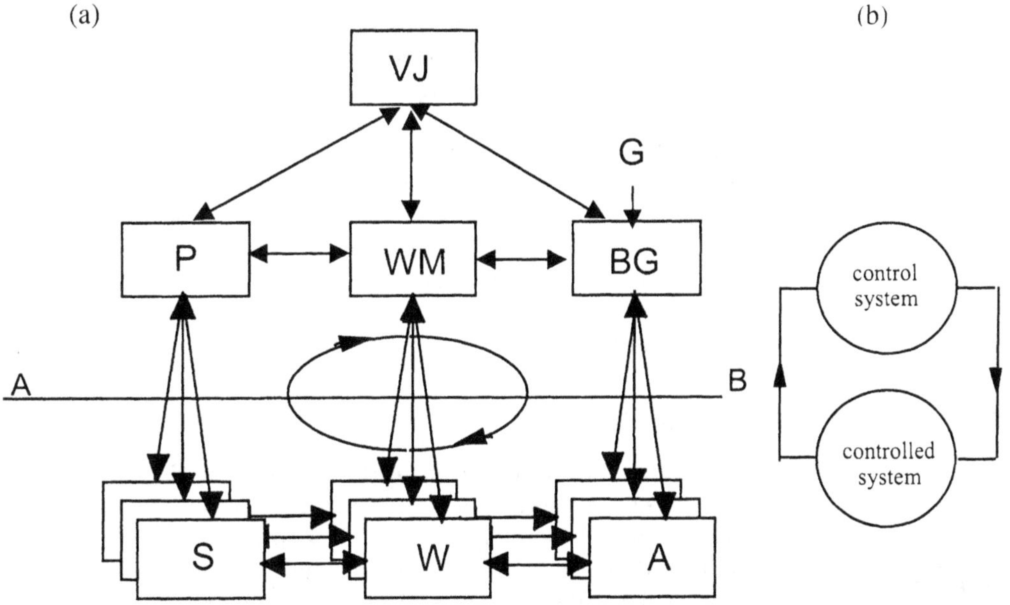

Figure 4.8. An elementary loop of functioning with multiple controlled subsystems.

[4]This feature of multiresolutional systems is similar to the "multirate control" feature.

ONE SET OF SP-WM-BG CAN HANDLE MANY SETS OF S-W-A

The arrows between the module WM and the set {W} indicate a virtual correspondence between the real world, W, and its representation in the world model, WM. These links are not channels of communication; they exist virtually. The knowledge in the WM is the system's best estimate of the real state of the world (W). The virtual links represent a noisy channel through which the system perceives the world (W). To the extent that the perception is a correct and relevant representation of reality, behavior is more likely to be successful in achieving the goal. If the virtual perception is not completely accurate, behavior is less likely to be successful.

Figure 4.8*a* and *b* describes the elementary loop of functioning as consisting of two domains: the control system and the controlled system (Figure 4.8*b*). The number of controlled subsystems is not given. Once the control and controlled subsystems are separated from each other, the levels of RCS can be separated too. Each will be considered a node of the NIST-RCS architecture.

The line *AB* is also a divider between the domains of that which is to be controlled and that which is the system (or systems) of control. It denotes the fact that the modules above this line are components of the control system. The designer is free to change them at the design stage. The control engineer, together with programmers, are free to put different control algorithms and different knowledge representation in this part of the system. The modules below this line are the subsystems being controlled. Signals of several sensors are fused together within one P, and one BG can control several actuators. In the RCS hierarchy the lower level of the control system can be regarded as part of the system being controlled.

EACH LOWER RESOLUTION LEVEL OF SP-WM-BG CAN HANDLE MANY S-W-A OF HIGHER RESOLUTION

This simplified structure leads to the multilevel structure shown in Figure 4.9 where nine controlled units are unified in three groups, which can be considered as three machines each equipped with three electric motors.

The three machines together form a manufacturing cell. The functioning of this cell requires a control system (which is the upper level of control). This control system receives a directive (or goal) that specifies what should be manufactured and how the results should be distributed in time (for the total horizon of time, T_{cell}). The upper-level control system generates some general control assignments to the machines that compose the cell. Each machine is assigned what to manufacture and how to interact with the others within some horizon of time T_{mi}, where $i = 1, 2, 3$ (for the total horizon of time $T_{mi} < T_{\text{cell}}$). Each machine has its own control system which generates detailed control assignments to their actuators. These (detailed) control assignments prescribe the things to be done by each actuator. Often it is impossible to use one control system for three controlled actuators; it is possible only in very simple cases. In Figure 4.9*b* the system is simplified to show three integrated control nodes of the architecture that control a much larger number of controlled subsystems (the number of controlled subsystems is not important) with clearly determined levels.

4.7.4 Integrated NIST-RCS Modules

We introduced an integrated structure earlier when we considered a multiplicity of controlled systems to be a single integrated control system, or a set of controllers, forming a control system at one level. Figure 2.6*b* gives an example of this integration.

WHAT IS A CONTROL SYSTEM FOR THE LEVEL BELOW MIGHT BE A CONTROLLED UNIT FOR THE LEVEL ABOVE

From Figures 4.8 and especially 4.9 we can see that presenting the ELF diagrams in the form of detailed graphs can be cumbersome, since the number of levels can grow and extensively branch. This is why at each level of the hierarchy, we consider the whole system below it as the "controlled system" and regard that single level as the "reality to be controlled." In Figure 4.10*a* we show that the whole system at the "input-output

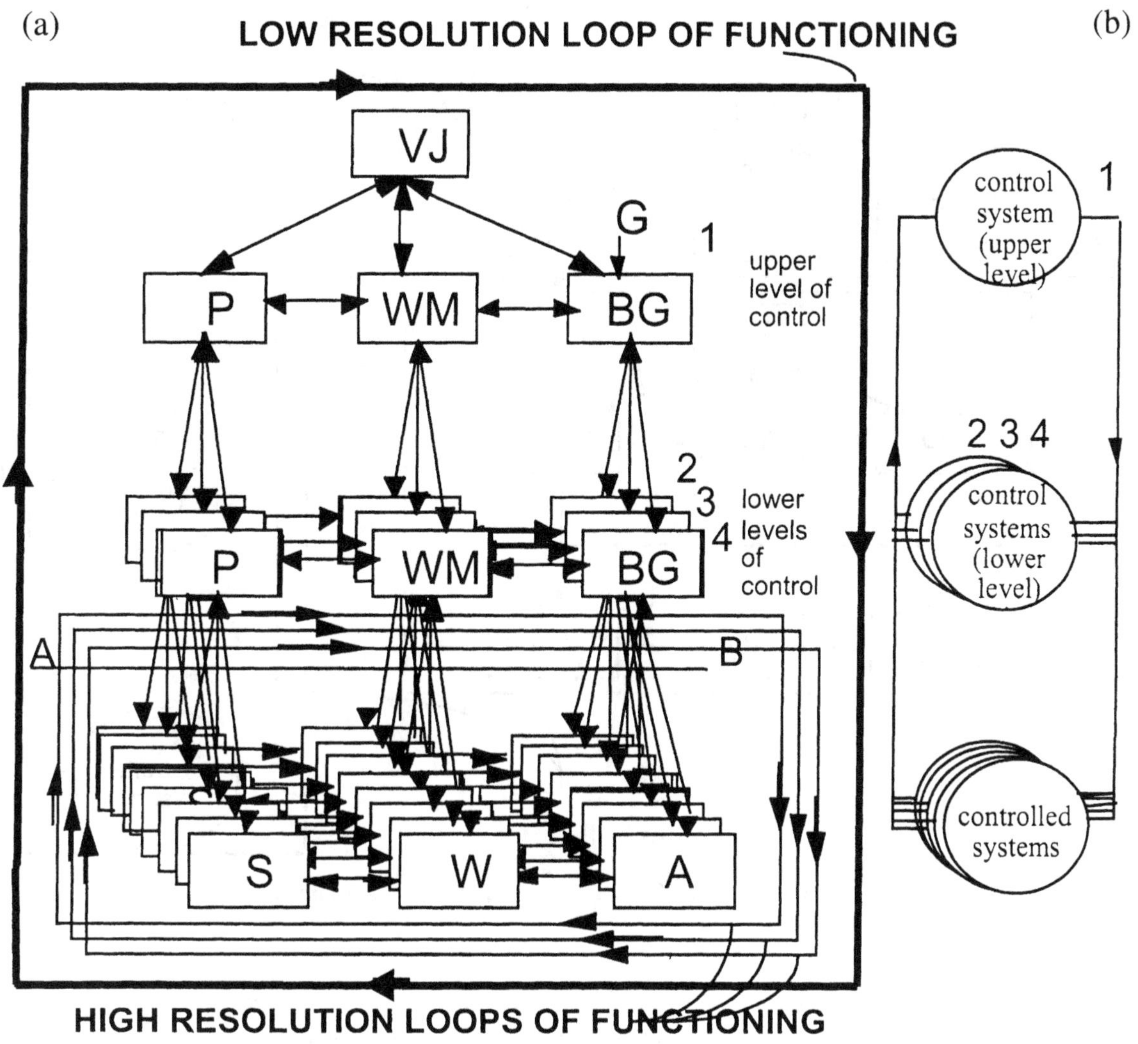

Figure 4.9. A two-level ELF, or NIST-RCS, with two control levels (nodes).

terminals" of the upper level can be considered a controlled system, as indicated by the rectangle. Similar integration is performed in Figure 4.10*b*.

The dividing line *AB* in Figure 4.10 works only for the first level of the ELF hierarchy; the second level has its own dividing line $A_V B_V$ which exists independently (in addition to *AB*).

Even at comparatively low branching (as in Figure 4.8, where the branching is 3) the graphical representation can be cumbersome. Therefore we can group the modules that emerge after task decomposition as shown in Figure 4.10. This is a form of shorthand we will use in our discussion of integrated modules, which we will treat as "group modules" or "vector modules."

A vector module can be interpreted as an integrated entity at a resolution level that does not preclude the designer and/or user from a successful pursuit of task decomposition. This is done by manipulating the knowledge transformation sets (K) and by using linguistic or vector algebra techniques. The hierarchy of task decomposition is a linguistic hierarchy, which demonstrates the relation of inclusion (nesting) that exists among

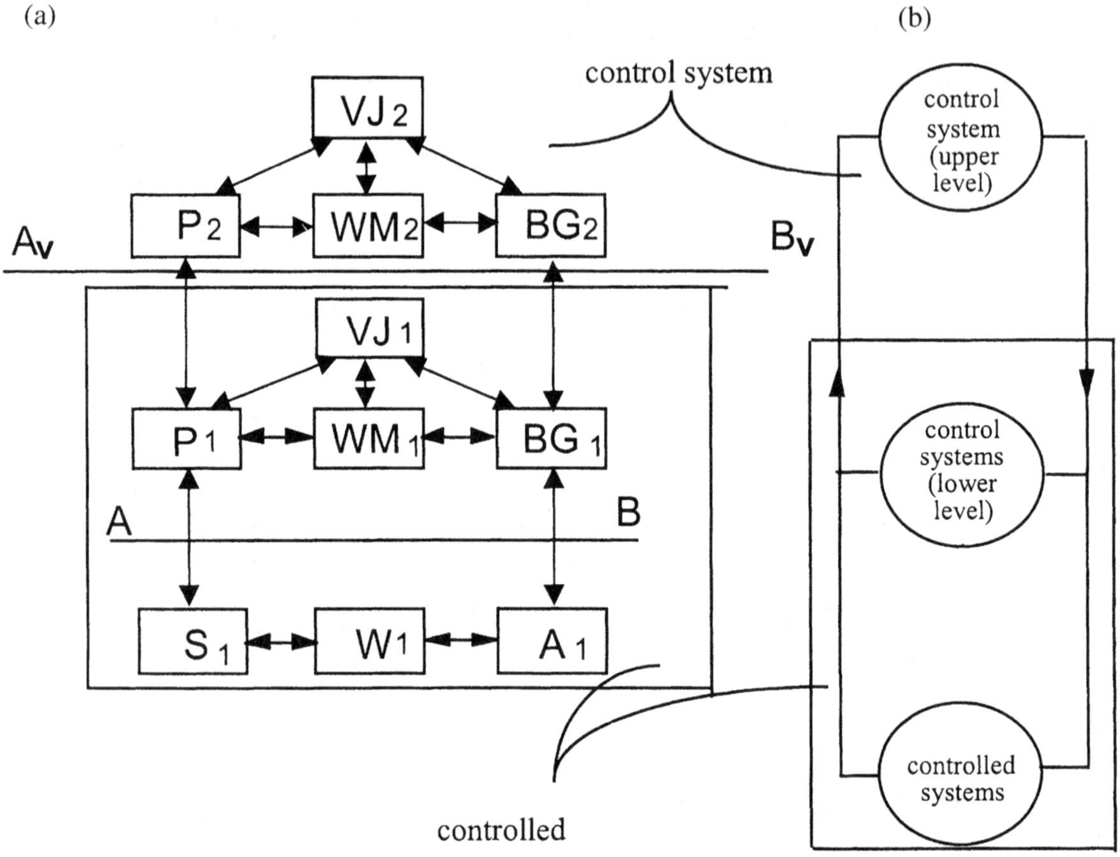

Figure 4.10. Modular two-level NIST-RCS structure in a compact form.

all corresponding sets K from different levels of resolution. This hierarchy $H(K)$ is partially induced by the physical hierarchy of the structural system decomposition. Its construction as an aggregate of subsystems represents the prior design efforts including the efforts to increase its efficiency by using different alternatives of task decomposition (e.g., experienced in the past, or intentionally synthesized).

4.8 AGENT-BASED REFERENCE ARCHITECTURES

4.8.1 Elements of Intelligent Software

The following key ideas are considered the domain of application for the development of software systems with elements of autonomy such as intelligent virtual enterprises and intelligent autonomous robots [48–50].

4.1 All agents are software units that are situated at a particular host and have inputs, outputs, and sets of rules. The software unit may be equipped with devices for interaction with a simulated "virtual world" (as shown in Figure 1.1*a*) or the physical world (as shown in Figure 1.1*b*).

- Agents, as software units ("bots"), may be equipped with hardware devices that replicate their output behavior in the real world and generate inputs depending on what is going on in the real world (customer invoices, sensor outputs).
- Hardware-equipped agents (e.g., ro-bots) are typically driven by a controller; if the controller contains a computer, the devices are embodied primarily in the software unit driving the computer. This creates an opportunity to decouple the software unit from the hardware unit; the essence of the "agent" is actually situated in the software unit.
- Software agents that can change the host are called "mobile agents" (usually, they are written in JAVA; examples are Aglets, Voyager, Mole, JATLite, Concordia).
- Agents that are spatially located in different physical locations, or located at different hosts while functioning as part of an external entity, are called distributed agents; their architectures are called "distributed architectures."
- Generally, agents act only reactively (the look-up tables determine the property of reactivity). Even if the agents are equipped with planners, the reaction is still considered a "reactivity to anticipation."

4.2 Intelligent systems are systems that include, or are composed of, intelligent agents. The concept of "intelligent agent" is introduced to distinguish these agents from regular agents that produce preprogrammed and/or reactive behavior. "Clients" are intelligent agents too; they react with the intelligent agent and affect its cost function.

- Intelligent agents are entities that are endowed with a goal, and as a result of having a goal, produce behavior that benefits themselves, or their clients that can be intelligent agents, too.
- Intelligent agents are equipped with extensive tools of "self-organization" and learning, can deal with situations with a high level of uncertainty, and function successfully under the circumstances.
- The concept of "benefit" is introduced to declare the existence of the cost-function associated with behavior, depending on environment, and built-in mechanisms that determine, using this cost-function, the alternatives during the decision-making process.
- Goal is a description of the state or states that should be achieved by the intelligent agent by generating output changes on the way, depending on the situation of the host, the prehistory of input-output couples of the agent under consideration, of other agents, clients, and so forth.
- Because relational networks develop into interactive and communicative networks, the sets of agents develop into *coalitions of agents*.

4.3 As we have already mentioned, the mechanisms of self-organization lead to the formation of *multiscale coalitions of agents*.

4.8.2 Functioning of the Agent-Based Level

First we have to establish what can happen at a level. Agents can be of the following types:

1. A simple input-output automaton with a fixed number of IF-THEN rules of transition and output mappings and single alternative of actions, or *simple agent* (SA).
2. A simple input-output automaton with a fixed number of IF-THEN rules of transition and output mappings and the ability to obtain from a list the alternatives of action, and the need to *select* one, or *simple agent with selector* (SAS).
3. A simple input-output automaton with a fixed number of IF-THEN rules of transition and output mappings and the ability to combine them into various sequences of action, and the need to *select* one, or *simple agent with combiner* (SAC).

Each of the three agents can be equipped with a capability to infer additional rules to enhance the transition function and output function look-up tables. This creates a family of learning agents.

4. A simple agent that can learn its experiences and infer additional rules, or *simple learning agent* (SLA).
5. A simple agent with selector that can learn its experiences and infer additional rules, or *simple learning agent with selector* (SLAS).
6. A simple agent with combiner that can learn its experiences and infer additional rules, or *simple learning agent with combiner* (SLAC).

As the number of rules grows, one can start searching for groups with similarities. This leads to the creation of the lower-resolution rules. Certainly, this can be done only via generalization of both antecedent and consequent parts of these rules, i.e., through the creation of new (generalized) input and output vocabularies and the new (generalized) list of states. In other words, this leads to the construction of a lower-resolution automata, or a new level of resolution (scale, granularity). An interesting feature of generalization is that it can be performed by a GFS-type operator that is equivalent to coalition formation. The latter presumes communication between the agents at the level.

7. A simple learning agent that can cluster similar rules and generalize on prior experiences, thus creating subsequent levels of lower resolution, or *generalizing learning agent* (GLA).
8. A simple learning agent with selector that clusters similar rules and generalize on prior experiences, thus creating subsequent levels of lower resolution, or *generalizing learning agent with selector* (GLAS).
9. A simple agent with combiner that can cluster similar rules and generalize on prior experiences, thus creating subsequent levels of low resolution, or *generalizing learning agent with combiner* (GLAC).

Generalizing learning agents are already multiscale agents: indeed, if the agent can generalize and build the second level out of the first, it is able to construct the third, fourth, and so on (see Chapter 10). With multiscale agents, the decision-making process is a multiresolutional one (top-down), while the process of generalizing the sensor information will progress similarly in the bottom-up direction. Another feature that further develops the agent architectures is the ability to simulate the alternatives of output before issuing it. Obviously, in the single level agent, this will lead to a delay in responses, therefore, this feature should be used at the lower levels of resolution where the rules are more general, and the "ticks" of the clock has lower frequency. This gives extra time for simulation preceding the decision making, which becomes a rudimentary "planning."

10. A generalized learning agent that is equipped with the ability to stimulate the alternatives of decision before submitting them at the output, or *multiscale planning agent* (MPA).
11. A generalizing learning agent with selector that is equipped with the ability to simulate the alternatives of decision before submitting them at the output, or *multiscale planning agent with selector* (MPAS).
12. A generalizing agent with combiner that is equipped with the ability to simulate the alternatives of decision before submitting them at the output, or *multiscale planning agent with combiner* (MPAC).

Finally the planning feature can be made more efficient by the ability to predict the processes by identifying their inner models. This gives additional kinds of agent architectures.

13. A multiscale planning agent that is equipped with the ability to predict the regular processes to improve the planing, or *multiscale predicting and planning agent* (MPPA).
14. A multiscale planning agent with selector that is equipped with the ability to predict the regular processes to improve the planning, or *multiscale predicting and planning agent with selector* (MPPAS).
15. A multiscale planning agent with combiner that is equipped with the ability to predict the regular processes to improve the planning, or *multiscale predicting and planning agent with combiner* (MPPAC).

It is important to state that representation is presumed in all types of agents: the very fact of having rules is equivalent to having a system of world representation. Later we will see (Chapter 10) that from a system of rules, a system of concepts can be deduced, and together both systems constitute the system of knowledge representation with all kinds of world model.

Recently a category of agents was introduced called *SciAgents*. This is one of the agent-based approaches to building multidisciplinary problem solving environments that are naturally parallel and highly scalable, and is convenient for a distributed high-performance computer environment. This environment allows for simulation of physical processes and design optimization using knowledge and computational models from multiple disciplines in science and engineering that are part of high performance computing and communication (HPCC) systems. The evolution of the Internet into the global information infrastructure (GII) makes it a rational tool for network scientific computing (NSC). NSC enables scientists to address the class of complex problems within the accelerated strategic computing initiative (ASCI) from DOE. In this paradigm, the design process operates at the scale of the whole physical system with a large number of components that have different shapes, obey different physical laws and manufacturing constraints, and interact with each other through geometric and physical interfaces. SciAgents are guided by this vision.

Such systems would allow the reuse and combine legacy software to solve the specific problem at hand and build a different MPSE for each composite problem they solve. One of the major goals of the MPSE concept is to allow low-cost and less time-consuming methods of building the software to simulate a complex mathematical model of physical processes. While working within this vision of problem-solving processes, this architecture requires the solvers to behave like agents (e.g., understand agent languages, use them to communicate data to other agents). This approach provides an agent wrapper for PSEs and other software modules, which takes care of the interaction with the other agents and with the other aspects of agent behavior.

The categories of agents that have been introduced determine the broad spectrum of architectures. The analysis of possibilities allows us to arrive at the following conclusions.

1. Architectures based on agents are equivalent to the NIST-RCS architecture. The existing applications of NIST-RCS are within the described spectrum of agent architectures. The terminology applied to NIST-RCS and the terminology of multiscale communities of agent are complementary. All other cognitive architectures, including the "subsumption architecture," are particular cases of NIST-RCS.
2. Elements of NIST-RCS are called "agents;" they can be of any type (from 1 through 13 in the list above).
3. Top-down functioning of NIST-RCS is called "global decomposition" and/or "tasks distribution."
4. Bottom-up functioning of NIST-RCS is called "goal formation" and/or "tasks formation." Both goal and tasks formation will be discussed later as part of the learning process (Chapter 10).
5. The highest potential of both intelligence and autonomy can be expected in the group of learning agents.

REFERENCES

[1] L. D. Erman, F. Hayes-Roth, V. R. Lesser, and D. R. Eddy, Hearsay-II speech understanding system: integrating knowledge to resolve uncertainty, *Computer Survey*, Vol. 23, June 1980, pp. 213–253.

[2] J. E. Laird, A. Newell, and P. Rosenbloom, SOAR: an architecture for general intelligence, *Artificial Intelligence*, Vol. 33, 1987, pp. 1–64.

[3] Honeywell Inc., *Intelligent Task Automation Interim Technical Report II-4*, Dec. 1987.

[4] J. Lowerie et al., Autonomous land vehicle, *Annual Report, ETL-0413*, Martin Marietta Denver Aerospace, July 1986.

[5] D. Smith and M. Broadwell, Plan coordination in support of expert systems integration, *Knowledge-Based Planning Workshop Proceedings*, Austin, TX, December 1987.

[6] J. R. Greenwood, G. Stachnick, and H. W. Kaye, A procedural reasoning system for army maneuver planning, *Knowledge-Based Planning Workshop Proceedings*, Austin, TX, December 1987.

[7] A. J. Barbera, J. S. Albus, M. L. Fitzgerald, and L. S. Haynes, RCS: The NBS real-time control system, *Proceedings Robots 8 Conference and Exposition*, Detroit, MI, June 1984.

[8] R. Brooks, A robust layered control system for a mobile robot, *IEEE Journal of Robotics and Automation*, Vol. RA-2, 1, March 1986.

[9] G. N. Saridis, Foundations of the theory of intelligent controls, *IEEE Workshop on Intelligent Control*, 1985.

[10] A. Meystel, Intelligent control in robotics, *Journal of Robotic Systems*, 1988.

[11] J. A. Simpson, R. J. Hocken, and J. S. Albus, The automated manufacturing research facility of the National Bureau of Standards, *Journal of Manufacturing Systems*, Vol. 1, No. 1, 1983.

[12] J. S. Albus, C. McLean, A. J. Barbera, and M. L. Fitzgerald, An architecture for real-time sensory–interactive control of robots in a manufacturing environment, *4th IFAC/IFIP Symposium on Information Control Problems in Manufacturing Technology*, Gaithersburg, MD, October 1982.

[13] J. S. Albus, System description and design architecture for multiple autonomous undersea vehicles, *National Institute of Standards and Technology Technical Report 1251*, Gaithersburg, MD, September 1988.

[14] J. S. Albus, H. G. McCain, and R. Lumia, NASA/NBS standard reference model for telerobot control system architecture (NASREM), *National Institute of Standards and Technology Technical Report 1235*, Gaithersburg, MD, 1989.

[15] A. Meystel, M. Montgomery, and D. Gaw, Navigation algorithm for a nested hierarchical system of robot path planning among polyhedral obstacles, *Proc. IEEE Int'l Conf. on Robotics and Automation*, Raleigh, NC, 1987, pp. 1616–1622.

[16] C. Isik and A. Meystel, Pilot level of a hierarchical controller for an unmanned mobile robot, *IEEE J. of Robotics & Automation*, Vol. 4, No. 3, 1988, pp. 244–255.

[17] A. Meystel, *Autonomous Mobile Robots: Vehicles with Cognitive Control*, World Scientific Publ., Singapore, 1991, p. 580.

[18] D. Apelian and A. Meystel, Knowledge based control of material processing: challenges and opportunities for the third millenium, in *Metallurgical Processes for the Year 2000* (H. Y. Sohn and E. S. Geskin, eds.), TMS, Warrendale, PA, 1989.

[19] C. Corson, A. Meystel, F. Otsu, and S. Uzzaman, Semiotic multiresolutional analysis of a power plant, ibid. in *Architectures for Semiotic Modeling and Situation Analysis in Large Complex Systems*, Proc. of the 1995 ISIC Workshop, Monterey, CA, 1995, pp. 401–405.

[20] A. Waxman et al., A visual navigation system for autonomous land vehicles, *IEEE Journal of Robotics and Automation*, Vol. 3, No. 2, 1987, pp. 124–141.

[21] B. Hayes-Roth, A blackboard architecture for control, *Artificial Intelligence*, 1985, pp. 252–321.

[22] G. A. Miller, The magical number seven, plus or minus two: some limits on our capacity for processing information, *The Psychological Review*, Vol. 63, 1956, pp. 71–97.

[23] A. Meystel, Theoretical foundations of planning and navigation for autonomous robots, *International Journal of Intelligent Systems*, Vol. 2, 1987, pp. 73–128.

[24] M. Minsky, A framework for representing knowledge, in *The Psychology of Computer Vision* (P. Winston, ed.), McGraw-Hill, New York, 1975, pp. 211–277.

[25] J. Albus and A. Meystel, A reference model architecture for design and implementation of semiotic control in large and complex systems, in *Architectures for Semiotic Modeling and Situation Analysis in Large Complex Systems* (J. Albus, A. Meystel, D. Pospelov, and R. Reader, eds.), Proc. of 1995 ISIC Workshop, Monterey, CA, 1995, pp. 33–45.

[26] E. D. Sacerdoti, *A Structure for Plans and Behavior*, Elsevier, New York, 1977.

[27] R. C. Schank and R. P. Abelson, *Scripts Plans Goals and Understanding*, Lawrence Erlbaum Associates, Hillsdale, NJ, 1977.

[28] D. M. Lyons and M. A. Arbib, Formal model of distributed computation sensory based robot control, *IEEE Journal of Robotics and Automation in Review*, 1988.

[29] D. W. Payton, Internalized plans: a representation for action resources, *Robotics and Autonomous Systems*, Vol. 6, 1990, pp. 89–103.

[30] A. Sathi and M. Fox, Constraint-directed negotiation of resource reallocations, CMU-RI-TR-89-12, *Carnegie Mellon Robotics Institute Technical Report*, March 1989.

[31] V. B. Brooks, *The Neural Basis of Motor Control*, Oxford University Press, 1986.

[32] J. Piaget, *The Origins of Intelligence in Children*, International Universities Press, New York, 1952.

[33] J. S. Albus, A theory of cerebellar function, *Mathematical Biosciences*, Vol. 10, 1971, pp. 25–61.

[34] P. D. MacLean, *A Triune Concept of the Brain and Behavior*, University of Toronto Press, Toronto, 1973.

[35] J. S. Albus, Mechanisms of planning and problem solving in the brain, *Math. Biosciences*, Vol. 45, 1979, pp. 247–293.

[36] S. Grossberg (ed.), *Neural Networks and Natural Intelligence*, Bradford Books, MIT Press, 1988.

[37] J. S. Albus, The cerebellum: a substrate for list-processing in the brain, in *Cybernetics, Artificial Intelligence and Ecology* (H. W. Robinson and D. E. Knight, eds.), Spartan Books, 1972.

[38] J. J. Hopfield, Neural networks and physical systems with emergent collective computational abilities, *Proceedings National Academy of Sciences*, Vol. 79, 1982, pp. 2554–2558.

[39] B. Widrow and R. Winter, Neural nets for adaptive filtering and adaptive pattern recognition, *Computer*, Vol. 21, No. 3, 1988.

[40] M. Minsky and S. Papert, *An Introduction to Computational Geometry*, MIT Press, Cambridge, MA, 1969.

[41] M. Ito, *The Cerebellum and Neuronal Control*, Chapter 10, Raven Press, NY, 1984.

[42] H. M. Huang, R. Hira, and R. Quintero, Submarine maneuvering system demonstration based on the NIST real-time control system reference model, *Proc. of the 8th IEEE Int'l Symposium on Intelligent Control*, Chicago, IL, 1993.

[43] F. Proctor and J. Michaloski, Enhanced machine controller architecture overview, *NISTIR 5331*, NIST, Gaithersburg, MD, December 1993.

[44] M. Nashman, W. Rippey, T. Hong, and M. Herman, An integrated vision-touch probe system for dimensional inspection tasks, *NISTIR 5678*, June 1995.

[45] K. Stouffer, J. Michaloski, R. Russell, and F. Proctor, ADACS—an automated system for part finishing, *NISTIR 5171*, NIST, Gaithersburg, MD, April 1993.

[46] R. Quintero and A. J. Barbera, A software template approach to building complex large-scale intelligent control systems, *Proc. of the 8th IEEE Int'l Symposium on Intelligent Control*, Chicago, IL, 1993.

[47] A. Newell, *Unified Thoeries of Cognition*, Harvard U. Press, Cambridge, MA, 1990.

[48] K. Sycara, *Distributed Intelligent Agents*, IEEE Expert, 1996.

[49] A. Haddadi, *Communication and Cooperation in Agent Systems: A Pragmatic Theory*, Springer Verlag, Lecture Notes in Computer Science, 1996.

[50] D. McKay, J. Pastor, R. McEntire, and T. Finin, An architecture for information agents, in *Advanced Planning Technology* (A. Tate, ed.), The AAAI Press, Menlo Park, CA, 1996.

PROBLEMS

4.1. Choose one system from list A and one from list B:

List A: manufacturing plant, autonomous mobile robot, army regiment, human being.

List B: vending machine, programmable disk player, personal computer (e.g., IBM PC), automobile.

Represent each of these systems as an ELF. Decompose the system into subsystems, including sensors, sensory processing, world model, behavior generation, value judgment, and actuators. Try to find at least two levels of hierarchy in each of the subsystems. Draaw a diagram of a hierarchical ELF. What is the difference between ELF hierarchies constructed for the systems of your choice A and B?

4.2. Draw the hierarchies obtained in Problem 4.1 in a treelike form similar to the one shown in Figure 4.3, or 4.8, or 4.9, depending on the number of levels you are dealing with. Describe functioning of the system that you have obtained at each level of the hierarchy. Specify for each of the subsystems its input and output variables. Demonstrate the phenomenon of nestedness for SP, WM, and BG.

4.3. For the description of functioning that you have received as a result of solving Problem 4.2, find a list of variables for each level of resolution. Organize them in the following groups:

(a) variables of the external world that are perceived by the sensors of the system,

(b) variables of sensory processing that are obtained after transformation and organization of the perceived variables,

(c) variables that are used in the world model,

(d) variables that are used for behavior generation,

(e) variables that are used for describing actuation.

To organize these groups of variables, it can be recommended to use input and output variables that you received as a result of solving Problem 4.2.

4.4. Construct a system of the top-down task decomposition and build a timing diagram for it similar to the one shown in Figure 4.5.

4.5. Use the same examples of systems A and B, analyze the results of solving Problem 4.2, and determine what can be called "control system" and "controlled system" as shown in Figures 4.7 and 4.8.

4.6. Unify the "control system" and "controlled system" of the highest level of resolution into a "virtual controlled system" as perceived from the lowest level of resolution (as shown in Figure 4.9). What is the merit of such construction of a virtual level?

CHAPTER 5

MOTIVATIONS, GOALS, AND VALUE JUDGMENT

5.1 INTERNAL NEEDS VERSUS EXTERNAL GOALS

Intention is considered an intrinsic property of an intelligent agent. A *strive to the goal* is in its core. If we have a behavior, we can figure out what is its implicit or explicit goal.

Since intelligent systems are goal-oriented, as we saw in Chapter 4, their functioning requires them to know, at each step, what is better and what is worse. The concepts of motivation, goals, and value judgment are treated always within the framework of living creatures, namely, human beings. But we will try and determine how these concepts can fit into a more general paradigm of intelligent systems, which includes not only natural systems but the constructed ones too. This chapter provides an introductory exploration of this issue [1–4]. This will give us more freedom in the subsequent parts of this book.

5.1.1 Neurophysiological Models

Instinctive Behavior

Instinctive models are based on implications of cause and effect and are constructed if and when possible. To call a behavior *instinctive* is straightforward and therefore appealing. Instinctive behaviors are easier to explain, and they probably play a substantial role in our understanding of motivational mechanisms.

DEFINITION OF INSTINCT

An *instinct* is an involuntary response by an animal to an external stimulus, resulting in a predictable and relatively fixed behavior pattern.

The simplest form of instinctive behavior is reflex action, or the response of an agent to a stimulus (see Chapter 3). Many living creatures are proven to be governed by chain reflexes. In millipedes, for example, the motion of one leg stimulates the motion of another. Even more complicated living creatures have recognizable models of "gait" in walking and running. This is very appealing for a programmer or for an engineer developing machine automation with fixed action patterns in which several body systems participate. Some of these fixed patterns may lead to a goal, but others may have several functions [5]. This can be simulated by an automaton, although the input and output vocabularies as well as the transition and output functions are often complicated [6].

DEFINITION OF REFLEX

A *reflex* in biology is a type of action consisting of comparatively simple segments of behavior that usually occur as direct and immediate responses to particular stimuli uniquely correlated with them.

Both instincts and reflexes are automatisms of various scales and therefore can be considered agents working at different levels of resolution. Reflex actions have widespread occurrence among complex animals. Many reflexes of mammals appear to be innate. They are inherited and are the common property of the species, and often of the genus. They include such simple acts as swallowing, the blink reflex, the knee jerk, and the scratch reflex, stepping, standing, and others. More complex patterns of coordinated muscular actions can be synthesized out of elementary reflex-related motion. Reflexes are at the core of much instinctive behavior in animals [7].

Humans have many innate reflexes which are directed toward optimization of the overall behavior. Reflexes drive the body's distance receptors (the eye and the ear); they are actively involved into orienting parts of the body in spatial relation to each other. Apparently intelligent systems are supposed to have a rich network of "reflexes." The following innate reflexes can be listed just for the eyes:

1. Paired shifting of the eyeballs, coordinated with turning of the head, to vary the focus of attention in the field of vision.
2. Contraction of the intraocular muscles to adjust the focus of the retina for the viewing of near or far objects.
3. Constriction of the pupil of the eye depending on illumination of the scene.
4. Blinking due to intense light or touching of the cornea.

This list gives an idea of how many innate reflexes can and should be built-in within the intelligent systems such as autonomous mobile robots just at the stage of design.

REFLEX ARC

In its simplest form, a reflex is viewed as a result of the existence of a *reflex arc* (see Chapter 3). The reflex arc consists of the sensory-nerve cell (or receptor) that receives the stimulation which is connected to another nerve cell that activates the muscle cell (or effector) which performs the reflex action eventually. The reflex arc contains also additional nerve cells that communicate with other parts of the body (beyond the receptor and effector). As a result of the integrative action of the nervous system in higher animals, the behavior of such organisms is more than the simple sum of their reflexes; it is a unitary whole that exhibits coordination among many individual reflexes and is characterized not by inherited, stereotyped set of stored *automatisms*, but by adaptable complex behavior.

Many automatic, unconditioned reflexes can thus be modified by conditioning of reflex responses (see Chapter 10). The experiments of I. Pavlov (1849–1936), for example, showed that if an animal salivates at the sight of food while another stimulus, such as the sound of a bell, occurs simultaneously, the sound alone can induce salivation after several trials. The animal's behavior is no longer limited by fixed, inherited reflex arcs but can be modified by experience and exposure to an unlimited number of stimuli.

REWARD AND PUNISHMENT

The reflex concept led to premature attempts to develop a psychology based on reflexes. Pavlov's results were understood by many as foundation of a new domain: physiology of behavior. The behaviorists E. R. Guthrie, C. L. Hull, and B. F. Skinner proposed principles to explain all actions as conditioned or learned responses to external and internal stimuli [8–10]. These principles were based in part on earlier reflex notions and upon the model of Pavlov's conditioned reflex. They alluded to an ability to shape the emergence of reflexes by repetitive rewards and punishments.

Gradually it became clear that the reflex linkage between stimulus and response is not as simple as was implied. The use of the conditioned reflex as a model for learning in classical conditioning experiments isolated important components of the total learning process in intelligent systems. Alone it is incomplete and cannot properly analyze and persuasively explain the complex interactions characteristic for the behavior of intelligent systems (see Chapter 10). The need in analysis of motivations became clear.

Motivational research has also progressed through discoveries made in the field of physiology. The discovery of separate nerve fibers for sensory and motor information, first suspected by the Greek physician Galen and separately confirmed by the English anatomist Sir Charles Bell in 1811 and the French physiologist François Magendie in 1822, led naturally to the development of the stimulus–response approach to motivation, which has become fundamental to the field.

Agents with Fixed Patterns of Behavior

Agents with fixed action patterns may be modified by learning processes. An example of such agent is the one with nest-building behavior. Canaries, even ones that have been reared in artificial cloth nests, have an innate ability to choose the appropriate materials and to line the nest with feathers. There exist many examples of fixed patterns in behavior generation.

OBSTACLE AVOIDANCE

An interesting robotics-oriented case is avoidance behavior (only weakly related to *obstacle avoidance* in robotics). Vision is the sense that most often is a part of the agent with avoidance behavior. Many small birds, for example, will react to the gaze of an owl or, more particularly, to its characteristic wide-set eyes. This reaction is innate; birds raised in laboratories have an avoidance reaction at the sight of an artificial owl with prominent eyes. The reaction of many people to the sight of snakes can be interpreted as an instinctive avoidance.

An animal that detects a predator can warn its fellows with a cry that will produce in them instinctive avoidance responses even before the enemy is visualized. The instinctive agent of one species can be an agent acquired by learning by another; a mixed colony of birds will react to the alarm cry of any of its constituent species. The chemical response of some individuals can produce avoidance behavior in others of the same species. Ants produce and emit a hormonelike substance into the air that signals to other members of the nest that they should take up defensive positions or flee.

BEHAVIOR AS STRING OF REFLEXES

The following fairly sophisticated agents are considered to be innate ones. As the signal of warning is received, a variety of avoidance agents can be activated, ranging from rapid flight to simulating death in expectation that the predator will be interested only in living prey. A squid has a sophisticated agent that will propel itself away from the source of danger, leaving behind an inky cloud to hinder pursuit. Noises and visual signs that serve to keep a group in contact will also be suppressed during flight. An antelope in flight will fold its tail to conceal the normally visible fluffy white patch. The agents are not of a simple stimulus-response nature. They have their sub-agents, equipped with feedbacks. When an animal cannot run away from danger, it may turn to face the predator with a threatening posture. A monkey, for example, will grimace, showing as many teeth as possible. Our intention is to demonstrate that this is not a voluntary behavior but a string of reflexes.

All animals have the agents of instincts. In general, the higher the animal form, the more flexible the behavior. Among mammals, agents with learning often prevails eventually over fixed action patterns, as when a young rat learns through experience to adjust its instinctive heaping and smoothing motions to the appropriate stage of nest building. In humans, behavior is heavily influenced by the socialization agents. The issue of heredity versus environment continues to be a controversial subject, since no theory of multiple agent functioning yet exists.

5.1.2 From Instinct to Motivation to Drive and to Emotion

The examples above of instinctive behavior need a more constructive explanation than that given by labeling the phenomenon with this term "instinct." We will develop more constructive models of mechanisms determining behavioral patterns observed in intelligent systems.

Instincts as Motivators

At the start of the twentieth century, instincts were believed to be the prime motivators of human behavior. W. McDougall (1871–1938) led the theorists who emphasized the power of motivation over perception and emotion. He claimed that one perceives what one's instincts motivate one to perceive, and the appropriate object, once perceived, generates a surge of emotions that inspire action. Freud had also based much of human behavior on irrational instinctive urges. Freud had attributed human motivation to two basic instincts: *eros* (the life or sexual instinct) and *thanatos* (the death instinct).

Drives

W. B. Cannon introduced the concept of *homeostasis* in 1915 to explain what he and others (including Freud) considered the main function of motivation—to regulate the body into stable existence [11]. He postulated that humans do not act until their internal conditions cross a certain threshold of imbalance. Nonbiological drives became known as learned drives and were thought to gain their motivational power through association with biological drives. Later theorists specified that the drive itself is a homeostatically conceived but non-goal-oriented.

R. S. Woodworth's thesis [12] was that instincts do not determine action, but rather that action occurs when the internal conditions of a homeostatic system exceed some threshold. *Drive* is what emerges in order to initiate the motivational changes. The lack of some substance within the body instigates the drive, which results in behavioral change until the drive is reduced. Lack of energy leads to a hunger drive, which leads to food-seeking behavior.

An elaborate theoretical model of drive was developed by C. Hull (1884–1952) in the 1940s [13]. He considered drive to be general in nature and producing various motives—hunger, thirst, and sex. In Hull's model, drive instigates behavior and thus increases activity. Hull introduced the notion of *drive stimuli* which direct the drive and are different for different motives. He believed that learning depends on adequate drive. Responses are strengthened by drive-stimulus reduction. When it is not reduced, learning does not happen.

Emotions

In 1884 and 1885, respectively, W. James [14] and C. Lange [15] had independently proposed a theory of physiological mechanisms of emotions. The theory argued, for example, that experiencing a dangerous event leads to bodily changes such as increased breathing and heart rate, and a surge of adrenaline. These physical changes are detected by the brain, and the emotion appropriate to the situation is experienced. Further developments questioned such a causal link: there is much evidence that emotional responses emerge as a result of the stored, or pre-simulated information.

Agents of emotions demonstrate very complicated structure. The brain, on receiving information from the senses, interprets an event as emotional while at the same time preparing the body to deal with the new situation. The emotional responses and changes in the body were thus seen to be preparations for dealing with a potentially dangerous emergency situation.

The Schachter-Singer cognitive-physiological theory of the emotion agent [16] proposed that both bodily changes and a cognitive label are needed to experience emotion completely. The bodily changes are assumed to occur as a result of situations that are experienced, while the cognitive label is considered to be the interpretation the brain makes about those experiences. In this view, one experiences anger as a result of perceiving the bodily changes (increased heart rate and breathing, adrenaline production, etc.) and interpreting the situation as one in which anger is appropriate or would be expected.

The Yerkes-Dodson law of the U-curve [17] states that as the excitement increases, performance improves, but only to a point; beyond that point any increase in excitement leads to a deterioration in performance. The search for a biological mechanism capable of altering the excitement level of an individual led to the discovery of a group of neurons (nerve cells) in the brain stem that also relate to an individual's attention level.

Clearly, all these theories invoke agents that exist for generating *goals* and *willpower*, but these topics were cautiously avoided because they bordered philosophical inquiry. We will include these concepts in our discussion but return to them later. For a designer of intelligent machines, the generation of goals and willpower is an important issue: It is integral to the designs of lookup tables and feedback loops, and of subsystems capable of will and goal generation.

5.1.3 Motivation

DEFINITION OF MOTIVATION

Motivation factors are the scales of values that guide the decision making processes of the agent toward the goal; these scales are affected both by the hierarchy of goals existing at all levels of resolution and by the needs at a particular level.

In 1938 H. A. Murray published a list of primary (innate) and secondary (learned) needs [18]. The needs are determined by the two main strives: to reach success and to avoid failure. These needs, he claimed, prove human behavior to be goal oriented. He raised the question of whether motivation is primarily the result of internal needs or external goals. Psychologists had presented many conflicting theories that intend to explain why individuals act at all, why they select the actions they do, and why some people are more motivated than others and succeed where others having talent and ability fail. Both internal and external stimuli had been analyzed. The basic motivations were studied themselves in order to distinguish innate needs from learned orientations. There was an interest to determining whether motivation directs behavior toward a particular goal, or whether it amplifies behavior driven by habits. Since our theory is supposed to be applicable to the coalitions of agents, we will focus upon motivations of a single agent but will put its scales of values in correspondence not only with the agent needs but also with the goals that have been developed for the agents at the levels of lower resolution.

NEEDS PRODUCE MOTIVES

After Darwin's theories were applied to psychological adaptation to the environment in the beginning of the twentieth century, researchers began to apply them to the concept of motivation. The view was that on the one hand, humans, like animals, are partially governed by their instincts for food, water, sex, and so on; on the other hand, the undeniable capacity for motivation must serve an evolutionary purpose too. Most behaviors were called instinctive, but what was often considered innate behavior was demonstrated to be modifiable by learning from experience.

Some motivational theorists noticed that motivation can be an aversive state: one to be avoided. After 1905, Freud's view of motivational processes could block sexual energy and displace into acceptable behaviors; that is, sexual energy as a motivation can be aversive. It was found that frequently motivation produces behavior that leads to increases in future motivation.

In the 1960s A. Maslow developed a theory of self-actualization [19, 20]. It was based on the observation that human motivation is influenced by a need for competence or control. Human behavior could be motivated by a need to become as much as one could possibly achieve. Maslow demonstrated that human motivation results from a hierarchy of needs. Among these are basic physiological demands and needs of safety, belonging, and progress toward self-actualization. The levels of motivation evoked stages. As lower-level needs are met, the motivation to meet the higher-level needs becomes activated. During this development it becomes more difficult to successfully fulfill the needs of each higher level. Nevertheless, as Maslow showed, not many reach the level of self-actualization. (This theory belongs to the set of cognitive theories; see Section 5.1.7.)

5.1.4 From Motives to Goals

From the preceding discussion we see that the analyses of behavior have all led to the fact that there is no motivation without a goal. The more strongly one needs or desires a goal, the more successful one will be in attaining it. Behavior therapists stress three factors of a person's attitude toward a goal that influence motivation:

1. The degree of ambivalence a person feels about his object of desire.
2. The ability to visualize one's goal clearly.
3. The ability to break the goal down into accomplishable smaller tasks.

The word "motivation" is derived from the Latin term *motivus* ("a moving cause"), which suggests the activating properties of the processes involved in motivation. The study of motivational forces helps to explain observed changes in behavior that occur in an individual. Motivation of the agent is not typically measured directly but rather inferred as the result of behavioral changes in reaction to internal or external stimuli; it is a performance variable. That is, the effects of changes in motivation are often temporary. An individual highly motivated to perform a particular task because of a motivational change may later show little interest for that task as the result of a further change in motivation. This suggests that value judgments of a system are constantly changing.

To sum up, there are primary, or basic, motives, that are unlearned and common to both animals and humans; secondary, or learned, motives, can differ from system to system. Primary motives are thought to include hunger, thirst, sex, avoidance of pain, and perhaps aggression and fear. Secondary motives include achievement, power motivation, and numerous other specialized motives.

Motives sometimes are grouped into "pushes" and "pulls." The first group deals with internal changes that evoke specific motive states. The second group represents external goals. Most motivational situations are combinations of push and pull conditions. So the listing of them is a difficult task. Some of the components can be best understood from a biological viewpoint; other components are learned and determined by the complexity of human activities. There are, of course, motives influenced by the cognitive processes. Future motivations are determined by our interpretation of the events around us. Basic motives are often influenced by a variety of elements. For example, hunger is determined by energy needs, by stress or anxiety, by depression, by loneliness, by social influences (seeing other people eating), and by desires to taste certain foods. This interaction of many factors creates difficulty in understanding even basic motivations. Indeed, one can expect determining motivations for constructed intelligent systems to be as difficult and uncertain as it is for determining motivations of living creatures.

5.1.5 Various Approaches

Nomothetic versus Ideographic Approaches

For studies of motivation it is still unclear whether it is better to focus on groups of individuals and attempt to draw general conclusions (the *nomothetic approach*) or on individualized cases of behavior (the *idiographic approach*). Both approaches have helped in understanding motivational processes; however, the nomothetic approach is the dominant one because statistical methods are considered more scientific than the construction of individualized models. One can expect that the scientific community will favor benchmark testing of multiple intelligent systems in multiple situations before the judgment about their behavior can be made.

Innate versus Acquired Processes

An important debate among psychologists concerns the degree to which motivational processes are innate (genetically programmed) versus acquired (learned). This debate echoes the discussion, within the area of machine learning on the balance between the amount of knowledge to be programmed at the designed stage and the knowledge to be learned (in both the supervised and non-supervised setting; see Chapter 10). In the 1920s, the idea that all patterns of behavior were learned replaced the hereditary approach. By the 1960s, the research results demonstrated that both positions are correct. Some motives, in some species, are innate; other motives, such as achievement motivation, seem to evolve via learning [21].

Internal Needs versus External Goals

Another area of the motivational research is whether motivation is primarily the result of internal needs or external goals (which amounts to understanding the differences between push and pull motives). Research suggests that some motives are best classified as internal (push motives), while others develop from external goals (pull motives).

In practice, we always deal with a combination of both internal and external motives [22, 23]. This is an important issue for constructed intelligent systems because in autonomous robotics there is a need for mechanisms that can generate internal needs. In Chapter 8 we will consider to some extent how internal needs are generated by *planning* processes.

Mechanistic and Behavioristic versus Cognitive Approaches

The last important area of debate is whether motivational information is mechanistic or cognitive. The first of these assumes that motivational processes are created by some automaton as a result of formal synthesis, so the intelligent system does not question or try to influence these processes. Researchers favoring the mechanistic point of view focus primarily on internal needs and genetically programmed behaviors.

The cognitive approach focuses on external and acquired motives. It emphasizes the importance of cognition in motivational processes. Cognitive motivational approaches assume that the active processing of information has important influences on motivation. Given the complexity of motivational processes, most theorists feel safe in assuming that some motive states are relatively mechanistic while others are more cognitive [24, 25].

The behaviorists, as we noted earlier, rejected theories of both instinct and will; they emphasized the importance of learning in behavior. This group determined behavior to be a reaction or response (R) to changes in environmental stimulation. As a result, in the 1920s, the concept of instinct was harshly criticized and fell into disrepute. Behaviorism came to dominate the thinking of motivational theorists, and a new motivational concept—drive, which was congenial to behaviorism's S-R approach—was born. The concept of *drive* was used to refer to obtaining specific items from the environment. Behaviorism remained dominant in motivational research until the 1960s despite arguments by the psychological community for the existence of a more active processing of information in both humans and animals.

Cognitive psychologists opened the way for other researchers to examine motivations concerning expectations of future events, choices among alternatives, and attributions of outcomes. Cognitive research also considered issues of learning. Studies of learning shed new light on motivation; significantly learning research showed that individuals are capable of learning new motives. It was demonstrated that new motives are acquired as a result of three learning techniques: classical conditioning, instrumental experimentation, and observational learning.

Genetics and Ethology

As we saw above, the ideas of James and McDougall, that some motivated behaviors are the result of innate programs manifested in the nervous system, met with a certain dissatisfaction as it became evident that it is not possible to discriminate between instinctive and learned behaviors; categorizing an observed behavior as instinctive did not explain why the behavior occurred. In the 1970s the evolutionary significance of animal behaviors appealed to theorists interested in the genetic basis of behavior. K. von Frisch (1886–1982), K. Lorenz (1903–1989), and N. Tinbergen (1907–1988) were awarded the Nobel Prize in 1973 for developing a new field of study: ethology. Ethology consists in studies of the behavior patterns of animals in their natural habitat. Ethologists began to find that the evolutionary significance of a particular behavior can best be understood after a taxonomy of behaviors for that species has been developed as a result of observations in nature. The most common techniques of ethologists are naturalistic observation and field studies.

Ethologists have demonstrated that some animal behaviors emerge in an automatic and mechanical fashion when conditions were appropriate. These behaviors are known as fixed-action patterns. They have several salient characteristics:

- They are specific to the species under study.
- They occur in a highly similar fashion from one occurrence to the next.
- They do not appear to be appreciably altered by experience.

Furthermore the stimulus that releases the genetically programmed behaviors is usually highly specific, such as a particular color, shape, or sound.

Although the largest number of studies conducted by ethologists has been on nonhumans, some ethological researchers have applied the same analyses to human behavior. One can expect that this approach may be productive in the analysis of the behavior of autonomous intelligent systems.

5.1.6 Development of the Concept of "Goal"

The concept of free will as proposed by Aristotle and others was a widely accepted philosophical position until it was generally rejected in favor of determinism. The latter holds that every behavior is the *effect* of some antecedent *cause*. Motivation is an antecedent to particular behaviors. Regardless of the explanation for a particular behavior, one never assumes that the system behaves randomly.

Aristotle's thought that the mind is a blank slate at birth and that, as a result of experiences, behaviors are learned. Descartes proposed the concept of mind-body dualism; it implied that human behavior could be understood as a result both of a free, rational soul and of automatic, nonrational bodily processes. The idea that behavior could be motivated by nonrational, mechanistic processes under some circumstances brought into being the notion of instinct. The mechanistic component of Descartes's dualism can be seen as the precursor of the study of genetic components of motivation, and his view of rational choices is a predecessor of modern cognitive approaches to motivation.

DEFINITION OF TELEOLOGY

Teleology (from Greek telos, *end*; logos, *reason*) gives explanations by reference to some purpose or end; namely it describes a final causality, in contrast to explanations by efficient causes only.

Human conduct is typically associated with the ends pursued or alleged to be pursued. The anthropomorphic tendency is to explain the behavior of other things in nature based on this analogy. The account of teleology was given by Aristotle when he declared that a full explanation of anything must consider not only the material, the formal, and the efficient causes but also the final cause—for whose purpose the thing exists or was produced.

Mechanistic explanations of natural phenomena appeal only to efficient causes. Were teleological explanations to be used, they would take the form not of saying (as in Aristotelian teleology) that things develop toward the realization of ends internal to their own natures but that all things, even biological organisms, are like machines ingeniously devised by an intelligent being.

I. Kant addresses teleology in his "Critique of Judgment" (1790). He cautioned that teleology can be only a regulative and not a constitutive principle; it is a guide to the conduct of inquiry rather than to the nature of reality.

In the late nineteenth century, the focus of discussions was on the issue of whether the phenomena of growth, regeneration, and reproduction characteristic of living organisms could be explained in purely mechanistic terms. H. Driesch revived the Aristotelian concept of entelechy or immanent agency, which was postulated for every organism. It became important to understand whether biological processes could be explained in purely physicochemical terms, or whether the structure, function, and organization of systems carried within themselves some kind of teleology. Organismic conceptions, proclaimed in the mid-twentiethth century by L. von Bertalanffy, opened a new perspective for these issues.

Since Aristotle, vitalists claimed that evolution involves a deterministic finalism, or directedness toward an end. A similar view was promulgated by those who subscribed to a Lamarckian view of evolution involving the inheritance of acquired traits. (These views were formulated in the constructive form by G. Mendel in 1865 and A. Weismann in 1885.) Most evolutionists believed that natural selection was nonrandom in evolution, that it gave evolution direction, so no concept of a final goal was required. Some evolutionists argued that the chance factors in mutation and selection, in addition to the unpredictability of environmental change, made it impossible to formulate deterministic laws. Similar considerations by others led to the claim that evolutionary biology is a paradigm of an after-the-fact exploratory science and that the course of evolution can never be predicted.

J. Locke (1632–1704) contributed to the motivational theory when he emphasized the importance of the sensory experience that underlies external stimulation. There was the conjecture that goals become valuable to us because of the sensory experience associated with these goals. Locke introduced the important mechanism of association in producing new and complex ideas, and this showed how nonmotivating experiences can become motivating. For example, if one pairs a nonmotivating stimulus with a highly motivating object several times, the formerly neutral stimulus begins to motivate behavior. The associative mechanism was the basis of Pavlovian classical conditioning.

Research on incentives proved fruitful in the study of goal formation. Incentive motivation is concerned with the way goals influence behavior. It is often assumed that the stimulus characteristics of the goal are what produce the goal's motivating properties. Unlike drives, which were thought to be innate, incentives are usually considered to be learned. An individual is not born preferring one goal over another, but rather these preferences develop as new goals are experienced.

Incentive motivation exceeds the primary motives of hunger, thirst, sex, or avoidance of pain. One of the most important aspects of this type of motivation is that any goal one seeks can motivate behavior. Goals serving as incentive motivators do not even need to physically exist at the time they activate behavior. The imaginary goals are as powerful motivators as those existing in reality.

Today the concept of goal seems to dominate the area of behavior generation. It is not a metaphor anymore. It is a reality of CNS functioning.

5.1.7 Cognitive Theories of Goal Formation and Comparison

Cognitive theories of goal formation and comparison explain behavior as a result of the active processing and interpretation of information into units of goals and incentives. The goal is understood as a state to be achieved, while incentives are understood as the costs associated with the goal state. Motivation is not seen as a mechanical or innate set of processes but as a purposive and persistent set of behaviors based on the information available. Expectations, constructed in the brain in the form of the desirable state in the future, serve to direct behavior toward particular goals.

Theories of goal generation include expectancy-value theory, attribution theory, cognitive dissonance, self-perception, and self-actualization.

Expectancy-Value Theory

Expectancy-value theory suggests that behavior is a function of the expectancies (E) one has and the value (V) of the goal toward which one is working; this is expressed as $B = f(E, V)$. When alternative behaviors are possible, the behavior chosen has the largest value of the combination function B. Expectancy-value theory is frequently useful in the explanation of social behaviors, goal achievement motivation, and work motivation.

Goal achievement was initially recognized as an important source of human motivation by H. Murray in the late 1930s. After 1940, D. McClelland and J. Atkinson developed a tool for measuring differences in goal achievement motivation. A theoretical model was developed based on the fundamental concepts of expectancy and goal value [26, 27].

The expectancy-value model of achievement motivation proposes that in general, the tendency to achieve the goal in a particular situation depends on two stable motives: the desire for success and a desire to avoid failure. It is possible to evaluate the probability of success in these terms. The desire for success is believed to result from prior goal achievement. A person with a history of successful goal achievement has a high value of orientation toward goal achievement. The motive to avoid failure is also determined by prior instances of unsuccessful goal achievement.

Since most realistic intelligent systems have experienced both successes and failures, the theory assumes that each person has different levels of motivation for success and motivation to avoid failure. The probability of success in a particular situation is an important factor in this achievement theory.

This theory makes several predictions which are interpretable as far as the general concept of intelligent systems is concerned:

1. Systems with extensive experiences of successful goal achievement will tend to set the alternative goals judged to be moderately difficult.
2. Systems with extensive experiences of failures will tend to choose goals that they judge to be either very easy or extremely difficult.
3. The choices made by systems with extensive experiences either of successful goal achievement or of failures differ because of the differing value of easy, moderate, and difficult goals for these two types of systems.

The model mathematically predicts that goals that require moderate effort to achieve will have the greatest value for systems with extensive experiences of success. Systems with extensive experiences of failure believe they are likely to be unsuccessful. For this reason the theory predicts that they prefer easy goals where success is likely or tasks so difficult that little embarrassment would ensue if they fail. Of course, not everything is so easily translatable into a general paradigm of "systems" like the term "people," which we have substituted by the term "systems" everywhere. It is harder to do this with the term "embarrassment."

Attribution Theory

Another important approach to goal achievement and its corresponding motivations rejects the expectancy-value formulation and analyzes instead the *attributions* that people make about achievement situations. This theory is much harder to apply to the systems different from people because it requires self-diagnostics (although the property of self-diagnostics can be required from a contemporary autonomous mobile robot). Attribution theory analyzes how people make judgments about someone's (or their own) behavior. It was discovered that people typically attribute behavior either to stable personality characteristics or to the situations at the time the behavior occurred [28].

The most obvious recommendations of this theory can be interpreted for the constructed systems, like this: "when a system is successful at a goal achievement and attributes that success to ability, that system is likely to approach successfully the new goal achievement situations in the future." Similarly, "if the success of a system was attributed to an intense effort, future goal achievement efforts would depend on a positive attitude of a system toward expending such effort in the future."

Theory of Cognitive Dissonance

A popular cognitive approach to the study of motivation is the theory of cognitive dissonance. L. Festinger's theory, proposed in the 1950s, states that people attempt to maintain consistency among their beliefs, attitudes, and behaviors. A motivational state called "cognitive dissonance" is produced whenever beliefs, attitudes, and behaviors are inconsistent. Then mechanisms are initiated that tend to return the system of cognition to the consistent relationships, such as in the relationships between attitudes and behaviors. If behavior is inconsistent with one's beliefs, it will result in modifying those beliefs. Behavior cannot be changed because it has already occurred; belief, on the other hand, can be changed. Dissonance theory is based on this prediction of the attitude of an intelligent system. Several studies have supported this view [29].

Self-perception Theory

Self-perception theory was developed as an alternative approach. It suggests that all individuals analyze their own behavior and make judgments about why they are motivated to do what they do. This is a pervasive approach which descends from ancient Indian philosophies and emphasized by J. Locke. Dissonance and self-perception theories are not necessarily mutually exclusive; both processes can and do occur but under different conditions [30].

Self-actualization Theory

This is the name for Maslow's hierarchy of needs (see Section 5.1.3). This book will focus on processes of computing the hierarchy of goals for an intelligent system.

5.2 VALUE JUDGMENTS

5.2.1 General Definitions

Value judgments provide the criteria for making intelligent choices. Value judgments evaluate the costs, risks, and benefits of plans and actions, and the desirability, attractiveness, and uncertainty of objects and events. Value judgment modules produce evaluations that can be represented as value state-variables. These can be assigned to the attribute lists in entity frames of objects, persons, events, situations, and regions of space. They can also be assigned to the attribute lists of plans and actions in task frames. Value state-variables can label entities, tasks, and plans as good or bad, costly or inexpensive, as important or trivial, as attractive or repulsive, as reliable or uncertain. Value state-variables can also be used by the behavior generation modules for both planning and executing actions. Value judgments provide the criteria for decisions about which course of action to take.

Value state-variable parameters may be overlaid on the map and egosphere regions where the entities to which they are assigned appear. This facilitates planning. For example, approach-avoidance behavior can be planned on an egosphere map overlay defined by the summation of attractor and repulsor value state-variables assigned to objects or regions that appear on the egosphere. Navigation planning can be done on a map overlay whereupon risk and benefit values are assigned to regions on the egosphere or world map.

DEFINITION OF EMOTIONS

Emotions are biological value state-variables that provide estimates of goodness (what is good, what is bad, and what are the degrees of goodness and badness).

Emotion value state-variables can be assigned to the attribute lists of entities, events, tasks, and regions of space so as to label these as good or bad, as attractive or repulsive, and so on. Emotion value state-variables provide criteria for making decisions about how to behave in a variety of situations. For example, objects or regions labeled with fear can be avoided, objects labeled with love can be pursued and protected, and those labeled with hate can be attacked. Emotional value judgments can also label tasks as costly or inexpensive, risky or safe.

DEFINITION OF PRIORITIES

Priorities are value state-variables that provide estimates of importance.

Priorities can be assigned to task frames so that BG planners and executors can decide what to do first, how much effort to spend, how much risk is prudent, and how much cost is acceptable for each task.

DEFINITION OF DRIVES

Drives are value state-variables that provide estimates of need.

Drives can be assigned to the self-related frame, to indicate internal system needs and requirements. In biological systems, drives indicate levels of hunger, thirst, and sexual arousal. In mechanical systems, drives might indicate how much fuel is left, how much pressure is in a boiler, how many expendables have been consumed, or how much battery charge is remaining.

5.2.2 Limbic System

In animal brains, value judgment functions are computed by all subsystems that conduct and/or support comparison of alternatives and choice among them. However, there is one subsystem whose role is determining preferences—the limbic system. State-variables produced by the limbic system include emotions, drives, and priorities. In animals and humans, electrical or chemical stimulation of specific limbic regions (i.e., value judgment modules) has been shown to produce pleasure and pain as well as more complex emotional feelings such as fear, anger, joy, contentment, and despair. Fear is computed in the posterior hypothalamus. Anger and rage are computed in the amygdala. The insula computes feelings of contentment, and the septal regions produce joy and elation. The perifornical nucleus of the hypothalamus computes punishing pain, the septum pleasure, and the pituitary computes the body's priority level of arousal in response to danger and stress [31].

The drives of hunger and thirst are computed in the limbic system's medial and lateral hypothalamus. The level of sexual arousal is computed by the anterior hypothalamus. The control of body rhythms, such as sleep–awake cycles, are computed by the pineal gland. The hippocampus produces signals that indicate what is important and should be remembered, or what is unimportant and can safely be forgotten. Signals from the hippocampus consolidate (i.e., make permanent) the storage of sensory experiences in long-term memory. Destruction of the hippocampus prevents memory consolidation [32].

In lower animals, the limbic system is dominated by the sense of smell and taste. Odor and taste provides a very simple and straightforward evaluation of many objects. For example, depending on how something smells, one should either eat it, fight it, mate with it, or ignore it. In higher animals, the limbic system has evolved to become the seat of much more sophisticated value judgments, including human emotions and appetites. Yet even in humans, the limbic system retains its primitive function of evaluating odor and taste, and there remains a close connection between the sense of smell and emotional feelings.

Input and output fiber systems connect the limbic system to sources of highly processed sensory data as well as to high-level goal selection centers. Connections with the frontal cortex suggests that the value judgment modules are intimately involved with long-range planning and geometrical reasoning. Connections with the thalamus suggests that the limbic value judgment modules have access to high-level perceptions about objects, events, relationships, and situations, for example, the recognition of success in goal achievement, the perception of praise or hostility, or the recognition of gestures of dominance or submission. Connections with the reticular formation suggests that the limbic value judgment modules are also involved in computing confidence factors derived from the degree of correlation between predicted and observed sensory input. A high degree of correlation produces emotional feelings of confidence. Low correlation between predictions and observations generates feelings of fear and uncertainty.

The limbic system is an integral and substantial part of the brain. In humans the limbic system consists of about 53 emotion, priority, and drive submodules linked together by 35 major nerve bundles [31].

5.2.3 Value State-Variables

It has long been recognized by psychologists that emotions play a central role in behavior. Fear leads to escape, hate lends to rage and attack, joy produces smiles and dancing, and despair produces withdrawal and despondent demeanor. All creatures tend to repeat what makes them feel good, and avoid what make them feel bad. All attempt to prolong, intensify, or repeat those activities that give pleasure or make the self feel confident, joyful, or happy. All try to terminate, diminish, or avoid those activities that cause pain, or arouse fear, or revulsion.

Emotions provide an evaluation of the state of the world as perceived by the sensory system. Emotions tell us what is good or bad, what is attractive or repulsive, what is beautiful or ugly, what is loved or hated, what provokes laughter or anger, what smells sweet or rotten, what feels pleasurable, and what hurts.

It is also widely known that emotions affect memory. Emotionally traumatic experiences are remembered in vivid detail for years, while emotionally nonstimulating everyday sights and sounds are forgotten within minutes after they are experienced.

Emotions are popularly believed to be something apart from intelligence—irrational, beyond reason or mathematical analysis. The theory presented here maintains the opposite. In this model, emotion is a critical component of biological intelligence, necessary for evaluating sensory input, selecting goals, directing behavior, and controlling learning.

It is widely believed that machines cannot experience emotion, or that it would be dangerous, or even morally wrong to attempt to endow machines with emotions. However, unless machines have the capacity to make value judgments (i.e., to evaluate costs, risks, and benefits, to decide which course of action, and what expected results, are good, and which are bad), machines can never be intelligent or autonomous. What is the basis for deciding to do one thing and not another, even to turn right rather than left, if there is no mechanism for making value judgments? Without value judgments to support decision making, nothing can be intelligent, whether or not it is biological or artificial.

Some examples of value state-variables are listed below, along with suggestions of how they might be computed. This list from [33] is not complete.

Goodness is a positive value assignable to any state-variable. It may be assigned to the entity frame of any event, object, or person. It can be computed as a weighted sum, or spatiotemporal integration, of all other positive value state-variables assigned to the same entity frame.

Badness is a negative value assignable to any state-variable. It can be computed as a weighted sum, or spatiotemporal integration, of all other negative value state-variables assigned to an entity frame.

Pleasure is a physical internal positive value assignable to any state-variable that can be assigned to objects, events, or specific regions of the body. In the latter case, pleasure may be computed indirectly as a function of neuronal sensory inputs from specific regions of the body. Emotional pleasure is a high-level internal positive value state-variable that can be computed as a function of highly processed information about situations in the world.

Pain is a physical internal negative value assignable to any state-variable that can be assigned to specific regions of the body. It may be computed directly as a function of inputs from pain sensors in specific regions of the body. Emotional pain is a high-level internal negative value state-variable that may be computed indirectly from highly processed information about situations in the world.

Success_observed is a positive value assignable to any state-variable that represents the degree to which task goals are met, plus the amount of benefit derived therefrom.

Success_expected is a value assignable to any state-variable that indicates the degree of expected success (or the estimated probability of success). It may be stored in a task frame, or computed during planning on the basis of world model predictions. When compared with success that has been observed, it provides a baseline for measuring whether goals were met on, behind, or ahead of schedule; at, over, or under estimated costs; and with resulting benefits equal to, less than, or greater than those expected.

Hope is a positive value assignable to any state-variable produced when the world model predicts a future success in achieving a good situation or event. When high hope is assigned to a task frame, the BG module may intensify behavior directed toward completing the task and achieving the anticipated good situation or event.

Frustration is a negative value assignable to any state-variable that indicates an inability to achieve a goal. It may cause a BG module to abandon an ongoing task, and switch to an alternate behavior. The level of frustration may depend on the priority attached to the goal, and on the length of time spent in trying to achieve it.

Love is a positive value assignable to any state-variable produced as a function of the perceived attractiveness and desirability of an object or person. When assigned to variables of the frame of an object or person, it tends to produce behavior designed to approach, protect, or possess the loved object or person.

Hate is a negative value assignable to any state-variable produced as a function of pain, anger, or humiliation. When assigned to variables of the frame of an object or person, hate tends to produce behavior designed to attack, harm, or destroy the hated object or person.

Comfort is a positive value state-variable produced by the absence of (or relief from) stress, pain, or fear. Comfort can be assigned to variables of the frame of an object, person, or region of space that is safe, sheltering, or protective. When under stress or in pain, an intelligent system may seek out places or persons with entity frames that contain a large comfort value.

Fear is a negative value assignable to any state-variable produced when the sensory processing system recognizes, or the world model predicts, a bad or dangerous situation or event. Fear may be assigned to the attribute list of an entity, such as an object, person, situation, event, or region of space. Fear tends to produce behavior designed to avoid the feared situation, event, or region, or flee from the feared object or person.

Joy is a positive value assignable to any state-variable produced by the recognition of an unexpectedly good situation or event. It is assigned to the variables of self-object frame.

Despair is a negative value assignable to any state-variable produced by world model predictions of unavoidable, or unending, bad situations or events. Despair may be caused by the inability of the behavior generation planners to discover an acceptable plan for avoiding bad situations or events.

Happiness is a positive value assignable to any state-variable produced by sensory processing observations and world model predictions of good situations and events. Happiness can be computed as a function of a number of positive (rewarding) and negative (punishing) value state-variables.

Confidence is an estimate of probability of correctness. A confidence state-variable may be assigned to variables of the frame of any entity in the world model. It may also be assigned to variables of the self frame, to indicate the level of confidence that a creature has in its own capabilities to deal with a situation. A high value of confidence may cause the BG hierarchy to behave confidently or aggressively.

Uncertainty is a lack of confidence. Uncertainty assigned to variables of the frame of an external object may cause attention to be directed toward that object in order to gather more information about it. Uncertainty assigned to variables of the self-object frame may cause the behavior generating hierarchy to be timid or tentative.

It is possible to assign a real nonnegative numerical scalar value to each state-variable. This defines the degree, or amount, of that value assigned to the state-variable. For example, a positive real value assigned by "goodness" defines how good; if

$$e := \text{"goodness"} \quad \text{and} \quad 0 \leq e \leq 10, \tag{5.1}$$

then $e = 10$ is the "best" evaluation possible.

Some value state-variables can be grouped as conjugate pairs. For example, goodness–badness, pleasure–pain, success–fail, and love–hate. For conjugate pairs, a positive real value means the amount of the good value, and a negative real value means the bad value.

For example, if

$$e := \text{"goodness–badness"} \quad \text{and} \quad -10 \leq e \leq +10, \tag{5.2}$$

then the variables with these values assigned must be considered

with $e = 5$ as good,	$e = -4$ as bad
with $e = 6$ as better,	$e = -7$ as worse
with $e = 10$ as best,	$e = -10$ as worst
with $e = 0$ as neither good nor bad	

Similarly, in the case of pleasure–pain, the larger the positive value, the better it feels. The larger the negative value, the worse it hurts. For example, if

$$e := \text{"pleasure–pain"} \tag{5.3}$$

then the variable with these values assigned must be considered

with $e = 5$ as pleasurable	$e = -5$ as painful
with $e = 10$ as ecstasy	$e = -10$ as agony
with $e = 0$ as neither pleasurable nor painful	

The positive and negative elements of the conjugate pair may be computed separately, and then combined.

5.2.4 VJ Modules

Values for the state-variables are computed by value judgment functions (VJ) residing in modules. Inputs to VJ modules describe entities, events, situations, and states. VJ functions compute measures of cost, risk, and benefit. VJ outputs are value state-variables.

Theorem The VJ value judgment mechanism can be defined as a mathematical or logical function of the form

$$\mathbf{E} = \mathbf{V}(\mathbf{S}), \tag{5.4}$$

where

E is an output vector of value state-variables,

V is a value judgment function that computes **E** given **S**,

S is an input state vector defining conditions in the world model, including the self.

The components of **S** are entity attributes describing states of tasks, objects, events, or regions of space. These may be derived either from processed sensory information or from the world model.

The value judgment function **V** in the VJ module computes a numerical scalar value (i.e., an evaluation) for each component of **E** as a function of the input state vector **S**. **E** is a time-dependent vector. The components of **E** may be assigned to attributes in the world model frame of various entities, events, or states.

If time dependency is included, the function $\mathbf{E}(t + dt) = \mathbf{V}(\mathbf{S}(t))$ may be computed by a set of equations of the form

$$e(j, t + dt) = \left(k\frac{d}{dt} + 1\right) \sum_{i} s(i, t) w(i, j), \tag{5.5}$$

where

$e(j, t)$ is the value of the jth value state-variable in the vector **E** at time t,

$s(i, t)$ is the value of the ith input variable at time t,

$w(i, j)$ is a coefficient, or weight, that defines the contribution of $s(i)$ to $e(j)$.

Each individual may have a different set of "values," namely a different weight matrix in its value judgment function **V**.

The factor $(k\, d/dt + 1)$ indicates that a value judgment is typically dependent on the temporal derivative of its input variables as well as on their steady-state values. If $k > 1$, then the rate of change of the input factors becomes more important than their absolute values. For $k > 0$, need reduction and escape from pain are rewarding. The more rapid the escape, the more intense, but short-lived, the reward.

Formula (5.6) suggests how a VJ function might compute the value state-variable "happiness":

$$\begin{aligned} \text{happiness} = \left(k\frac{d}{dt} + 1\right) &(\text{success-expectation} \\ &+ \text{hope-frustration} \\ &+ \text{love-hate} \\ &+ \text{comfort-fear} \\ &+ \text{joy-despair}), \end{aligned} \tag{5.6}$$

where

success, hope, love, comfort, joy are all positive value state-variables that contribute to happiness,

expectation, frustration, hate, fear, and despair are all negative value state-variables that tend to reduce or diminish happiness.

In this example, the plus and minus signs result from $+1$ weights assigned to the positive-value state-variables and -1 weights assigned to the negative-value state-variables. Of course, different brains may assign different values to these weights.

Expectations are listed in formula (5.6) as negative state-variables because the positive contribution of success is diminished if success_observed does not meet or exceed the success expected. This suggests that happiness can be increased if expectations are lower. However, when $k > 0$, the hope reduction that accompanies expectation downgrading may be just as punishing as the disappointments that result from unrealistic expectations, at least in the short term. Therefore lowering expectations is a good strategy for increasing happiness only if expectations are lowered very slowly or are already low to begin with.

Figure 5.1 gives an example of how a VJ module might compute pleasure–pain. Skin and muscle are known to contain arrays of pain sensors that detect tissue damage. Specific receptors for pleasure are not known to exist, but pleasure state-variables can easily be computed from intermediate state-variables that are computed directly from skin sensors.

The VJ module shown in Figure 5.1 computes pleasure–pain as a function of the intermediate state-variables of softness, warmth, and gentle stroking of the skin. These intermediate state-variables are computed by low-level SP modules. Warmth is computed from temperature sensors in the skin. Softness is computed as a function of pressure and deformation (i.e., stretch) sensors. The gentle stroking of the skin sensation is computed by a spatiotemporal analysis of skin pressure and deformation sensor arrays that is analogous to image flow processing of visual information from the eyes. Pain sensors go directly from the skin area to the VJ module.

In the processing of data from sensors in the skin, all of the computations preserve the topological mapping of the skin area. Warmth is associated with the area in which the temperature sensors are elevated. Softness is associated with the area where pressure and deformation are in the correct ratio. Gentle stroking is associated with the area in which the proper spatiotemporal patterns of pressure and deformation are observed.

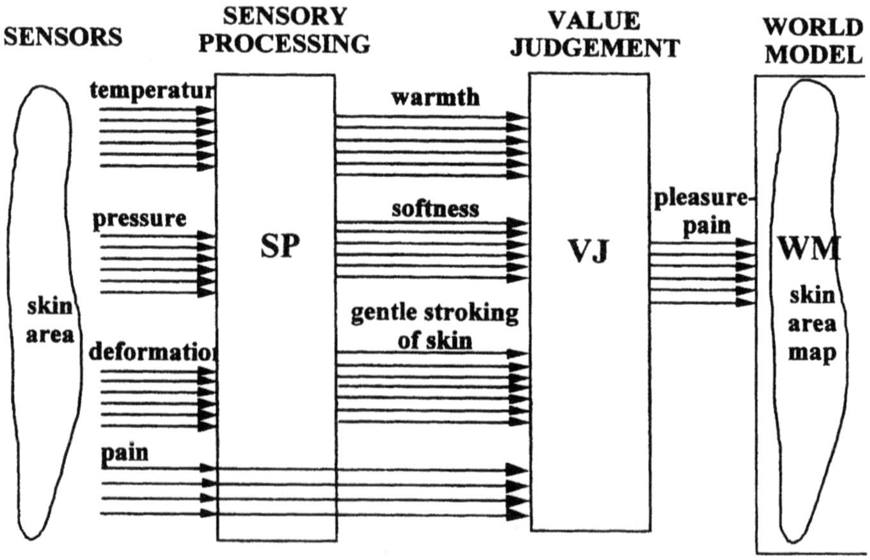

Figure 5.1. VJ module.

Pain is associated with the area where pain sensors are located. Finally pleasure–pain is associated with the area from which the pleasure-pain factors originate. A pleasure–pain state-variable can thus be assigned to the knowledge frames of the skin pixels that lie within that area.

5.2.5 Value State-Variable Map Overlays

When objects or regions of space are projected on a world map or egosphere, the value state-variables in the frames of those objects or regions can be represented as overlays on the projected regions. When this is done, value state-variables such as comfort, fear, love, hate, danger, and safe will appear overlaid on specific objects or regions of space. BG modules can then perform path planning algorithms that steer away from objects or regions overlaid with fear, or danger, and steer toward or remain close to those overlaid with attractiveness, or comfort. Behavior generation may generate attack commands for target objects or persons overlaid with hate. Protect, or care-for, commands may be generated for target objects overlaid with love.

Projection of uncertainty, believability, and importance value state-variables on the egosphere enables BG modules to perform the decision-making computations necessary for acting appropriately and for manipulating sensors and focusing attention. Replanning can be done directly on the egosphere and the rate of repetition can be very high.

Confidence, uncertainty, and hope state-variables may also be used to modify the effect of other value judgments. For example, if a task goal frame has a high hope variable but low confidence variable, behavior may be directed toward the hoped-for goal, though cautiously. On the other hand, if both hope and confidence are high, pursuit of the goal may be much more aggressive.

The real-time computation of value state-variables for varying task and world model conditions provides the basis for complex situation dependent behavior.

5.3 ACHIEVING THE GOAL: OPTIMIZATION VIA THE CALCULUS OF VARIATIONS

5.3.1 Notation and Basic Premises

In order to provide a rigorous framework for the ensuing discussion, it is necessary to make certain assumptions about the class of systems considered here. The suggested methods can be applied to many problems besides the mathematical descriptions given in formulas (5.7) to (5.16). Likewise, the algorithms presented here are designed to be implemented in a way that allows a system identification in which few assumptions are made about the structure. In order to establish the validity of these algorithms, we therefore begin by making the following assumptions: First, there exists a plant (object of control) belonging to the class described by the vector differential equation

$$\dot{\mathbf{x}} = \mathbf{f}(\mathbf{x}, \mathbf{u}, t), \tag{5.7}$$

where the state $\mathbf{x} \in R^n$ and the controls are $\mathbf{u} \in R^m$ for which partial derivatives $\partial\mathbf{f}/\partial\mathbf{x}$, $\partial\mathbf{f}/\partial\mathbf{u}$, and $\partial\mathbf{f}/\partial t$ exist for all values of $\mathbf{x}$, $\mathbf{u}$, and t inside the constrained region

$$\mathbf{x} \in X \subset R^n, \quad \mathbf{u} \in \Omega \subset R^n, \quad t_0 \leq t \leq t_f \tag{5.8}$$

(and X are closed convex sets containing the origin and X is compact) except, perhaps, for a finite number of points inside the boundaries of the region. The output of the plant is given by the function

$$\mathbf{y} = \mathbf{g}(\mathbf{x}, t) \quad \text{for } \mathbf{y} \in R^p \tag{5.9}$$

all of whose partial derivatives exist within the region defined by (5.8). For this plant a goal is defined as a required final position

$$\mathbf{x}(t_f) = \mathbf{x}_f, \tag{5.10}$$

where the final time may not be known a priori. This implies that a required trajectory

$$x_r(t), \qquad t_0 \leq t \leq t_f,$$

should be synthesized. In order to have a basis for constructing this required trajectory, a performance function

$$J(\mathbf{x}, \mathbf{u}, t)|_{\mathbf{x}_0, \mathbf{x}_f} \tag{5.11}$$

is required, for which there exists such a $\mathbf{u}^*(t)$ that

$$J(\mathbf{x}, \mathbf{u}^*, t)\Big|_{\substack{\mathbf{x}_0 \\ \mathbf{x}_f}} \leq J(\mathbf{x}, \mathbf{u}, t)\Big|_{\substack{\mathbf{x}_0 \\ \mathbf{x}_f}} \quad \forall t_0 \leq t \leq t_f, \forall \mathbf{u}. \tag{5.12}$$

It may be assumed that the plant is approximated by a set of difference equations

$$\mathbf{x}[k+1] = \mathbf{f}(\mathbf{x}[k], \mathbf{u}[k], k), \qquad 0 \leq k \leq K, \tag{5.13}$$

and

$$\mathbf{y}[k] = \mathbf{g}(\mathbf{x}[k], k). \tag{5.14}$$

In most applications, the discrete time values $t_0, t_1, \ldots, t_k, \ldots, t_K$, are uniformly distributed, and it may be assumed that

$$\Delta t_k = t_{k+1} - t_k = \text{constant}. \tag{5.15}$$

A performance index that is commonly encountered (subject to other constraints) is the time optimal cost function

$$J_K = \sum_0^K \Delta t_k \tag{5.16}$$

which can be evaluated piecewise at any given level of temporal resolution.

What is presented here in (5.7) through (5.16) does not allude to any particular representation or an analytical solution. It is a symbolic story that fits within any paradigm of design and control.

5.3.2 A Linearized Third-Order Plant: Model of a DC Motor

The maximum time problem involves the transition of a system from an arbitrary initial state to the goal state in a way that minimizes the time required for the transition. We write

$$\dot{\mathbf{x}} = \mathbf{A}\mathbf{x} + \mathbf{B}u,$$
$$y = c\mathbf{x}, \tag{5.17}$$

where the state vector **x** is

$$\mathbf{x}(t) = \begin{bmatrix} \theta(t) \\ \omega(t) \\ i(t) \end{bmatrix} \triangleq \begin{bmatrix} x_1(t) \\ x_2(t) \\ x_3(t) \end{bmatrix}.$$

The corresponding **A** and **B** matrices are

$$\mathbf{A} = \begin{bmatrix} 0 & 1 & 0 \\ 0 & -\frac{B}{J} & \frac{k}{J} \\ 0 & -\frac{k}{L} & -\frac{R}{L} \end{bmatrix} \quad \text{and} \quad \mathbf{B} = \begin{bmatrix} 0 & 0 \\ 0 & -\frac{1}{J} \\ \frac{1}{L} & 0 \end{bmatrix}. \tag{5.18}$$

The numerical values of these matrices are

$$\mathbf{A} = \begin{bmatrix} 0 & 1 & 0 \\ 0 & -1.4449 & 501.1407 \\ 0 & -15.6718 & -361.4744 \end{bmatrix} \quad \text{and} \quad \mathbf{B} = \begin{bmatrix} 0 & 0 \\ 0 & -3802.3 \\ 118.9 & 0 \end{bmatrix}. \tag{5.19}$$

In order to make the solution well-posed, bounds are specified for each input in the form

$$|u_i| \leq u_{i_{\max}}.$$

The solution of this problem provides

- The state trajectories of the plant (beginning at the initial state and terminating at the goal state, as function of time) which, when followed, lead to minimum time operation
- The input functions which, when applied to the plant in the given initial state, cause it to follow these optimal state trajectories.

In this example the goal state was chosen at a position offset 180 degrees relative to a zero initial state, with final velocity and current at zero. The solution is expected to provide velocity and current trajectories for the duration of control together with the corresponding voltage input.

5.3.3 Optimization via the Calculus of Variations

It would be interesting to explore similarities between the theory of optimum control and motivation, goals, and value judgment. Optimization is goal-oriented control; it is generated by using some techniques of providing *value judgment* about means that we use to reach the *goal* in the best possible way: These are mechanisms of *motivations*.

The classical theory of optimization—in which the tool is the calculus of variations and the dynamic programming is the search techniques of this tool—provides a powerful set of methods for controller design. It is not a universal tool: only certain realizations of system models are under consideration. Specifically, the model is frequently presented as a deterministic or stochastic systems described by not necessarily linear dynamical equations. A mathematically exact global minimum of a multiple-objective quadratic cost functional can often be shown to exist in this case. A system that strives to reach this minimum can be implemented as a possibly time-varying on-line feedback control policy.

These tools therefore encompass the entire paradigm of controller design for such systems, to the extent of specifying the expected final cost of control and verifying the stability of the control policy, as well as allowing on-line optimal identification of system parameters in some cases. Less general results have also been developed for other classes of cost functionals, and for systems described by certain partial differential equations. However, many questions remain unanswered such as:

- Can the calculus of variations paradigm be demonstrated and proved for automata representations? (We will admit a positive answer, since the automata have been shown equivalent to differential equations in a tessellated world.)
- In the stochastic system, would the process of optimization drive the system to the global optimum, local optimum, or optimum in average? (We could explore techniques that lead to the "optimum in average"; however, we hope that a procedure can be assigned leading to the global optimum.)
- Should the form of the cost-functional matter? (We hope, it does not).

We anticipate that following the process of optimization in a feedforward mode is obvious: The feedback mode could raise some problems of realization, but it does not. Let us discuss a complete solution of the minimum time problem with a practical bound on the input. This problem does not admit to a closed form solution because the Euler-Lagrange formulation leads to the generation of a two-point boundary value problem that must be solved numerically. There are no general algorithms for the solution of this problem, and various kinds of search techniques are usually applied in order to receive approximate solutions.

The key idea underlying the calculus of variations approach is actually very simple. In a nutshell, optimization involves constant *value judgment*: evaluation (by some method) of a function or functional $J(\cdot)$ and satisfaction of a set of inequality constraints. This process is to be followed by the selection of the largest of these as a solution by searching for its extrema. In a variety of cases with constraints or with inconvenient structure, the optima are non-unique, and results inconclusive (i.e., it may not be clear whether the minimum obtained is local or global).

In the case of dynamical systems, however, smooth constraints are introduced via the use of an *intuitive introduction of weights* with the aid of Lagrange multipliers. The weights technique is used regularly in quantitative psychology, operations research, and optimum control theory when several goals need to be addressed simultaneously (some cost functions or functionals are minimized, and some shown to satisfy inequality conditions). Indeed, it cannot be argued that the sum of the important factors for all motivating conditions is 1. Since the weights reflect our *motivations*, they do not remain constant during the process of control but must somehow be normalized.

YERKES-DODSON LAW IN ROBOTIC POSITIONING

To take an example, consider the process of moving a system from one point in space to another (positioning). At the beginning of this process, we are usually concerned with minimizing the time of positioning so that the strategies of control are close to the *dominant motivation*: Move as soon as possible. From our knowledge of the stochastic factors, we are aware that high-speed movement is associated with large errors. But it would not be prudent for us to think about final error from the beginning of our motion, so we place importance on the maximum value of the time factor. Only as we approach the *goal state* do we get more and more concerned with accuracy and even reduce the speed of motion; the importance of the accuracy factor rapidly grows in value. Thus toward the end of the positioning the importance of time is reduced drastically as the importance of the error value starts rising. Interestingly this can be regarded as a form of *motivation*: Initially we are concerned with time minimization, whereas toward the end we are concerned with energy minimization. It definitely reminds us of the Yerkes–Dodson law from Subsection 5.1.2!

The elegant result that emerges is that since there is essentially no difference between minimizing the *value judgments* $J(\cdot)$ and $J(\cdot) + 0$, the state equations of a system may be adjoined to the cost functional as follows: Since the model of our system can be represented in the form

$$\dot{\mathbf{x}} = \mathbf{f}(\mathbf{x}, \dot{\mathbf{u}}, t) \Leftrightarrow \dot{\mathbf{x}} - \mathbf{f}(\mathbf{x}, \mathbf{u}, t) = \mathbf{0}, \tag{5.20}$$

we have

$$J(\cdot) = j(\cdot) + \lambda^T(\dot{\mathbf{x}} - \mathbf{f}(\mathbf{x}, \mathbf{u}, t)), \tag{5.21}$$

where the *importance factor* λ is a Lagrange multiplier. The Lagrange multiplier can, in general, be a vector function of the system variables including time, and it is used to generate a scalar from the system equations. In this manner the *constrained* minimization of $J(\cdot)$ subject to remaining on the natural trajectories of the dynamical system is replaced by an extended *unconstrained* minimization problem.

Using the description of dynamical systems developed above, a general form of the cost is given by

$$J(t_0) = \Phi(\mathbf{x}(T), T) + \int_{t_0}^{T} L(\mathbf{x}(t), u(t), t)\, dt, \tag{5.22}$$

where $[t_0, T]$ is a time interval of interest. The functional essentially expresses the value judgment (cost of system operation) in terms of the accumulated *penalties*, or *losses* during the interval of operation, as well as the cost of the final outcome at time T (the final weighting function).

The general optimal control problem may be stated as the determination of a $u^*(t)$ on the interval $[t_0, T]$ that drives a system along a trajectory $x^*(t)$ such that the cost functional is minimized and some terminal function of the states $\Psi(\mathbf{x}(T), T)$ is brought to zero. The final state constraint allows the explicit inclusion of terminal state constraints and is not to be confused with $\Phi(\cdot)$, which is simply a weighting function.

HAMILTONIAN AS A FUNCTION OF MOTIVATION

Now, we have to unify all our motivation factors into a single function which will be called the Hamiltonian. If the extended cost functional is rewritten as in (5.21), with the Hamiltonian function defined as

$$H(x, u, t) \triangleq L(x, u, t) + \lambda^T(t) f(x, u, t), \tag{5.23}$$

then

$$J^f = \Phi(\mathbf{x}(T), T) + \nu^T \Psi(\mathbf{x}(T), T) + \int_{t_0}^{T} (H(\mathbf{x}, u, t) - \lambda^T(t)\dot{\mathbf{x}}(t))\, dt. \tag{5.24}$$

The expression above includes losses and constraints together with the importance factors λ, with the possibility to determine them quantitatively without relying on human estimation.

To find the minimum of the functional, a variational technique is employed. This technique is analogous to the method of function minimization using derivatives. Instead of determining the values of the function where all derivatives are equal to zero, however, we employ Leibniz's rule for small variations of $J(\cdot)$ as a function of corresponding variations in x, λ, ν, u, and t. This rule states that if $\mathbf{x}(t) \in R^n$ and if $J(\cdot)$ and $h(\cdot)$ are real scalar functions of $\mathbf{x}(t)$, then for

$$J(x) = \int_{t_0}^{T} h(\mathbf{x}(t), t)\, dt \tag{5.25}$$

increments of $J(\cdot)$ are

$$dJ = h(\mathbf{x}(T), T)\, dT - h(\mathbf{x}(t_0), t_0)\, dt_0 + \int_{t_0}^{T} [h_x^T(\mathbf{x}, t), t)\, \delta x]\, dt, \tag{5.26}$$

where $h_x \triangleq \partial h / \delta x$ and dJ is essentially

$$dJ = J(x + \delta x) - J(x). \tag{5.27}$$

For 5.24, dJ^f can be derived in terms of each of the variations δ as

$$\begin{aligned} dJ^f = {} & (\Phi_x + \Psi_x^T \nu)^T dx\Big|_T + (\Phi_t + \Psi_t^T \nu)^T dt\Big|_T + \Psi_\nu^T\Big|_T d\nu \\ & + (H - \lambda^T \dot{x}\, dt\Big|_T - (H - \lambda^T \dot{x})\, dt\Big|_{t_0} \\ & + \int_{t_0}^{T} \left[H_x^T \delta x + H_u^T \delta u - \lambda^T \delta \dot{x} + (H_\lambda - \dot{x})^T \delta\lambda \right] dt. \end{aligned} \tag{5.28}$$

Two simplifications can be made here. The first is the elimination of the term in $\delta \dot{x}$ by the use of integration by parts. Thus

$$-\int_{t_0}^{T} \left[\lambda^T \delta \dot{x} \right] dt = -\lambda^T \delta x \Big|_{t_0}^{T} + \int_{t_0}^{T} \left[\dot{\lambda}^T \delta x \right] dt. \tag{5.29}$$

The second simplification results from the approximation of the variation δx by the remainder in the Taylor series expansion

$$dx = \dot{x} \cdot dt + \delta x. \tag{5.30}$$

Substituting (5.30) into (5.29), and the result into (5.28), the complete form of the variations in $J^f(\cdot)$ are

$$\begin{aligned} dJ^f = {} & (\Phi_x + \Psi_x^T \nu - \lambda)^T dx\Big|_T + (\Phi_t + \Psi_t^T \nu + H)\, dt\Big|_T \\ & + \Psi_t^T\Big|_T d\nu - H\, dt\Big|_{t_0} + \lambda^T dx\Big|_{t_0} \end{aligned} \tag{5.31}$$

$$+ \int_{t_0}^{T} \left[(H_x + \dot{\lambda})^T \delta x + H_u^T \delta u + (H_\lambda - \dot{x})^T \delta \lambda \right] dt.$$

The variations of $J^f(\cdot)$ must be zero for an optimum solution. This leads to several necessary conditions which may be culled from (5.32) by setting the coefficients of each variational term to zero. The terms in dt_0 and $dx(t_0)$ are all eliminated by the assumption that the starting time and state are fixed.

For the system described in (5.20), the performance index in (5.21), and the Hamiltonian in (5.22)

$$\dot{x} = \frac{\partial H}{\partial \lambda} = f(\mathbf{x}, \mathbf{u}, t), \qquad t \geq t_0, \tag{5.32a}$$

$$\dot{\lambda} = -\frac{\partial H}{\partial x} \qquad t \leq T, \tag{5.32b}$$

$$\frac{\partial H}{\partial u} = 0, \tag{5.32c}$$

$$(\Phi_x + \Psi_x^T \nu - \lambda)^T \, dx(T) + (\Phi_t + \Psi_t^T \nu + H)\Big|_T dT = 0. \tag{5.32d}$$

Equation (5.32d) is explicitly allowed to remain unresolved so that the final time, T, can remain a variable as in the minimum time problem. If T is fixed, the second term in the equation is removed, and the boundary condition reverts to the coefficient of $dx(T)$. If, on the other hand, the final state is fixed, $dx(T)$ becomes zero, and the boundary condition is completely specified by the second term.

Equation set (5.32) is thus a list of necessary conditions to be fulfilled in order for a control to be optimal in the sense of the performance index of (5.23). There is an important point to be made here, which is that this general formulation does not say anything about the implementation of the optimal control. This may be determined simply as a function of time on the interval $[t_0, T]$. The basic method for doing so consists of solving (5.32c) for u and substituting the result into (5.32a) and (5.32b). The resulting $2n$-dimensional system of equations is solved using the boundary conditions that are determined from the value of $\mathbf{x}(t_0)$ and from (5.32d). The resulting boundary conditions are distributed equally between the initial and the final time. This situation is known as a two-point boundary value problem, and it is typically resolved using indirect numerical methods that can be quite complex.

By a happy coincidence, if the term inside the integral, $L(\cdot)$, is of the form

$$L(\mathbf{x}, \mathbf{u}) = \mathbf{u}^T \mathbf{R} \mathbf{u} + \mathbf{x}^T \mathbf{Q} \mathbf{x}, \tag{5.33}$$

then it may happen that the optimal control has the form

$$\mathbf{u}^*(t) = -\mathbf{k}\mathbf{x}(t), \tag{5.34}$$

where k may or may not be a nonconstant gain vector, depending on the structure of the plant and the choice of boundary conditions. This result provides a feedback solution; it is a useful development that has been investigated at great length by many authors.

Unfortunately, the application of this result depends on a number of conditions that do not necessarily hold in specific cases. In no particular order, these are as follows:

- The existence of practical performance criteria which are reproducible in the form of the performance index in (5.33). This condition holds for cases where energy is involved or a variation of error is of importance, which is frequently; real cases have a much broader set of forms for L.
- Nonheuristic a priori knowledge of the "correct" weighting of the **R** and **Q** matrices. Since this performance index trades off control energy against error, there exists an internal or implicit problem of balancing the multiple criteria represented by the matrices.
- Although the stability of the idealized version of this control law can be determined, its performance depends on the ability to supply the specified control energy. The introduction of constraints must be handled in an indirect manner, which lays open to question the utility of the approach in cases where such constraints exist.

These questions are largely peripheral to our purposes here because the performance index is not under investigation. It should nevertheless be clear that in general, the solution of the optimal control problem via the calculus of variations does not generate a feedback solution.

If the optimal input is constrained to lie within some limits, the earlier conditions still hold but one further condition, known variously as Pontryagin's maximum or minimum principle, is added. This says that

$$H(\mathbf{x}^*, \mathbf{u}^*, \lambda^*, t) \leq H(\mathbf{x}^*, \mathbf{u}, \lambda^*, t). \tag{5.35}$$

Thus the value of the Hamiltonian must be minimized over all admissible $\mathbf{u}$ for optimal values of the state $\mathbf{x}$ and costate λ.

For a linear system such as our example, $\dot{\mathbf{x}} = \mathbf{A}\mathbf{x} + \mathbf{B}\mathbf{u}$, and if the criterion of optimization is to minimize time of operation with constrained controls, it is possible to choose the performance index

$$J(t_0) = \int_{t_0}^{T} 1 \cdot dt \tag{5.36}$$

with T free.

The control problem is then to determine $|u_i(t)| \leq u_{\max}$, $1 \leq i \leq m$ and $t_0 \leq t \leq T$, where m is the control dimension that minimizes $J(\cdot)$ and drives the state of the system from $\mathbf{x}(t_0)$ to $\mathbf{x}(T)$.

The need for Pontryagin's principle is illustrated by the use of the direct approach as follows. The Hamiltonian for this problem is

$$H = L + \lambda^T f(x, u, t) = 1 + \lambda^T AX + \lambda^T Bu, \tag{5.37}$$

from which we have

$$\frac{\partial H}{\partial u} = \lambda^T B = 0. \tag{5.38}$$

Equation (5.38) is independent of u because the Hamiltonian is linear with respect to this variable. This means that the method of substitution of u in equation (5.32) cannot be used in this instance. Instead, the substitution (5.37) in (5.35) gives

$$1 + (\lambda^*)^T Ax^* + (\lambda^*)^T Bu^* \leq 1 + (\lambda^*)^T Ax^* + (\lambda^*)^T Bu \tag{5.39}$$

for

$$(\lambda^*)^T Bu^* \leq (\lambda^*)^T Bu. \tag{5.40}$$

Now the right-hand side of the inequality can be rewritten as

$$(\lambda^*)^T Bu = \sum_i u_i(t) b_i^T \lambda(t). \tag{5.41}$$

The sum in (5.41) is minimized (made closest to zero) when each $u_i(t)$ is opposite in sign and maximum in magnitude in comparison with the product $b_i^T \lambda(t)$. This suggests that

$$u_i^*(t) = -u_{\max} \mathrm{sgn}(b_i^T \lambda(t)). \tag{5.42}$$

Thus, using Pontryagin's maximum principle, it is possible to state that the controls for the minimum time solution of the stated problem, if it exists, require a sequential application of maximum values of input in opposite directions ("bang-bang"). The inputs are switched from $u_{i_{\max}}$ to $-u_{i_{\max}}$, and vice versa, occuring until the required state transition is accomplished. This ensures that the bang-bang controls that cause the required state transition uniquely specify the minimum time state trajectory for the linear system.

Naturally the operation of such controls depends on a number of circumstances. For a linear system it is sufficient that the plant be stable, though the controls could have a finite duration period during which the optimal value is undefined for some of the components. These singular cases can result when a decoupled part of a system achieves its final state earlier than the rest of the system.

Bang-bang controls are characterized by *switching instants*; these are the moments when the sign of the input reverses abruptly. It has been shown for systems with real poles that the number of switching instants is always less than the order of the system and that they constitute a unique set. It is therefore possible to solve for these instants directly in determining $\mathbf{u}^*(t)$. Unfortunately, the determination of switching instants is predicated on the approximate solution of the boundary value problem in some form, graphical, numerical, or both.

In doing so, it is necessary to determine the initial conditions for the *motivational importance factors* λ which cause (5.42) to evolve in a manner that places the input switching points at such a time that the goal state is achieved by the plant. This requirement may be seen to be equivalent to solving the augmented state equation

$$\begin{bmatrix} \dot{x} \\ \dot{\lambda} \end{bmatrix} = \begin{bmatrix} A & 0 \\ 0 & -A^T \end{bmatrix} \begin{bmatrix} x \\ \lambda \end{bmatrix} + \begin{bmatrix} B \\ 0 \end{bmatrix} u, \tag{5.43}$$

where u is bang-bang, $x(t_0)$ and $x(T)$ are completely specified, but neither the final time T nor the initial or final conditions for λ are known. The conventional way to solve this problem is to use *search*, in the form of a gradient method or another successive approximation technique. A more primitive way would be to try all possible combinations of the two switching times for the third-order example (exhaustive search; see Chapter 8).

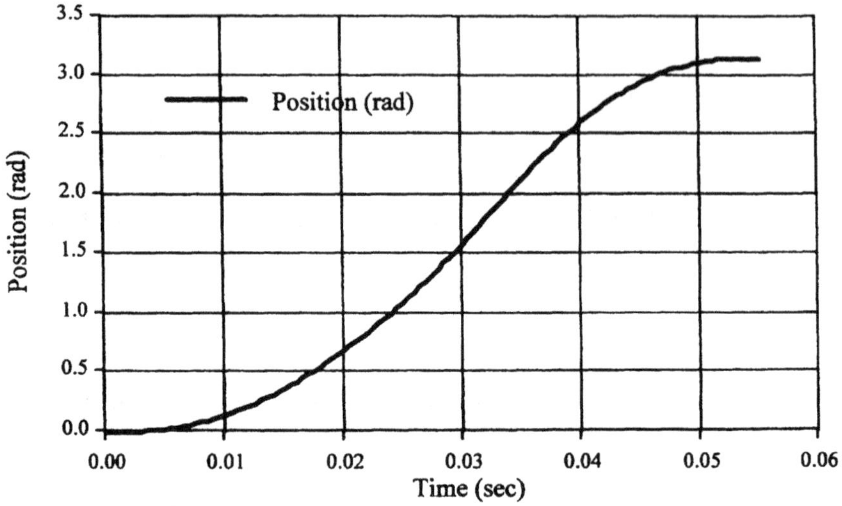

Figure 5.2. Optimal curve of position in relation to time.

5.3.4 Results and Discussion

For our example with dc motor, the (approximate) time optimal state, control, and adjoint trajectories are shown in the series of Figures 5.2 to 5.5. Evolution of the path in time is shown in Figure 5.2. The velocity–time curve does not take the form of a triangle as expected because the voltage constraints make it impossible to maintain the maximum acceleration. These peculiar curves for the applied torque (current) and velocity are determined by the law of optimum input demonstrated in Figure 5.5.

In our case the switching points for the input were determined by solving the linear system of equations in terms of eigenvalues and eigenvectors and by looking for changes of sign in the adjoint equations which placed the terminal state near $(\pi, 0, 0)$. A linear search was then performed at a 10 μs resolution to find the switching points that bring

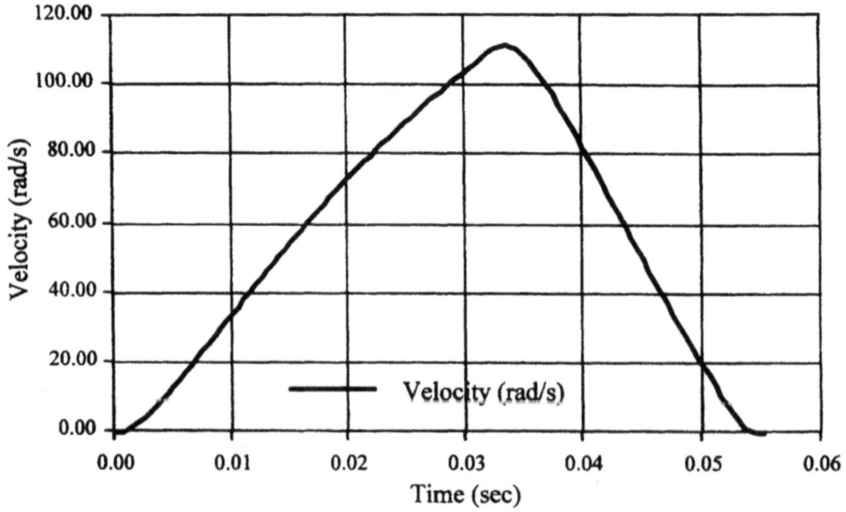

Figure 5.3. Optimal curve of velocity with relation to time.

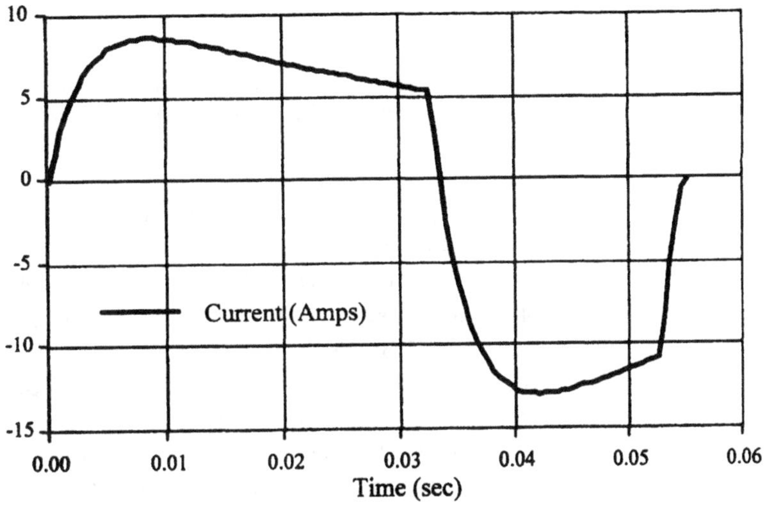

Figure 5.4. Optimal curve of current with relation to time.

the state of the system close to the desired state. Notice that even in this very simple case, one cannot complete the problem-solving process without *combinatorial search.*

Because of the simple structure of this problem, the solution can thus be determined relatively easily. In general, when the system model is not so convenient, or when Pontryagin's principle is not directly applicable, the search can be decidedly more complex, and it should be started at the early stages of the problem solving.

The foregoing example reaffirms our position on the significance of motivation, goals, and value judgment in problem solving. The situation gets substantially more complicated when the problem-solving methods are not prescribed by the library of solutions and must include the personality of the intelligent system.

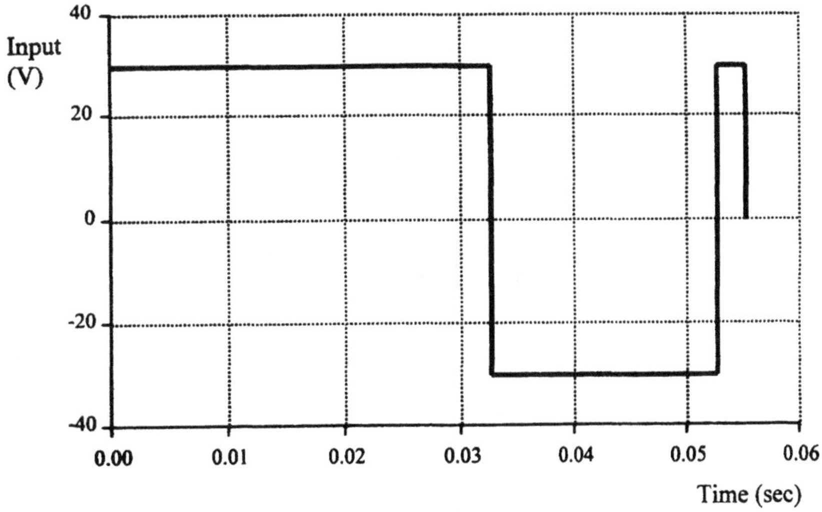

Figure 5.5. Optimal curve of input voltage with relation to time.

REFERENCES

[1] M. Frese and J. Sabini, Action theory: an introduction, in M. Frese and J. Sabini, eds., *Goal Directed Behavior: The Concept of Action in Psychology*, Lawrence Erlbaum, Hillsdale, NJ, 1985.

[2] M. Silver, Purposive behavior in psychology and philosophy: A History, ibid., pp. 3–17.

[3] C. R. Gallistel, Motivation, intention, and emotion: goal directed behavior from a cognitive neuroethological perspective, ibid., pp. 48–65.

[4] C. N. Cofer and M. H. Appley, *Motivation: Theory and Research*, Wiley, NY, 1964.

[5] C. Sherrington, *The Integrative Action of the Nervous System*, Scribners, NY, 1906.

[6] S. J. Beck and H. B. Molish, eds., *Reflexes to Intelligence*, Free Press, Glencoe, IL, 1959.

[7] R. Harper, C. Anderson, C. Christensen, and S. Hunka, eds, *The Cognitive Processes*, Prentice Hall, Englewood Cliffs, NJ, 1964.

[8] E. R. Guthrie, *The Psychology of Learning*, Harper and Row, NY, 1952.

[9] C. L. Hull, *A Behavior System*, Yale U. Press, New Haven, CT, 1952.

[10] B. F. Skinner, *Science and Human Behavior*, Macmillan, NY, 1953.

[11] W. B. Cannon, *The Wisdom of the Body*, W. W. Norton, NY, 1932.

[12] R. S. Woodworth, *Experimental Psychology*, Holt, Rinehart and Winston, NY, 1938.

[13] C. L. Hull, *Principles of Behavior*, Appleton-Century-Crofts, NY, 1943.

[14] W. James, What is emotion? *Mind*, Vol. 9, 1884, pp. 188–204.

[15] C. Lange (1995), *The Emotions*, Baltimore, 1922.

[16] S. Schachter and J. Singer, Cognitive, social, and psychological determinants of emotional state, *Psychological Review*, Vol. 69, 1962, pp. 379–399.

[17] R. M. Yerkes and J. D. Dodgon, The relation of strength of stimulus to rapidity of habit formation, *J. Comp. Neurol. Psychol.*, Vol. 18, 1908, pp. 459–482.

[18] H. A. Murray, *Explorations in Personality*, Oxford U. Press, NY, 1938.

[19] A. H. Maslow, *Toward a Psychology of Being*, Van Nostrand Rheinhold, NY, 1968.

[20] A. H. Maslow, *Motivation and Personality*, Harper and Row, NY, 1970.

[21] C. T. Morgan and R. A. King, *Introduction to Psychology*, McGraw-Hill, NY, 1966.

[22] J. Reykowski, Social motivation, *Annual Review of Psychology*, Vol. 33, 1982, pp. 123–154.

[23] M. Zuckerman, *Sensation Seeking: Beyond the Optimal Level of Arousal*, Erlbaum, Hillsdale, NJ, 1979.

[24] B. Schwartz, *Psychology of Learning and Behavior*, Norton, NY, 1978.

[25] Ed. L. Weiskrantz, *Analysis of Behiavioral Change*, Harper and Row, NY, 1968.

[26] D. C. McClelland, J. W. Atkinson, R. A. Clark, and E. T. Lowell, *The Achievement Motive*, Appleton-Century-Crafts, NY, 1952.

[27] J. W. Atkinson, *An Introduction to Motivation*, Van Nostrand Rheinhold, Princeton, NJ, 1964.

[28] H. H. Kelley and J. L. Michela, Attribution theory and research, *Annual Review of Psychology*, Vol. 31, 1980, pp. 457–501.

[29] L. Festinger, *A Theory of Cognitive Dissonance*, Harper and Row, NY, 1957.

[30] R. C. Wylie, *The Self-Concept*, U. of Nebraska Press, Lincoln, Nebraska, 1961.

[31] A. C. Guyton, *Organ Physiology: Structure and Function of the Nervous System*, Saunders, Philadelphia, PA, 1976.

[32] W. B. Scoville and B. Milner, Loss of recent memory after bilateral hippocampal lesions, *J. Neurophysiol. Neurosurgery Psychiatry*, Vol. 20, No. 11, 1957, pp. 11–29.

[33] J. Albus, Outline for a theory of intelligence, *IEEE Transactions on Systems, Man, and Cybernetics*, Vol. 21, No. 3, 1991, pp. 473–509.

PROBLEMS

5.1. Using examples of systems introduced in Problem 4.2 (from the Problems to Chapter 4), determine the goals of functioning. Determine the goal that enters the BG-module at each level of resolution. What kind of motivation has generated these goals?

5.2. Analyze the rules used in BG-modules of systems considered in Problem 5.1 and try to determine whether they are similar to *instincts* or *reflexes*. How would you characterize the motivators using the similar phenomena from the human activities?

5.3. Follow the evolutionary development of the systems under consideration (from both lists A and B). Demonstrate the development of motivations. Do this for all nodes in the tree of task decomposition.

5.4. Analyze the tree of motivations corresponding to the tree of task decomposition obtained in the solution for Problem 5.3. Construct the tree of cost-functions corresponding to this tree of motivation.

5.5. How would you apply these cost-functions for conducting value judgment for the subsystems of SP, WM, and BG. What are the value state variables? How are they related to the state-variables of this particular ELF?

5.6. Find anthropomorphic analogies for the value variables in the terms of the vocabulary presented in Subsubsection 5.2.3.

5.7. For the systems under consideration, apply reasoning from Subsection 5.3 targeting optimizing your system by using calculus of variations. Do you have all relevant information? If not, what information is lacking? What would be your cost-functional applied for optimization purposes.

CHAPTER 6

SENSORY PROCESSING

Within each node of the RCS architecture, sensory processing (SP) computations are conducted on the information that the sensors extract about the world and that thus generates successful behavior. We show in this chapter a hierarchy of SP modules in which input information of all levels is processed together with multiresolutional information of world model (WM). SP processes provide WM processes with the information necessary to maintain the knowledge database (KD) as a current, accurate, and relevant model of the world.

DEFINITION OF SENSORY PROCESSING

Sensory processing is a set of functional processes that operate on data from sensors to extract information about the world to create and/or update the world model.

6.1 IN-LEVEL AND INTER-LEVEL PROCESSES

Within each SP module, there are two kinds of processes:

Stage 1. Formation of the generalized entities from the elementary information units not yet organized into entities (inter-level processes). These are primarily various forms of using GFS (GFACS) on arriving information. The procedures include searching, focusing attention, and grouping applied to the input in various combinations.

Stage 2. Recognition and identification of the entities formed at the preceding stage (in-level processes). These are procedures of labeling the entities discovered at stage 1 and preparing them for use in WM.

Actually, these two stages require for dealing with the two adjacent levels of resolution simultaneously. The level of elementary information units is the higher-resolution level. The GFS operator is applied to these elementary units and generalizes them into the lower-resolution units by attention focusing, grouping, and combinatorial search (see Chapter 2). Grouped units get transformed into entities of lower resolution and enter the level above. At the lower-resolution levels, grouped elementary units of information exist in the form of generalized lower-resolution entities, and they are structured in an entity-relational network. The lower-level resolution receives additional information from its real or virtual sensors, and it groups its information into more generalized units in order to place them in the adjacent level right above by applying GFS-operator, and this process is repeated over and over again.

BOTTOM-UP DISAMBIGUATION

As the lower-resolution entities are created and supplemented by new information that arrives from the sensors, all the information is subjected to the GFS-inverse (GFS^{-1}) and is sent to the higher-resolution level (top-down) for verification and re-grouping, if necessary.

TOP-DOWN DISAMBIGUATION

At the stage 1 the GFS-operator is applied; it performs attention focusing, grouping, and combinatorial search. Also some basic processing functions are applied: (1) attention focusing (the FA part of GFS), (2) creation of grouping hypotheses (G and CS parts of GFS), (3) computation of attributes for grouping hypotheses (G part of GFS), (4) selection of group attributes and confirmation of grouping hypotheses (CS part of GFS), and (5) classification of the created entities, recognition of the created entities, and organization of grouped entities and events into an entity-relational network.

6.1.1 Focusing of Attention (F of GFS)

The regions of space and the intervals of time over which the sensory processes will operate on sensory inputs are selected by attention focusing. All other input can be masked or ignored. The shape, position, and duration of spatial and temporal masks and windows of attention are determined by an attention function that defines regions in the image requiring attention. The attention function selects regions that are important for one of two reasons:

1. They are relevant to tasks being addressed by the BG process.
2. They contain entities or events with attributes that fall outside some expected norm (e.g., they are evaluated as surprising or dangerous).

Windowed regions of space and time can be assigned priorities and allocated computational resources in proportion to their relative importance. Figure 6.1 shows an example of applying windows of attention to a front view of the intelligent mobile autonomous system (IMAS) developed at Drexel University in 1984–1987. This is what was visualized by a CCD camera installed at the vehicle. The image was scanned by sliding windows to determine singular entities as described in Chapter 2. As a result, the singularities were covered by rectangular windows of uniform density. A rule base was deciphering them into the concept of "obstacle–human."

WINDOWS OF UNIFORM DENSITY

However, it would take the LADAR from the further development of the autonomous vehicle (see Chapter 11) to recognize that one of two singularities caught by windows is just a shadow of the "obstacle–human."

Focusing of attention (FA) is a method for reducing the volume of information to be computed during the process of grouping (G). However, it can also be used as a tool for combinatorial search (CS). The well-known mechanism of "sliding windows" is an example of FA used in conducting CS.

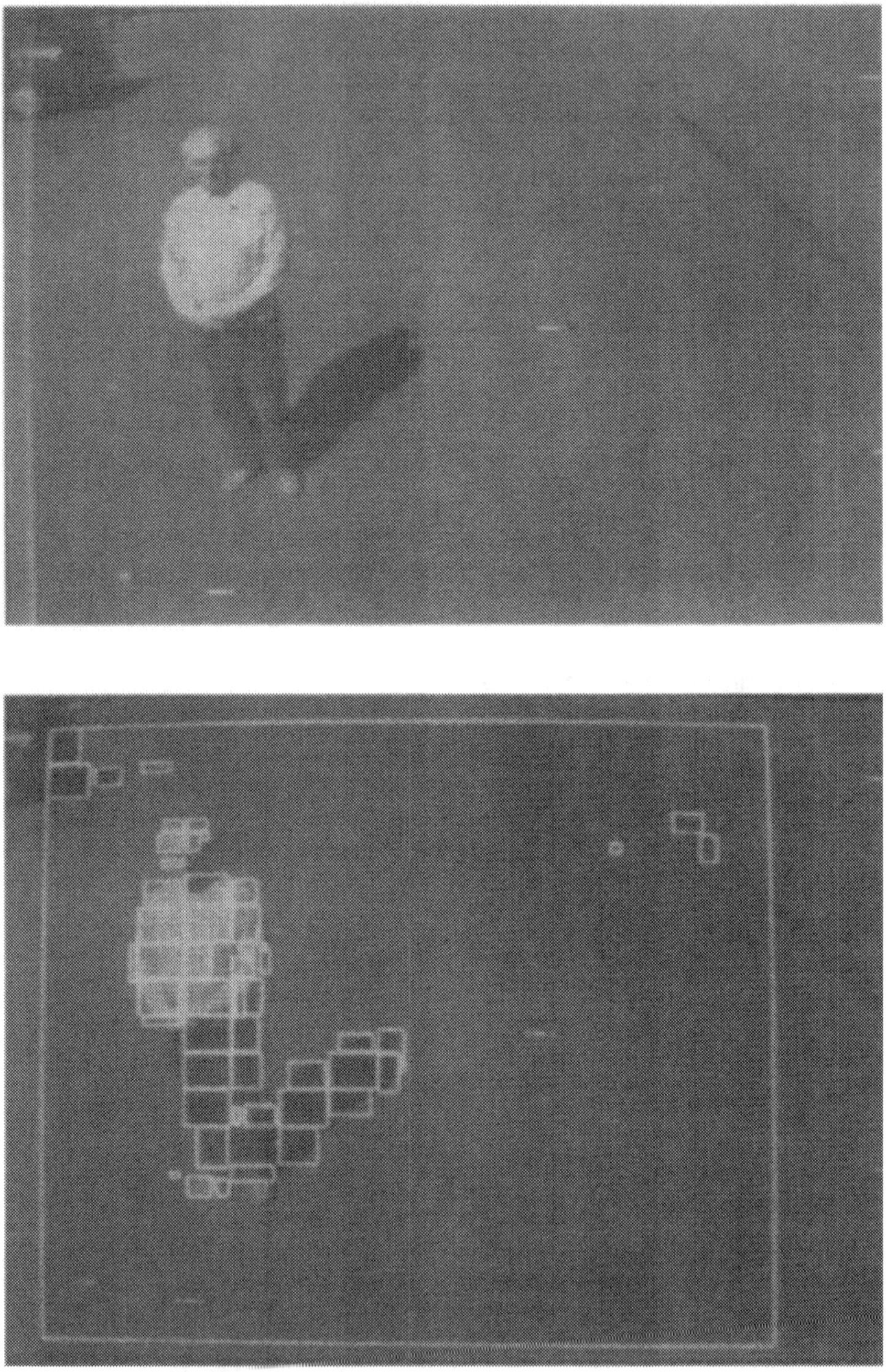

Figure 6.1. Typical attention-focusing images selected by cameras mounted on a vehicle.

6.1.2 Creation of Grouping Hypotheses (G and S of GFS)

Elementary information (potential subentities) is organized or integrated into entities—and the potential subevents into events—by creation spatial, or temporal, or joint spatio/temporal hypotheses. Likewise, temporal groupings segment larger time intervals into shorter intervals that can be associated with event labels or names. This decomposition should be seen as temporal patterns of signals incoming from each of three sensors. The signals are grouped by subevents along the time line, and a pattern of subevent strings forms the event. Each event, which consists of subevents that have been grouped together, has some duration and is bounded by points on the time line.

PROCESS DECOMPOSITION

Each of the lower-resolution events is being decomposed into higher-resolution events as demonstrated earlier in Figures 3.1 and 4.2. In Figure 6.2 this set of events (processes) is partially shown in a more convenient notation.

In signal processing, temporal grouping corresponds to establishing event boundaries along the time line. These may be bit, word, or frame boundaries. In Figure 6.2 each event on the time line has a name that points to an event frame with attributes of the event, such as its duration, spectral components, or intensity. Besides the list of attributes, each frame contains a set of pointers to subevents that belong to it, and a pointer to the higher-level event to which it belongs. The pointers are established by grouping processes.

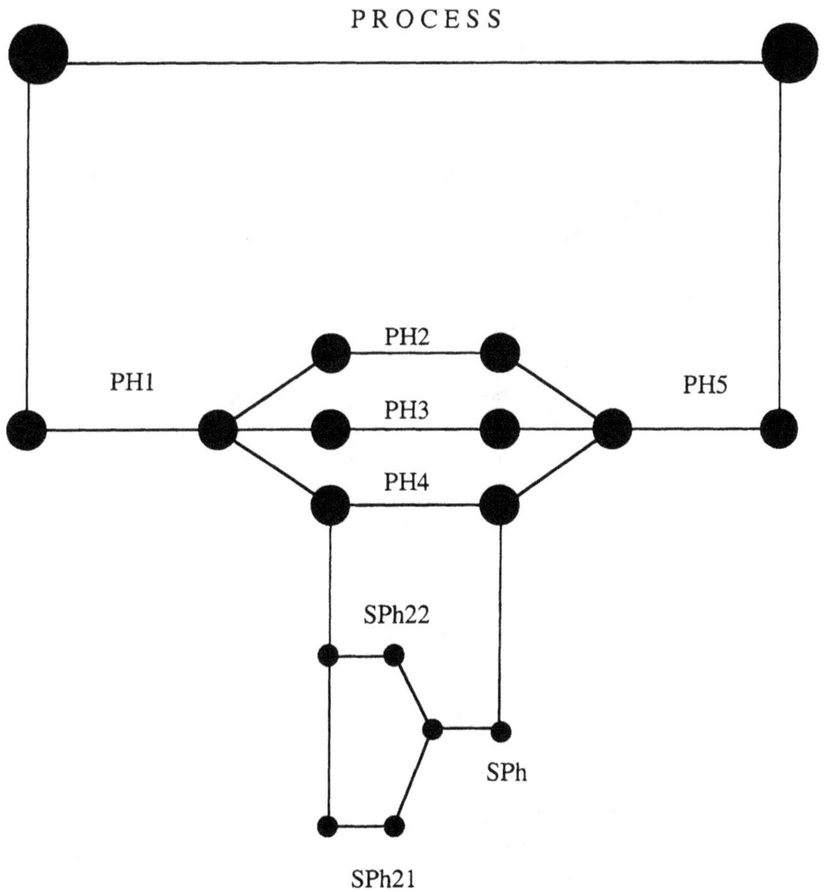

Figure 6.2. Part of the events from Figure 3.1 in the designer's notation.

In image processing, spatial grouping segments images into regions that can be assigned entity labels or names. Each pixel in the labeled region carries the name of the entity to which it belongs. This creates an entity image consisting of pixels labeled with entity names. For each region in an entity image, there is an entity frame that contains the attributes computed for the entity. For example, each pixel in the surface entity image has a pointer to the name (or location) of the surface entity frame to which it belongs. In the surface entity image, pixels are grouped into surface entities. In the object entity image, pixels are grouped into the object entity. Pointers in the entity frames describe the relationship between entities and provide links back to entity images.

Thus each pixel in the object entity image has a pointer to the name of the object entity frame to which it belongs. Each surface entity frame has a pointer back to the surface entity image, and each object entity frame has a pointer back to the object entity image. Each surface entity frame also points to the object entity frame to which it belongs, and the object frame has a set of pointers to the surface entity frames that are part of it. All of these pointers result from the grouping operation that collects pixels and organizes them into entities.

TOOLS OF GESTALT

The grouping of entities or events by hypothesis is based on the rules of gestalt, or gestalt heuristics (more on this in Chapter 10). One frequently exercised gestalt rule is that of spatial and temporal *proximity* (subentities are situated close together in an image, or subevents are situated close together along the time axis). Another important sign of belonging to the same entity is *similarity* (subentities have attributes such as color, texture, range, or motion, or subevents share the same spectral properties or sequential relationships). Both proximity and similarity can be observed within the heuristics of *continuity* (subentities have directional attributes that line up or lie on a straight line or on a sufficiently smooth curve) and *symmetry* (subentities are evenly spaced or are symmetrical about a point, line, or surface).

There is, of course, no guarantee that any grouping hypothesis for entity or event in SP produces an internal entity or event that has any correspondence to an entity or event in the real world. Usually more hypotheses are created than a reasonable amount of entities allows. For example, two pixels that are in close proximity in an image may lie on completely different entities in the world. Two pixels of the same color and intensity may lie on different objects, and two pixels with different colors and intensities may lie on the same object. Edges that line up in an image may or may not lie on the same edge in the world. There is also no guarantee that any SP event grouping hypothesis produces an internal event that corresponds to an external event. Two events that are next to each other on the time line may derive from completely different phenomena in the world. Grouping hypotheses need to be tested, and confirmed or denied, by observing how well predictions based on each grouping hypothesis match subsequent observations of sensory data over time under a variety of circumstances.

6.1.3 Computation Attributes for Grouping Hypotheses (G of GFS)

Event attributes such as waveform, frequency components, duration, and pattern of motion can be computed by integrating over the duration of the event. Entity attributes such as position, velocity, orientation, area, shape, and color can be computed by integrating subentity attributes over the region in an image covered by the hypothesized entity. For example, the brightness of each pixel entity is computed by integrating the photons within a spectral energy band falling on a photo-detector in the image plane during an interval of time. The color of a pixel is computed from the ratio of brightness of registered arrays of pixels in three different spectral bands. The spatial or temporal gradient of brightness or color at a pixel can be computed from spatial or temporal differences between adjacent pixels in space or time. Second, third, or fourth derivatives of brightness or color can be computed by convolution with spatial or temporal filters over larger neighborhoods or in registered lower-resolution images.

The length of an edge entity in an image can be computed by counting the number of pixels along the edge. The area of a surface entity can be computed by counting the number of pixels contained in it. The cross-sectional area of an object entity can be computed by multiplying the area of its projection in the image by the square of the ratio of its range to the focal length of the camera. The lateral velocity of an object entity can be computed by computing the angular motion of its center of gravity in the image multiplied by its estimated range. The radial velocity can be computed from range-rate measurements. The orientation of an edge can be computed by regression of edge pixels onto a tangent line segment. The curvature of an edge can be computed by regression of edge pixels on a curved line segment. For surface entities, attributes such as area, texture, color, shape, and orientation can be computed. For surface boundary entities, attributes such as length, shape, and type (i.e., intensity, color, or range discontinuity) may be computed. For object entities, attributes such as size, shape, color, texture, and state including position, orientation, and velocity can be computed by integrating pixel attributes over the region occupied by the entity. Entity attributes and states can be represented in the list of attribute-value pairs contained in the entity frames.

6.1.4 Selection and Confirmation (S of GFS)

Selection of group attributes and confirmation of grouping hypotheses (S of GFS) are processes that require measures for reducing the noise and enhancing the signal quality. The quality of computed values for entity attributes and states can be improved by using GFS operations with grouping signals with the help of time-windows (as in the well-known procedures of recursive estimation) that are actually another embodiment of GFS package (it has a focusing attention sampling tool, a computational tool for generalization, etc.).

Recursive estimation accomplishes two purposes: (1) It computes a "best estimate" (by grouping signals over a window of space and time) of entity attributes and states, and (2) it generates statistical "estimators" such as confidence factors for observed and estimated attributes and state values. Recursive estimation means that a new "best estimate" is computed with every new sensory observation. Each sensory measurement adds new information to what was previously known about entities and events in the world. Recursive estimation operates by comparing a prediction based on the current "best estimate" with an observation based on sensory input. The variance between an observation and the prediction is used to update the best estimate and to compute a confidence in that best estimate.

Confidence in the best estimate can be used to confirm or deny the grouping hypothesis that created the entity. When variance between observed and predicted attributes is small, confidence in the grouping hypothesis is increased. When the variance between the observed and predicted attributes is large, confidence is reduced. When the confidence factor for the entity attributes rises above a confirmation threshold, the grouping hypothesis that generated the entity is confirmed. When the confidence falls below a denial threshold, the grouping hypothesis is rejected and a new grouping hypothesis must be selected.

Recursive estimation uses the GFS concept, and therefore it employs the principle of hypothesizing and consecutive testing of hypotheses. At each point in time, the best estimate is a hypothesis about the state of the world. Based on that hypothesis, WM generates a prediction of what will happen next in the world and how that will affect the next sensory input. SP tests the WM hypothesis by comparing WM predictions with sensory observations. If predictions correlate with observations over a window of space and time, the WM best estimate hypothesis is confirmed within that window. If not, the best estimate is discredited, and a new hypothesis is advanced to account for the sensory input.

6.1.5 Classification, Recognition, and Organization of Entities

Recognition is a process that establishes a match between confirmed entities and entity classes stored in the system's knowledge database. It is based on similarities between the attributes and structure of a confirmed entity and an entity class. A variety of similarity measures can be used, including distances between, or the dot product of, the two attribute vectors. If the degree of similarity exceeds a recognition threshold, the entity is recognized as a member of the class.

Recognition generates for each geometric entity frame a pointer that contains the name of the entity class to which that frame belongs. Recognition also generates for each pixel in the corresponding entity image, a pointer that contains the name of the entity class to which it belongs. Thus, for each recognized entity, the entity image is labeled with the name of the entity class. Thus image and map overlays can be constructed with color-coded labels for roads, buildings, bridges, targets, friendly forces, and enemy positions.

Once images and maps are labeled with entity names, additional overlays can be generated that contain the attributes of the recognized entity class. By this means, images and maps can have regions labeled with attributes and values that cannot be directly observed such as names of rivers, depths of water, likelihoods of rain, or values of objects to be attacked or defended. Maps with such labels are what is needed by the BG hierarchy for making long-range plans and for developing strategies, tactics, and maneuvers to accomplish those plans.

Detection establishes a match between the attributes of a confirmed event and stored event classes in the system's knowledge database. Detection is based on the similarity between the attributes of a confirmed event and attributes of an event class. A series of detected events generates a historical trace of labeled events with event frames that describe the recent past and provide the basis for planning future actions. The historical trace of labeled events may be stored as a queue in short-term memory so that events can be grouped into higher-level events and attributes of higher-level events can be computed.

HOW GFS WORKS AT A LEVEL

Figure 6.3 is one of the possible data flow diagrams of the set of the processing functions that make up SP functionality and as described above. The input image is windowed by an attention function (FA) that selects what part of the image is important. FA decides what is important based on the task and makes a VJ assessment of the object and events detected in the sensory data stream within the context of a task assigned to a BG process at the same level. The current task being planned and executed in the BG hierarchy requests from the WM the prediction function that helps to search (CS) and selects the gestalt hypotheses (grouping, G). The WM prediction also determines the higher-resolution FA windows for regions of the image that are important to the task and therefore need to be processed. Regions that are irrelevant to the task are masked out and ignored. The task context also selects from a library of entity and event class frames in long-term memory a set of expected entities, events, and situations that are

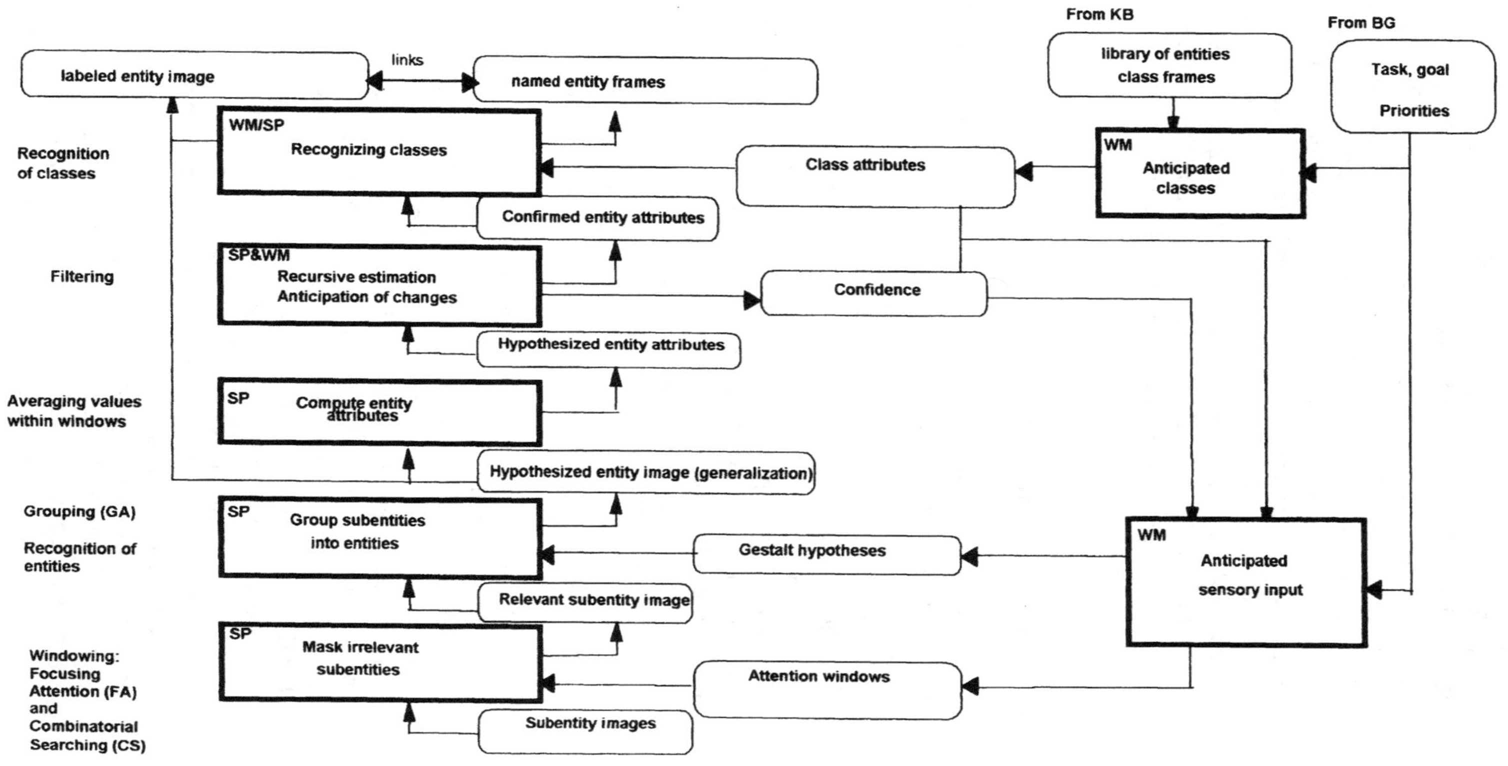

Figure 6.3. GFS functioning as a set of image processing procedures.

relevant to the task. This narrows the search for a match between observed entities and entity classes.

The FA constructs windows that contain relevant information. The grouping function (G) distributes the subentities to lower-resolution entities: The gestalt hypotheses should be created and used depending on the goal of the task. The attributes and states of the groups are computed and processed by recursive estimation (higher-resolution GFS). This process generates statistics for assessing preferences of hypotheses. If the grouping hypothesis corresponds to physical relationships in the real world, it will be confirmed. Otherwise, other grouping hypotheses will be tested. Finally entity attributes and states of confirmed groups are compared with attributes and states of entity classes stored in the knowledge database. When a match is recognized between a confirmed entity and a stored entity class, the class name is assigned to the confirmed entity frame and the corresponding region in the entity image is labeled with the class name.

6.2 SENSORY PROCESSING AS A MODULE OF THE LEVEL

The structure of ELF was introduced in Chapter 1, developed in Chapter 2, and extended to the multiresolutional case in Chapter 4. The hierarchy of SP is substantially interwoven with WM and BG hierarchies. The relationships and interactions among BG, WM, KD, SP, and VJ in a typical node of the RCS architecture are shown in Figure 6.4. The behavior generation (BG) processes contain the job assignment (JA), scheduling (SC), plan selector (PS) functions and executor (EX) subprocesses. The planner (PL) module overlaps BG, WM, and VJ processes. Planning involves the JA, SC, and PS subprocesses, the WM simulator, and VJ plan evaluator. The PS subprocess selects the best plan for execution by the EX subprocesses. The world modeling (WM) process supports the knowledge database (KD) that contains both long-term and short-term symbolic representations and immediate experience images. The WM contains the plan simulator, where the alternatives generated by JA and SC are tested and evaluated, as well as mechanisms for generating predicted images that can be compared with observed images. Sensory processing (SP) contains windowing, grouping, and filtering algorithms for comparing predictions generated by the WM process with observations from sensors. SP also has algorithms for recognizing entities and labeling entity images. The value judgment (VJ) process evaluates plans and computes confidence factors based on the variance between observed and predicted entity attributes.

A task command from level $i + 1$ specifies the goal and object of the task. The goal specification selects an entity from long-term memory and moves it into short-term memory. The planner generates a plan that leads from the starting state to the goal state. The EX subprocesses execute the plan using predicted state information in the KD short term memory. SP operates on sensory input from level $i - 1$ to update the KD within the node, and sends processed information to level $i + 1$.

A NODE OF RCS Each node of the RCS hierarchy closes a control loop. Input from sensors is processed through SP and used by the WM to update the knowledge database (KD). This provides a current best estimate $\underline{X}$ and a predicted state X^* of the world. A control law is applied to this feedback signal arriving at the EX subprocess. The EX subprocess computes the compensation required to minimize the feedback error between the desired state X_d and the predicted state X^* that emerges as a result of the SP/WM filtering process. The predicted state X^* is also used by the JA and SC functions and by the WM plan simulator to perform their respective planning computations. Labeled entity images

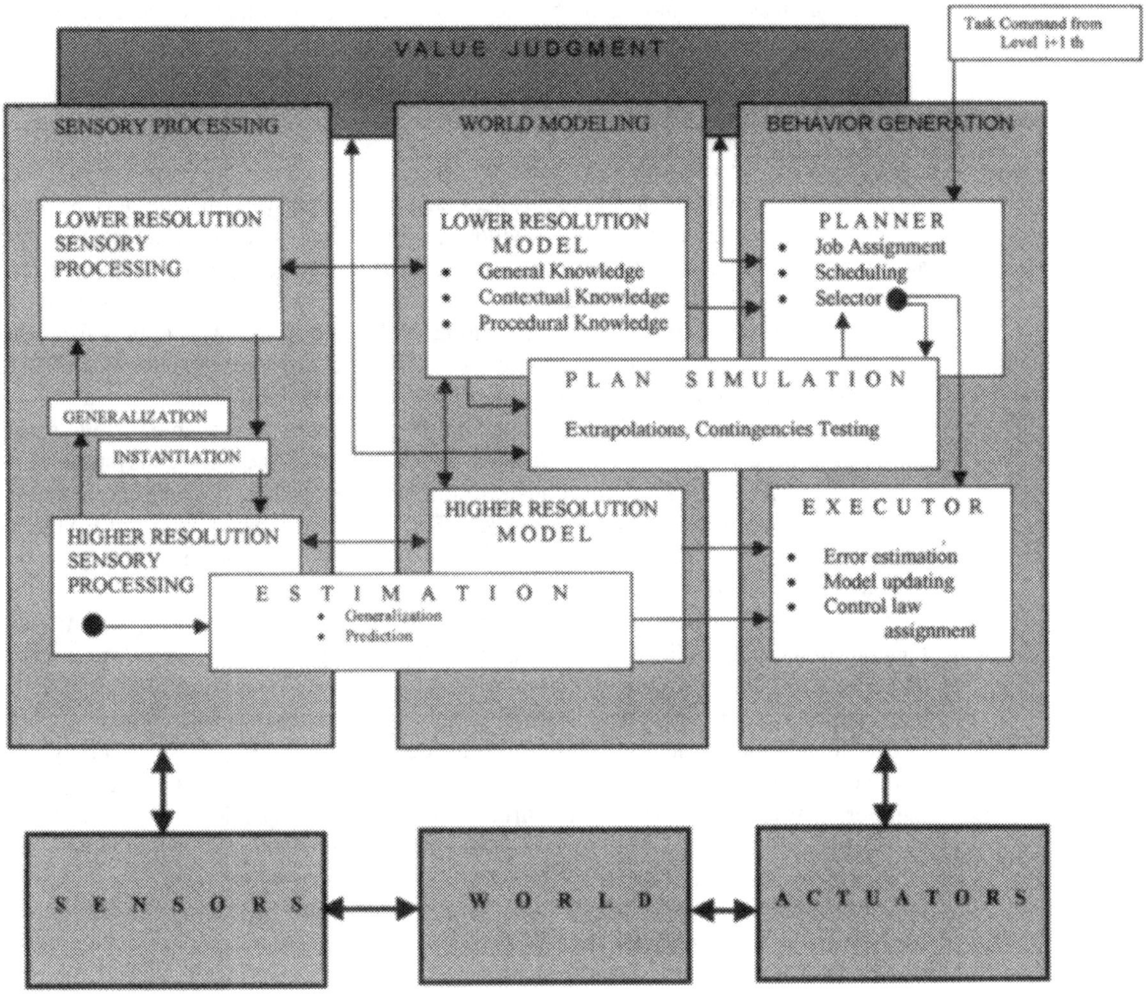

Figure 6.4. Relationships within a typical node of the RCS architecture.

may also be used to generate labeled maps (not shown in Fig. 6.4) for path planning by the BG process.

Within each node, windowing, grouping, computation, filtering, and recognition takes place. Within each node, planning and control functions are based on the knowledge maintained in the knowledge database.

Within each node, lower-level entities are recognized and grouped into higher-level entities. A model of the world is maintained that enables planning and control functions relevant to that node. Within each node, an internal model of the world enables SP to analyze the past and BG to predict the future so as to maximize that node's ability to achieve commanded behavioral goals.

SP is task dependent and goal directed. As shown in Figure 6.4, the task command to BG specifies the object on which the task is to be performed. This causes the WM to select from long-term memory the frame containing class attributes of the task object. This task frame is then moved into short-term memory where it can be used to generate expectations and grouping hypotheses for the SP process. The task command also specifies the task goal and other parameters such as priorities that can be used by the attention functions in SP to establish windows and masks.

6.2.1 Information Sources

Input to sensory processing is provided by sensors. Each sensor produces a signal that varies in time with the measurable physical phenomena that vary in time, too. SP processes operate on sensory signals to window, group, filter, recognize, detect, classify, and interpret them as entities, events, and situations that correspond in a meaningful and useful way to entities, events, and situations in the real world. Outputs from SP processes are used by the WM processes to keep the KD up to date. Outputs from the WM may be returned to the SP subsystem to facilitate knowledge-based windowing, grouping, filtering, comparison, recognition, and interpretation of sensory input.

Intelligent systems typically have large numbers of sensors that enable them to sense the environment. Sensors may be grouped into a variety of sensory subsystems. For example, a vision subsystem may consist of one or more cameras, including black-and-white or color TV, forward looking infrared (FLIR), and laser range imaging (LADAR) cameras. TV and FLIR cameras subsystems may consist of monocular, stereo pairs, or nested sets or cameras with different magnification.

SENSORS

A typical camera consists of a two-dimensional array of sensors that track the intensity of incoming radiation in the visual or infrared spectrum. The black-and-white cameras contain photo-detectors that integrate the intensity across the entire visual spectrum. Color cameras typically have three arrays of sensors that are differentially sensitive in the red, green, and blue regions of the spectrum. An intelligent system may also have acoustic sensors that detect sounds in the environment. These may be arranged in pairs or arrays so as to monitor the direction of incoming acoustic energy. Radar subsystems transmit beams of microwaves and detect reflected energy from objects in the environment. LADAR cameras transmit and detect reflected laser energy. Sonar subsystems transmit and detect reflected sound energy. Tactile sensors measure touch, vibration, force, or pressure. Other types of sensors measure position, velocity, acceleration, force, torque, temperature, pressure, magnetic fields, electric voltage or current, nuclear radiation, and presence of chemicals in the air or in fluids.

An intelligent vehicle system may have accelerometers, tilt meters, gyros, and global positioning satellite (GPS) receivers. Odometers and speedometers are used to estimate position, velocity, and heading. Internal vehicle sensors provide information about fuel levels, engine temperature, rpm, oil pressure, and vibration.

An intelligent manufacturing system may have sensors that measure position and orientation of parts; velocity, force, and vibration of tools or fixtures; or temperature, hydraulic pressure, and operating state of machines. Touch probes measure dimensions of parts and part features. Vision sensors measure properties of surfaces, the transparency of liquids, the number of parts, and the condition of materials. Sensors read bar codes and identification tags. Chemical plants, refineries, and power plants typically have sensors that measure temperature, pressure, flow rate, volume, and state of chemical reactions.

Output from the sensors provides input to the sensory processing subsystems (SP). In each node of the RCS hierarchy, the SP goal is to extract information from the sensory data stream that are relevant to the behavioral task being planned and executed in that node. As sensory data are processed, they are filtered, windowed, and combined or grouped with data from other sensors into entities, events, and situations with attributes and states that can be matched against known classes for recognition and classification.

The goal of a vision-based SP on an intelligent vehicle is to detect obstacles, to track targets, and to recognize objects, events, and situations. Camera platform encoders measure the pointing direction of cameras relative to the vehicle, and inertial sensors enable the camera platforms to stabilize images and determine absolute camera pointing direction and tracking velocity. Then the SP subsystem is used to determine the position and motion of objects in the world relative to the vehicle. Characteristic size, shape, color, thermal signature, and motion enables the detection and recognition of objects, such as roads, ditches, fences, rocks, trees, bushes, dirt, sand, mud, wire, smoke, fire, buildings, bodies of water, cars, trucks, and people.

The goal of an acoustic SP subsystem is to track and recognize acoustic signatures such as voices, laughter, bird songs, dogs barking, sounds of cars, trucks, airplanes, and footsteps. Acoustic SP functions detect frequency components, compare time and frequency patterns with predicted acoustic events, measure correlation and differences, close phase-lock loops, and control tracking filters.

As sensory data are processed, they are filtered, windowed, and combined or grouped with data from other sensors. Attributes and states are computed that can be matched against known classes for recognition and classification. As a result signals are transformed into symbols, images are transformed into maps, entities and events are detected, and identified as belonging to classes. An internal world model is constructed that is the intelligent system's best estimate of what exists in the world. This world model is used by BG processes to plan and execute behavior designed to maximize the probability of success in achieving behavioral goals.

6.2.2 State-Space Tessellation: Sampling

Outputs from sensors are typically in the form of analog voltages or currents. Analog signals can be transmitted and processed by analog computers or used directly in analog feedback loops. However, analog signals are subject to noise and may experience dissipation and distortion over long distances because of random variation of circuit parameters during transmission. Therefore analog sensor signals are typically sampled and digitized before transmission.

In biological systems, analog sensory signals represented by cell membrane voltages are sometimes used locally for analog computations, but more often they are transformed into impulse rates (or interpulse intervals) of action potentials for transmission long distances over axons. In artificial systems, voltages from sensors are typically sampled and digitized by analog-to-digital converters into binary numbers before being transmitted over long distances or processed in a digital computer.

The rate at which sensory signals are sampled establishes the resolution along the time line and hence constrains the bandwidth of signals that can be represented. The Nyquist criterion states that the highest-frequency signal that can be represented by sampling must be less than one-half the sampling rate. In practice, signals are typically filtered to eliminate frequencies above one-tenth of the sampling frequency.

Tessellation (e.g., quantization) of sensory signals into digital levels establishes the amplitude in the resolution. Thus the precision with which measurements can be represented is limited. Signals from photo-detectors typically are quantized with eight bits (256 levels) of resolution. Pressure, force, or tension sensors may be quantized with 10 bits (1024 levels) of resolution. Position sensors may be quantized with 12 to 16 bits (4096 to 64,000 levels) of resolution.

Arrays of sensors sample the incoming energy from the environment at many different points in space. The spacing of sensors in a camera, or in the retina, or on the skin, establishes the spatial resolution of the sensor array, and hence the level of detail in, or magnification of, the image that can be represented. Low-resolution image arrays may contain 128 × 128 (about 16,000) pixels or less. High-resolution images may contain 1024 × 1024 (about 1,000,000) pixels or more. Variable resolution may be used to produce high spatial resolution in dense regions of the sensor array and low resolution in sparse regions of the sensory array. For example, the eyes of many biological species have closely spaced photoreceptors in the fovea, and widely spaced sensors throughout the remainder of the retina. In the human eye, the density of vision sensors decreases roughly as the logarithm of the distance from the optical axis. This produces a high-resolution, highly magnified image in the fovea, with lower resolution and lower magnification in the periphery. This is called *foveal/peripheral vision*. A similar effect can be achieved in artificial systems through a variety of mechanisms such as log-polar CCD arrays, optical lenses with nonuniform magnification, zoom lenses, or multiple cameras with different focal length lenses.

Different sensors may be tuned (i.e., maximally sensitive) to different portions of the frequency spectrum of the incoming energy. For example, red, green, and blue cones in the retina are sensitive to energy in three different bands of the visible part of the spectrum. Rods are much more sensitive than cones to low-light levels but generate only gray-scale images. In the ears, the cochlea is a tuned cavity designed so that vibration sensors at different places in the cochlea are tuned to specific frequencies of sound. In the skin, different kinds of sensors are designed to detect different tactile sensations such as pressure, vibration, temperature, and pain. In the smell and taste systems, sensors are tuned to detect chemical properties of molecules with which they come in contact.

6.2.3 Noise, Uncertainty, and Ambiguity

In most cases sensors do not directly provide the information necessary for controlling behavior. Only in a few cases related to high-resolution levels, such as an encoder or tachometer mounted directly on a motor output shaft, does the sensor directly measure the world state that the BG system seeks to control. In the case of the lower-resolution levels of the BG hierarchy, where the goal may be to capture prey, to avoid a predator, or to attract a mate, very little about the state of the world can be directly measured. The state of objects in the world must be estimated and the identity of prey objects must be inferred from sensory data that are highly ambiguous and often intentionally deceptive. For example, prey animals use camouflage and deceptive behavior to confuse SP and WM subsystems in the predators.

Visual data can be especially ambiguous. For example, both the photo-detector in a camera and the retina of the eye measure only the intensity of light. The visual signal is completely ambiguous with respect to how far away the object is from where the light originated. The visual signal from a single photo-detector can also be ambiguous about what the object is and in what direction it is moving. SP functions must deduce this kind of information by interpreting sensory data in the context of signals from other sensors and information from the world model.

Sensory data are frequently confounded by noise. For example, images from cameras typically are noisy, particularly under low-light levels. Acoustic energy arriving at the ear typically is the result of a combination of sounds originating from a wide variety of sources, most of which are irrelevant to the behavioral task at hand. SP subsystems must extract from this cacophony of sounds information that is useful for understanding what is going on in the world.

6.2.4 Creation of Hypotheses and Testing

The ability of an intelligent system to extract the information needed to construct a reliable world model despite ambiguity and noise in the sensory input is due to a process of hypothesis making and testing. Intelligent systems make plausible hypothetical interpretations of the sensory input and then test how well these interpretations are confirmed by subsequent observations. Except for the most primitive senses of pain, smell, and taste, most perception involves formulating a hypothesis and then testing it. Clearly, the ability to recognize and analyze complex situations in the world depends on the ability of the intelligent system to make inferences, to reason about these inferences, to propose theories, and to see how well they stand up to further experimental observations. This process has been developed in human history as the scientific method. However, it exists in more primitive form in virtually every intelligent system.

DISAMBIGUATION

Recursive estimation is a form of hypothesis and testing. Recursive estimation is employed at every level of the SP/WM hierarchy to filter noise and interpret sensory input. At every sampled point in time, the WM generates a hypothesis in the form of a prediction based on its current best estimate of the state of the world. The SP unit then tests this hypothesis by comparing WM predictions with sensory observations.

A schematic of a simple hypothesis-test interaction between the SP and WM is shown in Figure 6.5. The WM makes a prediction based on the set of estimated state variables that reside in the knowledge database. WM predictions are compared in the SP with sensory observations. The sensory observations come directly from sensors or from lower-level SP subsystems. The correlation between sensory observations and WM predictions indicates the degree to which WM predictions are correct. If the correlation exceeds the threshold, the hypothesis is verified, or confirmed. When the hypothesis is verified, the focus of attention can be narrowed, and residual difference values can be used to update estimated state-variables to improve the hypothesis. Differences or correlation offsets indicate error between the WM prediction and SP observation. If errors rise above a certain threshold, the hypothesis is rejected, the focus of attention is broadened, and a new hypothesis is generated.

This is, in fact, recursive estimation. It tightly couples observations with predictions to extract a reliable estimate of the state of the world. By hypothesis and test, SP and WM work together to establish and maintain a correspondence between the real world and a model of the world in the knowledge database. The SP subsystems combine observations from sensors and lower-level SP functions with predictions based on knowledge already in the KD. Knowledge at time t is a combination of prior knowledge at time $t-1$ plus current observations from sensors and control signals sent to the output at time t.

The world model uses the current best estimate of the state of the world stored in the knowledge database to predict the current sensory input. This prediction is compared with sensory observations in sensory processing. Estimates of correlation and values of difference are returned to the world model to update the knowledge database. This, again, is a recursive estimation process. It can be expressed as follows:

$$KD(t) = KD(t-1) + \text{observation}(t) \tag{6.1}$$

where $KD(t)$ is the current best estimate of the state of the world at time t, and $\text{observation}(t) = \text{sensory input}(t) + \text{control output}(t)$.

The system records the incoming light intensity from a certain field of view at a fixed sampling rate. By this imaging process the information flow is discretized in two ways:

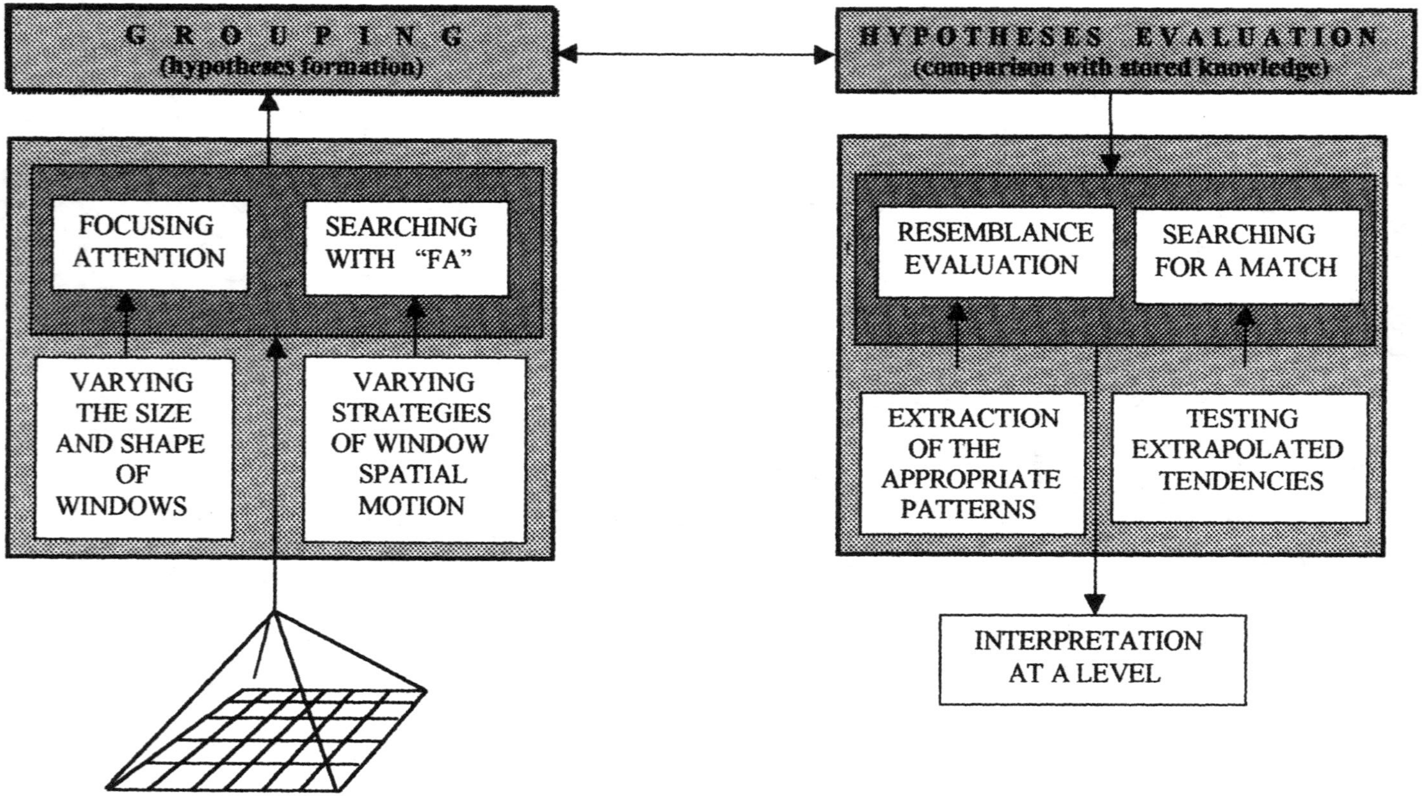

Figure 6.5. Hypothesis making and testing in an intelligent system.

there is a limited spatial resolution in the image plane determined by the pixel spacing in the camera, and a temporal discretization determined by the scan rate of the camera.

Instead of trying to invert this image sequence for 3D scene understanding, a different approach of analysis through synthesis has been selected, taking advantage of the available recursive estimation scheme after Kalman. From previous experience, generic models of objects in the 3D world are known in the interpretation process.

This approach comprises both 3D shape, recognizable by certain feature aggregations given the aspect conditions, and motion behavior over time. In an initialization phase, starting from a collection of features extracted by low-level image processing (lower center left in Fig. 6.6), object hypotheses including the characteristic conditions and the motion (transition matrices) in space have to be generated (upper center left in Fig. 6.6). They are installed in an internal "mental" world representation intended to duplicate the outside real world. This is sometimes called world 2, as opposed to the real world 1.

Once the grouping of objects has been instantiated in the world 2, exploiting the dynamical models for those objects allows the prediction of object states for that point in time when the next measurements are going to be taken. By applying the *forward* perspective projection to the visible features, using the same mapping conditions as in the CCD sensor, a model image is generated that should duplicate the measured image if the situation has been understood properly. The situation is thus "imagined" (right and lower center right in Fig. 6.6). The advantage of this approach is that the

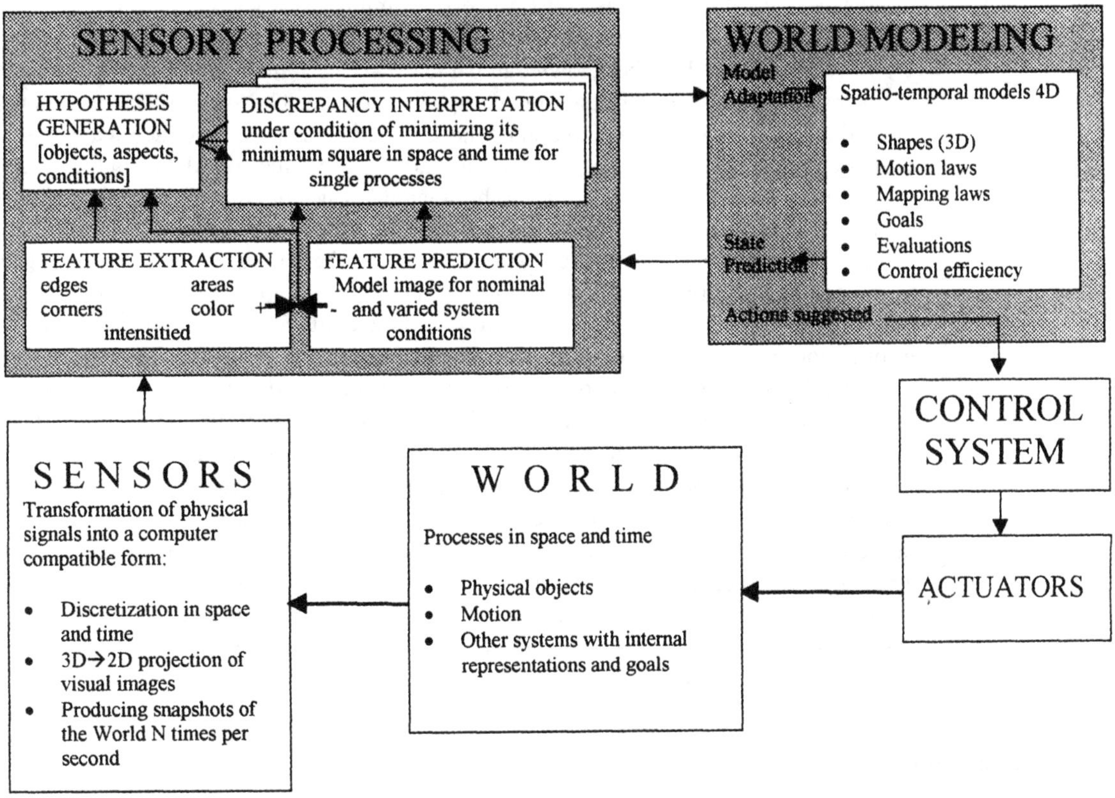

Figure 6.6. Basic scheme for 4D image sequence understanding by prediction error minimization.

internal 4D model [33] allows not only the actual situation at the present time to be determined but also the Jacobian matrix of the feature positions and orientations with respect to all state component changes (upper block in center right, lower right corner). This need not necessarily be done by analytical means but maybe achieved by numerical differentiation exploiting the mapping subroutines already implemented for the nominal case.

This rich information is used for bypassing the perspective inversion via recursive least squares filtering through the feedback of the prediction errors of the features. More details are available in Dickmans and Graefe [38].

A precise mathematical statement of the recursive estimation process is

$$\underline{x}(t|t) = x^*(t|t-1) + K(t)(y(t/t) - y^*(t)), \tag{6.2}$$

where

$x(t|t)$ = estimated state of the world at time t after a measurement at time t,
$y(t)$ = observed sensory input at time t,
$y^*(t|t-1)$ = predicted sensory input at time t based on estimated state at time $t-1$ plus control output at time $t-1$,
$K(t)$ = inverse measurement model and confidence factor.

The SP module receives observed sensory input $y(t)$ directly from sensors or from a lower-level SP subsystem. The WM processes simultaneously generate predicted sensory input $y^*(t|t-1)$ based on the previous best estimate of the state of the world $\underline{x}(t-1|t-1)$ stored in the knowledge database. The SP module compares observed atttibutes with predicted attributes and computes variance. The WM processes then use the variance $y(t) - y^*(t|t-1)$ to update the knowledge database, producing a new best estimate $\underline{x}(t|t)$. Statistics are kept on the variance to assess the confidence in the model.

Recursive estimation (RE) is one of the standard issues of control theory and theory of stochastic processes [43, 44]. A version of RE accepted in 4D-RCS [33] is described in [45]. The entity frame estimate at time $k-1$ after the measurement at $k-1$ is projected forward in time by the matrix $A(k-1)$, and combined with the effect of the control input $u(k-1)$ to provide a prediction of the entity attributes at time k based on information from measurements up to and including time $k-1$. The entity attributes are then transformed into image coordinates through the perspective projection $F(k)$, resulting in a predicted attribute image $y^*(k)$. This is compared with the observed attribute image $y(k)$. This comparison may consist of correlation or difference operations. The result is an error signal $y(k) - y^*(k)$. This error is then transformed back into attribute coordinates through the inverse Jacobian $K(k)$. The forward Jacobian can be computed numerically from the forward perspective projection $F(k)$ as described by Dickmanns. The transformed error is then added to the predicted entity attribute vector $x^*(k|k-1)$ to produce the new estimated entity attribute vector $x^*(k|k)$. Figure 6.7 illustrates how the recursive estimation process developed by Dickmanns can be incorporated into the RCS architecture [33].

6.2.5 Initialization

A recursive estimation loop requires initialization. There must be an initial estimate at time $t = 0$ in order to make an initial prediction. In simple cases the initial estimate can

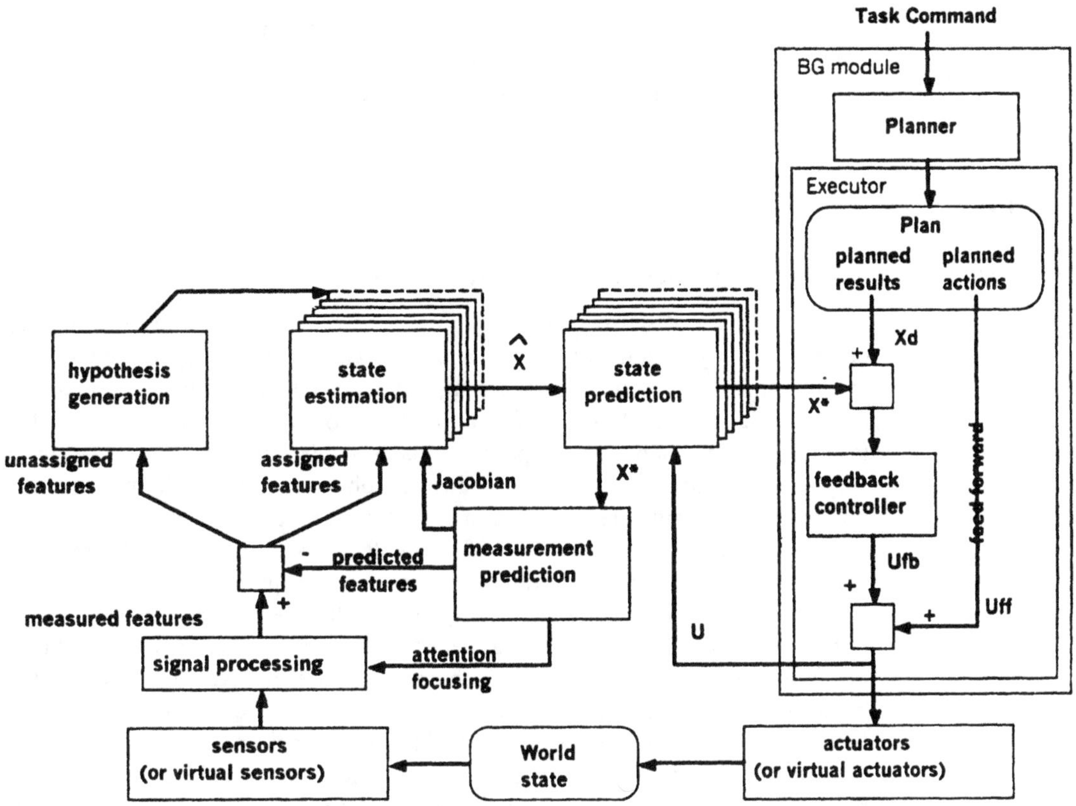

Figure 6.7. Distillation of the recursive estimation concepts showing how the 4D approach interfaces with the RCS concept of task decomposition in a behavior generation module.

be zero. This will produce an initial prediction of zero and an initial error equal to the negative of the initial observation. Thus the observation at $t = 0$ becomes the estimate at time $t = 1$. The recursive estimation process then proceeds to refine the $t = 1$ estimate with each new measurement of the state of the world.

In more complex cases, part of the initial estimate can be provided top-down by a priori information. Additional information can be derived through a series of hypothesis and testing procedures designed to build up a reliable model of the situation that exists in the world. For example, when an intelligent system first awakes from an unconscious state, it must first initialize itself to a known starting state and then initiate procedures that answer such basic questions as "Where am I?" "What is the situation in my immediate vicinity?" "Why am I here?" "What should I do next?"

An intelligent system must always have answers to these basic questions. If it does not, it must invoke procedures designed to discover the answers. For example, to discover where it is in the world, the system typically must orient itself with respect to gravity. It should have visual, tactile, and acoustic sensors with which to explore its immediate environment. Is it light or dark? Where is the ground? What is the ground made of? What is its slope? Are there objects? If so, where? What are their attributes? Are any objects moving? What are they? Where is North?

To know what to do next, an intelligent system must have a goal—even if that goal is to do nothing. If the system has no goal, it must make one up, or pick a goal from a library of goals, or receive a goal from an external source.

Once the initialization process is complete, the system knows where it is, where it is going, what it is doing, and what is going on in the world around it. It then can decide what is important, and begin to plan tasks and actions to accomplish its goals.

6.3 HIERARCHY OF SENSORY PROCESSING

6.3.1 Naturally Emerging Hierarchies of Representation in SP

The mechanism of sensory processing employs GFS, and thus creates hierarchically organized entity-relational networks in a natural way. The state of the world and the current situation are inferred via learning process introduced by GFS recursion (it is applied to its own results). GFS functions via hypothesis generation and its subsequent testing. The space of possible hypotheses that is required for interpreting the current sensory input is enormous even in the industrial robots with vision. In more sophisticated outdoor robots designed for long-term autonomous functioning, the visual and auditory sensory data contain so many cases of incompleteness, paradoxes, and ambiguity that it is impossible to formulate and test all the possible interpretations.

One way to reduce the number of hypotheses that need to be considered is to partition the space of possibilities into a hierarchy of levels so that the range and resolution of possibilities can be limited in space and time at each level. This reduces the combinatorial explosion of possible interpretations. A second way to reduce the number of hypotheses is to apply the hypothesis and test the paradigm within the context of a priori knowledge and information about tasks and goals in the BG hierarchy. This narrows the focus of attention, limits the possibilities that need to be considered, and suggests grouping hypotheses that are likely to be successful.

In many ways the SP, WM, KD, and VJ functions mirror the planning and control algorithms in the BG hierarchy. Just as BG/WM/VJ/KD generate behavior by decomposing tasks into subtasks for a number of agents over an extended period of time in the future, so too do SP/WM/VJ/KD generate perception by integrating information from a number of sensors over an extended period of time in the past. Computations in the BG hierarchy are primarily top-down, decomposing high-level conceptual representations of goals and tasks into low-level actions by a large number of actuators. Presently computations in the SP hierarchy are primarily bottom-up, integrating low-level signals from a multitude of sensors into high-level perceptions of situations and concepts; however, the need in two waves of processing (both bottom-up with GFS and top-down with GFS^{-1}) can be foreseen for the near future.

Nevertheless, the BG hierarchy receives important bottom-up inputs via WM from SP that enable reactive or reflexive behavior. Similarly SP receives important top-down inputs via WM from BG that enable focusing of attention, masking of unimportant signals, and selective application of sensory processing algorithms. Just as there are different kinds of planning and control requirements at different levels of the BG hierarchy, so there are different kinds of computational algorithms and data representations at different levels of the SP hierarchy. Just as the BG hierarchy is a hierarchy of tasks, the SP hierarchy is a hierarchy of event and entity classes.

The BG hierarchy requires many different levels of range and resolution from the knowledge database to plan and control behavior at many different levels of control. Thus a hierarchy of SP subsystems extracts information at many different levels of abstraction. At all levels, SP and WM processes work together to construct and maintain a distributed multiresolutional world model.

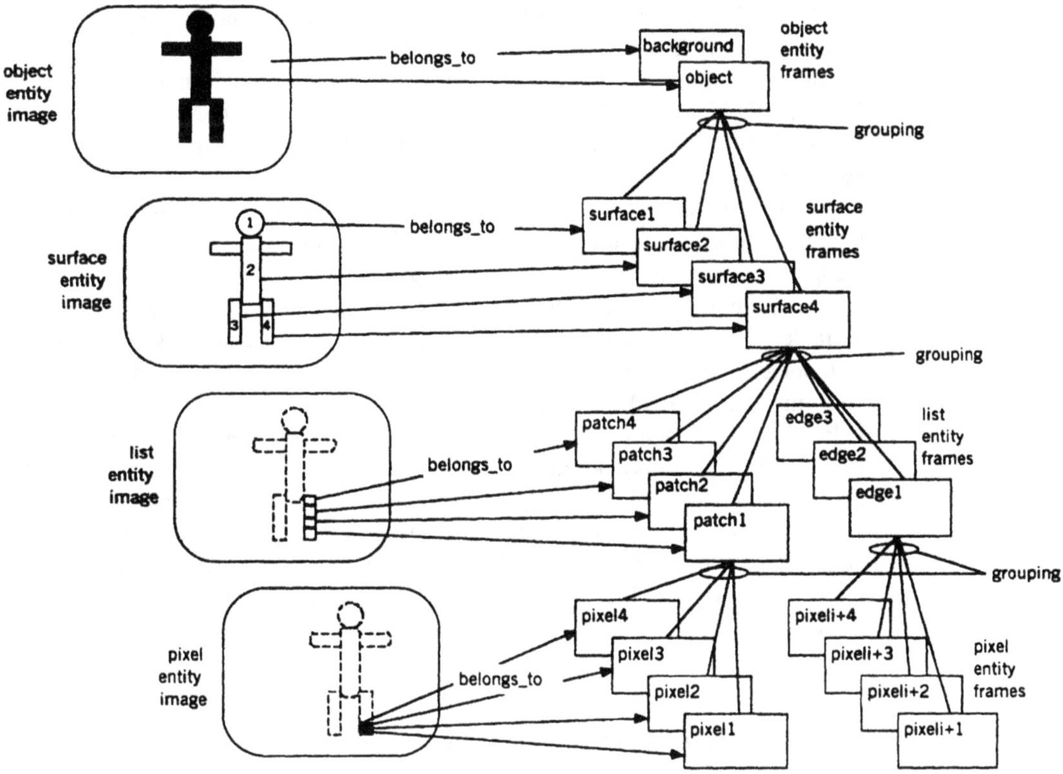

Figure 6.8. Example of hierarchical image and entity frames emerging from GFS-based bottom-up processing.

The transition between levels of abstraction results from the grouping of processes within SP at each level. Each grouping process collects entities and events into higher-level entities and events. For example in Figure 6.8, pixel entities are grouped into list entities, list entities are grouped into surface entities, and surface entities are grouped into object entities. As a result of this series of grouping processes, pointers are established between entity images and entity frames at each level.

The levels of abstraction in the WM hierarchy are intimately related to, and to a large extent defined by, the requirements of planning and control in the BG hierarchy. The knowledge database (KD) is organized as a hierarchy of entities, events, classes, images, and maps that are designed to provide the BG processes with the knowledge of the world at the level of resolution that is required to accomplish task goals at that level.

At each level, interactions between WM and SP functions can generate a variety of windowing, masking, grouping, predictive filtering, detection, and model-matching processes that enable recognition, segmentation, scene understanding, and learning. At each successively higher level, information from different sensory inputs is integrated and grouped into higher level more general entities and situations. Note that the acoustic and vision systems do not interact with each other or the BG hierarchy at the servo level.

For the SP functions to compare observations with predictions, it is necessary for observed attributes be comparable with predicted attributes. This means that hypothesized entities in the SP hierarchy must be comparable with the classes of entities in the WM and KD hierarchy. It also means that the observed entities in SP must be expressed in the same coordinate system as the predicted attributes from the WM.

The SP, WM, KD, VJ, and BG hierarchies are interconnected in the nodes of the RCS heterarchy. Each node is built around a BG process, with the SP, WM, KD, and VJ functions necessary to support that BG process. At each hierarchical level in the BG/WM/VJ hierarchy, there are different kinds of planning and control requirements. Similarly at different levels of the SP/WM/VJ hierarchy, there are different kinds of computational algorithms and data representations.

The task goals for the various operational units at different levels of the RCS hierarchy are different, as are the task objects. For example at the servo level, task objects are the state-variables, each of which can be computed from one or more sensor signals at a single point in space and time. Task goals are the desired values of these state-variables. State-variables are in sensor and actuator coordinates. At the primitive level, task objects are the edges, vertices, or trajectories that represent linear features in space and time such as road edges and obstacle edges. Task goals are the desired positions, velocities, or orientations of features in a vehicle coordinate frame. At the subsystem level, task objects may be the surfaces of obstacles or terrain features. Task goals are the desired relationships between the vehicle and task objects expressed in task object surface coordinates. At the vehicle level, task objects may be real world objects such as vehicles, buildings, trees, and bushes. Task goals are the desired relationships between the vehicle and task objects in task object coordinates. Typically tasks and world model knowledge are expressed in a different coordinate system at each different hierarchical level.

6.3.2 Processing at the Two Adjacent Levels of the Highest Resolution

Computations in the BG hierarchy are primarily top-down (see Chapter 7), decomposing high-level abstract representations of goals and tasks into low-level and hypothesizing plans. Computations in the SP hierarchy are primarily bottom-up, integrating low-level signals from a multitude of sensors into high-level perceptions of situations and concepts. There are, however, important bottom-up inputs to BG/WM from SP that enable reactive reflexive behavior. Similarly there are important top-down inputs to SP/WM from BG that enable focusing of attention, masking of unimportant signals, and selective application of sensory processing algorithms.

Let us consider functioning of SP at the two adjacent levels of the highest resolution. This analysis should be considered as an example: In all cases the SP result might be different.

LEVEL 1

Individual Sensor Signals Arrival and Organization

Inputs to level 1 of the SP hierarchy are signals from individual sensors integrated over a single sample period. An input set may represent the position, velocity, or force of a single actuator integrated over a single sample period. An input may represent the brightness measured by a single photo-detector in a CCD or FLIR camera, or the sound pressure measured by a single acoustic sensor. An input set may also represent a single range measurement from a LADAR camera or radar antenna. It may represent a single measurement by an accelerometer, gyro, speedometer, or GPS receiver.

Focusing Attention and Searching at Level 1 At level 1 the window of attention defines the region of space and time that is sampled by each sensor. An attention window may move about over the egosphere, and an attention schedule may define how frequently each sensor is sampled. For acoustic or microwave sensors, windowing may be performed by pointing directional microphones or antennae at high priority targets. For position, orientation, velocity, acceleration, and force sensors, windowing may consist of selecting those signals that are relevant to the current servo task command for each actuator. For vision sensors, windowing includes pointing and focusing a camera on the region of the egosphere that is most worthy of attention.

There may be several cameras of different types with different resolutions, fields of view, and frame rates. The attention system will focus the highest resolution cameras on the most important region in the image while using lower-resolution cameras to observe less important regions. The attention window may track the highest priority entity of attention, or saccade quickly from one high-priority entity to the next as the priority list is updated.

Generation of Grouping Hypotheses at Level 1 At level 1 the grouping simply means that each sensor integrates, or groups, all the energy impinging upon it during a spatial and/or temporal sample interval. For example, each photo-detector in a CCD camera integrates the incoming visible radiation over the spatial region occupied by that photo-detector and over the temporal interval of an exposure time period.

Computing the Attributes and Variables at Level 1 A number of attributes can be directly measured for each sensor. For an acoustic sensor, the amplitude of the acoustic signal can be computed. For a black-and-white TV camera, the intensity (I) of radiation at each pixel is measured. In a color TV camera, the intensity of red, blue, and green light (*rgb*) at each pixel is measured. In a laser range imager (LADAR), or in an imaging radar, a value of range (r) is measured at each pixel. For a forward-looking infrared (FLIR), the temperature at each pixel is measured.

Additional attributes for each pixel in an image can be computed from measured attributes in the local neighborhood. For example, the x- and y-intensity gradients (dI/dx, dI/dy) may be computed at each pixel by calculating the difference in intensity between adjacent pixels in the x- and y-directions, or by applying a gradient operator such as a Sobel operator on a neighborhood about a pixel. The temporal intensity gradient (dI/dt) can be computed by subtracting the intensity of a pixel at time $t - 1$ from the intensity at time t, or by applying a temporal gradient operator. Second-, third-, or fourth-order derivatives can be computed by convolution with spatial or temporal filters over larger neighborhoods. Differences in range at adjacent pixels can be used to compute spatial range gradients (dr/dx, dr/dy) and surface roughness, or texture (tx). Spatial range gradients can also be computed from shading (dI/dx, dI/dy) when there is knowledge of the incidence of illumination. Temporal differences in range at a point can yield range rate (dr/dt).

For a pair of stereo images, disparity can be computed at each pixel by a number of algorithms. When combined with camera vergence and knowledge of camera spacing, disparity can be used to compute range for each pixel at every point in time. If the camera image is stabilized (or camera rotation is precisely known), spatial and temporal gradients can be combined with knowledge of camera motion derived from inertial and speedometer measurements to compute image flow vectors (dx/dt, dy/dt) at each pixel (except where the spatial gradient is zero. In this case the flow rate sometimes can be inferred from the flow of surrounding pixels.) For stationary objects, time to contact (*ttc*) can be computed from image flow. Under certain conditions, range at each pixel can also be computed from image flow. Each attribute that can be measured or computed at each pixel forms an attribute image. The set of attributes that can be measured or computed at each pixel forms a pixel frame for each pixel.

Recursive Estimation at Level 1 Each sensor signal can be filtered in order to reduce or eliminate noise and to improve signal to noise ratio. Each pixel in an image can be filtered if image motion can be eliminated or accounted for. Figure 6.3 is an example of how recursive estimation might be used to filter pixel attribute images. In this example, each observed (i.e., measured or computed) attribute image is compared with a predicted attribute image. The prediction process uses an estimate of the image flow rate at each pixel to predict where a pixel attribute observed in an image at time t will occur in the next image at $t + 1$. Pixel signals from cameras are used to compute observed pixel attribute images. These are compared with predicted pixel attribute images and the difference is used to update estimated pixel attribute images. Perturbed attribute images are used to compute a correlation function over a set of image offsets. The peak of the correlation function is used to compute errors in the estimated pixel flow. This is combined with higher-level estimates of entity motions to estimate pixel flow at each pixel. Predicted pixel attribute vectors are compared with pixel class attribute vectors to classify pixels and generate a labeled pixel class image. The estimated flow rate at each pixel combines information from two sources:

1. Estimated image flow rates ($\underline{dx/dt}$, $\underline{dy/dt}$) and range rate ($\underline{dr/dt}$) that describe the current motion of entities in the image.
2. Commanded camera motions and other actions that affect motion of entities in the image.

The computation of image flow at a single pixel at a single point in time is notoriously unreliable and noisy. However, filtering over an interval in time by recursive estimation has been shown to provide considerable improvement in signal-to-noise ratio. Integration over a group of points comprising an entity further improves the signal to noise. The computation of image flow from a combination of entities at several different levels, each of which is recursively estimated, can provide a robust estimate of image flow at each pixel. The estimated flow rates $\underline{dx/dt}$ and $\underline{dy/dt}$ at each pixel are computed from a combination of estimates as follows:

1. From the correlation offset between a pixel in the predicted attribute image and the corresponding pixel in the observed attribute image.
2. From the level 2 entity flow estimate for the list entity to which the pixel belongs.
3. From the level 3 entity flow estimate for the surface entity to which the pixel belongs.
4. From the level 4 entity flow estimate for the object entity to which the pixel belongs.

In equation form this can be expressed as

$$\begin{aligned}\underline{dx/dt}(x, y, t-1) &= \mathrm{c1}^*\underline{dx/dt}(\mathrm{pixel}) + \mathrm{c2}^*\underline{dx/dt}(\mathrm{list}) + \mathrm{c3}^*\underline{dx/dt}(\mathrm{surface}) \\ &\quad + \mathrm{c4}^*\underline{dx/dt}(\mathrm{object}), \\ \underline{dy/dt}(x, y, t-1) &= \mathrm{c1}^*\underline{dy/dt}(\mathrm{pixel}) + \mathrm{c2}^*\underline{dy/dt}(\mathrm{list}) + \mathrm{c3}^*\underline{dy/dt}(\mathrm{surface}) \\ &\quad + \mathrm{c4}^*\underline{dy/dt}(\mathrm{object}),\end{aligned}$$

where estimated variables are underlined, and

c1 = confidence in the pixel level flow computation for the pixel at x, y,
c2 = confidence in the list level flow computation for the pixel at x, y,
c3 = confidence in the surface level flow computation for the pixel at x, y,
c4 = confidence in the object level flow computation for the pixel at x, y.

The correlation offset at each pixel generates a measure of pixel flow error. This is added to the predicted flow rate to generate an estimate of pixel flow rate. The pixel flow rate is then combined with higher-level estimates of list entity flow, surface entity flow, and object entity flow to compute a best estimate of flow at each pixel.

Once a reliable estimate of image flow is known, the attribute of a pixel at position (x, y) at time $t - 1$ can be predicted to appear at position $(x + dx, y + dy)$ at time t. This can be expressed as

$$I^*(x + dx, y + dy, t) = I(x, y, t - 1)$$

Comparison between predicted and observed images produces correlation and difference values. The attribute difference between observed image $I(x, y, t)$ and predicted (with zero offset) $I^*(x, y, t)$ is used to update the attribute estimate. Correlation of the observed image with a set of predicted images with different offsets produces a correlation function. The offset values that produce the maximum correlation value indicate the amount of position error (Δx and Δy) between the observed and predicted attribute images at each pixel. These position errors divided by the sample period yield the flow error at each pixel.

The combination of image flow estimates suggests how recursive estimation can provide a means for combining information about a single estimated pixel attribute from several sources. Each level in the hierarchy produces an estimate of image flow for each pixel. The higher levels estimate the flow rate of entities consisting of many pixels. Each higher level computes flow estimates for entities with more pixels. Thus, combining flow estimates from multiple levels produces good estimates of individual pixel flow rates at the lowest level.

Classification, Recognition, and Organization of Entities at Level 1 The final SP function performed at each level is the recognition of classes. The set of attributes for each pixel defines a pixel attribute vector for each pixel. At level 1, recognition is accomplished by comparing the estimated pixel attribute vector with pixel class attributes from a library of pixel classes. When the inner product between the estimated pixel attribute vector and the class attribute vector exceeds threshold, the pixel is recognized as belonging to the class.

The pixel class library contains a set of geometric classes. These include brightness-discontinuity, color-discontinuity, range-discontinuity, range-gradient-discontinuity, and surface-continuity. Discontinuity class attributes may include orientation, size, position, and velocity of the discontinuity. Discontinuity classes are important because pixels belonging to these classes tend to lie on the edges of objects, or on the boundaries of surfaces. Surface-continuity class attributes may include surface orientation, color, texture, temperature, position, and velocity. Surface-continuity classes are important because pixels belonging to these classes have attributes that describe the surfaces and objects to which they belong.

The pixel class library may also contain a set of generic classes. These may include obstacle, road, grass, rock, dirt, sand, water, sky, foliage, or moving object. Generic class attributes include surface orientation, color, texture, temperature, position, and velocity. Thus, pixels classified into surface-continuity classes may be further classified into generic classes.

Upon recognition, each pixel acquires one (or more pointers) that is(are) the name(s) of the pixel class(es) to which it belongs. The classification process thus generates one or more new pixel entity images where each pixel in the entity image(s) carries the name(s) of the pixel class(es) to which the pixel belongs.

Output from level 1 is thus a set of pixel attribute images plus one or more pixel entity images wherein each pixel contains the name or names of the class(es) to which it belongs. When there exists a range attribute for each pixel, attribute images may be transformed into a map of the ground that can be seen by the camera. This map may then be transformed into several maps at different levels in the RCS hierarchy with range and resolution at each level appropriate to the planning and control requirements at that level.

Summary of Level 1

1. A portion of the egosphere is windowed onto each sensor.
2. The energy falling on each sensor is integrated over a sample interval in space and time.
3. Observed values are computed for a set of attributes for each pixel.
4. Estimated and predicted values are computed for each attribute of each pixel by a recursive estimation filter process.
5. Pixels are classified as belonging to one or more classes, and one or more pixel entity images are formed in which each pixel has a pointer (or pointers) to the class (or classes) to which it belongs.

LEVEL 2

Entities and Events

At level 2 of the vision SP hierarchy, groups of a few (on average about 10) classified level 1 regions (pixels or events) can be grouped, processed, and analyzed as level 2 regions (list entities or events). For example, a group of pixel entities with contiguous brightness or color gradients might be grouped, processed, and analyzed as a list entity such as an edge, vertex, or surface patch. Also at level 2 the temporal strings of filtered attributes from a single sensor can be grouped, processed, and analyzed as level 2 events. For example, a temporal signal, or string of intensity attributes, from an acoustic sensor might be grouped as a tone, a frequency, or a phoneme.

Focusing Attention and Searching at Level 2 Windowing consists of placing windows around regions in the image that are designated as worthy of attention. At level 2 the placement of windows depends on the goal and priorities of the current level 2 task. The placement of windows also depends on the previous detection of higher level entities. For example, the placement of windows on the edge of the road or on the vehicle depends on previous detection and recognition of the road edge and vehicle respectively. The size and shape of each window is determined by the size and shape of recognized entities. The size of each window is also determined by the confidence factor computed by the level 2 recursive filtering process. If the confidence value for the entity in a window is high, the window will be narrowed to only slightly larger than the set of pixels in the entity. If the confidence value is low, the window will be significantly larger than the hypothesized entity. Until there exists a set of recognized list entities, level 2 windows are set to their maximum size.

Generation of Grouping Hypotheses at Level 2 Grouping causes an image to be segmented, or partitioned, into regions (or sets of pixels) that correspond to higher order entities. At level 2 the grouping is a process by which neighboring pixels of the same class with similar attributes can be grouped to form higher level entities. Information about each pixel's class is derived from the recognized pixel entity image. Information about each pixel's attributes is derived from the pixel attribute vector defined by the predicted pixel attribute images. Grouping is performed by a heuristic algorithm based on gestalt principles such as contiguity, similarity, proximity, pattern continuity, or symmetry. For example, contiguous pixels in the same geometric or generic class with similar attributes of range, orientation, and flow rate might be grouped into level 2 regions. A typical level 2 group consists of about 10 pixels.

A grouping is a hypothesis that the pixels in the group all lie on the same entity in the world. The grouping process thus generates a hypothesized list entity image in which each group is assigned a label and a frame in which a set of attributes can be stored. A grouping hypothesis can be tested by a recursive estimation filtering algorithm applied to each of the hypothesized entities. If the recursive filtering process is successful in predicting the observed behavior of a hypothesized group in the image, then the grouping hypothesis for that entity is confirmed. On the other hand, if the recursive filtering process is not successful in predicting the behavior of the hypothesized entity, the grouping hypothesis will be rejected, and another grouping hypothesis must be selected. If the computational resources are available, a number of grouping hypotheses may be tested in parallel so that the grouping hypothesis with the greatest success in predicting entity behavior can be selected.

The choice of which gestalt heuristic to use for grouping may depend on an attention function that is determined by the goal of the current task. It may also depend on the confidence factor developed by the recursive estimation process as well as on attributes of previously recognized list entities.

Computing the Attributes and Values of Variables at Level 2 Each of the regions defined by the grouping hypotheses has attributes that can be computed. Entity attributes differ from pixel attributes in that entity attributes are integrated over the entire group of pixels comprising the entity. For example, edge entities have attributes such as edge-length, edge-curvature, edge-orientation, position and motion of the edge center-of-gravity. Surface-patch entities have attributes such as area, texture, average range, average surface gradient, average color, position and motion of the surface center-of-gravity.

For each hypothesized level 2 grouping, computed entity attributes fill slots in an observed list entity frame.

Recursive Estimation at Level 2 At level 2 a recursive estimation process compares observed list entity attributes with predicted list entity attributes. Predicted entity attributes are generated from estimated entity attributes. Of particular importance for prediction are state attributes of position, orientation, and motion. Entity motion in an image can arise from three sources: (1) estimated dynamic attributes such as image flow rates ($\underline{dx/dt}$, $\underline{dy/dt}$) and range rate ($\underline{dr/dt}$) that describe the motion of entities in the image, (2) estimated motion of the camera platform, and (3) commanded actions that affect motion of entities in the image.

Comparison of observed entity attributes with predicted entity attributes produces correlation and difference values. Difference values are used to update the estimated entity attribute values. Estimates of image flow rates of entities at level 2 are generated from a combination of correlation offsets generated at level 2, plus estimates of flow attributes from higher levels. A confidence level for each estimated entity attribute value is computed as a function of the correlation and difference values. If predictions based on estimated entity attributes successfully track the behavior of observed entity attributes, the confidence level rises. When the confidence level of the recursive estimation filter rises above threshold, the hypothesized list entity grouping is confirmed. If the confidence level of the recursive estimation filter falls below threshold, the hypothesized level 2 grouping is rejected, and another grouping hypothesis must be selected.

Classification, Recognition, and Organization of Entities at Level 2 The set of attributes for each list entity define a list entity attribute vector. At level 2, recognition is accomplished by comparing the estimated list entity attribute vector with list entity class attributes from a library of list entity classes. When the inner product between the estimated list entity attribute vector and the list entity class attribute vector exceeds threshold, the list entity is recognized as belonging to the class.

The list entity class library contains a set of geometric classes. These include brightness-edge, color-edge, range-edge, range-gradient-edge, brightness-vertex, color-vertex, range-vertex, range-gradient-vertex, and surface-patch. Edge and vertex class attributes may include orientation, size, position, and velocity of the edge or vertex. Edge and vertex classes are important because pixels belonging to these classes tend to lie on the edges or corners of objects, or on the boundaries of surfaces. Surface-patch class attributes may include surface orientation, color, texture, temperature, position, and velocity. Surface-patch classes are important because pixels belonging to these classes have attributes that describe the surfaces and objects to which they belong.

The list entity class library may also contain a set of generic classes. These may include the top, bottom, or sides of an obstacle, the near edge of a ditch, the far edge of a ditch, the crest of a hill, the edge of a tree trunk, the side of a building, the edge of surface of a road, a grassy area, a rock, a patch of dirt, sand, or water, the sky, foliage of a tree or bush, or a moving object. Generic class attributes include surface orientation, color, texture, temperature, position, and velocity. Thus a list entity classified into surface-patch classes may be further classified into a generic class.

Upon recognition, each pixel acquires one or more additional pointers that contain the name (or names) of the list entity class (or classes) to which the pixel belongs. Output from level 2 is thus one or more list entity images wherein each pixel points to a list entity frame that contains the attributes of the list entity. Each list entity frame also has a pointer back to the list entity image that contains the pixels that belong to it as illustrated in Figure 6.2.

Summary of Level 2

1. Regions of the image containing list entities of attention are windowed.
2. Pixels with similar attributes are tentatively grouped into level 2 regions that comprise hypothesized list entities.
3. The attributes of each hypothesized list entity are computed.
4. The attributes of hypothesized list entities are filtered by recursive estimation, and each grouping hypothesis is either confirmed or rejected.
5. The attributes of each confirmed list entity are compared with the attributes of a set of list entity classes. Those that match are labeled with list entity class names.

Pixels in the classified regions are given pointers to the list entity frames to which they belong. Each frame also points back to the list entity image.

6.3.3 What Happens at the Levels Above?

LEVEL 3

Surfaces

At level 3 of the SP hierarchy, groups of list entities with geometric attributes that correspond to a surface entity class can be grouped, processed, and analyzed as level 3 regions called surface entities. For example, a set of contiguous surface patch list entities with similar range and motion might be grouped, processed, and analyzed as a surface entity. A set of edge and vertex entities that are contiguous along their orientation might be grouped, processed, and analyzed as a surface boundary entity. Temporal strings of level 2 events with attributes that correspond to level 3 events can be grouped into level 3 events such as sonograms, or 2-dimensional surfaces in frequency and time. For example, a temporal string of phonemes might be grouped as a word, or a temporal string of outputs from a set of frequency filters as a note or phrase in an acoustic signature.

Focusing Attention and Searching at Level 3 At level 3 the windowing consists of placing windows around regions in the labeled list entity image that are designated as worthy of attention. The selection of which regions to window is made by an attention function that depends on the goal and priorities of the current level 3 task, or on the detection of noteworthy attributes of estimated surface entities, or on inclusion in a higher-level entity of attention. The shape of the windows are determined by the recognized regions in the surface entity image. The size of the windows are determined by the confidence factor associated with the degree of match between the predicted surface entities and the observed surface entities.

Generation of Grouping Hypotheses at Level 3 At level 3 the recognized list entities in the same class are grouped into level 3 regions called surface entities based on gestalt properties such as contiguity, similarity, proximity, pattern continuity, and symmetry. Surface patches in the same class that are contiguous at their edges and have similar range, velocity, average surface orientation, and average color may be grouped into surface entities. Edges and vertices in the same class that are contiguous along their orientation with similar range and velocity may be grouped into boundary entities. The grouping is based on a hypothesis that all the pixels in the group have the common property of imaging the same physical surface or boundary in the real world. This grouping hypothesis will be confirmed or rejected by the level 3 filtering function.

Computing the Attributes and Variables at Level 3 Observed level 3 regions have attributes such as area, shape, position and motion of the center-of-gravity, range, orientation, surface texture, and color. Surface boundary entities have attributes such as boundary type, length, shape, orientation, position and motion of the center-of-gravity, and rotation. Boundary types may include intensity, range, texture, color, and slope boundaries. For each level 3 region the computed entity attributes fill slots in an observed surface entity frame.

Recursive Estimation at Level 3 The recursive estimation process compares the observed surface entity attributes with the predicted surface entity attributes generated from estimated surface entity attributes, planned results, and predicted attributes of parent entities. The comparison produces correlation and difference values. The difference values are used to update the estimated surface entity attributes. The correlation values are used to update estimates of entity motion. A confidence level associated with each estimated entity confirms or rejects the surface entity grouping hypothesis that created it.

Classification, Recognition, and Organization of Entities at Level 3 The recognition process establishes a match between the estimated surface entities and a generic or specific class of surface entities in the world model knowledge database. Typical generic surface entity classes are the ground, tree foliage, tree trunk, side of building, roof of building, road lane, fence, or lake surface. As a result of recognition, a labeled surface entity image is formed in which each pixel has a pointer to (or the name of) the surface entity to which it belongs.

At level 3 the surface entity image can be transformed into a map of the ground in front of the vehicle. In 4-D/RCS [33] this map is a 50 by 50 meter map in vehicle coordinates with a 10 centimeter resolution. In some cases this map may be used to update the map provided by the digitized battlefield database.

Summary of Level 3

1. Portions of the image containing surface entities of attention are windowed.
2. List entities with similar attributes are tentatively grouped into level 3 regions, or surface entities.
3. The attribute values of each topological surface and boundary entity are computed.
4. Attributes of each hypothesized surface entity are estimated by recursive estimation, and each grouping hypotheses is either confirmed or rejected.
5. The attributes of each confirmed surface entity are compared with the attributes of a set of surface entity classes, and those that match are assigned to surface entity classes. Each pixel in the classified regions are given pointers to the surface entity frames to which they belong, and each frame has a pointer to the surface entity image containing the pixels that belong to it.
6. A map in vehicle coordinates is generated wherein each pixel carries the name of the generic surface class to which it belongs.

LEVEL 4

Regions: Entities and Events

At level 4 of the SP hierarchy, groups of level 3 entities with attributes such as range, motion, orientation, color, and texture that correspond to an object entity class are grouped, processed, and analyzed as level 4 regions, or object entities. For example, a group of surfaces with coincident boundaries and similar or smoothly varying range and velocity attributes might be grouped into an object such as a building, a vehicle, or a tree. Attributes of surface entities comprising each object entity are combined into object entity attributes. Also at level 4, temporal strings of level 3 events such as words might be grouped into a sentence, or acoustic signatures might be grouped into a level 4 event.

Focusing Attention and Searching at Level 4 At level 4 the windowing consists of placing windows around regions in the image that are classified as worthy of attention.

The selection of which regions to so classify depends on the goal and priorities of the current level 4 task, or on the detection of noteworthy attributes of groups of level 3 regions. The shape of the windows are determined by set of pixels in the recognized object entity image. The size of the windows are determined by the confidence factor generated by the level 4 recursive estimation filter.

Generation of Grouping Hypotheses at Level 4 At level 4 the recognized surface entities in the same class are grouped into level 4 regions, or object entities, based on gestalt properties such as contiguity, similarity, proximity, pattern continuity, and symmetry. Surfaces in the same class that are contiguous along their boundaries and have similar range and velocity may be grouped into object entities. Boundaries that separate surfaces with different range, velocity, average surface orientation, and average color are used to distinguish between different objects. Level 4 grouping is based on a hypothesis that all the pixels in the region have the common property of imaging the same physical object in the real world. This grouping hypothesis will be confirmed or rejected by the level 4 filtering function.

Computing the Attributes and Variables at Level 4 For each of the hypothesized object entities, entity attributes such as position, range, and motion of the center-of-gravity, average surface texture, average color, solid-model shape, projected area, and estimated volume can be computed. For each hypothesized object entity, computed attributes fill slots in an observed object entity frame.

Recursive Estimation at Level 4 The recursive estimation process compares the observed object entity attributes with predictions based on the estimated object entity attributes. Included in the prediction process are estimated motion of object entities, and expected results of commanded actions. The comparison of observed entity attributes with predicted entity attributes produces correlation and difference values. Difference values are used to update the attribute values in the estimated object entity frames. A confidence level is computed as a function of the correlation and difference values. If the confidence level of the recursive estimation filter rises above a threshold, the object entity grouping hypothesis is confirmed.

Classification, Recognition, and Organization of Entities at Level 4 The recognition process compares the attributes of each confirmed object entity with attributes of the object entity classes stored in the knowledge database. A match causes a confirmed geometrical object to be classified as a generic or specific object entity in the world model knowledge database.

For example in the case of a ground vehicle performing a typical task, a list of generic object classes may include the ground, sky, horizon, dirt, grass, sand, water, bush, tree, rock, road, mud, brush, woods, log, ditch, hole, pole, fence, building, truck, tank, and the like. There may be several tens, hundreds, or even thousands of generic object classes in the KD. A list of specific object classes may include a specific tree, bush, building, vehicle, road, or bridge. As a result of recognition, a new object entity image is formed in which each pixel has a pointer to (or the name of) the generic or specific object entity to which it belongs.

At level 4 the object entity image can be transformed into a map of the ground in the vicinity of the vehicle. In 4-D/RCS [33] this map is a 500 by 500 meter map in vehicle-centered world coordinates with a 1 meter resolution. In some cases this map may be used to update the world map provided by the digitized battlefield database.

Summary of Level 4

1. Portions of the image containing object entities of attention are windowed.
2. Surface entities with similar attributes are tentatively grouped into level 4 regions, or object entities.
3. The attribute values of each hypothesized object entity is computed.
4. Attributes of each hypothesized object entity are estimated by recursive estimation, and each grouping hypotheses is either confirmed or rejected.
5. The attributes of each confirmed object entity are compared with the attributes of a set of object entity classes, and those that match are assigned to object entity classes. Pixels in the classified regions are given pointers to the object entity frames to which they belong, and each frame has a pointer to the object entity image containing the pixels that belong to it.
6. A map in coordinates of an intelligent system is generated wherein each pixel carries the name of the generic object class to which it belongs.

LEVEL 5 "Interactive Teams" of Entities and Events

At level 5 of the SP hierarchy, groups of object entities with similar range, motion, and other attributes are grouped, processed, and analyzed as a group1 collection or assemblage of objects. Attributes of object entities comprising each group1 entity are combined into group1 entity attributes. Also at level 5, temporal strings of level 4 events are integrated into level 5 events.

Focusing of Attention and Searching at Level 5 At level 5 the windowing consists of placing a window around regions in the image that are classified as group1 entities of attention. The selection of which regions to so classify depends on the task, and on the detection of entity attributes that are worthy of attention. The shape of the windows is determined by the collection of pixels in the recognized group1 entity image. The size of the windows is determined by the confidence factor associated with the level 5 recursive estimation process.

Generation of Grouping Hypotheses at Level 5 Object entities in the same class are grouped into level 5 regions, or group1 entities, based on gestalt properties such as contiguity, similarity, proximity, pattern continuity, and symmetry. Objects in the same class that are near each other and have similar range and velocity may be grouped into group1 entities. These grouping hypotheses will be confirmed or rejected by the level 5 filtering function.

Computing the Attributes and Variables at Level 5 For each of the hypothesized group1 entities, attributes such as position, range, and motion of the center-of-gravity, density, and shape can be computed. These computed attributes become observed entity attributes in an observed group1 entity attribute frame.

Recursive Estimation at Level 5 The recursive estimation process compares observed group1 entity attributes with predictions based on estimated group1 entity attributes and expected results of commanded actions. The comparison of observed group1 entity attributes with predicted entity attributes produces correlation and difference values. The difference values are used to update the attribute values in the estimated squad entity frames. A confidence level is computed as a function of the correlation and difference values. If the confidence level of the recursive estimation filter rises above a threshold, the group1 entity grouping hypothesis is confirmed.

Classification, Recognition, and Organization of Entities at Level 5 The recognition process compares the attributes of each confirmed group1 entity with attributes of group1 entity classes stored in the knowledge database. A match causes a recognized geometric group1 entity to be classified as a generic or specific group1 entity in the world model knowledge database.

For example, in the case of a ground vehicle, a list of generic group1 classes may include a woods, fields, groups of vehicles, groups of people, and clusters of buildings. There may be several tens, hundreds, or even thousands of generic group1 classes in the KD. A list of specific group1 classes may include a specific group of humans, trees, bushes, buildings, vehicles, or the intersection of two or more roads. As a result of recognition, a new entity image is formed in which each pixel has a pointer to (or the name of) the generic or specific group1 entity to which it belongs.

At level 5 the recognized group1 entity image can be transformed into a map of the ground in the vicinity of the vehicle. In a particular implementation such as an autonomous vehicle, this map might be a 5 by 5 kilometer vehicle centered map with a 3 meter resolution. In some cases this map may be used to update the map provided by the digitized battlefield database.

Summary of level 5

1. Portions of the image containing group1 entities of attention are windowed.
2. Object entities with similar attributes are tentatively grouped into level 5 regions, or group1 entities.
3. The attribute values of each observed group1 entity are computed.
4. Attributes of each hypothesized group1 entity are estimated by recursive estimation, and each grouping hypotheses is either confirmed or rejected.
5. The attributes of each confirmed group1 entity are compared with the attributes of a set of group1 entity classes, and those that match are assigned to group1 entity classes. Pixels in the classified regions are given pointers to the object entity frames to which they belong, and each frame has a pointer to the object entity image containing the pixels that belong to it.
6. A map in the lower resolution coordinates of an intelligent system is generated wherein each pixel carries the name of the generic object class to which it belongs.

6.4 MULTIRESOLUTIONAL NATURE OF SENSORY PROCESSING

The goal of this chapter is to demonstrate that the methods and techniques, developed for the intelligent systems as a whole, are applicable for each of its subsystem, for each module of ELF. Here it is demonstrated for sensory processing. Later we will demonstrate similar things for the world model and behavior generation.

The following components are inescapably present within the inner structure and processes of functioning of each of these modules, including sensory processing:

- *Multiresolutional representation of information.* In the case of vision, the advantages of multiresolutional processing are especially beneficial because the amount of information is very large and the benefits in complexity reduction are profound. At each level of resolution, the stored information is organized as an entity-relational network (ERN). Within ERN of each level, a grammar exists based upon causal and syntactic rules.

ERN

- *Using GFS for consecutive generalization of information.* Actually, GFS is being used for the bottom-up and GFS^{-1} for the top-down processing. As GFS produces generalized entities and populates the level of lower resolution with singularities found (new entities), it is being applied again to these entities in order to create the next lower level of resolution. The viability of the entities created as a result is determined by evaluating their consistency with the existing world model, or with a set of primary rules (interpretation within the context and within a paradigm).

GFS/GFS^{-1}

CLOSURE

- *Providing for information processing closure.* Semiotic closure provides for the eventual convergence of computational procedures at a level of resolution.

Of course, all these play a substantial role for hearing, touch, and other sensing systems. However, vision is probably the most active and even dominating sensory domain in the practice of the emerging intelligence. Thus, the results of the evolution of sensory processing and its rules and laws are very instructive as far as learning from the living examples of intelligence.

Present generation of professionals in computer-based sensory processing for intelligent systems learned from the original sources written by J. Gibson, D. Hubel, and D. Marr [1–3]. These sources established fundamentals for the subsequent wave of references [4–6] related to the overall system of visual perception. Then the micro-events became of interest and the interest to processes of interpretation was stimulated by [7–9]. It was early understood that the domain is driven by such general laws like "gestalt" [10, 11] and the laws of generalization and attention came under scrutiny of research community (e.g., [12]). The processes of search were of special interest to researchers in perception [13, 14], thus all elements of GSF happened to be encompassed by research.

At the same time, the high-level components required for interpretation were actively attempted, too. R. Duda and P. Hart provided several generations of scientists with a source that covered exhaustively for the current needs in the area of sensory processing [15]. Fundamental issues of representation of solid shapes in vision [16] and using experience-based rules and models [17, 18] were an important and powerful step in the development of this domain. Analysis of motion was the next crucial step in the development [19, 20]. Then the time came for high-level activities of the vision system including components and packages of complex inferences and solving the problem of visual recognition [21, 22].

Multiresolutional processing was looming in the early works by P. Burt, E. Riseman, A. Hanson, A. Rosenfeld, and others. P. Burt noticed the intrinsic laws of multiresolutional representation [23]. Multiresolutional integrated vision system "Visions" was reported in 1978 (we refer here to recent literature [24, 25]). The inevitability of using the concept of multiresolutional representation (MR) for the analysis of visual perception resulted in a wave of research in this area, and this helped to obtain numerous important advances [26–28]. Drexel University research in multiresolutional image processing has been developed to address the issue of intelligent systems [29–32]. National Institute of Standards and Technology has accepted MR approach in their activities [33]. MR approach was reflected both in theoretical textbooks [34] and in a multiplicity of practical results (e.g., [35–42]).

This little survey was not intended to be complete; rather, its goal is to underscore the directions of research that have surfaced during the last two decades and are linked with the research in the area of intelligent systems.

REFERENCES

[1] J. Gibson, *The Perception of the Visual World*, Houghton-Mifflin, Boston, MA, 1950.

[2] D. Hubel and T. Wiesel, Receptive fields, binocular interaction, and functional architecture in the cat's visual cortex, *Journal of Physiology*, Vol. 160, 1962, pp. 106–154.

[3] D. Marr, *Vision*. W.H. Freeman and Company, San Francisco, California, 1982.

[4] J. J. Koenderink, The structure of images, *Biological Cybernetics*, Vol. 50, 1984.

[5] S. Grossberg, Neural dynamics of surface perception: Boundary webs, illuminents, and shape-from shading, *Computer Vision, Graphics, and Image Processing*, Vol. 37, 1987, pp. 116–165.

[6] M. Fischler and O. Firschein, *Intelligence: The Eye, the Brain, and the Computer*, Addison-Wesley, Reading, MA, 1987, p. 168.

[7] D. Hubel, *Eye, Brain and Vision*, W. H. Freeman, NY, 1988.

[8] I. Biederman, Higher-level vision, In *Visual Cognition and Action, Vol. 2*, (D. Osherson, S. Kosslyn, and J. Hollerbach, Eds.) MIT Press, Cambridge, MA, 1990.

[9] G. Francis, S. Grossberg, and E. Mingolla, Cortical dynamics of feature binding and reset: Control of visual persistence, *Vision Research*, Vol. 34, 1994, pp. 1089–1104.

[10] W. Kohler, *Gestalt Psychology*, Liveright, London, 1929.

[11] K. Koffka, *Principles of Gestalt Psychology*, Harcourt, Brace, New York, 1935.

[12] S. Kosslyn, Mental imagery, In D. N. Osherson, S. M. Kosslyn, and J. M. Hollerbach (eds.), *Visual Cognition and Action, Vol. 2*, MIT Press, Cambridge, MA, 1990.

[13] A. L. Yarbus, *Eye Movements and Vision*, Plenum Press, 1967.

[14] D. Sparks, Translation of sensory signals into commands for the control of saccadic eye movements: Role of the primate superior collicullus, *Physiol. Rev.* Vol. 66, 1986, pp. 118–171.

[15] R. Duda and P. Hart, *Pattern Classification and Scene Analysis*, Wiley, NY, 1973.

[16] J. Koenderink and A. Van Doorn, The internal representation of solid shape with respect to vision, *Biological Cybernetics*, Vol. 32, 1979, pp. 211–216.

[17] M. Brady, Computational approaches to image understanding, *ACM Computing Surveys*, Vol. 14, 1982.

[18] T. Binford, Survey of model based image analysis systems, *International Journal of Robotics Research 1*, 1982, pp. 18–64.

[19] E. Adelson and J. Movshon, Phenomenal coherence of moving visual patterns, *Nature*, Vol. 30, 1982, pp. 523–525.

[20] A. Georgopoulos, R. Caminiti, J. Kalaska, and J. Massey, Spatial coding of movement: A hypothesis concerning the coding of movement direction by motor cortical populations, *Exp. Brain Res. Suppl. 7*, 1983, pp. 327–336.

[21] M. Brady and A. Yuille, An extremum principle for shape from contour, In M. Arbib and A. Hanson (eds.) *Vision, Brain, and Cooperative Computation*, Bradford Books, MIT Press, Cambridge, MA, 1987, pp. 285–328.

[22] S. Ullman, *High-level Vision: Object Recognition and Visual Cognition*, MIT Press, Cambridge, MA, 1996.

[23] P. Burt and E. Adelson, The laplacian pyramid as a compact image code, *IEEE Transactions of Communication*, COM-31, 1983, pp. 532–540.

[24] M. Hansen, Video processing technologies and the VFE-200 vision system: Sarnoff pyramid processing technologies and a roadmap for advanced video processing systems, *Sarnoff Corporation Internal Presentation*, Princeton, NJ, March 1998.

[25] E. Riseman and A. Hanson, A methodology for the development of general knowledge-based vision systems, In M. Arbib and A. Hanson (eds.) *Vision, Brain, and Cooperative Computation*, Bradford Books, MIT Press, Cambridge, MA, 1990, pp. 285–328.

[26] K. Chaconas and M. Nashman, Visual perception processing in a hierarchical control system: level 1, *National Institute of Standards and Technology Technical Note 1260*, June 1989.

[27] F. Corbetta, F. Miezin, S. Dobmeyer, G. Shulman, and S. Petersen, Attentional modulation of neural processing of shape, color, and velocity in humans, *Science*, Vol. 248, 1990, pp. 1556–1559.

[28] D. Van Essen and E. Deyoe, Concurrent processing in the primate visual cortex, In M. Gazzaniga (ed.) *The Cognitive Neurosciences*. Bradford Books, Cambridge, MA, 1995, pp. 383–400.

[29] A. Meystel, On the phenomenon of high redundancy in robotic perception. In J. Tou and J. Balchen (eds.), *Highly Redundant Sensing in Robotic Systems*, Springer-Verlag, Berlin, 1990, pp. 177–250.

[30] I. Rybak, S. Bhasin, M. Meystel, and A. Meystel, Multiresolutional stroke sketch adaptive representation and neural network processing system for gray-level image recognition, In D. Casasent (ed.), *Intelligent Robots and Computer Vision XI*, Proc. SPIE, Vol. 1826, Boston, MA 1992, pp. 261–279.

[31] M. Meystel, I. Rybak, S. Bhasin, and A. Meystel, Top-down/bottom up algorithm for adaptive multiresolutional representation of gray-level images, Abstracts of papers "IS&T/SPIE's Symposium on Electronic Imaging: Science & Technology, Jan. 31–Feb. 4 1993, San Jose, CA.

[32] A. Meystel, Multiresolutional system: Complexity and reliability, In O. Kaynak, G. Honderd, and E. Grant (eds), *Intelligent Systems: Safety, Reliability, and Maintainability Issues*, Computers and Systems Sciences, Vol. 114, NATO ASI, Series F, Springer-Verlag, Berlin, 1994, pp. 11–22.

[33] J. Albus, *4D/RCS: A Reference Model Architecture for Demo III, Version 1.0, NISTIR 5994*, Gaithersburg, MD, 1998.

[34] A. Rosenfeld and A. Kak, *Digital Picture Processing*, Academic Press, NY, 1976.

[35] G. J. VanderBrug, J. S. Albus, and E. Barkmeyer, A vision system for real-time control of robots, *Proceedings of the 9th International Symposium on Industrial Robots*, Washington, D.C., 1979.

[36] R. N. Nagel, G. J. VanderBrug, J. S. Albus, and E. Lowenfeld, Experiments in part acquisition using robot vision, *Proceedings of Autofact II, Robots IV Conference*, Detroit, MI, October 29–November 1, 1979.

[37] L. Matthies and A. Elfes, Sensor integration for robot navigation: Combining sonar and stereo range data in a grid-based representation, *Proceedings of the 26th IEEE Decision and Control Conference*, Los Angeles, December 9–11, 1987.

[38] E. Dickmanns and E. Graefe, a) Dynamic monocular machine vision, b) Application of dynamic monocular machine vision, *Journal of Machine Vision & Applications*, Nov. 1988, pp. 223–261.

[39] A. Meystel and L. Holeva, Interaction between subsystems of vision and motion planning in unmanned vehicles with autonomous intelligence, Proc. of SPIE, *Applications of Artificial Intelligence* (edited by J. Gilmore), Arlington, VA, 1984.

[40] H. Schneiderman and M. Nashman, A discriminating feature tracker for vision-based autonomous driving, *IEEE Transactions on Robotics and Automation*, December 1994.

[41] T. Hong, S. Legowik, and M. Nashman, *Obstacle Detection and Mapping System*, NISTIR 6213, August 1998.

[42] E. Dickmanns, An expectation-based, multi-focal, saccadic (EMS) vision system for vehicle guidance, *Proceedings 9th Int. Sump. on Robotics Research (ISRR '99)*, Salt Lake City, October 1999.

[43] F. L. Lewis, *Optimal Estimation*, Wiley, NY, 1986.

[44] C. W. Therrien, *Discrete Random Signals and Statistical Signal Processing*, Prentice Hall, Englewood Cliffs, NJ, 1992.

[45] E. Dickmanns, Machine perception exploiting high-level spatio-temporal models, *AGARD Lecture Series 185: Machine Perception*, Hampton, VA; Munich; Madrid, 1992.

PROBLEMS

6.1. Considering examples of systems that you have selected in Problem 4.2 of Chapter 4, analyze the functioning of their sensors and SP-module. What are the generalized entities that are formed from the elementary entities as a result of functioning of SP-module? What are the objects and events that are being *recognized* and what are the objects and events that are being *identified*?

6.2. Consider the same systems as in Problem 6.1. Where and how are the procedures of focusing of attention performed? What are the FA-algorithms utilized in these systems? Can you suggest different ideas about focusing attention for these cases?

6.3. Consider the same systems as in Problem 6.1. Where and how are the procedures of grouping and/or combinatorial search performed? Describe the possible alternatives of G and/or CS-procedures utilized in these systems. What are the attributes that have been used for constructing hypotheses, and how are the verification of hypotheses done? (See Subsubsections 6.1.3 and 6.1.4.)

6.4. Consider the same systems as in Problem 6.1. For the recognized objects, construct their description. Use Figure 2.1 for guidance about the construction of the object and event specifications ("frames").

6.5. Consider the same systems as in Problem 6.1. For the recognized events, construct the description of these events as changes that happen to the objects (construct a "frame" similar to the one constructed in Problem 6.4 for objects).

6.6. Consider a computer vision system. Using an image of your choice (a scene, possibly visualized by this system) and an algorithm of image processing of your choice follow this algorithm of image processing and illustrate it by using the chosen image. Show how the sequence of windowing (focusing attention and search), grouping, computing entity attributes, and recursive estimation end up with classification of attributes and labeling the entity in the image.

6.7. Consider the same systems as in Problem 6.1. For each of these systems, explain how the space tessellation is performed.

6.8. Consider a computer vision system using a particular camera and particular image processing software. Explain how the space tessellation is performed.

6.9. Explain the basic scheme of 4D image sequence understanding (use Fig. 6.6).

6.10. Using MATLAB, simulate a recursive estimation loop for a simple system of your choice.

6.11. Follow Subsubsection 6.3.2 and describe functioning of two levels of a mobile robot equipped with computer vision for a concrete visual scene of your choice.

6.12. Describe how the results of processing obtained in the solution of Problem 6.11 are used by the subsystem of behavior generation. Follow Figure 6.7, however, apply it for a concrete visual scene of your choice.

6.13. Construct an example of recursive estimation for ultrasonic and infrared sensors. Illustrate expressions (6.1) and (6.2).

6.14. Interpret recursive estimation at levels 4 and 5.

CHAPTER 7

BEHAVIOR GENERATION

7.1 PRELIMINARY CONCEPTS OF MULTIRESOLUTIONAL BEHAVIOR GENERATION

The NIST-RCS architecture described in Chapter 4 embodies the principles of behavior generation inferred from an abundant number of human activities and the experience of engineering science. It is constructed as a nested multiresolutional coalition of ELFs that communicate with the help of GFS and GFS^{-1} algorithm of generalization and instantiation of information between adjacent levels. The format of the architecture allows the jobs in the modules to be performed by a human or by a machine depending on available resources. The algorithms of operation and their arrangement follow the arrangements delineated by the architecture regardless of the performer.

The general framework of the RCS architecture as an organizational framework for intelligent systems is concerned with the general laws of knowledge representation or general laws of symbolic representation of the world and its processes rather than with a limited set of particular applications. It is related to the general laws of thinking, engineering thinking in particular, and this is why the results of this chapter are equally important at both the stage of systems design and the stage of analysis of the system's functioning. The implication is that the algorithms of behavior generation reflect and reproduce mechanisms of thinking. They reproduce it to the degree the procedures could be possibly reconstructed from the experimental data interpreted within our theoretical paradigm. The results can be related to nonautomated as well as to automated systems, including autonomous robots [1].

7.1.1 Definitions

The module of behavior generation is the next unit of ELF following the subsystem of world representation. The behavior generation (BG) module of the intelligent system receives a goal (presumably, from the upper level of the hierarchy of ELFs; see Chapter 2), retrieves relevant knowledge in the world model, and creates assignments for the lower level of resolution in the form of strings of tasks sufficient to compute the input control commands for the actuators or virtual actuators of the lower level. The BG module at the adjacent lower levels of the hierarchy considers these tasks to be its "goals." Our presentation is based on the following axioms and definitions:

DEFINITION OF BEHAVIOR

Behavior is the ordered set of consecutive-concurrent changes (in time) of the states registered at the output of a system (in space). In a goal-oriented system, behavior is the result of executing a series of tasks.

The *mathematical representation of behavior* is a time-tagged trajectory in an enhanced state-space. This trajectory is frequently called *motion*. A space-time (spatiotemporal) representation presumes the process of motion to be a sequence of time-tagged states (temporal sequence) in which each state is a vector in the enhanced state-space which is built upon coordinates corresponding to all variables of the process (including input, output, and inner state-variables).

DEFINITION OF BEHAVIOR GENERATION

Behavior generation is the process of planning and execution (including feedforward and feedback control) of actions designed to achieve behavioral goals.

DEFINITION OF BEHAVIORAL GOAL

A *behavioral goal* is a desired result that an action is designed to achieve or to maintain.

Behavior generation accepts task commands with specified goals and priorities, formulates and/or selects plans, and develops controls for the actions. Behavior generation develops or selects plans by using a priori task knowledge and value judgment functions combined with real-time information provided by real world modeling in order to find the best assignment of tools and resources to agents and to find the best time schedule of required actions (i.e., the most efficient feedforward plan to get from an anticipated initial state to a goal state; see Chapter 8). The module of behavior generation involves actions of the system by producing both a feedforward string of control commands and combining it by an additional component: feedback error compensation. Both feedforward and feedback control commands are assigned according to the control law chosen within the system.

DEFINITION OF CONTROL LAW

A *control law* is a model that computes a control signal given the predicted, desired, and measured states. This model can be represented by a set of equations, a set of logical rules, a table, an automaton, and so on. In all cases this model materializes as a computational algorithm or an architecture (an interrelated set of algorithms).

The model of control law can be symbolically represented in the form

$$\mathbf{u} = \{\mathbf{u}_{\mathrm{ff}}[\mathbf{x}_{\mathrm{d}}, \mathbf{x}^*], \mathbf{u}_{\mathrm{fb}}[\mathbf{x}_{\mathrm{d}}, \mathbf{x}^*, \mathbf{x}]\}, \tag{7.1}$$

where

$\mathbf{u}$ = input control signal (a string of commands evoked by achieving particular moments in time or coordinates of space),

$\mathbf{u}_{\mathrm{ff}}$ = feedforward control action (from a plan),

$\mathbf{x}_{\mathrm{d}}$ = desired world state (from a plan),

$\mathbf{x}^*$ = predicted world state (from the world model),

$\mathbf{u}_{\text{fb}}$ = feedback compensation computed according to the accepted principle of compensation,

$\mathbf{x}$ = measured world state.

A common form of the control law is additive:

$$\mathbf{u} = \mathbf{u}_{\text{ff}} + \mathbf{G}(\mathbf{x}^*, \mathbf{x}_{\text{ff}}, \mathbf{x}), \tag{7.2}$$

where $\mathbf{G}$ is a function that defines the additive feedback compensation depending on the desired, predicted, and actual (measured) state of the world.

Selection of the control law depends on the models of uncertainties accepted in a particular system and on the strategies of decision making contemplated and intended for dealing with the uncertainties. Among these strategies, the following are of frequent consideration:

- Compensation with minimizing the variance of deviations (e.g., the Kalman filter).
- Compensation with minimizing instantaneous, predicted, and cumulative errors (e.g., the PID controller in the multiplicity of its forms, where the "proportional" part corresponds to instantaneous error compensation, "integral" reflects the concern about cumulative component of the error, and "derivative" is the rough prediction of what will happen right away).
- Fast compensation unequivocally robust about the disturbances (variable structure control).

AXIOM 1

The computational control model treats the functional elements and knowledge database of an intelligent system as if they can be distributed over a set of computational nodes, and be represented within each computational node by a set of processes interconnected by a communication system that transfers information between them.

AXIOM 2

Any intelligent system contains knowledge required to determine the *action* necessary and sufficient to perform at least one set of tasks.

DEFINITION OF TASK

A *task* is a component of work to be done, or an activity to be performed. It can be described as a data structure representing the assignment. Any task leads to a *goal* state and can be performed by executing a sequential-parallel set of *actions* (processes).

DEFINITION OF ACTION

An *action* is an effort generated by the actuator producing changes in the world.

DEFINITION OF GOAL

The *goal* a state to be achieved or an objective toward which task activity is directed (e.g., a particular event). A goal can be considered an event that successfully terminates a task.

DEFINITION OF TASK COMMAND

A *task command* is an instruction to perform a named task. This is an assignment presented in the code pertaining to a particular module of the system. A task command may have the form:

DO<Task_name(parameters)>AFTER<Start State (or Event)>UNTIL
<Goal State (or Event)>

DEFINITION OF TASK FRAME

A *task frame* is a data structure in which task knowledge can be stored. It is a data structure specifying the *task knowledge*.

DEFINITION OF TASK KNOWLEDGE

Task knowledge is all knowledge required to accomplish a task. As the task frame is invoked, all task knowledge becomes available.

PRELIMINARY DEFINITION OF PLAN

A *plan* is a set of subtasks and subgoals that are designed to accomplish a task, a job, or a sequence (or a set of sequences for a set of agents) of subtasks intended to generate a series of subgoals leading from the starting state to the goal state. (More definitions are given in Chapter 8.)

DEFINITION OF JOB

A *job* is the description of one or many actions (activities) required to perform the task and formulate it in terms of the resources to be used.

Figure 7.1 illustrates a task plan. The main task is decomposed into three jobs for three agents. Each job consists of a string of subtasks and subgoals. The entire task plan consists of a set of subtasks and subgoals to be used in accomplishing the task. In this example, the task plan consists of three parallel job plans. Each job plan can be represented as a state graph where subgoals are nodes and subtasks are edges that lead from one subgoal to the next. A job plan can also be represented as a set of waypoints on a map such that the waypoints are subgoals and the paths between waypoints are subtasks.

A plan may be represented as a path through state-space from an anticipated starting state to a goal state. A plan may consist of a sequence of intended steps, or actions to be done, or subtasks to be performed by a single agent (a job plan) or by coordinated agents (task plan). A plan may be represented as an augmented state graph, a Pert chart, a Gantt chart, or a set of reference trajectories through a state-space. Alternatively, a plan can be represented as a program, or a flowchart, that generates a sequence of subtask commands for a number of agents so as to define a set of actions that drives a controlled system through the state-space from a start state to the goal state while using resources and satisfying constraints.

In general, a task plan consists of the following:

1. An assignment of job responsibilities to a set of agents in a form, for example, of a state graph. This assignment is developed by the agents of the ith level to the agents of $(i-1)$th level.
2. An allocation of resources to the agents.
3. A set of (possibly coordinated) schedules that defines the sequence of activities for the agents designed to achieve or maintain the goal state while satisfying a set of constraints.

The three job-plan state graphs in Figure 7.1 are shown as simple linear lists of subtasks and subgoals. In practice, each job-plan state graph can have a number of branching conditions that represent alternative actions that may be triggered by objects or events detected by sensors. For example, each job plan might contain emergency action sequences that could be triggered by out-of-range conditions.

Note that in practice of some social systems, agents of the ith level are called "superior" relative to the agents of $(i-1)$th level who are labeled as "subordinate" ones. In the terminology of multiresolutional systems theory, these anthropomorphic terms are meaningless.

Coordination among agents can be achieved by making state transitions in the state graph of one agent conditional upon states or transitions in states of other agents, or

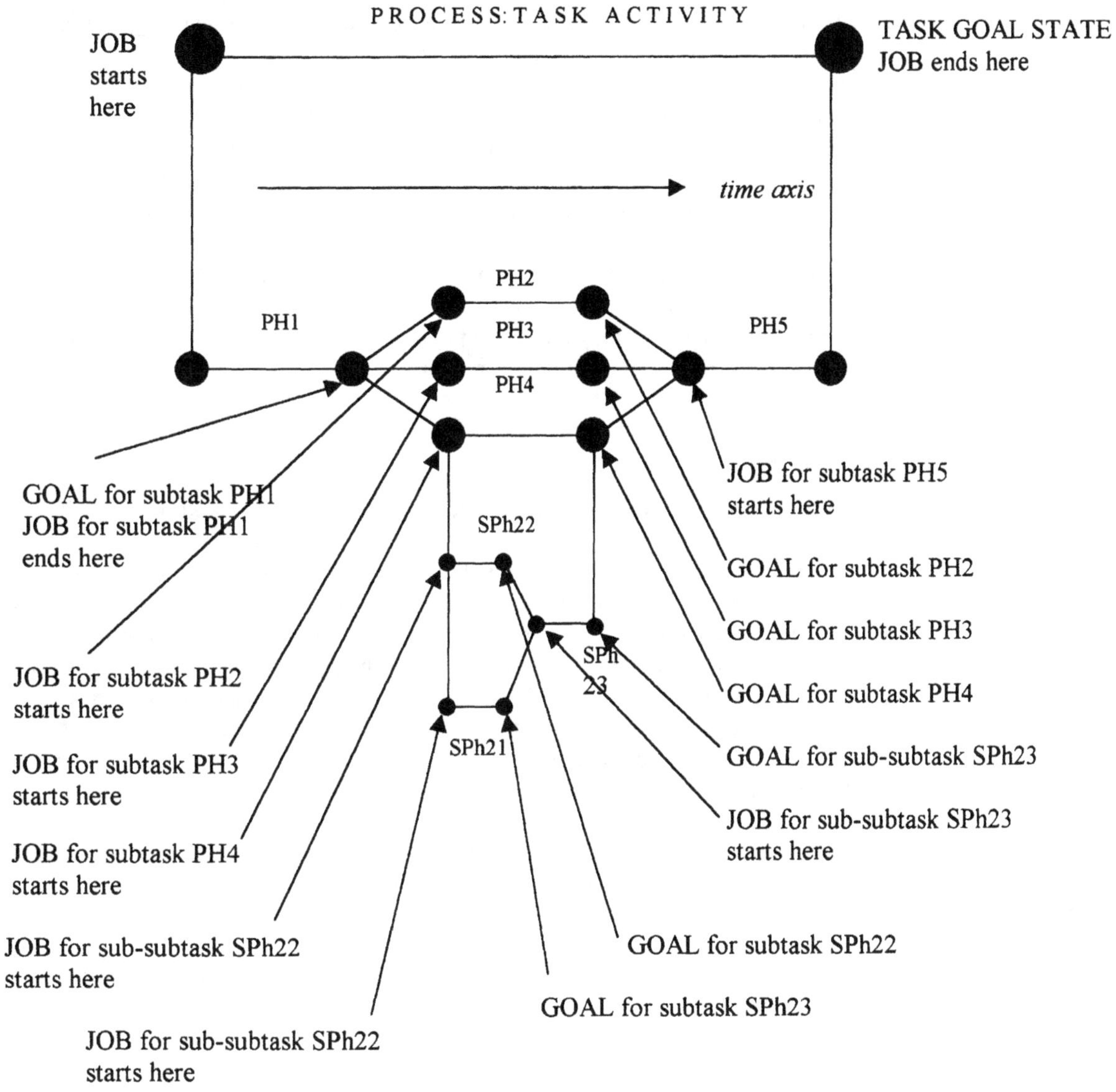

Figure 7.1. Task plan for a hierarchy of agents (see also Figures 3.1 and 6.2).

by making state transitions of coordinated agents dependent on states of the world. In the latter case, each of the agents must have access to information from sensors that measure the states of the world.

DEFINITION OF SCHEDULE

A *schedule* is the timing specification for a plan. A schedule can be represented as a time-labeled, or event-labeled, sequence of activities or events.

Task Frame

In systems where task knowledge is explicit, a task frame [24] can be defined for each task in the task vocabulary. The task vocabulary is the set of task names assigned to the set of tasks the system is capable of performing. For creatures capable of learning, the task vocabulary is not fixed in size. It can be expanded through learning, training, or programming. It may shrink from forgetting or program deletion.

Typically a task is performed by one or more "agents" on one or more objects. The performance of a task can usually be described as an activity that begins with a start-event and is directed toward a goal-event. Task knowledge is knowledge of how to perform a task, including information concerning which tools, materials, time, resources, information, and conditions are required, plus information regarding which costs, benefits, and risks to expect.

Task knowledge may be expressed implicitly in fixed circuitry, either in the neuronal connections and synaptic weights of the brain or in algorithms, software, and computing hardware. Task knowledge may also be expressed explicitly in data structures, either in the neuronal substrate or in a computer memory.

Task Knowledge

For any task to be successfully accomplished, there must exist task knowledge that describes how to do the task. Task knowledge may be acquired through learning or exists in the form of instinct. It may be provided as schema or algorithms designed by a programmer, or discovered by a heuristic search over the space of possible behaviors. Task knowledge may include a list of the tools and materials necessary to accomplish the task, a list of conditions necessary to start or continue the task activity, a set of constraints on the operations that must be performed, and a set of information, procedures, and skills required to successfully execute the task, including error correction procedures and control laws that describe how to respond to error conditions. This task knowledge, together with information supplied by the task command, can be represented in a task frame.

WHAT SHOULD BE IN THE TASK FRAME

A task frame is essentially a recipe that specifies the materials, tools, and procedures for accomplishing a task. A task frame may include the following:

1. *Task name* (from the library of tasks the system knows how to perform). The task name is a pointer or an address in a database where the task frame can be found.
2. *Task identifier* (unique id for each task commanded). The task identifier provides a method for keeping track of tasks in a queue.
3. *Task goal* (a desired state to be achieved or maintained by the task). The task goal is the desired result to be achieved from executing the task.
4. *Task objects* (on which the task is to be performed). Examples of task objects include targets to be attacked, objects to be observed, sectors to be reconnoitered, vehicles to be driven, weapons or cameras to be pointed.
5. *Task parameters* (that specify, or modulate, how the task should be performed). Examples of task parameters are priority, speed, time of arrival, and level of aggressiveness.
6. *Agents* (that are responsible for executing a task). Agents are the subsystems and actuators that carry out the task.
7. *Task requirements* (tools, resources, conditions, state information). Tools may include sensors and weapons. Resources may include fuel and ammunition. Conditions may include weather, visibility, soil conditions, daylight or darkness. State information may include the position and readiness of friendly forces and other vehicles, position and activity of enemy forces, and stage in battle.

8. *Task constraints* (upon the performance of the task). Task constraints may include visibility of sectors of the battlefield, timing of movements, requirements for covering fire.
9. *Task procedures* (plans for accomplishing the task, or procedures for generating plans). There typically will exist a library of plans for various routine contingencies and procedures that specify what to do under various kinds of unexpected circumstances.
10. *Control laws and error correction procedures* (defining what action should be taken for various combinations of commands and feedback conditions). These may be developed during system design and modified through learning from experience.

Explicit representation of task knowledge in task frames has a variety of uses. For example, task planners may use it for generating hypothesized actions. The world model may use it for predicting the results of hypothesized actions. The value judgment system may use it for computing how important the goal is and how many resources to expend in pursuing it. Plan executors may use it for selecting what to do next.

Task knowledge is typically difficult to discover, but once known, it can be readily transferred to other tasks. Task knowledge may be acquired by trial and error learning, but more often it is acquired from a teacher or from written or programmed instructions. For example, the common household task of preparing a food dish is typically performed by following a recipe. A recipe is an informal task frame for cooking. Gourmet dishes rarely result from reasoning about possible combinations of ingredients, still less from random trial and error combinations of food stuffs. Exceptionally good recipes often are closely guarded secrets that, once published, can easily be understood and followed by others.

Making steel is a more complex task example. The human race took many millennia to discover how to make steel. However, once known, the recipe for making steel can be implemented by persons of ordinary skill and intelligence.

In most cases the ability to successfully accomplish complex tasks depends more on the amount of task knowledge stored in task frames (particularly in the procedure section) than on the sophistication of planners in reasoning about tasks.

7.1.2 Behavior Generation as a Recursive Synthesis of Instantiations from Generalizations

As you remember, the primary information flow in SP was bottom-up: from instances toward generalities. Unlike the SP-hierarchy, the hierarchy of behavior generation is busy looking for the best strings of instantiations that fit within the general tasks descriptions arriving top-down.

The goal of this subsection is to demonstrate that with representation organized as a recursive hierarchy, the process of behavior generation can be described in a form of a recursive algorithm of synthesis. This algorithm will take care of the *planning/control* commands generation including task formation and decomposition at all resolution levels. It is shown above that the hierarchy is retained even when only one actuator is considered. This would be a *hierarchy of decision making* processes over a hierarchy of world representation where each of the levels can be characterized exhaustively by the value of resolution of knowledge representation. Later we will see that there exists also a bottom-up process of generalizing the emergent instantiations. Thus, the overall BG process is a result of joint "top-down/bottom-up" computation.

Task Vocabulary

Thus, behavior generation is inherently a hierarchical process. At each level of the behavior generation hierarchy, tasks are decomposed into subtasks that become task commands to the next lower level. At each level of a behavior generation hierarchy, there exists a task vocabulary and a corresponding set of task frames. Each task frame contains a procedure state graph. Each task name at the next level of higher resolution must correspond to the state transition function at the level under consideration. Therefore, each node in the state graph must correspond to a task name in the task vocabulary at the next lower level.

TASKS: A HETERARCHY OF OPPORTUNITIES

Behavior generation consists of both spatial and temporal decomposition. Spatial decomposition partitions a task into jobs to be performed by different subsystems. Spatial task decomposition results in a tree structure, where each node corresponds to a BG module, and each arc of the tree corresponds to a communication link in the chain of command (see Fig. 7.1).

In a plan involving concurrent job activity by different subsystems, there may be requirements for coordination, or mutual constraints. Some tasks may require concurrent coordinated cooperative action by several subsystems. For example, a start-event for a subtask activity in one subsystem may depend on the goal-event for a subtask activity in another subsystem. Both planning and execution of subsystem plans may thus need to be coordinated.

There may be several alternative ways to accomplish a task. Alternative task decompositions can be represented by an AND/OR graph in the procedure section of the task frame. The same is related to the hierarchy of jobs invoked by a particular decomposition tree. The decision as to which of several alternatives to choose is made through a series of interactions between the BG, WM, SP, and VJ modules. Each alternative may be analyzed by the BG module hypothesizing it, WM predicting the result, and VJ evaluating the result. The BG module then chooses the "best" alternative as the plan to be executed.

"Behavior" is the set of output activities of the system. Behavior generation is an important function of control systems. It is especially important when we talk about complex systems. In the subsequent material we consider systems that are controlled by NIST-RCS created for complex (and sometimes, large) systems.

"Behavior" is a term incorporated by the area of intelligent control only recently. It is sufficient to talk about "motion" in its general meaning: temporal development of the output coordinates, or "motion trajectory." It is convenient to cluster similar trajectories by their resemblance or by their goal. Then we receive more generalized patterns of maneuvers or motion including such behaviors as "pursuit," "avoidance," and "wall following." In the evolution of NIST-RCS architecture, the roots of behavior generation are found in the definition of task decomposition.

Task Decomposition

Behavior is generated by a task decomposition system that plans tasks by decomposing them into subtasks, and scheduling them to achieve goals. Goals are selected and plans generated by a looping interaction among task decomposition, world modeling, and value judgment functions. The task decomposition system hypothesizes plans, the world model predicts the results of those plans, and the value judgment system evaluates those results. The task decomposition system then selects the plans with the best evaluations for execution. Task decomposition also monitors the execution of task plans and modifies existing plans whenever the situation requires.

Task decomposition is a part of planning that is equivalent to designing the system's configuration. It can be done for a system with known components that can be arranged in various ways. The problem then of system decomposition into a set of interrelated subsystems is already solved at the design stage. Certainly this affects the way the task is decomposed. On the other hand, in many cases, task decomposition and the system design are done simultaneously. This leads to more appropriate results for both. This chapter relates both cases: behavior generation in a previously designed system and in a system that is being designed but whose design is not yet completed.

CLOSURES OF TASK DECOMPOSITION

Task decomposition results in a hierarchy of tasks that corresponds to the hierarchy of subsystems. In a goal-oriented hierarchical system, behavior is generated via special process of task decomposition. The latter includes processes of planning at each level of the hierarchy, and other process components of hierarchical decision making, including feedback compensation, task distribution, and the like. Task decomposition is a *closed-loop process*. Planning in turn, consists of several components including job assignment, scheduling, simulation, evaluation, and selection. After the best plan is selected, it is supposed to be executed; this presumes allocating it with a proper subsystem of actuation, monitoring the process of actuation and compensation for the deviations that occur.

The processes of functioning can be adequately decided by using the set of *loops of functioning*. Each loop contains and circulates all information required to generate a particular loop behavior. The need in the loops of functioning is determined by the need of behavior generation and the need to provide *consistency of functioning* within the loop.

In the next section we outline the quantitative analysis of the structures. We include a discussion of the tools dedicated to representing the systems in a hierarchical multiresolutional fashion and solving the control problems. It is demonstrated that substantial computational advantages can be gained when the control solutions are also obtained similarly: as a multiresolutional hierarchical system. The system of control in this case can be designed by minimizing the total complexity of the system. We evaluate the complexity of computations quantitatively and make recommendations for the design of knowledge heterarchies.

7.2 BG ARCHITECTURE

The place of behavior generation in the reference architecture determines the need in the constant interaction of the BG modules with the rest of the system. This can be modeled in the form of virtual loops.

7.2.1 Virtual Loops

Now we come to the matter of considering the relationships among modules at different levels. An example is illustrated in Figure 7.2 for the second level of control and the first level of control system together with the controlled system; the boxed elements are equivalent to the set of its virtual A–W–S (see Chapter 4).

In the manufacturing cell example (see Section 7.4.3), the level assignment (the set of parts) and the manufactured cell (the target object) consist of several machines; each machine contains several actuators. At this level of resolution we can consider the "cell" to be a performing unit or a control system. The description of the job to be performed is done in a generalized fashion as a group; it usually requires some attention to time,

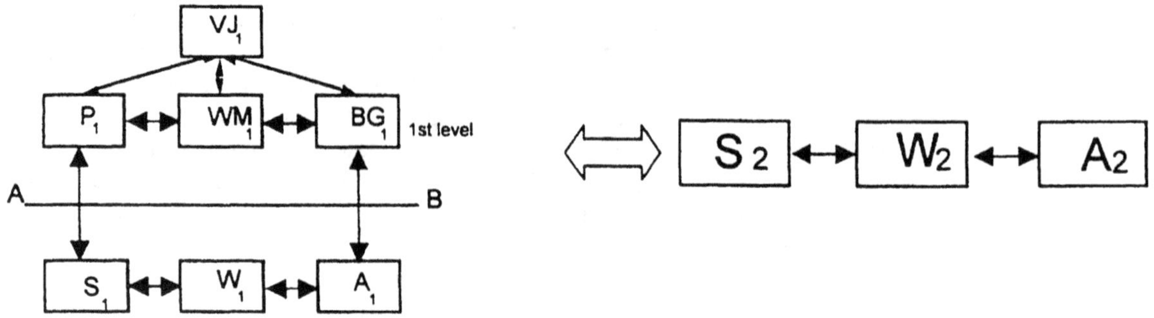

Figure 7.2. Equivalence between the system for the level 1 and a string $S_2W_2A_2$.

materials, and expense. At this level of resolution we can avoid talking about the details of machining, or the voltage required by the actuators; we regard the level (the RCS node, the control system) to be capable of handling several cells in coordinating their activities using the same input-output (interface) language.

The system for level 1 includes the whole ELF at this level: that is, both the controlled system $[S_1W_1A_1]$ and its control system $[P_1WM_1BG_1VJ_1]$. Jointly they constitute the imaginary (virtual) controlled system $[S_2W_2A_2]$ that exists only in "imagination" the second level of control. If one considers the first level of the control hierarchy (the lowest-abstraction level, or the highest-resolution level, or the low end controller), then it can be represented as it is shown in Figure 7.3*a*. In this figure the accurate parameters of the world model are given for the high-frequency clocks (the clocks of time discretization). The correctness of the physical model holds only for a small vicinity of the present state.

The overall physical model of the system can therefore be simplified before using it for the control loop. The order of the dynamic model can be reduced; it can be linearized and decoupled within an interval of time in the vicinity of the present state. When the second level of resolution is considered, the external system of S–W–A will include the first loop as a component (see Fig. 7.3*b*). For this reason the paradigm of execution for level 2 is not the real world S_1–W_1–$A_1$of the first level but rather the *virtual* set including $\{VJ_1, P_1, WM_1, BG_1\}$ together with the system triplet $\{S_1–W_1–A_1\}$.

YOUR MODEL MIGHT DEPEND ON THE LEVEL YOU ARE AT

The boxed region labeled "system controlled by the second level" is the new virtual set S_v–W_v–A_v which consists of virtual sensors, the virtual world, and virtual actuators. This world is denoted $\{S_v–W_v–A_v\}$; it is also real but its reality is valid only for level 2. Its discrete time is larger, though the frequency of its clock is slower. Its physical model can be described as functioning within a more extended time-span (planning horizon) around the present state. However, this model has a lower resolution in all its variables.

Therefore let us define the virtual loop as the loop of functioning as it is visualized by an observer from the viewpoint of a particular resolution level. This internal observer is not interested either in a further decomposition of the virtual aggregated view or in other levels of resolution and their scope of view (scope of attention). This position allows us to introduce a level vocabulary that gives the most efficient representation of the level.

For the second level system to be controlled, the virtual set of (ACTUATORS + WORLD + SENSORS)$_2$ is a set $\{P_1WM_1BG_1VJ_1, S_1W_1A_1\}$ that requires considering the output of the virtual sensor S_2 as a result of the corresponding generalization from S_1 combined with the activities of the set $\{P_1WM_1BG_1VJ_1\}$. This is the meaning of the box "system controlled by the second level" in Figure 7.3.

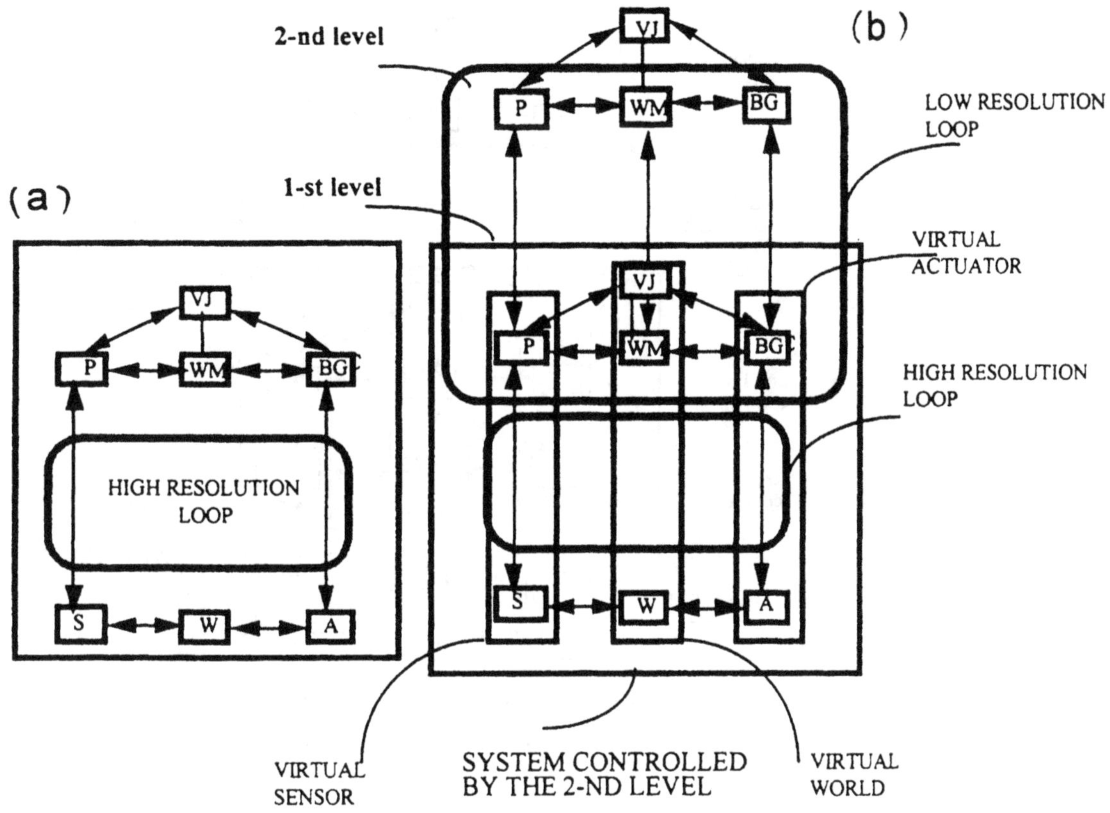

Figure 7.3. Representation of "output" at different levels.

The third level of resolution has the second-level control loop as the substitute for the real external world (Fig. 7.3). The time period within which the original physical model can be simplified is longer than in the second level of resolution. The intervals of discrete time are longer, the frequency is lower, and the accuracy is determined at a lower-frequency sampling with the numerical value of the variables averaged over the increased period of sampling.

It is clear from Figures 7.2 and 7.3 that both the system controlled by the second level (Fig. 7.2) and the system controlled by the third level (Fig. 7.3) are transformed into a set of virtual modules S_V–W_V–A_V for the corresponding level of control. The control loop "visualized" by the third level of control (its virtual control loop) works with the system of sensors–world–actuators and includes all levels. Each component of this virtual triplet constructed for the third level ($S_{V3}W_{V3}A_{V3}$) has a model and parameters that represent the whole hierarchy of sensors and actuators within the ontological hierarchy of the world at the third level.

CONTROL BY VIRTUAL MODULES

In other words, each level of the RCS architecture visualizes all higher levels of resolution underneath as a virtual world with its virtual actuators receiving the control commands and virtual sensors submitting new information of changes that happened.

On the other hand, for the lowest-level controlled system, the whole hierarchy of controllers with its three-level hierarchy of sensory processing, world model, and behavior generation, appears to be a single-level virtual control system as shown in Figure 7.5. Figure 7.5 and Figure 7.4 do not contradict each other; they give complementary views.

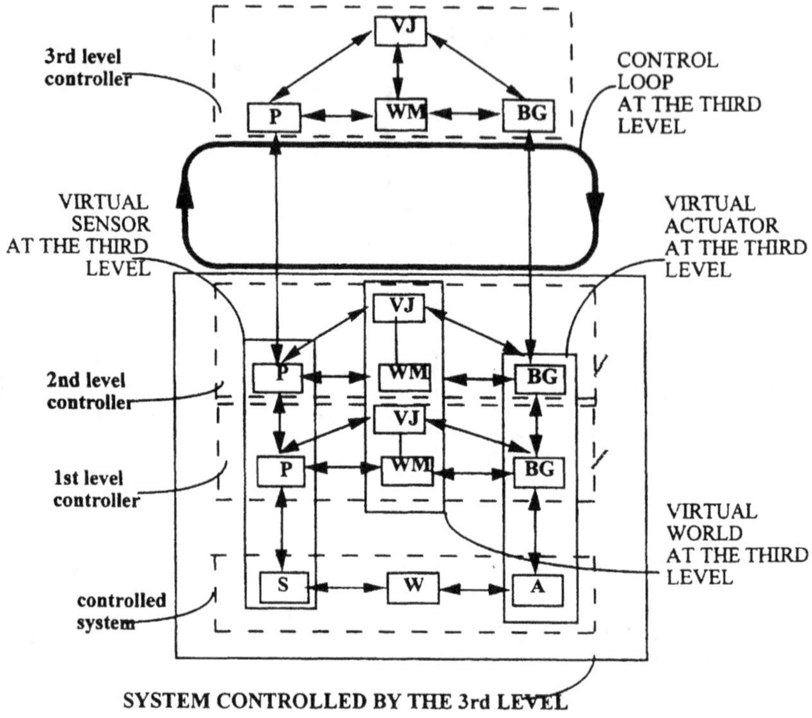

Figure 7.4. Virtual triplet of the controlled system actuator–world–sensor for the third level.

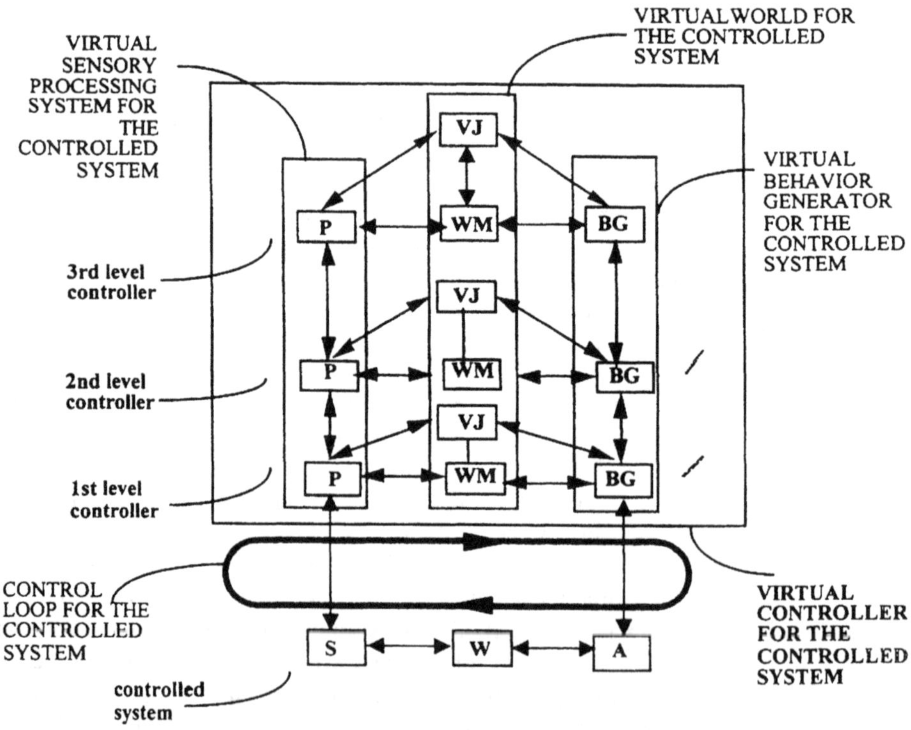

Figure 7.5. Three-level control system as visualized by the "observer" within the controlled system.

The intermediate level of the control system can take any of the following forms in its "virtual" representation:

- It can be simultaneously a controlled system for the above virtual control and the control system for the below virtual control. In the first case, the system might be the third level of the overall control system; in the second case, the system might be the first level of the overall control system linked with the lowest-level controlled system.
- It can be a part of the upper-level virtual control system.
- It can be a part of the lower-level virtual controlled system.

These conditions of belonging are meaningful at the design stage during modeling and simulation as well as in developing loops. In the Figure 7.6 we present two types of loops, unified and separated.

As Figure 7.6 shows, unlike conventional controllers which are represented by space-time dimensions, the multilevel (multiresolutional) control architecture is represented by *scale-space-time dimensions*. Thus, while all of these loops have the same world as a part of their structure, they have a different vocabulary, and the world is represented at different resolutions. Figure 7.6*a* shows the intermediate step which leads to Figure 7.10*b*. In a real system we design and plan for each level separately, and the consis-

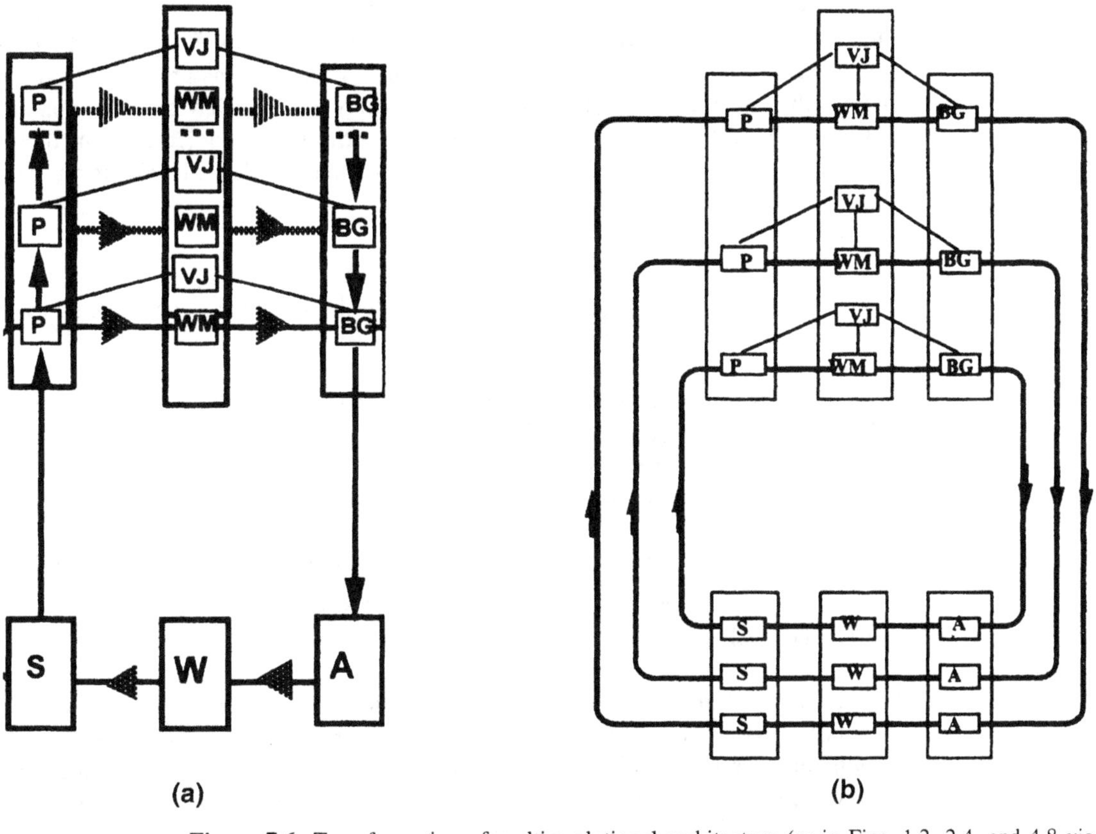

Figure 7.6. Transformation of multiresolutional architecture (as in Figs. 1.2, 2.4, and 4.8 via constructing Fig. 2.7*b*) into set of multiple virtual ELFs.

tency of the results of this level ELF functioning can be achieved if we have satisfactory conditions determined by the actual rules of decomposition for the level above into the adjacent level below, or the rules of aggregation of the levels below into the adjacent level above. In the case of Figure 7.6*a*, this has been done for the levels (control systems), but it is not clear what will happen with these loops at the bottom of the diagram (within the controlled system). In Figure 7.6*b* the loops are separated from each other by the decomposition of the sensor-set, actuator-set, and even the world (its ontology, of course.) There are three different virtual worlds driven by three different virtual actuators and perceived by three different sets of virtual sensors. (The real elements are parts of the hierarchy. The virtual ones are parts of the consistency check.)

In a discussion of the same world, these descriptions are easily connected. Their connection acts as a consistency check. However, in each loop of the resolutional level, the flow of knowledge is completely different from the flows of knowledge in other loops. The *virtual existence* of these different loops is the artifact of RCS control systems.

TWO CONSISTENCIES

In practice, the loops of different resolution levels are often considered and used anyway despite the fact that the consistency conditions are not satisfied. Frequently neither the vertical "between-the-loops" consistency is checked on a regular basis, nor the horizontal "within-the-loop" consistency. Representing the loops consistently internally and externally is one of the major problems in developing the world model.

7.2.2 Real-Time Control and Planning: How They Are Affected by the Sources of Uncertainty

Recursive behavior generation is one of the fundamental components of hierarchical architectures and computational schemes. It determines both the design and the functioning of the NIST-RCS architecture. The fundamentals of this process are presented here with a consideration of further applications of recursive behavior generation for the RCS architecture of large systems (e.g., manufacturing). As is indicated in the literature [1, 50, 52], in general, intelligent task decomposition takes the following form:

1. Reasons about space and time.
2. Alludes to geometry and dynamics.
3. By an implicit demand, formulates, synthesizes, and/or selects plans based on values such as cost, risk, utility, and goal priorities.

Task planning and execution are done, however, in the presence of uncertain, incomplete, and sometimes incorrect information that circulates in the functioning loop. The subsystem of behavior generation allows for task decomposition in the presence of the following sources of uncertainty simultaneously:

1. The implicit statistical experience embodied in the structure of the system.
2. The approximate models stored in the knowledge/databases.
3. The stochastic data available from sensors.

The danger that must be avoided is that the relations among the loops not be violated.

For the system to succeed in a dynamic and unpredictable world, its task decomposition (and therefore its planning) at the high-resolution levels must be accomplished in real time, for the dynamics and unpredictability of the real world at these levels occurs in real time. Either one should properly predict these factors online, or properly respond to them online, or preferably, both. This term "online" reflects different aspects of NIST-

RCS functioning. It means that NIST-RCS operates while the system functions, that is, simultaneously with the system it is designed to control.

The levels of low resolution do not require any fast, real-time decision making. Because the functioning loops of the different levels (as perceived by NIST-RCS) operate at different time scales, the phenomenon of being "simultaneous" does not necessarily mean to happen at the same moment of astronomical time. The coincidence of time instances is substituted by satisfaction of the conditions of inclusion for the sampling time units of corresponding resolution levels.

Second, the NIST-RCS alone has the same property of all modules at all levels operating simultaneously. (Some of the procedures are supposed to be done as the need arises. Other issues such as learning presume that the process of "preparation" is done off line while the results of this "preparation" are used online.)

RECURSIVE LINKAGE OF LEVELS

Third, in order to achieve online behavior generation via task decomposition for a sufficiently complex system, it is necessary to transform the planning problem into a hierarchy of planning problems pertaining to levels of resolution with different temporal planning horizons and different degrees of detail at each hierarchical level. This should be done to provide for a proper operation of the functioning loops associated with each of the levels.

When a level is sufficiently low in abstraction and high in resolution, the planning processes are performed in real time. They are considered "feedforward control," and this presumes a particular horizon of planning at each level of resolution. Once this is done, it is possible to employ multiple planners to simultaneously generate and coordinate plans for many different subsystems at many different levels.

7.2.3 Nesting of the Virtual ELFs

Hierarchical control always implies the phenomenon of nested loops. Nesting can be represented in the form of set of virtual ELF-loops shown in Figure 2.4. A situation in world W_i is measured by sensors S_i, and their signals enter the subsystem of perception P_i (which contains algorithms for processing sensor information.) The results of sensing undergo their primary organization. After this, they are incorporated (or rejected) by the knowledge base K_i where the world model WM_i is maintained (as knowledge sets and databases in a variety of representation forms).

After the goal G arrives to the subsystem of behavior generation BG_i, the latter makes decisions on task decomposition, planning, and execution. BG_i requests and receives from K the subset of knowledge required for the decision-making process. From BG_i, the decision about behavior arrives to the actuator A_i which develops changes in the world W_i.

This description relates to every level of the system, though each level operates at its own resolution and accuracy. It is important to realize that for proper functioning, the ith level should receive and submit information from and to its neighbor below, the $(i - 1)$th level, and should receive and submit information from and to its neighbor above, the $(i + 1)$th level. Each level contains knowledge (e.g., world models at its resolution). If a new model is required, it should be generalized upon the models held in the higher-resolution levels.

The property of nesting implies that the levels of decomposition (aggregation) are also levels of different resolution (granularity, or scale). The nesting of ELFs implies that each ith ELF nested within the $(i+1)$th ELF is an ELF of a higher resolution (finer granularity, smaller scale).

The property of nesting is implied by the very nature of the recursive NIST-RCS hierarchy (see Section 3.1). The purpose of the hierarchy of control is to achieve the goal of the system by generating efficient behavior. To do this, the number of control activities is reduced by unifying elementary units of the system into generalized units. This leads to the following *consistency conditions of nesting*:

1. The *goals* for the hierarchical levels are nested within each other: Each goal of the upper level contains goals of the lower level explicitly or implicitly. The goals of the higher resolution (HRGs) are represented within the lower-resolution goals (LRGs) ultimately as their components. On the contrary, each LRG is a generalized form of all HRGs which are nested within them.
2. The *world models* for the levels of hierarchy are nested within each other. The world models of the higher resolution (HRWMs) are represented within the lower-resolution world models (LRWMs) in a generalized form. Therefore a set of HRWMs is nested within each LRWM as its component.
3. The *behaviors* for the levels of hierarchy are nested within each other. The behaviors of the higher resolution (HRBs) are represented within the lower-resolution behaviors (LRBs) in a generalized form. Therefore a set of HRBs is nested within each LRB as its components. This leads to the similar statements concerned with components of the behaviors: plans, control, and actions.
4. The *plans* for the level of hierarchy are nested within each other. The plans of the higher resolution (HRPLs) are represented within the lower-resolution plans (LRPLs), ultimately as their components. On the other hand, each LRPL is a generalized form of all HRPLs which are nested within them.
5. The *actions* for the level of hierarchy are nested within each other. The actions of higher resolution (HRAs) are represented within the lower-resolution actions (LRAs) in a generalized form. Therefore a set of HRAs is nested within each LRA as its components.

PREMISES OF NESTING

These conditions suggest the following:

a. Any process of the world at ith level of resolution directly consists of the processes of the $(i-1)$th resolution level which in turn contain all processes of the $i-2$ level. Both systems and their processes can be decomposed in the resolution levels.
b. The virtual sensor module of the ith level can be regarded as a generalization of several real sensors at the $(i-1)$th level. Thus these sensors of the $(i-1)$th level are directly nested within the virtual sensor of the ith level. Sometimes a rough sensor can be installed (physically) at a lower level of resolution. Then it can be regarded as a generalization upon nonexistent (virtual) high-resolution sensors.
c. Consideration b can be repeated for the actuators. The virtual actuator of the ith level ("motion producer") can be a generalization of the real actuators performing several motions at the $(i-1)$th level. The A_i module should contain a direct representation of the set of the $\{A_{i-1}\}$ modules, contain all representations of A at all subsequent resolution levels.
d. The world model of the ith level contains elements of the world models of each of the several units of the $(i-1)$th level (in a generalized form). The values of parameters of these components of the overall model are constantly updated. This is why the set of $\{WM_{j,i-1};\ j=1,\ldots\}$ is directly nested within the WM_i.

e. All operations of ELF are performed taking into account the property of nesting in a multiresolutional model. For example, the BG_i module performs planning. Any planning presumes contemplation of future events and therefore contains simulations as a component. To judge the consistency of the results of planning at the ith level, a condition should be checked to ensure that the plans of the $(i-1)$th level modules are contained within the best alternative of the plan at the ith level. The operation of BG_{i-1} should be contained within the operation of BG_i.

f. The same should be repeated for the subsystem of SP_i. Recognition processes at the ith level require permanent interaction with recognition processes of the $(i-1)$th level. Recognition processes (e.g., in the area of computer vision) generally follow this recommendation, though it is often neglected or even contradicted.

7.3 STRATEGY OF MULTIRESOLUTIONAL CONTROL: GENERATION OF A NESTED HIERARCHY

7.3.1 Off-line Decision-Making Procedures of Planning-Control

In general, the controller can be represented as a box with three inputs and only one output. These inputs can be specified as follows (see Fig. 7.7*a*):

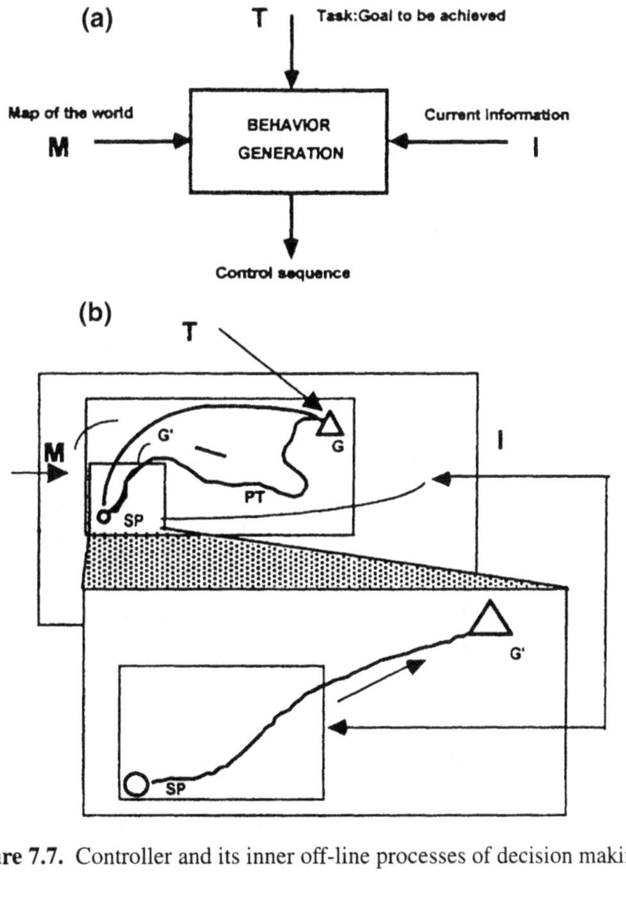

Figure 7.7. Controller and its inner off-line processes of decision making.

- *Task.* The goal G is to be achieved from the starting position SP (and conditions to be satisfied including the parametrical constraints, the form of the cost function, its value, or its changes).
- *Description of the "exosystem."* This is a map of the world (M) which includes numerous items of information to be taken in account during the process of control. The map of the world is often incomplete, and sometimes it is deceptive.
- *Current information* The current information (I) relates the vicinity of the working point delivered by the sensors in the beginning of the control process, and continuing to be delivered during the process of system's operation.

The control process is described by the trajectory of "working point" moving in the state-space. The processes within the controller are illustrated in Figure 7.7*b*.

A part of the planned trajectory PT in the immediate vicinity of SP can be determined with high accuracy (we will call it PT*). The selection of PT nevertheless is changed when the input I updates the map M in such a way that PT is no longer the best trajectory. It is sent to the level of higher resolution for refinement. The PT* part of the plan is not changed if no new information is expected. This simple consideration includes the following components of the processes of planning/control:

Off-line Planning/Control Algorithm

Step 1. Find the optimum plan PT based on the map M, and do the task formulation (SP, G, cost function, constraints) as follows:

a. Search for the alternatives of PT.

b. Compare the alternatives found.

c. Select the preferable alternative (which is accepted as a plan to be executed).

Step 2. Update the map information M in the vicinity of SP by using the sensor information I.

a. Analyze and interpret set I.

b. Compare M and I.

c. Decide on the required changes.

d. Create a new map in the vicinity of SP with the required deletions and additions.

Step 3. Refine the path planned within the updated zone of the map.

a. Determining the subgoal G′ (e.g., a point at the intersection of PT and the boundaries of the updated portion of the map).

b. Find the optimum plan PT* (repetition of all procedures listed in step 1).

Step 4. Track (follow) the optimum path PT*.

Step 5. Upon arrival to the new point of the selected trajectory (distinguishable from the initial point), loop to the step 2.

Both planning system and execution controller are intended to work together to solve the tracking problem. The similarity between our "positioning control" and our "tracking the target" will become more obvious after we realize that the tracking trajectory is being constantly recomputed during the plan computation process [3].

7.3.2 Nested Hierarchical Information Refinement during On-line Decision Making

The on-line process of consecutive information refinement is shown in Figure 7.8. At the top of the diagram the subgoal G'_i is found. Let us consider finding the refined plan PT* as a separate problem in which the new refined information I_{i+1} can be delivered for a part of the map M_{i+1} in the vicinity of the initial point SP. This consecutive process generates a nested hierarchy of computations which is characterized by an important precondition for each level: The motion starts if the subgoal is determined at the upper level, and as the new information updates map in some vicinity, the new subgoal should be determined for the lower level (level of higher resolution) as a result of the postmotion path planning at a given level.

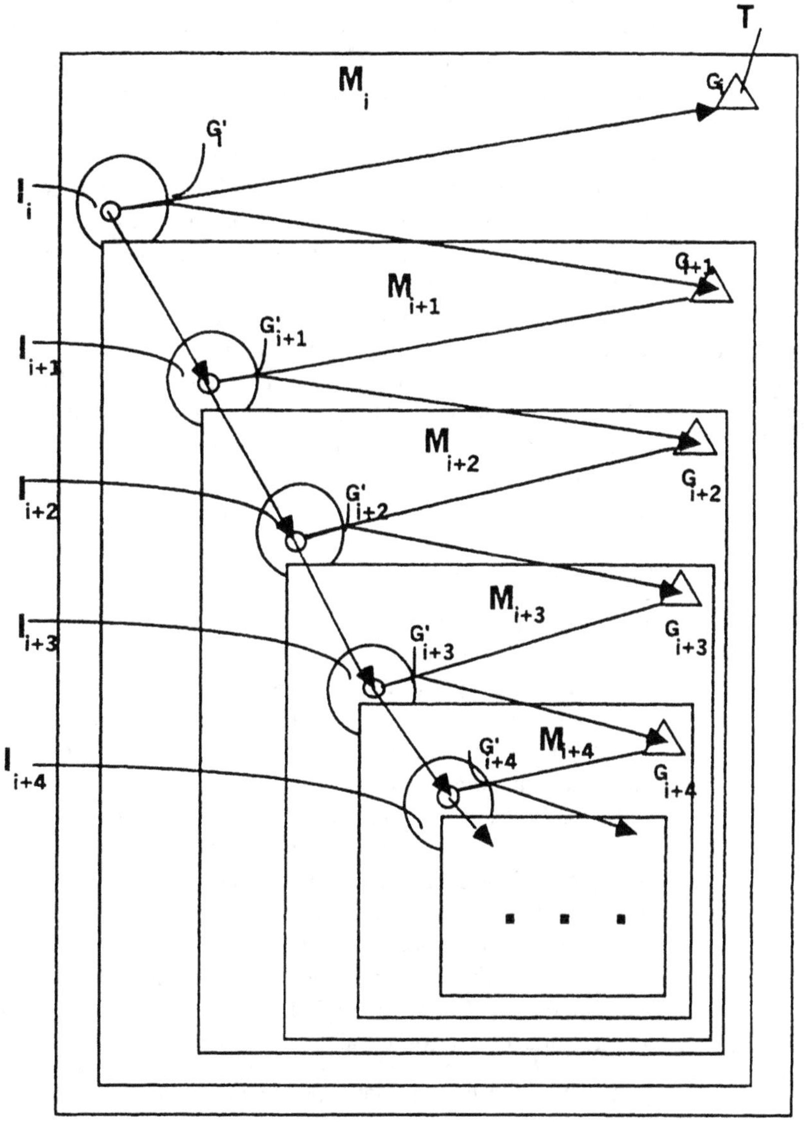

Figure 7.8. Nested hierarchical refinement of information during decision making.

The best planned trajectory can be considered a predicted trajectory. At each level of the system shown in Figure 7.8, the circled part of the trajectory is a part of the plan called PT*, and the rest is just a predicted trajectory. However, for the next consecutive lower level the situation becomes different. Within the plan assigned by the upper level, a plan is being refined, and the rest is becoming just a prediction. Since part of PT is considered a prediction anyway, the corresponding information is to be updated in the future. This generates such topics as sorting alternative plans, the evaluation of predictions by some probability measure, and syntheses of contingency plans.

The solution of the problem of generalized controller design is illustrated by Figure 7.9. A generalized controller consists of two main parts: an open-loop controller (OLC), or *feedforward controller*, and a closed-loop controller (CLC), or *feedback compensation controller*. The OLC is designed to generate a control input applied to the *plant* (G) to be controlled. The operation of planning (S) finds the output trajectory leading to the desirable final state with minimum cost of this operation. The only way to determine the *required input* to the plant is to construct the *inverse* of the plant (G^{-1}) and then apply to the input terminals of G^{-1} the *desired output* of the plant. Then, at the output terminals of G^{-1}, the required input to the plant is obtained so that $G \cdot G^{-1} = I$. Whatever we submit to the input (desirable output trajectory) must appear at the output (actual output trajectories) to the degree of accuracy of our knowledge of G.

BOTTOM-UP AND TOP-DOWN WAVES OF REFINEMENT

The computational structure G^{-1} is obtained on the assumption that the model G of the plant is known adequately. This presumes knowledge of the external environment, which is never the case. Thus, when the computed required input is applied, one must carefully compare the actual output with desired output; the difference (error) is used as input to another computational structure. It compensates the feedback and forms immediately a feedback compensation loop. It has been proven that in order to minimize the output error, CLC should be also based on the G^{-1} computational structure. (The following subtlety should be considered before using these recommendations: the computational structure G^{-1} should be determined only after G is stabilized (if unstable); in rare cases when G is a minimum phase, the structure G^{-1} is unstable and should be stabilized in the OLC computation is to be done online).

Consistency of the Control Theory

The ELFs and the multiresolutional ELFs associated with signs and symbol formation reflect the tendency to introduce control concepts into semiotics. The introduction of the concepts and premises of control theory, as well as the consistent usage of corresponding terminology in this discourse, is a result of striving for consistency. It allows for a theory to be built that explains and predicts at a level of consistency that satisfies the most demanding scientific standards.

In speaking about intelligent systems, which nervous systems definitely are, M. Minsky expressed a belief that in these kinds of systems, no consistency is required for their functioning. However strange it may sound, an entire generation of logical philosophers had wrongly tried "to force their theories of mind to fit the rigid frames of formal logic" [4]. In the science of local well-posed problems, the lack of consistency is a "scientific taboo." Consistency is consistency is consistency: We do not have anything like "gradually increasing," or "evolving" consistency. However, in the area of neurobiological architectures of control, we always deal with a contradiction in the following factors:

1. Our need to formulate specifications of the system, conditions of the test, and premises of the derivation with a level of scientific rigor typical for the control community.

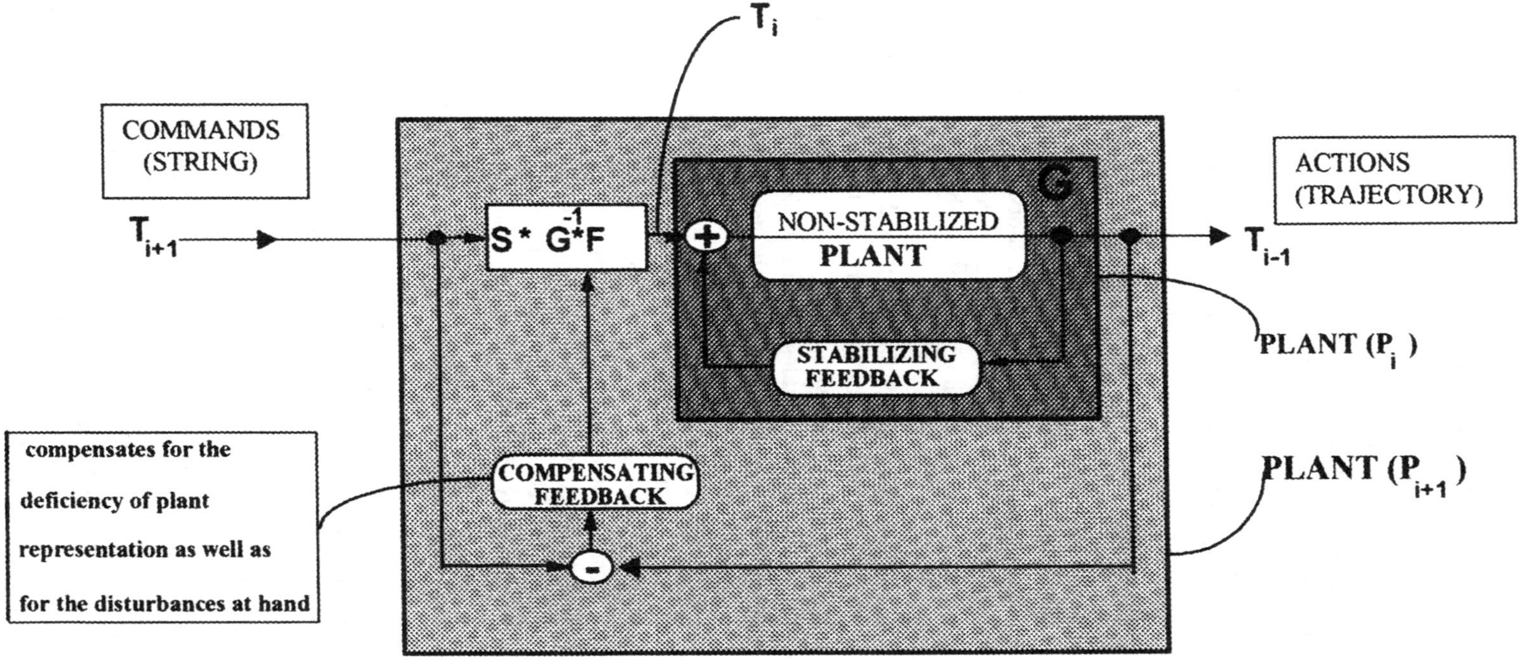

Figure 7.9. Block-diagram of the generalized controller.

2. The impossibility to provide sufficient statistics for a proof, given the interference of personal views and interpretations, and the like.

It is not easy to discuss the mechanisms of how the nervous system functions as a whole. Too much commonsense reasoning is involved, too much of a multidisciplinary blend is required in which no authority can approve the line of reasoning. The most intimate subtleties of the functioning of the CNS often seem to be illogical: How can they be explained within the framework of conventional logic? One can see this in any discussion concerning the definition of *intelligence*. In fact, it may happen that no scientific unity of results will ever be achieved in this area. Nevertheless, the theory of hierarchical automatisms persistently generalizes the nervous system and its parts into a set of automatisms. When this is done with boxes of generalized subsystems, it becomes control theory. Undoubtedly, somewhere at the top, there is intelligence to consider.

Control Structures

In Figure 7.9 the single-level architecture of control is given in its most general form. The plant is the system to be controlled (e.g., muscles and the external objects they act upon), the command generator is the feedforward controller (e.g., a subsystem of the nervous system) which develops a series of commands based on its knowledge of the plant model (it uses a procedure of inverting the desired motion, or has a look-up table stored).

It takes some computational effort to find the inverse of the desired trajectory. In order to simplify the feedforward controller, instead of inverting, we can use a library of stored command strings which can be considered elementary automatisms. This is equivalent to the storage of rules. If our knowledge of the plant is perfect (which never happens), no other control is required; the feedforward controller is sufficient, and our control system is the open-loop controller (OLC).

Since the real plant is different from our knowledge of it, the expected and real results of functioning will differ. The difference must be computed and compensated by the feedback (closed loop) controller. To be properly estimated, the output motion must include sensor information and the delay between the cause and effect. We must not forget that the output appears a bit later than when the control command is issued (with some delay Δt). Thus it is useful to compute the *predicted* output. As the command is issued at time t, a prediction is made about the output value at time $(t + \Delta t)$. As S. Grossberg [5] puts it, "Perceptions are matched against expectations." It turns out that there is a long history behind this important insight [6, 7].

The feedforward–feedback (OLC–CLC) architecture of the single-level control is widely discussed in the literature. M. Arbib uses it for analysis of the CNS (it is given in [8] with a reference to his book of 1981). Note that the need for a feedforward controller is frequently overlooked, and even within the control community many people rely only on feedback compensation. It is possible to demonstrate that controllers with both OLC and CLC are more efficient and less complex than controllers employing only feedback.

The error (the difference between the expected and real signals) has interesting characteristics. It is usually much smaller in magnitude than the desired signal. Its spectral density (the pack of signal frequencies it contains) is shifted toward higher values on a frequency scale. Clearly, instead of using the plant model which contains relevant information required for all frequencies, one can use two models together: a simplified model for the lower frequencies of the desired signal and another simplified model for the frequencies of higher bandwidth.

FEEDBACKS AS FEEDFORWARDS OF THE NEXT LEVEL BELOW

Since the sampling frequency of the compensation loop must be higher than the sampling frequency of the feedforward channel, and since the spatial resolution of the compensation loop must be higher than the spatial resolution of the feedforward channel, a conjecture can be made that the compensation process (for the feedforward channel) belongs to the level of resolution higher than the resolution of the feedforward channel. At this level the compensation commands for level 1 can be considered a feedforward control command for level 2. (The plant models for these channels differ in their bandwidth.) As soon as we start treating level 2 independently, our analysis can be recursively repeated, and a multiresolutional (multigranular, multiscale) hierarchy of control is obtained. This is but one way to introduce control hierarchies, based on the temporal and spatial resolution of compensation commands as opposed to feedforward commands.

Control Hierarchies

Another way of designing control hierarchies is by top-down task decomposition. Each task is broken down into its components as described in [9]. The task components are always spatially smaller and temporally shorter. Thus, in this case, the temporal and spatial resolution of the elementary actions (automatisms) increases top-down from level to level of the control hierarchy. In Figure 9.33 a system of signals is shown for a realistic model of a multiresolutional control hierarchy. One can see that the size and shape of signals confirm our characterization of them (see Chapter 9).

Control hierarchies were anticipated a long time ago [10, 11]. The rationale for them was clear: Task decomposition, gradual focusing of attention, and an increase of resolution were a powerful source of reducing the complexity of processing [12]. This hypothesis was later confirmed quantitatively [13]. Hierarchical controllers became a legitimate part of the intelligent control theory [14, 15]. A hierarchy of motor biological control system was described in [16, 26]: a and g motoneurons are above muscles, spinal interneurons are above motoneurons, the brainstem is above the spinal interneurons, the motor cortex is above the brainstem with the premotor cortex even higher. Higher up, the following levels can be found bottom-up: the thalamus, then the basal ganglia with the cerebellum. Finally, we arrive at the subcortical areas with association cortex at the top. Although a top-down hierarchy was admitted in control distribution, there was no such hierarchy in sensory processing. Feedforward control connections are given as if no feedback exists from each lower level to an adjacent level above [16, p. 413].

Planning includes the procedures of search, both for the desired output as well as for the input that should be applied to the plant when the desired output of the plant is given. Another term for this input is *feedforward control.* If we compare the real output with the desired output, the error of OLC can be found and the compensation can be computed and applied. We call the output of the feedback loop *compensation.* The compensation feedback controller is therefore a *closed-loop control*, or CLC. It is clear that planning can be done off-line as well as online, in advance as well as in real time. The notations of Figure 7.9 are related to the arbitrary level of any control system: T_{i+1} the task from the level above, G the transfer function of plant P_i (stabilized), T_i the task for plant P_i; this task can be obtained from the task of the level above T_{i+1} by applying the operator of planning/control $P/C = S^*G^{-1*}F$, where S is the generator of the string of input commands, G^{-1} is the inverse of the plant's transfer function, F is the feedback operator (compensation). The plant of the level of resolution together with the planning control operator P/C can be considered the plant of the upper level P_{i+1}.

The execution controller, or the first level of control (Fig. 7.10) is a machine for producing sequences of preassigned output states. The next adjacent level above, or the trajectory generator (second level), has at its input the *trajectory of motion required to perform a particular maneuver* (TMPM). The maneuver is understood as a recognizable sequence of elementary trajectories. Then the entire diagram of Figure 7.10 (the first level) is considered the machine for executing TMPM (e.g., in the set of PID execution controllers for turning all four wheels of our robot, all operators of CLC for G_1 are assumed to be multi-input–multi-output, or MIMO). MIMO TMPM machine (shown in Fig. 7.10*a* for a particular application) has at its output not the motion of a single wheel, and not the set of motions of the four wheels, but a trajectory of motion of the robot during some unspecified particular maneuver which is unknown at this level. In other words, these are the *actual time profiles* of speed, position, and orientation of the robot. In order to receive this output the desired TMPM has been obtained using $(S^*G^{-1*}F)_1$ operator.

The OLC is not expected to work properly because the actual load torques are not equal to the expected ones. A number of nonlinearities contaminate the picture such as the nonlinearity of the friction, the slipping factor between the wheels and the ground, and the high-order components of the Taylor expansion that were neglected in the transmission model. Thus the actual time profiles of position, speed, and orientation differ from the expected ones.

The maneuver controller has at its input a call for a particular maneuver (Fig. 7.10*b*). The MIMO maneuver generating machine has at its output not the trajectory of robot motion as a time profile but rather a description of some particular maneuver that is part of the motion schedule unknown at this level.

In other words, the input is presented as a verbal (logical) description in the terms of standardized behavior which is expected from a machine. This OLC is also not expected to work properly because the actual environment differs from the typical environment presented in the lookup table of the available TMPM goals. Thus the actual sequence of the maneuver primitives can differ from expected ones. This difference is being measured and used to generate the CLC part of the controller at this level. It compensates for the deficiency of plant representation as well as for such disturbances as "the terrain surface on the left is of different quality than expected." All hierarchical structures, applied in practice of intelligent control, follows the above scheme of assignments and actions decomposition [18–22].

7.3.3 Nested Modules

We have already introduced the phenomenon of nesting and the concept of nested module earlier (see Figs. 2.7b and 7.8).

All modules within the nested one are expected to interact. Let us focus attention on this phenomenon of interaction. The concept of "level" which works as a provisional idea for describing the hierarchy is incomplete because it obscures the functional essence of the system. The processes of functioning can be described only if we consider a loop in which all the ELF diagram processes can be recognized, and all causal links recorded and explained. This includes the modules in which the processes occur: that is, all sensors, perception, knowledge base, decision making, and actuators, since all these subsystems are interacting with the world. Discussion of the level of resolution, or hierarchical architecture, actually refers to the loop of the ELF diagram that can be substantiated at that level. The levels are nested in each other in the same manner as the modules are nested in each other.

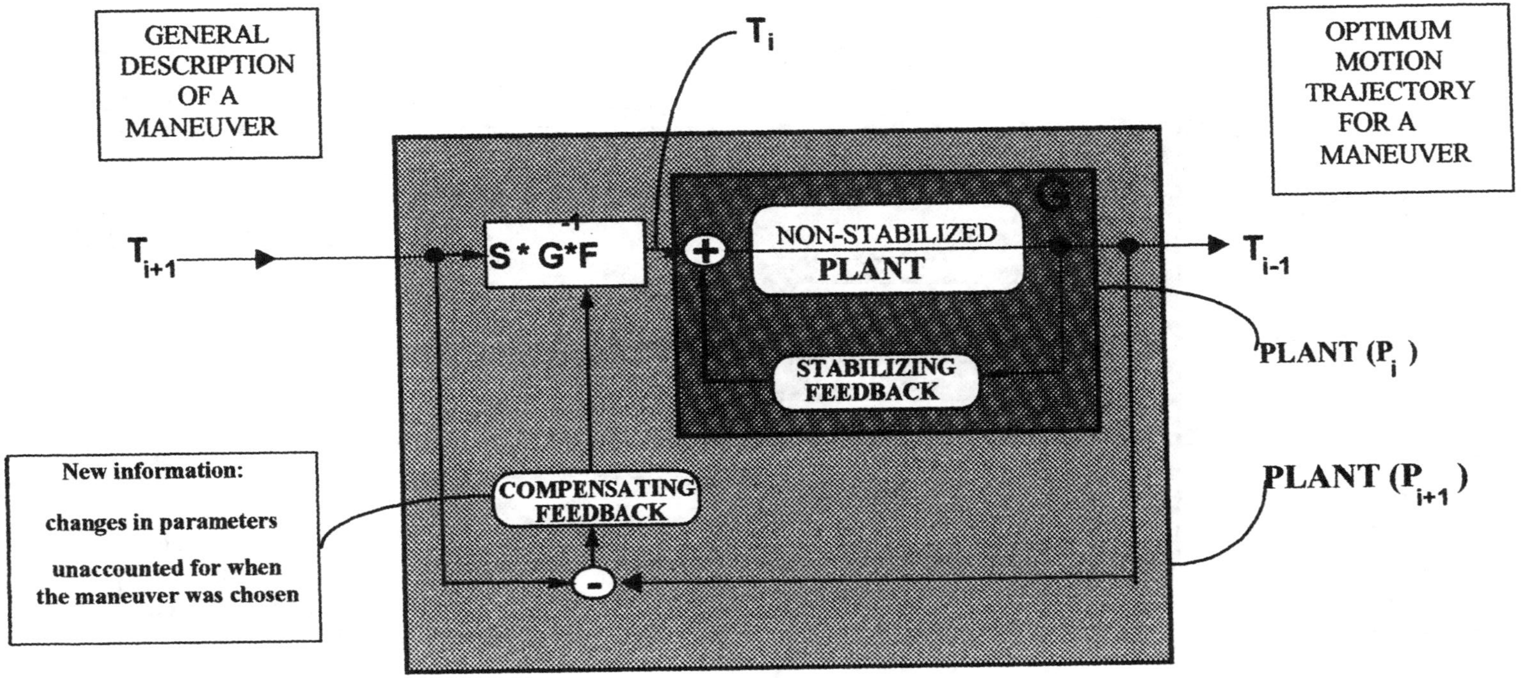

Figure 7.10a. Controller at the level of motion optimization.

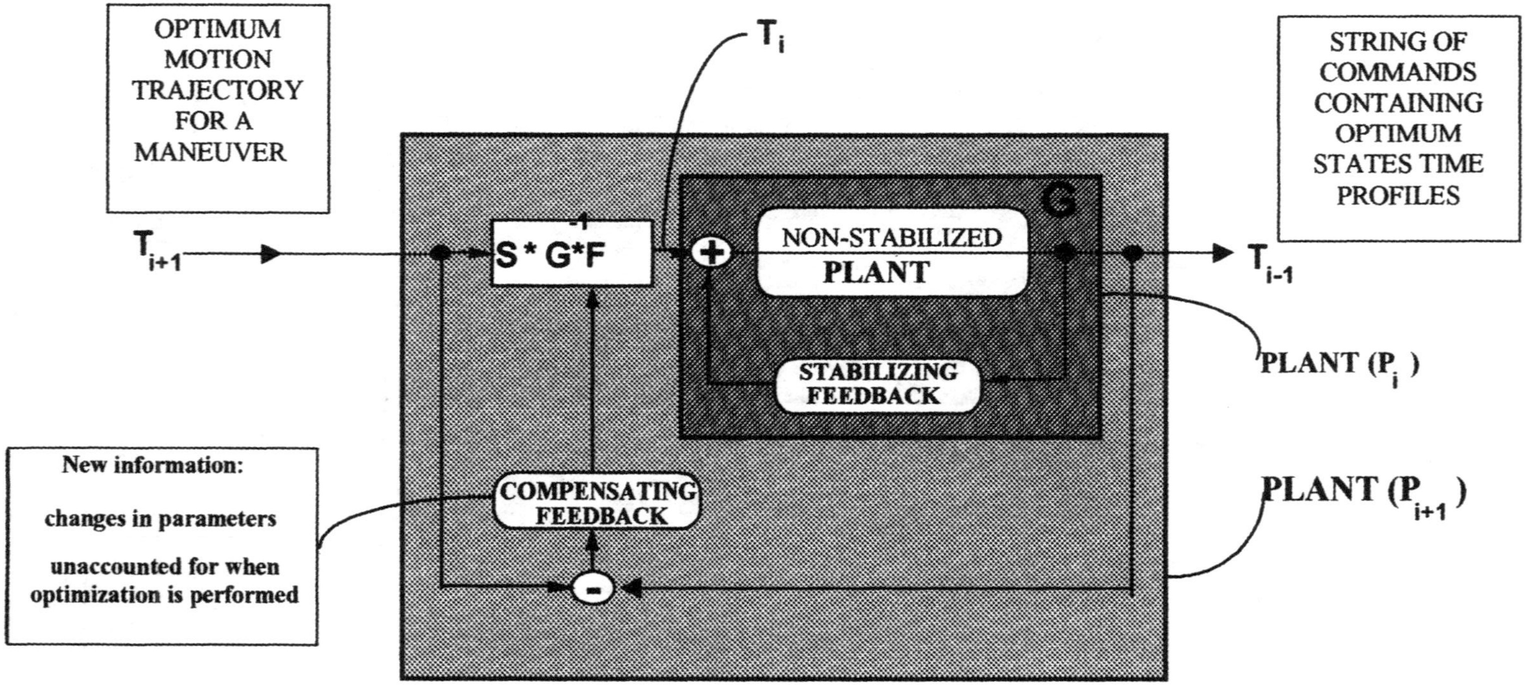

Figure 7.10b. Controller at the maneuver generation level.

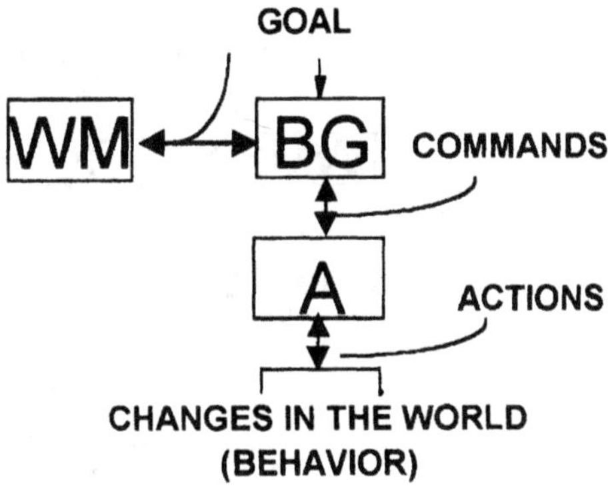

Figure 7.11. Fragment of the elementary loop of functioning: WM, world model; BG, behavior generation; A, actuator.

This phenomenon of a "nested module" must be considered in more detail. Let us take a module (WM, BG, or A) at a particular level of resolution; associated with this module are the models of the system, the snapshots of the world, the plans, and the actual trajectories of motion will be described with a particular resolution. This means that all of the higher-resolution levels of information are supposed to be fully represented within each of the boxes of interest. In this section, we are interested primarily in the following part of ELF shown in Figure 7.11.

The previous section showed that if the level of behavior generation has been introduced, then the other levels can be recursively derived from the description of a single level. We consider a loop of the level to be the unit that performs the original G. Saridis's triplet of the intelligent operations: organization, coordination, and execution [7]. Analysis shows that each of these three functions is being performed *at each level of resolution,* within ELF of this level, rather than being distributed top-down over the entire architecture of the system in the following manner: upper levels for organization only, middle ones for coordination only, and lower levels for execution only.

As we decompose the subsystems into next level sub-subsystems we immediately arrive at the phenomenon of nesting. If the subsystem can be decomposed, the sub-subsystems are the inner parts of it. To describe the functioning of the system, one should refer to the functioning of the sub-subsystems. Then, what is the purpose of having this separate consideration of subsystems and their parts? It is difficult to talk about functioning of a subsystem if the reference to all its sub-subsystems components is required. The property of nestedness (which is a direct result of decomposition/aggregation) allows us to simplify representation, information channels, and communication.

Let us make an experiment, and decompose all modules of the system shown in Figure 7.11. Figure 7.12 shows the decomposition of all the modules in the system, which is extremely cumbersome. It demands demonstration of a multiplicity of loops of functioning, and the complexity of this effort can grow rapidly.

Instead of dealing with the cumbersome hierarchy shown in Figure 3.5, we demonstrate that the relation of "belonging to" or "consisting of" can be substituted by using a kindred relation "nested in." This addresses an important issue of treating all objects of results of decomposition and all processes of the subsystems as components of the processes belonging to the systems that undergo decomposition. In Figure 7.11 only one module of the next level below is shown that is nested within the module of the concrete level. In fact, branching is a typical phenomenon, and several modules can be nested

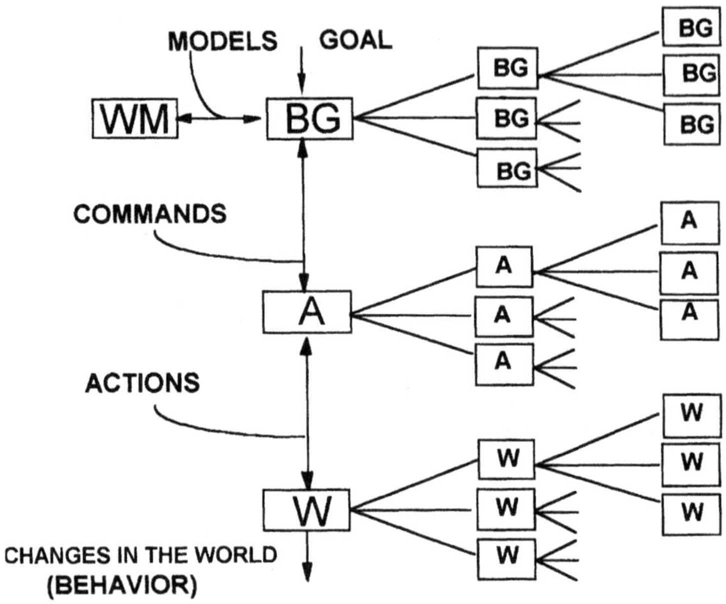

Figure 7.12. ELF fragment (see Fig. 7.11) with hierarchical decomposition.

within each module of the architecture. We will refer to all these factors as measures for the reduction of computational complexity.

Figures 7.10 and 7.11 demonstrate the fact that high-resolution level information is contained at lower-resolution level models and give descriptions and processes. This condition is a condition of consistency. If this condition does not hold, any operation produced by the system will be erroneous. This is important because the conditions of nesting might be different in different parts of the process. The latter can even include words of the vocabulary.

THE INSIGHT OF NESTING

For these reasons the concept of nesting must be clear on the scale of the overall system (see Fig. 7.13): All levels of the hierarchy within each subsystem (K, BG, etc.) form a *nested system*. This means that the operation of the inner box is a *part of functioning* of the external box and *not a separate operation*. Functioning of the lower-resolution box (see Fig. 7.11) cannot be separated from functioning of the higher-resolution boxes which are parts of the lower-resolution boxes, as shown in Figure 7.13. Inner boxes present the *symbol grounding* for the external boxes, which is their *consistency check*. The consistency check is a mandatory component of any implementation of the NIST-RCS system. This check should be considered as a part of the communication process.

Each inner module is obtained from the external module as a result of refinement. Each inner box considers the processes of its corresponding external box at higher resolution with more detail, enhanced vocabulary, and smaller unit of time scale. The totality of all processes of functioning of the inner modules is equivalent to the processes of functioning of the corresponding external module, but it is represented in this external module in generalized form with lower resolution.

The changes of time scale may be explained by the fact that at higher spatial resolutions, the processes are faster than at lower resolution. This means that at each higher resolution the frequency band shifts toward the next higher frequency.

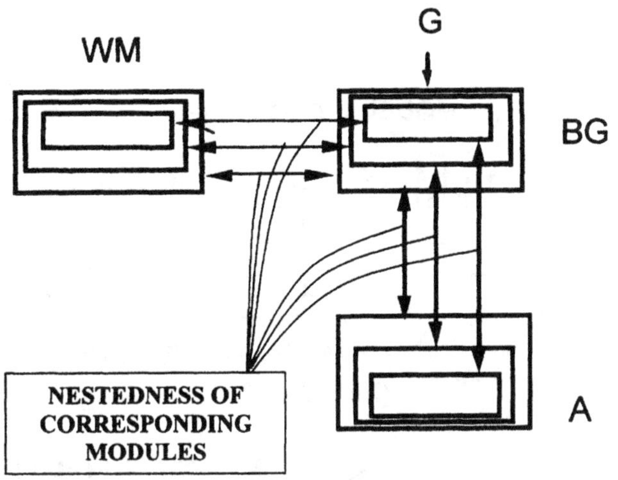

Figure 7.13. ELF fragment (see Figs. 2.7b and 7.11) as a nested structure for three levels of the system.

The phenomenon of nested modules allows for a better organization of the control system, and thus convenient software organization and automatically guided procedures of the NIST-RCS design.

7.4 OVERALL ORGANIZATION OF BEHAVIOR GENERATION

7.4.1 Main Concept of Computing Behavior

The main concept of behavior generation presumes that to achieve some goal (a state in the future), the controlled system must execute some behavior. The module of behavior generation consists of two submodules: one for behavior planning and one for behavior execution. These two submodules perform two distinct functions of the control system at a level that includes feedforward and feedback control. The structure of BG-module follows from the definition of its function.

Let us consider a more informative version of Figure 7.3 (see Fig. 7.14). This small segment of ELF (WM–BG–A–W) can be characterized by the following information flows between the subsystems of ELF:

1. A goal is submitted externally, for example, from an adjacent level of the NIST-RCS above the level under consideration. The goal is shown as a desired value of the output vector at the time T_{ph} which will be called later the *horizon of planning*.
2. Some models are transmitted from WM to BG. These models will be used to generate the required BG output.
3. Based on the desired goal, a set of commands (or a time function $U_{\text{v}(t)}$ of the command vector) is generated at the output of BG. It is given as a string (c_1, c_2, c_3, c_4).
4. The changes in the world are sampled by sensor functioning. This generates a string of signals.
5. As a result of the string of commands, a set of actions (or a time function of the action vector) is generated at the output of A. This appears as a string (a_1, a_2, a_3, a_4).

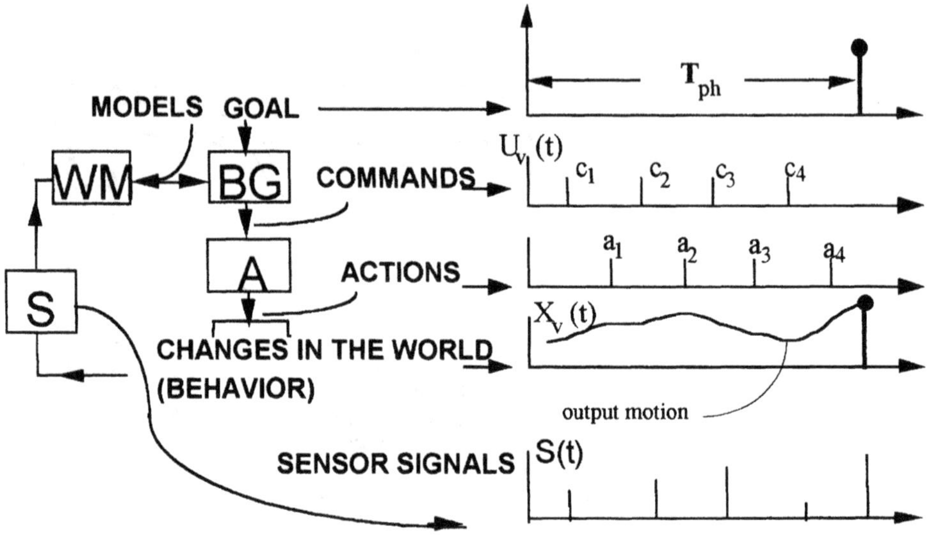

Figure 7.14. Detailed version of Figure 7.13.

6. The movement of output begins within world W; the results achieve the goal at the time T_{ph} as was prescribed by the level above.

Even if all these variables are determined, the process of behavior generation cannot be considered finished. The following additional activities must yet be performed:

First, because we are talking about virtual actuators and virtual world, the starting command c_1 or a string of commands shorter than the full string determined by BG, must be submitted to the BG module of the adjacent level below. A new behavior generation is initiated with higher temporal and spatial resolution. This is separate from the different groups of components that exist within the output vector X_v, and it presumes that various actuators are listed at the level below. The commands allow for decomposition into components and their distribution among the multiple virtual actuators existing at the adjacent level below, as shown in Figure 7.15.

Second, the models obtained from the WM have limited accuracy. This problem can occur, even though the models were accurate in the beginning. The models become inaccurate in time because of yet unknown changes in the world. The string of commands computed is also inaccurate. It will be quickly discovered that the virtual trajectory obtained as a result of functioning differs from the virtual trajectory expected. It is therefore prudent to introduce corrections to the set of commands computed in the expectation that these corrections will reduce the deviation in the movement that is observed.

A set of time functions is added to these two types of activities performed by the behavior generation subsystem that allows the system to achieve the desired goal. The time functions include the output motion trajectory, the string of actions that produces this motion trajectory, and the set of commands that generates these actions. Together these time functions are called *a plan*. The plan contains decisions about job distribution and scheduling, as well as all packages of coordinated schedules. It allows for the required decomposition into components for the virtual actuators of the adjacent level below. The process of finding these decomposable time functions is called *planning* and the sub-subsystem that produces it will be called the *planner*.

Figure 7.15 shows a part of the functioning loop determined for the ith level. The output of BG_i is submitted to the virtual actuators as determined for this level. Next the process of transforming plans into a set of output commands to the adjacent level below is called the *execution*. This process includes the opportunity to actively correct the plan online (compensate for the deviations, perform local predictions). The sub-subsystem performing this function is called the *executor*.

A plan of the level must exist before any execution starts, since the plan provides the program of functioning. Measures are taken to make the executed behavior as close to the plan as possible so that the goal can be achieved at the lowest possible cost. The cost is understood as the value of losses required to achieve the goal. It should include losses of energy, losses of time, expenses linked with manpower, cost of the tools and material, and wear of the devices required. The goal is presumed to be generated externally, and the goal entails the matter of cost functions to be applied in each particular case.

Each ELF uses its BG to feed the actuators A with its output. However, the output of A arrives to world W, where the motion is supposed to occur. This transition from the input commands to the motion of the world is the most fundamental process in the system because it is the main resource-consuming process, where the resource can be interpreted as time, energy, human power, and the like. In consideration of the cost of these sources, we always allocate a separate coordinate system for the action and the output trajectory. The actuators have commands as the input, actions as the output, and the desired motion in the external world as a result of this output.

The input to virtual actuators demonstrated in Figure 7.15 is an input to the BG-module of the adjacent level below. However, BG module of the ith level does not

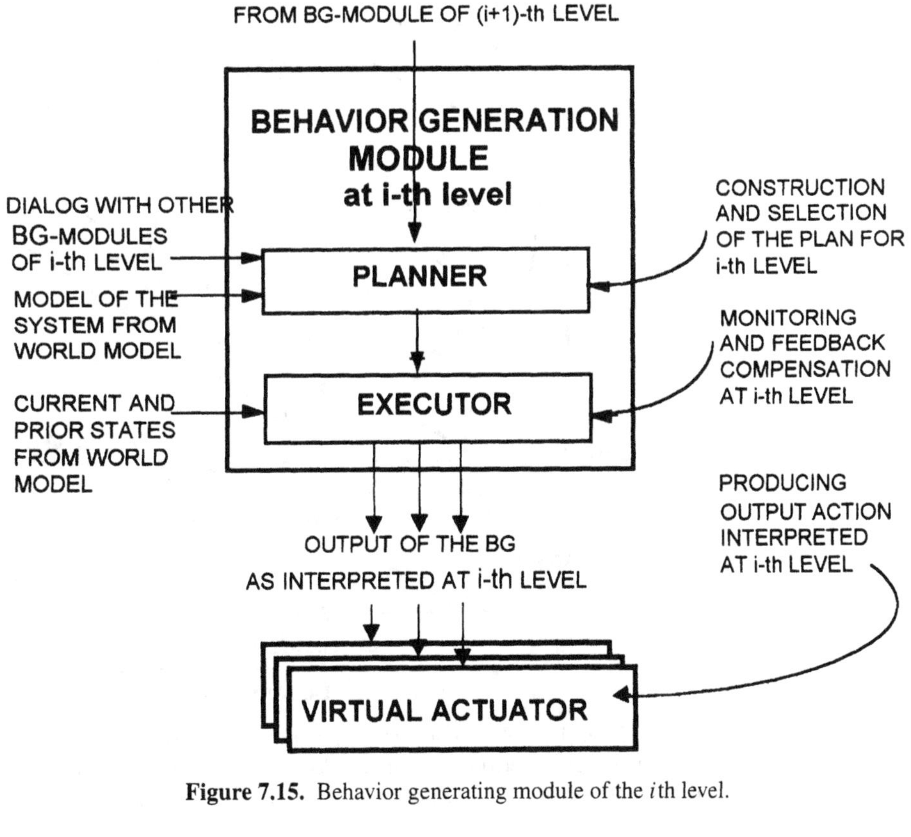

Figure 7.15. Behavior generating module of the ith level.

realize that the command is submitted to the BG module of the $(i-1)$th level. The executor of the ith level does what it is supposed to do. It submits its output to the actuator. It has the parameters of this actuator (the set of virtual actuator parameters) and receives the information about the execution process at this virtual level.

7.4.2 BG Modules

In the control architecture each level of the hierarchy contains one or more BG modules. At each level, there is a BG module for each controlled subsystem. The function of the BG modules is to decompose task commands into subtask commands.

Input to BG modules consists of commands and priorities from BG modules at the next higher level, plus evaluations from nearby VJ modules, plus information about past, present, and predicted future states of the world from nearby WM modules. Output from BG modules may consist of subtask commands to BG modules at the next lower level, plus status reports, plus "What is?" and "What if?" queries to the WM about the current and future states of the world.

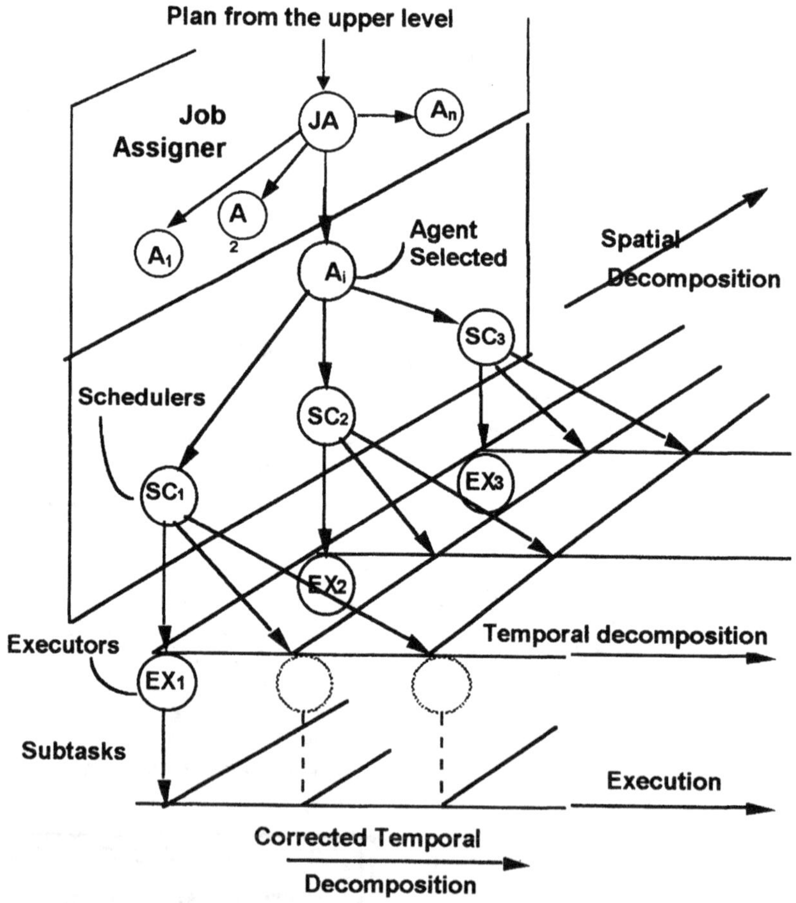

Figure 7.16. The job assignment (JA) module performs a spatial decomposition of the task. The schedulers (SC_i) perform a temporal decomposition. The executors (EX_i) correct the temporal decomposition and execute the plans generated by the planners.

Temporal decomposition partitions each job into sequential subtasks along the time line. The result is a set of subtasks, all of which when accomplished, achieve the task goal, as illustrated in Figure 7.16. The term "spatial decomposition" should be understood as representation in a coordinate system in which each coordinate represents a particular variable of the process.

Planners in turn consist of two components: job assigner (JA) and scheduler (SC). The job assigner performs the spatial decomposition among the coordinates (i.e., among the actuators that will perform the job). For each task decomposition ("spatial" decomposition) generated by the JA, the temporal distribution of all the subtasks is done by the scheduler. After a number of such tentative spatiotemporal distributions, the best of them is selected, and this concludes the process of planning. This separation of JA and SC should be understood as a theoretical tool; in algorithms of planning JA and SC are always intertwined.

Job Assignment Sublevel

The JA submodule is responsible for spatial task decomposition. It partitions the input task command into N spatially distinct jobs to be performed by N physically distinct subsystems, where N is the number of subsystems currently assigned to the BG module. The JA submodule may assign tools and allocate physical resources (e.g., arms, hands, legs, sensors, tools, and materials) to each of its subordinate subsystems for their use in performing their assigned jobs. These assignments are not necessarily static. For example, the job assignment submodule at the individual level may, at one moment, assign an arm to the manipulation subsystem in response to a <use tool> task command and later assign the same arm to the attention subsystem in response to a <touch/feel> task command.

The job assignment submodule selects the coordinate system in which the task decomposition at that level is to be performed. In supervisory or telerobotic control systems such as defined by NASREM [18], the JA submodule at each level may also determine the amount and kind of input to accept from a human operator.

Scheduler Submodule

For each of the N subsystems, there exists a scheduler submodule SC(j). Each scheduler submodule is responsible for decomposing the job assigned to its subsystem into a temporal sequence of planned subtasks.

JA AND SC ALWAYS WORK INTERACTIVELY

Scheduler submodules SC(j) may be implemented by case-based planners that simply select partially or completely prefabricated plans, scripts, or schema [19, 22] from the procedure sections of task frames. This may be done by evoking situation/action rules of the form, IF(case_x)/THEN(use_plan_y). The planner submodules may complete partial plans by providing situation-dependent parameters.

The range of behavior that can be generated by a library of prefabricated plans at each hierarchical level, with each plan containing a number of conditional branches and error recovery routines, can be extremely large and complex. For example, nature has provided biological creatures with an extensive library of genetically prefabricated plans, called instinct. For most species, case-based planning using libraries of instinctive plans has proved adequate for survival and gene propagation in a hostile natural environment.

Scheduler submodules SC(j) may also be implemented by seach-based planners that search the space of possible actions. This requires the evaluation of alternative hypothet-

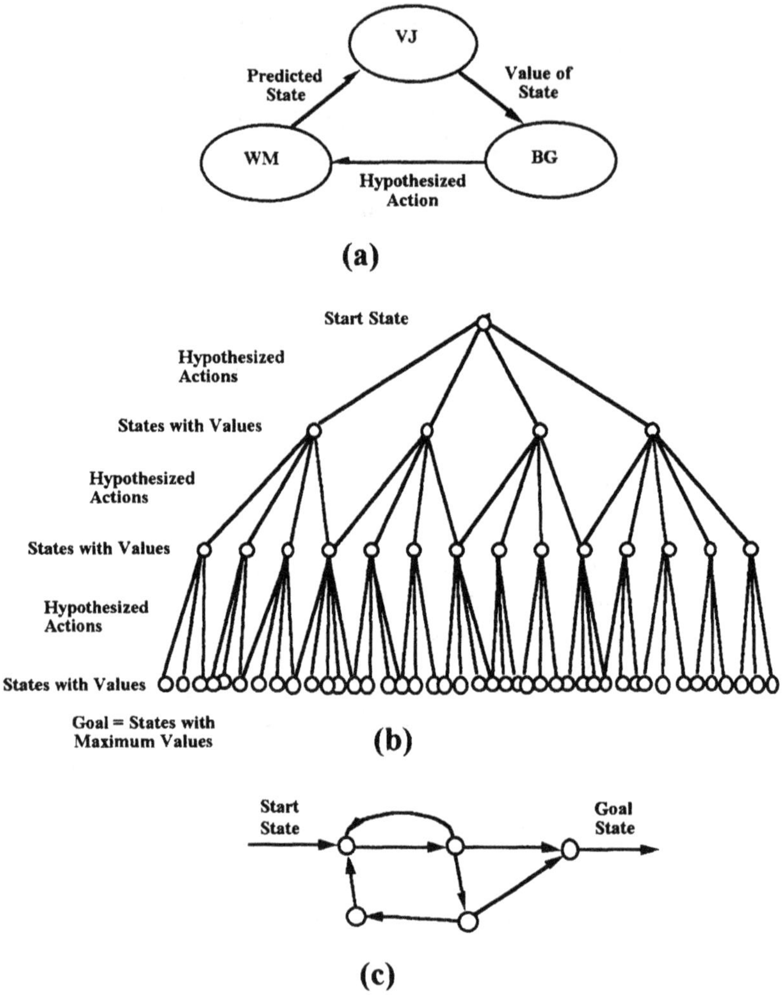

Figure 7.17. The planning loop (*a*) produces a game graph (*b*). A trace in the game graph from the start state to a goal state is a plan that can be represented as a plan graph (*c*). Nodes in the game graph correspond to edges in the plan graph, and edges in the game graph correspond to nodes in the plan graph. Multiple edges exiting nodes in the plan graph correspond to conditional branches.

ical sequences of subtasks, as illustrated in Figure 7.17 [1, 2]. Each planner SC(j) hypothesizes some action or series of actions, the WM module predicts the effects of those action(s), and the VJ module computes the value of the resulting expected states of the world, as depicted in Figure 7.17*a*. This results in a game (or search) graph, as shown in Figure 7.17*b*. The path through the game graph chosen by the planner's "selector," and leading to the state with the best value, becomes the plan to be executed by EX(j). In either case-based or search-based planning, the resulting plan may be represented by a state-graph, as shown in Figure 7.17*c*. Plans may also be represented by gradients, or other types of fields, on maps [23–25], or in configuration space (see Chapter 8).

Job commands to each planner submodule may contain constraints on time, or specify job-start and job-goal events. A job assigned to one subsystem may also require synchronization or coordination with other jobs assigned to different subsystems. These constraints and coordination requirements may be specified by, or derived from, the task frame. Each scheduler's SC(j) submodule is responsible for coordinating its schedule with schedules generated by each of the other $N - 1$ schedulers at the same level, and checking to determine if there are mutually conflicting constraints. If conflicts are found, constraint relaxation algorithms [26] may be applied, or negotiations conducted between SC(j) (cooperating schedulers) until a solution is discovered. If no solution can be found, the schedulers report failure to the job assignment submodule, and a new job assignment may be tried, or failure may be reported to the next higher level BG module.

Executor Sublevel

There is an executor EX(j), or a group of executors for each planner PL(j). The executor submodules are responsible for successfully executing the plan state graphs generated by schedulers within their respective planners. At each tick of the state clock, each executor measures the difference between the current world state and its current plan subgoal state, and issues a subcommand designed to null the difference. When the world model indicates that a subtask in the current plan is successfully completed, the executor steps to the next subtask in that plan. When all the subtasks in the current plan are successfully executed, the executor steps to the first subtask in the next plan. If the feedback indicates the failure of a planned subtask, the executor branches immediately to a preplanned emergency subtask. Meanwhile its planner begins work selecting or generating a new plan that can be substituted for the former plan which failed. Output subcommands produced by executors at level i become input commands to job assignment submodules in BG modules at level $i - 1$.

Planners PL(j) operate on the future. For each subsystem there is a planner that is responsible for providing a plan that extends to the end of its planning horizon. Executors EX(j) operate in the present. For each subsystem there is an executor that is responsible for monitoring the current ($t = 0$) state of the world and executing the plan for its respective subsystem. Each executor performs a READ–COMPUTE–WRITE operation once each control cycle. At each level, each executor submodule closes as a reflex arc, or servo loop. Thus executor submodules at the various hierarchical levels form a set of nested servo loops. Executor loop bandwidths decrease on average about an order of magnitude at each higher level [28–39].

7.4.3 Realistic Examples of Behavior Generation

At some level of aggregation, a turning machine tool cuts a piece of metal to properly shape it; this is our goal. The relative motion of the cutter and the piece of metal is the output motion that leads us to the goal (consider this to be Example 1). One can interpret the actuator at another level of resolution as assembly activities where different parts are to be put together. Then, the ordered attachment of the parts will be the desired motion, and the assembled object is the goal (consider this Example 2). Finally one can interpret the actuator as a manufacturing shop that gives as output several sets of manufactured parts according to a time schedule. The goal is to have the output of these parts at this shop in stock according to the required schedule (consider this Example 3). In all three examples the goals and the output motions are different from the input commands introduced at the input of the actuators.

Example 1

METAL CUTTING

The goal is a trajectory of relative motion of the blank and the cutter. Several alternatives can be considered typical for the metal-cutting machines: the spindle and the turret, or the cross-feed carriage (in the turning machine), the mill and the table (in the milling machine). This relative motion is a shape-forming factor. The spindle rotates with a particular speed, which is determined by a spindle motor equipped with mechanical devices. This motor should receive a command "on" at the beginning and "off" at the end, which is the feedforward control, FFC: It can be interpreted as a feedforward schedule of commands. Because the blank metal resists cutting, there are resistance forces on the cutting surface.

The force developed by the actuator must overcome the resistance force. As a result of these forces, deformations emerge: The blank bends and the cutter wears. A torque is created on the shaft of each electrical motor, and the current increases in its windings. The first-order differential equation of the electrical circuit equilibrium for dc motor under load is written as follows:

$$u(t) - e(t) = i(t) \cdot r + L\frac{di}{dt}, \tag{7.3}$$

where

$u(t)$ = voltage applied;
$e(t)$ = value of the counter-EMF equal to $k_1\omega$, which gives

$$e(t) = k_1\omega(t), \tag{7.4}$$

k_1 = a constant value,
$\omega(t)$ = velocity measured on the shaft of the motor,
$i(t)$ = armature current,
r = armature resistance.

The first-order differential equation also holds for the equilibrium of mechanical system:

$$T_{\mathrm{m}}(t) - T_{\mathrm{c}}(t) = J\frac{d\omega}{dt}, \tag{7.5}$$

where

$T_{\mathrm{m}}(t)$ = value of torque developed by the motor as a result of current $i(t)$ so that

$$T_{\mathrm{m}}(t) = k_2 i(t), \tag{7.6}$$

k_2 = a constant value
J = value of inertia on the shaft,
T_{c} = value of torque developed by the black resisting to the process of cutting.

Equations 7.4 through 7.6 are obtained for the spindle actuator; a similar system of four equations can be written for the feed actuator.

The value of torques $T_{\mathrm{cs}}(t)$ and $T_{\mathrm{cf}}(t)$ for the spindle (s) and feed (f) actuators correspondingly depends on many variables such as the depth of cutting (d), the type of cutter (c), the velocity of spindle actuator $\omega_{\mathrm{s}}(t)$ and the velocity of the feed actuator $\omega_{\mathrm{f}}(t)$. The world can be characterized by the function F of the cutting process, which can be written in the form of an equilibrium equation of cutting. One possible form of this equation is

$$F[d, c, T_{cs}(t), \frac{dT_{cs}(t)}{dt}, T_{fs}(t), \frac{dT_{fs}(t)}{dt}, \omega_c(t), \frac{d\omega_c(t)}{dt}, \omega_f(t), \frac{d\omega_f(t)}{dt}] = 0. \quad (7.7)$$

The system of equations (7.3) through (7.7) has an infinite number of possible solutions. In order to perform planning, a set of constraints must be added and a condition that minimizes (or constrains) the cost function. The cost function C for the electrical motor can be interpreted as consisting of time, energy, materials, and the like, separately or in a combination:

$$C = \sum_i C_i \rightarrow \text{min.} \quad (7.8)$$

To find the process of output motion, equations (7.3) through (7.8) are solved jointly or simulated. The simulation is organized so that the overall effort can be distributed between the actuators.

We have intentionally selected the simplest case with only two equations which are linear and first order. In reality there are more equations, and they are often nonlinear and have higher order. But the three requirements remain unchanged in all cases: The equilibrium equation of the processes in the actuator must hold, the equilibrium equation of the processes in the world must hold, and a cost function minimization condition (or constraint) must be added.

By solving the equations under different conditions, or by simulating the process of cutting under different conditions, search is performed. This allows us to determine the command sequence, called the feedforward control, that is supposed to generate a "reference trajectory" for subsequent tracking. When applying this control, the motion deviates from the required (reference) trajectory unless the speed (and/or torque) is measured and its deviations are compensated by proper varying of the input voltage (feedback compensation control, FBC).

Note that during the planning process we have determined the time functions for all components of the plan at once: the output motion, the current trajectories (actions), and the set of control commands. This was possible because we have comparatively simple problem.

The electrical motor represented by equations (7.3) through (7.8) is a virtual one: All parameters and dependencies are generalized and conceal a higher-resolution layer. This does not violate the general approach that we have introduced here. We can envision that after this stage of planning is performed, the actuators we dealt with will turn out to be virtual actuators. The next stage of planning (with higher resolution) is performed when the next decomposition is done. Indeed, the turret will follow the prescribed trajectory. The latter is performed in a plane by the motion of two electrical motors. Their trajectories are assigned by a constantly varying voltage (feedforward control, FFC), and the trajectories' deviations are compensated for (feedback compensation control, FBC).

Example 2

ASSEMBLING

The goal is to manufacture an object by assembling it from a set of parts. The output trajectory is a best set of the parallel/sequential strings of activities that are required to achieve the assembly of an output trajectory with minimum cost (the output schedule). The input vector trajectory is contained within the schedule of commands that prescribes the activities to be performed.

Several alternatives that relate to the assembly operation can be considered: manual assembly, robotic assembly, partially automated assembly, and so on. Within each of these alternatives, the relative motion of the parts is to be performed so that they match and attach to each other. Although the rules of attachment are prescribed by the drawings of the assembly, the relative motion can be performed in many ways. The fitting couples should be determined and attached to each other if it does not contradict the subsequent formation of the more complex group of parts.

Then the performing actuators receive proper commands in the order determined by the plan of assembly. The plan can be interpreted as a feedforward control, FFC; it consists of the reference trajectory to be tracked at the output and the schedule of commands to be applied at the input. Each of the relative motions linked with the search of matching positions and subsequent fastening procedures requires maneuvering the parts in their relative motion. The process resists our effort to speed it up in the same way as the difficulty of maneuvering the moving device increases in the cluttered environment when we try to increase the speed of its motion. Because of inertia one has to expend more energy in avoiding collision, and as the accuracy of motion reduces, the time of machining instead increases.

The actuator force must overcome the force required for quick maneuvering. Motion errors occur as a result of these forces. The oscillatory component increases in the torque on the shaft of each electrical motor, heating its windings. If the assembly is performed by a single gripper equipped by three (X, Y, Z) actuators, the first-order differential equations of the electrical circuit equilibrium for dc motor under load is written as

$$\left\{\left[u(t) - e(t) = i(t) + L\frac{di}{dt}\right], [e(t) = k_1\omega(t)]\right\}_j, \qquad j = 1, 2, 3, \tag{7.9}$$

where j is the number of the actuator. The first-order differential equation holds also for the equilibrium of mechanical system:

$$\left\{\left[t_{\mathrm{m}}(t) - T_{\mathrm{fr}}(t) = J\frac{d\omega}{dt}\right], [T_{\mathrm{m}}(t) = k_2 i(t)]\right\}_i, \qquad i = 1, 2, 3, \tag{7.10}$$

where $T_{\mathrm{fr}}(t)$ is the value of torque developed by the friction and weight of the part. The set of equations (7.9) and (7.10) is obtained for the X-actuator. Similar system of four equations can be written for the Y and Z actuators. Let us denote this addition set $(7.9)_i$ and $(7.10)_i$. For the overall set of actuators we will receive a set of equations (7.9) and $(7.10)_i$.

The world is characterized by the initial configuration of parts and configuration space where all possible trajectories of motion can be found by the consecutive *simulation of elementary moves to form pairs* by *searching*.

The system of equations (7.9) and $(7.10)_i$ has an infinite number of possible solutions. To perform simulation and/or search, a set of constraints are added, and an equation that minimizes (or constrains) the sum of all cost functions C_i, $i = 1, 2, \ldots,$ is computed for all actuators involved. As in the previous case, it includes components that depend on time, energy, materials, and so on, separately or combined:

$$\sum C_i \to \min. \tag{7.11}$$

To perform the search, equations (7.9) through (7.11) are jointly simulated: Their analytical solution at each step of the search is either not easily available or more expensive computationally than the simulation. The simulation is organized to allow for an eventual distribution of the assembly operation between the actuators.

We have intentionally selected the simplest case with only two equations which are linear and first order. In reality we have more equations, they are often nonlinear and have higher order. But three things hold: (1) the equilibrium equation of processes in the actuator, (2) the equilibrium equation processes in the world, and (3) the condition that a cost function minimization (or constraint) should be added.

Only for the simple assemblies is this joint search-simulation computationally affordable. In most cases the output motion is simulated without a simultaneous simulation of the transient processes in the actuators. The search is performed by minimizing the cost function, which is artificially simplified by generalizing its components. After the output trajectories are planned, they are inverted to the input of the actuators by using different analytical or computational techniques.

The reality of the process of assembly differs from our expectations, and feedback compensation control is required. Indeed, as we apply the chosen plan of assembly, the motion might deviate from the required trajectory unless the speed (and/or torque) is measured and its deviations are compensated by properly varying the input voltage (feedback compensation control, FBC). These corrections might lead to a need to replan, since under new circumstances better results could be obtained as a result of repeating the process of search from an intermediate stage of the assembly.

Example 3

MANUFACTURING SHOP

The goal coincides with the output motion; it is the output schedule of sets of manufactured parts. The only difference that might be envisioned is that the goal might include a constraint on expenses allowed. The goal will be achieved if the input schedule of assigning these parts to the particular machines and manpower is determined. Among the many possible time and space assignments, there is a set of preferable assignments that provides the output schedule required with the best or assigned losses.

The output motion trajectory can be obtained if a best set of the parallel/sequential string of activities is assigned to the virtual actuators (cells, and/or machines, together with a particular manpower). Then the goal will be achieved with minimum losses (the output schedule). The input vector trajectory is the input schedule of commands that allows these activities to be performed. An activity is understood as an assignment of a particular part to a particular machine operated by a particular person. Information is available about productivity of the machines with particular parts to be manufactured (sometimes the productivity is affected by the worker doing the job). Also, as in the case with assembly, a rule base that contains a set of constraints that demand particular sequencing of jobs (precedence rules) is available.

Several alternatives related to manufacturing shop functioning can be considered. Prior grouping of the machines, parts, and manpower would reduce the expected volume of search during the planning process. Proper grouping would create intermediate levels of hierarchy in manufacturing so that in distributing part groups among the machines, general matches can be attained. This can ensure that more general cost functions are minimized. Although the rules of precedence are prescribed, the assignment can be performed in a vast number of ways. The ways of improving the planning process become more complicated because the number of cost functions is usually large and not properly ordered.

The role of actuators is played by the machines with the manpower distributed among them. The world can be characterized by the initial configuration of machines, and the totality of parts to be manufactured. All possible trajectories of motion can be found by consecutive *simulation of elementary assignments to form noncontradictory strings*, namely by *searching* in the configuration space. The cost function C can be interpreted separately or as combination of time, energy, materials, and so on. The search is organized so that the overall effort allows for the eventual distribution of the assembly operation between the actuators.

The reality of the manufacturing shop can differ from our planning expectations. The feedback compensation control should be introduced as the schedule corrections.

7.4.4 Generalization upon Realistic Examples: A Sketch of the Theory

We can see that after our image of the output motion in the world has been found, as a result of the search, it should be inverted to the input of the actuator to determine the required commands.

At each level of resolution of the NIST-RCS hierarchy, the image of the virtual output motion is determined (the motion $X_v(t)$ in the virtual world, W_v). Then this motion is inverted to the input of the virtual actuator (A_v) and the function $F(uv)$ of the virtual input control should be computed. If the virtual actuator has a transform function T_{av} then the following holds for the input trajectory $U_v(t)$:

$$U_v(t) = X_v(t)^* T_{av}^{-1}. \tag{7.12}$$

Given T_{av}^{-1}, we worry about receiving both the desired virtual output motion $F(uv)$ and the required input control commands $X_v(t)$. These two functions in most cases are not the same. Goal arrives to the level of NIST-RCS in the form of the set of states $\{X_{v1}, X_{v2}, \ldots, X_{vn}\}$ which should be achieved at the moments of time $t_1, t_2, \ldots, t_n$ where n is the horizon of planning and the minimum difference between two consecutive milestones is never less than Δt, which is the time-scale interval at the level.[1] We can see that the goal at a level is assigned as a set of the time-tagged milestones of the output trajectory (of the motion) within the state-space.

If this trajectory is found in the form $X_v(t)$, and the properties of the system are known in the form of transformation function T_{av}, the input trajectory (program of control) can be found $U_v(t)$ by inverting the output to the input terminals of the system to be controlled. (Eventually $U_v(t)$ is transformed into the discrete string of commands $\{U_{v1}, U_{v2}, \ldots, U_{vn}\}$; the minimum difference in time should be equal Δt, which is the time-scale interval at the level.) Therefore *plan* can be understood as a symbolic and/or numerical description for a couple $\{X_{v(t)}, U_{v(t)}\}$ which includes both the desirable (output) behavior of the system, and the output control trajectory which is required to obtain this output trajectory.

IMAGINING THE FUTURE

Consequently a plan can be found by performing two steps of operation:

Step 1. Given the output milestones, simulate the processes and search for possible combinations of the output functioning that form the admissible set of virtual output trajectories of motion.

Step 2. Find the input commands to be submitted to the virtual actuator at the level.

In some realistic cases, step 1 can be skipped over, and only step 2 is required. Often $T_{av} = I$, and to find output is the same thing as to find the input. Other variations are also

[1] Δt is the "atomic" (indivisible) time interval at a level.

possible; however, the complete and adequate description of the operations required for planning includes both steps of the operation. In most of the NIST-RCS systems, finding the desirable output functioning nevertheless is the main problem. This problem is solved by using the main NIST-RCS algorithm of planning based upon multiresolutional search algorithm.

We tag the output trajectory of the motion and the input commands that are supposed to produce this motion in elementary units of time pertaining to the particular level of resolution under consideration.

It is typical to decompose a schedule into a consecutive set of milestones (intermediate goals) on the way to the final goal. Task decomposition transforms the goal of the system into a hierarchy of the subgoals (which are the goals for the subsystems). The subgoals contain the set of assignments including determining inputs to the subsystems, feedback control laws and gains of the compensation controller.

In this subsection we have described how the goal state stimulates planning the output behavior that generates the goal states for subsystems. In the same way, the initial input goal submitted to our system has emerged as a result of BG process at a level above (a lower-resolution level). This concept has a broad basis, and many examples can be given to show that in practice goal creation and behavior generation occur in such a way.

7.4.5 Algorithm of Multiresolutional Hierarchical Planning (NIST-RCS Planner)

Let us consider a space Ω_i in which the world model, WM, is represented at a particular level i with resolution ρ_i (for both space and time). From the $(i-1)$th level a goal

$$G_i(T_i, SP_i, FP_i, \rho_i, J_{i+1.1} < J_{i+1} < J_{i.2}, J_{i.1} < J_1 < J_{i.2}) \tag{7.13}$$

arrives that demands task T_i to be performed having the starting and final points SP_i and FP_i as a part of the task. It contains a description of the job to be done which can be performed at the ith level of NIST-RCS for a particular set of virtual actuators. The motion trajectory from SP_i to FP_i is to be found with the value of final accuracy ρ_i, and the cost constraints of the upper level $J_{i+1.1} < J_{i+1} < J_{i+1.2}$ are added to the cost constraints of the ith level $J_{i.1} < J_1 < J_{i.2}$. The condition of job constraints for the ith level can be substituted by the condition of minimizing some of the components of the J_i vector.

We will introduce the following three operators:

WHO MIGHT ACT?

1. *The operator of the job assignment (JA).* This operator determines the virtual actuators of the $(i+1)$th level of resolution to be involved in behavior generation and the groups of coordinates in which jobs to be performed are represented at the ith level of resolution. JA assigns coordinates to these groups of job labels. In other words, the JA operator introduces the alternatives of spaces in which the output motion should be described at this level of resolution in order to choose one of them. This operator maps the task (T_{i+1}) that has arrived from the $(i+1)$th level of NIST-RCS into the space of output covered by the virtual actuators of the ith level of NIST-RCS. Computationally this is done via combinatorial mapping from the set of tasks into the set of virtual actuators under constraints introduced from the prior experiences.

$$\text{JA: } (T_{i+1}, \Omega_i, \rho_i) \to \{\{A_v\}_{i,j}\}_k, \quad \text{or} \quad \{\{A_v\}_{i,j}\} = \text{JA}(T_{i+1}, \Omega_i, \rho_i),$$
$$j = 1, 2, \ldots, n, \; k = 1, \ldots, m, \tag{7.14}$$

where JA is the combinatorial grouping operator which synthesizes the groups that perform the job assignment. It does this by mapping from the task description into a possible combination of the coordinates of the state-space Ω_i at the resolution ρ_i. $\{\{A_v\}_{i,j}\}_k$ is the set of meaningful distributions of the task among the n virtual actuators of the ith level of resolution, j is the number of an actuator, and k is the number of alternatives of job assignment.

The result of this search gives a map that determines the density of the subsequent search graph. This map will be called *alternatives of the job assignment* and the best time schedule will be computed for each of these alternatives.

HOW WOULD THEY ACT?

2. The *operator of state-space search for the admissible set of time schedules of the output trajectory (SC)*. This operator provides for a combinatorial grouping of the points of the output space into a set of strings for the subsequent consideration of these strings as alternatives of the output trajectory. The input trajectory can also be $U_v(t)$ obtained during the process of planning (scheduling). Otherwise, it a separate computation (7.12) can be performed to find the input commands from the description of the required output results.

$$\text{SC: } (\{\{A_v\}_{i,j}\}_k, SP, FP, J) \rightarrow \{X(t)\}_1, \quad \text{or}$$

$$\{X(t)\}_1 = \text{SC} \rightarrow (\{\{A_v\}_{i,j}\}_k\}_1, SP, FP, J) \tag{7.15}$$

where

SC = the string concatenating, or scheduling operator,

$1 = 1, 2, \ldots, a$ is the number of alternatives of the time schedules for the output trajectory retained for the subsequent comparison,

$\{X(t)\}_1$ = the set of admissible strings (output motion trajectories) connecting the starting point SP and the finish point FP and providing the string belonging to an interval delineated by the cost constraints of the upper level $J_{i+1,1} < J_{i+1} < J_{i+1,2}$ added to the cost constraints of the ith level $J_{i,1} < J_i < J_{i,2}$ correspondingly, where indexes 1 and 2 denote the lower and upper bounds of the desirable cost.

CHOOSING THE COURSE AT A LEVEL

3. *The operator of selection and focusing of attention (PS)*. This operator determines the only trajectory (the best one), together with the subset of the output space (which is considered to be at a higher level of resolution as the subset of the output space for the particular solution refinement)

$$\text{PS: } (\{X(t)\}_1\ \{J_p\}) \rightarrow \text{ENV}_{i-1}(X_i^*, w), \quad \text{or } ENV_{i-1}(X_i^*) = \text{PS}(\{X(t)\}_1\ \{j_p\}), \tag{7.16}$$

where

$\{X(t)\}_1$ = set of all admissible output trajectories,

X_i^* = The best output trajectory (minimizing the sum of costs under consideration),

$\{J_p\}$ = the totality of all cost measures including not only those mentioned above, but also the results of simulation if the latter are available,

w = the parameter of the envelope (e.g., the "width" of the envelope),

ENV_{i-1} = the result of surrounding the best trajectory by a zone of the space called the *envelope* for the subsequent search procedures at the next $(i-1)$th level of resolution.

Joint functioning of these three operators gives the algorithm of planning: PLANNER = JA*SC*PS. Clearly, by selection of the X_i^*, we simultaneously determine the best schedule, the required set of actuators, and the optimum set of inputs. PLANNER = JA*SC*PS is equivalent to a single operator of multiresolutional searching for the best output trajectory.

The multiresolutional search for the best output trajectory can be concisely described as follows: For $k = 1, \ldots, m$, do the following string of procedures:

Step 1. $\mathrm{JA}(T_{i+1}, \mathrm{ENV}_{i-1}(X_i^*), \rho_i)$.
Step 2. $\mathrm{SC}(\{\{A_v\}_{i,j}\}_1, SP, FP, J)$,
Step 3. $\mathrm{PS}(\{X(t)\}_1\{J_p\})$.

MULTIRESOLUTIONAL OPTIMIZATION

The algorithm of control can be represented as a diagram

$$\begin{array}{ccccccc} & & T_{i+1}, \rho_i & & SP, FP, J & \{J_p\} & \\ & & \downarrow & & \downarrow & \downarrow & \\ \mathrm{ENV}_i(X_{i+1}^*, w_{i+1}) & \rightarrow & \mathrm{JA} & \longrightarrow & \mathrm{SC} & \longrightarrow \mathrm{PS} & \rightarrow \mathrm{ENV}_{i-1}(X_i^*, w_{i+1}) \\ & & \{\{A_v\}_{i,j}\}_1 & & \{X(t)\}_1 & & \end{array} \tag{7.17}$$

or a recursive expression

$$\mathrm{ENV}_{i-1}(X_i^*) = \mathrm{PS}(\mathrm{SC}(\mathrm{JA}(\mathrm{ENV}_i(X_{i+1}^*), w), \rho_k)SP, FP, J). \tag{7.18}$$

Equations (7.17) and (7.18) could be considered the NIST-RCS algorithm of BG planning, but the planning is not sufficient for behavior generation. The feedforward control must be supplemented by feedback compensation. Thus PLANNER in the BG module is later supplemented by a feedback compensation sub-subsystem called EXECUTOR.

Algorithm of Planning

The planning algorithm specifies search in the state-space as a means of finding the desirable state-space trajectory. This procedure entails a complete set of required results (spatial plans with job distribution among the agents and the temporal schedules). The fundamental issue relevant to the algorithm is how the goal is assigned and represented. In the vast multiplicity of cases, the goal is considered to be a known state in the future.

In many problems this state is unknown before its vicinity has been reached. Then, this vicinity is regarded as a known state at the lower level of resolution. For example, the goal can be assigned as a "safe position on the northern slope of the mountain Z." After the slope is achieved, the procedure of planning should be performed again, since the goal state is supposed to be known.

Often, we deal with problems of control in which the goal is assigned not as a state but as a condition to be satisfied (and this condition can be satisfied in a variety of states). For example, the goal can be assigned as "being in the visibility range from the moving objects of a particular type (in a particular region)." Then the low-resolution goal is to achieve this particular region. The high-resolution position this region is defined based on other criteria. After a search for moving objects is initiated, areas of high probability to detect moving objects are recognized and found. The next high-resolution planning will occur upon detection of a moving object.

After the present state is known, after the goal state is assigned, after the cost function is clarified at the ith level of resolution, the algorithm of planning consists of two steps:

Step 1. Find the alternatives plans recommended by the library of stored solutions. At this particular level, compare them under conditions of the assignment and select one that is "the best."

If the solution is found, exit successfully.

Otherwise go to Step 2.

Step 2. Search in the state-space for the sets of feasible trajectories for achieving the goal at the ith level of resolution.

The search is done both in the state-space and in time within the search envelope of the ith level. It is determined whether achieving this goal requires any cooperation of the subsystems existing at the adjacent level of higher resolution. Different alternatives of combination of the subsystems of the $(i-1)$th level of resolution are usually available. Judgment about the preferable combination of the systems component is made based on the results of the planning. If there exists a system (or systems) of the ith resolution level that plan their operation for themselves, and our system is cooperate, it should interact with these systems during the search.

Implications of Step 1 First, the search procedure should be prepared and executed. This search will look for a trajectory in the state-space that leads to the goal from the initial state. It minimizes a cost function or keeps the process within some specified boundaries (instead of minimizing the cost function) or both. Selection of the states is determined by mapping from the space of problem into the space of components of the solution. The latter also affects the alternatives related to the cooperative agents of the next level of higher resolution which should be selected for performing the job.

Second, it would be inefficient to repeat the search for a particular set of conditions if they are encountered again. Storing the prior "good" solution is presumed. It is an example of learning from experience. Many of the situations can be simulated off-line and the results of simulation with a sufficient degree of belief can be stored. This is an example of learning from the simulation results. Finally similar situations may be experienced in other systems of this type. Some of the results with a sufficient degree of belief may be included in the storage. This is an example of learning from an expert. The procedure of search in the state-space can be supplemented by the procedure of searching in the library of successful solutions known from the prior searches and from previous experience with functioning, simulation, or expertise judgment.

Step 3. Evaluate these trajectories by their "goodness" and the probability of success. "Goodness -of-fit" evaluations should be based on the cost-criteria important for the system including the processes of cooperation with other systems.

Step 4. Store in a list the limited number of trajectories with the best goodness-of-fit values. The length of this list may vary depending on the environment, the system under consideration, and the computing power available.

Step 5. Within the best trajectory, assign a limited scope of it (horizon in time and the envelope in the state variables) for the consecutive refinement (i.e., translating the results of planning into the language of the next higher-resolution level). The stochastic reality factors lead to the phenomenon that the degree of belief for the results of search (that this is "the best" trajectory) gradually decreases along the time axis for the trajectory received as a result of planning, since our belief was based on an open-loop execution during the selection process.

Step 6. Apply the subset of the planned trajectory (within the limited time horizon) to the virtual actuator and execute it using feedback information for online compensation. If this was the first level of the NIST-RCS, consider the job finished.

Step 7. Submit the following information about the trajectory after performing the compensation to the adjacent level of higher resolution (or, which is the same, the $(i-1)$th level below):

a. The final "point" of the trajectory designated by the limited scope assignment from step 5. (This point is actually an area of the indistinguishability zone.)

b. The boundaries of the state-space "stripe" confine the search envelope at the higher-resolution level.

This is performed using translation of the results of planning into the language of the next higher-resolution level.

Step 8. Execute at the $(i-1)$th level steps 1 through 7. Continue until the schedule of operation demands execution. Then submit the output of level $(i-1)$ to its actuator. Continue computation at all levels simultaneously until the system shuts down. Figure 7.18 illustrates functioning of the algorithm [1, 31].

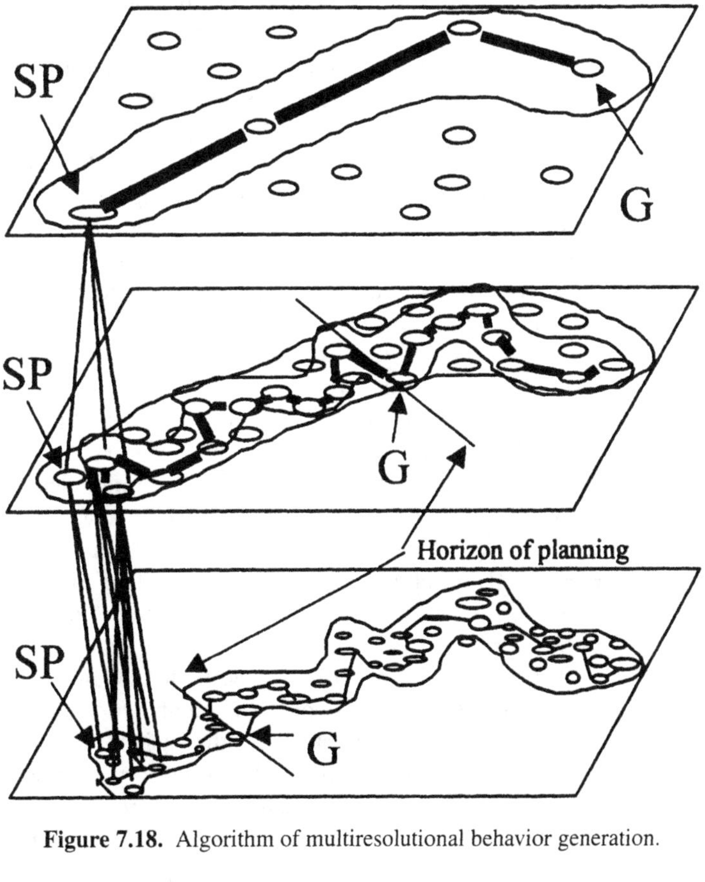

Figure 7.18. Algorithm of multiresolutional behavior generation.

7.4.6 BG Module: An Overview

The BG module is considered in two interrelated aspects: as a part of the overall generic processing node of the NIST-RCS, and internally as a *generic behavior-generating module*.

Generic Processing Node

EXTERNAL DESCRIPTION

From the preceding subsections we learned that all subsystems allow for representation in the form of ELF hierarchies, which can be considered systems of nested ELFs. A few additional comments can now be made in the view of our interest in the processes within the BG module. In any set of ELF subsystems we want to focus on the processing entity SP–WM–(VJ)–BG. We will call this ELF entity a *generic processing node* (of NIST-RCS).

The sensory processing (SP) module contains filtering, detecting, and estimating algorithms, plus mechanisms for comparing predictions generated by the WM module with observations from sensors. The SP has algorithms for recognizing entities and clustering entities into higher-level entities. The value judgment (VJ) module evaluates plans and computes confidence factors based on the variance between observed and predicted sensory input. SP communicates actively with WM (and its knowledge storage) in order to be able to hypothesize about clusters detected and to make converging the process of multiresolutional image recognition.

The world-modeling (WM) module contains, or has an access to, the knowledge/database (KD), with both long-term and short-term symbolic representations and short-term iconic images. In addition, WM contains a simulator for testing the hypothesis that emerges during the process of knowledge organization. This simulator is employed by BG for testing the alternatives generated by JA and SC search algorithms for the subsequent comparison in the plan selector (PS) [32–44].

Finally the behavior-generating (BG) module contains submodules of the planner (PL) and the executor (EX), as shown in Figure 7.19. The planner has sub-submodules of job assignment (JA), scheduling (SC), and plan selector (PS). Before the final selection of the best plan happens, the planner sends the alternatives to WM for modeling of the alternatives within a larger picture. One can argue whether or not the planner should send the alternatives for simulation back to the WM module or should request the model from WM and perform simulation within BG. It seems that in different systems this can be done in a different way.

Each node of the NIST-RCS hierarchy is a control system and it closes a control loop (ELF). Functioning of this ELF is demonstrated in Figure 7.19 [52]. Sensor inputs are processed through sensory processing (SP) modules and used by the world modeling (WM) modules to update the knowledge database (KD). This provides a current best estimate (see Fig. 7.19) of the state of the world to be used as a feedback signal to the executor (EX) submodule. The EX submodule computes the compensation required to minimize the difference between the planned reference trajectory and the current state of the world.

The estimate $\hat{x}$ is also used by the JA and SC functions and by the WM plan simulator to perform their respective planning computations. The vector $\hat{x}$ is used in generating a short-term iconic image that forms the basis for comparison, recognition, and recursive estimation in the image domain in the sensory processing module.

The structure demonstrated in Figure 7.19 is built for a system when it is convenient to search for the alternatives of the schedule of commands by performing a specific

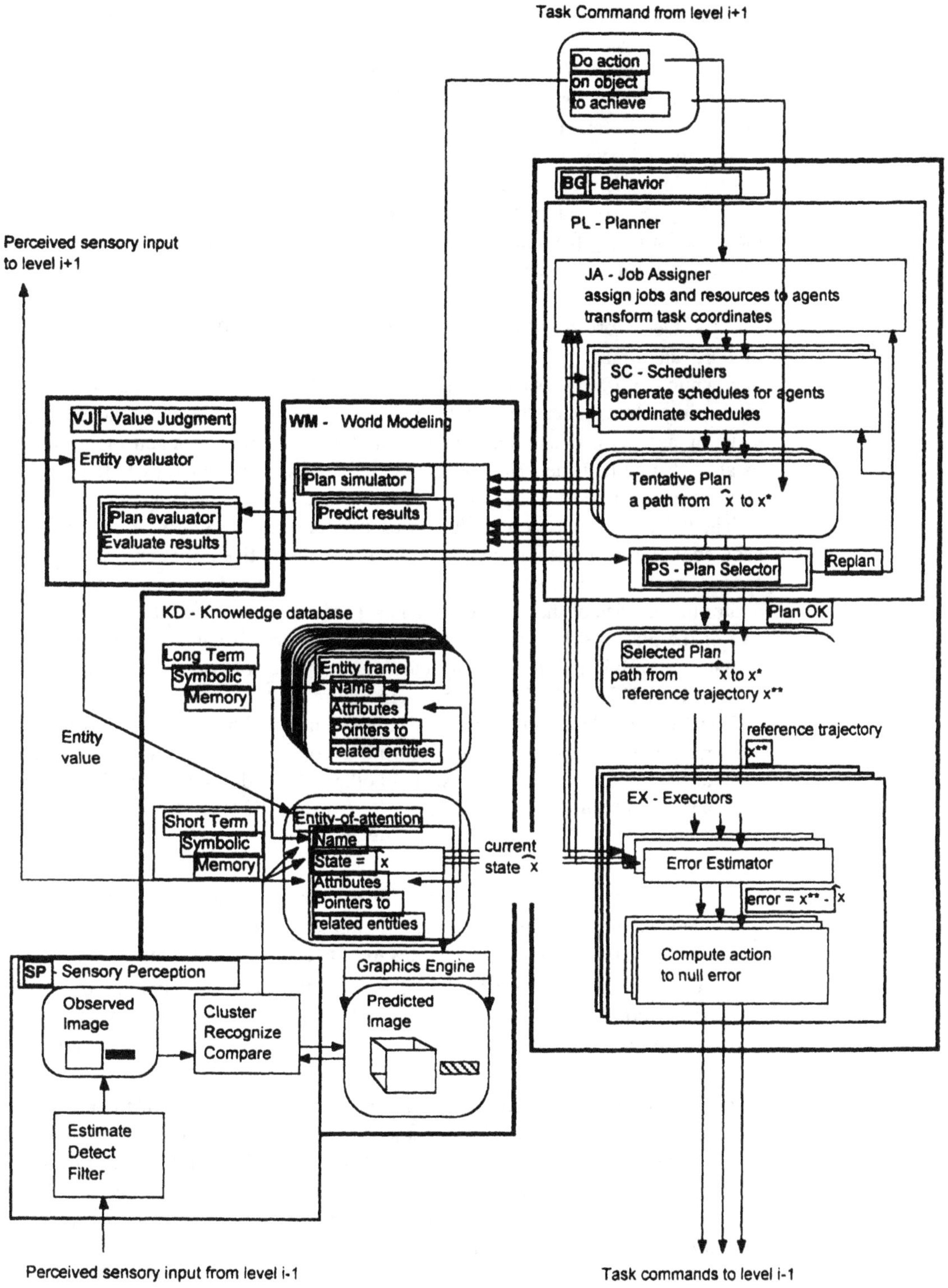

Figure 7.19. Relationships within a single node of the NIST-RCS.

search. First, we introduce the alternatives of job assignment and then compute schedules for these particular alternatives of job assignment. We will later introduce a more general planner for which the structure in Figure 7.22 is just a particular case. At this stage we intend to discuss the functioning of a BG subsystem within a generic node of the NIST-RCS system.

Figure 7.22 gives a level of detail that allows us to apply it to a single level of control (a single loop ELF) or interpret it as an arbitrary level of the multiresolutional hierarchy of NIST-RCS. In the latter case the existence of lower-resolution levels can be seen in the "goal" arriving from above, while all lower levels are represented by an imaginary (virtual) controlled system.

Internal Description of the BG Module Let us start with a description of the full BG module functioning within the ELF of a particular level. It is realistic to assume that we have a number of sensors including those measuring values of some particular parameter (and using a transducer "physical variable-symbolic variable") and those dealing with visual images. It is convenient to deal with these two groups of sensors separately. Let us also assume that the WM module is equipped with a rich knowledge database that allows for full support all sensory processing and decision making.

The behavior generation module at each level of the NIST-RCS hierarchy performs such cognitive activities as construction of the alternatives of solutions in time and in space. For example, it suggests trajectories of the output motion, suggests schedules and/or trajectories of the input commands, performs selection of the preferable alternative of the plan, its decomposition, and the monitoring of its execution by the virtual actuator. The BG module is constructed as a sequence of two submodules: PLANNER (PL) and EXECUTOR (EX), as shown in Figure 7.19.

The BG module communicates with world model from which it receives the relevant mappings of the system to be controlled (computational model), and the results of simulation. It also communicates with all other BG modules at its level of resolution.

A particular solution of the PLANNER shown in BG (Fig. 7.22) consists of the sub-submodules JA, SC, and PS which operate as follows:

- JA forms tentative alternatives of distribution for the jobs and resources to subagents, and transforms coordinate systems from task to subtask coordinates (e.g., from endpoint, or tool, coordinates to joint actuator coordinates).
- SC computes a temporal schedule of subtasks for each alternative of job distribution contemplated by JA and coordinates the schedules between cooperating subagents (e.g., coordinate joint actuator trajectories to generate desired endpoint trajectories).
- PS accounts for the cost of the alternatives, for the results of their simulation by WM and evaluation of these results by VJ, and for the number of replannings in a particular situation. PS selects a subset of the plan for sending it to the higher-resolution adjacent level of the NIST-RCS hierarchy. Since JA and SC operate in the language of the next adjacent higher level of resolution, *the task turns out already to be decomposed.*

Together, the assignment of jobs and resources to subagents, the transformation of coordinates, and the development of a coordinated input command schedule, as well as the output motion trajectory for each subagent constitute a plan. Therefore the output from JA and SC is a set of tentative alternatives of a plan. JA and SC may generate several tentative plans. Each is sent to the world model where the expected results are simulated.

The BG module communicates with the world model from which it receives the state information. The results of simulation are sent to the value judgment (VJ) module where the cost/benefit evaluation for each alternative is performed. The evaluation is returned to the plan selector (PS) sub-submodule for a decision as to the best plan of action (see Fig. 7.20).

This process can be iteratively repeated for different zones of the state-space where the possible plans are generated. The PS performs its function of selecting one plan with the best results of the VJ's evaluation. With the proper hardware architecture, plans can

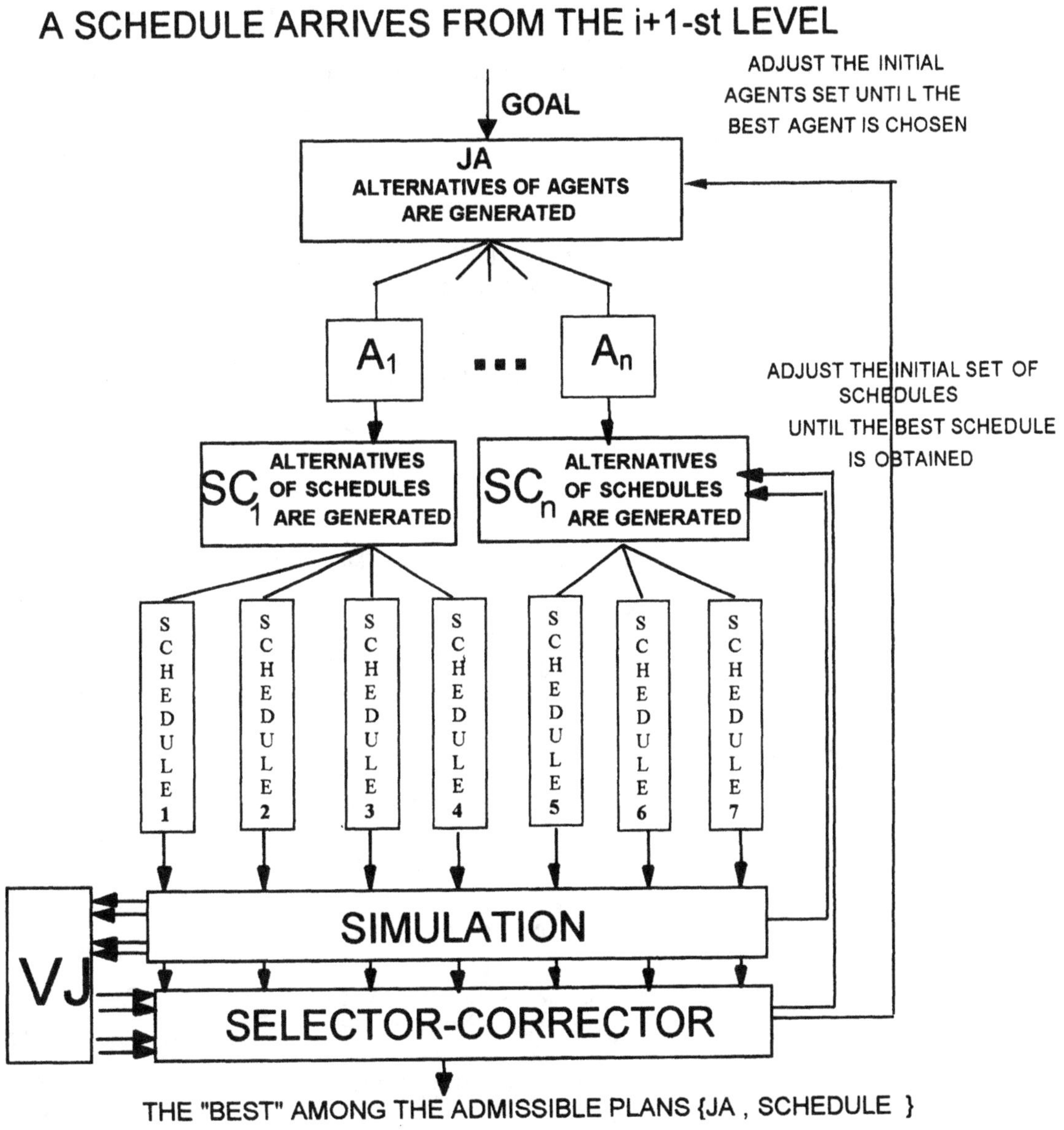

Figure 7.20. Diagram of the planning process.

be developed and evaluated in parallel. From the set of tentative alternative plans, the best plan is submitted for execution. The detailed description of the planner is given in Section 7.5.

At each hierarchical level, plans are expressed in a vocabulary of subtask commands that can be accepted as input by the executor (EX) submodules at that level. For each executor the string of planned subtask commands constitutes a reference trajectory through the subtask space. Since the real trajectory will differ from the planned one, the value of the error should be estimated. There are many algorithms of estimation that will be discussed in Chapter 9. However, one thing is important: Regardless of how the estimation is performed, the action computed should null the error. This process is called *compensation*.

The output of the BG module is applied to the input of the virtual actuator (VA). At each level, the results of planning have to be transformed in action. However, the phenomenon of "action" is interpreted at each level in a level-specific way. After the process of decision making is finished, the "decision" is formulated in the form of plan. The "plan" contains information of the goal to be achieved, virtual actuators to participate, and motions they should execute. This plan is applied to the system as it is visualized at this particular level, and the result of planning is obtained in the form understood at this particular level.

How the BG Module Operates In Figure 7.21 a menu of operations and an arsenal of available techniques are presented. These can be applied in various technical solu-

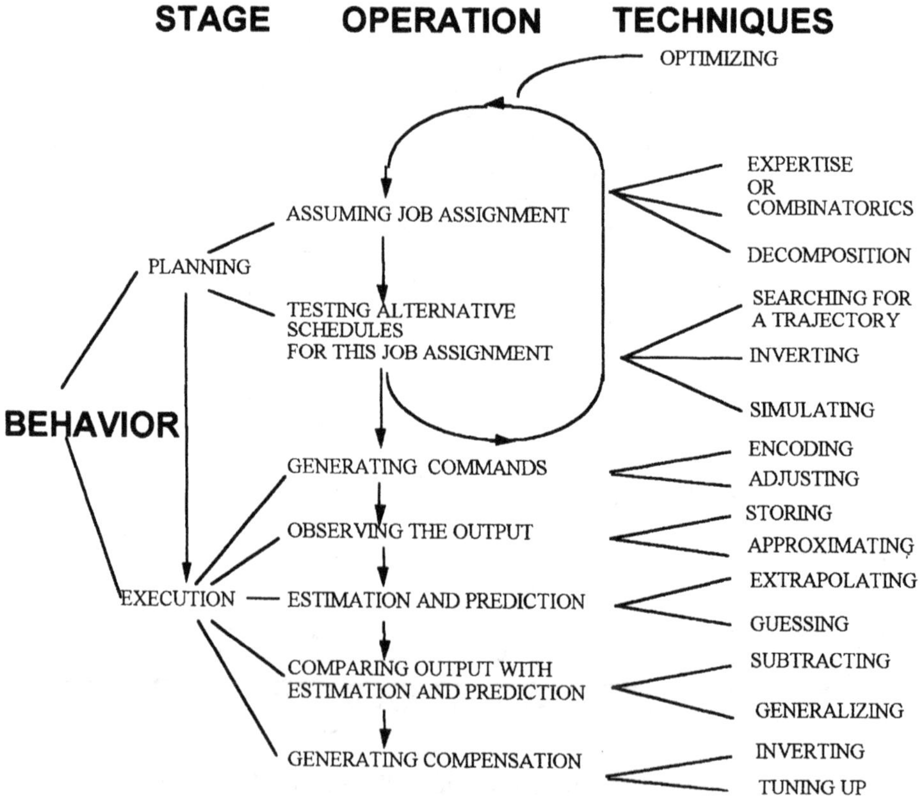

Figure 7.21. Stages, operations, and techniques in the BG module.

tions of BG module. The BG module receives information necessary for planning from the world model and from other BG modules at the level of consideration. The world model updates regularly the state information (or the string of states if the algorithm of planning requires). The world model provides the computational support of the search procedures performed by the BG module if there are no appropriate plans in storage.

These two consecutive stages (planning and execution) are performed using the following operations:

1. At the stage of planning
   a. the set of participating agents is selected,
   b. the schedules of motion are designed for the participating agents,
   c. the input signal (command) strings are computed for feedforward control (FFC).
2. At the stage of execution
   a. the sequence of commands for FFC is encoded and adjusted,
   b. the error of functioning is evaluated,
   c. the compensation component of the input is computed for feedback control (FBC).

In performing these jobs the PLANNER makes a combinatorial optimization and searches for the best combination of agents and best trajectory of motion. Since in most cases systems are defined via their experimental data, selection of both the set of agents and the best trajectory descends to a recursive procedure of joint search and simulation. This can be performed off-line.

In the meantime the EXECUTOR works primarily online in real time. The error (deviation of the real trajectory from the plan) is measured and compensated for. Both the PLANNER and the EXECUTOR are discussed in the subsequent sections in more detail.

The functioning of BG module is illustrated in Figure 7.22. It starts by assuming different possible job assignments (or the distribution of the work to be done among the available agents). Alternative job assignments are chosen for subsequent analysis and comparison based on the available combinations of feasible assignments. The list of feasible assignments is based on the BG module's experience in generating similar behavior in the past.

Suppose that we have two feasible job assignments. Each generates a different schedule (for each agent taking part in this job assignment). The two schedules are received after the desirable motion for each agent is obtained and the output is inverted to determine the input schedule. Then each schedule is checked via simulation. The simulator is contained within WM module. The schedule enters the simulated string of actuator–world–sensors. The result of simulated sensing are processed in the perception module. The quality of the particular schedule is evaluated within the VJ module. After testing all schedules, the result of this testing determines the best plan for the subsequent execution.

The PL submodules must communicate with each other if the results of their computation are to be coordinated. They also receive from the world model a system at a particular level of resolution that can be used for joint searches. The EX submodules at each level also communicate with one another. They do not need to have a model of the system, but they need to know each other's model in order to synchronize their execution.

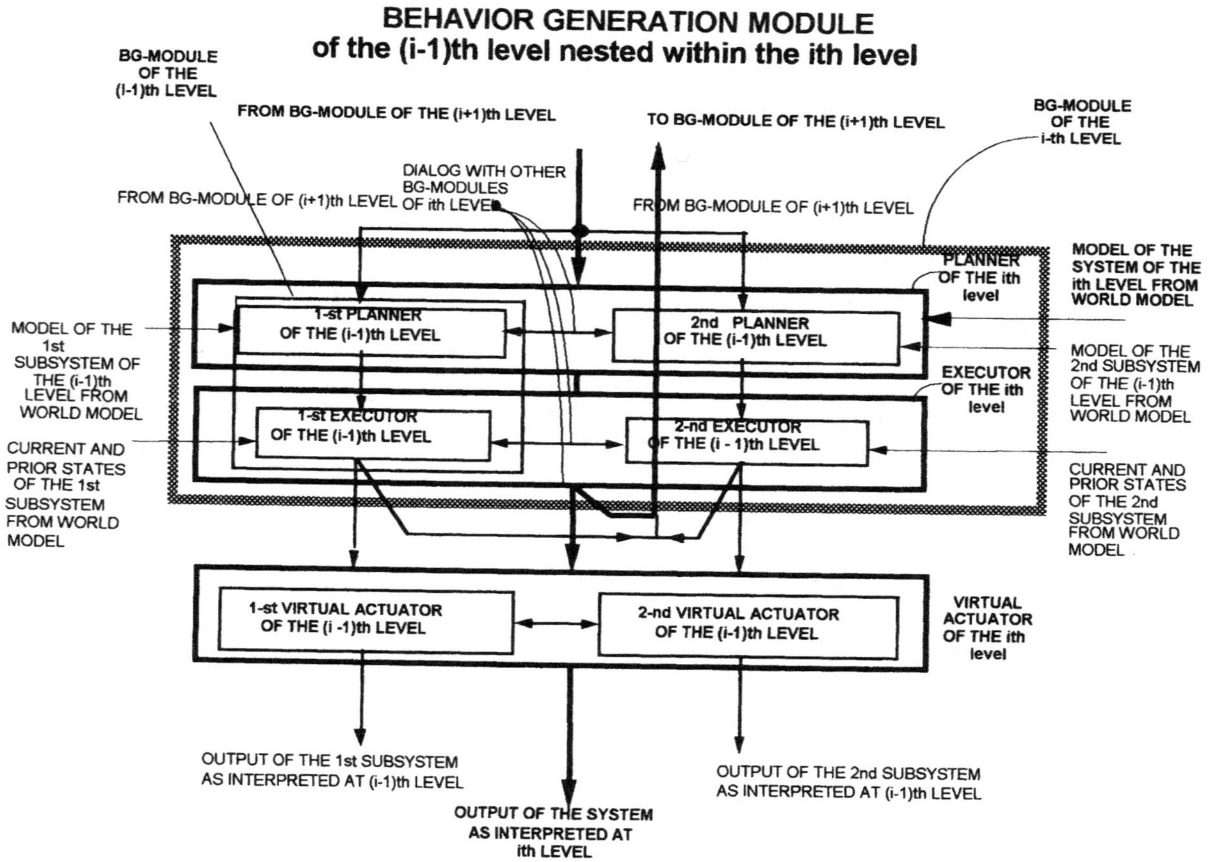

Figure 7.22. More than one BG module at a level works simultaneously (both nesting and concurrency phenomena are observed).

The entire top-down planning/job assignment process is performed before the actual execution starts. This can only take place if the concurrent operation of all PL submodules can be represented as a nested system. Sometimes the algorithms of BG allow for starting the operation with the current best plan.

We will focus on the phenomenon of "nesting" because it is an important component of the BG module's functioning. Indeed, the plan at each level cannot be considered complete without incorporating the results of planning from the higher-resolution levels. The procedures of high-resolution planning is therefore considered a part of the planning procedure at any level of consideration. This phenomenon applies in both top-down and bottom-up processes.

As Figure 7.25 shows, the plan proceeds as a sequence of commands from the (i+1)th level down to the ith level. It is developed in a twofold manner. Before the execution started, a command string with no changes is introduced by the EXECUTOR, since no compensation is yet present. In this period the compensation function is temporarily disabled. As soon as the PLANNER of the ith level has accomplished its operation, it submits the results both to the $(i - 1)$th level and to the $(i + 1)$th level.

It is important to note that planners communicate as independent units. Their communication encompasses their inputs and their outputs. However, their inner (sub-subsystems) functioning is conducted in a totally independent way.

Sending the plan to the $(i - 1)$th level allows the $(i - 1)$th level to look ahead, since the PLANNER module of the $(i - 1)$th module is at the input of the VIRTUAL ACTUATOR of the $(i - 1)$th level; that is, the start of functioning for the level above means the functioning of the actuator. Otherwise, without feedback from the PLANNER module of the $(i - 1)$th level, the PLANNER module of the ith level would not be able to complete its mission. Sending the results from PLANNER of the ith level to the PLANNER of the $(i + 1)$th level does the same thing: Without this communication, its operation would not be complete. Indeed, it verifies at the $(i + 1)$th level that the plan is admissible, or that it must be corrected and resubmitted.

Here we return to the concept of nestedness mentioned earlier. Just as each BG module of the lower level is nested within its adjacent level from above, each PLANNER of the ith level is nested within the PLANNER of the $(i+1)$th level. A similar phenomenon can be expected for their sub-submodules too.

7.5 PLANNER

In this section we delineate specific details of the PLANNER as a subsystem of the BG module. Planning is a popular topic in the area of intelligent systems. However, in AI research planning was separated from the process of control. It is either considered a component of knowledge-based task formation or a component of the general decision-making process—never as belonging in a control process. In control theory, planning has become only recently a subject of discussion, after supervisory control became a legitimate theoretical topic. The PLANNER as a component of a multiresolutional control hierarchy, is a new topic, and many details of it will be discussed in this section.

7.5.1 Computation Process of Planning

In our discussion, a plan is understood to be a set of data that includes the following components:

- The output time-trajectory of motion. This can be represented, for example, as a time schedule[2] for the components of the output vector. Generation of this output trajectory is assigned to the set of virtual actuators of the virtual control system at the output of the node under consideration. This trajectory is considered the best one out of all possible alternatives. It is denoted X_i^* in Section 7.4.
- The trajectory of the action vector. This can be represented, for example, as a time schedule of actions. It is denoted $\{\{A_v\}_{i,j}\}_k$ in Section 7.4.[3]
- The time-trajectory of the input control vector. This can be represented, for example, as a time schedule of the control vector (or the vector of control commands).

These components of a plan cover the control problem completely. The feedforward control solution is satisfied by the plan. However, if the model we use for planning differs from the reality, the output computed as part of plan will differ from the desirable output. The input prescribed by the plan will entail even larger mistakes in the output. This is why the WM is supposed to constantly update the set of models employed for planning. This is also why the BG module is equipped by an EXECUTOR to perform the online feedback compensation (see Chapter 9).

It is presumed (and it can be proved for many practical cases) that there exists an optimum plan which maximizes the cost-function for a particular case. Certainly there exists the best plan which provides for a sufficient value of the cost-function for a particular case or which is enclosed in the "envelope of desirability."[4]

A plan can be decomposed in the same way as the objects, actions, and tasks. Multiple subplans are possible in the general case when different combinations of actuators are available. Each has its own plan with different schedules for performing tasks in a time interval (each as an individual plan) that can be assigned to the actuators. It is presumed (and it can be proved for many practical cases) that there exists a best plan that maximizes the cost-function for each case.

PLANNING/ CONTROL CONTINUUM

In this section we find that there is no difference between planning and feedforward control. Planning and control problems are interrelated; they do not have much meaning one without the other. This is why we use the term *planning/control* where possible to underline the inseparable character of these two operations. Planning is equivalent to feedforward control at each level of resolution, although, for the adjacent level of the higher resolution (the adjacent level below), the plan is associated with the goal that came from above. It differs from the control computed at that level by some inherent *lookahead* connotation.

7.5.2 Epistemology and Functions of the PLANNER

Planning is interpreted as a design of required activities, and as a design system it functions in the domain of knowledge representation. If the system is built (and this was our assumption in the beginning in order to simplify our presentation), this design is provided for a system that already exists. However, for much higher levels of generalization (lower levels of resolution), planning means providing a design for both required activities and a system that is a platform for the realization of these activities.

[2]A *time schedule* is comprised of two ordered lists with links of correspondence between them. The first list is a list of vectors, and the second one is a list of time instances.

[3]In this notation, i is the level of resolution, j is the number of an actuator, and k is the number of the alternative of job assignment.

[4]H. Simon call these plans "satisficing" plans.

The planner has as input the string of commands (the assignment) from the upper level, which entails the results of models obtained from the world representation of the upper level. It also has models obtained from the WM at the level under consideration, and it has information about current states, and information about planning activities of other planner submodules of other BG modules at this particular level of resolution. The plan is the output of the planner and is a string of commands to obtain the desirable motion of all of the subsystems under its control at a particular resolution level; this is submitted to the higher-resolution levels together with the description of the desirable motion at the output.

Planning has its own time constraints. We will try to avoid unnecessary search and reduce the space search as much as possible. Any reduction of the search envelope after some particular limit can lead to a loss of the best plan. We will consider the envelope reduction as increasing the efficiency of planning and reducing its reliability. This is the trade-off between the replanning frequency and optimality. If the plan is recomputed frequently enough, it needs only to direct the process properly at the point of consideration while the future details are less important.

The planner uses knowledge of the functioning loop, since this knowledge is represented in the world model. The latter is the result of learning and reflects our long-term WM knowledge of the system. This knowledge has two types of errors:

1. *Large errors*. These errors are due to unmodeled variability of the objects represented in the world. These are produced by the prior operations of generalization that synthesized these objects from the higher-resolution objects of the level below, where errors determine the interval of certainty about each parameter.
2. *Small errors*. These errors are due to the minimum discrete increments of representation at the particular level of resolution.

Large errors can emerge during the learning process, and they determine the variable costs of the alternative plans. Small errors affect the lower bound of a plan's error tolerance. Small errors can be reduced only at the higher-resolution levels (if they exist). Large errors lead to alternative trajectories formed by the PLANNER. Small errors interfere with the functioning of the EXECUTOR.

A plan is developed based on the model WM_{i+1}, and it is applied at the ith level which uses a different world model (WM_i) that is more narrow and more precise. The functions of the PLANNER can be listed as follows:

- The PLANNER's main function at each level is to find the best set of plans at that level. The PLANNER's main function as a multiresolutional nested system is to find the multiresolutional planning/control system of input commands.
- The PLANNER of the ith level receives the assigned subset of a PLAN from the $(i + 1)$th level (the higher adjacent level) which is called the TASK. This TASK is finalized by the BG module of the $(i + 1)$th level only after corrections are introduced by the EXECUTOR online ("online" here refers to the virtual ELF of the $(i+1)$th level). The PLANNER of the ith level uses this TASK and the model submitted by WM to create tentative combinations of the plan distribution JA among the virtual actuators[5] (VA), and then it contemplates which of all possible output trajectories performing the TASK is the best one.

[5]The PLAN is determined for the virtual actuator, not for an agent. We use the term "agent" as an equivalent for ELF because in the literature, agents have elements of intelligence similar to that of ELF.

- The PLANNER determines the appropriate alternatives (hypotheses) of the desirable combinations of VAs to perform the task. These combinations can be known in advance for the simple case. However, in general, they are found by making all possible combinations of VAs and testing all of them (combinatorial search).
- The PLANNER computes the best trajectory of motion for each hypothesis of the combination of VAs. In some cases this trajectory can be computed analytically. However, generally, the output trajectory can be obtained also as a result of search. All meaningful strings are created (combinatorial search). One or more strings regarded as the best are inverted into the input commands space (all this is done for each combination of VAs which has been determined by selecting the coordinates of the input space). So two consecutive search operations are performed; otherwise, the PLAN cannot be obtained.
- After this double search is done, the PLANNER distributes the best trajectory of motion among the VAs and determines the best trajectory of each VA's motion.
- The PLANNER transforms the best trajectory of the VA's motion into the VA schedule (the task's time distribution, the input command's time distribution, the remaining actions' time distribution). Note that with the PLANNER's output we receive a set of schedules which, when selected, was already distributed among the virtual actuators.
- The PLANNER transfers a set of schedules to the EXECUTOR (see Section 7.6).
- The PLANNER under consideration of the higher-resolution level (HRL) reports inconsistencies and singularities to the BG of the lower-resolution level (LRL).
- The PLANNERs of the HRL are considered a nested component of the PLANNER of the level under consideration. The PLANNER corrects its PLANs after receiving from the HRL PLANNER its responses concerning the inconsistencies and singularities. However, it does not communicate with PLANNERs of other levels of resolution (only with the adjacent level below).
- The PLANNER communicates with other planners of the same resolution level to coordinate the work processes, for example, shares resources in situations that arise due to uncertainties at that stage of planning.
- The PLANNER of the ith level receives from the PLANNERs of the submodules of the $(i-1)$th level their final results of refined planning and evaluates the overall performance. If replanning is required, it replans and resubmits the new plan to the levels of higher resolution.
- The PLANNER requests a new planning alternative if its adjacent LRL PLANNER responded with a result that is not satisfactory.
- The PLANNER makes the final decision on the final alternative package of schedules.

At the highest level of resolution no job distribution is required: the ACTUATOR is not a virtual actuator anymore. It introduces changes in the real world. (Of course, the metaphor of a virtual actuator is continued into the domain of reality, but we are not going to do this).

7.5.3 Planning in Multiresolutional Space versus Planning in Abstraction Spaces

In this book we depart from the usual planning paradigm as it is treated in artificial intelligence. The problem of planning has been traditionally treated in the AI area following the STRIPS, ABSTRIPS, NOAH, MOLGEN, and SIPE paradigms. Planning is considered to be a theory of reasoning about actions, and meta-planning was introduced to make this process more efficient (PANDORA paradigm[6] [35, 46, 47].) Of course, anything can be regarded as reasoning. But AI reasoning amounts to manipulating collections of schemata-like stimulus–response couples[7] [37]. The conventional AI paradigm of planning can be presented formally with the help of automata theory[8] [48]. Our paradigm is broader and cannot be satisfied by simple automata formalism.

NO IMAGINATION WITHOUT REPRESENTATION

In all prior planning paradigms (including AI) the advantages of multiresolutional (multigranular, multiscale) systems such as the NIST-RCS were not fully appreciated[9] [35]. The commonalities of planning techniques were not understood, since they were not visualized within the NIST-RCS paradigm. The commonalities between the planning and feedforward control (FFC) were not noticed at all. This is especially clear from the way multigranular (multiresolutional) planning is treated in the literature. In this book we contend that there are the similarities between planning and FFC. The differences among the levels of different granularity are simply in the length of lead time and the frequency component in the plan.

Although many results are related to planning in abstraction space, the AI authors give it an entirely different interpretation than the one we convey in the NIST-RCS[10] [1, 18, 19, 30, 49–58]. Indeed, the multiple resolution levels of the NIST-RCS hierarchy are not the "levels of abstraction" associated by many AI researchers with hierarchical systems. The processes of aggregation/decomposition used in the NIST/RCS are rather associated with generalization/instantiation and not with abstraction/specialization as AI authors often assume.

Unlike "abstraction spaces," our levels of multiple resolution can be legitimately associated with frequency or scale decomposition similar to those known from fractal or wavelet theories [59, 60]. Planning in the NIST-RCS involves solving a control problem within separate "frequency domains." What we call "planning in generalization levels," or in "multiscale levels," or in "levels of different resolution," could be rather called "finding controller for spectral regions" or "multiple bandwidth controllers" if indeed the spectral density of the system looks as it does as shown in Figure 7.23.

If planning is a feedforward control, why not treat it as a control and why not look for an analytical solution? This question is based on apparent confusion. There really is no methodology for finding the output trajectory in a feedforward control that satisfies conditions such as the set of constraints and/or delivers a minimum cost-function. The solution is known only for extremely simple, trivial cases. In control theory, searching for a feedforward control is frequently avoided by determining inputs online as a function of outputs or other state-variables of the system. This strategy is accepted based on the prevailing opinion that it is better than determining inputs as functions of time.

[6]Reasoning implies using logic for inference. The planning paradigm proposed in this book corresponds to the logic of multiresolutional control systems. In this way we supplement the reasoning typical for the predicate calculus of the first order (without contradicting it).

[7]M. Arbib's theory of schemata is a powerful tool for applying concepts of automata theory to the domain of intelligent systems.

[8]Recall that at each level of resolution, the automata representation has an equivalent representation in the language of differential and integral calculi, and vice versa.

[9]E. Sacerdoti uses multilevel planning. However, generalization supplemented by an explicit change of granularity could be of high advantage to the planning process.

[10]"All classical AI planners use distinct and well-defined planning levels. A new planning level is created by expanding each node in the plan with one of the operators that describe actions. These levels are often referred to as hierarchical which implicitly associates them with abstraction levels. In fact they are independent of any abstraction level: a new planning level may or may not result in a new abstraction level depending on which operators are applied."

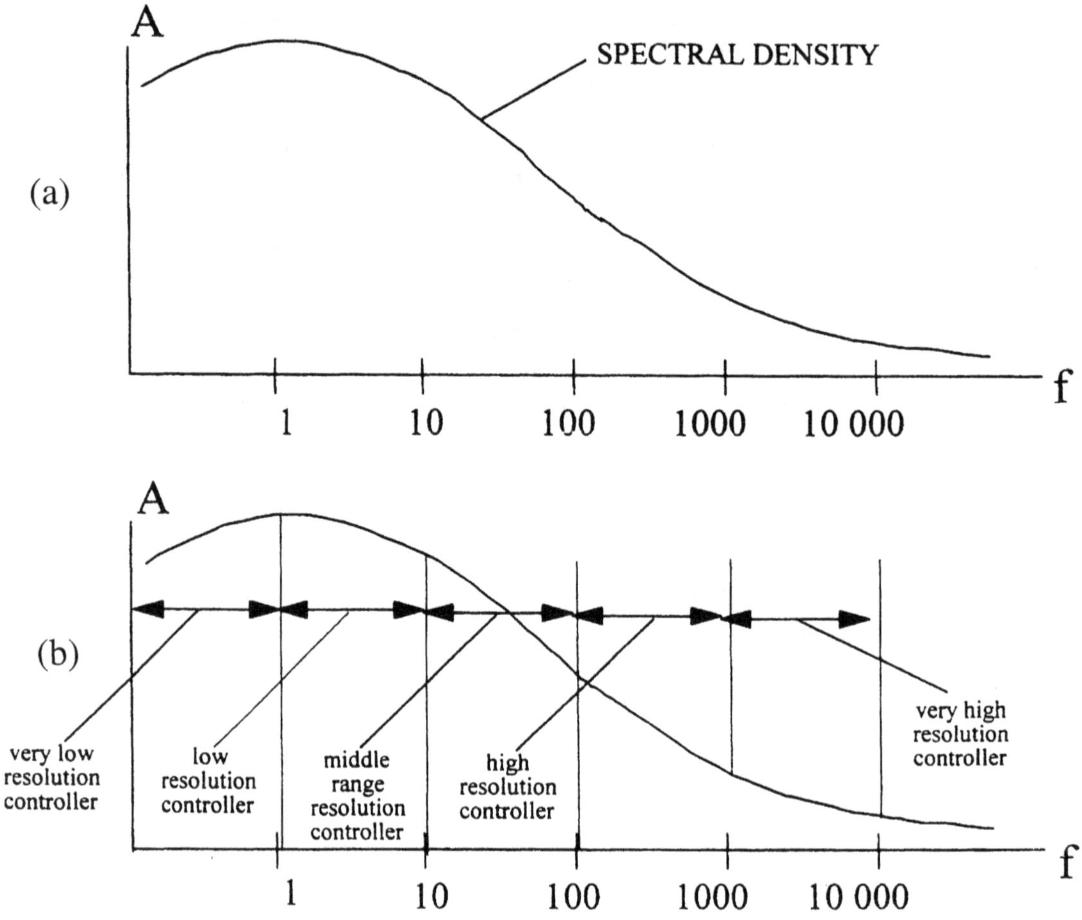

Figure 7.23. Concept of spectral region controllers.

This is often true, but it is also true that combining these two principles can give better results. This is exactly the procedure employed in the NIST-RCS: determining the most significant part of the input as a function of time (PLANNER) and determining the compensatory part of the input by taking into account a direct measure of the output (or/and other variables). There is a lot of evidence that if this principle is applied consistently throughout the entire hierarchical system, it can give the best results in comparison with all other possible decisions.

For most cases feedforward control requires search. We treat planning at each level as a procedure of search (first in the storage, then in the state space.) Other authors do not visualize the uniformity of the search processes and address the issues of motion planning as if they are different from the task planning[11] [61, 63, 65, 67, 68]. We do not distinguish between these because we address planning at all levels as a search in the state-space.

[11]The conceptual barrier existing between continuous systems and DES can be easily avoided in the case of planning.

7.5.4 Planning in the Task Space versus Motion Planning

Planning is equivalent to feedforward control. Planning involves finding a desirable trajectory of motion (including both control and output variables) using the available knowledge at the time of planning. Task space planning is associated with *discrete events*, while motion planning is usually associated with *continuous motion*.

GRANULATION MAKES ALL THE DIFFERENCE

We see the difference between these two planning problems only in the resolution by which the state space is discretized.[12] The world models, the state information, and the control sequences in NIST-RCS are discretized anyway. In the task space, the space is naturally discretized because of topological discontinuities, which are also called *morphological discontinuities* and *catastrophes*. These are the places and moments of time when the cutting tool enters the workpiece or finishes the process of cutting, or when the gripper of a manipulator leaves the initial position and arrives at the final point, or when the process of obstacle avoidance starts and ends.

From the last example involving an obstacle, we can see that the event of "starting the process of obstacle avoidance" is a fuzzy event. Its discrete time cannot be assigned with precision. One can easily deduce that between the two domains—one is discrete events and the other is the domain of continuous motion—there is an area of fuzzy transition. Therefore, between the domains of "task planning" and "motion planning," there is a continuum of different state-space tessellations. This is why in NIST-RCS, we will not distinguish between the principles and techniques of planning that are applied at different levels of resolution.

As the hypotheses about available JOB ASSIGNMENTs (and the initial decomposition of the work to be done) are completed in JA, the motion trajectory (including its output and input time-trajectories) should be found in SC which allows for the ACTION REQUIRED. This is the first stage of PLANNING. Preplanned trajectories can be stored[13] and browsed through when the need arises, or a search for the desirable trajectory of motion can be conducted.

Since the result is assigned to the agents in the form of a final state as required (under cost-function conditions and constraints), PLANNING proceeds with developing schedules, or which is the same, finding the output time-profiles for the agents. The results of planning include the trajectories <OUTPUT> to be executed (FFC) and the time-profiles or trajectories <INPUT> to be applied to execute the output trajectories. The results of planning can be interpreted as time-tagged strings of commands <TASK TO THE NEXT LEVEL>. The time-profile <OUTPUT> is found by solving the optimization problem, that is, by minimizing the cost-function S:

$$\langle \text{OUTPUT} \rangle \longleftarrow S_{\min F}[T_k, R_k(\text{WM}, \text{JA}), \rho_{ik}, \Delta t_k], \tag{7.19}$$

where

- T_k = task submitted to the BG module at the kth level of NIST-RCS,
- $R_k(\text{WM}, \text{JA})$ = representation admitted within the paradigm of the WM and JA functioning,

[12]The principles of representation including choice of the combination "scope of representation" and "minimal distinguishability zone" in space and in time are described in [1].

[13]The double-searching during planning activities is not only the source of a particular result but also a source of meaningful alternatives that can be stored in the library of "alternative solutions." This is one of the first entrances into the future RCS system with learning. The system can learn not only from its real experiences but also from the imaginary situations it has encountered during search procedures in its "imagination."

- ρ_{ik} = minimal distinguishability zone admitted at the particular spatial resolution in the ith coordinate at the kth level of resolution,
- Δt_k = minimal time interval admitted at the particular temporal resolution at the kth level of resolution.

As the PLANNING of motion trajectory is occurring for each of the subsystems at a level, the cooperation of these subsystems is negotiated, and the FINAL JOB DISTRIBUTION is performed. The second part of PLANNING executes the desired motion trajectory. The time-profile of the input to the lower-(higher-) resolution level is found. This is done by an inverse procedure that results in the sequence of the output EXECUTION commands. If the level operator is presented in the form

$$\langle \text{OUTPUT} \rangle \longleftarrow T(\langle \text{INPUT} \rangle), \tag{7.20}$$

where T is the transformation function of the controlled system (a plant operator, a model of the system); the inverse procedure can be expected (the "input" is computed from the "output" prescribed) and is written as

$$\langle \text{INPUT} \rangle \longleftarrow T^{-1}(\langle \text{OUTPUT} \rangle). \tag{7.21}$$

Substituting for <OUTPUT> the right-hand side of (7.19), we obtain

$$\langle \text{INPUT} \rangle \longleftarrow T^{-1*} S_{\min F}, \tag{7.22}$$

where $P = T^{-1*} S_{\min F}$ is the planning function, and $F = F[T_k, R_k(\text{WM, JA}), \rho_k, \Delta t_k]$. The activities of the BG module include planning and execution with corresponding cooperation among the agents, and this should be reflected in the interfaces of all modules. The output trajectory and the corresponding INPUT, as well as COMPENSATION, can be based on different existing models within the system of representation in the WM module (entity-relational graph, object-oriented representation) of the machine.

7.5.5 Reactive versus Deliberative Decision Making

Decision making reflects the epistemological characteristics of the system. Deliberative decision making is possible if a model exists, and there is sufficient time to simulate different hypotheses and select one of them. However, when the models are not available and/or there is no time to build the hypotheses and test them, reactive decision making is the only choice. We have to have on hand a menu of predetermined responses that are considered to be beneficial in most cases.

The issue of reactive planning arose during research at SRI on combining SIPE [65] (System for Interactive Planning and Execution Monitoring) and PRS [66] (Procedural Reasoning System) to control an indoor mobile robot. In the AI community, it is thought that robots should be controlled in a reactive manner. As we have explained earlier in this book, this idea is insufficient for complex cases; no intelligent operation can be done without online (or off-line) planning.

Deliberate decision making is therefore considered a different issue from *reactive* decision making: "The ability to act appropriately over a broad range of situations without deliberation is called reactivity, and is an important measure of competence for robots controlling dynamic and unpredictable processes" [67]. In the literature there are many myths and fantasies related to reactivity. Many authors believe that all agents should be reactive agents (apparently due to the lack of knowledge of predictive control systems).

Also many authors do not realize that reactive control is in fact "feedback control"; some authors even propose "slow planners embedded into fast real-time control systems" [68].

SHORT TERM VS. LONG TERM

The NIST-RCS is based on the notion that a goal-oriented system must be both deliberative and reactive[14] in generating its behavior; this includes planning/control activities at each level of resolution. Behavior is deliberative to the extent of the existing knowledge of the model and the expected situation. Behavior is reactive with regard to unpredictable and unexpected factors; the factor of "unexpectedness" must be handled well. This combination of deliberative/reactive (or reflective/reactive) strategy of behavior generation holds at each level of resolution. Another interpretation of the behavior generation strategies and processes can be formulated as follows: Each level can be goal-deliberative and unknown-circumstances-reactive, as well as known-circumstances-deliberative and subordinate-deliberative. Most of this terminological discussion would be unnecessary if we represent the operation of NIST-RCS in the terms of feedforward/feedback control which are correspondingly computed using known and unexpected information.

7.5.6 What Is Inside the PLANNER?

The PLANNER is effectively a feedforward (FFC) submodule. Its function is to find the optimal FFC function that presumes both spatiotemporal motion design and motion distribution among the agents. This submodule performs three operations:

1. Spatial planning including job assignment among the virtual actuators.
2. Coordinate transformation from task to subtask.
3. Temporal planning (scheduling).

The structure of the PLANNER submodule contains the following set of submodules (see Fig. 7.24.) The operations of spatial and temporal planning are not separable in principle. They can be considered separately only for the sake of presentation and/or for very simplistic cases of the NIST-RCS functioning. More important, they are executed separately by existing computational systems because it is convenient algorithmically to interlace sequentially the procedures of spatial and temporal searching. On the contrary, the procedures of comparison and decision making are combined in a stand-alone set of operations that should handle a set of results obtained from the several cycles of spatial and temporal search for alternatives.

Sub-subsystems

The PLANNER does its job by generating and contemplating different alternatives (hypotheses) of distributing the future activities among the potential subsystems at the higher resolution level. The PLANNER determines them both in space and time. Combining hypothetical alternatives and analyzing them amounts to simulation. Planning presumes *simulation* of multiple hypotheses of the future. Interestingly all algorithms of search contain a simulation of the processes that have not yet happened. However,

[14] "Reactive"—based on response to a stimulus—usually alludes to responding to something not expected and/or planned. This definition contains some circling because if one knows how to respond to a particular stimulus, this means that one must have deliberated and prepared the response. An example of a reactive response is the avoidance maneuver around a sudden obstacle. The word "reactive" is often substituted by "reflexive" because "reflex" is regarded as "action in response to a stimulus."

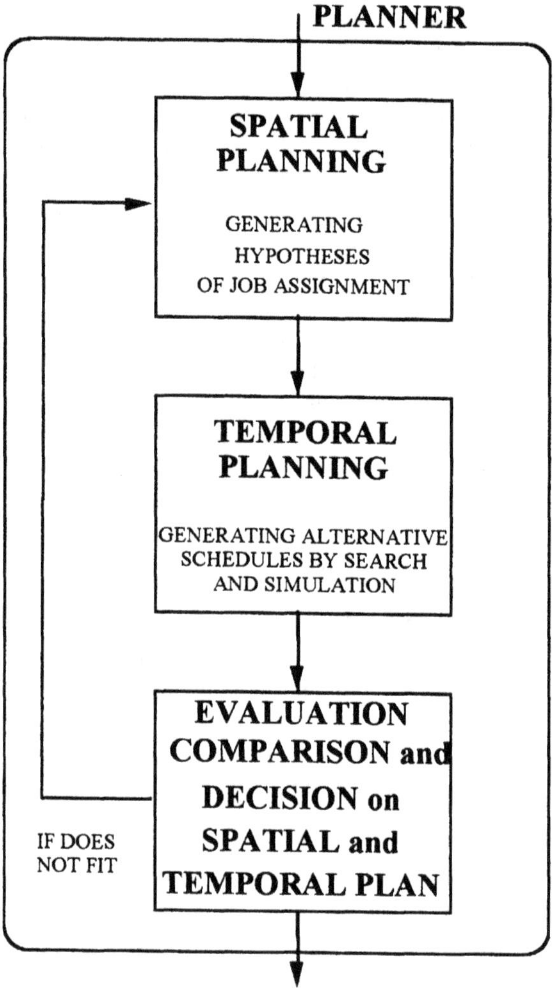

Figure 7.24. Sequential structure of the search for a spatial and temporal plan.

in planning we are interested in knowing "What if." This immediately creates a link between the concept of planning and the following concepts of intelligent information processing:

ALWAYS THE SAME: GENERALIZATION AND INSTANTIATION

1. *Grouping (G) of the units of knowledge based on their similarity.* We are interested in patterns of the state that emerge within the state when we contemplate its evolution into the future. The building of patterns demands a grouping of entities and/or other relations of similarity and recognition of similarity within emerging patterns.
2. *Focusing of attention (FA).* This is required; otherwise, the abundance of computational procedures to be performed would create substantial computational complexity.
3. *Combinatorial synthesis (CS).* This involves the creation of imaginary "possible worlds," or plausible alternatives of the state evolution.

This triplet of concepts (GFACS) [69] forms a basis for computing *plans*. The inner structure of the PLANNER is constructed according to its conceived functioning described in the sections above. This structure is shown in Figure 7.25 for a sequential processing (other versions are also possible.) The PLANNER comprises three compartments (Fig. 7.25):

GFS PRODUCES TEAMS

1. The JA subsystem generates tentative assignments of consecutive elements of behavior (jobs, operations, contemplated strings of input commands, other decomposition in the space representation) subject to subsequent simulation as a part of searching the best plan. These activities can be also interpreted as "team generation" resulting from the search/simulation. JA modifies the teams to correct the assumptions and improve the results of subsequent simulation. JA cannot make any corrections unless each of the "teams" generated a search for the best functioning in time. This search is performed by the scheduler (SC). Search procedures performed by JA and SC can be unified into a joint process of generating planning alternatives. Each team can be consisdered equivalent to what was called an "agent" in Figure 7.20.

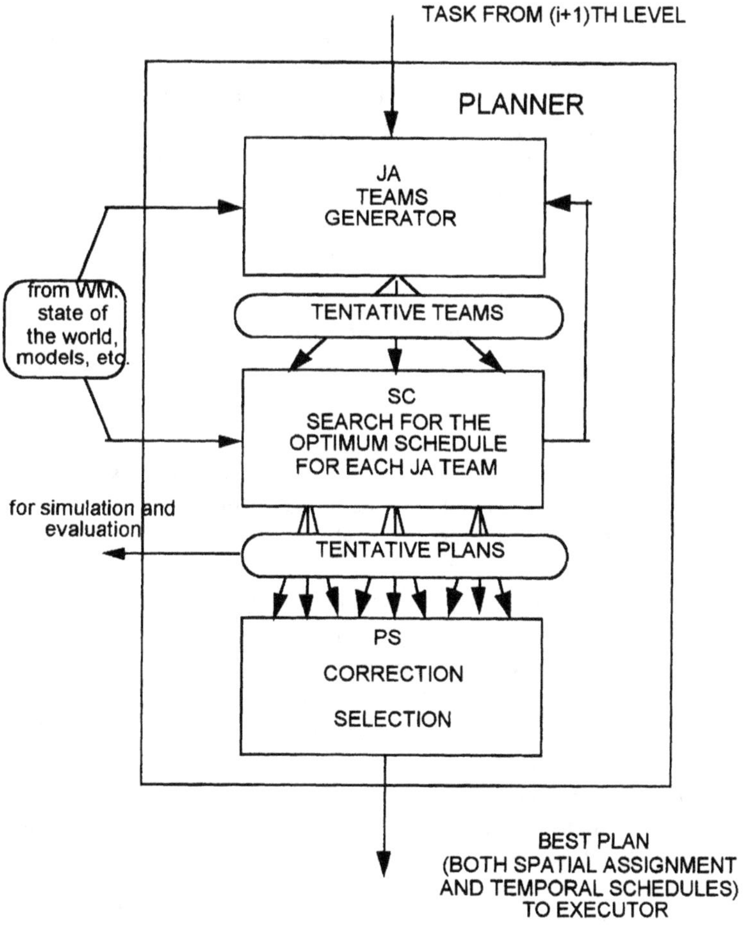

Figure 7.25. Inner structure of the PLANNER.

GFS^{-1} PRODUCES STRINGS OF ACTIONS

2. The SC subsystem searches for the optimum schedule for each of the teams postulated by JA. It performs the time decomposition of system functioning. In searching for a schedule, which can satisfy a set of requirements, it simulates the functioning process. Each search algorithm does this abstractly. Sometimes it is desirable to repeat a simulation in a less abstract manner with more detail in a broader paradigm. Then the limited number of alternatives is sent to WM for simulating them in more detail. As the results of search/simulation arrive, SC modifies the time distribution of jobs, and JA corrects the teams to improve the results of the subsequently repeated simulation.

PICK ONE!

3. The CORRECTION/SELECTION subsystem is supposed to make a final choice. There is a feedback loop from the results of search to the combinatorial process within the JA and SC. It allows for refining the prior coarse results of combinatorial synthesis, and searching for a more accurate fitting within the condition of "satisficing" and/or an "optimum."

Process of Planning

The PL submodules accommodate a variety of planning algorithms. These range from a sample lookup table of precomputed plans (or "scripts") to a real-time search in the state-space or configuration space, or even game-theoretic algorithms for cooperating multi-agent or competitive groups. However, regardless of how the plans are synthesized, a plan consists of a spatial decomposition of a task into a set of job assignments and of resource allocations to virtual actuators, plus a schedule of subtasks for each actuator, ordered along the time line. In many cases it is required that the schedules for the actuators be coordinated to produce coordinated actions.

In general, planning consists of the following steps:

Step 1. Generate a set of alternative plans including time and space distribution.

Step 2. Simulate the likely results of those plans by testing the alternatives.[15]

Step 3. Correct the alternatives by using the results of testing.

Step 4. Evaluate those results according to some cost/benefit criteria.

Step 5. Select the tentative plan with the best evaluation for execution.

The functional structure of the PLANNER shown in Figure 7.25 is that of the PL sub-submodule. The PL's function is to generate and/or select a plan in response to the input of a task from the next higher level BG module. Let us discuss this in more detail (see Fig. 7.26 from [52]).

It has already been mentioned that PLANNER submodules can be further decomposed into the sub-submodules of job assigner (JA), scheduler (SC), and plan selector (PS). These sub-submodules function as follows: JA assignments are applied for tentative scheduling. The assignment can be considered a "tentative team formation." For each tentative team, a schedule is computed by SC. SC intermediate results are constantly challenged by introducing a new assignment from JA. So the sequence of functioning looks like a string [JA–SC–JA–SC. . .]. Therefore JA and SC form a loop from which final versions of plans are chosen by the selector SC. Before the alternatives

[15]Frequently the search for alternatives by JA and SC is performed on a sufficiently complete model of the system, which validates it as a "simulation." A separate simulation stage is required only if the model employed by JA–SC is a simple one and discrepancies are not critical.

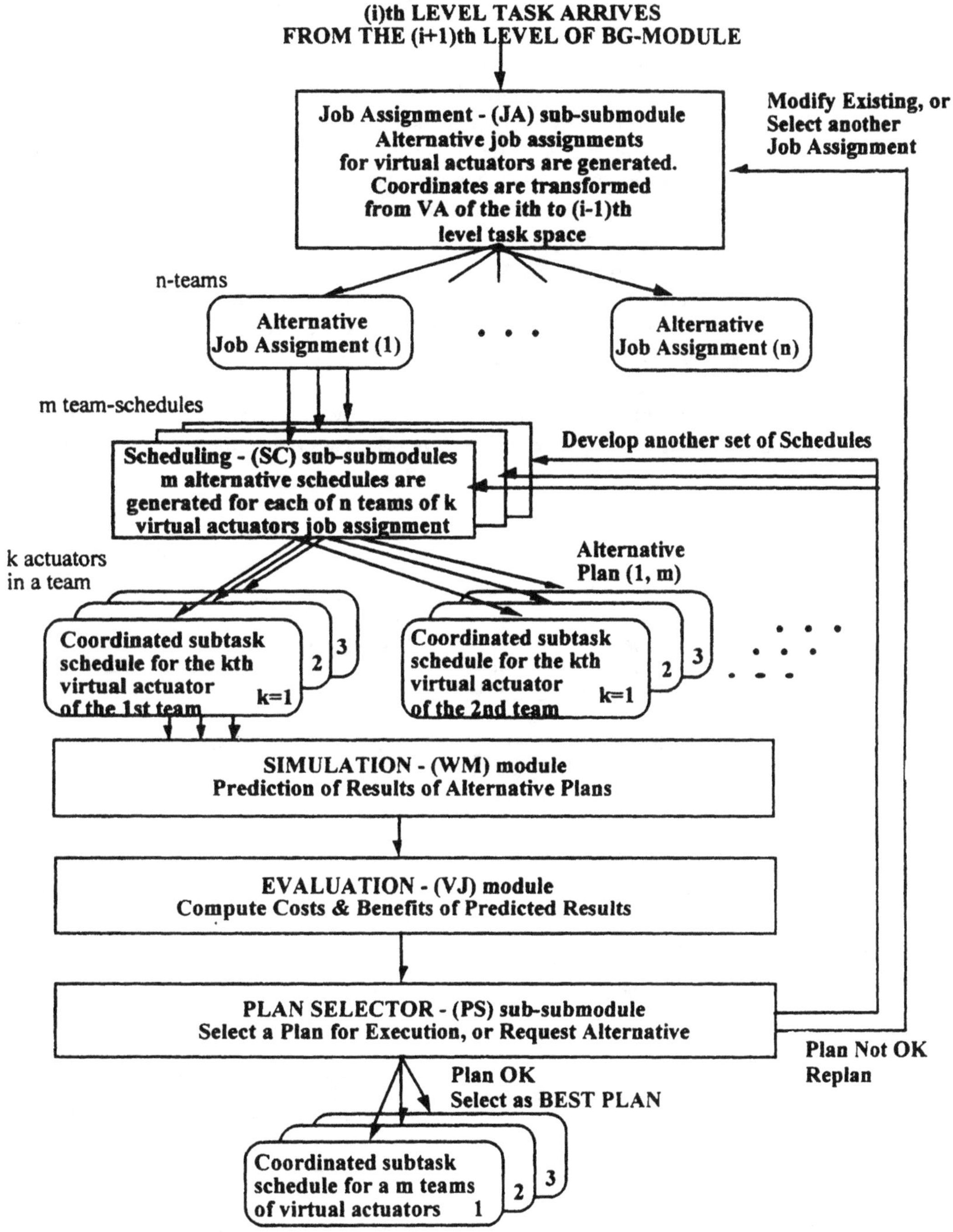

Figure 7.26. Diagram of the planning submodule.

submitted to PS for final selection, an inquiry is possible to WM and VJ modules for simulating the meaningful set of alternatives.

JA generates the alternative job assignments for the virtual actuators. It transforms the task from the level of virtual actuators of the $(i + 1)$th level of resolution (the adjacent level from above) into the language of virtual actuators of the ith level (the level under consideration). After the translation is complete, JA applies the operator of "tentative teams formation" (which is a synthesis of the combinations under constraints) and generates a set of n "actuator teams" of the ith level of resolution. Each of the teams is considered an alternative of job assignment.

Each alternative of job assignment is considered an input to the scheduler, SC, together with the world model pertaining to the case. The SC generates m schedules for each of the n teams and, for each of these, k schedules for all the single actuators of m schedules of n teams. The output is voluminous. Every opportunity to reduce ("prune") the search must be considered.

The schedules are simulated in WM and evaluated in VJ outside of BG (Fig. 7.27). In a well-structured case with extensive prior experience the stage of external simulation

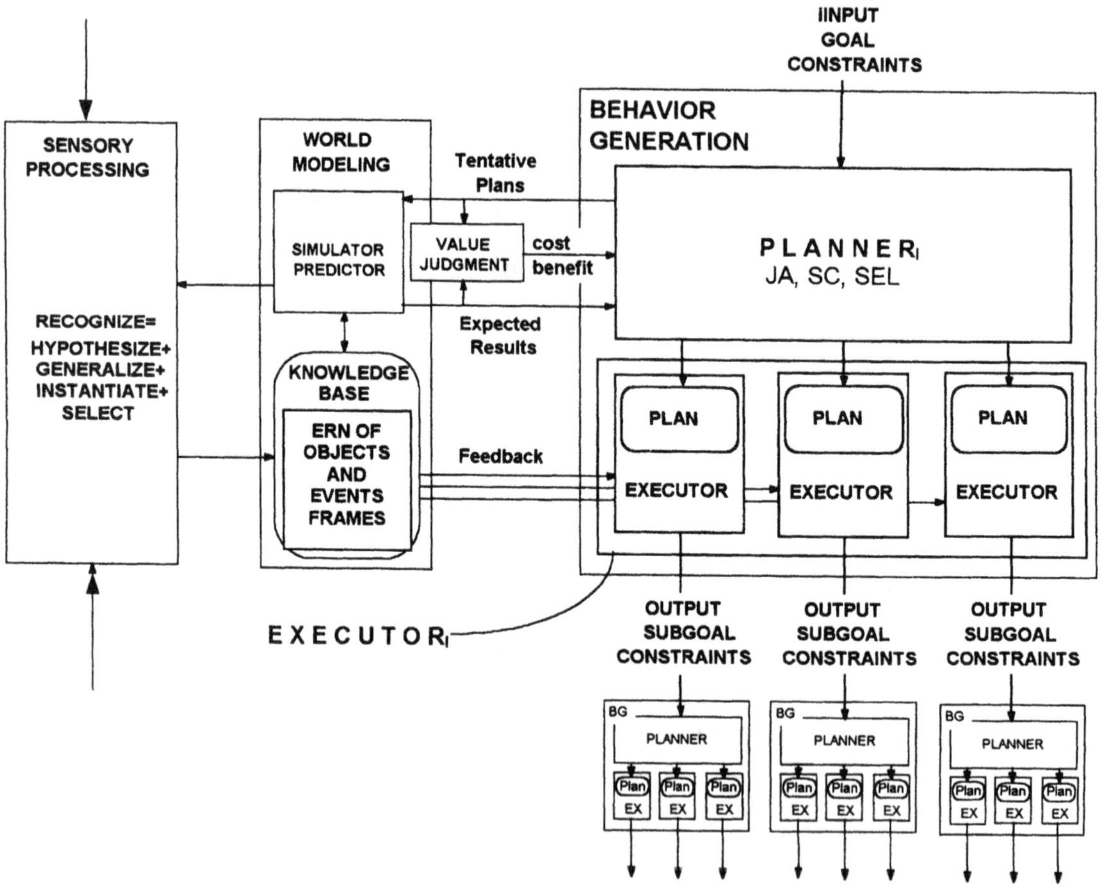

Figure 7.27. Interface of BG with WM for the imaginary case with branching "3" from level to level. Task command inputs are decomposed into plans that become subtasks for subordinate BG units. In the WM-module the "simulator" part is emphasized for better understanding of the process of functioning.

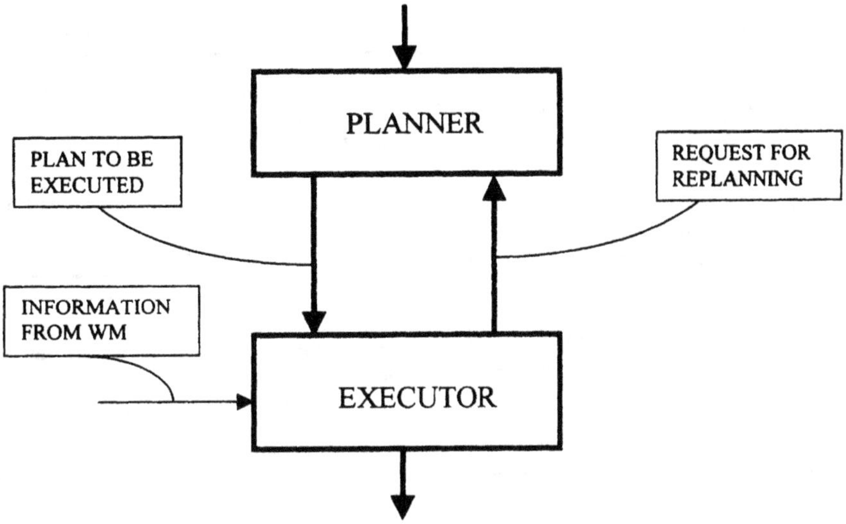

Figure 7.28. The EXECUTOR-PLANNER feedback is demonstrated, which is a command for REPLANNING. If the real trajectory differs from the PLAN substantially, i.e., leaves the boundaries of the planning envelope, replanning is requested. Otherwise, the process of replanning is performed regularly according to the accepted schedule.

can be skipped. All schedules are evaluated and compared in PS, after which one set of plans is selected as the best plan. For this plan the functions of "output," "action," and "input commands" are prepared for further communication within a smaller horizon (for the $(i-1)$th level of resolution) and then submitted for execution (Fig. 7.28).

Tentative Teams of JA

The JA subsystem is based on a mechanism that forms combinations among the available actuators and tentatively distributes their share in support of the overall motion. Most intelligent systems have redundancy, which allows for a great many possible combinations. The total job, J_T, should be a sum of the actuators jobs J_i ($i = 1, 2, \ldots, n$, where n is the total number of available actuators) with different relative contribution determined by the share coefficient s_i as in the equation of job distribution (Fig. 7.29):

$$J_T = s_1 J_1 + s_2 J_2 + \cdots + s_n J_n, \qquad s_1 + s_2 + \cdots + s_n = 1. \qquad (7.23)$$

Eventually job assignment provides an example of spatial planning (the same way we talk about space in which an assignment is performed). Spatial planning aims toward obtaining the best combination of the agents (virtual actuators) that are intended to collectively perform a job. This combination of agents is obtained by forming plausible hypotheses and testing them using the model of the system (in simulating the motion of the system).

Spatial planning starts with selection of a subgoal based on the schedule submitted from above. Then plausible hypotheses are formed for the job assignments. These hypotheses are evaluated and ranked. Information is to be organized as shown in Fig. 7.29*b*.

Job assigning (or job distribution) starts with a target point selected according to the plan submitted from above. It consists in forming hypotheses of job assignment. In

(a)

Job Assignment module
Name: JA(#)
Function:
Read input buffers
Compute conditions
Find match in table
Compute function
Compute output
Wait for trigger, then Go to Read

Input = (Task Name, Goal, Object, Parameters, Mode, Command ID)
State = {s0, s1,...,sN}
World Knowledge = (availability of agents, error conditions, estimated time to completion of jobs, position of objects, state of completion of task, state of workplace)

Mode - {Manual, Automatic, Mixed}
Conditions = (f(Input + World Knowledge) + State + Mode)
Compute function = compute the alternatives of the set of agents, and rank them, given the conditions
Compute output = place the set of Job Assignment Alternatives in output

(b)

Job Assignment Module

Input Buffers

Input Command from EX(i,j,)
Input Command from Operator
World Knowledge(i,j)

Input Condition	**Output Assignment**
Condition_1	Assignment_1
Condition_2	Assignment_2
Condition_3	Assignment_3
....	....
Condition_N	Assignment_N

Output Buffers

List of Job Assignments Hypotheses
Status of Job Assignment computation
Query of World Model for World Knowledge(i,j)

Figure 7.29. Job assignment subsubmodules for spatial planning.

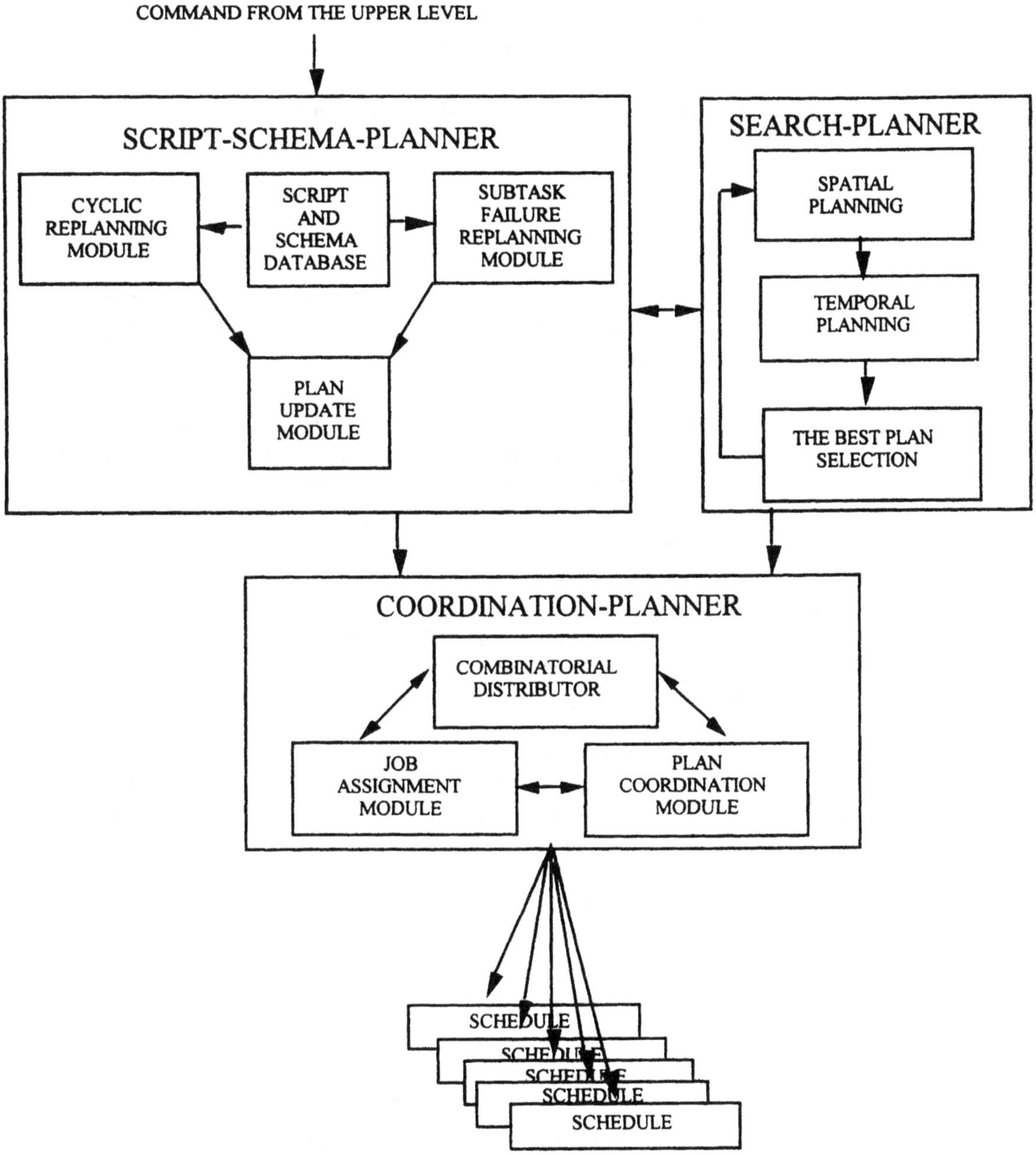

Figure 7.30. A variety of methods by which the PLANNER can be implemented.

order to do this, it must produce a number of tentative decompositions of the corrected plan (both before and after compensation) into the sequence of commands and then verify the results of planning at the next resolution level. The decomposition is done such that the total sequence is resolved as a cooperative effort developed by several agents (subsystems of the adjacent level of higher resolution). After the higher levels of resolution are through with their BG process, the PLANNER aggregates the results and decides whether they satisfy the constraints. If not, it starts the process of replanning.

In applications, synthetic PLANNERs seem to be beneficial. In the system shown in Figure 7.30 three concepts are put together:

1. Using "schemata" of rational activities taken from experience.
2. Search, as described above.
3. Set of coordination rules.

Instead of storing the analytical model, one can store numerous prior experiences and utilize them in forward testing.[16]

The algorithms of search by forward testing of the model provide the opportunity to eliminate any critical theoretical and computational problems. The following important benefits can be considered:

1. The need for implementation of an inverse system for planning is removed.
2. The difficulties associated with the variational techniques are avoided (in most cases these difficulties preclude optimum planning).
3. The computations are simplified because approximations can be introduced instead of searching for precise solutions.

In using forward search, the first approximation, which must be used, is the conversion of a typically continuous time system into its discrete time equivalent. Since nearly all practical problems involve sampled data systems, this is not actually a restriction. By selecting an appropriate sampling period, it is possible to vary the structure of the discrete time equivalents of the continuous system and thereby change some of their properties. The effect of this approximation on the computed optimum state-space trajectory and its inverse will be that the desired function and the output due to the computed inputs will agree only at sampling instants. This approach requires that the invertible mappings among the states and the labels of tasks be explicated.

We do not differentiate between the combinatorial search in the space of tasks and the combinatorial search for the motion trajectory. We believe these processes are different only in the level of discretization and are unified under the label "searching for the output trajectory."

SC Searches for Best Schedule

Evolution of simulated systems functioning in time generates the schedules that are submitted subsequently to the agents at the higher levels of resolution. Scheduling is performed by searching for the minimum-cost trajectory of motion in the state-space. The cost of all simulated schedules is computed by the submodule VJ. The search for the alternative schedules is performed via combinatorial construction (synthesis) of the output trajectories of motion at the selected level of discretization. This construction is equivalent to the process of simulating the system's motion under hypothetical conditions. As a result of this search, a set of two to three near-minimum-cost trajectories of the output motion are obtained.

After this, the set of near-minimum-cost output trajectories is inverted using the dynamic model of the system. As a result we receive a set of control schedules that allows

[16]SCRIPTS and SCHEMATA databases are obtained from prior experience. The technique of building these databases is outlined in two separate reports "Planning" and "Learning" (see Chapters 8 and 10).

us to check the convenience of the control to be applied. These schedules are simplified, linearized, and discretized according to the specifics of the system.

Certainly all the planning processes at all the levels of resolution cannot be initiated simultaneously. Later a "lead time" will be introduced and demonstrated. The essence of JOB ASSIGNER/scheduler functioning will become clear from the following description.

The state-space for the future searching of the best motion trajectory is bound from above and from below at each level of resolution by the scope of attention (any area beyond the scope of attention won't be required for searching) and by the smallest distinguishable element of the space (determined by the scale, granularity).

Synthesis of the state-space trajectory results with a time-tagged spatial curve that can be easily recomputed into the time axis for each subsystem of interest (scheduling). The upper bound of space is translated into the planning horizon (in time). When the output trajectory and the input command sequences are found, schedules are obtained for both the inputs and the outputs. They must be made consistent with the list of precedence conditions which is an essential list for any particular environment. If one or more conditions from this list are violated, the constraints are introduced into the state-space of the search, and the search is repeated ("replanning").

PS Makes Final Choice or Initiates Correction

Regular updating of models and correction of parameters are required; since the state-space is discretized, parameters contain a substantial stochastic component (possible error), and the tree of search is frequently pruned to reduce complexity of computations. After two to three near-minimum-cost schedules are selected, the coefficients should be varied in the "equation of job distribution" (7.23). This allows for refinement of the optimum trajectory. The refined command schedule, together with the output expected trajectory, is submitted to the output as the best plan, which includes (1) a set of time schedules (of feedforward control sequences of commands) for the selected set of actuators (virtual actuators) and (2) a set of corresponding anticipated output trajectories.

Functioning of the PLANNER Submodule

Two additional capabilities should be added that use prior experience of planning and coordinating parallel plans. After the command arrives from the upper level, the submodule performs the following operation (see Fig. 7.31):

1. Browsing in the library of solutions (e.g., SCRIPTS and SCHEMATA databases are presumed);

If the results of browsing are unsatisfactory, a search in the state-space is initiated:

2. Search in the state-space (by the SEARCH-PLANNER).

Since the desirable trajectory of motion is found at a level of an intelligent system, it is considered a plan at this level. On the other hand, given all uncertainties of the initial information, newly arriving sensor data, and intervals of discretization (inaccuracies inflicted by resolutions), this plan is actually no more than a fuzzy suggestion for the adjacent level with higher resolution of the subspace where the next search is to be conducted. This trajectory of motion is regarded as a "pipe" or a "strip," and it is a

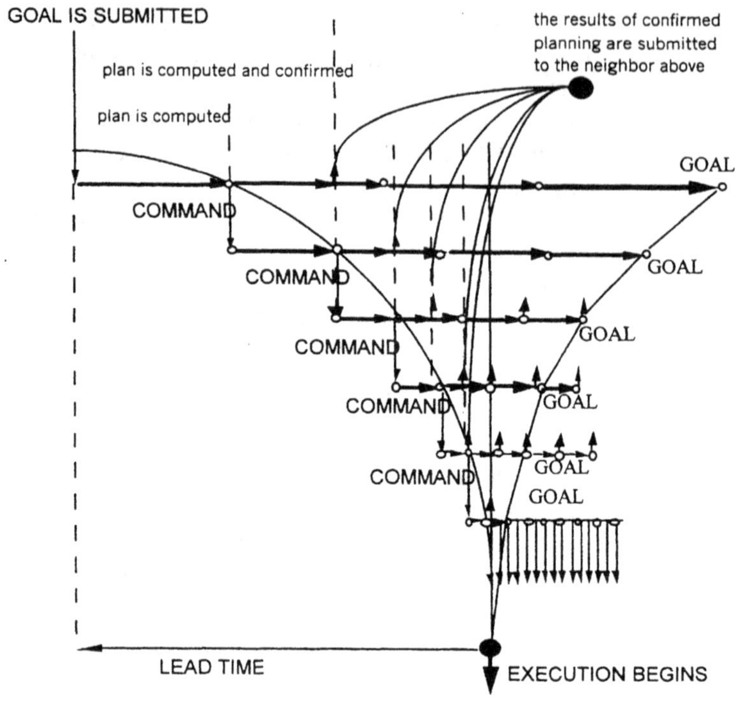

Figure 7.31. Evaluation of the lead-time required for planning.

narrow enough space for the subsequent search. If the motion is performed within this strip, it is already a "satisficing" solution. This is why we refer to this plan as the "strip of satisficing motion."

The boundaries of this strip are the constraints. If the real motion is executed within these constraints, no replanning would be required.

3. Coordination of the schedules, which entails the results of spatial planning.

The results of browsing are based on the existing solutions menu. If the appropriate solution does not exist, or has a low goodness, the search is initiated. The results of the search are accepted if they have a higher goodness-of-fit value. In this case they are also stored in the solutions library. The selected planning solution is submitted to the next level for exploration, where a similar sequence of activities is performed. The results from the next level confirm or reject, the results of the planning at the level under consideration. Usually, the vocabulary of the lower level is richer and contains information about the agents' performances. The initial stage of coordination is performed by selecting the best possible alternative of the plan. The final stage of coordination occurs during the stage of execution.

Lead-Time Diagram

The PLANNER receives the plan from the upper level (after its decomposition by the JA of the upper level) and determines the trajectory of desirable motion in the state-space. This means that the PLANNER has some cost-function and is capable of searching for the best alternative motion to be executed. Each level of higher resolution is supposed to initiate its planning operation only after the upper level has completed these procedures.

The starting points of the planning processes at all levels of resolution are illustrated in Figure 7.31. As this figure shows, the process has two waves: top-down and bottom-up. The search is done by constructing combinations of possible strings with the consecutive comparison of the results of this synthesis. Since other planners at a level are doing the same thing as part of their functioning, the search procedure must be done in cooperation with other planners. All planners exchange the current information about the intermediate results of planning among themselves. Of course parallel processing is assumed.

Replanning

From the lead-time diagram, we can see that the process of planning at the ith level is completed after the plan is confirmed at the level below. While the level below is operating, a concurrent process of planning ahead can start if the fraction of rejected plans is not too high. The confirmation is submitted to the upper level. If there is no confirmation, then a diagnosis is performed to assess the processes of browsing or search in the state-space.

Replanning is done after corrections are made. The replanning can be done in a smaller subset of the state-space. Therefore it is expected to be faster than the initial planning. Even if the initial plan was accepted, the search for the better alternatives should continue. The search in the state-space admits the best of available alternatives. The reduced envelope of the state-space will speed up the search, so the best solution might still be missing because of our desire to narrow the envelope as much as possible, and the process of replanning must be run for all zones of the alternatives.

Planning as a Time-Related Process

The planning at the upper (low-resolution) levels ends well before the actual motion starts. This is explained by the fact that all levels must complete their planning processes before any motion starts. The time delay of real-time motion is equal to

$$\sum t_d = \sum t_{\mathrm{pl}.i}, \qquad i = 1, 2, \ldots, n,$$

where $t_{\mathrm{pl},i}$ is the planning time at the ith level of the BG hierarchy and n is the number of levels in the hierarchy of control.

Just as the upper-level planning is completed, the next level starts its process of planning. Completion of the lowest-level planning initiates the real-time control, which generates control assignments to the lowest-end controller with the maximum frequency available. At each control level, new planning is initiated. This may begin immediately after the previous planning is completed. New planning may also be initiated if a warning indicates that the real motion at the level below is not executed as expected, and the boundaries of the "strip of satisficing motion," or the "strip of reliable planning," were crossed.

The time span between the two curves in Figure 7.31 characterizes the period of uncertainty in the level of control. The flow of regularly arriving information about motion execution is absorbed by the system only as it is learning. Otherwise, the planning is regular and starts right as the previous planning is completed, and any replanning occurs only if an alarm has been sounded. The curve of the planning horizon determines the time tag predicted by the planning algorithm. The upper (lowest-resolution) level of the NIST-RCS hierarchy has the largest horizon; the low end controller has the value of horizon determined by the frequency of control commands:

$$\text{horizon} = \frac{A}{\text{frequency}},$$

where A is a number depending on the domain of application.

7.6 EXECUTOR: ITS STRUCTURE AND FUNCTIONING

7.6.1 Processing the Results of Planning

The planner works on assignment. Its function is to distribute the tasks among the virtual actuators (job assignment, or spatial planning) and schedule them (temporal planning). This gives the most efficient functioning of the system. The system searches among all possible alternatives by using models from the WM. However, these models become obsolete very quickly. Even if nothing changes, the models may have errors because of erroneous discretizations, generalizations, or assumptions.

The models and the discretizations may also be flawed because of generalizations and a limited resolution. Then the real situation (or "virtual situation" at intermediate levels) will depart from the expected one.

The results of spatial and temporal planning (job assignments and schedules) are formulated in terms of outputs. Task decomposition generates the tasks. Each task is described in terms of the desired result and its relation to other tasks. If the machine is the triplet actuators–world–sensors, the results of planning will describe what should happen in the world rather than what should be the input of the actuator. The triplet actuators–world–sensors can of course be replaced by a single plant (which is considered a "virtual plant"). The plant input is computed to handle two needs: (1) to determine the input to the virtual actuator at the level of consideration, and (2) to compensate for the imperfectness of the model at hand.

Inverse Plan

The input to the virtual actuator is satisfied by inverting the output of plant. This is a JA procedure that uses either the analytical inverse expression P^{-1} or operates a forward search if the system is represented in a noninvertible form. If the projected input was not previously submitted with the desired output of the plan, the plan should be inverted.

Figure 7.32 shows the feedforward controller (FFC) submodule. In this figure, FBC takes place of the planner (as described in Section 7.5). The decomposed and inverted trajectory of motion is received as a result of the joint computations JA+SC. The inverse and decomposition computations are sent to the plant model for simulation and to the plant, or next level (higher level) of resolution for execution.

Error Compensation

The feedback mechanism serves to compensate for model imperfections. The actual motion is compared with the expected results. The expected results are part of the plan, or more accurately, are part of the simulation results obtained from WM after simulation (see Fig. 7.32). Since the world model is constantly updated, the plant representation at the moment of "execution" might differ from the plant representation during the

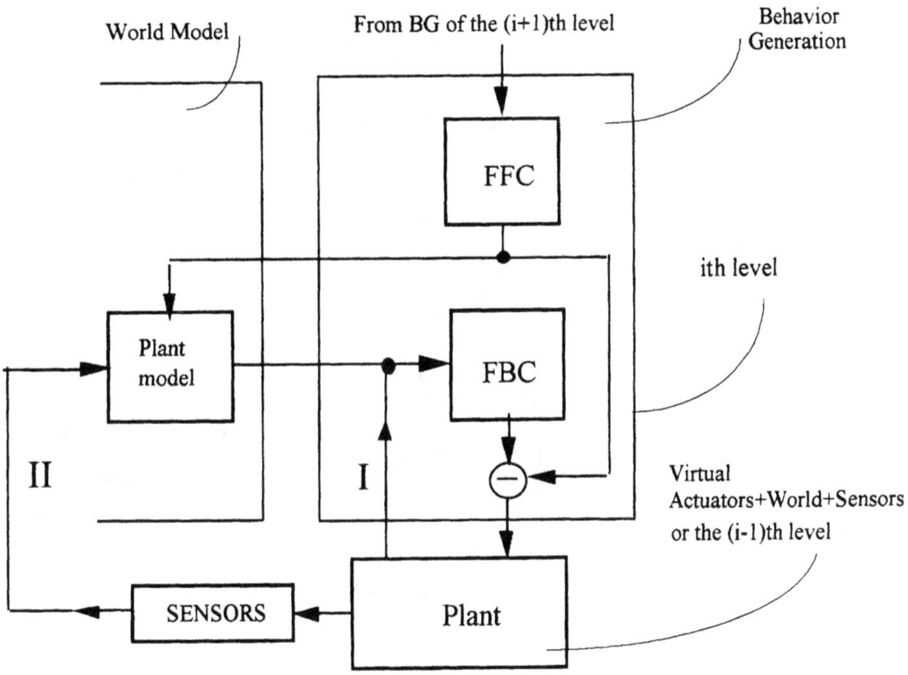

Figure 7.32. General structure of EXECUTOR.

process of planning. This is why one might be willing to simulate the output motion using the updated model of plant and compare the results of simulation with the actual motion. The simplest possible comparison is shown in Figure 7.32. The information on the state arrives through one of two possible paths: directly from the plant (I) or from the plant model which is constantly updated via loop II. Both loops I and II are subsystem channels for state and output estimation (in the reality, the NIST-RCS levels are equipped with subsystems of "recursive estimation").

The difference between the plan and the actual values of the state variables ("error") is computed within the submodule FBC as feedback compensation and added or subtracted from the plan (the output of FCC, the feedforward controller). The compensation depends on the nature of the system and the level of resolution at which this compensation is performed.

We can see that EXECUTOR must adjust the planning results to the reality of system functioning *online*. Feedback compensation allows it to correct the control command quickly, even in advance. No off-line activities can reduce the deviations that occur in reality. The EXECUTOR's function is to deal with unavoidable predicaments. One possible solution is to employ the concept of "prediction." In this way an error factor may be built into the forecast. Often this can be performed with high reliability.

The EXECUTOR translates the PLAN into a sequence of commands, which is submitted to the actuators, or the *virtual* actuators, for the level under consideration, and introduces a compensation command. Its role of reactive online control is typical for conventional control theory, since there are preselected control rules.

The EXECUTOR works from a menu of control rules that are applicable in a wide number of cases. This menu presents the most desirable compensation processes for the system at hand. The choice of compensation arrives from the planner must recognize

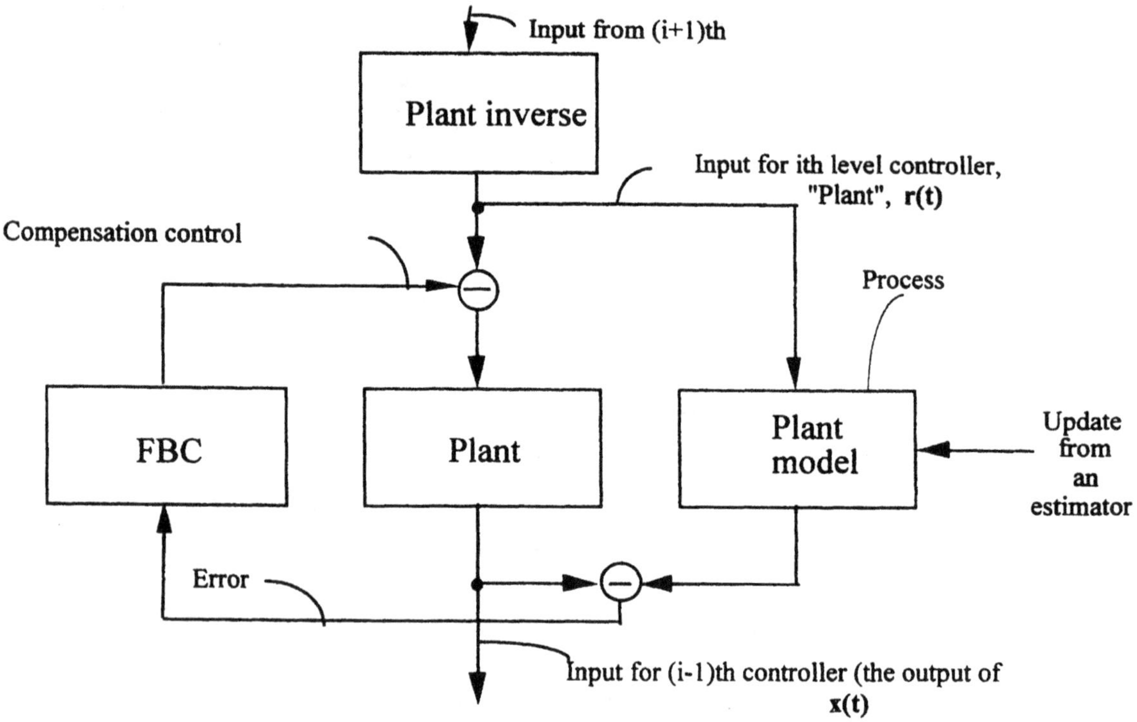

Figure 7.33. Interpretation of the PLANNER and the EXECUTOR in terms of control theory.

the required compensation rule in accordance with the plan and the real conditions of the world. The process of compensation requires proper distribution among the agents, although the complexity is not as high as in the case of feedforward plan development.

Figure 7.32 can be better understood if we compare it with a representation that is common in the area of control theory shown in Figure 7.33. The differences are due to our allowances for factors beyond control theory, such as the aging of the model and its quality, which is evident by its adequacy and its ability to represent the system well.

The EXECUTOR also depends on information contained in the TASK FRAME. For this reason the EXECUTOR acts on the errors of planning detected as undesirable deviations by the sensor. When the deviation exceeds some predetermined boundaries, the EXECUTOR informs the PLANNER that a replanning procedure must be performed. Finally, to introduce compensation in a timely fashion, the EXECUTOR is equipped with a built-in predictor of any deviation development process. This reduces the amount of final errors in ELF functioning.

7.6.2 Structure of EXECUTOR

The function of the EXECUTOR is to provide for online support of the operation system by introducing corrections into schedules. The inner structure of the EXECUTOR is built according to the account of its functioning described in Figures 7.32 and 7.33. This structure is shown in Figure 7.34. It is presumed in the diagram that the PLANNER has submitted both the set of required output trajectories and the set of recommended input schedules (input commands). Current results of the process simulation have arrived from the WM.

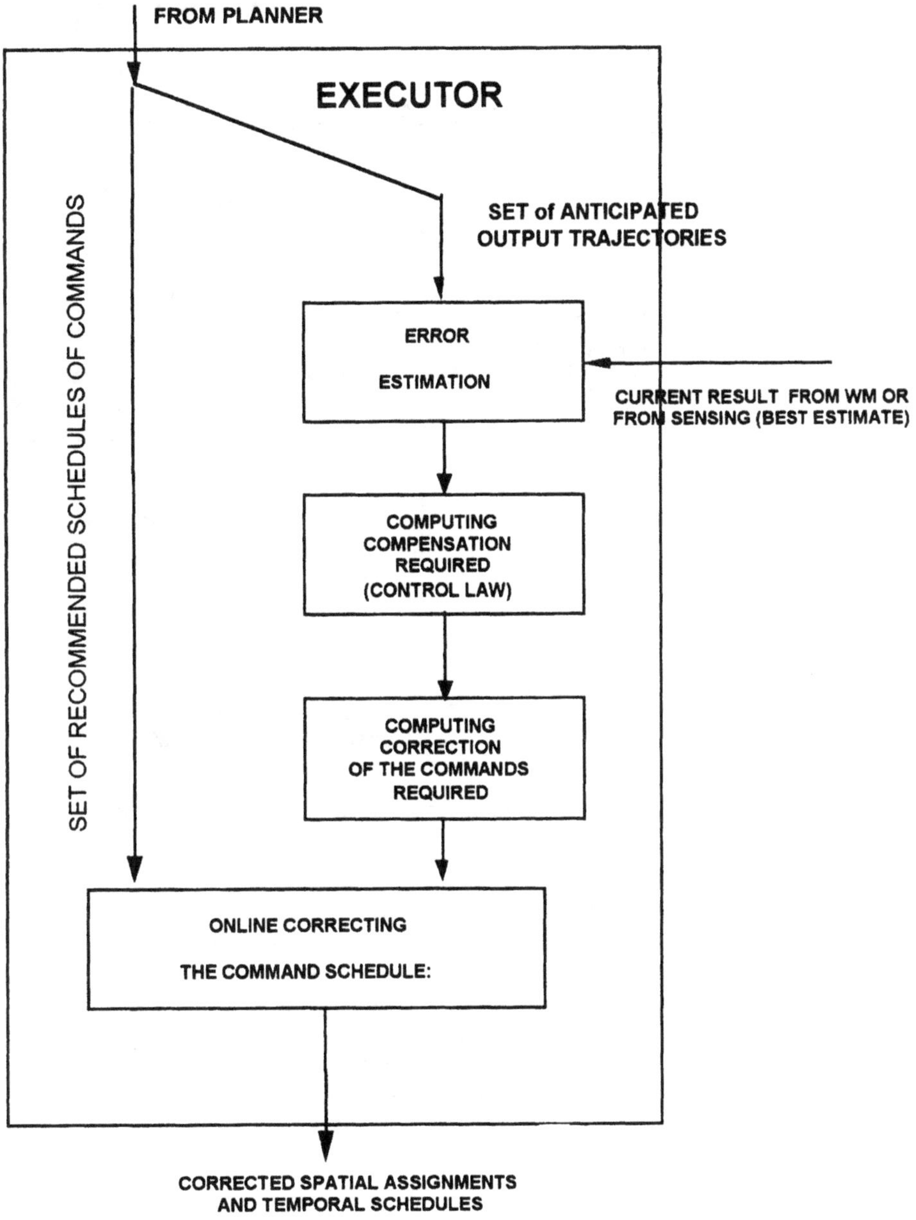

Figure 7.34. Inner structure of the EXECUTOR, type 1.

The anticipated output trajectory then is compared with the current functioning, and the difference is estimated (as an error). The error may in fact not be just the difference between the anticipated trajectory and the current measurements; the submodule of error estimation can include various types of prediction.

Based on a menu, a control rule is chosen, and a correction to the predicted feedforward control command is computed. This correction is used to supplement the control commands that arrive from the planner together with the anticipated output trajectories, or it can be computed from the anticipated output trajectories (the plan) by using the in-

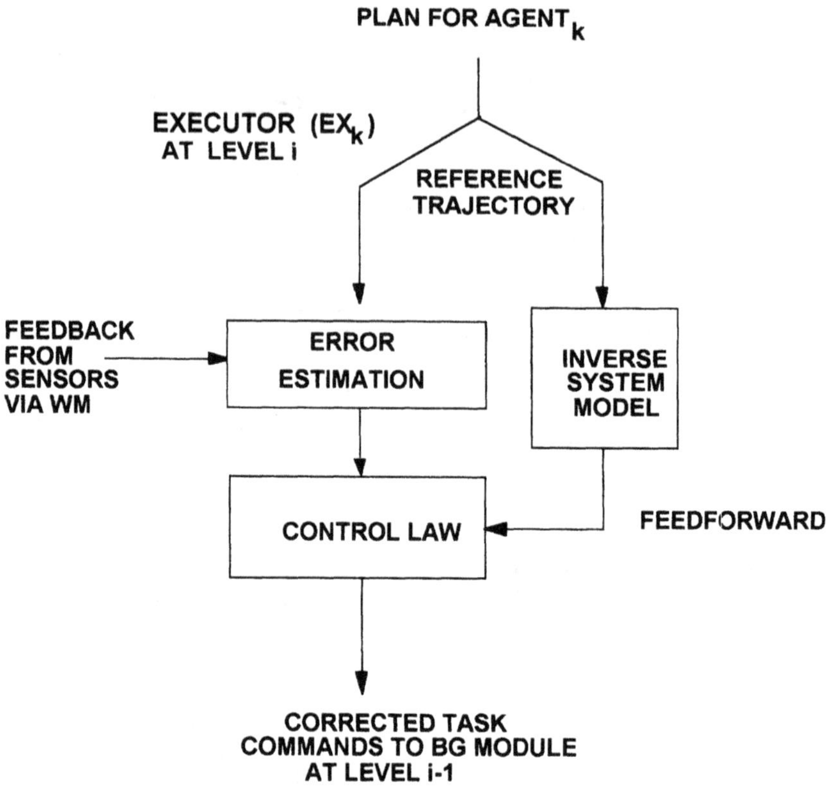

Figure 7.35. Inner structure of the EXECUTOR, type 2.

verse submodule (shown in Fig. 7.35). Among the present types of EXECUTORS, those based on multiresolutional variable structure controllers and multiresolutional Kalman filters have a good prospect for future industrial applications of NIST-RCS systems. Another type of EXECUTOR with an inverse submodule is shown in Figure 7.35. Here the input commands are not obtained from the PLANNER. They are restored in the EXECUTOR by the inverse operator. In this case both the error estimate and the control sequence arrive to the "control rule" submodule where the corrected control sequence is computed.

7.6.3 Operations of the EXECUTOR

Now we are ready to discuss the functioning of the EXECUTOR in more detail and the way its subsystems operate.

Error Estimation

The anticipated output trajectory (the plan) is compared with the actual result of sensing. There are many techniques of estimation, and under different circumstances some of them are more preferable than others. The sequence of the estimation process is illustrated in Figure 7.36. The ESTIMATOR gives an *anticipated error* at its output.

The key element of the ESTIMATOR is the submodule EXTRAPOLATOR. In all cases of estimation the measurements are compared with their expectations based on a

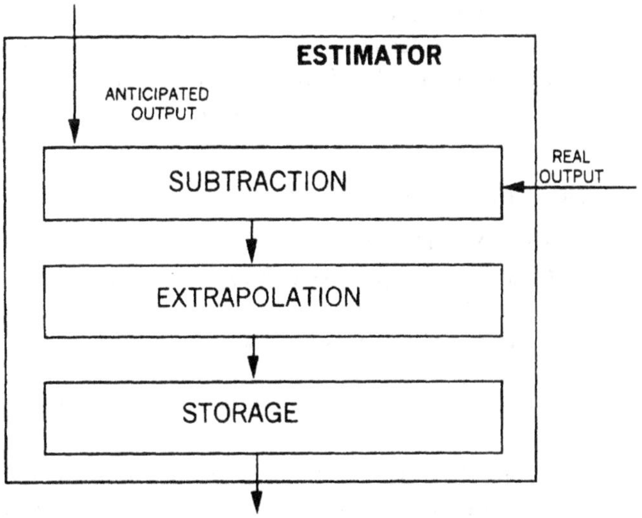

Figure 7.36. Structure of the ESTIMATOR.

particular theory of process development or an algorithm of prediction. The expectation of the computed error depends on the actual system. The required correction is predicted by the control command according to statistics of measurement and assumptions about particular properties of the system: its ability to respond to particular statistical characteristics of the error function. Since all estimation results can be assigned some degree of belief, the predicted control command can only be trusted to a particular degree too. It has been demonstrated that in a hierarchical system, the total prediction is more adequate than if estimation is done at a single resolution level. Various schemes of recursive estimation seem to be appropriate for many resolution levels.

Computing the Required Compensation

Depending on our anticipation of error in our system working in a particular environment, a particular control rule can be suggested. This rule would allow us to compute adequately the value of control compensation. (There are good reasons to expect that in the case of RCS, the *equivalent control rule for variable structure* makes some preferences among the alternatives.) Variable structure controllers are applicable in several ways: for example, as "bang-bang" controllers for minimum-time systems and as fuzzy controllers with a hierarchy of fuzzy convergence to the assignment. The estimator is not involved in computing the complete control command. It computes only part of it, the error compensation.

On-line Correction of Command Schedules

On-line correction is a sub-subsystem of a subsystem of the EXECUTOR (see Figs. 7.35 and 7.36), and it is introduced to the flow of command signals sent to the system of actuators, or (virtual actuators). As a result the final version of the schedules arrives at the corresponding control system right on time, at the moment when it is ready to be applied.

This operation can be considered similar to the reactive response of the intelligent agent, as presented in the literature on the reactive control of robotic intelligent agents with preassigned "behavior." Thus the EXECUTOR along with the ACTUATOR at its output can be considered the analogue of the intelligent agent.

7.6.4 EXECUTOR as a TASK GENERATOR

The EXECUTOR is a submodule for accommodating plans for particular actuators. Central to this process is error compensation. As the real-time control is performed, the results of planning are corrected by the feedback compensation (FBC) mechanism which is similar to the inverse operator and takes the form

$$\langle \text{COMPENSATION} \rangle \longleftarrow F(\langle \text{OUTPUT}_\text{p} \rangle - \langle \text{OUTPUT}_\text{r} \rangle),$$

where F is a feedback operator that represents the accepted strategy of feedback and $\langle \text{OUTPUT}_\text{p} \rangle$ and $\langle \text{OUTPUT}_\text{r} \rangle$ are planned and real (observed) output trajectories correspondingly.

Selection of a precise strategy F of feedback control depends on the context and the requirements of the user. There are many laws of control that reflect different strategies. If the processes are fairly slow, proportional feedback can be good enough. It can be improved by adding an operation of prediction, which helps in general. Often a rough prediction can be made by straightforward differentiation. Numerous experiences have shown that even for dynamic processes a simple derivative of the process is sufficient; in this way a proportional-derivative strategy is obtained (PD control). To account for cumulative error, designers add to the feedback signal a component that is proportional to the integral of the error. Integration enables proportional-integral-derivative feedback, which is the well-known PID controller. If the designer is concerned about the statistical probability of error, the strategy of control should include a computation of the square of this error.

Finally the equation of the real-time operation of the ith level is written as

$$\text{TASK} \longleftarrow C_{\text{rt}}(\langle \text{INPUT} \rangle,\ \langle \text{COMPENSATION} \rangle),$$

where C_{rt} is the operator of real-time control, and it is used, for example, in the direct addition

$$C_{\text{rt}}(\langle \text{INPUT} \rangle,\ \langle \text{COMPENSATION} \rangle) = \langle \text{INPUT} \rangle + \langle \text{COMPENSATION} \rangle.$$

Now the ASSIGNMENT GENERATION can be considered a convolution

$$\text{ASSIGNMENT GENERATION} \longleftarrow [\text{S}^*\text{P}^{-1}]^* F,$$

which is the knowledge inverse operation (K_{IO}) for the behavior generator.

This process refers to the system as a whole. However, it is the same for every level of resolution. Obviously other subsystems can be attached to this picture. We focus upon a small section only because the BG module can be used to determine the whole functioning of the system; other subsystems provide support for the BG module. The fundamentals of productivity, efficiency, and effectiveness are contained in this module. Its connection to the system connects it to all other levels.

7.7 CONCLUSIONS: INTEGRATING BG IN THE INTELLIGENT SYSTEM

- In a nested multiresolutional system (e.g., NIST-RCS) all levels of resolution are considered and its elementary loop of functioning (ELF) system, which includes the world, sensors, perception (sensory processing), the world model, behavior generator, and actuators.
- The behavior of multiresolutional intelligent systems is generated as a result of the joint functioning effort of the perception, world modeling, value judgment, and behavior generation modules which together perform the planning and control of a system.
- Behavior generation, as a subsystem of the NIST-RCS hierarchy, is based on the functioning of its inner submodules responsible for planning and for control of the resolution levels.
- The planner consists of three sub-submodules for job assignment, scheduling, and selection of the best plan. The planner acts as a feedforward controller: It computes sets of commands that are valid if the knowledge of the system is adequate. Planning starts by assigning a job; this assignment outlines the alternative spatial distributions of the job among the participating agents. The scheduler searches for the minimum cost plan for all alternatives of the job assignment. The selector chooses the best schedule among the alternatives.
- The EXECUTOR consists of the sub-submodules for inverse computation, error estimation, and error compensation. The executor provides the feedback control system.
- In all cases the process of developing the NIST-RCS architecture should start with its behavior generation module, which includes the planning and executor since they affect most of the performance of the system.
- All subsystems of the ELF support the behavior-generating subsystem. The development and functioning of other subsystems (including the sensors and actuators) is determined by the way the behavior-generating structure is designed and operates.
- Behavior can be generated only in a loop (elementary loop of functioning). Analysis of the systems starts with identification of existing loops. This is a nontrivial procedure, since finding a loop can be done only by hypothesizing: by constructing them tentatively and exploring whether the loop functioning does not contradict any experimental data. In turn the hypotheses depend on our familiarity with the systems of signs that can be identified within the environment of interest.
- The NIST-RCS system has as many loops as it has levels of resolution. The number of levels of resolution depends on the complexity of the computations in addition to symbol grounding, namely finding a correspondence between the hypotheses and the existing experimental data.
- All loops should be treated separately. This means that their "languages" should be developed and utilized according to specifications written for the loop. Translation of a language at one level of resolution into a language at an adjacent level is considered a behavior generation process. Each loop has to be checked for satisfactory conditions of inclusion (related to their knowledge flows).
- All loops have their own flow of knowledge (information). The rules of consistency and the laws of conservation are formulated and checked for each loop. All loops

have to be checked for consistency by verifying conditions of nesting top-down and bottom-up in each module of the ELF.

- At the core of the behavior generation subsystem is the concept of recursive nested hierarchies. All NIST-RCS results and applications are determined by the concept and implicit formalisms of recursive nested hierarchies. The concept of a hierarchy in the RCS is different from the mathematical concept of a decision tree because the nodes at each level are interrelated and form a network.
- The formalisms of recursive hierarchies are based on a fundamental decision-making procedure (determined by the triplet of intelligence with its components: generalization, focusing of attention, combinatorial search) applied to the data (knowledge) structures typical for the area of application.
- The algorithms of recursive hierarchies are problem invariant. The result of NIST-RCS design are determined by the context information to which the formalisms of recursive hierarchies are applied.
- All integrated complex systems allow for effective modeling by the formalisms of recursive hierarchies.
- The property of nesting requires that the hierarchies satisfy additional rules of inclusion.

REFERENCES

[1] J. Albus and A. Meystel, Behavior generation in intelligent systems, *NIST Technical Report*, Gaithersburg, MD, 1996.

[2] J. Albus, Outline for a theory of intelligence, *IEEE Transactions on Systems, Man, and Cybernetics*, Vol. 21, No. 3, May/June 1991, pp. 473–509.

[3] A. Meystel, Nested hierarchical control, in *An Introduction to Intelligent and Autonomous Control*, (P. Antsaklis and K. Passino eds.), Kluwer Academic Publishers, Boston, 1992, pp. 129–162.

[4] M. Minsky, A framework for representing knowledge, in *The Psychology of Computer Vision*, (P. Winston, ed.), McGraw-Hill, New York, 1975, pp. 211–277.

[5] S. Grossberg (ed.), *Studies of Mind and Brain*, Reidel Publishing Co., Holland, 1982.

[6] S. Grossberg, *Neural Networks and Natural Intelligence*, Bradford Books, MIT Press, 1988.

[7] J. J. Hopfield, Neural Networks and physical systems with emergent collective computational abilities, *Proceedings National Academy of Sciences*, V. 79, 1982, pp. 2554–2558.

[8] M. Arbib, *The Metaphorical Brain*, New York, Wiley, 1972.

[9] J. S. Albus, *Brains, Behavior, and Robotics*, BYTE/McGraw-Hill, Peterborough, NH, 1981.

[10] S. N. Salthe, *Evolving Hierarchical Systems*, Columbia University Press, New York, 1985.

[11] B. Ziegler, *Object-Oriented Simulation with Hierarchical Modular Models: Intelligent Agents and Endomorphic Systems*, Academic Press, San Diego, CA, 1990.

[12] J. S. Albus, Mechanisms of planning and problem solving in the brain, *Math. Biosciences*, Vol. 45, 1979, pp. 247–293.

[13] Y. Maximov and A. Meystel, Optimum architectures for multiresolutional control, *Proc. IEEE Conference on Aerospace Systems*, May 25–27, Westlake Village, CA, 1993.

[14] G. Saridis and C. S. G. Lee, Approximation of optimal control for trainable manipulators, *IEEE Trans. on Systems, Man, and Cybernetics*, Vol. 8, No. 3, 1979, pp. 152–159.

[15] J. Albus and A. Meystel, A reference model architecture for design and implementation of semiotic control in large and complex systems, in *Architecture of Semiotic Modeling and Situation Analysis*, (J. Albus, A. Meystel, D. Pospelov, and T. Reader, eds.), Proc. of the IEEE Workshop at the 10th ISIC, Monterey, CA, 1995, pp. 33–45.

[16] M. Ito, *The Cerebellum and Neuronal Control*, Chapter 10, Raven Press, NY, 1984.

[17] G. N. Saridis, Foundations of the theory of intelligent controls, *IEEE Workshop on Intelligent Control*, 1985.

[18] J. S. Albus, H. G. McCain, and R. Lumina, NASA/NBS standard reference model for telerobot control system architecture (NASREM), *National Institute of Standards and Technology, Technical Report 1235*, Gaithersburg, MD, 1989.

[19] A. J. Barbera, J. S. Albus, M. L. Fitzgerald, and L. S. Haynes, RCS: The NBS real-time control system, *Proceedings Robots 8 Conference and Exposition*, Detroit, MI, June 1984.

[20] Honeywell, Inc., Intelligent task automation interim technical report II-4, Dec 1987.

[21] J. E. Laird, A. Newell, and P. Rosenbloom, SOAR: an architecture for general intelligence, *Artificial Intelligence*, Vol. 33, 1987, pp. 1–64.

[22] D. Smith and M. Broadwell, Plan coordination in support of expert systems integration, *Knowledge-Based Planning Workshop Proceedings*, Austin, TX, December 1987.

[23] C. Isik and A. Meystel, Knowledge based control for an intelligent mobile autonomous system, *Proc. 1st IEEE Conf. on Artificial Intelligence Applications*, Denver, CO, Dec. 2–4, 1984.

[24] R. Chavez and A. Meystel, Structure of intelligence for an autonomous vehicle, *Proc. IEEE Int'l Conf. in Robotics*, Atlanta, GA, March 13–15, 1984, pp. 584–593.

[25] J. Lowerie et al., Autonomous land vehicle, *Annual Report, ETL-0413*, Martin Marietta Denver Aerospace, July 1986.

[26] A. Sathi and M. Fox, Constraint-directed negotiation of resource reallocations, *CMU-RI-TR-89-12*, Carnegie Mellon Robotics Institute Technical Report, March 1989.

[27] V. B. Brooks, *The Neural Basis of Motor Control*, Oxford University Press, 1986.

[28] J. R. Greenwood, G. Stachnick, and H. S. Kaye, A procedural reasoning system for army maneuver planning, *Knowledge-Based Planning Workshop Proceedings*, Austin, TX, December 1987.

[29] A. Meystel, M. Montgomery, and D. Gaw, Navigation algorithm for a nested hierarchical system of robot path planning among polyhedral obstacles, *Proc. IEEE Int'l Conf. on Robotics and Automation*, Raleigh, NC, 1987, pp. 1616–1622.

[30] C. Isik and A. Meystel, Pilot level of a hierarchical controller for an unmanned mobile robot, *IEEE J. of Robotics & Automation*, Vol. 4, No. 3, 1988, pp. 244–255.

[31] A. Meystel, *Autonomous Mobile Robots: Vehicles with Cognitive Control*, World Scientific Publ., Singapore, 1991.

[32] D. Apelian and A. Meystel, Knowledge based control of material processing: challenges and opportunities for the third millennium, in *Metallurgical Processes for the Year 2000* (H. Y. Sohn, and E. S. Geskin, eds.), TMS, Warrendale, PA, 1989.

[33] C. Corson, A. Meystel, F. Otsu, and S. Uzzaman, Semiotic Multiresolutional analysis of a power plant, ibid. in *Architectures for Semiotic Modeling and Situation Analysis in Large Complex Systems*, Proc. of the 1995 ISIC Workshop, Monterey, CA, 1995, pp. 401–405.

[34] B. Hayes-Roth, A blackboard architecture for control, *Artificial Intelligence*, 1985, pp. 252–321.

[35] E. D. Sacerdoti, *A Structure for Plans and Behavior*, Elsevier, New York, 1977.

[36] R. C. Schank and R. P. Abelson, *Scripts Plans Goals and Understanding*, Hillsdale, NJ, Lawrence Erlbaum Associates, 1977.

[37] D. M. Lyons and M. A. Arbib, Formal model of distributed computation sensory based robot control, *IEEE Journal of Robotics and Automation in Review*, 1988.

[38] D. W. Payton, Internalized plans: a representation for action resources, *Robotics and Autonomous Systems*, Vol. 6, 1990, pp. 89–103.

[39] L. A. Suchman, *Plans and Situated Actions: The Problem of Human-Machine Communication*, Cambridge University Press, Cambridge, UK, 1987, p. 203.

[40] H. Samet, The quadtree and related hierarchical data structures, *Computer Surveys*, 1984, pp. 16–22.

[41] P. Kinerva, *Sparse Distributed Memory*, MIT Press, Cambridge, 1988.

[42] J. S. Albus, A new approach to manipulator control: the cerebellar model articulation controller (CMAC), *Transactions ASME*, September 1975.

[43] J. S. Albus, Data storage in the cerebellar model articulation controller (CMAC), *Transactions ASME*, September 1975.

[44] E. Kent and J. S. Albus, Servoed world models as interfaces between robot control systems and sensory data, *Robotica*, Vol. 2, 1984, pp. 17–25.

[45] E. D. Dickmanns and T. H. Christians, Relative 3D-state estimation for autonomous visual guidance of road vehicles, *Intelligent Autonomous System 2* (IAS-2), Amsterdam, December 1989, pp. 11–14.

[46] E. Sacerdoti, *A Structure for Plans and Behavior*, Elsevier, 1977.

[47] D. Wilkins, *Practical Planning: Extending the Classical AI Planning Paradigm*, Morgan Kaufmann, 1988, pp. 46–50.

[48] M. Arbib, *Theories of Abstract Automata*, Prentice Hall, Englewood Cliffs, NJ, 1969.

[49] J. A. Simpson, R. J. Hocken, and J. S. Albus, The automated manufacturing research facility of the National Bureau of Standards, *Journal of Manufacturing Systems*, Vol. 1, No. 1, 1983.

[50] J. S. Albus, C. McLean, A. J. Barbera, and M. L. Fitzgerald, An architecture for real-time sensory-interactive control of robots in a manufacturing environment, *4th IFAC/IFIP Symposium on Information Control Problems in Manufacturing Technology*, Gaithersburg, MD, October 1982.

[51] J. S. Albus, System description and design architecture for multiple autonomous undersea vehicles, *National Institute of Standards and Technology, Technical Report 1251*, Gaithersburg, MD, September, 1988.

[52] J. Albus and A. Meystel, A reference model architecture for design and implementation of semiotic control in large and complex systems, in *Architectures for Semiotic Modeling and Situation Analysis in Large Complex Systems* (J. Albus, A. Meystel, D. Pospelov, and T. Reader, eds.), Proc. of 1995 ISIC Workshop, Monterey, CA, 1995, pp. 33–45.

[53] A. Meystel, Multiscale Systems and Controllers, *Proceedings of the IEEE/IFAC Joint Symposium on Computer-Aided Control System Design*, Tucson, AZ, 1993, pp. 13–26.

[54] H. M. Huang, R. Hira, and R. Quintero, Submarine maneuvering system demonstration based on the NIST real-time control system reference model, *Proc. of the 8th IEEE Int'l Symposium on Intelligent Control*, Chicago, IL, 1993.

[55] F. Proctor and J. Michaloski, Enhanced machine controller architecture overview, *NISTIR 5331*, NIST, Gaithersburg, MD, December 1993.

[56] K. Stouffer, J. Michaloski, R. Russel, and F. Proctor, ADACS - an automated system for part finishing, *NISTIR 5171*, NIST, Gaithersburg, MD, April 1993.

[57] R. Quintero and A. J. Barbera, A software template approach to building complex large-scale intelligent control systems, *Proc. of the 8th IEEE Int'l Symposium on Intelligent Control*, Chicago, IL, 1993.

[58] R. Bostelman, J. Albus, N. Dagalakis, and A. Jacoff, Applications of the NIST robocrane, *Proc. of the 5th International Symposium on Robotics and Manufacturing*, Maui, HI, 1994.

[59] Mary Beth Ruskai et al. (eds.), *Wavelets and Their Applications*, Jones and Bartlett Publishers, Boston, 1992.

[60] B. Mandelbrot, *The Fractal Geometry of Nature*, W. H. Freeman and Co., 1977.

[61] R. Wilensky, *Planning and Understanding*, Addison-Wesley, 1983.

[62] E. Dynkin and A. Yushkevich, *Controlled Markov Processes*, Springer, 1975.

[63] J.-C. Latombe, *Robot Motion Planning*, Kluwer, 1991.

[64] A. Meystel and S. Uzzaman, Planning via search in the input-output space, *Proc. of the 8th IEEE Int'l Symposium on Intelligent Control*, Chicago, IL, 1993.

[65] D. Wilkins, *Practical Planning: Extending the Classical AI Planning Paradigm*, Morgan Kaufmann, 1988, pp. 25–39.

[66] M. Georgeff et al, Reasoning and planning in dynamic domains: an experiment with a mobile robot, *Technical Note 380*, SRI, Menlo Park, CA, 1986.

[67] T. Dean and M. Wegman, *Planning and Control*, Morgan Kaufmann, 1991, p. 206.

[68] L. Kaebling, An architecture for intelligent reactive systems, in *Reasoning about Actions and Plans*, (M. Georgeff and A. Lansky, eds.), Proc. of the 1986 Workshop, Morgan Kaufmann, 1986, pp. 395–410.

[69] Learning algorithms generating multigranular hierarchies, in *Mathematical Hierarchies and Biology*, (B. Mirkin, F. R. McMorris, F. S. Roberts, and A. Rzhetsky, eds.), DIMACS Series in Discrete Mathematics, Vol. 37, American Mathematical Society, 1997.

[70] J. Mendel, *Lessons in Digital Estimation Theory*, Prentice Hall, 1987.

[71] L. Kelmar, Manipulator servo level world modeling, *NIST Technical Note 1258*, NIST, Gaithersburg, MD, 1989.

PROBLEMS

7.1. Make a list of control laws that you know from a course on control theory. Present them in an analytical form (if feasible). For each of these control laws demonstrate the form of tasks that will be obtained as a result of task decomposition entailed by a particular control law. Use the task frame as suggested in pages 261–263.

7.2. Consider a case of a mobile robot (within a particular scene under a particular assignment) is to conduct the maneuver of obstacle avoidance. Consider a case when the robot learns more about the obstacle while executing the maneuver. Perform task decomposition for this maneuver. How did you make this decomposition? Where did you get the list of tasks that should be consecutively executed?

7.3. For each of the tasks obtained as a result of solving Problem 7.2, construct the complete task frame. Use the task frame as suggested in pages 261–263.

7.4. Construct a two-level RCS Architecture for a mobile robot utilized for solving Problem 7.2. Demonstrate control level at low resolution, control level at high resolution. Demonstrate the virtual loop consisting of the control level at low resolution and controlled system that is "perceived" by the control system of low resolution (see Subsubsection 7.2.1).

7.5. Use the mobile robot from Problem 7.2 with task decomposition from Problem 7.3 and architecture from Problem 7.4. Plan and execute several increments of the obstacle avoidance maneuver following the reasoning demonstrated in Figure 7.7. Consider PLANNING and EXECUTION as separate activities.

7.6. Represent generation of the obstacle avoidance maneuver in a nested way as described in Figures 7.9 and 7.10. Construct a temporal diagram of activities following Figure 7.14. Demonstrate commands, actions, changes in the world, sensor signals.

7.7. Construct a temporal diagram of joint activities of actuators of steering and propulsion following the diagram Figure 7.16. Construct equations in the form similar to (7.3) through (7.8).

7.8. Describe the process of planning at each step using the model for description shown in Figure 7.17.

7.9. Consider the mobile robot from Problem 7.2 with task decomposition from Problem 7.3 and architecture from Problem 7.4. Plan and execute several increments of the obstacle avoidance maneuver by using multiresolutional hierarchical planning described in Subsubsection 7.4.5.

7.10. For the robot from Problem 7.9, construct the generic processing node following Figure 7.19.

7.11. For the robot and planning process from Problem 7.9, describe the process of planning in detail following Figure 7.26. Demonstrate the steps of the formula (7.17).

7.12. Construct the lead time diagram for the case of Problem 7.9. (Follow Fig. 7.32).

7.13. For the robot from Problem 7.9 and the plan developed as a result of solving this Problem, describe the process of execution in detail following Figure 7.33.

7.14. Demonstrate all algorithms of EXECUTOR that should be used in solving Problem 7.13.

CHAPTER 8

MULTIRESOLUTIONAL PLANNING: A SKETCH OF THE THEORY

8.1 INTRODUCTION TO PLANNING

We will treat any *behavior* as *motion*. Thus we will treat any behavior generation as a motion control process. Each task planning is just another component of the more general paradigm in which the control of motion starts with a contemplation of the desired output motion for the system.

The process of contemplation presumes searching for alternatives of the desired motion in a way that the cost-function is minimized while all constraints are satisfied. These alternatives of motion are constructed and tested in the *imagination* of the intelligent system. The imagination for computer-based intelligence amounts to modeling and simulation.

Intelligent systems exercise multiresolutional processes of control. Therefore planning in an MR system takes a multiresolutional form, too. In this chapter, multiresolutional (multiscale) planning is introduced as a nested system of the state-space search processes, and this provides exceptionally efficient computational algorithms. It is demonstrated that search in the state space allows for efficient planning/control procedures. The results of two-level nested hierarchical planning are given as a representative simulation example. It is demonstrate that multiscale searching achieves results similar to those obtained by human strategies of planning.

8.1.1 Overview of the Key Results in the Area of Planning

The planning of the motion trajectory for any system is a natural step to its functioning. Planning, in terms of the more general paradigm of *motion control*, consists of two components: motion planning and motion execution. Motion planning is presumed to be performed in advance, while motion execution is to be done online. In this book, we demonstrate that at each level of resolution, there is a part of "in advance" control, and a part of "online" control.

Starting in the 1960s, the interest in motion planning began to grow and spread to various domains of application. Motion planning is a central topic both in artificial in-

FINDING ROUGH PLAN SHOULD PRECEDE CONTROL

telligence and in intelligent control. Intelligent control is at the intersection of three related scientific fields: operations research, artificial intelligence, and control theory. OR emerged in the 1940s and instigated the analysis of queues, graph theory, and methods of optimization. The study of planning was extended to AI research in the 1960s, on corresponding processes of human cognition, and the first explicit effort on planning algorithms was related to human thought simulation [1]. A. Newell, H. A. Simon, N. Nilsson, and other prominent researchers in AI developed the fundamental algorithms for the existing results in robot motion planning. Traditionally "dynamics" had no place in AI planning; it was considered a prerogative of control theory.

In the 1970s, K.-S. Fu, G. Saridis, and their students initiated research of control systems that incorporated planning and recognition [2, 3]. This combination eventually brought to fruition the new discipline of intelligent control [4]. By definition, then, intelligent control incorporates OR, AI, and control theory. One interesting problem about it is the analysis of planning for robotics. Usually, the mainstream specialists in control theory developed the so-called reference trajectory, which referred to the input to control systems; this was considered a plan and computed as a part of the design process. However, the traditional control specialists considered everything not related to stability analysis to be extraneous.

STRIPS [5, 6], and A* [7] became classical elements of planning in AI and robotics. The subsequent development in the area of robot path planning branched enormously:

1. Problems of representation became critical.
2. Combinatorics of tasks and dynamics of systems became intertwined.
3. Planning developed hierarchically.
4. Computational complexity became the real limitation for the development of theories.

Some of the milestones in the evolution of motion and path planning are as follows:

- In 1966 J. E. Doran and D. Michie applied a graph-theoretic mechanism for path planning [8].
- In 1968 W. E. Howden introduced the "sofa problem" in treating the geometric problem of motion planning [9].
- In 1968 the A* algorithm was introduced by P. Hart, N. Nilsson, and B. Rafael [7].
- In 1971 the use of strips was presented by R. E. Fikes, P. Hart, and N. Nilsson [5, 6].
- In 1979 the concept of search was attempted for dealing with obstacles by T. Lozano-Perez and M. A. Wesley [10].
- In 1979 J. Albus introduced the method of task decomposition for hierarchical systems; later it became a part of the NIST-RCS methodology with nested planning processes at all levels of the control hierarchy [11].
- In 1981 T. Lozano-Perez applied "configuration space" to manipulator's planning [12].
- In 1982, a first textbook appeared edited by M. Brady, J. M. Hollerbach, T. L. Johnson, T. Lozano-Perez, and M. T. Mason [18].
- In 1983 M. Julliere, L. Marce, and H. Place developed a mobile robot with planning via tessellated space [13].
- In 1984 R. Chavez and A. Meystel [14] introduced the concept of searching in a space of various (nonuniform) traversability.

- In 1985 J. E. Hopcroft, D. A. Joseph, and S. H. Whitesides analyzed the geometry of robotic arm movement in 2D bounded regions [15].
- In 1986 A. Meystel demonstrated that the most efficient (least computational complexity) functioning of a multilevel learning/control systems with a search for use in planning can be provided by a proper choice of a ratio of lower-level/higher-level resolution [16]. This concept of a multiresolutional planning/control hierarchy provided theoretical support for the hierarchical architecture of intelligent system.
- In 1985 to 1987 M. Arbib's school of control via "schemata" came up with numerous schemes of "reactive" behavior. This gave birth to a multiplicity of robot control concepts that explored and exercised reactive behavior generation.
- During the period 1985 to 1995 many researchers associated problems of robotic motion planning with short-term (local) reactive behavior (e.g., obstacle avoidance). Nevertheless, the interest in the search within the state-space was perpetuated.
- In the meantime the primary focus of robotics shifted to systems that do not require planning (robotics with "situated behavior"). Thus the interest in planning diminished (R. Brooks, MIT, R. Arkin, Georgia Tech) and the curiosity of researchers shifted toward emerging phenomena in robots with rudimentary intelligence.
- In 1991 a comprehensive text was published by J. C. Latombe [17] that outlines most of the theories and experiences approved by the practice in a variety of applications.

Ten years of research and experience (1982 to 1991) has helped clarify an important maxim: The process of robot motion planning is equivalent to searching within the state-space for a trajectory of motion leading to a goal.

8.1.2 Definitions Related to Planning

The following definitions of a plan are typical for most dictionaries (e.g., *Merriam-Webster's Collegiate Dictionary*):

DEFINITION OF PLAN (NOUN)

Plan (as a noun) **1**: a drawing, or diagram drawn on a plane: as (a) a top or horizontal view of an object, (b) a large-scale map of a small area; **2**: (a) a method for achieving an end; (b) an often customary method of doing something; (c) a procedure: a detailed formulation of a program of action; (d) goal, aim; **3**: an orderly arrangement of parts of an overall design or objective; **4**: a detailed program (as for payment or the provision of some service).

DEFINITION OF PLAN (VERB)

Plan (as a verb) **1**: to arrange the parts of: design; **2**: to devise or project the realization or achievement of <a program>; **3**: to have in mind: intend: to make plans.

These definitions can be applied to any module of the intelligent system that receives a goal, retrieves relevant knowledge in the world model, and creates strings of tasks for the actuators (or the similar modules below in the hierarchy, which the latter consider their "goals").

The professional definitions for specialists, involved in the planning and control of robots, are recommended by the NIST research report on behavior generation in intelligent systems (including robots) [19]. The system of behavior generation is supposed to be constructed out of BG modules; each module equipped with a planner. Within this paradigm we have the definition:

DEFINITION OF PLAN GENERATED BY INTELLIGENT SYSTEM

Plan is the set of schedules for the group of agents that are supposed to execute these schedules as a cooperative effort and accomplish the required job (achieve the goal) as a result of this effort. To find this set of schedules different combinations of agents are tested, and different schedules are explored.

Plan is also defined as:

- The course of events determined within BG module and reproduced in the world to achieve the goal in the desirable fashion, or
- A description of the set of behaviors that leads to the goal in a desirable fashion (this description is represented as a set of schedules), or
- A state-space trajectory that describes the behavior of system leading to the goal and provides satisfaction of constraints and conditions on some cost function, or cost functional (whose conditions might include having the value of this cost function/cost functional within some interval, maximizing, or minimizing it).

Thus the plan controls the system. Its main components are the final state, which is achieved at the end of each planning interval, and a string of the intermediate states which are often supplemented by their time schedules and are bounded by a value of admissible error. The plan consists of task space/time decompositions such that subtasks are distributed in space and time. It may be represented as a PERT chart, a Gant diagram, a state transition graph, a set of schedules complemented by the account of resources required (e.g., a bill of materials, tools and manpower requirements, delivery schedules, and cost estimates). Each plan is characterized by its goal, a *time horizon*, a *set of agents (performers)*, and *its envelope*.

OPTIMAL PLAN

The *optimal plan* leads to the goal achievement while minimizing (or maximizing) a particular cost function or a cost functional. The optimal plan is found (synthesized) only by the comparison of all alternative feasible (admissible) plans.

ADMISSIBLE PLAN

Admissible plans are all meaningful plans that can be built within the specified constraints.

SATISFICING PLAN

The *satisficing plan* is one of the admissible plans within a narrowed set of constraints. It is one of the state-space trajectories that is constructed within the desirable boundaries specified by a customer. In other words, this is a sufficient, satisfactory, but not necessarily the best plan.

SPATIAL PLAN

The *spatial plan* is a state-space trajectory (in the enhanced state-space which includes inputs, outputs, and states of the system). The state-space trajectory is represented at the output of the planning sub-module as the result of selection of agents and jobs assigned to them, their responsibilities and criteria of their performance.

TEMPORAL PLAN

The *temporal plan* is a schedule of actions and/or events. See *Schedule*.

DEFINITION OF PLANNING

Planning is the design of a course of events determined within the BG module, the design of a desirable state-space trajectory, or the design of the feedforward control function. It is the intended future for the system. A planning is performed on the assumption that we know the agents (of the adjacent higher level of resolution) that will cooperate in the further delineation of the plan. Each assumption corresponds to one alternative plan. Another alternative has another assumption about performing agents and leads to another plan. The design of the desirable motion of the system entails that many supportive components of operation also should be planned: the algorithms of feedback compensation, inputs to the energy converters, the scope of sensing (focus of attention), and others.

PLANNING ENVELOPE

The *planning envelope* is a subset of the state-space that is considered during planning, together with a corresponding world model which is submitted to BG at the higher level of resolution for refinement.

PLANNING HORIZON

The *planning horizon* is the time interval within which a plan exists. The degree of belief for each future state of the plan falls off as time t grows large because the stochastic component of the operation affects the verifiability of the results. For some value of time in the future, the degree of belief is lower than the degree required for the decision-making process. This value of time is the maximum useful *planning horizon*.

PLANNING STRATEGY

The *planning strategy* is the set of rules that guides the algorithm of planning. It is the orientation toward receiving either the *optimal* or the *satisficing plan*.

REPLANNING

Replanning is the process of planning that is performed to assure that the chosen plan is still valid, or if the top-down and bottom-up processes of plan propagation did not converge. The need in replanning can emerge (1) if the initially selected version of plan distribution failed, (2) if the prescribed conditions of compensation fail to keep the process within the prescribed boundaries, (3) if the world model has changed, and (4) or if the goal has changed.

DEFINITION OF RESOURCES

Resources are material things and units of information usually taken into account in constructing the cost function; for example, time, energy, materials, remaining life-span of the system, the degree of fault-tolerance, and money.

DEFINITION OF RESOLUTION

Resolution is the property of the level of hierarchy which limits the distinguishability of details at that level; these details will emerge as objects at the levels of higher resolution. *Synonyms*: scale, granulation, coarseness.

DEFINITION OF SCHEDULE

Schedule is another term for the "temporal plan"; it is the description of the development of a process in time. A schedule is obtained by computing the state-space trajectory within the time domain. The schedule should give the start and the end events and provide for coordination, reduced queues, and elimination of the "bottlenecks." Schedule can be also defined as a job-time event-gram.

DEFINITION OF SCHEDULING

Scheduling is the process of outlining the temporal development of the motion trajectory.

In sum, the definitions above all point to the critical value of planning, which can be restated more comprehensively:

GENERAL DEFINITION OF PLANNING

Planning is a process of searching for appropriate future trajectories of motion leading to the goal. Searching is performed within the system of representation.

8.1.3 Planning as a Stage of Control

A general control diagram is shown in Figure 8.1. The "desired motion trajectory" ("reference trajectory") is applied at the input of the control diagram. As a result, the output

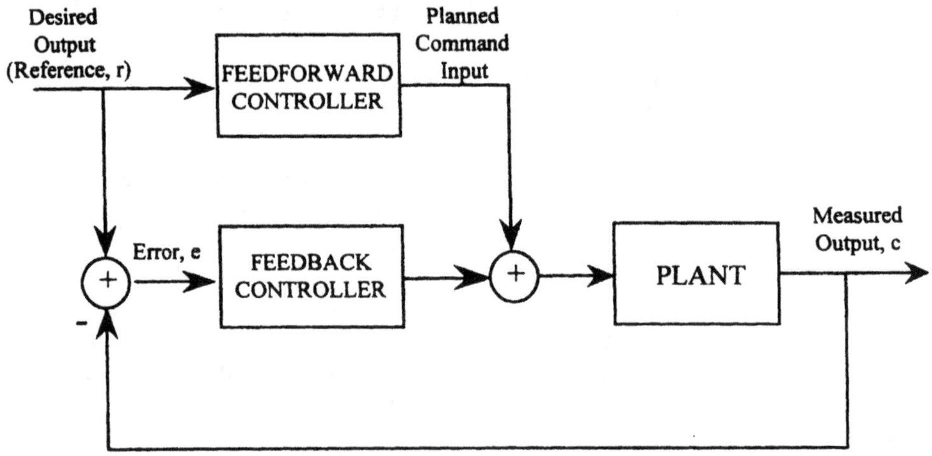

Figure 8.1. Combined feedforward/feedback control architecture.

motion is obtained. If the measured output differs from the desired motion, the difference between them enpowers the feedback controller to perform the compensation. It turns out that the feedforward role of the control systems is partially that of a planner.

Intuitively the reaction to error must occur after the error is detected. Therefore it can be expected to be relatively slow compared to the predictive correction that is available via the feedforward channel. Because feedforward exists outside the immediate scope of the feedback loop, the feedforward controller can inject a priori a particular bias into the operation of that loop. This bias may consist of a nominal command applied by an expert to effect the continuous operation of a machine in a factory, or it may be a linearizing or decoupling torque generation scheme for a robotic manipulator; in either case, the planned command input is produced according to an analysis of the system model in some form—mathematical, linguistic (or both)—in order to improve the performance of the overall system.

If the reference trajectory for the system has been synthesized, then it is necessary for the feedforward controller to be an inverted dynamical model of the plant. Then, the only task for the feedback would be to cancel the effects of the *unmodeled* dynamics and disturbances. The transfer function can be simplified to unity if the plant and feedforward controller are exact inverses of each other. In fact most of the existing results in the area are based on this assumption. The work of Brockett [20] is generally acknowledged as the first formal treatment of the problem of the inversion of multivariable linear time invariant systems and references [21–24] list extensions of those results to nonlinear, time-varying, and discrete-time cases. Using inverse for finding the control law was discussed and important results were obtained in [25].

The general approach to planning as a synthesis of feedforward commands is to implement an algorithmic procedure (for approximate inversion) to determine the nominal input function or trajectory that leads to an approximate tracking of the reference trajectory.

DEFINITION OF APPROXIMATE INVERSION

Approximate inversion is an algorithmic procedure for planning an input function in some admissible input set which, when applied to a plant, causes its outputs to follow a trajectory that minimizes the deviation of those outputs from a prescribed trajectory over some closed interval of time. An input, so determined, is referred to as an *approximate inverse* of the reference trajectory over the input set.

The full planning process is involved finding the reference trajectory. One way is to optimize the output motion of the system. Specifically, the optimal control of systems suggests using the calculus of variations and Pontryagin's principle which, in theory, will provide both the reference trajectories for the system and the inputs required to generate them. In practice, optimal solutions are hard to generate for all but the simplest problems. In many other cases, such as those involving systems with large, distributed parameters, and computer-based models, it is not possible to apply the classical theory at all.

Optimization is typically performed by using the tools of searching. Search should be performed within some envelope around the desired trajectory. Thus an envelope around the desired trajectory is submitted to the input as a primary assignment. This envelope contains the initial and the goal points. This envelope encloses the space for the subsequent search of the optimum trajectory.

8.2 EMERGING PROBLEMS IN PLANNING

8.2.1 Generic Problems of Behavior Generation

Robotics is an integrated domain that provides for (a) blending the goals of functioning and (b) testing the means of achieving these (i.e., blended) goals; it is a domain that requires close planning. In 1983 T. Lozano-Perez introduced the idea of search in a configuration space. Experience over the years in using this search has made it clear that an exhaustive search is computationally prohibitive if the configuration space is tessellated with the accuracy required for motion control. But the theory has made one important thing obvious: Planning is equivalent to synthesizing admissible alternatives and searching for the trajectories entailed by these alternatives. This development has helped us realize that planning must combine an exhaustive (or meaningfully thorough) search off-line as a part of the algorithm of off-line control.

It was in the early 1980s that engineers stopped talking about control of actions and introduced the more balanced term "behavior generation." Behavior generation became a codeword for the joint processes of testing the alternatives within the mechanism of planning (open loop, feedforward control) blended with an online sorting of the feedback alternatives for error compensation (closed-loop control, or execution).

Behavior generation alludes to many mechanisms of planning and execution. At the present time, these mechanisms still are not fully understood, and a general theory of planning does not exist. However, there is some merit in discussing a subset of problems in which the goal is determined as attainment of a particular state.

Behavior generation [19] can involve many different mechanisms of planning and execution. These mechanisms are not well known. We will discuss a subset of problems in which the goal is defined as an attainment of a particular state. Other types of problems can also be imagined: In chess, for example, the goal state is clear (to win), but this goal cannot be achieved by simply reaching a particular position in a space (even in a descriptive space). Most of the problems related to the theory of games, and thus with pursuit and evasion, are characterized by a similar predicament and are not discussed here.

8.2.2 Structural Sources of Problems

Any problem of planning is associated with the actual existence of the present state (PS), the actual or potential existence of the goal state (GS), and the knowledge of values for all or some states as far as a particular goal is concerned (KS). The cumulative cost of trajectories to a particular goal (or goals) can be deduced from this knowledge. On the other hand, the knowledge of costs for the many trajectories traversed in the past can be obtained, which is equivalent to knowing cumulative cost from the initial state IS to the goal state GS (from which the values of the states can be deduced).

In other words, any problem of planning contains two components: The first is to refine the goal (i.e., bring it to the higher resolution); the second is to determine the motion trajectory to this refined goal. These two parts can be performed together or separately. Frequently we treat them separately. Then they are formulated as follows: (1) given PS, GS, and KS (all paths) find the subset of KS with a minimum cost, with a preassigned cost, or with a cost in a particular interval; (2) given PS and GS from the lower-resolution level and KS (all paths) find the GS with a particular value.

In [26] two important issues are introduced for the area of planning: controllability and recognizability. The controllability issue arises when the number of controls is smaller than the number of independent parameters defining the robot's configuration. The recognizability issue occurs when there are errors in control and sensing: how well can the robot recognize goal achievement. Both issues can affect the computational complexity of motion planning. The set of controllability, recognizability, and complexity is especially important to the development of autonomous robots.

8.2.3 Representation and Planning

Planning in a Representation Space with a Given Goal

MENTAL REHEARSAL OF UPCOMING ACTIVITIES

The world is assumed to be judged by using its state-space (or the space of representation) which is interpreted as a vector space with a number of important properties. Any activity (motion) in the world (space of representation) can be characterized by a trajectory of motion along which the working point or present state (PS) is traversing this space from one point (initial state, IS) to one or many other states (goal states, GS). The goal states are given initially from the external source as a "goal region" or a "goal subspace" in which the goal state is not completely defined in a general case.

From the point of view of planning, the state-space does not differ from the configuration space. Indeed, the upcoming behavior is represented as a trajectory in the state-space (and/or configuration space). One of the stages of planning (often the initial one) is defining where exactly the GS is within the *goal region* (which was the goal state at the lower resolution). In many practical problems the designer must focus on planning procedures in which one or many GS remain unchanged through all periods of their functioning (before they are achieved). Traversing from IS to GS is associated with consumption time or some other commodity (cost). So a straightforward exhaustive search allows for exploring all possible alternatives.

Planning as a Reaction to Anticipated Future

Researchers in the area of reactive behavior have introduced a method of potential fields for producing comparatively sophisticated obstacle-avoiding schemes of motion. Reactive behavior is considered to be the antithesis of planning. But this is not so. Motion based on planning can be called reactive too. The difference is that in reactive behavior, robots usually react to the present situation. In a system with planning they react to the anticipated future.

ANTICIPATORY REACTIVE BEHAVIOR

Thus planning can be considered an anticipatory reactive behavior. The difference is in the fact that anticipation requires richer representation than the simple reactive behavior requires.

Types of Representation Available for Planning

All representation spaces are acquired from external reality by learning processes. Many types of learning are mentioned in the literature (supervised, unsupervised, reinforcement, dynamic, PAC, etc.). Before classifying a need in a particular method of learning and deciding how to learn, we like to figure out what exactly we should learn. It is important in planning procedures to find out whether the process of learning can be separated into two different learning processes: that of finding the world representation and that of finding the appropriate rules of action. These two kinds of learning are just two interrelated sides of the same core learning process. Thus, both learning processes should be conducted jointly.

All subsequent knowledge should be consigned to the representation space. If no GS is given, any pair of state representations should contain implicitly the rule of moving from one state to another. In this case, while learning, we inadvertently consider any second state as a provisional GS.

We call a representation "proper" if it is similar to the mathematical function and/or field description: At any point the derivative is available together with the value of the function; the derivative can be considered an action required to produce the change in the value of the function. We call a representation "goal oriented" if the value of the action at each given point describes not the best way of achieving an adjacent point but the best way of achieving the final goal. Both proper and goal-oriented representations can be transformed within each other.

Artifacts of Representation Space

The representation of the world can be characterized by the following artifacts:

- Existence of states with its boundaries determined by the resolution of each state (which is presented as a tessellatum, or an elementary unit of representation, the lowest possible bounds of attention).
- Characteristics of the tessellatum which is defined as an indistinguishability zone (the resolution of the space shows how far the "adjacent" tessellata, or states, are located from the "present state" (PS) tessellatum).
- Lists of coordinate values at a particular tessellatum in space and time.
- Lists of actions to be applied at a particular tessellatum in space and time in order to achieve a selected adjacent tessellatum in space and time.
- Existence of strings of states intermingled with the strings of actions required to receive next consecutive tessellata of these strings of states.
- Boundaries (the largest possible bounds of the space) and obstacles.
- Costs of traversing from a state to a state and through strings of states.

Often the states contain information that pertains to a part of the world that is beyond their ability to reach; this part is called the *environment*. The part of the world to be controlled is the system for which the planning is performed. We refer to it frequently as *self*. Thus part of the representation is related to self, including knowledge about actions which this self should undertake in order to traverse the environment.

It is seen from the list of artifacts that all knowledge is represented at a particular resolution. Thus the same reality can be represented at many resolutions and a multiresolutional representation is presumed. The system of representation is expected to be organized in a multiresolutional fashion. As a result there is a need to apply a number of special constraints and rules. The rules of inclusion (aggregation/decomposition) are especially important.

8.2.4 Classification of Planning Problems in Intelligent Systems

Geometric Models

This is the domain of practical problems. Among a wide variety of famous theoretical problems is the "sofa" problem which has evolved into "piano-movers" problem. A thorough survey is given in [27]. An interesting geometric model based on Snell's law is presented in [28].

Collision-Free Robot Path

Most of the FINDPATH algorithms of the 1980s were based on searching for a minimum path string of vertices within a so-called visibility graph (a graph comprising all vertices of the polygonal objects connected with visibility lines [29–32]).

Nonholonomic Path Planning

Mobile robots can be considered single-body devices (carlike robots) or comprised of several separate bodies (tractors towing several trailers sequentially connected). These robots are known to be nonholonomic; that is, they are subject to nonintegrable equality kinematic constraints involving velocity. The number of controls is smaller than the dimension of the configuration space. The range of possible controls has additional inequality constraints such as due to the mechanical stops in the steering mechanism of a tractor. It is demonstrated for the nonholonomic multibody robots that the controllability rank condition theorem is applicable even when there are inequality constraints on the velocity in addition to the equality constraints [33–34].

Uncertainty and Probabilistic Techniques for Path Planning

Most of the techniques for searching the minimum-cost paths on a graph are deterministic ones, and introduction of uncertainty became a new source of challenge [35–37]. An approach to motion planning with uncertainty for mobile robots is introduced in [38]. Given a model of the robot's environment, a sensory uncertainty field (SUF) is computed over the robot's configuration space. At every configuration, the SUF is an estimate of the distribution of possible errors in the sensed configuration, and it is computed by matching the data given by the robot sensors against the model. A planner uses the SUF to generate paths minimizing the expected errors. The SUF has been explored for the classical line-striping camera/laser range sensor.

Planning relies on information that becomes available to the sensors during execution so that the robot can correctly identify the states it traverses. A set of states is chosen, a motion command is associated with every state, and the state evolution is evaluated.

The interdependence of these tasks can be avoided by assuming the existence of landmark regions in the workspace; these are considered "islands of perfection," where the position sensing and motion control are accurate [39].

Planning in Unknown or Partially Known Environment

Planning in an unknown environment is a problem that defies our purpose to derive a search process from precise knowledge of an environment. Indeed, a map of a maze might be unknown, but a strategy of behavior in a maze does exist. All that is required is to have a "winning" strategy of actions to use in uncertain situations. There is an area of research oriented toward finding the most general rules of dealing with different types of environment [56–58].

Planning in Redundant Systems

Nonredundant systems have a unique trajectory of motion from a state to a state. A redundant system is defined as a system in which more than one trajectory of motion is available from one state to another. It can be demonstrated for many realistic system-environments that (1) they have a multiplicity of traversing trajectories from a IS to a GS, and (2) these trajectories can have different costs.

These systems contain many alternative space traversals. Redundancy occurs when the system is a stochastic one. The number of available alternatives gets high when we consider also the many goal tessellata of a particular level of resolution which are due to the condition of assigning a goal at a lower-resolution level in multiresolutional systems (e.g., NIST-RCS).

In nonredundant systems a problem like this does not exist in planning: Only one trajectory of motion is available. Since the trajectory of motion to be executed is a unique one, the problem is to determine this trajectory and to provide tracking of it by an appropriate control system. Many research results demonstrate that redundancy can be considered an important precondition (1) for the need of planning, and (2) for performing planning successfully [59–62].

Figure 8.2 shows realistically the situation of redundancy that occurs for most planning problems. There are many paths from two geographical points in the 3D space depicted in the figure. If the only requirement is minimum time, a comparison of several paths will determine the chosen path. However, by introducing additional preferences

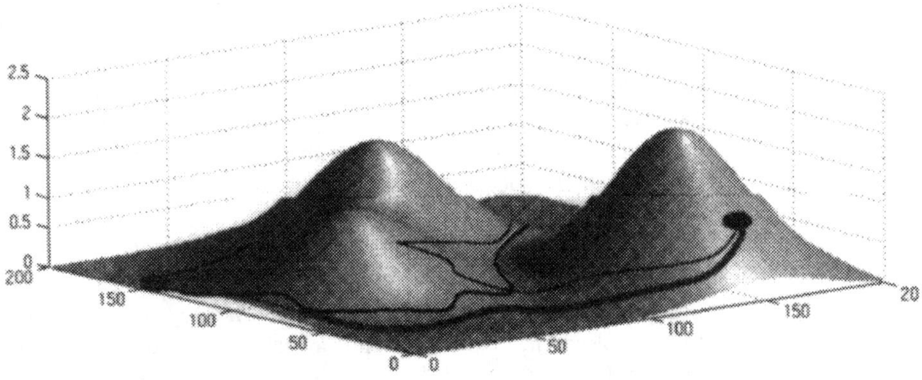

Figure 8.2. Multiple alternative plans.

and components of the cost-function, the redundancy can be effectively reduced and even eliminated.

8.3 PLANNING OF ACTIONS AND PLANNING OF STATES

8.3.1 Algorithms of Planning

Planning constructs the goal states and/or the preferable strings of states connecting the present state with the goal states. One of the successful techniques is associated with task decomposition [40]. Task decomposition is related to the consecutive refinement, namely consecutive increases of resolution for both actions and states. Planning involves both of these components (Fig. 8.3).

The first component of a planning algorithm is the translation of the goal state description from the language of low resolution to the level of high resolution. Frequently this means increasing the number of state-variables. In all cases it means increasing the scale of representation, which is the same as a reduction of the indistinguishability zone or of the tessellatum size associated with a particular variable.

The second component is the simulation of all available alternatives of motion from the initial state, IS, to one or several goal states, GS, for a selection of the best trajectory. Procedurally this simulation is performed as a search, that is, via combinatorial construction of all possible strings (groups). To make this combinatorial search for a desirable group more efficient, we reduce the space of searching by focusing attention; that is, we preselect the subset of the state-space for further searching.

Thus all planning algorithms consist of two components: a module for exploration of spatial distribution of the trajectory, and a module for exploration of the temporal distribution. No algorithm of planning is conceivable without these two components. The need in planning is determined by the multi-alternative character of reality. The process of planning can be made more efficient by using appropriate heuristics.

8.3.2 Visibility-Based Planning

The intelligent observer (IO) is introduced in [41] as a mobile robot that moves through an indoor environment while autonomously observing moving targets selected by a human operator. The robot carries one or more cameras that allow it to track the objects while at the same time sensing its own location. It interacts with a human user who issues task-level commands, such as indicating a target to track by clicking in a camera image. The user can be located far away from the observer, communicating with the robot over a network. As the IO performs its tasks, the system provides real-time visual feedback to the user. We have implemented a prototype of the IO which integrates basic versions of the four main components: localization, target tracking, motion planning, and robot control. We have performed initial experiments using this prototype that demonstrate the successful integration of these components and the utility of the overall system.

A particular problem of computing robot motion strategies is outlined in [42]. The task is to maintain visibility of a moving target in a cluttered workspace. Both motion constraints (as considered in standard motion planning) and visibility constraints (as considered in visual tracking) are taken in account. A minimum path criterion is applied. Predictability of the target is taken in account. For the predictable case, an algorithm that computes optimal numerical solutions has been developed. For the more

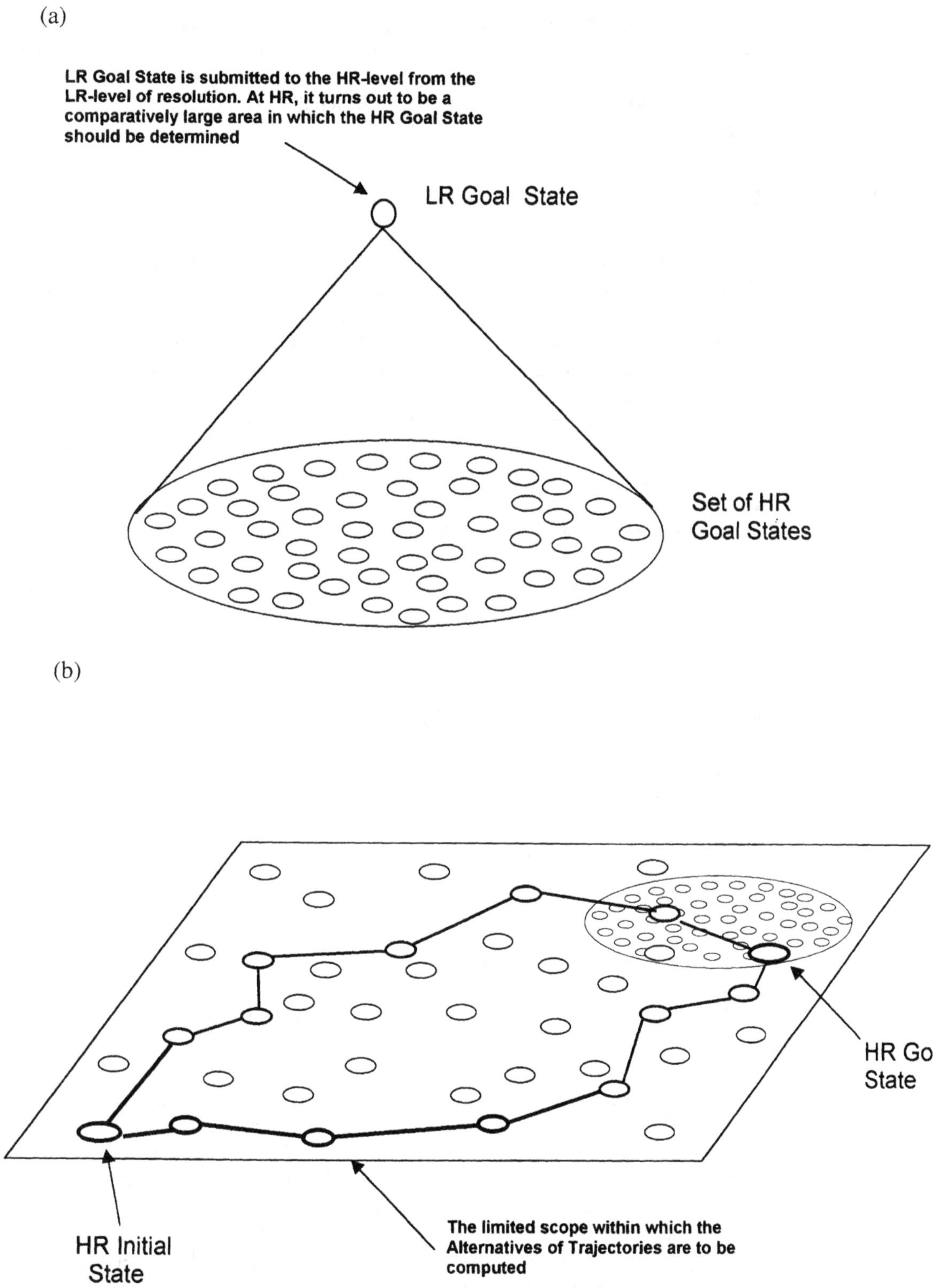

Figure 8.3. Two parts of the planning problem. LR, lower resolution; HR, higher resolution.

challenging case of a partially predictable target, two online algorithms have been developed that each attempt to maintain future visibility with limited prediction. One strategy maximizes the probability that the target will remain in view in a subsequent time step, and the other maximizes the minimum time in which the target could escape the visibility region.

8.3.3 Local Planning: Potential Field for World Representation

The most pervasive method for navigating with minimal planning effort is using potential field construction around the obstacles [43, 44]. The potential field presumes adding to the world representation such properties that will increase the cost of moving into particular directions. An approach to robot path planning is proposed in [45] consisting of building and searching a graph connecting the local minima of a potential function defined over the robot's configuration space. The planner based on this approach can solve problems for robots with many more degrees of freedom. The power of the planner derives both from the "good" properties of the potential function and from the efficiency of the techniques used to escape the local minima of this function. The most powerful of these techniques is a Monte Carlo technique that escapes local minima by executing brownian motions. The overall approach is made possible by the systematic use of distributed representations (bitmaps) for both the robot's workspace and configuration space.

Genetic search is one of the tools for local planning. In some environments it gives positive results and can be recommended for use [46, 47].

8.3.4 Global Planning: Search for the Trajectories

The most general way of planning is by global searching. It consists of the following stages:

1. Populate the world with the randomly assigned "points" that become vertices of the search graph.
2. Connect them in the vicinity.
3. Determine the cost of edges.
4. Run the graph search algorithm (e.g., Dijkstra algorithm or A*).

There are some problems that can be resolved in each particular case. Indeed, the "density" of future vertices of the search graph is to be selected. The concept of "vicinity" should be discussed, and the value of this vicinity should be properly evaluated. Different techniques of pruning the search-tree should be discussed. This area is explored in [48–52].

One of the most powerful results in the area of planning was based on space tessellation [10, 12]. Randomized space tessellation became a productive and efficient technique of performing space tessellation in the practice of planning. It can be applied not only for path planning but also for any dynamical problem of finding minimum cost trajectory in a state space (see an example in [53]).

STOCHASTIC OPTIMIZATION?

Several randomized path planners have been proposed [54]. They are recommended to a variety of robots. A general planning scheme is introduced that consists of randomly sampling the robot's configuration space. The choice of points candidates can be determined by a relation between the probability of failure and the running time. The running time only grows as the absolute value of the logarithm of the probability of failure that we are willing to tolerate. An interesting theoretical paper [56] can be useful in analysis of the methods of planning using randomized dissemination of points in the state space.

8.4 LINKAGE BETWEEN PLANNING AND LEARNING

8.4.1 Learning as a Source of Representation

Learning is defined as knowledge acquisition via analyzing the experiences of functioning. These experiences have their knowledge structure extremely dependent on the goal of an action that has generated an experience. Thus learning concerns the development and enhancement of the representation space under various goals. The representation can be characterized in the following ways (Fig. 8.4*a*):

- By a set of trajectories (to one or more goals) previously traversed.
- By a set of trajectories (to one or more goals) previously found and traversed.
- By a set of trajectories (to one or more goals) previously found and not traversed.
- By the totality of (set of all possible) trajectories.
- By a set of trajectories executed in the space in a random way.

One can see that this knowledge contains implicitly both the description of the environment and the description of the actions required to traverse a trajectory in this environment. Moreover, if some particular system is the source of knowledge, then the collected knowledge contains information about properties of the system that moved in the environment.

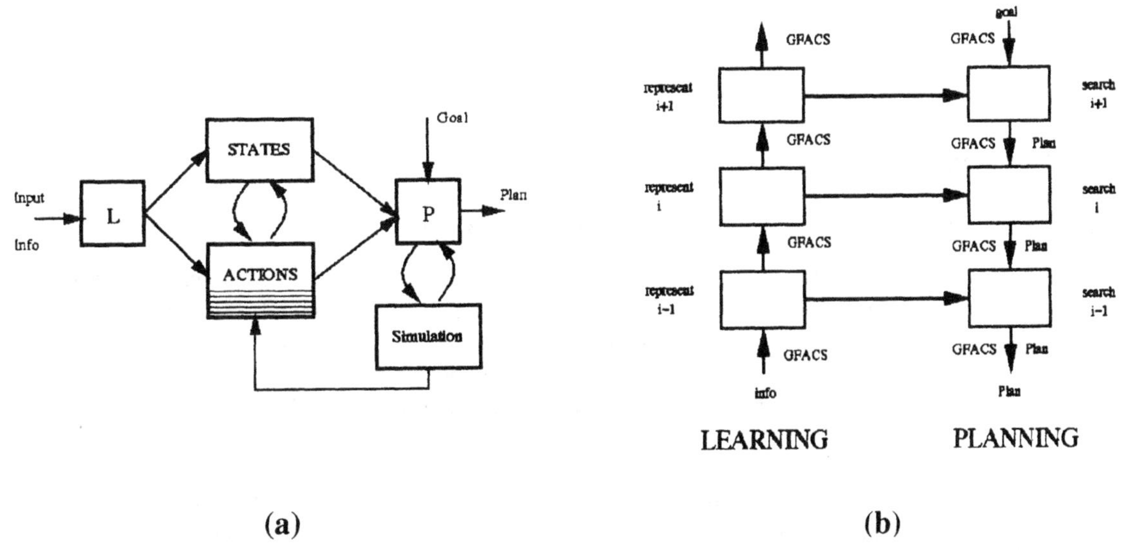

Figure 8.4. Functioning of GFACS in the joint learning-planning processes. (*a*) Performed both via new experiences and plans simulation. (*b*) Performed at all levels of resolution simultaneously.

All this information arrives in the form of experiences which record states, actions between each couple of states, and evaluation of the outcome. The collection of information obtained in one or several of these ways forms the knowledge space (KS).

If the information base contains all tessellata of the space with all costs among the adjacent tessellata, we usually call it the representation. Thus the representation can be considered equivalent to the multiplicity of explanations on how to traverse or how to move. In other words, all the kinds of learning mentioned above are equivalent.

8.4.2 Interrelations between Planning and Learning

Planning is learning from experience in the domain of imagination; searching in the state-space is exploration of these imaginary experiences. Planning is performed by searching within a limited subspace:

- For a state with a particular value (designing the goal).
- For a string (a group) of states connecting SP and GP satisfying some conditions on the cumulative cost (planning of the course of actions).

The process of searching is associated either with collecting the additional information about experiences or with extracting from KS the implicit information about the state and moving from state to state, or learning. In other words, planning is inseparable from and complementary to learning. Learning is a source of the multi-scale (multiresolutional, multigranular) representation. Figure 8.4b illustrates how the multi-scale representation emerges by a consecutive generalization of experiences. Planning, however, presumes a consecutive refinement of imaginary experiences. For both generalization and refinement a set of procedures is used including grouping (G), focusing of attention (FA) and combinatorial search (CS) which are denoted GFACS.

The planning-learning processes are always oriented toward the improvement of functioning in engineering systems (improvements of accuracy in the adaptive controller) and/or increasing the probability of survival (emergence of the advanced viruses for the known diseases that can resist various medications, e.g. antibiotics.) Thus both processes relate to single systems as well as to populations of systems, and they determine their evolution.

8.5 PLANNING IN ARCHITECTURES OF BEHAVIOR GENERATION

8.5.1 Hierarchical Multiresolutional Organization of Planning

An important premise for introducing multiscale algorithms of planning is organization of multiscale (multiresolutional, multigranular) world model. It is presumed that each system can be represented as a multiscale model, that is, as a hierarchy of models with differing details. Therefore planning and control are required at each level of resolution (see [19]). A multiscale world model as well as a multiscale system of planning/control modules requires consecutive bottom-up generalization of the available information. Levels of generalization and the overall multiscale representation, as discussed here, are considered to be characterizations of the same object with different degrees of accuracy. The preceding statement is given in mathematical form by applying concepts of the single level state-space representation for the (not necessarily linear time invariant) system [63–64]:

$$\dot{x}(t) = A(x, u, t)x(t) + B(x, u, t)u(t),$$
$$y(t) = C(x, u, t)x(t), \tag{8.1}$$

where the state, input, output, and time belong to their respective domains as follows:

$$x \in R^n, \quad u \in R^m, \quad y \in R^{p_1} \quad t \in R^+.$$

Thus it is possible to form a solution of these equations as mappings describing the state transition and output functions:

$$\Phi: R^n \times R^m \times R^+ \to R^n \times R^+,$$
$$\Psi: R^n \times R^m \times R^+ \to R^p \times R^+, \tag{8.2}$$

so for any input function u on the interval $[t_0, t_f]$ it is possible to determine the corresponding output function y on the same interval. If it can be shown that there exists a pair of functions

$$\Phi': R^{n'} \times R^{m'} \times R^+ \to R^{n'} \times R^+,$$
$$\Psi': R^{n'} \times R^{m'} \times R^+ \to R^{p'} \times R^+, \tag{8.3}$$

for which n' is strictly less than n, and for which the same input function u generates the output function y' such that the inequality

$$\left| \int_{t_0}^{t_f} (y'(t) - y(t))\, dt \right| < \varepsilon \tag{8.4}$$

holds for all admissible inputs in the input function space where ε is a value that depends on the level of resolution under consideration. Then it is claimed that

$$\{\Phi', \Psi\} \text{ is an } \varepsilon\text{-generalization of} \{\Phi, \Psi\}. \tag{8.5}$$

The strictness of this formulation may be relaxed by considering a stochastic measure for associating a confidence level with the generalization to construct the concept of ε-generalization *nearly everywhere*. Thus

ε-GENERALIZATION

$$P\left[\left\| \int_{t_o}^{t_f} (y'(t) - y(t)\, dt \right\| < \varepsilon \right] < \tau \tag{8.6}$$

is a statement of the belief that the constraint holds with a probability defined by the pre-assigned threshold τ.

This formulation can be extended to an ordered collection of epsilons $\{\varepsilon_1, \varepsilon_2, \ldots, \varepsilon_k\}$, thereby defining a hierarchy of models that describe the same input-output behavior with increasing degrees of accuracy. The necessity of considering all elements of the input and output vectors as time-varying functions may also be relaxed so that at some level i, $u_{k_i}[t_0, t_f]$ could be considered constant in the interval, whereas at some lower level (at higher resolution) the same input may be represented as a time-varying function.

The ability to formulate the world models with this hierarchical generalization will be shown in the following example to be an essential device for coping with the complexity associated with the planning of system operation in a combined feedforward–feedback controller.

8.5.2 Case Study: A Pilot for an Autonomous Vehicle

In order to begin planning the path of a vehicle in a space with obstacles, it is necessary to have a geometrical representation of the extent and the elements of the search space, including the vehicle. This job is accomplished by creating geometrical models of the workspace, vehicle, and obstacles. Templates for these objects are components of the simulation package SIMNEST developed by Drexel University with NIST participation for design and control of hierarchical systems. The package allows the customization of the workspace to accommodate new maps and vehicle parameters. However, the principles of representation and control are easy to apply in any simulation package.

Having constructed the iconic representation of the physical elements of the problem, the next step is to specify their behavior and interactions. In our case a kinematic and dynamic model of the vehicle will be required, and it will be necessary to be able to identify the occurrences of possible collisions between the vehicle and obstacles so as to successfully avoid them.

SUCCESSIVE GENERALIZATION

The state of the vehicle in Figure 8.5 is often described in the literature in terms of the following variables: (1) x-position, (2) y-position, (3) orientation, (4) speed, and (5) steering angle. We will consider this description to be the level of complete informedness, or maximum resolution (even though we can include more complex issues such as the effect of roll on cornering dynamics or skidding). This description is sufficient to illustrate the principle of *successive generalization* required for obtaining a representation that fits into our concept of planning/control algorithm of *consecutive refinement*. (In addition it is necessary to know whether the vehicle is rolling or sliding so that, depending on the case, velocity can be treated as a scalar.)

Depending on the acceleration of the vehicle which is accomplished by the propulsion and steering actuators, it is possible to formulate a model of the system in which the next position of the car, given its current state, is a function of the effective steering angle and velocity α and ν, during the interval of modeling.

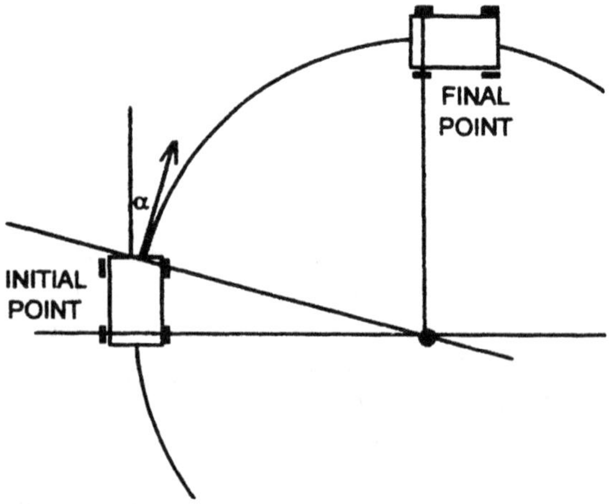

Figure 8.5. Construction of a system model for a four-wheel vehicle.

The center of rotation of the vehicle is given by the point of intersection of the lines shown in Figure 8.5, corresponding to the extensions of the lines joining the fixed rear axle and the perpendiculars to the planes of the front wheels. (It is clear that the single steering angle is an abstraction because, for these three lines to meet at a point, it is necessary that the two front wheels be rotated through different angles.)

The distance d moved on the circumference of the circles shown in Figure 8.6 can be determined from velocity and elapsed time to be

$$d = \int_{t_i}^{t_f} v\,dt. \tag{8.7}$$

Therefore the new coordinates of the vehicle with respect to the global xy-position can be determined by solving simultaneously the equation of the circles centered at the vehicle with radius 1 as shown, and centered at the center of rotation with radius r. The four solutions of the quadratic equations

$$\begin{aligned}(x - x_1)^2 + (y - y_1)^2 &= r^2,\\ (x - x_2)^2 + (y - y_2)^2 &= 1^2\end{aligned} \tag{8.8}$$

(where the subscripts refer to the indices of the centers of the two circles) correspond to the case of forward and reverse motion with positive (right turn) and negative (left turn) steer as shown in the figure. The correct global coordinates after moving the vehicle are restored according to the context in which motion occurs.

The orientation of the vehicle after motion occurs can be determined from the slope of the line joining the coordinates of the rear wheels of the vehicle, since this line is normal to the direction in which the vehicle is facing. The approach described is used to model the kinematics of the vehicle; the dynamics are a function of the actuators used in the vehicle.

Without loss of generality, the dynamics of the rolling vehicle can be modeled by the motion of an object with inertia subject to accelerating and braking forces. This model of motion is valid as long as (1) the vehicle does not slide because the force normal to its trajectory exceeds the reaction due to static friction and (2) the wheels of the vehicle do not lose traction.

Assuming that other models exist for describing these regimes of operation, we can proceed with the completion of our "comprehensive" model. The approach to system modeling at this stage begins to be influenced by the desire to solve the motion planning problem using the principles of consecutive refinement. This means that the vehicle

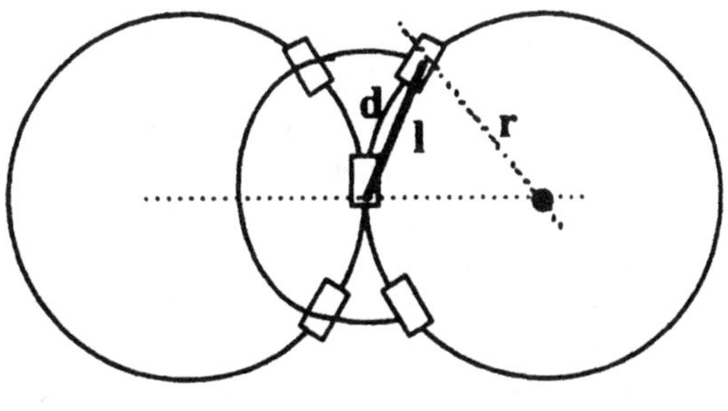

Figure 8.6. Position computation.

should be represented at more than one level of abstraction, and a rough plan should be created with an imprecise model before an attempt is made to synthesize the final plan. Since this example illustrates the use of two nested levels of planning, it will only be necessary to construct these two versions of the vehicle model.

First Level: Low-Resolution Planning (Trajectory Planning)

The technique of planning which is being exemplified here involves a top-down search with successive refinement. It is assumed that the results of this search will be considered open-loop control input for a particular level (a reference curve) with a feedback-tracking controller compensating for current errors that could not be taken care of at the stage of planning for this particular level. Thus, not only is it necessary to have a model of the environment for executing procedures of planning, a tool for performing the search in it must also be available. This tool is described in [63–65, 77]. It is also available as one of the options in the procedures library of NIST-RCS. This procedure performs the search from a graph representation of the workspace. The operation is fairly complex; it must incorporate successively more refined information in the two descriptions to be used for search.

In generating the graph of the first level, the model used is one that allows the inclusion of the least amount of detail that is considered sufficient to provide rough predictions of qualitative dynamical behavior. This is accomplished by representing the vehicle as an object whose velocity is conditioned by its change of direction. Thus, if the base velocity of the object is $v_{\max}$, then the velocity during motion requiring a change of direction of θ degrees is

$$v = v_{\max} \cos \theta. \tag{8.9}$$

In this rudimentary model only the resistance to change of direction is shown; other kinematic or dynamic issues are not represented. In order to utilize this model for planning, the first-level search procedure begins by tessellating the horizontal workspace of the vehicle. The rule of tessellation is novel in that a regular grid is not utilized. Instead the xy-plane is partitioned into regular subsets, and a random coordinate pair is used to represent each subset. This approach eliminates the idiosyncrasies associated with regular grids without utilizing a biasing heuristic. The number of points used to represent the space is a function of the average interpoint spacing and the narrowest passageways in the map. The influence of these parameters can be determined by experiment.

TESSELLATION IN THE VICINITY

The connectivity of the randomly tessellated version of the workspace is decided by the definition of vicinity at a given level of the representation. If the concept of vicinity is relaxed to include the whole space, the graph could hypothetically be fully connected. On the other hand, curtailing the measure of vicinity may lead to a graph that is not connected. This problem can be addressed by using heuristics, or by trial and error. There remains the issue of whether edges in this graph traverse forbidden regions in the workspace or whether the vehicle will intersect obstacles during such traversals. Such potential intersections are tested individually during the construction of the graph. The concept of the vehicle being represented as an expanded point is exercised and the degree of expansion is fixed by iterations beginning with the minimum dimensions of the vehicle.

The search minimizes the estimated time of traversal at each level of representation. At the first level of resolution, edge cost is computed using the expression:

$$t = \frac{d}{v_{\max} \cos \theta}, \tag{8.10}$$

where t is time (cost), d is the euclidean distance between neighboring points, $v_{\max}$ is the maximum velocity of the vehicle, and θ is the change in direction between consecutive points on the trajectories being tested. Thus the point-to-point traversal is assumed to occur at maximum velocity unless it involves a change in direction, which slows down the vehicle.

Upon completion of the graph representation of the workspace, a version of the best-first graph search algorithm (the so-called best-first algorithm) is invoked to determine a path between the start and the goal positions in the graph. The algorithm is not exactly the same as that described in [63] because the graph may conceivably be generated during search, the result of search is not necessarily (or even probably) globally optimal, and it is being applied to a different class of problems. However, the search process is at least similar to Dijkstra's algorithm.

The trajectory determined at the first level of the search, embodied by a string of xy-coordinates, is then provided to the next level for refinement.

Second Level: High-Resolution Planning (Maneuver Planning)

The second level refines the results of the first search by utilizing a similar search procedure, but it operates on a reduced space in the vicinity of the first (approximate) plan and the graph generation process takes into account the body configuration, kinematics, and dynamics of the vehicle in greater detail.

The vehicle model that is used at this second level includes the kinematic constraints of the vehicle as well as dynamical constraints on acceleration. Thus the model of the second level utilizes the assumption that the steering angle and velocity of the model remain constant in the interval of modeling to predict the new coordinates of the vehicle. The kinematic model of Figures 8.5 and 8.6 is thus utilized in its final form. Also a maximal rate of change of steering angle and velocity is used to describe the dynamics of steering and acceleration. A neighboring point is deemed to be unreachable from another if the transition requires a change of steering larger than allowed for a single transition.

In order to utilize the information provided by the rough plan of the first level, a neighborhood of this trajectory is constructed. The neighborhood is a closed subset of the workspace and is constructed as shown in Figure 8.7. The envelope is constructed as a string of roughly designated vicinities.

The trajectory, which is described by a string of points, is enclosed by sequence of cells whose width is a function of the characteristics of the trajectory itself. As can be

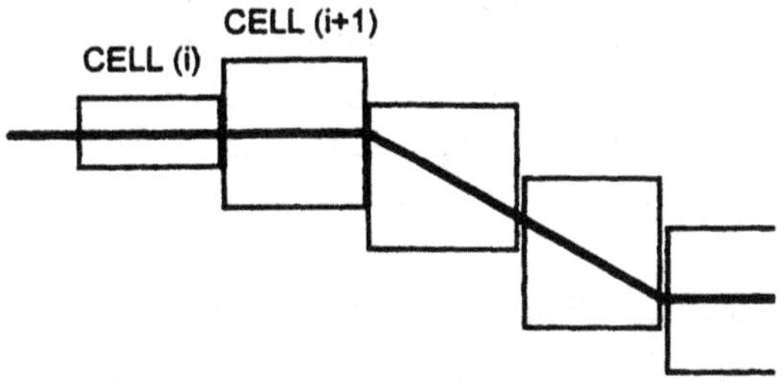

Figure 8.7. Construction of the vicinity of a rough trajectory.

seen in the figure, straight segments are enclosed in narrow cells such as cell i whereas a change in direction establishes the need for greater width, as in cell $i + 1$. The area of the cells must be increased if complex maneuvers are expected (e.g., turns or direction reversal) because new constraints will be included in the new level of description and the abrupt changes that can be made with the slack constraints of the rough description may no longer be possible. The selective refinement of the scope of search is thus a heuristic for guaranteeing consistency in this paradigm. One further heuristic is applied at this stage; the cell size is increased if the baseline cell intersects an obstacle, since avoiding the obstacle may involve departing significantly from the nominal trajectory.

The number of dimensions in the search space is increased at the second level, where velocity becomes an additional axis. Thus each cell of Figure 8.7 is decomposed into enough subcells to achieve the assigned average spacing of this second level and each randomly generated point has associated with it an x-position, a y-position, and a velocity.

SUCCESSOR VALIDATION

A method of successor validation is utilized to determine children for each parent in order to preserve the fidelity of the representation of this level. This process begins with an evaluation of the steering angle α which is required to attain each candidate successor point from each putative parent node. Inverse kinematic routines are used to determine this information as well as the corresponding center and radius of rotation. Constraints are applied in the following manner:

1. If the required steering angle exceeds the maximum steering angle, the successor is invalid.
2. If the required *change* in steering angle exceeds the maximum allowable value, the successor is invalid.
3. The time of transition is constrained by the maximum velocity allowed for the required radius of rotation. This value is determined from the maximal centripetal force which can be tolerated without slippage.

The edge cost is determined by the distance to be moved along the circumference of the circle and by the random velocity which is the third coordinate of the parent node. It has been determined experimentally that in order for the inverse kinematics routine to be numerically stable, the density of points in the xy-plane should be such that, on average, at least three points are located in a unit equivalent to the cross-sectional area of the vehicle.

One additional correction is to establish that the real volume of the vehicle, which is still represented by its center, does not intersect obstacles. For this computation an area corresponding to the size of the vehicle is marked off around each valid node and tested for intersection with forbidden regions such as obstacles. In this manner the graph representation of the second level can be modified to accommodate a realistic amount of detail in its model of the workspace.

Once more, upon completion of the graph representation of the workspace, the search algorithm is invoked to determine a path between the start and the goal positions in the graph. The resulting solution includes not only position trajectories but also velocity information and the steering angles required to follow the trajectory.

Results of Planning the Path of the Vehicle

Figure 8.8 depicts the process of planning via snapshots of the screens presented to the user during planning. The process of search is shown in the left panel of Figure 8.8. The

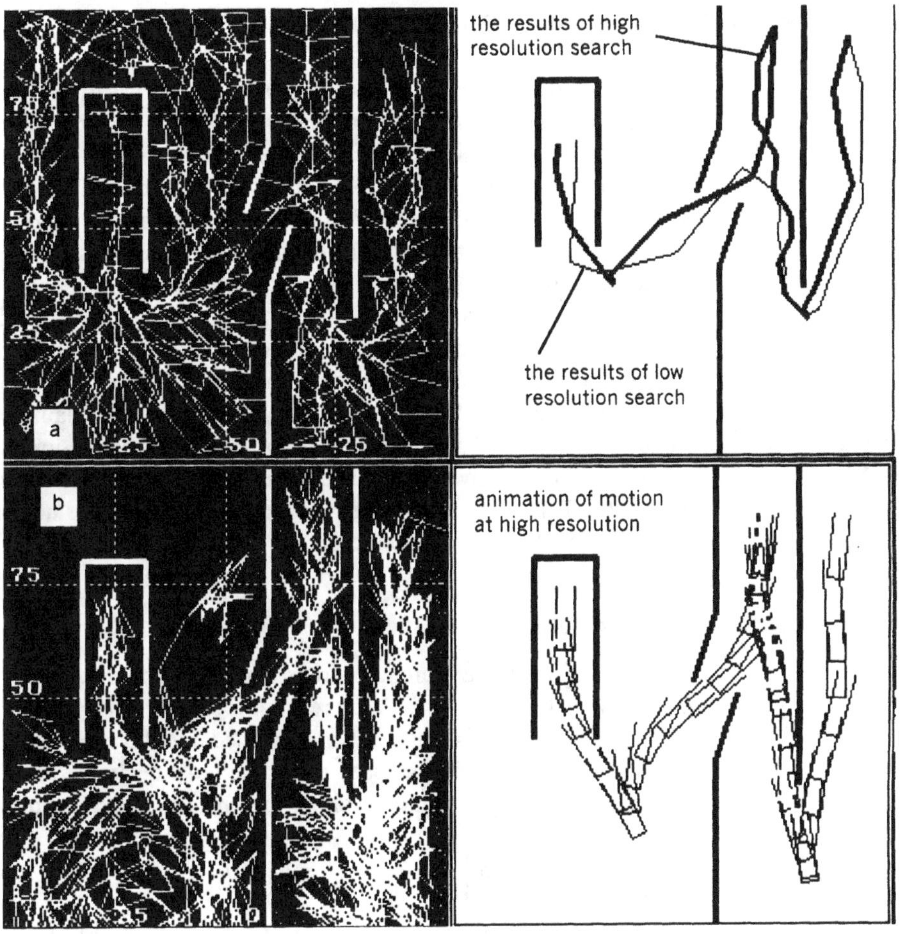

Figure 8.8. Planning via search in a multiresolutional state-space: (*a*) Search in the whole space at low resolution; (*b*) Search in the reduced space at high resolution.

upper left panel shows the search in full space at low resolution. The lower left panel shows search in a reduced search space but at higher resolution. The final trajectory of the vehicle is shown in the right panel of the figure for a workspace including a garage, wall, and gate.

The order of synthesis of this result can be seen beginning with Figure 8.5, which is a depiction of the search tree at a low level of resolution overlaid on a description of the workspace. The kinematics of the vehicle are clearly absent from this consideration as can be seen by the result of search at the first level in the upper right panel of the Figure 8.8 (thin trajectory lines) but they are evident in the search at the next level (bold trajectory lines) by which one can see the maneuvering of the vehicle. The search at this high-resolution level is depicted in the lower panels of Figure 8.8 where the reduced search tree of that level is shown.

This sequence of figures demonstrates that it is possible to synthesize complex maneuvers such as reversing and K-turns without using an expert rule base generated by a human being. Comparatively complex maneuvering is performed just by constructing a hierarchical representation of the system and searching for successive approximations to construct an ε-optimal solution of the problem at a particular level of resolution.

8.6 PATH PLANNING IN A MULTIDIMENSIONAL SPACE

The standard A* algorithm has been used in the past to find a minimum cost path in variable cost space [74]. Unfortunately, when this algorithm is presented with a number of equally minimal cost paths from which to choose, it fails to choose the straightest path. This problem is known as the trap of multiple, equal cost solutions. This anomaly is introduced by the discretization of the space to be searched and minimization of the deviation from the straight line. In [48] a modified cost function was proposed that could avoid this trap. This function was applied to evaluate every successor as the algorithm progressed. The multirule A* algorithm was proposed in [69]. This algorithm applies many rules in addition to the cost evaluation. The multirule A* search also solved this trap and proved to be more efficient. Multiple rules were not incorporated into the main cost-function. They were only applied when the main cost-function found two equally optimal paths. The MRA* algorithm (multiresolutional A*) provides a framework in which many evaluation functions can be combined into a multiresolutional hierarchy. This section discusses the generalization of this algorithm for application to n-dimensional state-space search.

MULTIRESOLUTIONAL A*

8.6.1 State-Space Representation

The state-space to be searched is represented by and discretized into an n-dimensional matrix of real numbers, with each real number representing what we call the set of desirability values, D. This real number is mapped into the interval [0, 1], with 0 representing the least desirable point in space and 1 the most desirable point in the state-space. An example of this representation for a 3-dimensional space is presented as follows:

$$\begin{bmatrix} 0.5 & 0.8 & 0.6 \\ 0.7 & 0.9 & 0.1 \\ 0.3 & 0.9 & 0.3 \end{bmatrix}_{z=0}, \quad \begin{bmatrix} 0.2 & 0.4 & 0.1 \\ 0.6 & 0.2 & 0.7 \\ 0.4 & 0.9 & 0.5 \end{bmatrix}_{z=1}, \quad \begin{bmatrix} 0.6 & 0.8 & 0.5 \\ 0.1 & 0.6 & 0.8 \\ 0.9 & 0.9 & 0.3 \end{bmatrix}_{z=3}. \tag{8.11}$$

Note that the space is simply a hierarchy of 2D matrices. For example, the position ($x = 0, y = 0, z = 0$) has the value 0.5 assigned to it. The position ($x = 1, y = 2, z = 3$) has the value 0.9 assigned.

This definition (with 1 representing maximal desirability) was chosen because of the original application of this algorithm to the problem of path planning of an autonomous vehicle [48, 69, 74]. In this application, cost is evaluated as the time of motion from tile to tile in the space. Desirability or traversability [48, 69, 74] is defined as

$$T_{\mathrm{L}} = k(dS/dt)_{\mathrm{L}}, \tag{8.12}$$

where

T_{L} = the traversability at location L,
$(dS/dt)_{\mathrm{L}}$ = the max velocity achievable at L,
k = some positive constant.

The traversability is defined at every point in the space by a function T, such as $T(x, y)$ in 2D space and $T(x, y, z)$ in 3D space. We can then calculate the cell to cell cost between any two adjacent cells as

$$C_{AB} = \frac{d(A, B)}{2T_A} + \frac{d(A, B)}{2T_B}. \tag{8.13}$$

More formally, the n-dimensional space to be searched can be represented by an undirected, acyclic graph, $G = (V, E)$, where V is the set of points in the discretized space and E is a set of cost values corresponding to each point in the space. The set E is calculated from the set of desirability values, D. Between any two *adjacent* points in the space, a distance metric, d, is imposed. From the definition of a metric space, d obeys the following:

$$d(x, x) = 0, \tag{8.14a}$$

$$d(x, y) = 0 \quad \text{if } x = y, \tag{8.14b}$$

$$d(x, y) = d(y, x), \tag{8.14c}$$

$$d(x, z) = d(x, y) + d(y, z), \tag{8.14d}$$

for all adjacent points x, y, z. The euclidean metric

$$d(x, y) = \sqrt{(x_1 - x_2)^2 + (y_1 - y_2)^2}, \tag{8.15}$$

where $x = (x_1, x_2)$, $y = (y_1, y_2)$ was used to find the optimal path in R^2. This euclidean metric can of course be generalized to R^n by

$$d(x, y) = \sqrt{\sum_{i=1}^{n} (x_i - y_i)^2}, \tag{8.16}$$

where $x = (x_1, x_2, \ldots, x_n)$, $y = (y_1, y_2, \ldots, y_n)$. Further generalizations in the metric itself are also possible:

$$d(x, y) = \sqrt{\sum_{i=1}^{n} |x_i - y_i|^p}, \tag{8.17}$$

which is the familiar L_p norm of a vector. In fact other metrics that do not even vaguely resemble the familiar euclidean metric can be used so long as (8.14a) through (8.14d) hold. Although this seems counterintuitive, it is a valid metric. Other less intuitive metrics include the city block or taxi cab metric,

$$d(x, y) = \sum_{i=1}^{n} |x_i - y_i| \tag{8.18}$$

and the supremum normal metric

$$d(x, y) = \sup_{i=1\ldots n} |x_i - y_i|. \tag{8.19}$$

The latter metric is actually employed in a popular computer operating system to determine the distance between two points in an RGB color space [48]. In this book the euclidean metric will be arbitrarily used, although one may discuss the results of using other metrics to guide the search of the state space if they prove interesting.

8.6.2 Expert Rules/Heuristics

Cost in a Multidimensional State-Space

As stated previously, the state-space is represented by an n-dimensional matrix of real numbers that are mapped into the interval [0, 1]. An element with a value of 0 corresponds to a most costly state and an element with a value of 1 corresponds to a least costly state. From this representation of the state-space and by using one of the previously discussed metrics, cost can be defined as follows:

$$C_{xy} = d(\mathbf{x}, \mathbf{y})/2S_x + d(\mathbf{x}, \mathbf{y})/2S_x, \tag{8.20}$$

where S_x is the state-space value at point x and S_y is the state-space value at point y. Note that

$$\lim_{s\to 1} \frac{d(\mathbf{x}, \mathbf{y})}{s} = d(\mathbf{x}, \mathbf{y}) \quad \text{and} \quad \lim_{s\to 0} \frac{d(\mathbf{x}, \mathbf{y})}{s} = \infty, \tag{8.21}$$

which corresponds to the definition that a point with a value of 0 is most costly and a point with a value of 1 is least costly. The $d(\mathbf{x}, \mathbf{y})/2$ appears because in moving from point $\mathbf{x}$ to point $\mathbf{y}$, it is assumed that half of the distance traversed is in the neighborhood of point $\mathbf{x}$ and the other half is in the neighborhood of point $\mathbf{y}$. Given that it is possible to calculate the cost between any two points in the state-space, $\mathbf{x}$ and $\mathbf{y}$, it is also possible to calculate the total cost of a given path, $\mathbf{P}$. Let a path $\mathbf{P} = (p_1, p_2, \ldots, p_n)$ where the p_i are the points along a given path $\mathbf{P}$. Therefore the total cost for path $\mathbf{P}$ is

$$\sum_{i=2}^{n} \frac{d(p_{i-1}, p_i)}{2S_{p_{i-1}}} + \frac{d(p_{i-1}, p_i)}{2S_{p_i}}. \tag{8.22}$$

Finally, f^*, which is assigned to the current node under consideration, is simply $f^* = g^* + h^*$. In traditional A* terminology, equation (8.22) includes the function g^* which is an estimate of the smallest possible minimal cost path from the start node to the current node under consideration. The heuristic function, h^* is an estimate of the cost from the current node under consideration to the goal node which is simply the distance (by one of the metrics previously described) from the current node under consideration to the goal node.

Selecting the Straightest Path

Because of the trap of multiple, equal cost solutions [69], it is desirable to select not only an optimal path from a set of equally optimal paths but to select that optimal path which is straightest. This trap is introduced as a by-product of the discretization of the state-space. Although fewer equal cost paths may exist in a continuous state-space, the discretization may introduce what appears to be many more equal cost paths.

To evaluate the relative straightness of a given path in n-dimensional space a straight line in n-dimensional space must be constructed. To accomplish this, first consider the two-point form [48] of a straight line in 2-dimensional space:

$$\frac{y - y_1}{x - x_1} = \frac{y_2 - y_1}{x_2 - x_1}, \tag{8.23}$$

where point $p_1 = (x_1, y_1)$ and $p_2 = (x_2, y_2)$. The two-point form [69] of a straight line in 3-dimensional space is

$$\frac{y - y_1}{x - x_1} = \frac{y_2 - y_1}{x_2 - x_1} = \frac{z - z_1}{z_2 - z_1}, \tag{8.24}$$

where point $p_1 = (x_1, y_1, z_1)$ and $p_2 = (x_2, y_2, z_2)$. For the problem of defining a path in n-dimensional space between two arbitrary points the equations above can be generalized as

$$\frac{p_1 - s_1}{e_1 - s_1} = \frac{p_2 - s_2}{e_2 - s_2} = \frac{p_3 - s_3}{e_3 - s_3} = \cdots = \frac{p_n - s_n}{e_n - s_n}, \tag{8.25}$$

where $\mathbf{S} = (s_1, \ldots, s_n)$ is the location of the starting point in the state-space, $\mathbf{E} = (e_1, \ldots, e_n)$ is the location of the ending point in the state space, and $\mathbf{P} = (p_1, \ldots, p_n)$ is an arbitrary point along the line between point **S** and point **E**. Given this equation and the points **S** and **E**, it is simple to evaluate the relative straightness between points along a path and the straightest possible path (a straight line) by solving for each point along the straight line. For example,

$$p_1 = \frac{(p_2 - s_2)(e_1 - s_1)}{e_2 - s_2} + s_1 \tag{8.26}$$

The relative straightness of a path, $\mathbf{V} = (v_1, \ldots, v_n)$, is then determined by the sum of the differences between a point on the straight line, **P**, and a point on the path, **V**, as follows:

$$\Delta = \sum_{i=2}^{n} |v_i - p_i|. \tag{8.27}$$

Therefore, when these are presented two equally "optimal" paths, the path of minimal deviation is chosen. This general equation for geometric straightness can be modified to include a different weight for each component of the vector. This will cause the algorithm to optimize differently for each component of the vector. The deviation Δ can be calculated as follows:

$$\Delta = \sum_{i=2}^{n} w_i |v_i - p_i|, \tag{8.28}$$

where w_i is the weight or emphasis to be placed on each component of the vector. Since we wish to minimize the deviation, with larger w_i we can emphasize or select against less desirable components of the vector. This weight is applicable to problems of control and pattern recognition in which each component is considered to be an element of a feature vector. Different components of the feature vector may be more desirable than others. Again, we hope to discuss such applications in future works if the results prove interesting.

Variance Minimization

Variance minimization as described in [48, 69] can be applied directly during the search. In terms of autonomous vehicle guidance, this technique can be used to select paths that contain roads. In other terms, this method selects paths that contain less high frequency components or minimizes the derivative of desirability along the path.

8.6.3 Techniques to Reduce Computation Time

Now we are ready to present some tools for reducing the computation time based upon heuristically evaluating the computational complexity of the A* algorithm.

Intelligent Successor Generation

In 2D space each node can have as many as 9 successors, and in 3D space each node can have as many as 27 successors. Generally, this is the number of adjacent points in the space, which can be calculated as 3^n where n is the number of dimensions in the space. This is derived from the fact that each point can differ from its neighbor at most by ± 1 (one tessellatum).

Since cycles directly back to the given node have a distance of 0 and therefore a cost of 0, they must be disallowed. Therefore the number of possible successors is reduced by 1 to $3^n - 1$.

The number of successors that needs to be generated can be further reduced by eliminating all paths that contain kinks, quirks, twists: strange moves that emerge as a result of choosing the successor. A path without any kinks is defined as follows: Let n be the successor of a parent node, p. Let g be the parent of the node p. Then, for a path without kinks, $d(p, n) = d(g, n)$ for all n, p, and g along a path. These nodes do not need to be generated because they are always more costly than the path from g to n directly. Therefore the number of nodes generated for a given successor can be reduced from 3^n to $3^n - 3^{n-1} - 1$. This is one of the many possible heuristics that constrain the motion and reduce the value of complexity.

Resolution

Choosing properly the set of levels of resolution to represent the space can greatly reduce the search time [52, 76]. Then the algorithm of search can be applied again at higher resolutions as a hierarchy of consecutive refinement when high-resolution sensory data are available.

Searching Open and Closed

The A* algorithm generally proceeds by removing that node with lowest f^* value from OPEN. The successors of this node are then generated and then this node is added to CLOSED. Every time that a successor is generated, OPEN and CLOSED must be searched to determine if a node(s) with the same location is already present. Therefore the following steps can be taken to decrease the time that is necessary to search and update OPEN and CLOSED:

Step 1. Maintain the list, OPEN, in nondecreasing order so that the time required to find the node with minimal f^* value is $O(1)$ instead of $O(n_{\mathrm{OPEN}})$. If access to OPEN is through a simple list maintained in nondecreasing order, the time required to find the node of minimal value (findmin time) is $O(1)$, but the time required to insert a node is $O(n_{\mathrm{OPEN}})$. When a concrete algorithmic solution is being selected, a measure is to be taken to increase the insert and delete times. Then findmin time becomes $O(1)$, $O(\log_{d^n})$ for insert, and $O(d \log_{d^n})$ for delete.

Step 2. The search of OPEN for a given successor s can be discontinued if a node n already on OPEN is found such that LOCATION(s) = LOCATION(n) and $f^*(n) < f^*(s)$. Further CLOSED does not have to be searched at all in this case, and the successor s can be discarded. In this case the search effort for OPEN and CLOSED is reduced from $O(n_{\text{OPEN}}) + O(n_{\text{CLOSED}})$ to $O(n_{\text{OPEN}})$.

Step 3. To further decrease the search time of OPEN for a node at a given location, an n-dimensional matrix of pointers to lists of nodes at a given location may be maintained. This reduces the search time of OPEN for nodes at a given location from $O(n_{\text{OPEN}})$ to $O(1)$. If it proves to be too memory-intensive, a hashing algorithm is used. The location is hashed to form the index into a 1D array of pointers to lists of nodes.

Step 4. If CLOSED is maintained as a list, the search time is $O(n_{\text{CLOSED}})$. This time can also be reduced to $O(1)$ if CLOSED is also maintained as an n-dimensional matrix of pointers to lists of nodes at a given location. If this proves to be too memory-intensive, the hashing algorithm can also be applied to CLOSED.

Finally note that the lists described above should be doubly linked to facilitate insertion and deletion.

8.6.4 Experimental Results

Figures 8.9 and 8.10 demonstrate the results of two searches in 3D space. The desirability assigned to each area is 0 for the polygons, 1 for areas that are not adjacent to the polygons, and slightly less than 1 for areas that are adjacent to the polygons.

To graphically present the results of searches of n-dimensional space, a series of 2D plots of all the dimension pairs is recorded. In 3D space, $x - y$, $x - z$, and $y - z$ are plotted. In 4D space, $x - y$, $x - z$, $x - u$, $y - z$, $y - u$, and $z - u$ are plotted.

Figure 8.9 demonstrates the algorithm's ability to find an optimal path in an area that is obstructed by polygons. When viewing the path from left to right, note that the algorithm has found a path that proceeds straight from the edge of the first polygon to the edge of the second polygon in an area assigned a desirability (D) value of 1. The path then proceeds straight along the face of the second polygon, rounds the corner, and heads straight toward the goal. Note that the diagonal descent of the path occurs in the area that is assigned a D value of 1. The more costly diagonals are avoided in the areas adjacent to the face of the polygons because the desirability value in these areas is lower.

Figure 8.10 demonstrates the straightness of the optimal path that was found. Visual inspection of Figure 8.10*b* shows that the areas of lower desirability adjacent to the edges of the polygons have been avoided while the overall straightness of the optimal path has been maintained. The diagonal descent, when viewing the path from left to right, occurs in areas where $D = 1$.

In conclusion, it is possible to modify the traditional A* algorithm to add a number of heuristics or rules that allow it to distinguish between a number of equal cost solutions. Further the search of 2-dimensional space using this algorithm can be generalized to allow it to determine optimal paths in n-dimensional space. Finally, significant steps

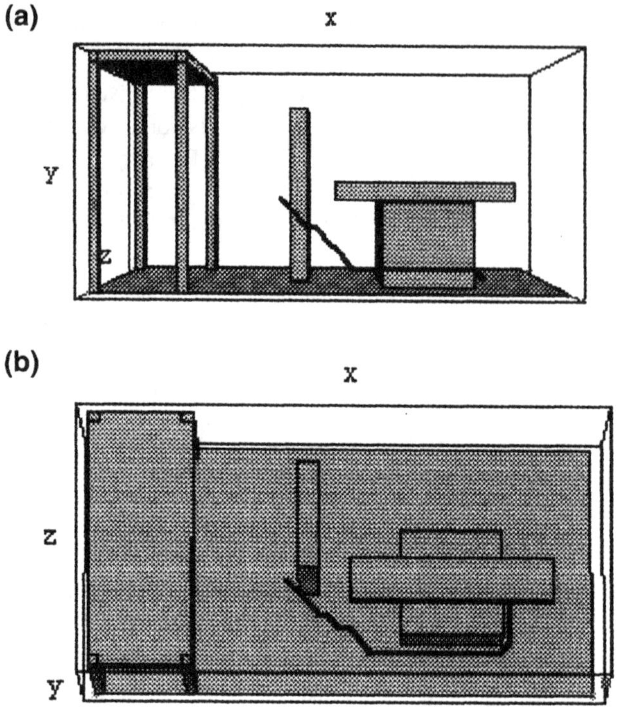

Figure 8.9. Example of search results: (*a*) Viewed in front; (*b*) viewed from above.

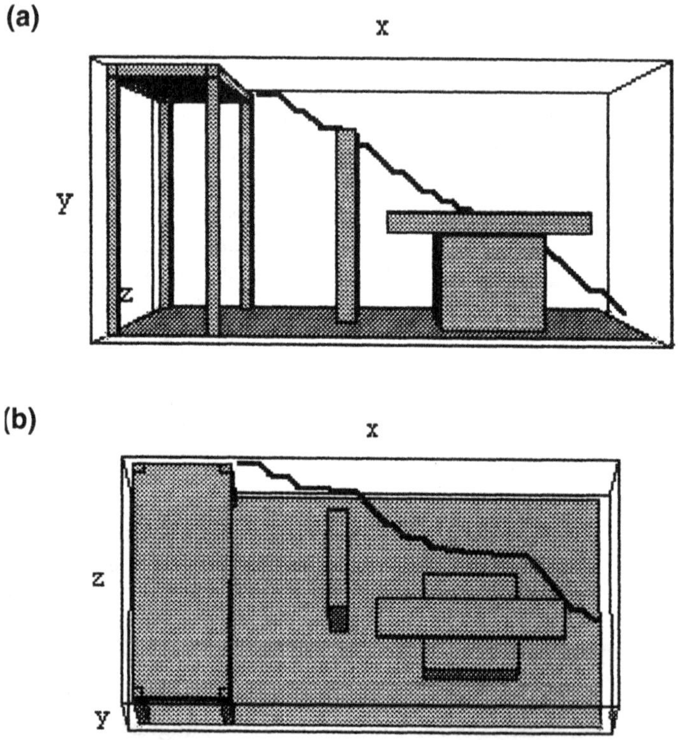

Figure 8.10. Another example of 3D search results: (*a*) Viewed from the front; (*b*) viewed from above.

can be taken to reduce the computational complexity of this algorithm by choosing appropriate data structures to represent OPEN and CLOSED.

8.7 MULTIRESOLUTIONAL PLANNING AS A TOOL OF INCREASING EFFICIENCY OF BEHAVIOR GENERATION

8.7.1 Multiresolutional Planning Embodies the Intelligence of a System

Planning consists of a job assignment and scheduling. The job assignment distributes the motion among the spatial coordinates. Scheduling distributes the motion along the time axis. Together they contribute to the search process. The search is performed by constructing feasible combinations of the states within a subspace (feasible means: satisfying a particular set of conditions). The search is interpreted as exploring (physically or in simulation) many possible alternatives of motion and comparing them afterward.

Each alternative is created by using a particular law of producing the group of interest (cluster, string, etc.) Usually grouping presumes exploratory construction of possible combinations of the elements of space (combinatorial search) and, as one or many of these combinations satisfy conditions of being an entity, substitution of this group by a new symbol with its subsequent treatment as an object (grouping).

The larger the space of search, the higher is the complexity of search. This is why a special effort is made to reduce the space of search. This effort is the focusing of attention, and it results in determining two conditions of searching, namely, its upper and lower boundaries: the upper boundaries of the space in which the search should be performed, and the resolution of the representation (the lower boundaries).

A fundamental procedure in both learning and planning is the formation of multiple combinations of elements (during the search procedure, S) satisfying required conditions of transforming them into entities (grouping, G) within a bounded subspace (focusing of attention, F). Since these three procedures work together, we will talk about them as a triplet of computational procedures that includes grouping, focusing of attention, and search (see GFS, or GFACS in Chapter 2). Notice that in learning this triplet creates lower-resolution levels out of higher-resolution levels (bottom-up), while in planning it progresses from the lower-resolution levels to higher-resolution levels (top-down).

The triplet of computational procedures is characteristic for intelligence of the system. Its purpose is the transformation of large volumes of information into a manageable form which ensures successful functioning. The joint functions of learning and planning explains the pervasive character of hierarchical architectures in all domains of activities.

REDUNDANCY IS A REQUIRED PROPERTY

The GFACS need to contain a great many alternative space traversals is stimulated by a property of knowledge representation (it is a property that allows representations to be redundant). The redundancy of representations meets the learning and planning need in GFACS; otherwise, the known systems would not be able to function efficiently (it is possible that redundancy of representations is a precondition for the possibility of life and intelligence).

Representations actually reduce the redundancy of reality. If we eliminate all redundancy, problems can be solved in a closed form (no combinatorics is possible and/or necessary). Sometimes a full reduction of redundancy is impossible, and combinatorial search is the only way of solving a problem. If a problem cannot be solved in closed form, we can reintroduce redundancy to allow the functioning of the GFACS.

At each level of resolution, planning is done as a reaction to slow changes in a situation that invokes the need for learning and planning and thus active interference in order to (1) take advantage of the growing opportunities or (2) take necessary measures before any negative consequences can occur.

Any deviations from a plan are compensated for by compensatory mechanisms also in a reactive manner. Thus both feedforward control (planning) and feedback compensation are reactive activities as far as interaction system–environment is concerned. Both can be made active in their implementation. This explains different approaches in control theory.

Examples

1. Classical control systems are systems with no redundancy they can be solved in a closed form. Thus they do not require any searching.
2. Any stochastics introduced to a control system creates redundancy and calls for either elimination of the redundancy and bringing the solution to a closed form or performing search.
3. Optimum control allows for a degree of redundancy that fulfills the need in searching.

In Figure 8.11 multiresolutional planning via consecutive search with focusing of attention and grouping is demonstrated for the control problem of finding a minimum-time motion trajectory. The space is learned in advance by multiple testing, and its representation is based on knowing that the distance, velocity, and time are connected by a simple expression which is sufficient for obtaining computationally the theoretically correct solution with an admissible error. Several methods can be applied in constructing the envelopes of attention.

Instead of solving the second-order differential equation analytically (which is trivial) we find the solution by multiresolutional search to better illustrate its application. One can see that initial search is done at low resolution: the whole space is discretized by sparsely populating it with random points. Then we increase the density of points but place them only within the envelope around the previous search results.

8.7.2 Multiresolutional Planning Reduces the Complexity of Computations

In [76] the phenomenon of computational complexity reduction was analyzed for a particular case of path planning. In this section we will enhance these results for a more general class of cases. We outline our goal of research and demonstrate that this goal is within reach [75].

The following three propositions introduce us to the situation characteristic for our system.

Proposition 1

a. For a stochastic process with a particular spectral density and a particular way of computing the one-step prediction, there exists a set of sampling rates such that if this method of computing the one-step prediction is applied (the BASE method) to each sampling, the results can be combined in such manner that the total prediction will have the same accuracy with a lower complexity of computations.
b. The highest (and subsequently all other) rates of sampling can be increased so that the complexity of the overall hierarchical computation is kept the same as in

(a)

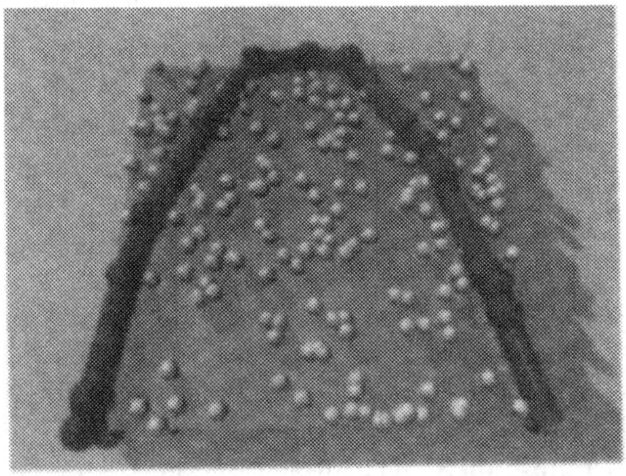

(b)

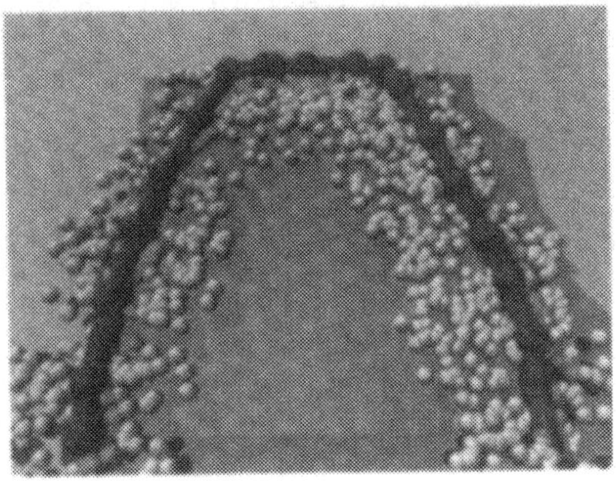

(c)

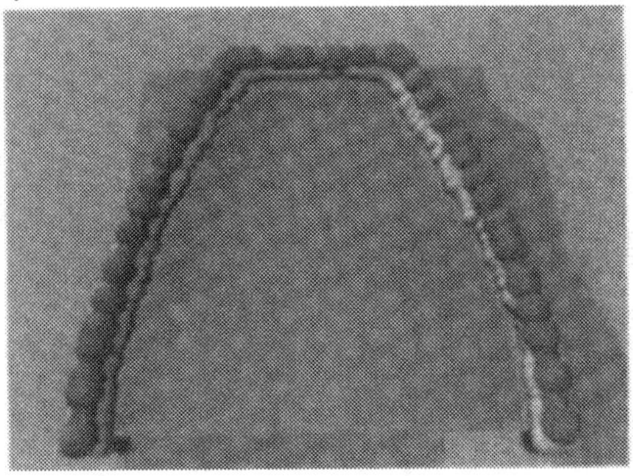

Figure 8.11. Solving a minimum-time control problem by multiresolutional planning with grouping and consecutive focusing of attention.

the BASE method, but the accuracy of the forecast will be increased. This can be formulated in a different way: The high resolution (and subsequently all other) rates of sampling can be increased so that the accuracy of the forecast will be increased while the complexity of the overall hierarchical computation will be kept the same as in the BASE method.

c. The temporal base for the processing and estimation (the minimum background interval required for the forecast) can be reduced with subsequent reduction of the complexity, and the accuracy of the forecast will be increased.

Proposition 2 If an arbitrary signal with spectral density $S(\omega)$ is channeled into n separate streams by n interval filters (0 through ω_1, ω_1 through ω_2, etc.), then the sum of spectral densities of these n channels is equal to the original $S(\omega)$.

Proposition 3 The strong version. If an arbitrary signal with spectral density $S(\omega)$ is channeled into n separate streams by n interval filters (0 through ω_1, ω_1 through ω_2, etc.), then the corresponding spectral densities of the channels can be obtained by "slicing" the original spectral density into the intervals determined by the filters.

Comments

1. In these three propositions (they seem to be equivalent) the reduction of complexity or the increase of accuracy is evaluated in an ordinal rather than in a cardinal manner. This means that we need to compare the variances and conclude that $\sigma_1 > \sigma_2$ always without making any real computations (the computations will be made as a confirmation that our theorem is indeed correct, a numerical example).
2. It remains unclear whether we should sample the same "curve" at all levels, or whether, before sampling at the next level (with higher sampling frequency and shorter BASE), we have to subtract the lower frequency approximation from the original curve, and consider the remainder a new curve for estimation.

Let $x_1(t)$ and $y_1(t), \ldots, y_Q(t)$ be stationary stochastic processes. Let in a multiresolutional system of control as need arises to estimate the future values of some cost-function $V_1(x(t)), \ldots, V_m(x(t))$. The future moments of time are denoted $t + T_{1,\max}, \ldots, t + T_{m,\max}$ correspondingly.

The value of cost-function $V_k(x(t))$ can be considered as a function of $y_1(t), \ldots, y_Q(t)$:

$$V_k(x(t)) \equiv W_k(y_1(t), \ldots, y_Q(t)). \tag{8.29}$$

If for some values of $q = q_1, \ldots, q_s$ the following inequality holds

$$T_{k,\max} \geq T_{yq_{i,\min}}, \tag{8.30}$$

then the expectation of the future value $W_k(y_1(t), \ldots, y_Q(t))$ in the moments of time $t+T_{k,\max}$ under condition that the values of $y_q(u)$ was observed at $u \leq t (q = 1, \ldots, Q)$ does not depend on the values of $y_{q_1}(t), \ldots, y_{q_s}(t)$ or this dependence is negligibly small.

Thus, when the condition (8.30) holds, the conditional expectation $E\{W_k(y_1, \ldots, y_Q)/ y_1(u), \ldots, y_Q(u), u \leq t\}$ is a function of a smaller number of variables than Q, which is considered within a larger time interval corresponding to resolutions at levels q (where $q \neq q_1, \ldots, q_s$). In this way the computational complexity of solving problems of estimation and/or control is reduced. The fast-changing component $x(t)$ is extracted from the stochastic process, and the time of decay of the correlation is smaller than the time of decay for slow-changing components that describe changes of the lower resolution.

To demonstrate how the computational complexity in a multiresolutional system can be reduced, we take the problem of extrapolating a stochastic process with variances $\sigma_1^2, \ldots, \sigma_m^2$ which are the required future moments of time. The maximum value of the time interval Δt_k, at which

$$D_{x,\max}(\Delta t_k) \le \sigma_k^2, \tag{8.31}$$

Now we extrapolate the stochastic process $x(t)$ with a predetermined variance σ_k^2. The algorithm for the extrapolation is robust, and does not require prior knowledge of the spectral density of the stochastic process $x(t)$. This algorithm uses a number of samples $x(s \cdot \Delta t_k)(s = \ldots, -1, 0, 1, \ldots)$, which can be chosen as follows:

$$N \le \begin{cases} \operatorname{int}\left[\dfrac{T_{x,\min} - T_{\max}}{\Delta t_k}\right] + 1 & \text{if } T_{x,\min} > T_{\max}, \\ 0, & \text{and vice versa.} \end{cases} \tag{8.32}$$

The int[·] gives the integer value of x. We can evaluate the computational complexity of the extrapolation as a function of the number of samples N.

If the extrapolation of the stochastic process $x(t)$ is performed by extrapolating its components processes $y_q(t)(q = 1, \ldots, Q)$, then the required quantity of measurements $y_q(s \cdot \Delta t_k)$ is equal to

$$N_{yq,k} \le \begin{cases} \operatorname{int}\left[\dfrac{T_{yq,\min} - T_{\max}}{\Delta t_k}\right] + 1 & \text{if } T_{yq,\min} > T_{\max}, \\ 0, & \text{and vice versa.} \end{cases} \tag{8.33}$$

If the extrapolation value does not require prior information on the process $x(t)$ (which can sometimes reduce the number of samples), then in the expressions (8.32) and (8.33) we take only the equality.

It is possible to demonstrate (see lemma 1) that

$$T_{yq,\min} \le T_{x,\min}. \tag{8.34}$$

LEMMA 1 Let $x(t)$ and $y_1(t), \ldots, y_Q(t)$ be stochastic processes in a multiresolutional system. For each q level of resolution, assign the values of variance σ_q^2 and σD_q defined above. It is natural to assume that the bands of frequencies for the filters are chosen in such a way that the stochastic process $y_q(t)$ is distinguishable at a q level resolution. This means that the following conditions exist: If $T_{x,\min}$, $T_{yq,\min}$, where $q = 1, \ldots, Q$, are intervals determined for the corresponding stochastic processes, then $T_{x,\min} \ge T_{yq,\min}$ for all $q = 1, \ldots, Q$, and at least one of these inequalities is closed to the equality. In other words, the computational complexity of extrapolation in multiresolutional systems that use measurement $y_q(t)$ with interval Δt_k does not exceed the computational complexity of extrapolation where samples of $x(t)$ are used with the same intervals Δt_k, with the same required variances σ_k^2, and computed for the stochastic process $x(t)$ to the horizon Δt_k.

LEMMA 2

Let us assume that a stochastic process can be represented as a sum of several stochastic processes $x(t) = x_1(t) + \cdots + x_m(t)$ and can be characterized by the intervals of decaying correlation

$$u_x = u_{x_1} \gg u_{x_2} > \cdots > u_{x_m}. \tag{8.35}$$

Then, by using m filters, we can obtain a decomposition $x(t) = \sum_{q=1}^{m} y_q(t)$ such that

$$u_x = u_{y1} \gg u_{y2} > \cdots > u_{ym}. \tag{8.36}$$

THEOREM OF MULTI-RESOLUTIONAL EXTRAPOLATION

If the stochastic process $x(t)$ allows for representation as a sum of slow and fast processes (see Lemma 1):

$$x(t) = x_1(t) + \cdots + x_m(t) \tag{8.37}$$

with time intervals of correlation decay that satisfy the inequality (see Lemma 2):

$$u_x = u_{x_1} \gg u_{x_2} > \cdots > u_{x_m}, \tag{8.38}$$

then there exist bands of frequencies $[0, \omega_1]$, $[\omega_1, \omega_2]$, ..., $[w_{m-1}, w_m]$, with intervals of time $\Delta t_q (q = 1, \ldots m)$, that determine the sampling rates of the stochastic process $y_q(t)$ at the output of a particular q-band filter, and the computational complexity of extrapolating the process $x(t)$ in a multiresolutional system that uses measurements $y_q(s \cdot \Delta t_q)(s = \ldots, -1, 0, 1, \ldots)$ is significantly smaller than the computational complexity of extrapolating $x(t)$ in a single-level system that uses only measurements $x(s \cdot \Delta t_0)(s = \ldots, -1, 0, 1, \ldots)$, with the same accuracy of extrapolation.

To prove the theorem we must demonstrate that if the stochastic process allows for representation of $x(t)$ as a sum of slow and fast processes (8.37) for which inequalities (8.38) hold, then the bands $[w_{q-1}, w_q]$, where $q = (1, \ldots, m)$, can be chosen to build filters that will produce processes $y_q(t)$ such that

$$x(t) = \sum_{q=1}^{m} q_q(t), \tag{8.39}$$

and

$$u_x = u_{y_1} \gg u_{y_2} > \cdots > u_{y_m}. \tag{8.40}$$

Then the extrapolation of the process $y_q(t)$ (for $q > 1$), with the time interval Δt_q satisfying the required accuracy of extrapolation, will have substantially lower computational complexity than an extrapolation within the same time interval Δt_q of the process $x(s \cdot \Delta t_q)(s = \ldots, -1, 0, 1, \ldots)$, since, as in (8.19) and (8.20), we have

$$N_{xq} = \begin{cases} \operatorname{int}\left[\dfrac{u_x - T_{\max}}{\Delta t_q}\right] + 1, & \text{if } u_x > T_{\max}, \\ 0, & \text{and vice versa,} \end{cases} \tag{8.41}$$

$$N_{uq,k} = \begin{cases} \text{int}\left[\dfrac{u_{uq} - T_{\max}}{\Delta t_k}\right] + 1, & \text{if } u_x > T_{\max}, \\ 0, & \text{and vice versa.} \end{cases} \tag{8.42}$$

For $q > 1$, $u_x \gg y_{yq}$, then

$$N_{yq,k} \ll N_{xk}. \tag{8.43}$$

Now we have an additional information about the process $y_1(t)$ with a large interval of the correlation time $y_1 = u_x$ and additional information that it is a slow-changing process. This additional information allows us to extrapolate the process $y_1(t)$ to the horizon Δt_k based on a smaller number of samples than the number $N_{y1,k}$ determined by expression (8.43).

For the slow-changing process $y_1(t)$, an accurate extrapolation can be obtained for a small horizon Δt by using the extrapolated value $y_{1,\max}(t + \Delta t_s)$ determined for the larger interval Δt_s. In this regard the following expression holds:

$$y_{1,\max}(t + \Delta t) = y_1(t) + \frac{\Delta t}{\Delta t_s}[y_{1,\max}(t + \Delta t_s) - y_1(t)]. \tag{8.44}$$

By expression (8.44), the computational complexity of extrapolating $y_1(t)$ to a smaller horizon Δt is reduced, while the accuracy of the extrapolation is close to the optimal accuracy $D_{1,\max}(\Delta t)$.

The computational complexity of extrapolating a stochastic process $y_1(t)$ to a larger horizon Δt_k can be evaluated by the expression

$$N_{y1,k} = \text{int}\left[\frac{u_{y1} - T_{\max}}{\Delta t_k}\right] + 1. \tag{8.45}$$

If the process $x(t)$ is extrapolated to a smaller horizon Δt without extracting the slow-changing component $x(t)$, the number of required samples is determined by expression (8.41):

$$Nx(\Delta t) = \text{int}\left[\frac{u_x - T_{\max}}{\Delta t}\right] + 1. \tag{8.46}$$

Thus, at all values $q = 1, \ldots, m$, the expression holds, $N_{x,k} \gg N_{yq,k}$, since at $q > 1$ we have $u_{yk} \ll u_x$ and at $q = 1$ we can use the extrapolation results with the interval of time $\Delta t_s \gg \Delta t_k$. Therefore

$$N_{y1} \ll \max_{l \le k \le m} N_{xk}. \tag{8.47}$$

These inequalities prove the assertion of our theorem that the reduction of computational complexity in the multiresolutional system is significant.

8.7.3 Applying the General S³-Search Algorithm for Planning with Complexity Reduction

The main overall feature of the S³-search planning algorithm is a reduction in the complexity of planning. In this section we address in more detail the source of complexity:

the tessellation process. We begin with the S^3-search algorithm of planning and focus on the role of tessellation in this algorithm. Many steps of this algorithm have been analyzed in [77].

This algorithm has been developed for finding the set of nested multiresolutional optimum motion trajectories in any state-space. As a result, we obtain a quasi-optimum motion trajectory within an efficient computational scheme.

S^3-Search Algorithm

Step 1. Select a bounded region in the domain of system operation (the state-space), within which to investigate alternatives. This step assigns the total scope of the search.

Step 2. Discretize this volume in a way which utilizes computational means in an optimal manner. For example, construct a grid and randomize the position of every grid node.

Step 3. Test the plant, within the specified vicinity of the initial point, using a set of sample inputs and a predefined strategy (or acquire this information from alternate sources, e.g., archived data). The output of this step is a constrained map of permissible local events (or transitions) and related costs, organized as a graph. Repeat this for each consecutive point of the search tree.

Step 4. Search the completed graph for the best solution(s) in the graph. At each consecutive point go back to step 3 to investigate the vicinity and determine the next node candidates. These solutions form the basis for the selection of a new, smaller envelope at the next level.

Step 5. If no solutions exist, return to step 1 with failure.

Step 6. If necessary precision has not been obtained, construct the envelope around the solution to limit the scope, increase resolution within this limited scope, and return to step 1 with success.

Step 7. Exit with solution.

The general algorithm may be understood as follows: It is assumed that there exists a maximally fine tessellation of the space of operation of a system that can be stored and analyzed with a limited amount of computational resources. A precise analysis can therefore be accomplished but must be fixed in advance. An imprecise candidate solution (or group of solutions) of the optimization problem can be obtained via search, provided that a local evaluation of alternatives has been performed.

Assuming that the evaluation has been completed (where completeness implies that a number of adequate trials have been attempted), the results of search can take three forms: (1) The precision of representation is sufficiently high to indicate that the solution is complete. (2) The solution is not sufficiently precise, so the envelope around the solution(s) must be evaluated at higher resolution. (3) No solution is found, either because none exists or because the envelope selected for refinement is not sufficiently large, so the scope of analysis must be re-evaluated.

The overall approach can thus be visualized as a recursive process in which the same functions are applied to the results of preceding function evaluations until a result with adequate precision is produced. One method for the implementation of the described functionality is described in the next few sections.

8.7.4 Tessellation

Development of a multiresolutional representation is performed via a series of tessellations. Tessellation of a closed subset of a state-space means partitioning this polytope into nonoverlapping subsets of tessellata (granules, boxes, or tiles). These subsets form an equivalence class in which all points within a tile are identified with a single label. The natural interpretation is that each of these labels represents the center of a box, plus or minus its measure of error in a manner similar to the tessellation used in digital systems. There are two major differences, however. The first is that the tessellation used in computers or analog/digital conversion is always finer than the final precision required of any result, so tessellation errors can be ignored. In the method proposed here, the desired final precision is achieved only at the lowest level, and all the preceding levels contain more roughly tessellated data.

The second difference is that the tessellation is based on an a priori estimation of the associated memory requirements which are calculated for all the data structures in terms of bytes/unit of information, taking into account all the required algorithms. Tessellation levels are thus constrained directly by computational resources, which is a practical way to approach the problem.

The purpose of tessellation is to provide a template into which the results of testing can be placed to form a knowledge base. The size of the tessellation determines the level of precision of the multiresolutional hierarchy at which the results will be incorporated.

The following is a list of definitions necessary to clarify some of the terms used above.

DEFINITION OF CLOSED COVER

Let $x \subset R''$. A family of S closed sets S_i is called a closed cover of X if $X = \cup_i S_i$.

DEFINITION OF TESSELLATION

Let $X \in R''$ and closed cover $S = \{s_i\}$. A tessellation Z of X is a set $Z = \{z_i\}$ containing exactly one element from each set S_i. This finite set may alternatively be described as an index set, or a set of *labels*.

DEFINITION OF TESSELLATUM SIZE

Given a tessellation Z of X the size of the tessellatum with respect to some norm, ρ, defined on X, is defined as

$$\frac{1}{s} = \min\{r : S_i \subseteq B(z_i, r) \forall i\},$$

where $B(\cdot)$ denotes the ρ-norm ball centered at z_i and with radius r. A tessellation with size s will be denoted as Z_s.

If the sets of the family S that defines a tessellation Z have pairwise disjoint interiors, that is, if $S_i \cap S_j = \emptyset, i \neq j$ (which will be a requirement in the remainder of this discussion), then S reduces to an equivalence relation in X as follows:

DEFINITION OF EQUIVALENCE

Consider a closed cover S of X with pairwise disjoint interiors, and two points $x_1, x_2 \in X$. x_1 and x_2 are equivalent modulo S if $\exists i$ such that $x_1, x_2 \in \text{int}(S_i)$. The points on the common boundary are assigned arbitrarily to either of the classes. The two points that are equivalent modulo S will be denoted $x_1 \equiv x_2$.

DEFINITION OF TESSELLATION OPERATOR

Consider a tessellation $Z_s\{z_i\}$ of a given set X. It follows from Definitions 8.7.4 and 8.7.4 that for any point $x \in X$ there exists an element $z \in Z_s$ such that $z \equiv x$. The operator that assigns $z \rightarrow x$ will be defined as the tessellation operator and denoted $Z_s(x) = z$.

In essence, the preceding definitions show that a system description can be formulated mathematically with precision. The requirements include the existence of a closed subset of the state-space of the target system, which is partitioned into nonoverlapping subsets that form an equivalence class. A tessellation operator maps any vector-valued state into a unique label that represents a locality, or a tile, in the space. Every point is indistinguishable from all other points within a tile. The number of elements in a tessellation can be assigned on the basis of the (expected) future storage and computation costs.

The tessellation is equivalent to the ε-net that forms the basis of complexity evaluation in Kolmogorov's interpretation. Thus the number of tiles used at a particular level of the multiresolutional representation can be mapped directly to epsilon entropy, and this number is determined by the tessellation size.

Assuming the existence of a closed polytope in the state space representing the working envelope of the system in question, or $X \subset R^n$, a tessellation $Z_s = \{z_i\}$ may be created by the following algorithm:

Tessellation Algorithm

Step 1. Select elements N in $Z_s = \{z_i\}$ based on previous computational and memory optimization.

Step 2. Distribute these elements among the n dimensions of the space in such a way that $\prod_{i=1}^{n} n_i \leq N$, where n_i is the number of intervals on each axis.

Step 3. Determine the size of the intervals, δ_i, on each axis by computing

$$\delta_i = \frac{x_{i_{\max}} - x_{i_{\min}}}{n_i}, \qquad 1 \leq i \leq n, \tag{8.48}$$

using prior knowledge about the bounds, $X_{i_{\max}}$ and $x_{i_{\min}}$.

Step 4. Locate the centers of the intervals on each axis, and collect them in sets as

$$C_j = \{c_i{:}c_i = x_{i_{\min}} + (i - 0.5)\delta_i, 1 \leq i \leq n_i\}, \qquad 1 \leq j \leq n. \tag{8.49}$$

Step 5. Form a set of n-tuples by taking the cross-product, $CN = C_1 \times C_2 \times \cdots \times C_n$, over all axes. Label each n-tuple with an integer in $[0, N-1]$. Both *CN* and the set of labels are now valid tessellations of X, in which tiles are defined by the weighted ∞-norm, $|(\alpha_1 x_1, \alpha_2 x_2, \ldots, \alpha_n x_n)||_\infty$, where the x_i signify distance from the center of a tile along the ith axis and the αs are used to make the measure commensurate.

Step 6. Because all points in a tile are equivalent (by Definition 8.7.4), a random shift is assigned to each element of *CN* such that it remains within its tile but is no longer located at its geometric center.

Step 7. The new elements of *CN* are now assigned as the individual elements of the tessellation $Z_s = \{z_i\}$, and the algorithm is complete.

The choice of labels is nontrivial because the tessellations are organized into a hierarchy in which the top level has the largest tile size, and so on, down to the final level which has the smallest tessellation (Fig. 8.12).

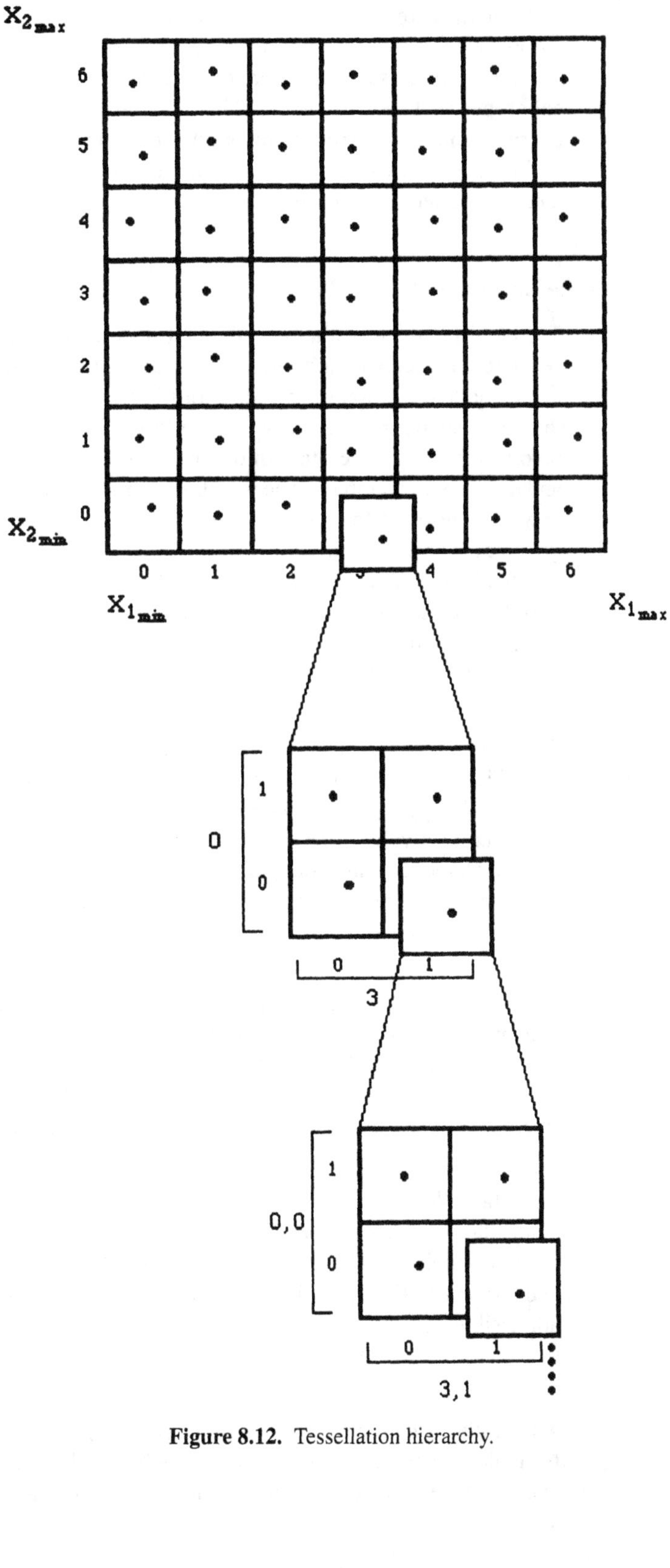

Figure 8.12. Tessellation hierarchy.

If the tiles are organized in a manner that takes into account the nesting of the tessellations, whereby each label can be decomposed into a set of labels at the next level, then the task of the tessellation operator is considerably simplified, and the inheritance of local properties becomes clearer. The resulting simplification allows both the tessellation operator and its inverse (for mapping from a label to its representative state marker) to be implemented using binary masking and shifting operations directly from the codes and knowledge of the base tessellata.

8.7.5 Testing of the Representation

The building of a tessellation is the first step in the process of determining the ε-optimal sequence of tiles which approximates the optimal trajectory. The next important stage is the characterization of the edges or tile-transitions which will be chosen to represent the local behavior of the tessellated system. The manner in which this is accomplished involves testing of the plant in order to document feasible state transitions, their costs, and the associated inputs. The results can be stored in the form of generalized transitions between tiles, and they naturally generate a graph that represents system behavior at a given level of precision.

This investigation of local behavior must be carefully organized because it is supposed to provide all of the information about the dynamics of the target system required for global ε-optimization. First, some mathematical preliminaries are dealt with in the following definitions (based on [78]).

DEFINITION OF DIRECTED GRAPH

A *digraph* G is a set $\{V, E\}$ of vertices and edges. The vertices are a set of labels for which any tessellation Z may be substituted. The edges E form a set of ordered pairs of the type (v_i, v_j), where $v_i, v_j \in V$. The edge set may be used to signify the existence of a valid transition or connection between the first and second members of its own elements. This nonsymmetric, transitive relation E may be expanded to include additional elements representing the cost and cause of each edge (or transition) in the form $(v_i, v_j, c_{ij}, u_{ij})$.

DEFINITION OF CTNC

Control Tessellated Null Controllability (CTNC) A dynamical system is control tessellated null controllable in $W \subseteq X$ if, for any open set $Y \subseteq X$ containing the origin, there exists a number $s_u(W, Y) \in R^+$ such that for all the tessellations Ω_s of Ω with $s \geq s_u$ there exists a sequence of tessellated controls $u[k] \in \Omega_s$ such that the system can be steered from any initial condition $x_0 \in W$ to Y without violating the state constraints.

DEFINITION OF STNC

State Tessellated Null Controllability (STNC) The system above is state tessellated null controllable if, in a region $W \subseteq X$ for any open set $Y \subseteq X$ containing the origin, there exists a number $s_x(W, Y) \in R^+$ such that for all tessellations Z_s of X with $s \geq s_x$ and for any initial condition $x_0 \in W$ there exist a finite number n and a sequence of admissible controls $u[k] \in \Omega, k = 1, 2, \ldots, n$ such that $z[k] = Z_s(x[k]) \in X, k = 0, 1, 2, \ldots, n$ and $z[n] \in Y$.

DEFINITION OF JTNC

Jointly Tessellated Null Controllability (JTNC) A system is jointly tessellated null controllable if there exists a number s_0 such that the system is state tessellated null controllable when the controls are restricted to a controllable tessellation Ω_s of Ω with size $s \geq s_0$.

In directed graph the edge set consists of ordered n-tuples, $(v_i, v_j, c_{ij}, u_{ij})$. These, for the sake of convenience, may be considered a parent, child, or successor vertex, the cost of the transition from the parent to the successor, and the associated input label.

In the case under consideration this information is equivalent to a generalization of the cost of moving from one tile to another, as opposed to the point-to-point motion usually investigated. To make matters worse, the imposition of input constraints partitions the continuous state-space into controllable and uncontrollable subsets, even if the original continuous or discrete system is controllable (reachable). In [78] Sznaier and Sideris conclude that all extensions always assume that the set of possible control laws is a dense subset of R^n and the initial condition of the system is perfectly known. So it is immediately clear what cost should be placed on a given tile-to-tile transition and also that the transitions are valid given the input constraints and the dynamics of the system. The remaining definitions are included to indicate the kind of information that is required to guarantee complete tessellated controllability. In general, this requirement is more stringent for pointwise state controllability (as opposed to controllability to the origin).

Since it is not clear how to establish correspondences between pairs in a given tessellation analytically, an experimental way to do this must be developed; it also establishes the cost of transitions and characterizes the local dynamics of the system under consideration. The use of an experimental approach provides several advantages in any case. The most attractive of these is that empirical data, which are actually obtained from real measurements, can be processed to provide the necessary information directly. In addition empirical evidence of the existence of controllability can be obtained directly from the system.

In other words, a testing-based strategy can be used in place of an analytical model, which may in any case be unavailable. Alternatively, a complex model can be simulated off-line to acquire the required information. Finally, if an analytical model is available, this too can be solved numerically to implement the testing strategy. The inability to establish links between tiles thus turns out to be an advantage after all.

The experimental approach works as follows (see Fig. 8.13). The trajectories of the system emanating from a tile in response to each of a set of tessellated admissible inputs of the system may be determined from an initial condition (within a tile) for a particular tessellated input. Neighboring tiles (in n dimensions) can differ from a parent tile in up to n coordinates. That is to say, each tile is surrounded by 8 immediate neighbors in two dimensions, or 26 in three dimensions, and so on. As can be seen in the two-dimensional example in the figure, the first distinguishable change in state will typically occur along a particular axis, the one corresponding to the state with the fastest dynamics. The change can be detected using the tessellation operator as soon as a tile to tile transition occurs. This is shown in the figure as the transition from tile (4. 2) to (5, 2).

Thereafter the trajectory evolves within the same tile, approaching and then receding from the state associated with the tile as a representative value in the tessellation algorithm. At the point of closest approach (by the infinity norm), the transition is considered to be complete, and the cost and input may be recorded in the graph.

Similarly the state continues to evolve across the boundary into tile (4, 3). This may be considered a unit change along the second axis, and recorded upon reaching the point of closest approach to the marker state for that tile. Unit increments along each axis, if and when they exist, for each input in the tessellated admissible set may be recorded in this way for all the tiles. The number of edges which will be generated for each parent

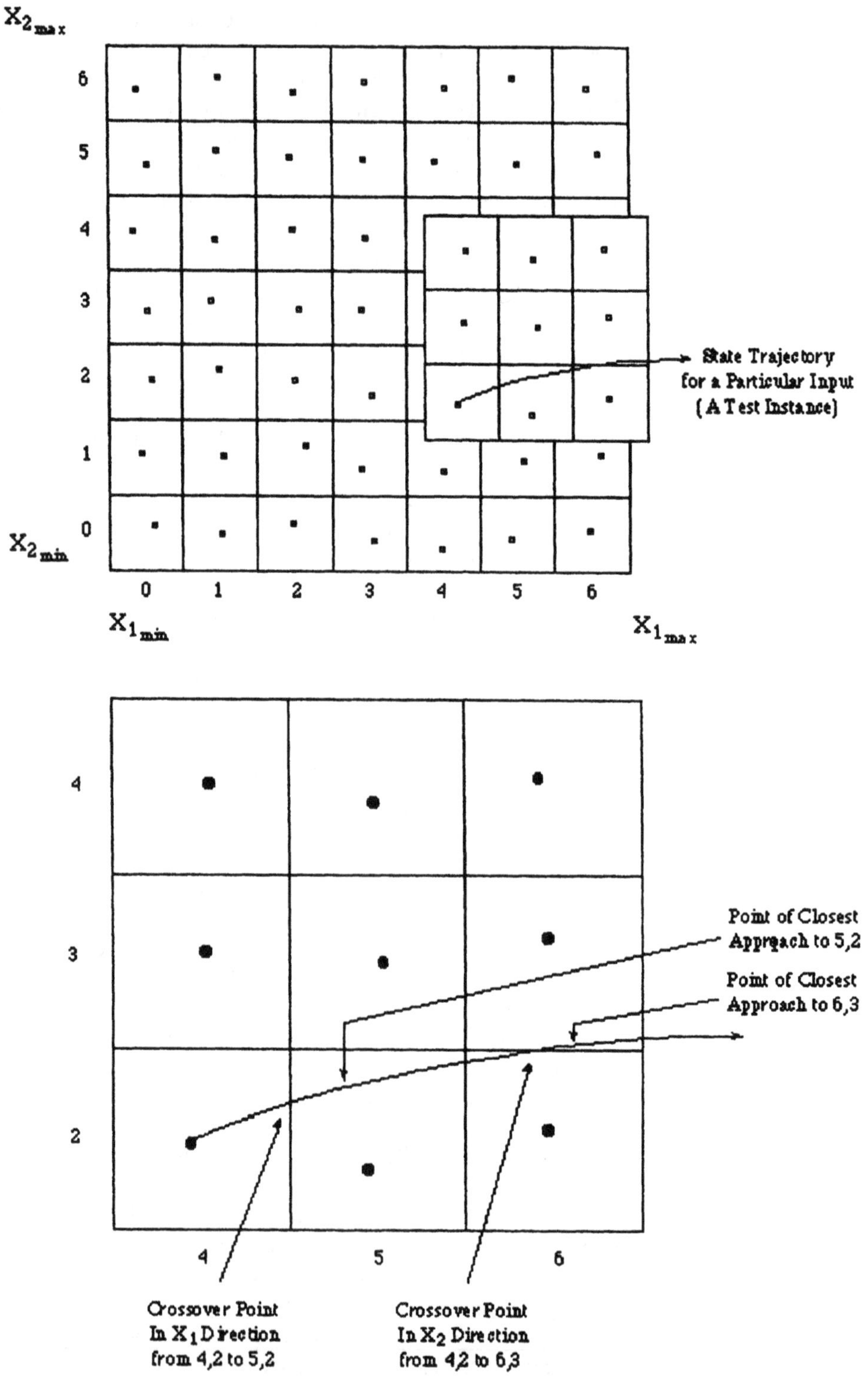

Figure 8.13. Strategy of testing.

using this method will be less than or equal to $n \times m$, where n is the order of the system and m is the number of input quanta.

If averaging is desired, a representative set of initial conditions may be selected, randomly or otherwise, for each test input, to achieve a generalization of the test result in terms of the averaged cost of the transition.

Connect Algorithm

Step 1. For each element z_i in Z_s with tile marker y_i

1.1 For each input, from $u_{\min}$ to $u_{\max}$,

1.1.1 For $j = 1$ to n,

1.1.2 *Either* solve $\dot{\mathbf{x}} = \mathbf{f}(\mathbf{x}, \mathbf{u}, t)$ in time with $x(0) = z_i$ *or* analyze measured data until $|(y_i)j - x_j| > \delta_j/2$, where j signifies the component of state under investigation, and δ_j is the discrete on that axis.

1.1.3 Record the tile characteristic point z_{i+1} of the tile that contains x_j.

1.1.4 *Either* continue to solve $\dot{\mathbf{x}} = \mathbf{f}(\mathbf{x}, \mathbf{u}, t)$ iteratively in time *or* analyze measured data until

$$|(z_{i+1}) - x||_{k+1} - ||(z_{i+1}) - x|_k > 0, \tag{8.50}$$

where the index k signifies the index of the iteration, and the formula represents the first time that the distance of the evolving trajectory from the point z_{i+1} increases. The point of closest approach is then x_k. Record the cost of reaching this point as the cost of the transition from z_i to z_{i+1}.

1.1.5 Next j.

1.2 Next input.

Step 2. Next element of tessellation.

⋮

Step 8. End.

8.7.6 Search

The goal of a search algorithm is usually that of identifying a particular solution that optimizes or satisfies a predetermined criterion. In this thesis, however, it is not a particular solution but an ensemble that is being sought.

It is for this reason that a variation on Dijkstra's algorithm (Chapter 1, Section 1.3) was designed for the task at hand. In doing so, the Dijkstra solution was preserved so that the result would continue to be an optimal one. The primary change is that in consonance with the new motivation, the information that is discarded in Dijkstra's algorithm is not immediately thrown away.

In order to demonstrate the changes from the original algorithm, the graph and search trees from Section 1.3 are duplicated in Figure 8.14. Dijkstra's algorithm is described as follows:

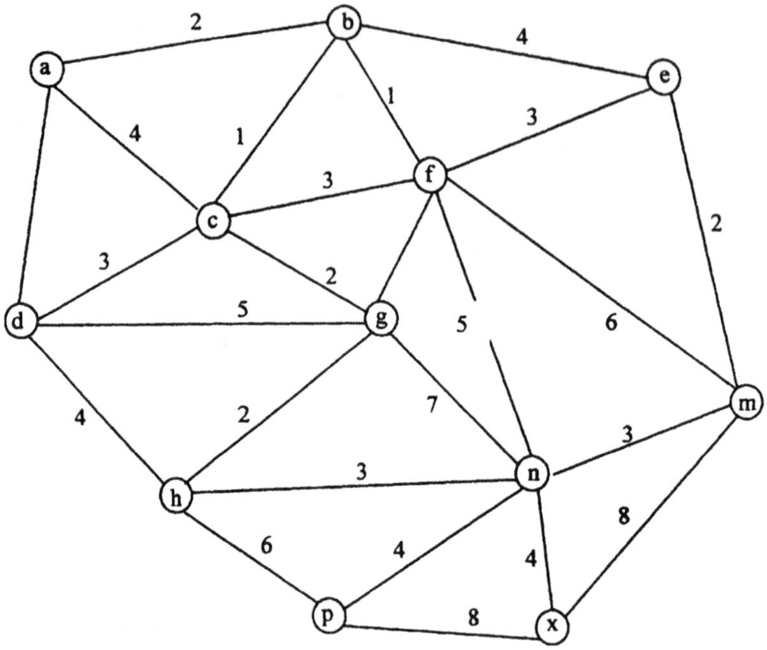

Figure 8.14. Example of an undirected graph.

Step 1. Initialize a search tree of size equal to the number of nodes in the graph with the edge from the start node to itself, with a transition cost of zero, and mark it "open" for future examination.

Step 2. Select the cheapest "open" node in the tree and mark it "closed" (investigated).

Step 3. If the goal node is achieved, traverse pointers to retrieve the solution path.

Step 4. "Expand" the current node, generating all of its successors (children) from the graph.

Step 5. For each child that has not been previously expanded

5.1 If the child is not in the "to" column of the tree, add the new edge, else:

5.2 If the existing cost to reach the child is higher than the newly determined cost, switch the source node and cost on the existing edge.

Step 6. Go to step 2.

The search tree is given in Figure 8.15 in which an optimal solution from node a to node x is identified as the sequence $\{a, b, f, n, x\}$. Now change step 5 of the algorithm as follows:

Step 5. For each child

5.1 If the number of occurrences of the child are less than the preassigned limit, add the new information to the tree, and mark the new child open.

5.2 If the new child has a lower cost than all existing ways of getting to that child

5.2.1 If the same {to, from} pair exists in the tree, replace it with the new information, and mark the new child open.

5.2.2 If two identical {to, from} pairs exists in the tree, replace the more expensive one with the new information, and mark the new child open.

5.2.3 Replace the most expensive {to, from} pair with the new information, and mark the new child open.

5.3

5.3.1 If the same {to, from} pair exists in the tree, and it is more expensive, replace it with the new information, and mark the new child open.

5.3.2 If two identical {to, from} pairs exist in the tree, replace the more expensive one with the new information.

5.3.2.1 If the replaced edge was more expensive, mark the new child open.

5.3.2.2 Add the incremental cost of the more expensive substitution to edges built from the previous node, if it has already been expanded.

5.3.3 If the new pair is less expensive, replace the most expensive {to, from} pair with the new information, and mark the new child open.

The effect of this change is that instead of collecting the unique optimal histories for each new node in the form of a search tree, a graph is obtained in which distinct suboptimal pasts are represented. The number of alternate histories is a parameter that can be selected on the basis of the number of acceptable computations during search.

TO	FROM	COST	OPEN
a	a	0	0
b	a	2	0
c	b	3	0
d	a	1	0
e	b	6	0
f	b	3	0
g	f	4	0
h	d	5	0
m	e	8	0
n	f	8	0
p	h	11	0
x	n	12	0

Figure 8.15. Final search tree for Dijkstra's algorithm.

The reason for elaborate set of checks in the reworked algorithm is that having received the graph or adjacency list from testing, it is entirely possible that many ways of getting from one node to another may have been examined. Consequently {to, from} pairs with more than one cost may have been recorded, in addition to the same *to* node having many possible *from* nodes (ancestors). It is the goal of the algorithm to record not only the best of these but also to remember a prespecified number of distinct alternate ancestors with increasing costs. Step 5 of the algorithm performs this task by ensuring that any pair added to the search graph replaces only repeated edges or higher-cost edges except in step 5.5.2.2. This is a special case of an edge that is discarded because it is a duplicate but is cheaper than the new edge that is added because it is a new distinct alternative, and not because it is cheaper.

Under these conditions, if the replaced node has already been expanded, all the future edges that have already been incorporated into the tree must be updated to correctly account for the cost of attaining the replaced *to* node. The need for this modification will be established in the section on envelope creation.

If the change is implemented for the example that was solved with Dijkstra's algorithm, and if the number of histories is set to two, the graph obtained is as shown in Figure 8.15.

As may be seen from the table, two ways of reaching each *to* node are remembered. The change does not, however, change the optimal solution found by the algorithm. This may be found in a manner similar to that of Dijkstra's algorithm by noting that the cheapest way to reach x is from n, to n from f or h, f from b, b from a, and a from a, resulting in the familiar path $a \to b \to f \to n \to x$. The immediate result of the change is that another optimal path $a \to d \to h \to n \to x$ is also represented in the search graph.

Thus the space requirements of the modified algorithm are a constant factor larger than Dijkstra's algorithm, which depends on the fanin to each node, and a constant average overhead is introduced in the number of operations required at each iteration. The complexity of the algorithm in the sense of Knuth remains $O(N^2)$. Finally the property of always finding the optimal solution is preserved. The reason for accommodating the increased information is explained in the next section.

8.7.7 Consecutive Refinement

Once the search has been repeated at several levels of resolution (top down) the reduced envelope of the last level will have a diameter less than or equal to the desired precision for the system, so no further search is necessary. The solution consists of a sequence of recommended state transitions, the accumulated costs, and the required inputs. If desired, smoothing may be performed in order to derive a continuous curve describing derived optimal behavior, which is accurate within the limits of precision of the final level of analysis.

At the first iteration of the overall algorithm, however, the search volume consists of the entire space of operation of the plant in question. Constraints on the space are in the form of limits of safe operation, and they constitute hard bounds that cannot be exceeded. The choice of the number of variables to be considered constitutes the second important aspect of the problem. Given a fixed amount of memory or storage space, an increase in the number of variables naturally limits the initial resolution of the tessellation (discretization) that is possible within the closed volume of investigation. It is possible, however, to introduce variables individually as their contribution to system dynamics becomes significant (at later iterations), when the precision of the representation is higher and the volume smaller. This restricts the growth of complexity and adds a degree of freedom to the options available to the planner. This possibility is deferred for future research. However, it should be clear that the partition of variables into a hierarchy of significance (or graded effect) is consistent with the desire to create an efficient hierarchical decomposition of system behavior.

Under normal circumstances, when the search algorithm for a level returns with success, the search volume is reduced by the algorithm for successive refinement. This is done so that the smaller search volume can be investigated at higher precision without a growth in complexity. Alternately, increases in the search volume may occur when the search algorithm, which is designed to make decisions based on empirical evidence as opposed to prior knowledge of system dynamics, is forced to operate on a reduced envelope that excludes parts of the space necessary to generate an approximate trajectory subject to the more precise description of system dynamics at a new level.

The natural desire to maximize the reduction of search volume from level to level precludes the possibility of deliberately overestimating this volume. Consequently failure in the search algorithm may be used as feedback to the refinement algorithm and the prior search may be repeated with a larger envelope at the same resolution if it was because of the inadequacy of the graph resulting from the previous reduction of the search space. Under these conditions of failure, algorithms for determining areas of inadequate analysis must be invoked to repeat the process at the same level of resolution as the previous iteration with a larger search volume.

Unfortunately, conventional measures on the state-space are inadequate for building an envelope because the distance between two points in this space is not a true indicator of the cost of the transition between them (unless the cost is expressed solely as a function of state). The only measure that can therefore be applied with any consistency is the cost-function itself; the question is, How?

In the method under consideration, it is suggested that an efficient way of doing this is to consider the output of the search algorithm to be not a particular trajectory but an ensemble. This ensemble may be parametrized by including all paths that differ in cost from the optimal cost by some constant factor at each level of resolution. Then the convex-hull of the locus of the ensemble can be expected to have a high probability of containing the optimal solution at the highest resolution.

Envelope Algorithm

Step 1. Initialize a table of size equal to the number of nodes in the search graph with the goal node and the cost of including in the optimal path to itself as $c_{\text{est}}(x, x) = 0$ and $c_{\text{min}}(x, a) =$ cost of the optimal path.

Step 2. For each open node, k, in the table

- 2.1 Compute an estimate for the cost of reaching the goal from each predecessor node, $k - 1$, using $c_{\text{est}}(x, k - 1) = c_{\text{est}}(x, k) + c(k, k - 1)$. Derive $c(k, k - 1)$ from the original graph, and determine, from the search graph, the optimal cost $c_{\text{min}}(k - 1, a)$ of reaching $k - 1$.
- 2.2 If the existing value of $c_{\text{est}}(x, k - 1)$ is less than currently in the table, replace it, or if $k - 1$ is not in the table add it, and mark the node open.

Step 3. Go to step 2. (If no open nodes remain, then exit.)

The algorithm works as shown in Figure 8.16. Its operation can be explained as follows: An estimate for the cost of passing through the goal node on the way to the

NODE	$c_{\min}(k, a)$	$c_{\text{est}}(x, k)$	$C(k)$	OPEN
x	12	0	12	1 (1) 0
m	8	8	16	1 (2) 0
n	8	4	12	1 (3) 0
e	6	10	16	1 (4) 0
f	3	14 (3) 9	17 (3) 12	1 (5) 0
h	5	7	12	1 (7) 0
b	2	14 (5) 10	16 (5) 12	1 (8) 0
g	4	10 (7) 9	14 (7) 13	1 (6) 0
c	3	12 (8) 11	14 (8) 14	1 (9) 0
d	1	11	12	1 (10) 0
a	0	12	12	1 (11) 0

Figure 8.16. Table for envelope algorithm.

goal is clearly the optimal cost determined from the search algorithm. This is marked in the first row of Figure 8.16.

The second pair of rows in the figure is generated by considering the ancestors of the only open node in the figure at the second iteration. Thus m and n are added to the table and marked open. The optimal costs to each of these nodes is 8, and the cost of going to x from them (according to the original graph) is 8 and 4, respectively. This information is added to the table. At the next iteration both m and n are open and are both examined in order. Now m has the ancestors e and f. Their optimal costs from the start are 6 and 3, respectively. An estimate for the cost to the goal is obtained by adding their direct costs to the estimate for m. This gives $2 + 8 = 10$ for e, and $6 + 8 = 14$ for f. Proceeding similarly for n, the cost-to-go is estimated as $5 + 4 = 9$ for f and $3 + 4 = 7$ for h. Since the new cost estimate for f is lower than the one in the table, this is replaced, and if f had already been closed, it would be reopened for a reexamination of its past.

The procedure just described is carried out until no open nodes remain, at which point the correct estimates for each node have been obtained. This result is shown in the figure, and it is clear that out of the 12 nodes eight can be included in an optimal path, two more can be traversed with an increase in cost of one or two units, and passing through the remaining two requires an increase in cost of four units.

Thus an envelope with a cost relaxation of 10% will contain nine nodes and one with a relaxation of 25% will contain ten, and so on. In the general case the nodes represent tessella or tiles, so their union defines the new envelope. If the envelope is required to be a connected set and is not, closure can be accomplished by using the same infinity-norm measure as in the tessellation algorithms. Naturally the size of the envelope should be such that a higher resolution analysis is possible within the constraints of computational resources. In the event that the volume of the envelope is prohibitively large for a selected relaxation parameter value, it may be reduced by using a geometrical measure; however, this may require several iterations in which the size of the envelope is adjusted to include relevant parts of the space that were inadvertently omitted.

8.7.8 Evaluation of Complexity

A quantification of the complexity of the overall algorithm of planning is relatively easy to determine using the Kolmogorov measure of epsilon entropy. It will be shown, however, that the results of this approach are counterintuitive. The usual measure of algorithmic complexity, as defined by Knuth, is then examined and is shown to provide a more realistic evaluation of performance.

Kolmogorov Complexity

Since the epsilon-entropy measure is additive and is based on the size of the epsilon net for a problem, it is easy to identify both the epsilon and the entropy for each level of the multiresolutional representation, and then to add them in order to arrive at a single number for the overall structure.

This may be done by recalling that a tessellation is equivalent to an epsilon net for the space of search, with epsilon equal to the tessellation size. Thus the entropy of a level of the hierarchy may be expressed as

$$H_\varepsilon(A_i) = \log_2(N_\varepsilon(A_i)), \qquad (8.51)$$

where A_i is the search volume at the ith level and $N_\varepsilon(A_i)$ is the number of tessella of size e required to cover the volume.

The composite entropy of the hierarchy may then be expressed as

$$H_\varepsilon(P) = \sum_{i=1}^{k} H_\varepsilon(A_i), \qquad (8.52)$$

where P denotes the problem under investigation, and encompasses all the levels of representation, ending with the highest-resolution level, the kth.

Conversely, the entropy of a single-level representation at the highest resolution may be computed from

$$H_\varepsilon(P_{\text{single level}} = \log_2(N_{\varepsilon_k}(A_0)), \qquad (8.53)$$

where A_0 is the volume of the full search space at the lowest level of resolution, and $N_{\varepsilon_k}(\cdot)$ is the number of tiles required to cover a given volume using ε_k, the tessellation size of the level of highest resolution.

Complexity in Terms of Polynomial Bounds

In terms of primitive computations, costs are incurred at each stage of the algorithm, including tessellation, testing, search, and refinement. Conveniently the upper bound on these computations may be derived from the complexity of the least efficient of the algorithms, which happens to be a variation on the well-documented search algorithm.

To see this, it is sufficient to first establish that the search algorithm is of $O(N^2)$ complexity and then to show that the rest are of lesser or equal complexity, which actually happens to be the case.

As was mentioned earlier [79], the complexity of Dijkstra's algorithm *in an abstract graph with N nodes* $O(N^2)$ in time. This evaluation is derived from the fact that a graph with N nodes can have no more than N^2 edges. Since each edge is examined no more than once in the application of Dijkstra's algorithm, the complexity is bounded from above by a quadratic function of N.

As was also noted earlier, the number of additional comparisons involved in the modified search algorithm is proportional to the fan in allowed in order to be able to apply the envelope algorithm. Since this is a fixed number, the complexity of the search algorithm is multiplied by a constant factor, and its complexity remains $O(N^2)$, although the number of node expansions is increased by a factor proportional to the fan in.

The complexity of the tessellation algorithm is negligible in practical terms because it is simply a process of assigning labels to N tiles at each level. Its complexity bound is thus $O(N)$. The testing algorithm assigns a number of edges to each of N nodes whose number is bounded by the number of input quanta. Since this is a fixed number, the number of tests is always less than or equal to kN, where k is the number of input quanta. Thus the complexity of this algorithm is also $O(N)$.

Finally the envelope creation algorithm expands each node back to its ancestors, computing an estimated cost for each node. In the worst case each expansion may cause the re-expansion of all other nodes because the estimated cost is too high at the previous iteration (this of course will never happen in practice). Thus the number of expansions is bounded from above by N^2 and the complexity can be no larger than $O(N^2)$.

Thus the overall complexity of the set of algorithms is also of order N^2. The trouble with this measure is that it does not actually indicate how long a series of computations it will take; it is just an indirect indicator of growth in computation time with problem size. Furthermore there exists no formal method of introducing hardware performance into this equation. For instance, a program for problem size $2N$ may run four times slower than one of size N, simply because of the failure of the processor's cache of optimization algorithms. Nevertheless, short of introducing empirical measures—for which a powerful argument can be made—this measure is still a good indicator of when an algorithm is truly inefficient or costly.

The measure of problem size, N, however, is a misleading indicator. For instance, it is easy to assume that the search algorithm is of acceptable complexity because the order with respect to the number of nodes is a graph. In fact the problem size should not be determined by the number of nodes at all but rather by the precision required of the solution of the search algorithm. This opens up a thorny issue.

Recall that the general idea is to use an e-sized tessellation of some n-dimensional state-space as the vertex set of the graph to be searched. Intuitively it is clear that for a given number of dimensions, the number of vertices will be inversely proportional to that power of the tessellation size. Specifically, consider the closed, bounded, n-dimensional space, and let the tessellation size be such that there are k intervals per dimension (or axis)

$$k_i = \frac{x_{i_{\max}} - x_{i_{\min}}}{\varepsilon_i}, \qquad 1 \leq i \leq n. \tag{8.54}$$

Assume that also the order of the k_i's is the same. Then the total number of vertices is

$$N \cong k^n = \left(\frac{1}{\varepsilon}\right)^n. \tag{8.55}$$

The complexity of search in the state-space with respect to tessellation size is thus

$$O(\varepsilon^{-2n}) = O(k^{2n}), \tag{8.56}$$

which differs significantly from the (correct!) $O(N^2)$ result for dimensions larger than two. What is known more precisely is that higher-order models or smaller tessella can automatically cause the algorithm to slow to an intolerable crawl—if memory requirements do not already cause it to become incomputable.

The use of multiple levels affects this figure substantially. The number of nodes to be examined at a time by the search algorithm is limited by the number of tessella at a given level. Instead of performing the search on a tessellation of the entire search volume at the highest resolution, it is then possible to perform the number of searches which is required to arrive at a particular precision using the same acceptable N at each level.

The resulting computations may then be calculated from the product k times N^2, where k is the number of levels and N is a much smaller number than it would be for a full tessellation at the highest resolution. Minimum complexity for a particular case of S^3 search was found in [76].

REFERENCES

[1] A. Newell and H. A. Simon, GPS: A Program that simulates human thought, in E. A. Feigenbaum and J. Feldman, eds., *Computers and Thought*, R. Oldenbourg K. G., 1963, pp. 279–293.

[2] K.-S. Fu, Learning control systems, in *Advances in Information System Sciences*, J. T. Tou, ed., Plenum Press, New York, 1969.

[3] G. Saridis, *Self-organizing Control of Stochastic Systems*, Marcel Dekker, New York, 1977.

[4] G. Saridis and A. Meystel, eds., *Proc. IEEE Workshop on Intelligent Control*, Troy, NY, 1985.

[5] R. E. Fikes and N. Nilsson, STRIPS: a new approach to the application of the theorem proving to problem solving, *Artificial Intell.*, **2** (1971): pp. 189–208.

[6] R. E. Fikes et al., Learning and executing generalized robot plans, *Artificial Intell.* **3** (1972).

[7] P. Hart, et al., A formal basis for the heuristic determination of minimum cost paths, *IEEE Trans. Sys. Sci. Cybern.* (1968): 100–107.

[8] J. E. Doran and D. Michie, Experiments with the graph-traverser program, *Proc. Royal Society* A (1966): 235–259.

[9] W. E. Howden, The sofa problem, *Computer Journal* **11** (1968): 299–301.

[10] T. Lozano-Perez and M. A. Wesley, An algorithm for planning collision free paths among polyhedral obstacles, *Commun. ACM* **22** (1979): 560–570.

[11] J. S. Albus, Mechanisms of planning and problem solving in the brain, *Math. Biosci.* **45** (1979): 247–293.

[12] T. Lozano-Pérez, Automatic planning of manipulator transfer movements, *IEEE Trans. Syst. Man Cybern.* **11** (1981): 509–681.

[13] M. Julliere et al., A guidance system for a mobile robot, *Proc. of the 13th Int. Symp. on Industrial Robots*, April 17–21, 1983 **2**: 17–21.

[14] R. Chavez and A. Meystel. Structure of intelligence for an autonomous vehicle. *Proc. IEEE Int. Conf. on Robotics and Automation*, 1984, pp. 584–591.

[15] J. E. Hopcroft et al., *SIAM J. Comput.* **14** (1985): 315–333.

[16] A. M. Meystel, Planning in a hierarchical nested controller for autonomous robots. *Proc. IEEE 25th Int. Conf. on Decision and Control*, 1986, pp: 1237–1249.

[17] J. C. Latombe, *Robot Motion Planning*. Kluwer Academic, Boston, 1991.

[18] M. Brady et al., *Robot Motion: Planning and Control*, MIT Press, Cambridge, 1982.

[19] J. S. Albus and A. Meystel, *Behavior Generation in Intelligent Systems*, NIST, Gaithersburg, MD, 1997.

[20] R. W. Brockett, Poles, zeros, and feedback: state space interpretation, *IEEE Trans. Auto. Cont.* **10** (1965): 129–135.

[21] F. Albrecht et al., Path controllability of linear input-output systems, *IEEE Trans. Auto. Cont.* **31** (1986): 569–571.

[22] H.-W. Wohltmann, Function space dependent criteria for target path controllability of dynamical systems, *Int. J. Cont.* **41** (1985): 709–715.

[23] U. Kotta, Right inverse of a discrete time non-linear system, *Int. J. Cont.* **51** (1990): 1–9.

[24] H. Seraji, Design of feedforward controllers for multivariable plants, *Int. J. Cont.* **46** (1987): 1633–1651.

[25] B. Widrow and E. Walcech, *Adaptive Inverse Control*, Prentice Hall, Upper Saddle River, NJ, 1996.

[26] J. C. Latombe, Controllability, recognizability, and complexity issues in robot motion planning, *Proc. 36th Annual Symp. on Foundations of Computer Science*, 1995, pp. 484–500.

[27] V. Akman, *Unobstructed Shortest Paths in Polyhedral Environments*, Berlin: Springer-Verlag, 1987.

[28] N. Rowe and R. Richbourg, An efficient Snell's law method for optimal path planning across multiple two-dimensional irregular hemogeneous-cost regions, *Int. J. Robot. Res.* (1990): 48–66.

[29] R. E. Tarjan, A unified approach to path problems, *J. Assoc. Comp. Mach.* **28** (1981): 577–593.

[30] A. Meystel and E. Koch, Computation simulation of autonomous vehicle navigation, *Proc. IEEE Int. Conf. on Robotics Automation*, 1984.

[31] D. Keirsey et al., Autonomous vehicle control using AI techniques, *IEEE Trans. Softw. Eng.* **11** (1985): 986–992.

[32] M. Montgomery et al., Navigation algorithm for a nested hierarchical system of robot path planning among polyhedral obstacles, *Proc. IEEE Int. Conf. on Robotics and Automation*, Raleigh, NC, 1987, pp. 1616–1622.

[33] J. C. Latombe, A fast path planner for a car-like indoor mobile robot. *Proc. 9th Nat. Conf. on Artificial Intelligence*, Anaheim, CA (1991): 659–665.

[34] J. Barraquand and J. C. Latombe, Nonholonomic multobody mobile robots: controllability and motion planning in the presence of obstacles, *Algorithmica* **10** (1993): 121–155.

[35] P. Burlina, D. DeMenthon, and L. S. Davis, Navigation with uncertainty: reaching a goal I a high collision risk region, *Proc. IEEE Int. Conf. on Robotics and Automation*, 1992, pp. 2440–2445.

[36] J. Borenstein and Y. Koren, Histogramic in-motion mapping for mobile robot obstacle avoidance, *IEEE Trans. Robot. Auto.* **7** (1991): 535–539.

[37] R. Cerulli et al., The auction technique for the sensor based navigation planning of an autonomous mobile robot. *J. Intell. Robot. Sys.* **21** (1998): 373–395.

[38] H. Takeda et al., Planning the motions of a mobile robot in a sensory uncertainty field, *IEEE Trans. on Pattern Anal. Mach. Intell.* **16** (1994): 1002–1017.

[39] A. Lazanas and J. C. Latombe, Motion planning with uncertainty: a landmark approach, *Artificial Intelligence* **76** (1995): 285–317.

[40] R. G. Simmons, Structured control for autonomous robots, *IEEE Trans. Robot. Auto.* **10** (1994): 34–43.

[41] C. Becker, H. Gonzalez-Banos, J. C. Latombe and C. Tomasi, An intelligent observer, in O. Khatib and J. K. Salisbury, eds., *Lecture Notes in Control and Information Sciences 223*, Experimental Robotics IV, Springer-Verlag, New York, 1997, pp. 153–160.

[42] S. M. LaValle et al., Motion strategies for maintaining visibility of a moving target, *Proc. IEEE Int. Conf. on Robotics and Automation*, 1997.

[43] K. S. Al-Sultan and M. D. S. Aliyu, A new potential field-based algorithm for path planning, *J. Intell. Robot. Sys.* **17** (1996): 265–282.

[44] J.-O. Kim and P. K. Khosla, Real-time obstacle avoidance using harmonic potential functions, *IEEE Trans. Robot. Autom.* **8** (1992): 338–349.

[45] J. Barraquand et al., Numerical potential field techniques for robot path planning, *IEEE Trans. Sys. Man Cybern.* **22** (1992): 224–241.

[46] C. Hein and A. Meystel, A genetic technique for planning a control sequence to navigate the state space, *Goddard Conf. on Space Applications of AI*, NASA (1994): 113–120.

[47] K. Sugihara and J. Smith, A genetic algorithm for 3-D path planning of a mobile robot, Tech. Report, Dept. of Information and Computer Science, University of Hawaii at Manoa, Sept. 1996.

[48] G. Grevera and A. Meystel, Searching in a multidimensional space, *Proc. 5th IEEE Int. Symp. on Intelligent Control*, vol. **2**, Philadelphia, 1990, pp. 700–725.

[49] C.-F. Liaw and C. C. White III, A heuristic search approach for solving a minimum path problem requiring arc cost determination, *IEEE Trans. Sys. Man Cybern.*, A**26** (1996): 545–551.

[50] R. Bhatt et al., A real-time guidance system for an autonomous vehicle, *Proc. IEEE Int. Conf. on Robotics and Automation*, Raleigh, NC, 1987, pp. 1785–1791.

[51] R. Bhatt et al., A real-time pilot for an autonomous robot, *Proc. IEEE Int. Symp. on Intelligent Control*, Philadelphia, 1987, pp. 135–139.

[52] A. M. Meystel, Theoretical foundations of planning and navigation for autonomous robots. *Int. J. Intell. Sys.*, **2** (1987): 73–128.

[53] A Meystel and A. Lacaze, Introduction to integrated learning/planning paradigm, A. Meystel, ed., *Proc. 1997 Int. Conf. Int. Systems and Semiotics: A Learning Perspective*. NIST Publ. 18, Gaithersburg, MD, 1997, pp. 523–528.

[54] A. Meystel, *Autonomous Mobile Robots: Vehicles with Cognitive Control*, World Scientific, Singapore, 1991.

[55] J. H. Tsitsiklis, Efficient algorithms for globally optimal trajectories, *IEEE Trans. Automatic Control*, Vol. 40, No. 9, 1995, pp. 1528–1538.

[56] T. Skewis and V. Lumelsky, Experiments with a mobile robot operating in a cluttered unknown environment. *Proc. IEEE Int. Conf. on Robotics and Automation*, 1992, pp. 1482–1487.

[57] K. Sun and V. Lumelsky, Path planning among unknown obstacles: the case of a three-dimensional cartesian arm, *IEEE Trans. Robot. Auto.* **8** (1992): 776–786.

[58] V. Lumelsky, S. Mukhopadhyay, and K. Sun, Dynamic path planning in sensor-based terrain acquisition, *IEEE Trans. Robot. Auto.* **6** (1990): 462–472.

[59] M. Kircanski and M. Vukobratovic, Contribution to control of redundant robotic manipulators in an environment with obstacles. *Int. J. Robot. Res.* **5** (1986): 112–119.

[60] A. A. Maciejewsky and C. Klein, Obstacle avoidance for kinematically redundant manipulators in dynamically varying environments. *Int. J. Robot. Res.* **4** (1986): 109–116.

[61] R. G. Roberts and A. A. Maciejewski, Nearest optimal repeatable control strategies for kinematically redundant manipulators, *IEEE Trans. Robot. Auto.* **8** (1992): 327–337.

[62] T. S. Wilkman and W. S. Newman, A fast, on-line collision avoidance method for a kinematically redundant manipulator based on reflex control. *Proc. IEEE Int. Conf. on Robotics and Automation*, 1992, pp. 261–266.

[63] A. M. Meystel, Multiscale systems and controllers, *Proc. IEEE/IFAC Joint Symp. on Computer-Aided Control System Design*, Tuscon, AZ, 1992, pp. 13–26.

[64] A. Meystel and S. Uzzaman, Planning via search in the input-output space, *Proc. IEEE Int. Symp. on Intelligent Control*, Chicago, 1993.

[65] J. Albus, A. Meystel, and S. Uzzaman, Nested motion planning for an autonomous robot, *Proc. IEEE Regional Conf. on Aerospace Control Systems*, Westlake Village, CA, 1993.

[66] P. J. Burt, Smart sensing, in Herbert, ed., *Machine Vision*, Academic Press, New York, 1988.

[67] B. S. Everitt, *Graphical Techniques for Multivariate Data*, North-Holland, Amsterdam, 1978.

[68] P. Graglia, *Path Planning in the Multi-Resolutional Traversability Space*, MSEE Thesis, Drexel University, Philadelphia, PA, 1988.

[69] G. J. Grevera and A. Meystel, Searching for an optimal path through pasadena. *Proc. IEEE Intl. Symp. on Intelligent Control*, Arlington, VA, 1988, pp. 308–319.

[70] *Inside Macintosh*, vol. 5, Addison-Wesley, Reading, MA, 1986, p. 156.

[71] G. McCarty, *Topology: An Introduction with Application to Topological Groups*, New York, McGraw-Hill, 1967.

[72] P. K. Pearson, Fast hashing of variable length text strings, *Commun. ACM* 33 (1990): pp. 677–680.

[73] R. E. Tarjan, *Data Structures and Network Algorithms*, Society for Industrial and Applied Mathematics, Philadelphia, 1983.

[74] P. Graglia and A. Meystel, Planning minimum time trajectory in the traversability space of a robot, *Proc. IEEE Int. Symp. on Intelligent Control*, Philadelphia, 1987, pp. 82–87.

[75] E. Khazen and A. Meystel, Why multiresolutional hierarchies reduce complexity of systems for estimation and control, *Proc. IEEE Int. Symp. on Intelligent Control*, Gaithersburg, MD, 1998.

[76] Y. Maximov and A. Meystel, Optimum design of hierarchical controllers, *Proc. IEEE Int. Symp. on Intelligent Control*, Glasgow, Scotland, 1992, pp. 514–520.

[77] S. Uzzaman, Multiresolutional search for the design of feedforward controls, PhD dissertation proposal, April 1995.

[78] M. Sznaier and A. Sideris, Feedback control of quantiser constrained system, *IEEE Trans. Auto. Cont.*, Vol. 39, No. 7, 1994, pp. 1497–1502.

[79] A. V. Aho, J. E. Hopcroft, and J. D. Ullman, *Data Structures and Algorithms*, Addison-Wesley, 1983.

PROBLEMS

8.1. In a field with several obstacles, assign the initial and the final points. Use a topographical map created for some of your realistic experiences. Find the best (minimum-time) path from the initial to the final points by using a visibility graph-based principle of planning.

8.2. Solve the same problem using potential field-based planning.

8.3. Compare the visibility graph-based planning, the potential field-based planning, the search-based planning for solving the problem of motion in the cluttered environment.

8.4. Construct a graph (15 nodes) with weighted edges. Use Dijkstra algorithm for searching the least expensive path from one point to another in this graph.

8.5. For a system of the first order (e.g., a mass moving by a force) with one degree of freedom, search for the optimum motion trajectory in coordinates speed = F(path). Apply S^3 search algorithm (see Subsubsection 8.7.3). Use random distribution of search nodes.

8.6. Consider XY coordinate system and a rectangular space represented in this coordinate system. Divide this rectangular space in several arbitrary segments and assign various traversabilities to these segments. Assign an initial point and the goal point. Use S^3 search to find the minimum-time trajectory of motion from the initial to the final points.

CHAPTER 9

MULTIRESOLUTIONAL HIERARCHY OF PLANNER/EXECUTOR MODULES

This chapter is dedicated to technical issues, including algorithms and procedures, relating to the nested hierarchy of PLANNER/EXECUTOR modules of an intelligent system. The discussion starts with a demonstration of the properties of the PLANNER/EXECUTOR hierarchy in a multiresolutional hybrid controller (Section 9.1); then the theoretical premises are outlined, and the notion of the canonical hybrid PLANNER/EXECUTOR module is introduced. The concept of optimal functioning is presented for this shell, and a theoretical solution is demonstrated for one of the shells by using classical optimization techniques as the optimal PLANNER/EXECUTOR Shell.

In many cases functioning of the shell can be secured by a simple production system that provides a quasi-minimum time functioning of PLANNER/EXECUTOR. Then a more complicated case is addressed: when should the minimum time functioning be achieved in the obstacles strewn state-space? Since the output trajectory has been obtained, its inverse is computed. We discuss the more frequent case where the approximate inverse is computed via forward searching. Finally the multiresolutional hierarchy of PLANNER/EXECUTOR modules is considered as a multirate hierarchical control system.

9.1 HYBRID CONTROL HIERARCHY

The PLANNER/EXECUTOR modules form a hierarchy of a hybrid control system. Hybrid systems are defined as systems in which a discrete event system (DES) is used to supervise the behavior of a continuous state system (CSS). A number of sources [1–5] treat hybrid control systems as a couple of DES and CSS connected via an "interface" that contains D/A and A/D converters and performs some auxiliary functions (Fig. 9.1*a*).

In a hybrid control system of the PLANNER/EXECUTOR modules, DES plays a role of a supervisory controller (the controller in Fig. 9.1*a*). The plant is the part

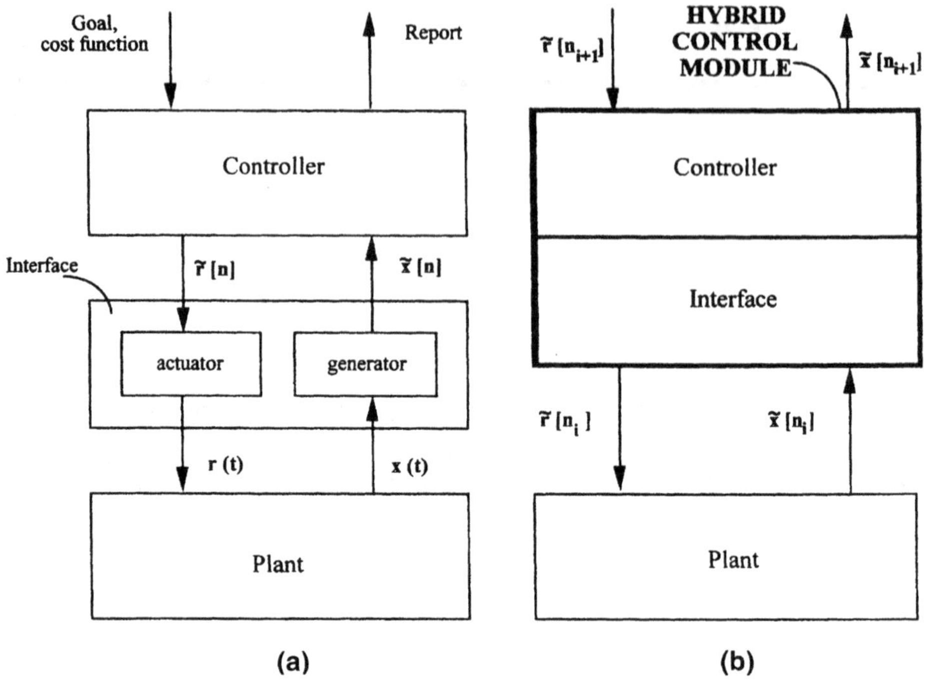

Figure 9.1. Models of the hybrid control systems of PLANNER/EXECUTOR modules: (*a*) Conventional structure of PLANNER/EXECUTOR modules; (*b*) canonical hybrid control module (PLANNER/EXECUTOR modules).

of the model that represents the entire continuous state (CSS) portion of PLANNER/EXECUTOR modules ([1], p. 334). Thus the plant is usually considered together with the continuous state controller: Both are treated together as PLANNER/EXECUTOR modules (i.e., control agents). The latter is analyzed for both single agent and multi-agent cases. This model is thoroughly investigated in [2–5]. It was demonstrated that the system can be considered a goal-generating system and therefore can qualify for being called an intelligent controller.

In the subsequent sections we propose to generalize the concept of hybrid systems and to define them as systems in which discrete event systems DES(n_i) of a sampling frequency (n_i) are used to supervise the behavior of the discrete event systems DES(n_j) with another sampling frequency (n_j). This is a theoretical framework for analysis and design of programs and processes for the hierarchy of PLANNER/EXECUTOR modules. Theoretically CSS are treated as DES with infinitely high frequency of sampling, or with high enough frequency of sampling and linear interpolation between the samples. Now we can consider the plant to be a classical plant from control theory, not a control agent. The plant is presumed to be modeled either with differential equations or by constructing its transfer function, or by constructing an automaton in which the transition and output functions are Lookup tables obtained from differential equations and/or transfer functions by approximating them with difference equations (or *rules*).

MULTIRESOLUTIONAL HYBRID SYSTEMS

Also we introduce the following modification to the model presented in Fig. 9.1*b*: We will demonstrate explicitly that the overall goal of control arrives to the controller externally, and that the results of control should be reported externally too. This external source of goals is the upper level of the PLANNER/EXECUTOR hierarchy, and the recipient of the computation results is the adjacent level below. After this, we are able to merge the interface with the control module. Thus we receive a hybrid control module that is called a canonical hybrid PLANNER/EXECUTOR module, since it is obtained as a generalization and extension of the results from [1–5].

The canonical PLANNER/EXECUTOR module can be applied to itself as many times as necessary, thus allowing the development of funnel (stemlike) and branching (treelike) recursive architectures. It turns out to be a convenient tool for analysis and design of different types of the hierarchical PLANNER/EXECUTOR modules that are introduced by the virtue of applying the algorithms of PLANNER/EXECUTOR modules recursively. Two types of PLANNER/EXECUTOR modules-hierarchies are discussed in this chapter: the funnel-hierarchy in which the sampling frequency of control gradually increases as the granularity of representation reduces and the subset of attended state-space becomes smaller, and the tree-hierarchy in which the branching recursion is applied and the flow of control information is being decomposed top-down. In the tree-hierarchy, each branching from the level above to the level below is associated with substituting one word of the vocabulary of the lth level by several words of the vocabulary of the $(l-1)$th level of the hierarchy.

The PLANNER/EXECUTOR module is implemented for the analysis and design of the NIST-RCS control system which is recommended for large and complex systems, including intelligent systems (autonomous robots, unmanned power stations, systems of integrated and/or intelligent manufacturing, etc.).

9.2 THEORETICAL PREMISES

The following premises are considered prerequisites for discussing the hierarchical hybrid control system of PLANNER/EXECUTOR modules:

Premise 1 *Representations of systems by automata and by differential equations are equivalent.*

This premise is essential for constructing and applying the concept of canonical PLANNER/EXECUTOR modules. The statement of this equivalence has been known in the literature on control theory since the 1950s, and it is a commonplace in the literature on hybrid systems. Nevertheless, we should reiterate it with some comments. The procedure of obtaining the plant automaton was proposed in [2] and is applicable in all practical purposes. From the point of view of the digital control module, the plant behaves as an automaton: It receives and understands the digital control commands, exercises some virtual transition and output functions, and when it returns sensor information, it arrives in a digital form. Introduction of the plant automaton is unavoidable for resolving the problem of plant controllability. The conditions of controllability were obtained based on the LaSalle invariance principle. Thus a procedure was introduced that permits a system to learn the realizable plant automaton.

DYNAMIC SYSTEMS ARE REPRESENTABLE BY AUTOMATA

One of the ways of representing continuous systems as virtual DES uses the operation of generalization[1] (see [5]).

Premise 2 *Representations both of the system and of its controller are granulated (tessellated).*

[1] A. Pury and P. Varaiya use the term "abstraction" in [5]. We prefer the term "generalization," since it better conveys the meaning of the procedures we apply and the results we obtain. More on this issue can be found in Chapter 2.

Even for the case of continuous control, the reality of space tessellation remains. It is determined by the threshold of sensitivity which can be explicated for all components of the system[2]. In the case of Fig. 9.1*b* the phenomenon of granulation becomes characteristic of all PLANNER/EXECUTOR modules and their resolution. It was suggested in [3] that the state space should be tessellated as follows:

REPRESENTATION IS TESSELLATED

- The cells (tessellata, granula) should represent all distinguishable states.
- Transitions should be placed between states that represent adjacent cells.
- Transitions should be labeled with the symbols generated by the hypersurfaces separating the associated cells.
- The design of the output function of the automaton should take into account the state-space tessellation and the transitions that it entails. This process of state-space tessellation is similar to the one described in [6] (and later refined in [7]).

Premise 3 *Each PLANNER/EXECUTOR module is characterized by the value of resolution for all variables (including the control variables) as well as for time.*

Resolution of the state-space is determined by the size of granula (cell, tessellatum) that is induced by both the value of limit in sensitivity and value of unmodeled component in the plant representation. Lower resolution corresponds to the larger granula, and vice versa. Resolution increases when the frequency of sampling increases because the change of the variables is smaller during the smaller interval of time. These matters are discussed in [3, 8].

Premise 4 *The results of state-space tessellation is the alternative model of the PLANNER/EXECUTOR module.*

BG-MODULE IS TESSELLATED AUTOMATON

The set of state-space-situated cells (the results of state-space tessellation) with the information of transitions from a state to a state (the edges of the corresponding graph) can be considered a model of the PLANNER/EXECUTOR module. Then any realizable string of edges constitute a state-space representation of the particular process of control. We will call such a graph the *state-space representation of PLANNER/EXECUTOR module*. Since all transitions are time processes, the PLANNER/EXECUTOR module is supposed to be associated with the time space as determined in [4].

SEARCH IS APPLIED FOR CONTROL FINDING

Premise 5 *Search is a general procedure for control finding and includes the analytical cases as the shortcuts.*

The trivial method of finding the control trajectory is a construction of the string of motion segments that connects the initial state with the final state. If the value of cost is associated with these segments, the optimum control can be found by any method of the graph-search which minimizes the total cost of the trajectory.[3] Since search procedures are time-consuming, analytical representations are often a useful tool of increasing the efficiency of computations at the stages of control or design. In the case of large and complex (including intelligent) systems, analytical expressions are frequently not available (or not reliable). Thus search algorithms might become the only way of finding the alternative of motion trajectory or other control recommendations for control. Search

[2]Actually any limit in sensitivity is equivalent to considering an *approximated discrete signal system* instead of talking about *continuous system*.

[3]It is obvious that the "optimum" trajectory found in this way could be only approximation of the optimum trajectory. Indeed, all information related to the single segments is obtained for the universal (general) use for possible constructing multiple alternatives of the optimum trajectory. Thus the validity of this information does not necessarily hold after the segment becomes a part of the particular trajectory.

in the state-space can be considered competitive with the variable-structure control system with the inevitable well-known phenomenon of chattering that happens when the system switches from one control invariant to another.

COMPLEXITY DEPENDS ON SEARCH ALGORITHM

Premise 6 *Algorithms of search in the state-space determine the computational complexity of a controller.*

There exists a variety of search methods, starting with the carefully guided search algorithms and ending with the exhaustive, or A*, search algorithms (as recommended in [6, 7, 9]). The general dependence is obvious between the resolution of the state-space representation (assigned as its granulation, tessellation, scales) and the complexity of the search: When the resolution is higher, the complexity of search grows. This is why there is a tendency not to have the resolution at a level of the PLANNER/EXECUTOR module higher (or the tessellation cells smaller) than some particular value depending on the total scope of control (or the volume of the state-space in which the motion of the system is expected). The latter is defined by the overall set of boundaries and constraints that are supposed to be provided as a part of design specifications.

ARCHITECTURE DETERMINES DESIGN

Premise 7 *Principles of control systems design determine the final architecture.*

The general case of controller design is relevant to the problems of control formulated for intelligent systems. In this case the following principles are held:

- A goal state (the desirable final state) and a set of relevant cost-functions are assigned (externally) as a major framework for any control system design.
- Within this general case, two particular cases are separated. First is the case where the reference trajectory is assigned (externally) to be obtained at the output of the plant. An input control sequence should be computed that ensures that this reference trajectory is followed with a particular accuracy before the final point is reached. In the second case, only the final state is assigned with no reference trajectory given. This requires that (1) the designer either compute the reference trajectory, or the control system (then the second case will become the first case) or that (2) the control command be somehow generated at each consecutive moment of time by using an evaluation of deviation of the present state from the final state in each consecutive current state.
- Three strategies of control are therefore possible:

    1. The control sequence is computed online by using a particular law of compensating the deviations from the reference trajectory (tracking with feedback control, FBC). We can use it as the output plan and compute the input plan by inverting the output plan (feedforward control, FFC).
    2. The desired output motion (the reference trajectory) is computed off-line by optimizing the motion from an initial state to the final. Then we are able to use the computed optimal output trajectory as a reference (the desired motion). After this, we can return to the first case and use it as the output plan, compute the input plan by inverting the output, and then perform tracking (1), with compensation control (FBC) only for deviations from the reference trajectory (FFC+FBC).
    3. No particular reference trajectory is favored. We can compute only the feedback control by compensating on-line the deviations from the goal state (positioning with feedback control, FBC).

Since the second framework is the most general one, in our further discussion we will consider a system in which both FFC and FBC are to be introduced.

9.3 CANONICAL HYBRID PLANNER/EXECUTOR MODULE

9.3.1 Basic Control Architecture

For constructing the basic control architecture we will follow [1–4]. The reference trajectory $\bar{x}_r(t)$ is assigned as a continuous trajectory, and a set of intermediate goals $G\{g_m\}$, $m = 1, 2, \ldots, f$. Thus the controller is chosen so as to ensure T-stable transitions between the goal states. The plant state equations can be written as

$$\frac{d\bar{x}}{dt} = f(\bar{x}, \bar{r}), \tag{9.1}$$

where $x \in R^n$, $\bar{r} \in R^m$, and $f : R^n \times R^m \to R^n$.

The architecture to be used is shown in Figure 9.1*a* and refined in Figure 9.2*a*. A hierarchical hybrid control system with a nested system of PLANNER/EXECUTOR modules uses a logical DES submodule of PLANNER to supervise the behavior of a continuous (actually, higher-resolution) data plant EXECUTOR. In this chapter we extend this case to a more general one where the controller DES(n_{i+1}) with sampling frequency n_{i+1} supervises the behavior of the plant DES(n_i) with a higher sampling frequency n_i (the continuous plant can be considered a particular case).

Each BG module receives its goal (or its reference trajectory) externally, namely from the upper adjacent level of the hierarchy of control and returns a report of performance as the goal is achieved (or at the intermediate stages of tracking). The control process is built on a general concept of a model following a state feedback control system. It uses an internal reference model which is represented by a stable minimum phase transfer function, $P^{\text{ref}}(s)$. This reference model computes an "ideal" trajectory, denoted $x_{\text{r}}(t)$, between the two goal sets corresponding to a T-stable transition. As the reference model state moves along $x_{\text{r}}(t)$, the controller that is used switches sequentially between N_g virtual EXECUTORs. The index ($m = 1, \ldots, N_g$) will be used to specify which control EXECUTOR is "active." The mth EXECUTOR uses output feedback through the linear system $F_m(s)$.

The reference model is designed in such a way that the reference trajectory does not pass too close to regions in the state-space associated with events other than initial event g_i and terminal event g_j. We will assume that the reference trajectory remains at least a distance ε (which is a granulation-induced value) from goal sets other than g_i and g_j. The value of ε is larger or equal to the indistinguishability zone[4] linked with the final accuracy of control, with the concept of ε-optimality introduced later, and with the granulation of state space which is produced in all DES systems.

The objective of the PLANNER/EXECUTOR module design is to find intermediate goal sets and associated control sequences that ensure transitions between intermediate goal sets, all of which are situated on the optimum trajectory. To accomplish this objective, we first generate the reference trajectory, $x_{\text{r}}(t)$. We will assume that the sample points are spaced closely enough so that the reference trajectory between consecutive setpoints will be interior to the union of two open spheres of radius γ centered at the setpoints. This will ensure that the reference trajectory is covered by the sequence of

[4]See Chapter 2.

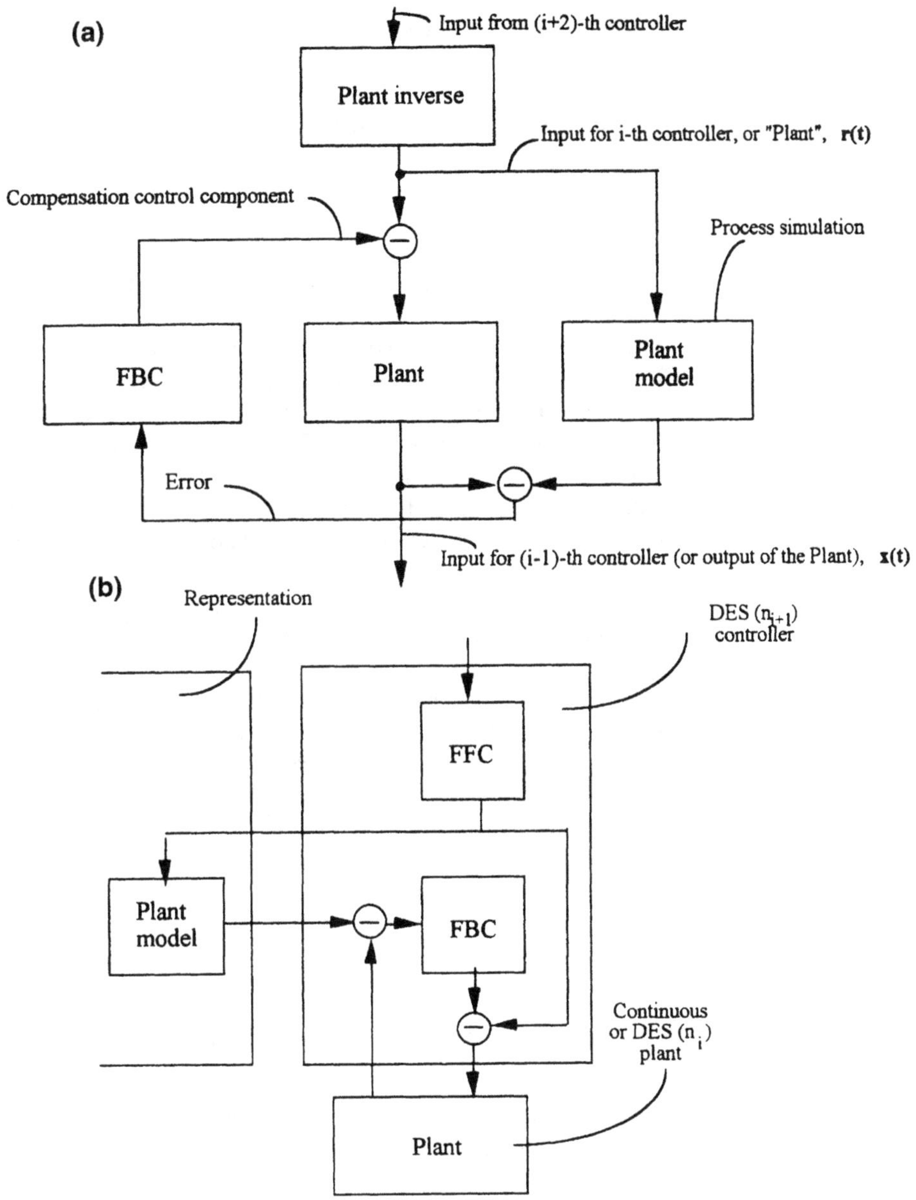

Figure 9.2. Hybrid control module: (*a*) Basic structure; (*b*) as implemented within the NIST-RCS architecture.

spheres centered at the setpoints. With the mth EXECUTOR, we will associate a subsequence of N_m contiguous setpoints denoted S_m. For each subsequence S_m a linear aggregated plant, $P_m^{\mathrm{a}}(s)$ will be constructed to satisfy

$$P\left(\frac{S}{x_{r,mi}}\right) = \left(I + N_m\left(\frac{s}{x_{rmi}}\right)\right) P_m^{\mathrm{a}}(s), \tag{9.2}$$

providing stability under multiplicative perturbation that ensures the desired accuracy of the aggregated plant. Both the concept of aggregation and the synthesis of control trajectories, goals, and EXECUTORs presume sequential interpretation. However, without loss of generality, most of the results remain valid for the parallel decomposition and aggregation of both controllers and trajectories of motion.

With the mth EXECUTOR, the mth intermediate goal set will be defined as

$$g_m = \bigcup_{x_r \in X_m} n(x_r, \gamma), \tag{9.3}$$

where $n(x_r, \gamma)$ is the special neighborhood set with the radius $\gamma > \varepsilon$ and the goal set g_m is within a distance γ of some point in X_m.

The controller represented by our BG module (including the PLANNER and the EXECUTOR) is a discrete event system (DES).[5] It receives as inputs a goal and a sequence $x_s[n]$ of symbols generated within the interface as a result of processing "sensor" information from the functioning plant. Thus these symbols can be considered a subset of the finite alphabet X of plant symbols. The supervisor outputs a sequence $r[n]$ of symbols drawn from a finite alphabet R of control symbols. The dynamics of the overall system is modeled by a deterministic finite automaton. The DES controller, C_d, for the hybrid system, is the ordered 4-tuple, $C = (S, X, R, \Phi, Q)$, where S is a finite alphabet of controller symbols, X and R are finite alphabets of plant state and control symbols, respectively. $\Phi : S \times X \to S$ is the controller's transition operator. $Q : S \to R$ is the controller output function.

Now let us assume that both parts of the system shown in Figure 9.1a are DES. Interface between the plant and supervisory DES, or between two DES of different frequency of sampling ($n_i < n_{i+1}$), works as follows: The bottom-up part of it (generator) transforms the sensor information of the plant's continuous time (or of higher frequency DES(n_i)) state trajectories, $x(t)$ (or $x(n_i)$), into a sequence of plant symbols, $x(n_{i+1})$ at the DES(n_{i+1}) lower frequency of functioning (n_{i+1}), and submits these symbols $x(n_{i+1})$ to DES(n_{i+1}) at the appropriate instants of time. The top-down part of the interface (actuator) converts a sequence $r[n_{i+1}]$ of control symbols into a plant input, ($r(t)$, or $r(n_i)$).

9.3.2 Recursive Application and Nesting of the Canonical Hybrid PLANNER/EXECUTOR Modules

The concept of nesting is connected with operations of generalization (abstraction) that presume inclusion and are connected with the properties of representation in the multilevel PLANNER/EXECUTOR modules [10]. Nesting is unavoidable in generalizing a continuous system into a discrete one, and it evolves in generalizing the high-resolution DES (with high-frequency sampling) into the lower-resolution DES (with lower-frequency sampling).

The approach is outlined in [5] where there are proposed two methods of generalized representation. With the first method, a differential equation is approximated by a simpler differential inclusion. Instead of analyzing the system with the inclusion $x \in f(x)$, one analyzes the simpler inclusion $x \in g(x)$ where $f(x) \in g(x)$. The second inclusion is a conservative generalization of the first inclusion, since the solutions of the first inclusion are also solutions of the second inclusion. The second inclusion may be simpler computationally; for example, it may be a piecewise constant inclusion.

The reach set of the abstracted (generalized) automaton is a conservative or upper bound on the reach set of the original hybrid automaton. The conditions of inclusion

[5] In the control literature, the PLANNER is frequently called the "supervisor."

must be satisfied for both space and time. A timed automaton is a hybrid automaton in which all the continuous variables are clocks. To conservatively generalize a differential inclusion, we look at its evolution, namely at the set of trajectories starting with the initial conditions at the same space tessellatum at the lower resolution. We obtain bounds on the evolution of the differential inclusion and "encode" this information with a clock. The relations between the differential equation and the clock are then made by a mapping. Each control location can be abstracted in this manner provided that we associate an initial set with it. Thus, whenever a transition is made into a new control location, the state must actually be within.

Let us consider a plant DES P which generates a language K. Assume that $K_{(i)}$ is the desired logical behavior which we want our hybrid control system to exercise. The ellipsoidal algorithm test can be used to decide in finite time whether or not a given arc of control is indeed T-stable. We can therefore construct a sequence of plant DES by successively testing all the arcs of control. As each testable arc is identified, we build up an ordered sequence of plant DES instantiations $P_{(1)}, P_{(2)}, \ldots, P_{(i)}$ whose associated languages are subsets of the specified language K^*. In this sense the languages $K_{(i)}$ generated by this sequence of DES plants represent better and better approximations of K^* which are guaranteed of being T-stable. In other words, we obtain the following sequence:

$$K^* \supset \ldots \supset K_{(i)} \supset \ldots \supset K_{(3)} \supset K_{(2)} \supset K_{(1)}. \tag{9.4}$$

The largest $K_{(i)}$ whose language is contained within K^* forms the supremal controllable sublanguage of the K^*. If it turns out that P_d is a testable plant DES, then the supremal controllable sublanguage is K^*. If $K_{(i)}$ turns out to be smaller than K^*, then this means that our formal specification cannot be realized by the stable hybrid system. We must either settle for the behaviors realized by $P_{(i)}$ or else modify the plant and interface until the formal specification can be achieved. Basically the procedure outlined above represents a way of constructing T-stable DES controllers for the hybrid system. The final plant DES obtained in this manner represents a modification of the original specification K^*, which is guaranteed of being a logically stable interpretation of the continuous-state plant's behavior.

If each language K is associated with a particular timed automaton, the latter will form a "funnel-hierarchy" in which the control processes will be nested (see Fig. 9.3*a*). The tree-hierarchy is demonstrated in Figure 9.3*b*. The controller of the upper $(i + 1)$th level controls a system that has been decomposed into three parallel subsystems (I, II, and III). Each is "perceived" by the supervisory controller of the $(i + 1)$th level as a plant. Thus the system shown in Figure 9.3*a* has three "virtual plants" of the ith level. Similar tree decomposition happens in the subsystem I at the next level. However, the subsystem II perpetuates top-down as a funnel-hierarchy. Finally the subsystem III does not perpetuate at all: It has a real plant at the next level top-down which has the appropriate granularity.

The phenomenon of building the "virtual plants" to satisfy the information needs of the supervisory controller is a unique one. We illustrate it in the Figure 9.4.

9.3.3 World Model: Maintenance of Knowledge

We can see that in the hierarchy of hybrid PLANNER/EXECUTOR modules, a number of new problems emerge that are not typical for the conventional control systems. Usually the controller is a carrier of the plant model in the appropriate form: in the form

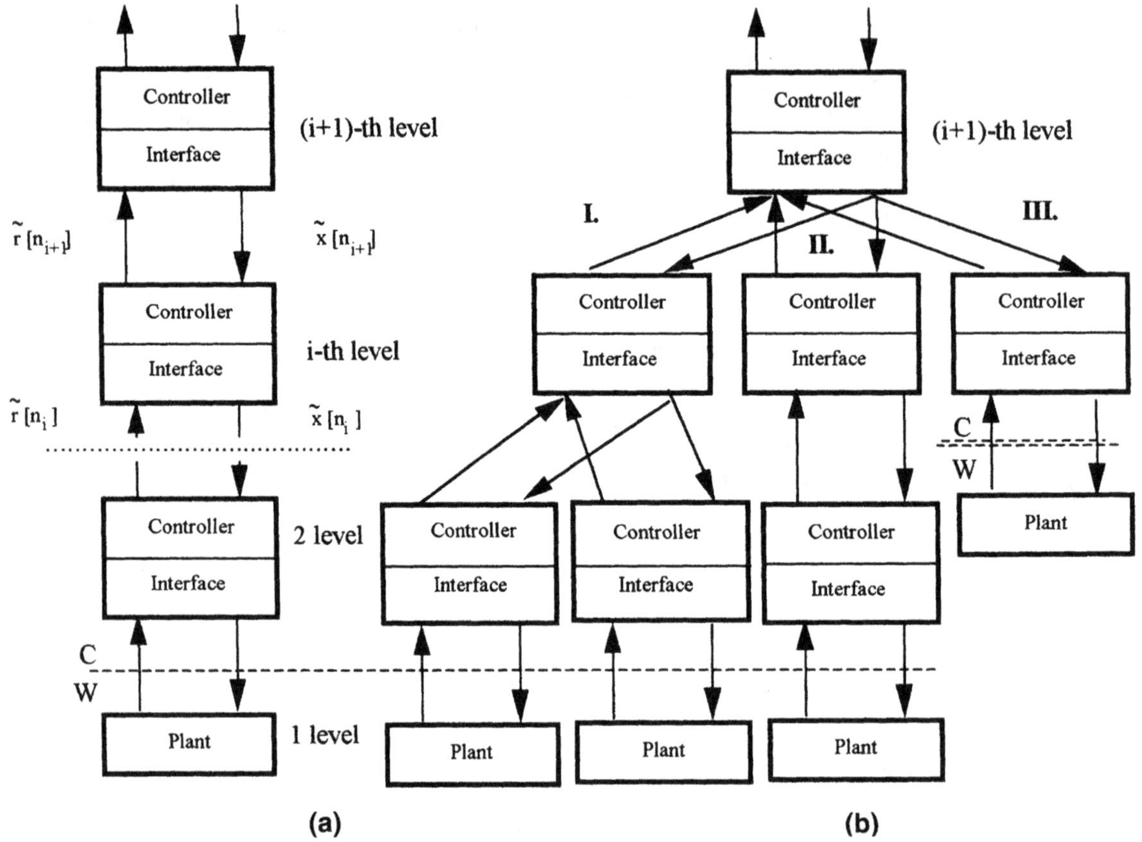

Figure 9.3. Different hierarchies formed of the multiplicity of PLANNER/EXECUTOR modules: (*a*) A funnel-hierarchy; (*b*) a tree-heirarchy. C, area of control system. W, area of the external world.

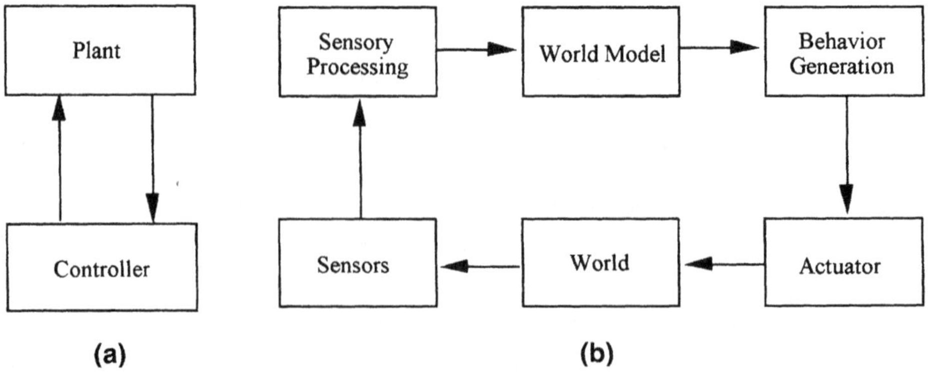

Figure 9.4. Change in the structuring of control systems: (*a*) Conventional structuring of a control; (*b*) elementary loop of functioning (ELF).

of differential equations, in the state-space equations form, and in the form of automata knowledge organization. It is not easy to have the model of a plant be part of the hierarchical control system: The hierarchy of control should handle a hierarchy of models, sometimes with complex relationships among them. This is why the tendency to separate the system of the world model (WM) from the system of behavior generation (BG) is now dominating in the area of large, complex (including intelligent) systems. The faculty of processing the sensor information is becoming a separate issue too, and the subsystems of sensory processing (SP) demonstrate a tendency to separation.

Obviously these three subsystems together (SP–WM–BG) are equivalent to the control system, but the architectures of these control systems become different in comparison with the architectures we are accustomed to from the 1960s and 1970s. This is why instead of lumping together the plant-controller as the control system (see Fig. 9.4*a*), the structuring now includes the sensors–world–actuators as the subsystems of the plant, and sensory processing (perception)–world model–behavior generation as shown in Figure 9.4*b* (see also ELF in Chapter 2).

While the subsystems have become more complex and independent, they still must interact, and their inner structures must be compatible. We will discuss in this chapter the close relationships between the architectures of the world model and behavior generation.

RECURSIVE HIERARCHIES

The hierarchy of the automata strongly depends on the hierarchy of the languages on which these automata are based. It is clear that in the hierarchy of the PLANNER/EXECUTOR modules (similar to those shown in Fig. 9.3), a nested system of languages is maintained as suggested by (9.4). The hierarchies of PLANNER/EXECUTOR modules can be called *recursive hierarchies* because the same set of algorithms is applied to the results of their application again and again. Recursive hierarchies have special relations between the nodes of the adjacent levels: all nodes of the set of m nodes $\{n_j\}_i$, (where $j = 1, 2, \ldots, m$, at the ith level of resolution counting levels from bottom to top), attached to a particular kth node $n_{k(i+1)}$ of the adjacent $(i+1)$th level from above, are obtained as a result of the special procedure called *decomposition* or *refinement*. This procedure presumes inversely that the properties of the node $n_{k(i+1)}$ can be obtained from the properties of nodes $\{n_j\}_i$ by a special procedure called *aggregation* or *generalization*, which is an integration over the j index. Decomposition alludes to the refinement of properties and functions, while aggregations are done via integration which is essentially a generalization.

Recursive hierarchies can be designed for processing information by changing the computational model from level to level so that an increase in resolution is accompanied by a reduction in the interval of computation. It can be shown that recursive computations always lead to nested structures of information and end up with a system of nested knowledge transformation sets $\{K\}$. The property of nesting allows for the levels to be decoupled functionally, although levels remain dependent on each other even after decoupling is done. Normal functioning still requires coordination of nesting conditions with adjacent levels from above and from below. This property ensures consistency when search is required instead of analytical methods, and when a change in resolution is accompanied by changes in vocabulary.

The multiresolutional hierarchy of knowledge can be represented graphically as a tree focused (by attention) at each level of resolution (see Fig. 9.5*a*). For comparison, we show a tree not truncated by attention (Fig. 9.5*b*).

This relationship entails the inclusion properties of the knowledge transformation sets: All global properties and functions of the system that can be represented by a recursive hierarchy contain all the properties of the lower levels of the hierarchy (higher-

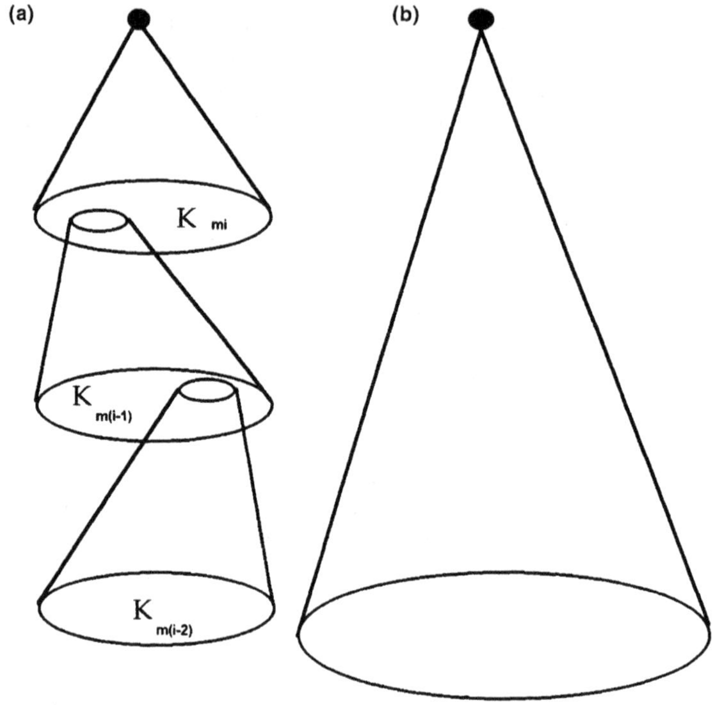

Figure 9.5. Focusing of attention on the tree of knowledge representation.

resolution levels) in an integrated (generalized) form. Thus the subsystems of the higher-resolution levels are not totally autonomous but are expected to carry out the assigned tasks assisted by other cooperating EXECUTORs at their own level in supporting their own parent systemic goals and objectives. Their autonomy is limited by the fact that they are contained within the adjacent lower-resolution level as its part in its pursuit of the global goal. The system of recursive hierarchy is thus built using the recursive modules.

As will be shown in the following example, the ability to formulate world models with this hierarchical generalization is essential for coping with the complexity involved in planning the operation of a combined system.

On the other hand, the inverse procedure can be done with a decomposition of the model into a set of multiple EXECUTORs (separate controllers). In other words, a decomposition can be assigned to satisfy the information needs of the system.

Let us discuss the architectures which emerge as a result of the hybrid control recursion with and without decomposition.

9.3.4 Nesting and Bandwidth Separation

Designers and users of hybrid hierarchies benefit from the following useful property of hybrid nested hierarchical controllers: bandwidth (time-scale) separation which emerges as a result of gradual reduction of frequency of sampling bottom-up. This property allows to decompose the control assignment into a number of frequency bandwidths, design and use separate controllers for each of the bandwidths, and superimpose the control outputs within the plant. Undoubtedly, each of the frequencies (time-scale) ranges is easier to control, since it is easier to satisfy the performance criteria within the limited bandwidth; each of the controllers deals with a simplified model of the plant (e.g., one that is meaningful for a particular time-scale range). Finally both the design and control processes will have reduced computational complexity or lead to a higher-quality result at the same computational complexity.

Bandwidth decomposition is a by-product of all hierarchical decompositions in the area of control. This fact was realized only recently in a number of papers on multiscale signal representation and wavelet decomposition. We will make explicit use of it in a systematic way in order to explore its potential for the design and functioning of control systems.

Our aim is to develop a method for the decomposition of a hybrid system and its controller into a real-time intelligent control hierarchy of uniformly constructed modules. In existing systems the decomposition is done first by using the system's semantics (organization, and natural disaggregation in parts) and then often intuitively narrowing the scope of attention (see Fig. 9.5). In nested hierarchies of PLANNER/EXECUTOR modules, the development of a recursive algorithm is possible for system representation and control at all levels of the hierarchy.

In existing hierarchical systems, separation of the overall system into appropriate hierarchical levels is done in a heuristic (knowledge based or expert based) way. The MIMO system separation into several decoupled transfer functions (by state feedback or precompensator) already belongs to classical control methods (see [11, 12]). Although the aggregated levels presented in the literature make control easier, they do not actually have hierarchical properties.

Time-scale decomposition of perturbed systems has been widely investigated in last 20 years [13], [14] on decomposition of regulator, near-optimal regulator for systems with fast and slow modes. Weak connections and time-scale aggregation of nonlinear systems was explored in [15]. Similar problems in discrete time case were discussed in [16] for multirate and composite control, and in [17] for optimal control. A systematic design method was outlined in [18]. Decomposed states observer design was the topic of [19].

MULTIPLE TIME SCALES CONTROLLER

The multiple time-scale approach expands the idea of perturbation-based separation (see [20]). A thorough analysis of multiple time-scale behavior of singularly perturbed linear systems was presented in [21], and an aggregation method for level production was provided as well.

The main idea of computational simplification produced by decomposition found its justification in many applications (e.g., [22] for aerospace guidance and [23] on nonlinear models for power systems).

In all the previously mentioned results, the decomposition is always induced by intrinsic, heuristically denoted system properties that reduce the separate application to small set of systems with natural time-scale properties. Since the work has been done mostly for the regulator problem, the presence of multiple time-scales implies the decomposition of the system transfer function into corresponding reduced-order transfer functions with distinct frequency bands that are superimposed.The decomposition not only simplifies the computations because of the reduced order of the partial models but also provides better control within the reduced bandwidths by the separate controllers than can be provided for the total frequency (time-scale) range by the initial single controller.

The linearity of the time-frequency transformations opens the possibility for decomposition of the plant model and control design. The decomposition may be induced by the frequency of the desired system output trajectory (trajectory tracking problem) or by the spectral density of the prescribed trajectory set (whose characteristics are given in the design stage).

The multiresolutional hierarchical algorithms introduced in the area of computational mathematics as domain decomposition methods or multigrid methods [24, 25] have distinct properties that deal with the different frequencies at different levels of resolution. The hierarchical aggregation of linear systems with multiple time-scales was discussed in [26]: This included a thorough mathematical treatment of nested hierarchical Markov controllers. Multiscale statistical signal processing was recommended in [27]. Recently an effort was made to formulate multiscale signal processing techniques and a multiscale systems theory methodology [28].

Multiresolutional control systems are discussed in [29, 30]. It is important to recognize that the multiresolutional controller emerges as the process of model aggregation proceeds (see Fig. 9.6). Both the high-resolution plant and its high-resolution DES are recognized by the middle-resolution level controller as the middle-resolution plant: But the middle-resolution interface cannot handle the high-resolution properties of the overall system. This process recurs at every next level above. The hybrid hierarchical system works as a natural bandwidth separator. Most of this effect is due to the granulation of the information at hand. And the purpose of this bandwidth separation is the reduction of the computational complexity of the control system.

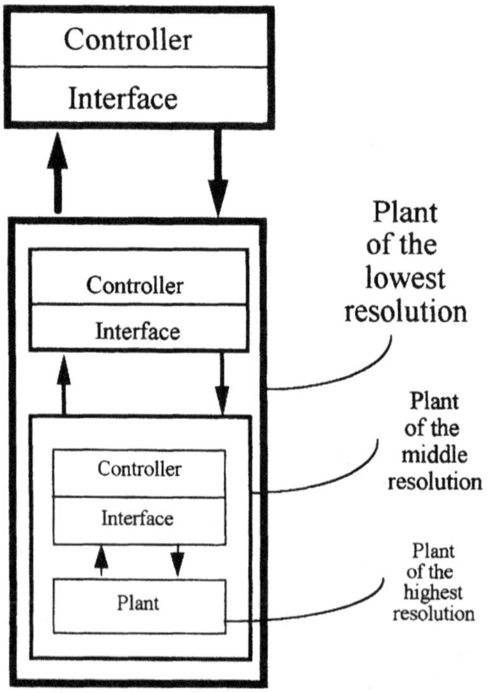

Figure 9.6. Generalization of the plant model in the hybrid funnel (nested) hierarchical controller.

9.4 QUASI-MINIMUM TIME FUNCTIONING OF PLANNER/EXECUTOR MODULES EQUIPPED WITH A PRODUCTION SYSTEM

Intelligent systems (IS) are nonlinear stochastic systems that are supposed to operate under conditions of constantly changing goal assignment as well as varying components of the mathematical model of the system and constraints. A method is proposed of minimum time control of mechanical motion for an IS in which control assignment can be changed during the command execution. The string of commands is based on the enhanced Bellman optimality principle for stochastic systems, so they are recomputed as soon as the new goal of motion is assigned. The problem of computational efficiency is solved by structuring the motion trajectory in such a way that new solution is obtained via a simple combining of the trajectory with a limited number of standard components given in the form of algebraic equations to be solved online. This allows a model-based production system to be used for synthesizing the control sequences in real time.

9.4.1 Assumptions about System Dynamics

The basic goal of the PLANNER/EXECUTOR at a level of the IS hierarchy is to find a minimum cost command sequence to have the system move a required predetermined distance while taking into account the assignment and the initial conditions to the system. The motion of the system is executed frequently in minimum time, and hence the control has to keep the system's constraint variables at their maximum limit. Since this is a stochastic system, and since the assigned goal of the motion can be reassigned before the motion is completed, the enhanced Bellman optimality principle is applicable [31] as was suggested in [32]. This principle states for stochastic problems that whatever the present information and the past decisions, the remaining decisions must constitute an optimal policy with regard to the current information set [31].

The PLANNER/EXECUTOR of an IS has at its output multiple actuators. Each actuation system with its PLANNER/EXECUTOR is considered to be independent from each other and can be modeled and computed individually. This assumption is attributed to the fact that even if actuators are coupled together, one could find a system of representation wherein they are decoupled or could apply one of many existing methods of decoupling [33].

JERK-CONTROLLED SYSTEM

Each of the independent actuator systems can be frequently modeled by a differential equation of the third order with each of the derivatives constrained. These constraints put a maximum and minimum value on the state-space variables. The input of the system is a "jerk" which is computed as the rate of change of this force, or as the time derivative of the force. Since, as was stated above, the goal is to obtain minimum time control, the jerk input is always selected at a maximum level. Hence the input alternatives are the maximum positive jerk, the maximum negative jerk, or no jerk. After a jerk is applied, a change in the current value of the force will appear depending on the direction in which the jerk is applied and the interval of time in which it was applied. The change in input force causes a direct change in the acceleration of the system. This in turn caused a change in the velocity and distance moved by the system. Each of these intermediate variables have constraints. As explained earlier, a jerk input has a maximum value. Its force and the velocity also have maximum positive and maximum negative values. The distance traversed is the output variable in the system of execution. Thus, for the third order, system equations are written for the state-space variables as follows:

$J(t)$ = jerk input (rate of change of force).
$F(t)$ = Force input.
Resistance(t) = Force of resistance.
$V(t)$ = Velocity.
$D(t)$ = Distance traversed.

Recall from Chapters 7 and 8 that the minimum time motion control is provided by applying the maximum possible input control (bang-bang control) and by keeping all variables of the system always at their maximum values. Hence the system of equations for the state-space variables is

$$F(t) = \int J(t)\,dt, \tag{9.5}$$

$$V(t) = \int \left[F(t) - \frac{\text{resistance}(t)}{\text{mass}} \right] dt, \tag{9.6}$$

$$D(t) = \int V(t)\,dt. \tag{9.7}$$

Even though equations (9.5) to (9.7) hold only for linear deterministic motion, they can be used for any type of actuator system. For example, they can be used for controlling angular motion, in applying a torque, or for finding the rate of change of the torque.

The system we are dealing with can be represented as a third-order dynamic system controlled by jerks (which is the first derivative of acceleration) [34, 35]. Theoretical diagrams for this type of control are shown in Figure 9.7. Simulated time trajectories

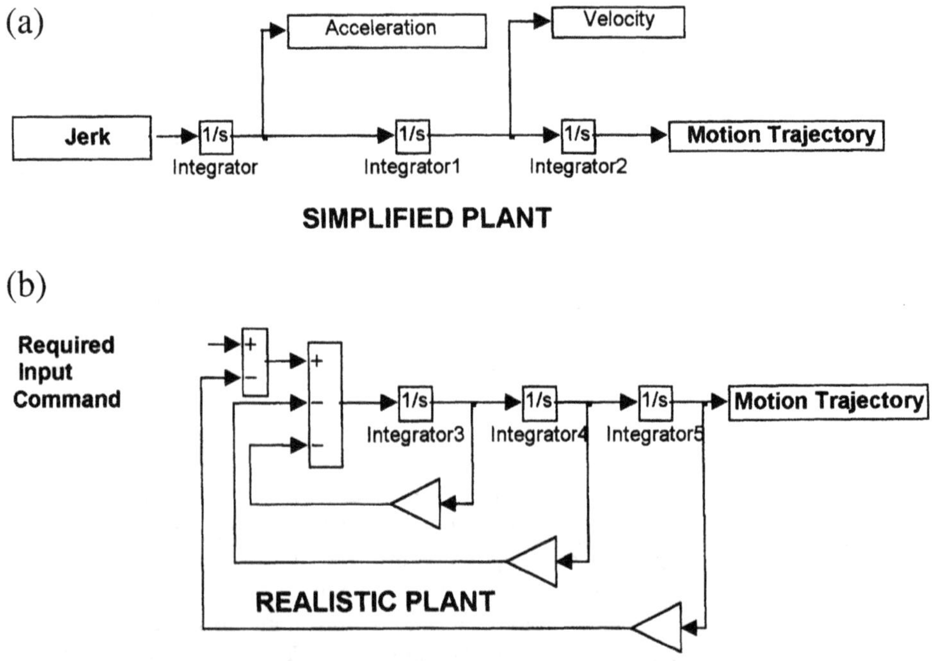

Figure 9.7. The structure of the jerk-controlled system. (*a*) Ideal; (*b*) realistic.

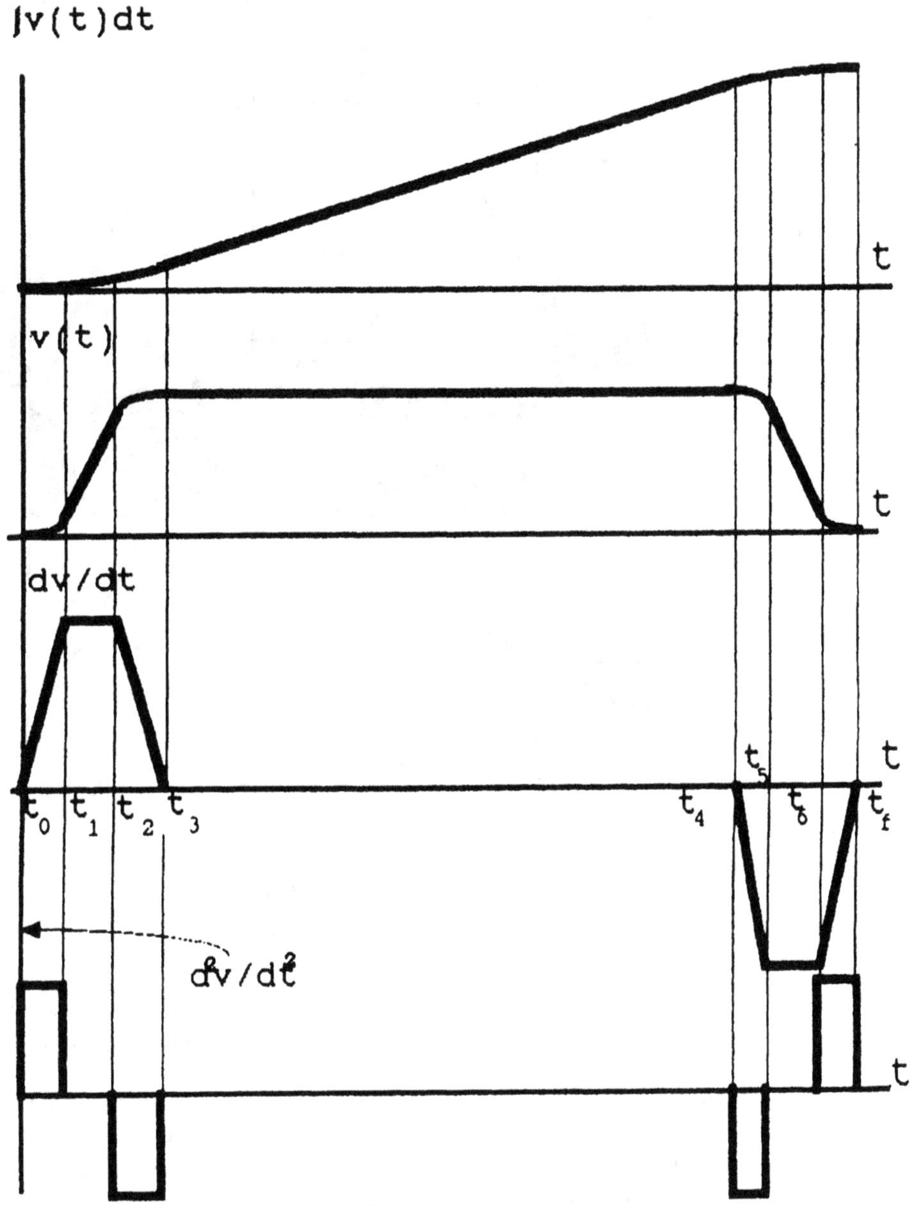

Figure 9.8. Theoretical diagram of an execution controller with jerk-control.

of all of the system controls and state variables if there were no constraints on any of the variables are shown in Figure 9.8. Each of the state-space variables $F(t)$ and $V(t)$ has constraints imposed on it. The constraints affect its motion, as described previously. First of all, the force which is shown to rise between period 0 and t_1 tends to reach a value that is greater than the maximum force and so is physically not possible. Second, the velocity tends to reach a maximum value at t_2 which again can become greater than the permissible limit, and hence the total force applied between 0 and t_2 has to be reduced. Thus we have two constraints imposed on the system of equations:

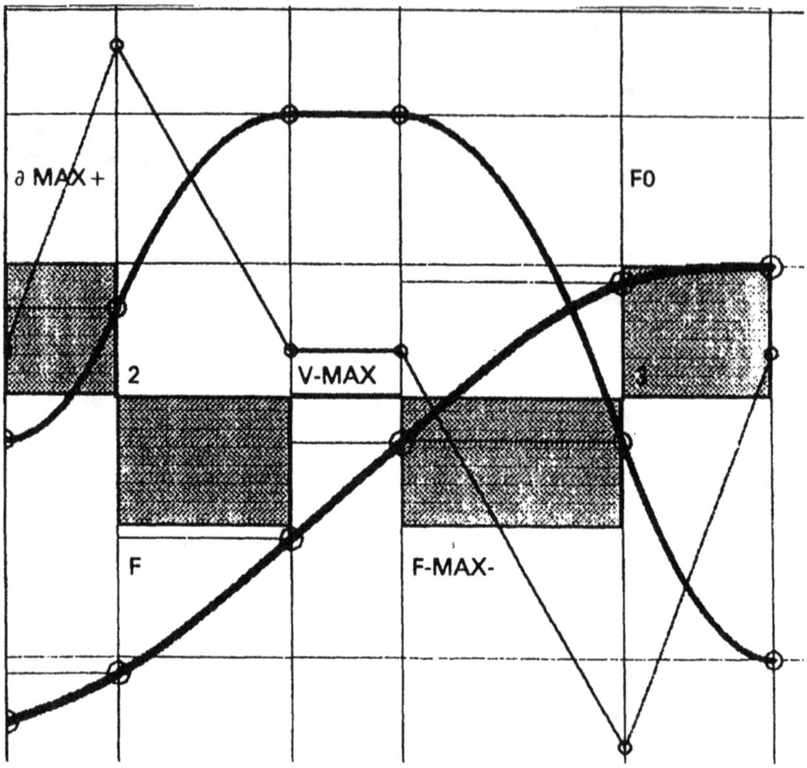

Figure 9.9. Theoretical motion trajectories with no constraints.

$$F(t) < \text{Force-max}, \; V(t) < \text{Velocity-max}.$$

We will observe the change in the control procedure by a stepwise application of each of these constraints. First is the case where the maximum velocity reached by the system is constrained. This is shown in Figure 9.9. As it can be observed from the graph, the jerk input has to be modified so as to remain within the velocity constraint. Then, in addition to constrained velocity, we will impose the constraint of maximum force both in the positive and negative directions. This is shown in Figure 9.10.

In order for us to solve the minimum time control requirements, we must determine the value of the time periods and type of jerk input during each period. As can be seen from Figure 9.11, all that is required to be changed is the timing of the jerks ("switching times") for different system configurations and their various initial conditions. The sequence in which these jerks will be applied remains the same. Therefore each period of a jerk application might be classified as a different control region. The control regions can grow or shrink depending on the system initial conditions and variables.

9.4.2 Assumptions about Propulsion Force Development

In (9.7) the input force is given as a summation of the rate of change of force, since the jerk input remains constant over the entire increasing range of the input force. This assumption is not valid for a wide range of cases. For example, in the braking force

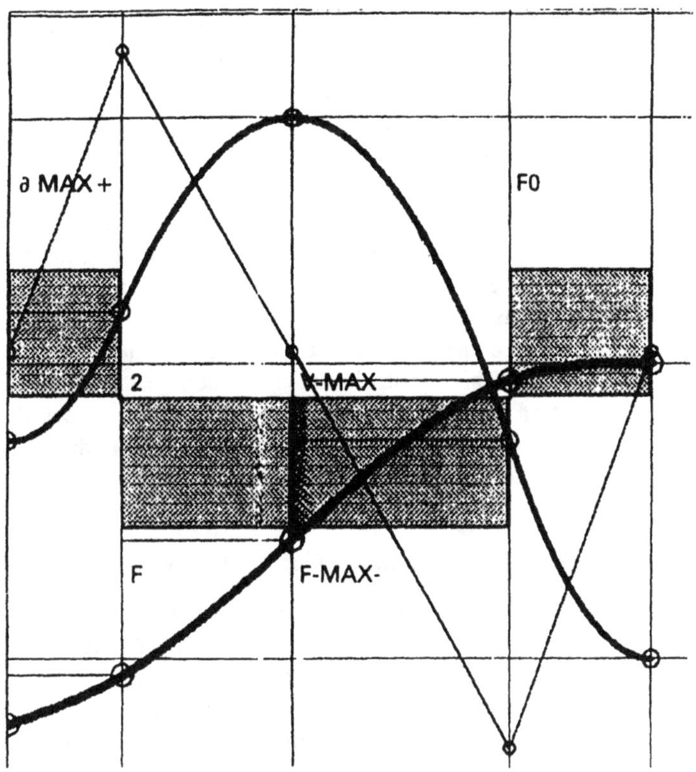

Figure 9.10. Theoretical motion trajectories with velocity constraints.

achieved by the introduction of excess resistance force (e.g., in propulsion drives with drum brakes), the jerk is not constant over the entire range of the input forces. To compensate for such nonlinearities, we add the component of frictional force which is dependent on the current magnitude of the force and of the jerk being applied.

An important set of active factors is omitted from the system of equations (9.6) to (9.8). This is the set of delays that exists in practice in most actuation systems. This is another obvious deficiency of the analytical model besides the set of continuous nonlinearities neglected in the model of motion. Thus from the beginning, we know that our equations contain substantial errors. We will assume that all of these sources of errors are included (fictitiously) in the force of resistance.

9.4.3 Assumptions about the Resistance Force

Resistance force can be visualized as a nonlinearity of the input force mechanism. This nonlinearity can be attributed to many different factors. The force generator may have an internal nonlinearity and not be able to give out constant output force that has its own stochastic effect for the system model. Delays in force generation were mentioned above. Three different simple cases of resistance force will be considered now:

1. Constant resistance force (slope).
2. Linear resistance force (velocity).
3. Dry friction resistance force.

These three different resistance forces do not cover all the sources of resistance forces, but they do in a sense cover the major practical applications. For example, a constant resistance force can be used in modeling motion against a constant gravitational force when the object is going up the slope and hence is storing potential energy. The linear resistance force model is applicable when viscous resistance is encountered and hence the friction force is proportional to velocity. Dry friction resistance is a combination of the first two types. In dry friction, the resistance force is constant, but when the input force is less than the resistance force, the resistance becomes equal to the input force; hence no motion can be achieved. This means that the resistance force is not active as was in Case 1.

These models of resistance forces are made more realistic by adding one more component. This is a stochastic component that has different expectations, variances, distribution laws, and the like. Later in this chapter, we will explore two types of distributions of the resistance force: uniform and Gaussian distribution at different mean values and different variances (as shown later in Fig. 9.14).

9.4.4 Determining Standard Components of the Control Cycle

In the constrained system mentioned before, each level is constrained, and hence the system has for a minimum time motion eight different regions within which the behavior of the system has to be controlled. It is easy to verify that under a broad variety of control conditions, the cycles to be performed consist of the same standard components. The different forms of these eight regions are as follows (see Fig. 9.8).

1. Propulsion force increases from initial value to the value of the resistance force.
2. Propulsion force increases from the level of equality with the resistance force to the maximum positive value.
3. Propulsion force is kept at a constant maximum value.
4. Propulsion force decreases from maximum positive value to the level of the resistance force.
5. Propulsion force decreases to a zero value.
6. Propulsion force continues to decrease to the maximum negative value (if the resistance force does not act in the opposite direction, which is valid for the case of positive value of velocity).
7. Propulsion force is kept constant at a negative maximum.
8. Propulsion force increases from the maximum negative value to the level of the resistance force.

As is obvious, the regions mentioned here define the sequence of switching times (or the time sequence of control commands) for the input actuation so as to obtain a desired response of the minimum time motion to the goal state. Each of these regions have the peculiar property of similarity in the sense of the variables controlled in each time zone. Also it can be seen from the form in which the solution is shown that these regions are interdependent, since the final goal of the system is to get a specific output form of the system.

Another major point to be made here is also about the system of constraints which contributes to the interrelationships between the regions. The actual effect of each region will be considered next; the interrelations will be discussed later. An analytical descriptive production system can be constructed to generate control sequences under all possible initial conditions and take into account all possible interrelations between the regions.

As can be observed from the description of the eight regions under consideration, there are only three types of input to the system. As these basic inputs combine, they give different forms of motion under different initial conditions. The set of inputs together with a tree of alternatives of possible motion models under various initial conditions and applicable constraints constitutes the production system for synthesizing the control sequence. These three inputs are as follows:

1. Positive jerk application.
2. No jerk application.
3. Negative jerk application.

Group 1: Positive Jerk Application

The force is increased from its initial value to the force of resistance by application of a positive jerk. In this region the velocity is acquiring acceleration. The distance covered is the integration of the increasing velocity with acceleration. The system of equations during this period is as follows:

$$\text{jerk-input} = \text{f-dot-max}^{+} \tag{9.8}$$

$$\text{f-input} = \text{f-dot-max}^{+} \times \text{time} \tag{9.9}$$

$$\text{f-friction} = \text{f-dot-max}^{+} \times \text{time} \tag{9.10}$$

$$\text{velocity} = \int \frac{(\text{f-input} - \text{f-friction})}{\text{mass}}\, dt = 0 \tag{9.11}$$

$$\text{distance} = \int \text{velocity}\, dt \tag{9.12}$$

Group 2: No Jerk Application

The force is kept at its initial value and held constant during the period of motion. There is no jerk applied in this region. The velocity grows linearly during this period as a result of constant acceleration. Also, if the initial value of the total force applied is zero, the velocity remains constant and hence will grow linearly with distance.

The system of motion equations during this period is presented as follows:

$$\text{jerk} - \text{input} = 0 \tag{9.13}$$

$$\text{f} - \text{input} = \text{force} - \text{initial} \tag{9.14}$$

$$\text{f} - \text{friction} = \text{const} - \text{friction} - \text{force} \tag{9.15}$$

$$\text{velocity} = \int \frac{(\text{f} - \text{input} - \text{f} - \text{friction})}{\text{mass}}\, dt \tag{9.16}$$

$$\text{distance} = \int \text{velocity}\, dt \tag{9.17}$$

Group 3: Negative Jerk Application

The negative jerk region is where the total input force is reduced from its initial value. This is the same as acceleration reduced from initial value. This causes the system to reduce the rate of growth of velocity. The system of motion equations during this period is as follows:

$$\text{jerk-input} = \text{force-dot-max} - \tag{9.18}$$

$$\text{f-input} = \text{force-dot-max} - \times \text{ time} \tag{9.19}$$

$$\text{f-friction} = \text{const-friction-force} \tag{9.20}$$

$$\text{velocity} = \int \frac{(\text{f-input} - \text{f-friction})}{\text{mass}}\, dt \tag{9.21}$$

$$\text{distance} = \int \text{velocity}\, dt \tag{9.22}$$

9.4.5 Production System for Combining the Control Cycle from Standard Components

Given the system of requirements for a particular distance to be traversed by the system, we want to find the timing for each region. Each region has a predetermined type of jerk input. First we need to list the vocabulary of the system constraints and their relative effects. The system constraints can be described as follows:

jerk-max$^+$ = maximum possible positive jerk of the system.
jerk-max$^-$ = maximum possible negative jerk of the system.
force-max$^+$ = maximum possible force generated by the system.
force-max$^-$ = maximum possible braking generated by the system.
force-resistance = resistance force (friction force).
velocity-max = maximum possible velocity of the system.
mass = mass of the system.

The set of initial conditions, listed below, also changes the behavior of its IS PLANNER/EXECUTOR modules. These initial conditions are imposed upon the intermediate variables of the system which have the constraints noted above.

force-init = initial value of the force at time zero.
velocity-init = initial velocity at time zero.
distance-required = distance that is to be moved by the system.

These regions appear to be in sequence. Nevertheless, the timing of each cannot be carried out in a sequence because of the interdependence between the regions. So we will show here the order in which each switching time is calculated and hence will determine the sequence of switching times and the values of the constrained variables during each time period.

A timing region is described by the following variables. They define the state of the system at every point during this time period.

region-left = pointer to the region left of the current one.
region-right = pointer to the region right of the current one.
jerk-input = type of jerk input to be given (jerk-max$^+$, jerk-max$^-$, 0).
time = time at which this region starts operation.
force = force value as it enters this region.
velocity = velocity value of the force as it enters this region.
distance = distance traversed by the system from the start of this region.
delta-time = time period for which this region remains active.
delta-force = change of force over this region.
delta-velocity = change of velocity over this region.
delta-distance = distance traveled during this region.

It was mentioned above that there are eight different regions (reg). These regions have different inputs and constraints. The equation for computing the state variables of each region are shown below.

1. Increase force from initial value to force of resistance.

$$\text{jerk input} = 1$$
$$\text{time} = 0$$
$$\text{force} = \text{initial value of force}$$
$$\text{velocity} = \text{initial velocity}$$
$$\text{distance} = 0$$
$$\text{delta-time} = \frac{(\text{force-resistance} + \text{force})}{\text{jerk-max}^+}$$
$$\text{delta-force} = (\text{force-resistance} + \text{force})$$
$$\text{delta-velocity} = 0$$
$$\text{delta-distance} = 0$$

2. Increase force from initial value to maximum positive.

$$\text{jerk input} = 1$$
$$\text{delta-time} = \frac{(\text{force-max}^+ - \text{force-resistance})}{\text{jerk-max}^+}$$
$$\text{delta-force} = (\text{force-max}^+ - \text{force-resistance})$$
$$\text{delta-velocity} = \frac{(\text{jerk-max}^+ \times \text{delta-time}^2)}{(2 \times \text{mass})}$$
$$\text{delta-distance} = \frac{(\text{jerk-max}^+ \times \text{delta-time}^3)}{(6 \times \text{mass})}$$

3. Keep force constant maximum.

$$\text{jerk input} = 0$$
$$\text{delta-velocity} = (\text{vel-max}^- \text{delta-velocity-reg1} + \text{delta-velocity-reg3})$$
$$\text{delta-time} = \frac{(\text{delta-velocity} \times \text{mass})}{(\text{force-max}^+ + \text{force-resistance})}$$
$$\text{delta-force} = 0$$

4. Decrease force from maximum positive to force of resistance.

$$\text{jerk input} = -1$$
$$\text{delta-time} = \frac{(\text{force-max}^{+} + \text{force-resistance})}{\text{jerk-max}^{-}}$$
$$\text{delta-force} = (\text{force-max}^{+} + \text{force-resistance})$$
$$\text{delta-velocity} = \frac{(\text{jerk-max}^{-} \times \text{delta-time}^{2})}{(2 \times \text{mass})}$$
$$\text{delta-distance} = \frac{(\text{jerk-max}^{-} \times \text{delta-time}^{3})}{(6 \times \text{mass})}$$

5. Keep the force constant at force of resistance.
6. Decrease force to maximum negative.
7. Keep maximum negative force constant.
8. Increase force from maximum negative to force of resistance.

This methodology enhances the results of [34], in which the deterministic force of resistance is presumed to be equal at the stages of acceleration and deceleration.

9.4.6 Simulation of a Real System

Once the design of the system is complete, the next step is to simulate and study the effects of sampling and of different types of noise added to the system. As was mentioned before, the type of noise we will be adding is resistance force (i.e., friction). This resistance force can have different forms of randomness as described above. In the simulated system we have the state variables of the system, which are as follows:

time = time at the instance of simulations.
force(t) = magnitude of force at time t.
velocity(t) = velocity at time t.
distance(t) = distance traversed by system at time t.
resistance(t) = resistance force executed by system at time t.

Depending on the type of region that the system is in, the jerk input is given for a finite time period. Hence the change in force is computed resulting from the jerk. Then changes in velocity and distance are computed, and a new state of the system is obtained. The equations for these computations are as follows:

$$\text{force}(t + \Delta t) = \text{force}(t) + (\text{jerk}(t + \Delta t) \times \Delta t)\,, \tag{9.23}$$

$$\text{velocity}(t + \Delta t) = \frac{\text{velocity}(t) + ((\text{force}(t + \Delta t) + \text{resistance}(t + \Delta t)) \times \Delta t)}{\text{mass}}\,, \tag{9.24}$$

$$\text{distance}(t + \Delta t) = \text{distance}(t) + (\text{velocity}(t + \Delta t) \times \Delta t)\,. \tag{9.25}$$

The state of the simulated system at each instance is found by using these equations. This simulation assumes that the increase of time, Δt, is small, and hence the assumption is that the system behaves linearly during this time. The results of the simulation [35] for different combinations of input conditions and parameters of the system are shown in Figure 9.11.

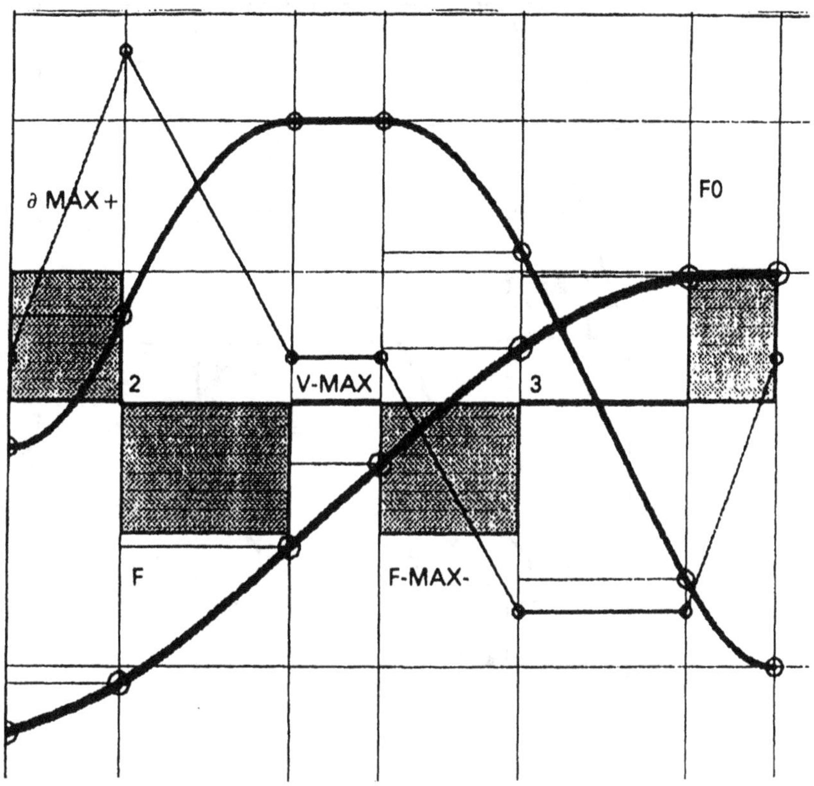

Figure 9.11. Theoretical motion trajectories with velocity and acceleration constraints.

Earlier in this section we learned that the resistance force has to be computed for each time period. This resistance force is a type of noise which we are adding to the system. We have two different models for simulating the noise in the resistance force over time: the Gaussian distribution and a uniform distribution. The expectation, variance, and the frequency of introducing the new random value can be varied. A simulated cycle of operation is shown in Figure 9.12*a*. The stochastic component of resistance is simulated in the form shown in Figure 9.13.

The stochastic resistance force illustrates the uncertainty that is typical for the real resistance force, and it incorporates some of the random factors and model deficiencies mentioned above. The resistance force has an average value that is constant, or that is changed according to a known deterministic law. The maximum dispersion amplitude is varied as a percentage of the average value. The value of the assigned variance of the resistance force is kept constant for a predetermined time period. This time period might remain constant or vary over the complete range of the system simulation. These parameters can be dependent on the type of region the system is in. Hence we have the resistance force varying from (resistance-force − amplitude) to (resistance-force + amplitude). This range can have different distributions. The simulation has two different ones, as shown in Figures 9.13 and 9.14.

From the equations for velocity and distance given above, we know that the noise added to the system will cause a corresponding component of noise in the velocity and distance parameters. Hence the error of motion can accumulate and cause error in the final state of the system. This error can occur in velocity or distance, or in both. By using

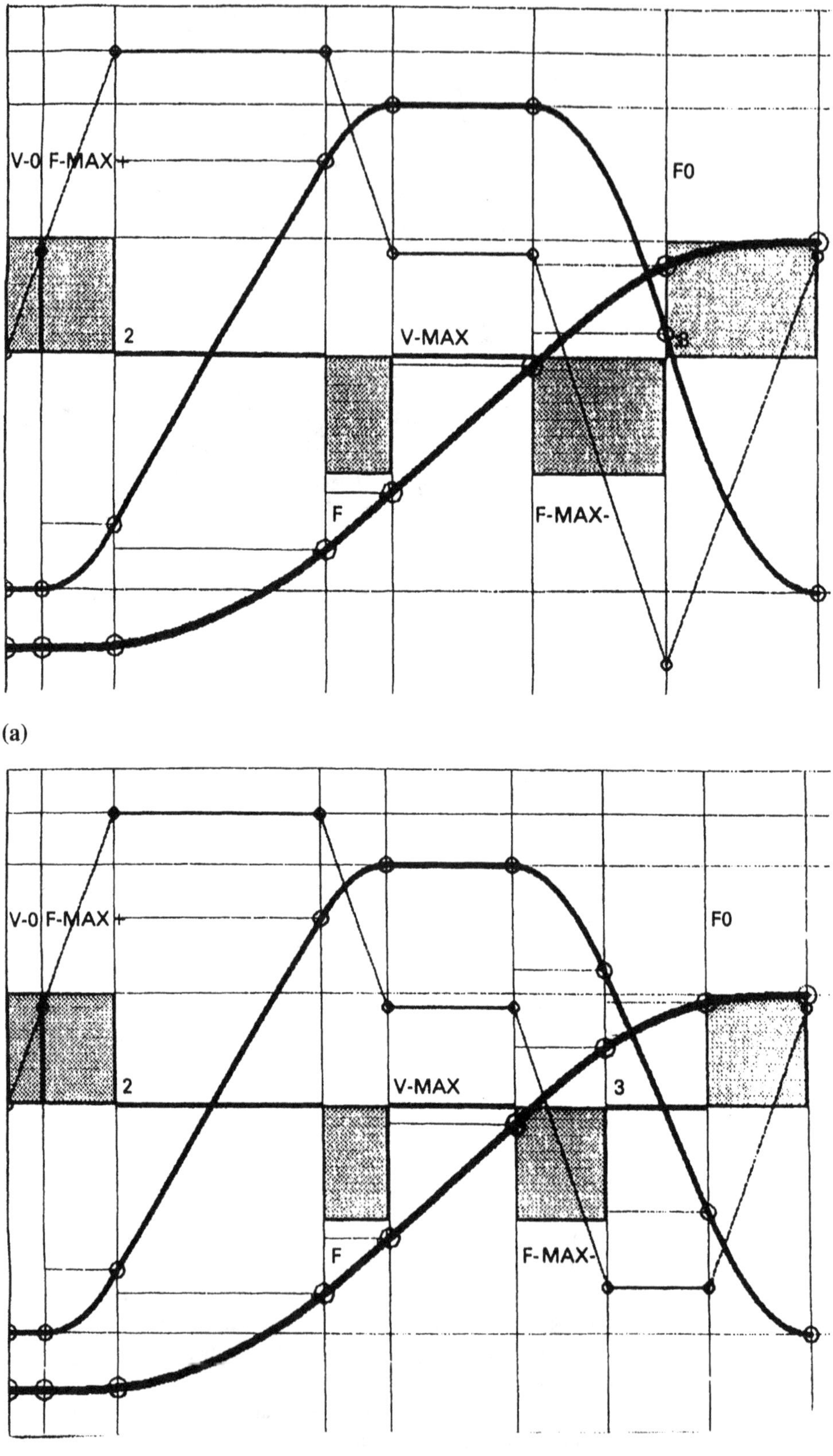

(a)

(b)

Figure 9.12. Synthesized cycles.

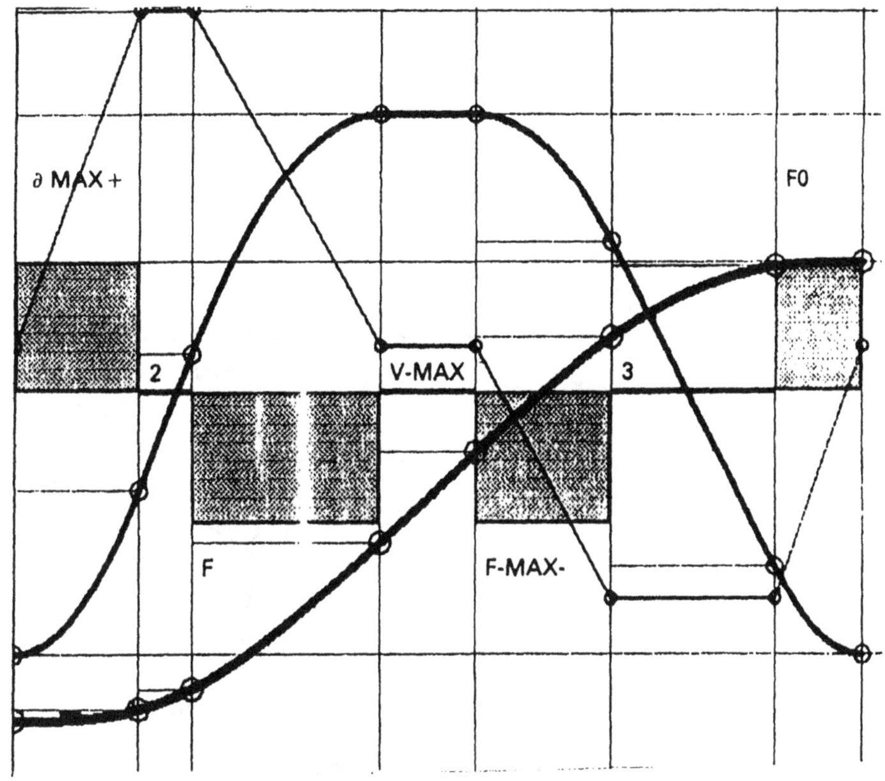

(c)

Figure 9.12. (*Continued*)

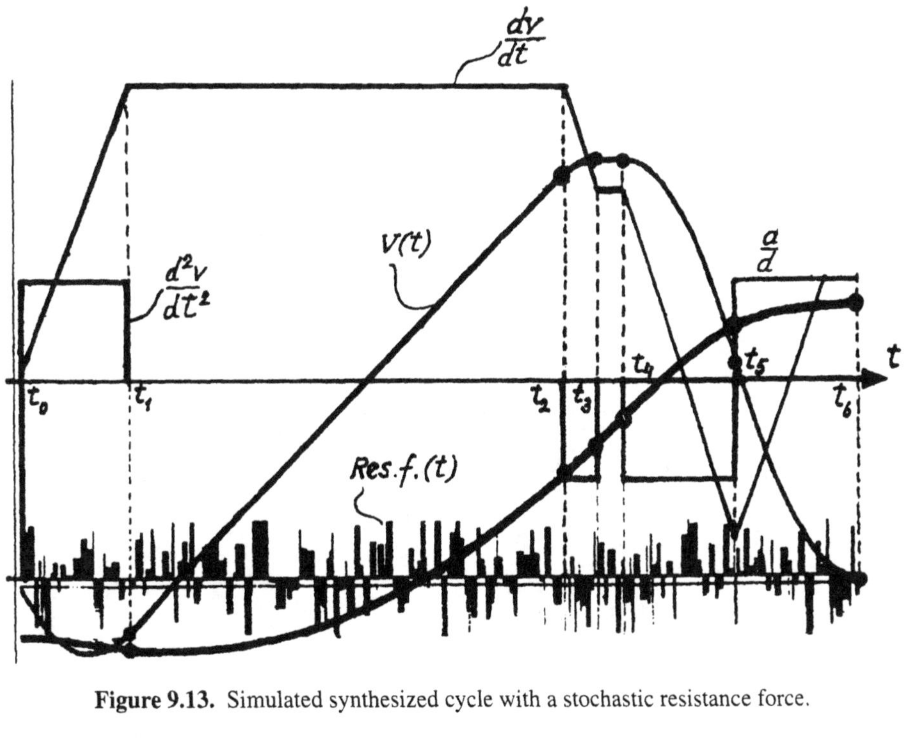

Figure 9.13. Simulated synthesized cycle with a stochastic resistance force.

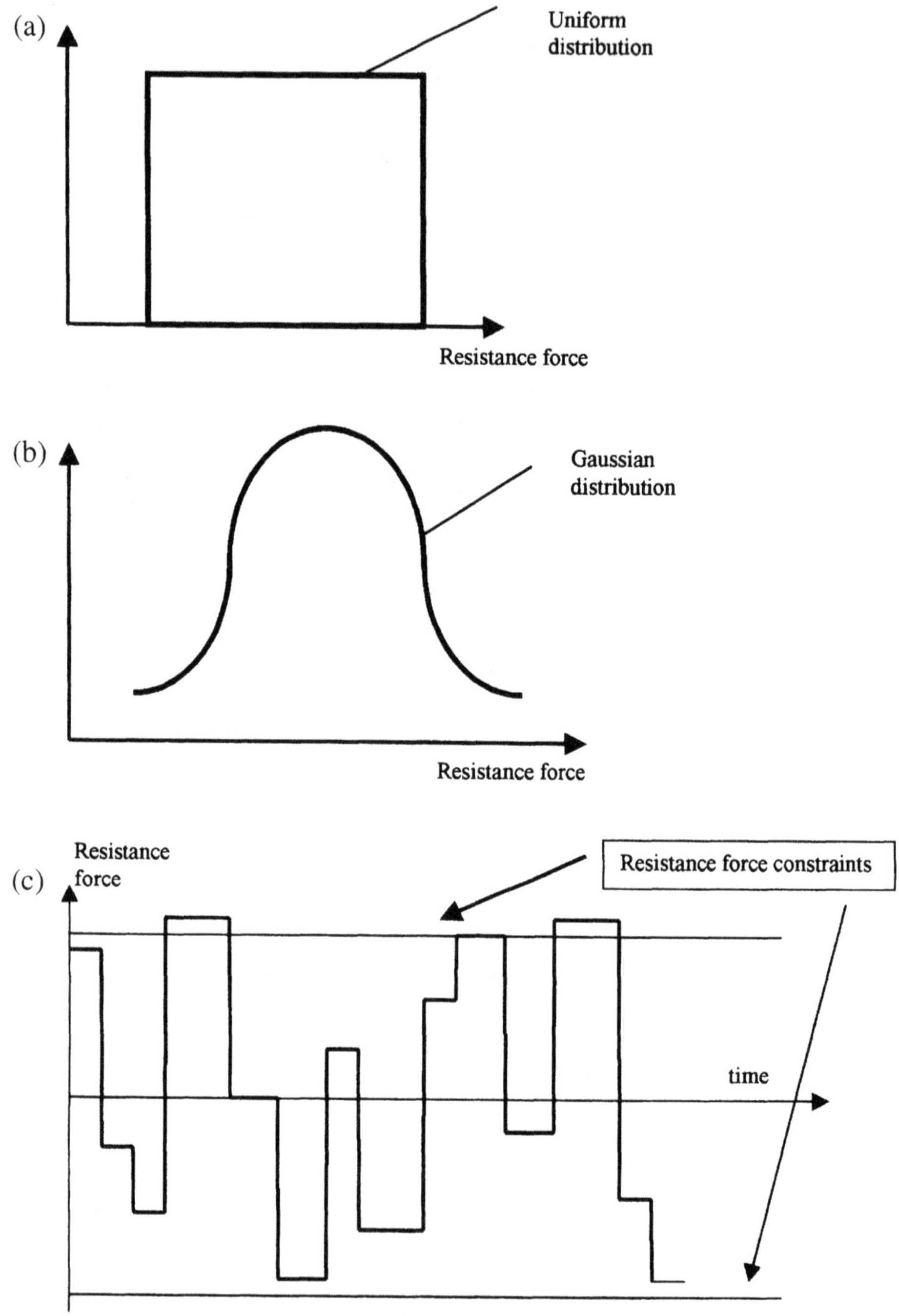

Figure 9.14. Characteristics of a resistance force: (*a*) Time-function, (*b*) uniform, and (*c*) Gaussian distributions.

this technique in a model of the noise of the resistance force, we can make an accurate error estimation.

As we have seen from the simulation results, the mechanism of a production system can be successfully utilized for an execution scheme that serves the PLANNER/EXECUTOR for IS submodules described in [35]. Different laws of distribution and other characteristics of the stochastic components lead to the quantitative results that are easily identified and used.

The simulation suggests that a feedback model is applicable in the production system of a BG module. After the stochastic component is identified, a set of corrections (tabulated prior to system operation) is introduced, and a new string of the switching times is submitted to the EXECUTOR.

9.5 APPROXIMATE INVERSE VIA FORWARD SEARCHING

The synthesis of feedforward commands is based on solving inverse problems which are frequently ill-posed. This section demonstrates that feedforward searching can provide the approximate solutions in such cases. The notion of approximate inversion is introduced, and a simple algorithm for performing this operation is described. The results are shown for SISO and MIMO cases, and some avenues for further research are suggested.

9.5.1 Introduction

Automated planning includes determining the desirable output trajectory as well as the feedforward commands that are applied to receive this output trajectory. In many cases the reference trajectory is predetermined by the design specifications. However, the system is noninvertible, and an analytical method for finding the feedforward commands is not known. As we will see, forward searching can be used to generate input commands that obtain planned output. Feedback control can be considerably reduced by a feedforward control synthesis, and the design of feedback controllers can be simplified as a result. It will be shown that since the synthesis of feedforward commands relies on the solution of inverse problems, and in many cases it does not necessarily have exact solutions, approximate solutions can be obtained by using forward searching. A definition of approximate inversion will be introduced, and a simple binary search algorithm will be described that synthesizes feedforward controls for practical cases. The simulation results will illustrate the utility of this approach.

The attempts to find an approximate inverse when it is not available analytically has allowed the use of planning controllers within a broad domain of application. The fundamental premise underlying design efforts in planning is that a joint feedforward/feedback (or open- and closed-loop) control architecture comprise the bulk of the actuation effort. This is supplied with a representative set of input–output alternatives to which corrections are applied on the basis of observations or measured deviations, design (or planning) as well as input function synthesis. By neglecting globally optimal performance, such sufficing algorithms of planning via search can be used to generate successively (not necessarily always in real time) more focused and, consequently, refined planned trajectories for implementation in conjunction with feedback corrections. The creation of such a hierarchical planning structure is additionally useful in the parallel implementation of control with multirate correction/sampling, whose operation is in conjunction with plans having different horizons of attention and accuracy.

The hierarchical planning approach to control design that can thus be recommended is one that builds successive approximations to a near-optimal result. It creates finite representations of (dynamic) system capabilities (and of the world of which the system is a part) and analyzes the optimality of strings of events that are shown to lead to the required end result. By focusing more closely on a subset of the world at each iteration, it is possible to converge to an arbitrarily precise approximation of optimal behavior. Although there may be loss of precision for a mathematical system model, the information obtained for the synthesis of actuation plans is the result of testing rather than a generalization of testing which is the characteristic analytical system model.

Closely associated with this approach to representing the system is the efficiency and complexity associated with information handling in various systems of knowledge representation. While these are not issues normally associated with control theory, they are an essential component in the development of automated planning for control, so a parallel, hierarchical organization of planning, as well as execution, is unavoidable.

9.5.2 Concept of Approximate Inverse Image of the Reference Trajectory

Let us consider again the concept of joint feedforward/feedback architecture of control. Figure 9.15 shows the usual structure of a control system that employs a feedforward controller to synthesize the input commands for the desired (planned) output of the plant.

For simplicity, we let the figure refer to a linear time-invariant system with output feedback, and we denote the plant and feedback controller by their transfer functions $P(s)$ and $F(s)$. If the feedforward controller does not contribute an input signal, then the overall transfer function of the system from reference to output can be written in the usual form as

$$\frac{C(s)}{R(s)} = \frac{PF(s)}{1 + PF(s)}. \tag{9.26}$$

However, for perfect tracking of the reference trajectory, this transfer function should be unity, which may seem impossible. Intuitively it is likely that the reaction to error will

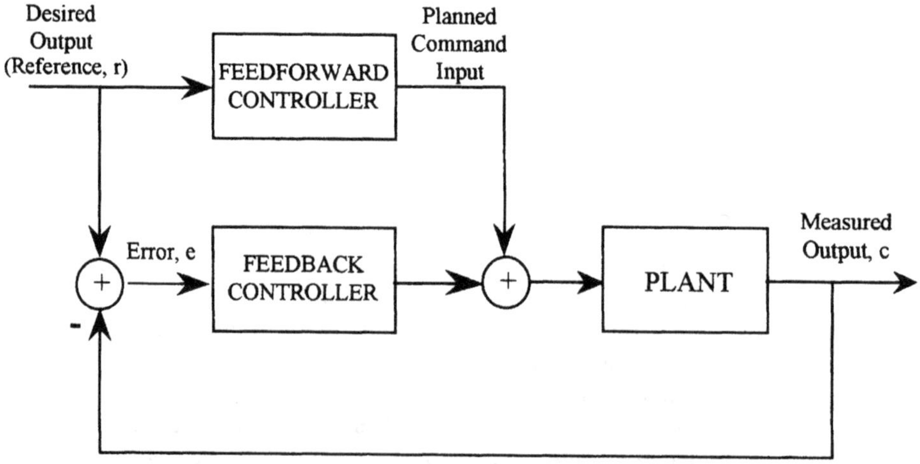

Figure 9.15. Combined feedforward/feedback control architecture.

be relatively slow compared to the predictive correction that is available via feedforward compensation. By virtue of its existence outside the immediate scope of the feedback loop, the feedforward controller injects a known bias into the operation of that loop. Regardless of whether this bias consists of a nominal command applied by an expert to effect the continuous operation of a machine in a factory or whether it is a linearizing or decoupling torque generation scheme for a robotic manipulator, the planned command input is produced on the basis of an analysis of the system model in some form—mathematical, linguistic, or both—in order to improve the performance of the overall system.

If a reference trajectory (or function) for such a system has been synthesized, then it may be necessary for the feedforward controller to be treated as the inverse dynamical model of the plant. Then the only task for the feedback component would be to cancel the effects of the unmodeled dynamics and disturbances. This can be shown by denoting the feedforward controller by $Q(s)$ and manipulating variables (in the overall system of Figure 9.15) to obtain:

$$\frac{C(s)}{R(s)} = \frac{PF(s)}{1 + PF(s)} + \frac{PQ(s)}{1 + PF(s)}. \tag{9.27}$$

The transfer function in (9.28) can be taken as unity if the plant and feedforward controller are exact inverses of each other. In fact, the bulk of the results in the area are based on this assumption. The work of Brockett [36, 37] is generally acknowledged as the first formal treatment of the problem of the inversion of multivariable LTI systems, and references [38–51] list extensions of these results to nonlinear, time-varying, and discrete-time cases.

The construction of minimal inverses and other realizations that are stable or robust in some sense has received attention in the past (e.g., [49]). The problem addressed here is related to the existence of an *inverse image* of a reference trajectory (rather than the inverse dynamical system itself). There are several reasons for this redirection of focus. First, although the inverse dynamical system has been depicted in the literature as a generator of input commands *based purely on the reference trajectory* (and not on the output of the plant), and although the differentiations inherent in such a dynamical inverse are potentially likely to require large (possibly infinite) command inputs, the emphasis has remained on determining a unique inverse image for the feedforward controller by actually implementing the inverse dynamical system. Second, in modern applications, controller implementation is digital, so there is considerable interest in a design that takes into account practical issues such as digitization, tessellation, and input saturation.

It is suggested here that the general approach to feedforward command synthesis is to implement an algorithmic procedure for the determination of a nominal input function or trajectory that leads to approximate tracking of the reference trajectory. To this end, we introduce the following definition:

DEFINITION OF APPROXIMATE INVERSION

Approximate inversion is an algorithmic procedure for determining an input function in some admissible input set U which, when applied to a plant, causes its outputs to follow a trajectory which minimizes the deviation of those outputs from a prescribed trajectory over some closed interval of time. An input, u^*, so determined shall be referred to as an Approximate Inverse of the reference trajectory y^* over the input set U.

The term "approximate inverse" has previously been used by Zames [52], although not in this context. It is in fact related to the complexity of other solution methods with related goals. But there is another reason for introducing approximate inversion. When the desired output is not specified, it should be found by using techniques of optimization. Specifically, the optimal control of systems via the calculus of variations and Pontryagin's principle should, in theory, provide both the reference trajectories for a system as well as the inputs required to generate them. In practice, optimal solutions are hard to generate for all but the simplest problems. In many other cases, such as those involving systems with large, distributed parameter, and computer-based models, it is not possible to apply the classical theory at all. Finally, even in cases with moderate computational complexity, for example, the path planning of manipulators, it is usual that a trajectory is prescribed, either on the basis of expert advice or by search, and it is the job of the control system to track this assignment. In these cases it is particularly desirable that a method such as approximate inversion be available to predict the feasibility and performance of a system subject to a synthesized reference trajectory.

9.5.3 Approximate Inversion via Search Procedures

There are many reasons for preferring forward (in time) search over traditional methods of inversion via implementations of the inverse dynamical system. First, it allows the determination of approximate inverses when exact dynamical inverses of the plant cannot be used to derive feedforward commands. Then there is the breadth of applicability of this method, which includes systems with large, complex models as well as those with combined linguistic-mathematical rule-based models. In both these cases, analytic inversion may not be feasible, nor is an exact inverse even of interest as long as performance criteria are met. It is possible, for instance, to conduct a search with a weighted cost functional for multivariable optimization of a MIMO system.

Another situation that may be conveniently avoided is the case where a desired output function is not physically realizable because of constraints, and working backward from the final state of the system leads to meaningless results if that state is only approximately reachable. Also, if planning is being performed online, with predictions being made up to a horizon of attention in the future, it may not be possible to plan behavior backward from the eventual goal of the system. Thus planning by forward testing may be seen to offer significant advantages over competing strategies of feedforward control.

In using forward search, the first approximation is the conversion of a typically continuous time system into its discrete time equivalent (which is done automatically by a numerical simulation). Since nearly all practical problems involve sampled data systems, this is not actually a restriction. In fact, by selecting an appropriate sampling period, it is possible to vary the structure of the discrete time equivalents of the continuous system and thereby change some of their properties (e.g., [54]). The effect of this approximation on the computed inverse will be that the given desired function and the output due to the computed inputs will agree only at sampling instants. The effects of changing the sampling frequency and harnessing multirate sampling is reserved for further research.

The simplest possible search technique that can be applied to a memoryless SISO, LTI dynamical system is a binary search. In the case described, the binary search is an exhaustive method that provides a guaranteed admissible solution if it exists, and it is accurate to an arbitrary nonzero accuracy and is based on the usual L_2 measure (euclidean distance). The method may be described in the context of a discrete time system defined by the following equations:

$$x[k+1] = \mathbf{A}x[k] + \mathbf{B}u[k], \tag{9.28}$$

$$y[k] = \mathbf{C}x[k], \tag{9.29}$$

where **A**, **B**, and **C** are $n \times n$, $n \times 1$, and $1 \times n$ dimensional constant matrix/vectors. Now let an output sequence $y^*[k]$ be defined for some finite k in the interval $0 \leq k \leq p$, and let the desired command tracking accuracy be defined as

$$||y^*[k] - y[k]|| \leq \varepsilon.$$

Also let the admissible set of inputs be defined as

$$u \quad \text{such that} \quad U_{\min} \leq u \leq U_{\max},$$

which is a statement of the usual physical constraints on inputs to real systems. Finally, recall that this algorithm is guaranteed to converge only in the case of memoryless systems, even though it may do so for a larger class of systems. Then the binary search is illustrated in Figure 9.16, and it involves the following algorithm:

Step 1. For each k in the interval $0 \leq k \leq p$, beginning at $k = 0$.
Step 2. Test $U_{\min}$ and $U_{\max}$ by applying them to the model of the system.
Step 3. Compute the magnitude $||\cdot||_2$ of the error given by

$$E_{\min} = y^*[k] - y_{\min}[k], \text{ and } E_{\max} = y^*[k] - y_{\max}[k].$$

Step 4. If $||E||$ is less than or equal to e, record $u[k]$ and continue with the next k (i.e., return to step 2).
Step 5. If $||E_{\min}|| < ||E_{\max}||$, set $U_{\max} = (U_{\min} + U_{\max})/2$ and go to step 2. Otherwise, set $U_{\min} = (U_{\min} + U_{\max})/2$ and go to step 2.

A further possible modification is to replace the termination condition of both these algorithms with a loop counter and to stop the refinement of the intervals of search after a fixed number of iterations instead of trying to achieve an arbitrary accuracy.

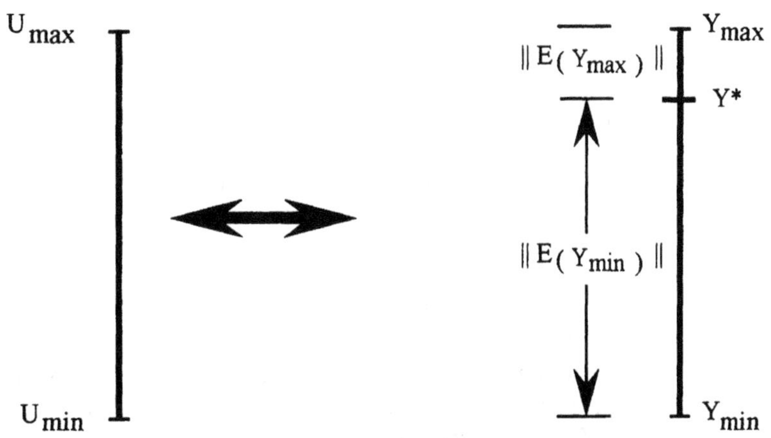

Figure 9.16. Illustration of a binary search.

This situation would arise if the allowable extent of error were larger than the desirable amount and the physically attainable level were between them and unknown a priori.

The reason for the choice of a euclidean norm in reducing the search space is that it appears to better extend the algorithm to more than one dimension. Before this purely intuitive approach, which has not yet been exhaustively analyzed, is described, it is useful to consider the proof of the euclidean algorithm in the case where the exact inverse lies in the admissible set of inputs.

For the system described above, the outputs due to two inputs u_1 and u_2 applied at time k and beginning from the same state x:

$$y_1[k+1] = \mathbf{CA}x[k] + \mathbf{CB}u_1[k] \tag{9.30a}$$

and

$$y_2[k+1] = \mathbf{CA}x[k] + \mathbf{CB}u_2[k] \tag{9.30b}$$

Now we can write

$$E_1[k+1] = \mathbf{CB}\left[u^*[k] - u_1[k]\right] \tag{9.31a}$$

and

$$E_2[k+1] = \mathbf{CB}\left[u^*[k] - u_2[k]\right]. \tag{9.31b}$$

Clearly, if $||E_1[k+1]|| \leq ||E_2[k+1]||$ and u^* is in the closed interval $[u_1, u_2]$, then

$$|u^*[k] - u_1[k]| \leq |u^*[k] - u_2[k]|,$$

and a new interval of search containing u^* may be defined by $[u_1, (u_1 + u_2)/2]$.

Unfortunately, the same argument does not carry over to the multidimensional case because of the interactions between the outputs of a system. A hypothetical situation is illustrated in Figure 9.17.

The selection of the upper right quadrant of the two-dimensional input space (indicated by the shaded region in Figure 9.17) on the basis of the smallest error from among the tested input/output pairs is not an approach that can be guaranteed. The reduced space may or may not include u^*, and

$$||E_1[k+1]|| \leq ||E_2[k+1]||$$

does not imply that

$$||u^*[k] - u_1[k]|| \leq ||u^*[k] - u_2[k]||$$

but just that

$$||\mathbf{CB}\left[u^*[k] - u_1[k]\right]|| \leq ||\mathbf{CB}\left[u^*[k] - u_2[k]\right]||,$$

which requires a special structure for the **C** and **B** matrices in order for the reduced space to be guaranteed to contain u^* (e.g., if their product is the identity matrix).

There remains one additional matter related to the determination of an approximate inverse of system outputs, which is that of the memory of the system. The system de-

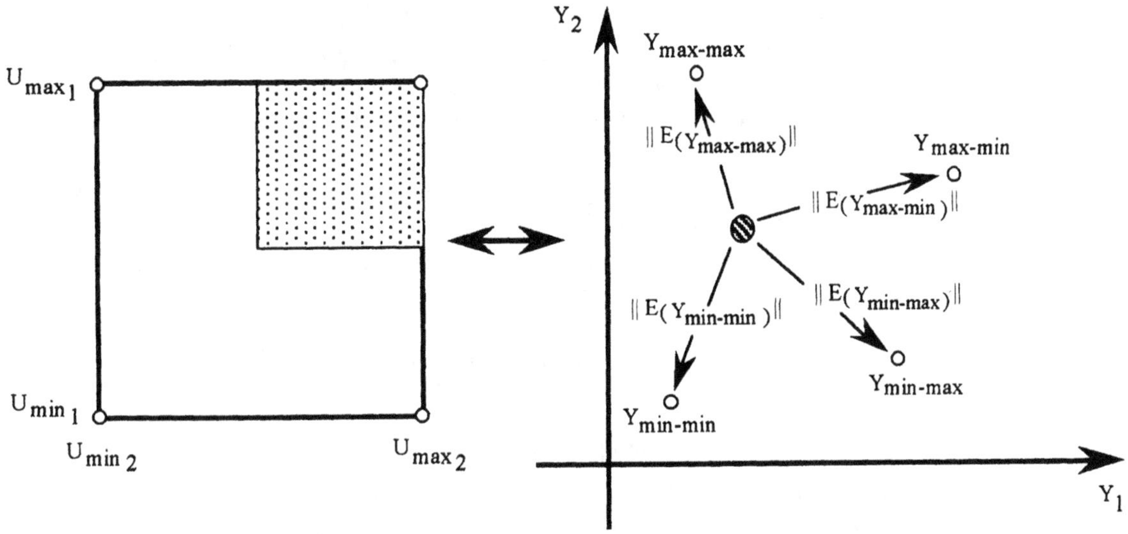

Figure 9.17. Using a euclidean measure in a two-dimensional I/O map.

scribed by equations (9.33) below is said to have memory if its output at time $k+1$ is a function of states or inputs at instants prior to k. Other contexts in which this property has been described are Markovian processes and information lossless automata (e.g., [41, 54, 55]). A simple example of a system with this property is described as follows: Let

$$x_1[k+1] = x_1[k] + x_2[k], \tag{9.32a}$$

$$x_2[k+1] = u[k], \tag{9.32b}$$

and

$$y[k] = x_1[k]. \tag{9.32c}$$

Then

$$y[2] = x_1[1] + u[0]. \tag{9.32d}$$

The last expression implies that the output at time k is dependent on the input at time $k-2$, and this can have significant implications for planning carried out on the basis of outcomes one time step ahead of the current state. The consequent possibility of planning finite chains of controls for systems with memory is investigated as part of further research which uses backtracking algorithms and model testing to generate epsilon-optimal reference trajectories as well as input functions.

9.5.4 Experimental Results

In order to investigate the outcomes of using the approaches described above, several dynamical models are analyzed. The examples include a single link manipulator with velocity as prescribed outputs and terminal voltage as controlled input, and a two link manipulator with a full, coupled, nonlinear model using the two joint velocities as prescribed outputs in conjunction with the two actuator terminal voltages as inputs

The models of the systems described above are used to plan the inputs required to generate output behavior belonging to several classes. These include:

1. Single objective planning, using a single input, of output functions that are known to have a physically realizable exact inverse.
2. Single objective planning, using a single input, of output functions that are known *not* to have a physically realizable exact inverse.
3. Dual objective planning, using two inputs, of output functions that are known *not* to have a physically realizable exact inverse.

Case 1: A Single-Link Manipulator (SLM)

A single-link manipulator was utilized to analyze the performance of a euclidean-based search algorithm whose results are presented in this section. The model used was

$$\begin{bmatrix} \frac{\theta(t)}{dt} \\ \frac{\omega(t)}{dt} \\ \frac{i(t)}{dt} \end{bmatrix} = \begin{bmatrix} 0 & 1 & 0 \\ 0 & -\frac{B}{J} & \frac{k}{J} \\ 0 & -\frac{k}{L} & -\frac{R}{L} \end{bmatrix} \begin{bmatrix} \theta(t) \\ \omega(t) \\ i(t) \end{bmatrix} + \begin{bmatrix} 0 & 0 \\ 0 & -\frac{1}{J} \\ \frac{1}{L} & 0 \end{bmatrix} \begin{bmatrix} v(t) \\ \tau(t) \end{bmatrix} \tag{9.33a}$$

and

$$\begin{bmatrix} y_1(t) \\ y_2(t) \end{bmatrix} = \begin{bmatrix} 0 & 1 & 0 \\ 0 & 0 & 1 \end{bmatrix} \begin{bmatrix} \theta(t) \\ \omega(t) \\ i(t) \end{bmatrix}, \tag{9.33b}$$

where

$$\theta(t) = \text{angular position of link,}$$
$$\omega(t) = \text{angular velocity of link,}$$
$$i(t) = \text{armature current,}$$
$$\tau(t) = \text{torque,}$$
$$v(t) = \text{voltage.}$$

By simple manipulation, the inverse of this system can be shown to be

$$v(t) = k\omega(t) + Li'(t) + Ri(t), \tag{9.34a}$$

$$\tau(t) = -J\omega'(t) - B\omega(t) + ki(t). \tag{9.34b}$$

An ideal (unachievable) voltage and current trajectory can be prescribed as in Figures 9.18*a* and *b*. From this, by applying equations (9.33), the corresponding voltage and torque may be derived. Figures 9.19*a* and *b* show the general form of these physically unrealizable results.

If, on the other hand, minimum time controls are determined for the plant in question, the requirements and results are as shown in Figure 9.19. The result of inverting the

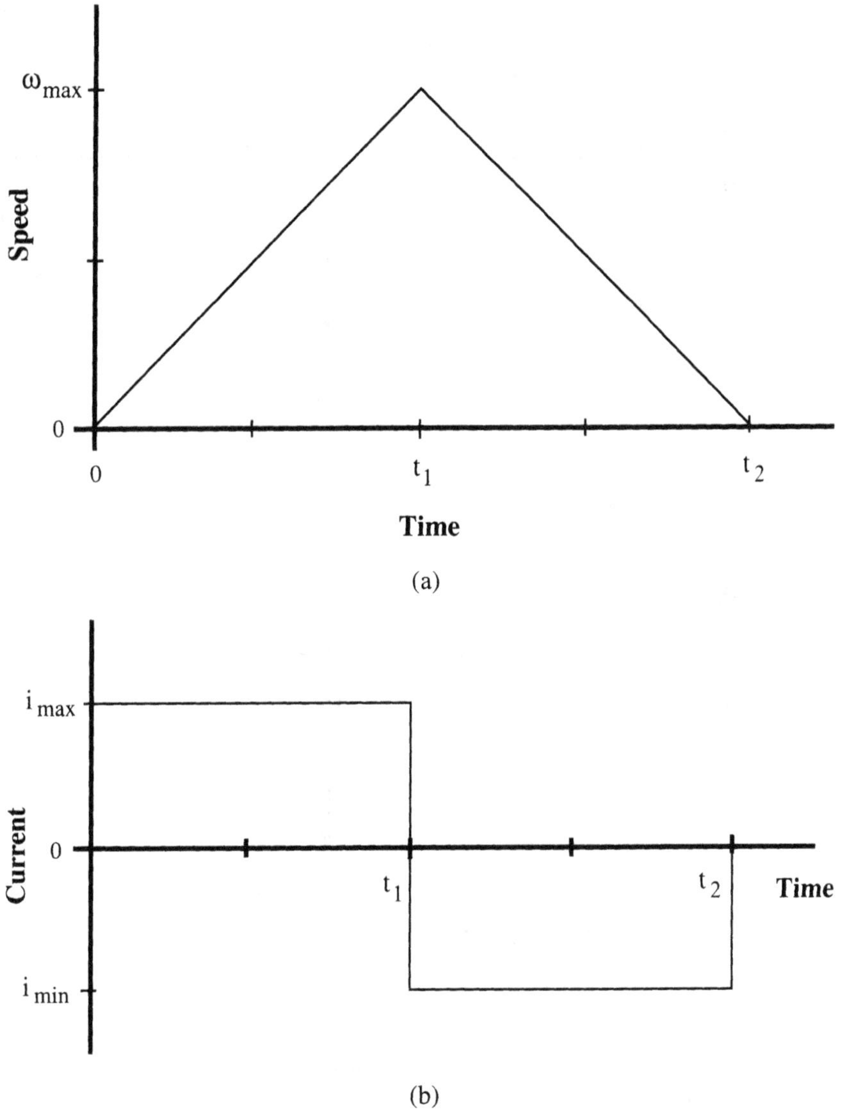

Figure 9.18. (*a*) Ideal velocity assignment for the SLM; (*b*) ideal current assignment for the SLM.

minimum time velocity trajectory by search is indistinguishable from that found by the numerical solution of the two-point boundary value problem associated with the minimum time problem.

Figures 9.18 through 9.23 depict the implementation of minimum time control using the results of Figures 9.18. The resulting error shown in these figures is due to an initial velocity error in the cases where feedforward control is used alone, simple constant gain proportional feedback control is used alone, or when both are used together (with the same feedback gains).

The desired triangular velocity trajectory in Figure 9.18*a* is noninvertible in the exact sense, yet approximate inversion is possible and is demonstrated by Figure 9.21. It is to be noted that in some cases the approximate tracking of an ideal, triangular velocity

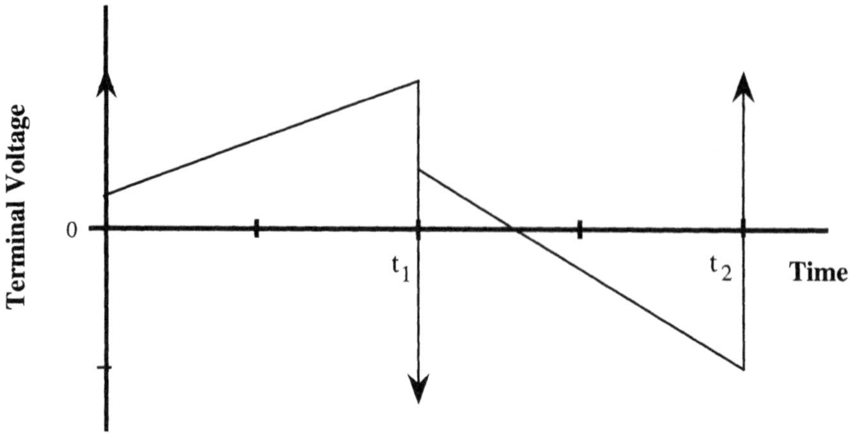

(a)

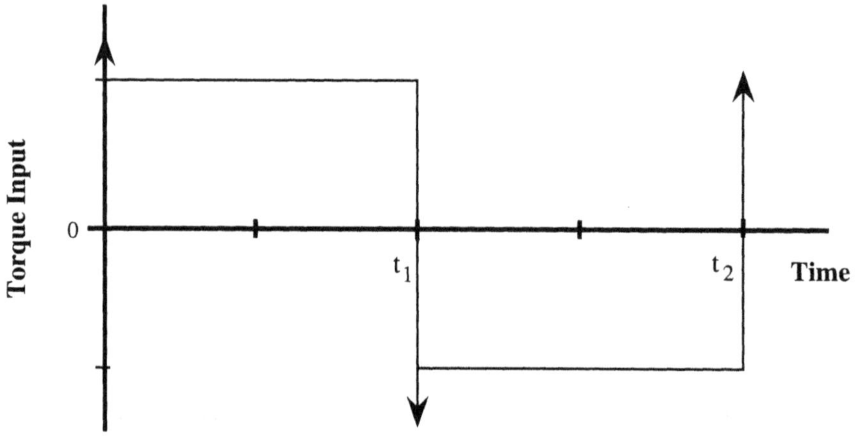

(b)

Figure 9.19. (*a*) Ideal voltage input found by analytical inversion of assignment; (*b*) ideal torque found by analytical inversion of assignment.

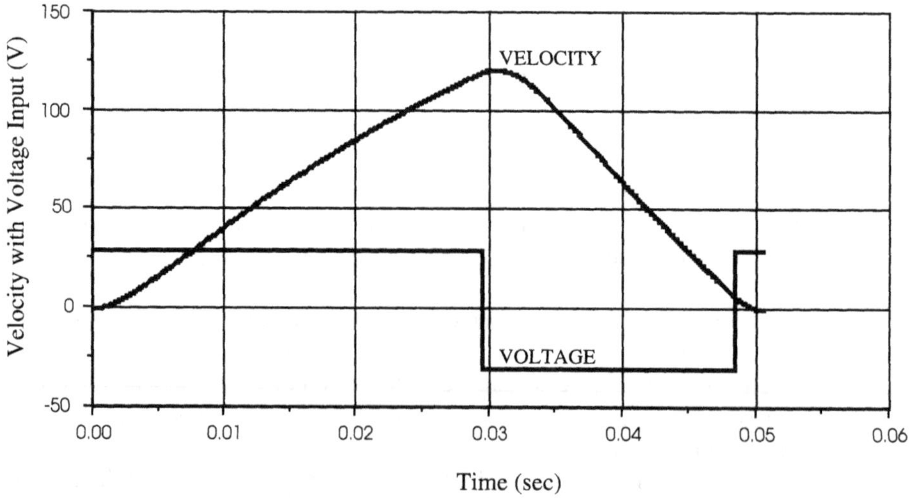

Figure 9.20. Actual and ideal (reference) velocity trajectories generated by the search.

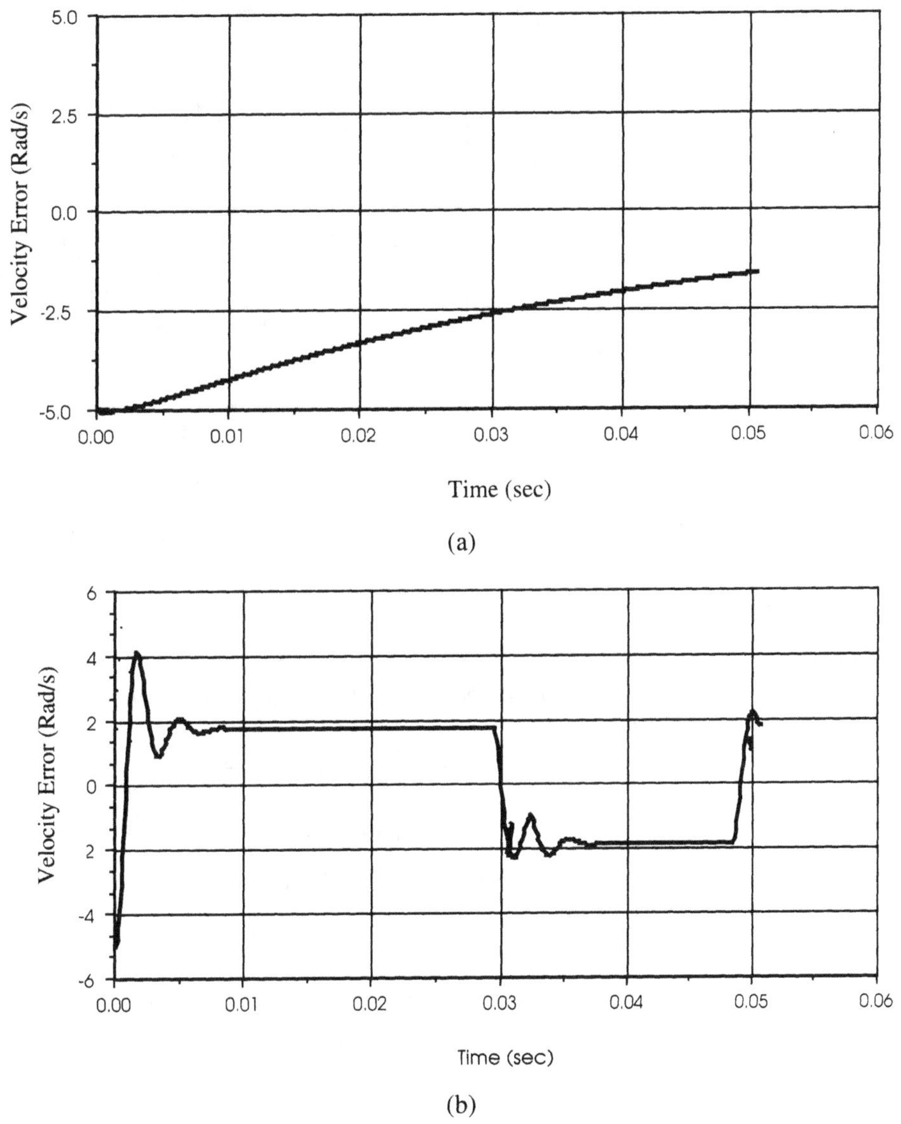

Figure 9.21. Error due to an ideal (unrealizable) reference trajectory.

trajectory may be a suitable substitute for the actual minimum time trajectory derived from optimization. The resulting error and inputs are shown in Figures 9.20 and 9.21. Note similarity between the inputs in Figures 9.19*a* and 9.21.

Case 2: A Two-Link Manipulator

The next example to be considered is that of a two-link manipulator whose inputs are the actuator voltages and whose outputs are the joint velocities. In order to construct this system in the form of a composite model, the equations for the SLM are duplicated to receive a six-dimensional state vector with indexes corresponding to joints 1 and 2. The dynamic effects of the links are included via their torque relations [56]:

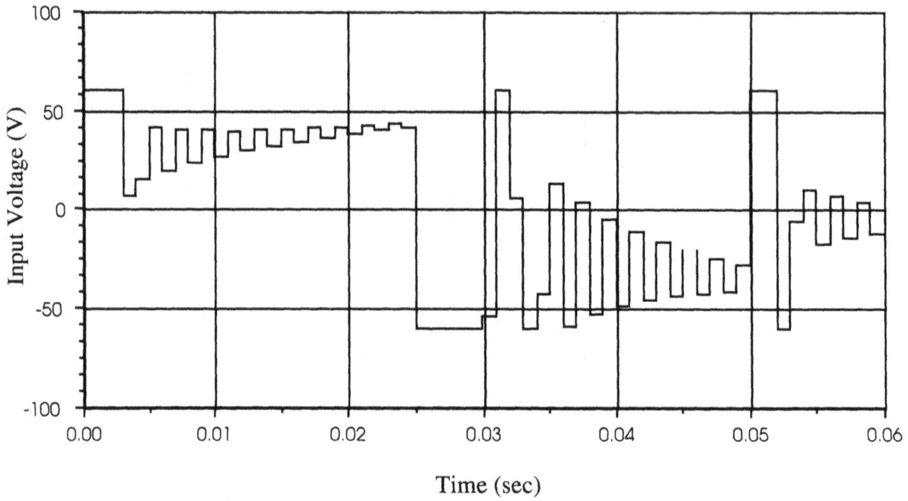

Figure 9.22. Input voltage determined by the search for approximately tracking the ideal reference.

$$\begin{aligned}
\tau_1(t) = {} & [(m_1 + m_2)d_1^2 + m_2 d_2^2 + 2m_2 d_1 d_2 \cos(\theta_2(t))]\frac{d\omega_1(t)}{dt} \\
& + [m_2 d_2^2 + m_2 d_1 d_2 \cos(\theta_2(t))]\frac{d\omega_2(t)}{dt} \\
& - 2m_2 d_1 d_2 \sin(\theta_2(t))\omega_1(t)\omega_2(t) - m_2 d_1 d_2 \sin(\theta_2(t))\omega_2^2(t) \\
& + (m_1 + m_2) g d_1 \sin(\theta_1(t)) + m_2 g d_2 \sin(\theta_1(t)) + \theta_2(t)), \\
\tau_2(t) = {} & [m_2 d_2^2 + m_2 d_1 d_2 \cos(\theta_2(t))]\frac{d\omega_1(t)}{dt} + m_2 d_2^2 \frac{d\omega_2(t)}{dt} \\
& - 2m_2 d_1 d_2 \sin(\theta_2(t))\omega_1(t)\omega_2(t) - m_2 d_1 d_2 \sin(\theta_2(t))\omega_1^2(t) \\
& + m_2 g d_2 \sin(\theta_1(t) + \theta_2(t)).
\end{aligned}$$

These coupled nonlinear torques are inserted in place of the external torque in equations 9.32.

The requirement from the system is to follow the trapezoidal velocity trajectories shown in Figure 9.23 together with the actual output. The joint accelerations are shown in Figure 9.24 and the results of search for the inputs are shown in Figures 9.25 and 9.26. It can be seen how the input to the first link reacts to the switch in the desired acceleration of the second joint, and vice versa toward the end of control.

In this example, the model parameters were chosen to produce an oscillatory step voltage response of the joints. This tendency, as well as joint coupling, is eliminated by the feedforward signals.

We have demonstrated that when the system is not invertible, the approximate solution can be obtained by using forward searching. The error of approximation can be reduced by varying the search parameters, which can be done off-line. Then the role of feedback is to compensate for the error of approximation, which reduces complexity of control and speeds up the online process of feedback compensation.

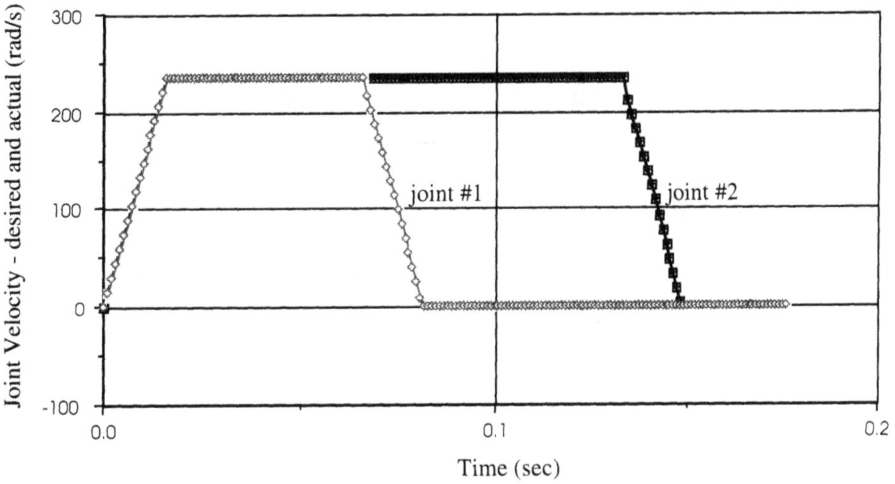

Figure 9.23. Reference and actual trajectories of a two-link manipulator.

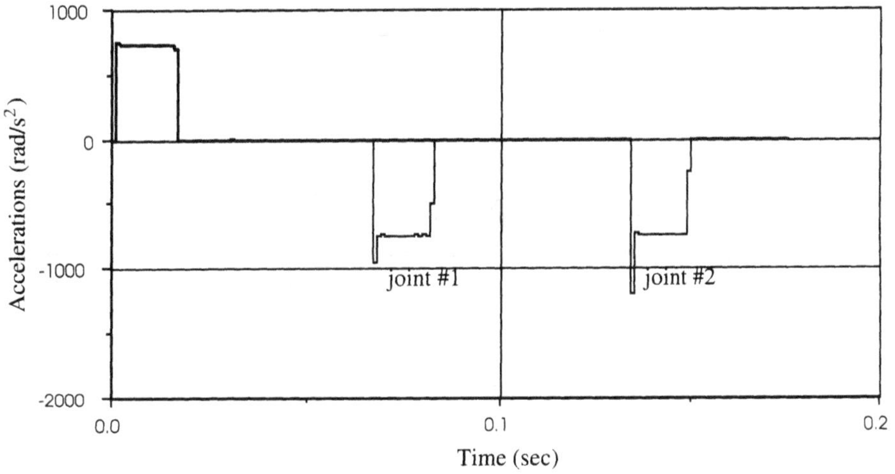

Figure 9.24. Accelerations experienced during the tracking of velocity curves in Figure 9.23.

Future work should focus on the necessity to represent the possible alternatives (which are themselves a finite collection representing infinite possibilities) in a nested hierarchy of a scale and scope in which strategies of control can be proposed and refined on the basis of increased focus of attention, rules of refinement, and implication models which are the source of the reference trajectory.

9.6 MULTIRATE HIERARCHICAL PREDICTIVE CONTROL SYSTEM

In Section 7.6 we mentioned that the methodology we presented for trajectory tracking in presence of various kinds of disturbances might have wider applicability. Our goal is to introduce a system that will exploit the opportunities given by computational advances in multiresolutional hierarchies and to create a structure that will eventually

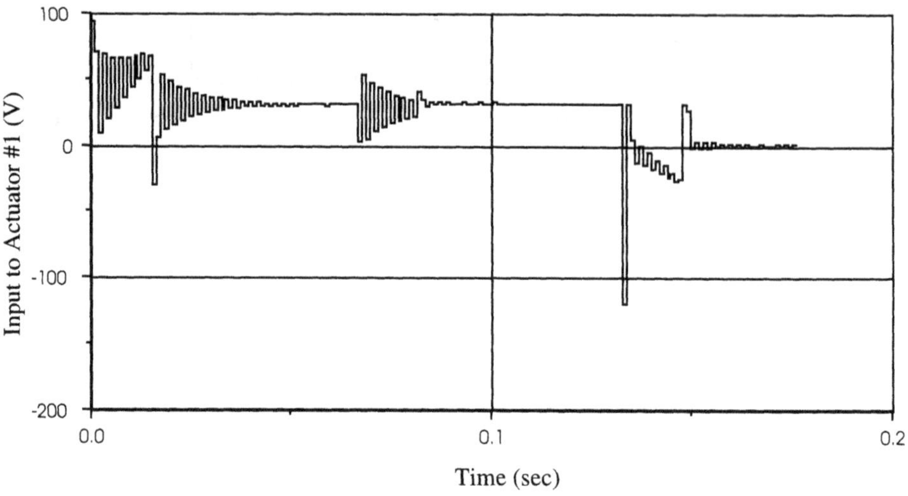

Figure 9.25. Input voltage trajectory for the first actuator of a two-link manipulator.

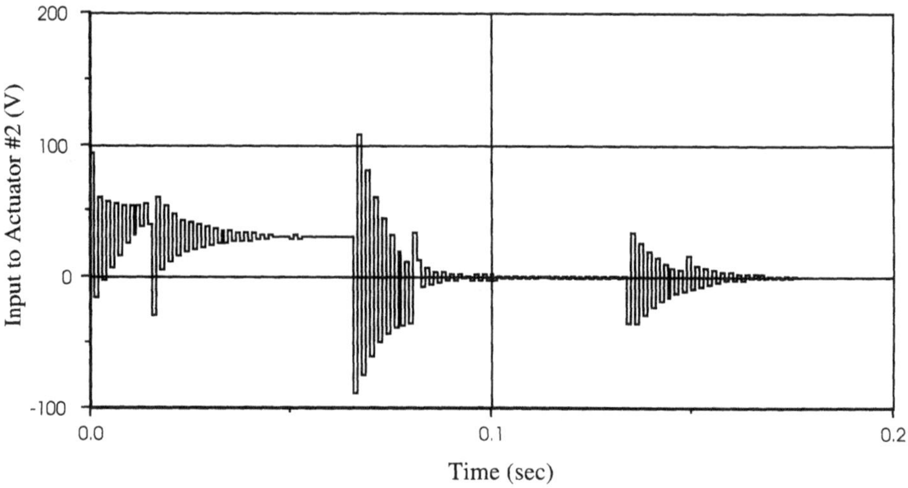

Figure 9.26. Input voltage trajectory for the second actuator of a two-link manipulator.

embrace intelligence features. We present a further development of the predictive recommendations from the Section 8.7. Nested feedback loops are derived from the multiresolutional representation of control variables and hierarchical prediction of the next sample error vector. This method leads to a simple design process and demonstrates the advantages over other approaches.

9.6.1 Introduction

In the existing hierarchical controllers, separation of the overall system into appropriate hierarchical levels is done in a natural, heuristic way. J. Albus and A. Meystel [57–59] gave a detailed theoretical elaboration of the nested hierarchical controllers principles. The time-scale decomposition of singularly perturbed systems has been widely investigated over the last 20 years. P. Kokotovic and R. Yackel [60] gave basic theorems

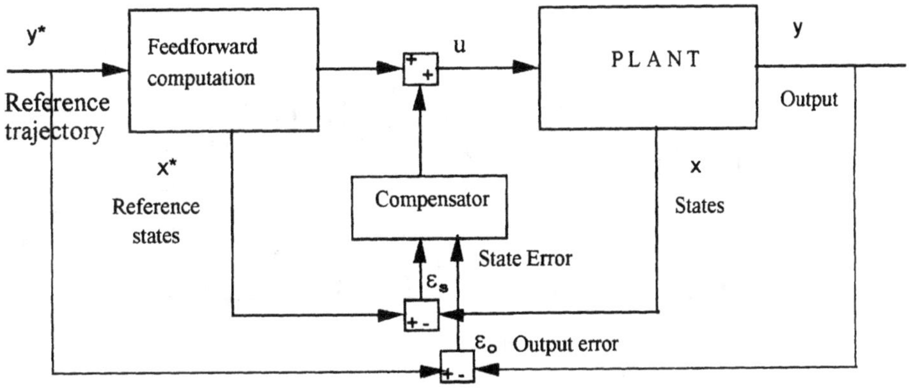

Figure 9.27. Complete control system.

for singular perturbations of linear systems, and P. Kokotovic and J. Chow [61, 62] have proposed the decomposition of the near-optimal regulator for systems with fast and slow modes. Multirate properties emerging from sampled data control systems were recognized by A. Chammas and C. Leondes [63], T. Hagiwara and M. Araki [64] and other works exploring these basic ideas.

The accepted fact is that trajectory tracking control cannot, in most cases, be achieved with only feedback control, and this leads us to the known [65] structure shown in Figure 9.27. Feedforward is presumed to be computed off-line, and the desired states trajectories are given in advance (these so-called reference curves are not always provided for all the states).

Our goal is to design a hierarchical compensator based on multiresolutional signal decomposition and tracking error prediction. This designed feedback system will be shown to lead to a significant improvement of tracking, an increase of control robustness, higher utilization of existing computational power and, above all, creating an evolutionary structure proven to be the best ground for adaptivity, learning, and implementation of other features of intelligent control.

This subsection will deal with the emerging need for prediction, averaging of wavelet (scaling function) choice, and a procedure for a hierarchical controller buildup and its evaluation in a classical sense. A comparison with PID control for a simple second order plant shows derived advantages. The conclusion will show the possibilities opened by the present research and will make projections of future advances in this area. The theory is presented for LTI systems; generalized and case characteristic conclusions are clearly indicated. The notation is explained online and the simplified examples should make this material accessible to a wide class of system engineers.

9.6.2 Hierarchical Structure with Averaging the Wavelet Algorithm

The known prediction algorithms rely either on a known error-generating model (recursion), which is an unrealistic assumption, or on an estimation based on the history of the process. The latter requires substantial memory and leads to increasing computational complexity which can jeopardize real-time operations. The hierarchical structure has the advantage of significant computational savings as well as intrinsic parallelism, as shown in [66].

The hierarchical controller architecture that we propose can be sketched as in Figure 9.28. It is introduced heuristically here; its derivation is the subject of the following

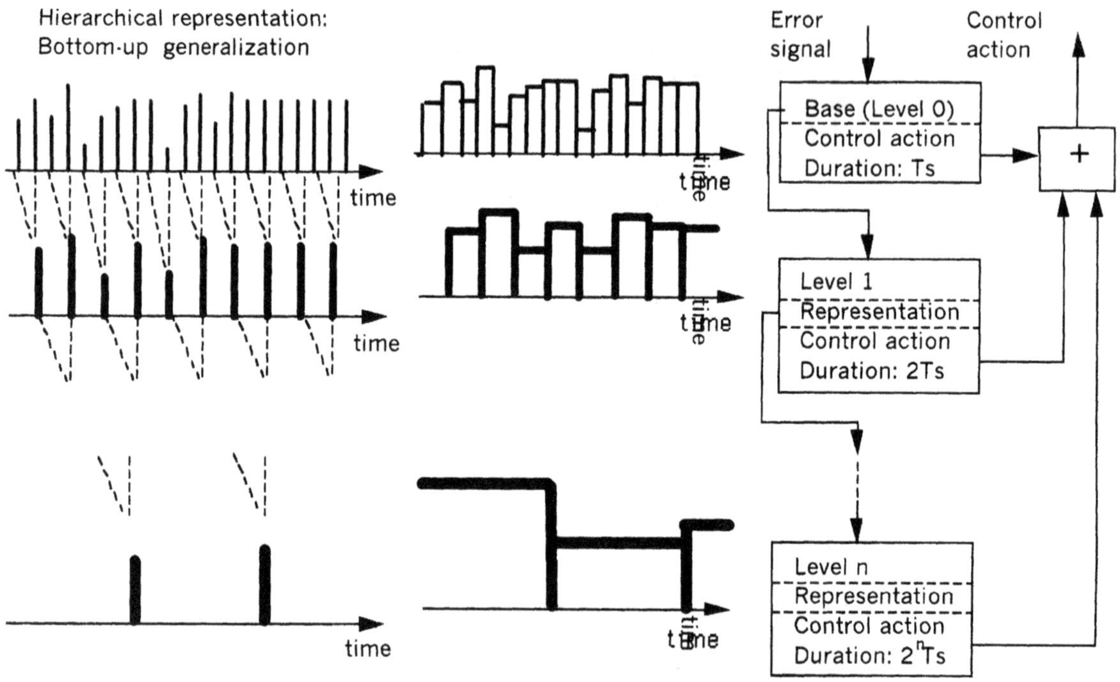

Figure 9.28. Representation of the hierarchical controller.

chapters. The error to be corrected presumably has a wide spectrum of frequency components. This fact requires a large memory for quality prediction at a single level. Let us suppose, for example, that our predictor uses N error samples and has $C = O(N^2)$ computational complexity. The multiresolutional representation of this signal can be done with the same number of samples and perfect reproducibility, with these (samples) distributed to m levels of frequency hierarchy [138]. Denoting number of samples per level by N_i, the overall complexity is

$$N = \sum_{i=0}^{m-1} N_i \rightarrow C_h = O\left(\sum_{i=0}^{m-1} N_i^2\right), \qquad (9.35)$$

which is reduced in comparison with the one level $C = O(N^2)$.

The multiresolutional representation allows for the prediction computation to be performed at each level of the resolution with less effort than making the implementation. Further computational savings result from the fact that the high-frequency prediction does not require all history of a signal but a finite width window, and therefore fewer samples per resolution level.

The heuristic treatment of the hierarchical and nested multiresolutional controller is known [128–130]; however, no formalism has been introduced for dealing with an analytical control design. We propose an analytical technique based on wavelet theory that allows for the selection of the number of resolution levels per controller and for the selection of resolution ratios between two consecutive levels. The derived analysis is suboptimal, since no computational complexity is involved in cost evaluation. In the subsequent presentation we will use the following terms:

- Higher or lower resolution. For levels with higher accuracy or lower accuracy. The accuracy will be determined by the value of the error representation at the level.
- Highest resolution. For a signal meant to be received as a result of subtracting two functions: the highest resolution of the reference curve and the output of the sensor (or of the highest resolution of the state observer).
- Signal. For the difference $\mathbf{x}(n) - \mathbf{x}^*(n)$, which actually is the control error, and it is reduced to the allowed specification level.

Taking a set of signal samples, we now want to find a systematic way of building one sample representation with the lower resolution required for the next adjacent level (see the appendix to this chapter for definitions of representation and resolution). Namely we want to find a constant to represent m samples of error. Since our design goal is to minimize the maximum error deviation, we can minimize the probability of exceeding a certain bound. The Tchebysheff inequality holds regardless of the density function that we use, so we can take the time sequence as a random variable. It yields

$$\mathbf{P}\{|\mathbf{x}(k) - \text{const}| \geq \text{bound}\} \leq \frac{\mathbf{E}\{|\mathbf{x}(k) - \text{const}|^2\}}{\text{bound}^2}. \tag{9.36}$$

Minimization of $\mathbf{E}\{(\mathbf{x}(k) - \text{const}^2\}$ shows that the best constant is the average of the $\mathbf{x}(k)$ sequence.

Multiresolutional signal representation is therefore done by averaging the wavelet transformation. Analysis is done following [67].

9.6.3 Hierarchical Controller Design

Suppose that the starting level of the signal representation corresponds to the measured error sample. How many samples should we use to build the next resolution level (the value of $\mathbf{s}_1 \in \mathbf{Z},\ \mathbf{s}_1 > 1$)? An optimal solution would take into account the computational cost as well as the performance goal. If we decide to pursue suboptimality, the problem will involve minimizing the function $\mathbf{J}$, the distance between the signal, and its representation:

$$\mathbf{J} = \max|\mathbf{f}(k) - \mathbf{D}^1\mathbf{f}_w|_{k\in(-\mathbf{s}_1+1.0)}, \tag{9.37}$$

$$\mathbf{J} = \max\left|\mathbf{f}(k) - \frac{\mathbf{f}(-\mathbf{s}_1+1) + \mathbf{f}(-\mathbf{s}_1+2) + \ldots + \mathbf{f}(0)}{\mathbf{s}_1}\right|_{k\in(-\mathbf{s}_1+1.0)} \tag{9.38}$$

Taking the worst case, where $\mathbf{f}(k)$ limited $(-\mathbf{M}, \mathbf{M})$ for some constant $\mathbf{M} > 0$ we obtain

$$\mathbf{J} = \left|\mathbf{M}\left(1 - \frac{1}{\mathbf{s}_1}\right) + \frac{(\mathbf{s}_1 - 1)}{\mathbf{s}_1}\mathbf{M}\right|_{k\in(-\mathbf{s}_1+1.0)}, \tag{9.39}$$

which is minimal for $\mathbf{s}_1 = 2$. Since similar argument can be applied for all consecutive levels, the suboptimal level generation is achieved by a wavelet transform with $\mathbf{s}_j = 2^j$.

The starting level control signal is therefore the state error feedback mentioned earlier. The first-level feedback controller predicts within its representation the error signal and computes the control action, which lasts two consecutive basic rate samples.

Next (second) level computes the action upon its prediction; thus the applied action lasts for two instances of the first level, four instances of the basic sampling rate. The procedure continues recursively until the point where the new level introduction does not improve performance. In order to determine this limit, a frequency domain analysis is made of the wavelet transform.

Tedious mathematical derivation can be avoided by recognizing that the averaging transform of a given signal to level j is equivalent to the sum of $\mathbf{s}_j$ signal shifts sampled at a rate $2^j T_s$, and divided by 2^j.

After averaging the multiresolutional representation, we take a Fourier transform of the frequency bandwidth nested controllers in the resulting hierarchy. For optimization we use $\mathbf{s}_j \neq 2^j$, which is subject to further research. Since each level action has its own signal frequency, the control signal is computed using a corresponding $2^j T_s$ sampled data plant model. A plant with a variety of natural frequencies makes model reduction at the upper levels possible. The real power of the correlation of prediction and design will become apparent in the hierarchical state observers and in the learning intelligence implementation based on this architecture. We will use the notation AWT for averaging wavelet transform.

The feedback compensator design procedure to be used in our example runs as follows:

1. Record the tracking error of a system with feedforward and feedback.
2. Find its next level of wavelet transform.
3. Evaluate the usefulness of its application to control.
4. Apply the same procedure to next level and regard previous results as a compact system.

9.6.4 Simulation Results

Consider a single-link manipulator for the plant in our simulation example. A second-order dc linearized system with no dry friction will be used in driving the link. Nonlinearity due to gravitation will be introduced as a source of error to be compensated by the control system:

$$x'(t) = \begin{bmatrix} -1.5 & 501.1 \\ -15.7 & -361.5 \end{bmatrix} x(t) + \begin{bmatrix} 0 \\ 118.9 \end{bmatrix} v(t), \quad y(t) = [\ 1 \quad 0\ ]x(t). \tag{9.40}$$

Now transfer the function (Laplace)

$$H_s = \frac{60{,}000}{s^2 + 363s + 8450} \tag{9.41}$$

The trajectory reach in the frequencies is chosen as our control goal. Any parametric disturbances will be checked by a change of plant parameters during the simulation control. A random number generator with uniform distribution will simulate an input and output white noise disturbance. In addition there is the influence of gravitation as a second component of the input noise.

A comparison is made with a PID controller, since it has shown the best experimental results. The PID gains are determined by an exhaustive search for values that minimize the tracking error. The prediction gain vector $\mathbf{p}$ for each level of hierarchy is found by minimization of **J**:

$$\mathbf{J} = \mathbf{w}_0^t(\mathbf{I} - \mathbf{bp})^t(\mathbf{I} - \mathbf{bp})\mathbf{w}_0 + \sum_{i=1}^{i=k} \mathbf{w}_i^t(\mathbf{A}^i\mathbf{bp})^t\mathbf{A}^i\mathbf{bpw}, \tag{9.42}$$

where vector $\mathbf{w}_i$ represents the weighting factor (equal to **c** in this case) $k = 5$ (the results of the **J** minimization are given in the appendix to this chapter). Matrices **A**, **b**, and **c** are computed from the continuous space-state system with different "sampling" times for different levels. The function **J** reflects our desire to have a perfect match, resulting in a zero error in the next step as well as minimizing the influence of this action in the future steps. Since the new control will be applied while $i > 0$ events occur, less significance is given to these members (but they must not be excluded since they guarantee stability). The choice of **w**'s and k can be made after further investigation.

The prediction problem involves making a forecast by using a simple exponential predictor:

$$\mathbf{x}(n+1) = \theta\mathbf{x}(n) + \theta(1-\theta)\mathbf{x}(n-1) + \theta(1-\theta)^2\mathbf{x}(n-2) + \ldots \tag{9.43}$$

The performance of the predictor is dependent on the choice of θ. The tracking boundaries are calculated for maximum absolute value of prediction and matrix norms (not least upper bound, LUB). Figure 9.29 gives the results of feedforward tracking when no disturbances are applied and the input is precomputed off-line (compare with Fig. 9.30).

Table 9.1 compares the results of the graphic illustrations and shows the superiority of the hierarchical AWT prediction controller over PID in all aspects of performance: gravitation as the input disturbance, parametric disturbance, and the combination of both with the addition of random noise. The sum of errors is compared with the maximum absolute value. Third controller is PID which has two additional levels; it shows that the hierarchical structure is not bound to prediction: It introduces a new way of dealing with a phenomenon that was hidden by the information overflow at one level.

A graphic representation of the tracking error is shown in Figure 9.31. All sources of error demonstrated in Table 9.1 have been applied. We can see that with only two levels of the hierarchy the magnitudes of the error are reduced in two to three times. The in-

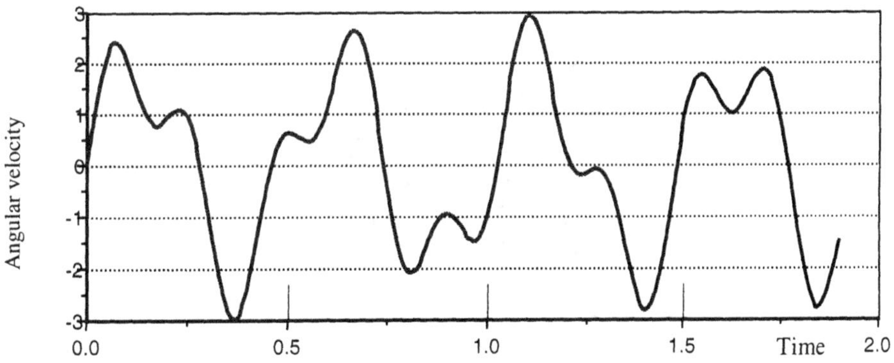

Figure 9.29. Perfect tracking with feedforward applied with no disturbances present (both curves coincide).

Table 9.1. Comparison of performances

Controller	Input Noise Gravitation	Parametric Disturbance	All Sources of Disturbances
PID			
Error sum	12.732	18.489	22.662
Max absolute	0.226	0.230	0.331
Hierarchy of two levels			
Error sum	2.49	7.02	9.64
Max absolute	0.216	0.108	0.147
PID with two additional levels			
Error sum	12.88	16.71	21.62
Max absolute	0.226	0.219	0.328

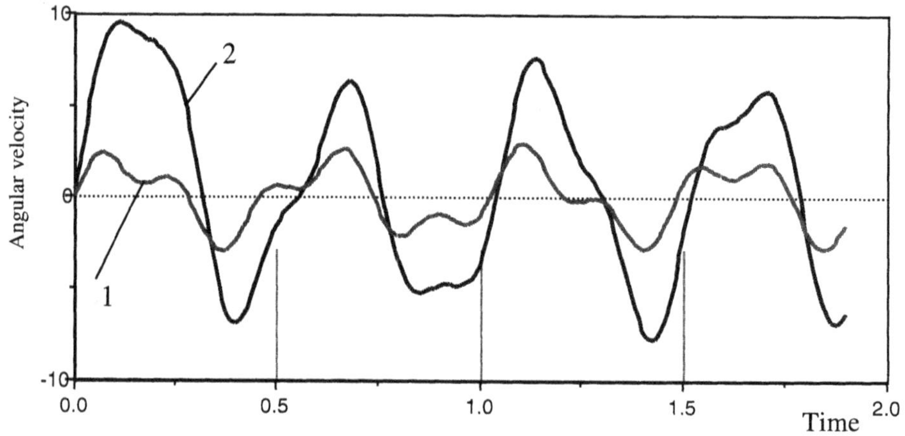

Figure 9.30. Desired (1) and produced (2) output with feedforward only, and with all disturbances present.

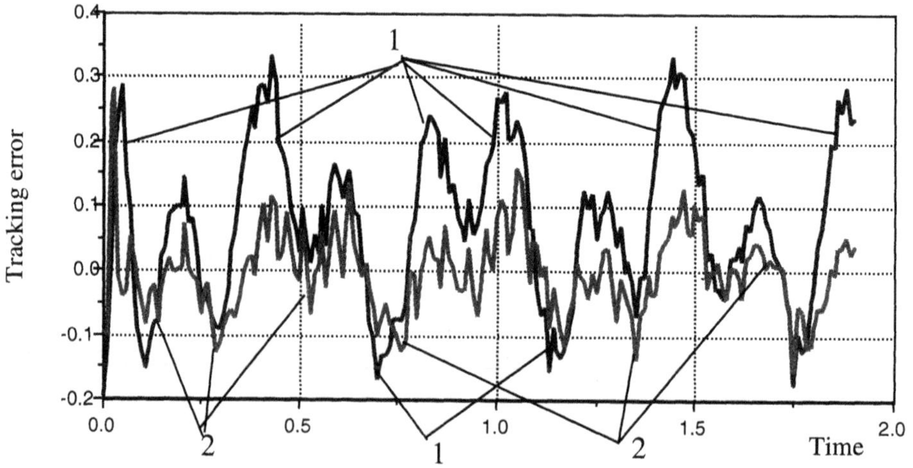

Figure 9.31. Comparison between the tracking errors of the PID (1) and the AWT prediction controllers (2).

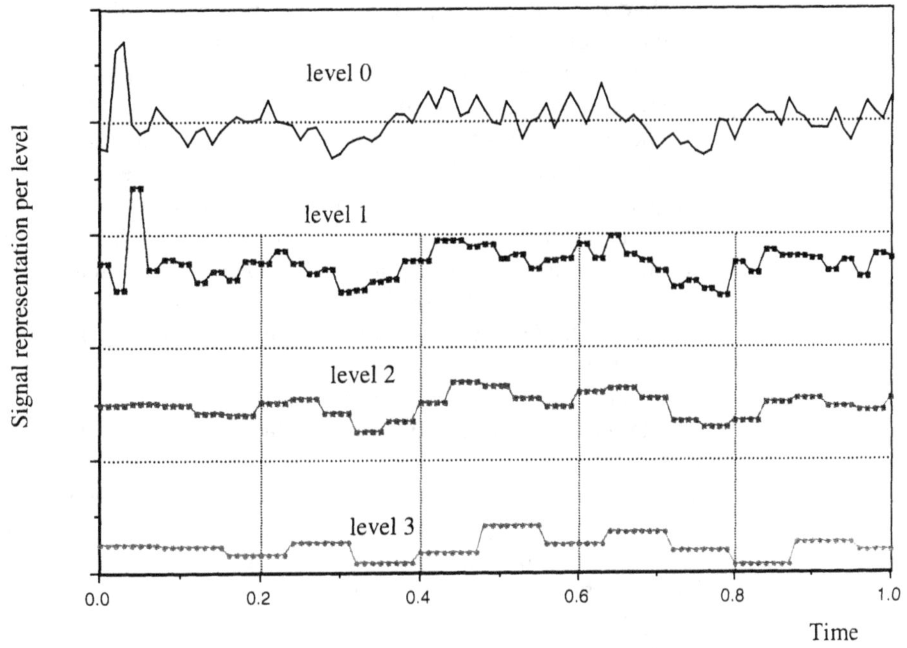

Figure 9.32. Signal representation of the AWT prediction controller.

teresting detail to note is that the PID controller ends up with a low frequency deviation from the zero level. This deviation does not exist in the AWT predictor controller case.

Figure 9.32 shows the signal representation for four levels of the AWT hierarchy. Each level has its own model of the plant according to the time between the two control actions:

$$\mathbf{A}_1 = \begin{bmatrix} 0.8354 & 1.1930 \\ -0.0374 & -0.0216 \end{bmatrix} \quad \mathbf{b}_1 = \begin{bmatrix} 1.1405 \\ 0.2865 \end{bmatrix}. \tag{9.44}$$

$$\mathbf{A}_4 = \begin{bmatrix} 0.1468 & 0.2186 \\ -0.0069 & -0.0102 \end{bmatrix} \quad \mathbf{b}_4 = \begin{bmatrix} 6.0379 \\ 0.0699 \end{bmatrix}. \tag{9.45}$$

It is readily apparent that higher-level models can be simplified due to differences of natural frequencies in their plants. This fact becomes more obvious for higher-order systems with a great variety of natural modes and a larger number of control levels.

9.6.5 Multirate Hierarchical Controller

The linearity of the time frequency transformations opens the possibility for decomposition of the plant model and control design induced by the frequency characteristics of the desired system output trajectory (the trajectory-tracking problem) or the spectral density of the prescribed trajectory set (characteristics to be given in the design stage).

The multiresolutional hierarchical algorithms introduced in computational mathematics as domain decomposition methods, or multigrid methods have distinct properties that deal with different frequencies at different levels of resolution.

Here we try to establish a direct link between the multiple time-scale signal representation and the nested multiresolutional principle of control by developing a theory and design techniques of building different levels of a hierarchical controller by bandwidth separation and decomposition.

The need for decomposition of the model, operation description, and the design and control procedures has become essential because of the increasing complexity of the intelligent control systems and the incompleteness of the available system model, its inadequate representation, and the insufficiency of the available sensor information. Indeed, separated design and control are easier and are implementable in a parallel manner; the complexity of computations is lower, and this opens the possibility for further control sophistication. Online feedforward computation, additional degree of design freedom, noise attenuation for aggregated levels can be explored. The time-frequency properties of the system and the desired trajectories can be considered a design tool in determining the levels of the control hierarchy, and thus they deserve a thorough investigation.

We are at a point in time when we can develop a theory of frequency bandwidth decomposition and propose methods of design that will allow for deriving the recommended laws of frequency (time-scale) range separation from design specifications. Our intention is to explore the advantages and disadvantages this separation entails. Our ability to decompose the system into bandwidth subsystems, to reduce its complexity, and to improve the dynamics and conditions for error compensation will be investigated.

The extension of existing methods of design of hierarchical controllers based on frequency bandwidth separation involves a number of issues. Their analysis is expected to yield new results in intelligent control. We are interested in the following three areas:

1. Formalization of a *method of hierarchical representation of systems via frequency decomposition* in such a way that frequency (time-scale) range separation can satisfy a criterion of goodness including complexity, dynamics, and error under conditions satisfying other design criteria. Our purpose of building a representation hierarchy via frequency decomposition is directly rooted in the requirements of controller design and operation. It is our observation that the roots of frequency decomposition are in the statistical representation of the assignments for functioning. This work involves determining the frequency (time-scale) ranges and their overlap, as well as the preferred number of levels.
2. Using a full system description including its dynamics effects which is different at different levels of hierarchy because of the different frequency bandwidth. It is our observation that real systems decompose into subsystems directly responsible for dealing with different frequency (time-scale) ranges anyway. In systems where the plant does not allow for physical bandwidth decomposition, the model can be decomposed to simplify the process of design. An investigation of the achievement (and definition) of frequency (time-scale) range decomposition that delivers near-optimal behavior will be carried out. Other methods that introduce reduced order models will be developed for the design of a controller with frequency bandwidth separation.
3. Experimental validation of theoretical predictive models using a system with a multilink manipulator that includes hierarchical planning and control. Some preliminary results are given for SISO case, and we sketch out a possible approach for the MIMO case.

Frequency Spectrum Decomposition of the Control Assignment

The linearity of the Fourier transformation allows us to regard any given signal and any transfer function as a sum of its distinct bandwidth components:

$$s(t) = \text{const}^* \int_0^{w_0} S(w)^* \exp(jwt)dw = \text{const}^* \sum_{w_0=0}^{w_{n-1}=w_0} \int_{w_i}^{w_{i+1}} S(w)^* \exp(jwt)dw.$$

$$s(t) = \text{const}^* \int_0^{w_0} \sum_{i=0}^{i=n-1} S(w_i)^* \exp(jwt)dw$$

so

$$S(w) = \sum_{i=0}^{i=n-1} S(w_i)$$

for $S(w_i) = 0$ everywhere but in its band $w_i - w_{i+1}$. This equation describes the control assignment as an instantiation of a deterministic function; it can be equally applied in stochastic functions when a number of realizations are expected to be performed so that their spectral density can be analyzed.

Controller Decomposition: Implementation Issues

If an assignment can be considered a spectrum, the controller is then a package of different controllers, each oriented toward its own bands of frequencies. In other words, any transfer function contains at the same moment multiple transfer functions each fit for its own package of operational frequencies and thus being prepared for its own subset of the assignment package.

An example of division into two bands will illustrate the basic ideas we want to investigate. Then each submodel will have its own controller which we will design for a subset of assignments. The presented control will be compared to the proportional output feedback. We will analyze this simple type of controller with the intention of expanding the theory to other types of controllers later.

Thus the error for the step function response is expressed as:

$$E_1(s) = \frac{s^2 + a_1^* s + a_2}{s^*(s^2 + a_1 + k^* b_1)^* s + a_2 + k^* b_2}.$$

If the ideal filters divide band 0–w_0 into two 0–w_f and $wf - w_0$, we have

$$E_2(s) = \frac{s^2 + a_1^* s + a_2}{s^*(s^2 + (a_1 + k_l^* b_1)^* s + a_2 + k_l^* b^2} + \frac{s^2 + a_1^* s + a_2}{s^*(s^2 + (a_1 + k_h^* b_1)^* s + a_2 + k_h^* b_2}.$$

Each is identically 0 outside its band. A comparison shows that the steady-state error E_2 gives the identical expression as E_1 with k_l in place of k. The choice of k_h can be made based on the characteristics of the stochastic signal in its band and the desired dynamical behavior (which is not arbitrary, since this is feedback output). The extra design freedom allows for separation of the design for steady state and the given noise dynamics trade-off (which is now local, i.e., within the band). Clearly, examples can be constructed for any number of submodels, and this brings us to the idea of infinite controller decomposition.

Infinite Decomposition

For a particular real (nonwhite) stochastic signal with a definite spectral density, the simple proportional feedback output can be evaluated with respect to this spectral density and some generalization of this density, such as its average. The division of the control problem into frequency bands improves the control, since we can select better the value of feedback gain to compensate for the noise in the more narrowband (increase of resolution). Asymptotically, in having an infinite number of bands, we can deal with the possible frequency of noise in a band and therefore can eliminate it without sacrificing dynamical characteristics.

The Problem of Filtering

The theoretical consideration of filtering is based on an implicit assumption that there exists a set of devices that distribute frequencies of the signal to corresponding channels. For the E_2 case, real (second-order) Butterworth filters give the same asymptotic behavior with respect to the time frequency. Further aspects of filter choice and dynamical influence will be studied as applied to real frequency (time-scale) range decomposition. The experimental results of a dc model (presented in the Appendix to Chapter 11) support our approach and encourage further investigation.

Sampled Data Systems

Computer control implies the use of sampled data systems. Discrete time sampling, which is our preliminary interest, provides a case of filtering that simplifies the hardware involved while introducing a number of disturbances because of the approximate character of filtering. The discrete case is of course not at all analogous to continuous time, and the opportunity to use samples in filtering creates some additional unsolved problems which we leave for further investigation. As is intuitively clear (and experimentally confirmed too), *if two proportional output gains are introduced, one for each sample (according to the Shannon theorem), and another for every nth sample (n is to be properly determined), a better tracking of the desired output can be achieved. The reason is that the slow gain gives a constant bias to the plant during periods of inactivity and eases the problem of error compensation for the fast gain.*

MIMO Systems

We want to expand the idea of frequency (time-scale) range decomposition (relatively easy understandable in the SISO case) to the MIMO domain. Using the results from the SISO case, we will base the generation of the hierarchy level in our bandwidth decomposition on input-output decoupling, time-scale system decomposition, and model decomposition with respect to the frequency characteristics of the desired output trajectories. The relationship between these components is investigated in order to produce the best performance.

The structure shown in Figure 9.33 can be generated to the point where all levels have an open-loop upper-level structure. Feedback compensation is actually applied only at the lowest level of the system. Our goal is to design a hierarchical compensator based

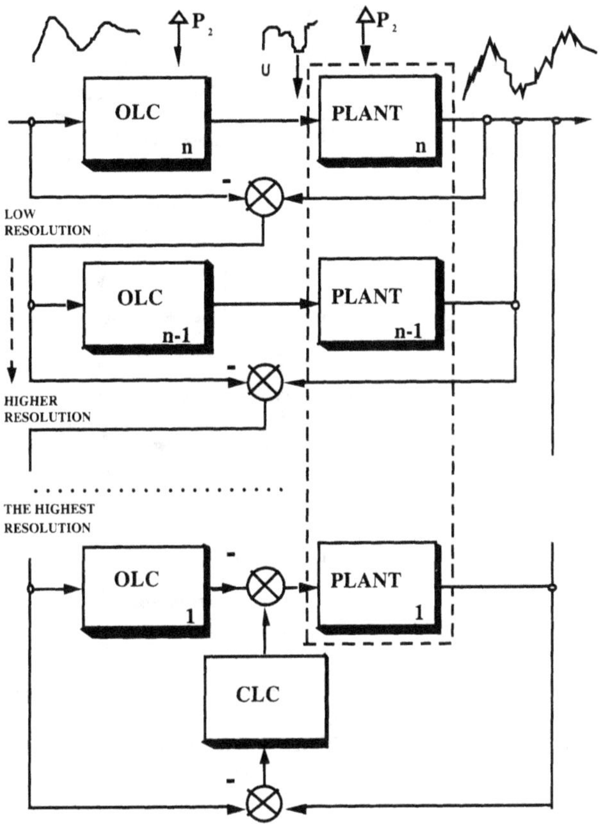

Figure 9.33. Hierarchical evolution by multirate decomposition of the OLC-CLC control system.

on multiresolutional signal decomposition and tracking error prediction. A feedback system, thus designed, will be shown to significantly improve tracking, increase control robustness, utilize higher existing computational power, and, above all, create an evolutionary structure with the best foundation for adaptivity, learning, and implementation of other features of the intelligent control. The theory of LTI systems (controllable and observable system or systems parts) is generalized here with case specific conclusions clearly indicated.

Multiresolutional representation allows the prediction computation to be performed at each level of resolution, thus there is much less effort in making the implementation realizable. Further computational savings result from the fact that a high-frequency prediction does not require all the history of a signal but a finite width window, so there are fewer samples per resolution level.

In the subsequent discussion we will consistently use the following terms:

- Higher or lower resolution. For the levels, higher or lower accuracy is determined by the value of the error representation at the level.
- Highest resolution of the signal. By this we mean the resolution received as a result of subtracting two functions: the highest resolution of the reference curve and the output of the sensor (or of the highest resolution of the state observer).
- Signal. For the difference $\mathbf{x}(n) - \mathbf{x}^*(n)$, the signal is actually the "control error," and it is reduced to the level allowed by the specifications.

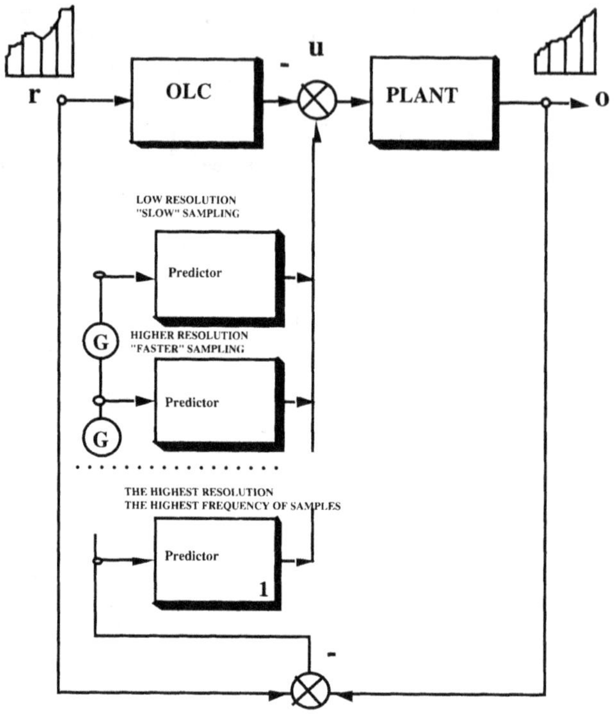

Figure 9.34. Hierarchical multiresolutional controller.

The starting level control signal is therefore said to be the state error feedback difference. The first-level feedback controller signal is a representation of original signal at a lower resolution (less the frequent samples). Each level predicts within its representation the error signal and computes the control action of its sampling time duration (as illustrated on Fig. 9.34).

The next level $(n + 1)$ computes the action upon its prediction (within its signal representation derived from signal representation at nth level), and thus the applied action has the duration of this representation sample time. The procedure continues recursively until the point where the introduction of a new level does not improve performance. A frequency domain analysis of the wavelet transform is used to determine this limit.

9.7 CONCLUSION

The multiresolutional systems of the PLANNER/EXECUTOR modules emerge as we develop multirate control systems. The mathematical descriptions of a control structure, multiresolutional signal representation, and the representation-resolution paradigm tend to formalize previously heuristically-treated phenomena. The treelike approach has left many unanswered questions (optimal level distribution, predictor structure, wavelet hierarchical observer, adaptivity, learning intelligence implementation) so we are challenged to make further investigations. It is evident so far that multiresolutional control can be approached both by multiresolutional decomposition of state variables and by multiresolutional decomposition of sampling schedules.

REFERENCES

[1] M. Lemmon and P. Antsaklis, A computationally efficient framework for hybrid controller design, *Proc. IEEE/IFAC Joint Symp. on Computer-Aided Control System Design*, Tucson, AZ, 1994, pp. 174–179.

[2] J. Stiver and P. Antsaklis, Extracting discrete event system models from hybrid control systems, *Proc. IEEE Int. Symp. on Intelligent Control*, Chicago, 1993, pp. 298–301.

[3] J. Stiver, P. Antsaklis, and M. Lemmon, An invariant approach to the design of hybrid control systems containing clocks, in R. Alur, T. Henzinger, and E. Sontag, eds., *Hybrid Systems III, Lecture Notes in Computer Science*, vol. 1066, Springer-Verlag, Berlin, 1996, pp. 464–474.

[4] H. Ye, A. Michel, and P. Antsaklis, A general model for the qualitative analysis of hybrid dynamical systems, *Proc. 34th CDC*, New Orleans, LA, 1995, pp. 1473–1477.

[5] A. Puri and P. Varaiya, Verification of hybrid systems using abstractions, Preprints of IFAC'96, vol. J, Identification II, Discrete Event Systems, San Francisco, 1996, pp. 467–472.

[6] A. Meystel, Planning in a hierarchical nested controller for autonomous robots, Proc. IEEE 25th CDC, Athens, Greece, 1986.

[7] A. Meystel, Architectures, representations, and algorithms for intelligent control of robots, In M. Gupta and N. Singha, eds., *Intelligent Systems: Theory and Applications*, IEEE Press, New York, 1996, pp. 732–788.

[8] A. Gollu, A. Puri, and P. Varaiya, Discretization of timed automata, *Proc. 33rd CDC*, Lake Buena Vista, FL, 1994, pp. 957–960.

[9] J. Albus, A. Meystel, and S. Uzzaman. Nested motion planning for an autonomous robot, *Proc. IEEE Conference on Aerospace Systems*, May 25–27, Westlake Village, CA, 1992.

[10] A. Meystel, Nested hierarchical control, in *An Introduction to Intelligent and Autonomous Control*, P. Antsaklis and K. Passino, eds., Kluwer Academic, Boston, 1992, pp. 129–162.

[11] F. Callier, W. Chan, and C. Desoer, I/O stability of interconnected systems using decompositions: an improved formulation, *IEEE Trans. Auto. Cont.* (1978).

[12] M. Hauts and M. Heymann, Linear feedback decoupling—transfer function analysis, *IEEE Trans. Auto. Cont.* **28** (1983).

[13] P. Kokotovic and J. Chow, Eigenvalue placement in two-time-scales systems, in *Proc. IFAC Symp. on Large-Scale Systems*, 1976, pp. 321–326.

[14] P. Kokotovic and J. Chow, A decomposition of near-optimum regulators for systems with fast and slow modes, *IEEE Trans. Auto. Cont.* **21** (1976).

[15] G. Peponides and P. Kokotovic, Weak connections, time scales and aggregation of nonlinear systems, *IEEE Trans. Auto. Cont.* **30** (1983).

[16] B. Litkouhi and H. Khalil, Multirate and composite control of two-time-scale discrete-time systems, *IEEE Trans. Auto. Cont.* **28** (1985).

[17] G. Blankenship, Singularly perturbed difference equations in optimal control problems, *IEEE Trans. Auto. Cont.* **26** (1981).

[18] A. Saberi and P. Sanuti, Time-scale structure assignment in linear multivariable systems by high-gain feedback, *Proc. 27th Conf. on Decision and Control*, December 1988.

[19] A. Saberi and P. Sanuti, Decentralized observers for a large scale systems with two-time scales, in *Proc. JACC*, San Francisco, June 1977.

[20] U. Ozguner, Near-optimal control of composite systems: the multi time scale approach, *IEEE Trans. Auto. Cont.* **24** (1979).

[21] M. Coderch, A. Willsky, S. Sastry, and D. Castanon, Hierarchical aggregation of linear systems with multiple time scales, *IEEE Trans. Auto. Cont.* **28** (1983).

[22] M. Buren and K. Mease, Aerospace plane guidance using geometric control theory, *Proc. American Control Conf.*, 1990.

[23] G. Peponides, P. Kokotovic, and J. Chow, Singular perturbations and time scales in nonlinear models of power systems, *IEEE Trans. Circuits Sys.* **29** 1982: 758–767.

[24] A. Brandt, in W. Hackbush and U. Trottenberg, eds., *Guide to Multigrid Development, in Multigrid Methods*, Lecture Notes in Mathematics, vol. 960, Springer-Verlag, Berlin, 1982.

[25] A. Brandt, D. Ron, and D. Amit, Multi-level approaches to discrete states and stochastic problems, in W. Hackbush and U. Trottenberg, eds., *Multigrid Methods II*, Lecture Notes in Mathematics 1228, Springer-Verlag, Berlin, 1986.

[26] S. G. Mallat, A theory for multiresolution signal decomposition: the wavelet representation, *IEEE Trans. Pattern Anal. Mach. Intell.*, **11** (1989): 674–693.

[27] M. Basseville, A. Benveniste, K. C. Chou, and A. S. Willsky, Multiscale statistical signal processing: stochastic processes indexed by trees, MTNS 89, Amsterdam, June 19–23, 1989.

[28] A. Benveniste, R. Nikoukhah, and A. S. Willsky, Multiscale system theory, Internal publication no. 518, IRISA, INRIA, February 1990.

[29] U. Ozguner and M. Dogruel, Multiresolutional automata theory, *Proc. of the 1995 ISIC Workshop*, Monterey, CA, 1995, pp. 55–59.

[30] P. Filipovich and A. Meystel, Hierarchical prediction feedback control, *Proc. American Control Conf.*, San Francisco, June 2–4, 1992.

[31] Y. Bar-Shalom, Stochastic dynamic programming: cautions and probing, *IEEE Trans. Auto. Cont.* **26** (1981).

[32] A. Meystel, Nested hierarchical controller for intelligent mobile autonomous system, *Intelligent Autonomous Systems, Proc. Int. Conf.*, Amsterdam, 1986.

[33] A. Meystel and A. Guez, Elements of the theory of design, Proc. 4th Int. Symp. on Large Engineering Systems, Calgary, Alberta, Canada, June 1982.

[34] A. Guez, Minimum time control of multilink manipulators, PhD Thesis, University of Florida, Gainesville, 1983.

[35] R. Bhatt et al., A real time pilot for an autonomous robot, *Proc. IEEE Int. Symp. on Intelligent Control*, Philadelphia, 1987.

[36] R. W. Brockett and M. D. Mesarovic, The reproducibility of multivariable systems, *J. Math. Anal. Appl.* **11** 1965: 548–563.

[37] R. W. Brockett, Poles, zeros, and feedback: state space interpretation, *IEEE Trans. Auto. Cont.* **10** 1965: 129–135.

[38] L. G. Birta and I. H. Mufti, Some results on an inverse problem in multivariable systems, *IEEE Trans. Auto. Cont.* **12** (1967): 99-101.

[39] F. Albrecht, K. A. Grasse, and N. Wax, Path controllability of linear input-output systems, *IEEE Trans. Auto. Cont.* **31** (1986): 569–571.

[40] M. K. Sain and J. L. Massey, Invertibility of linear time-invariant dynamical systems, *IEEE Trans. Auto. Cont.* **14** (1969): 141–149.

[41] J. L. Massey and M. K. Sain, Inverses of linear sequential circuits, *IEEE Trans. Comput.* **17** (1968): 330–337.

[42] L. M. Silverman, Properties and applications of inverse systems, *IEEE Trans. Auto. Cont.* **13** (1968): 436–437.

[43] L. M. Silverman, Inversion of multivariable linear systems, *IEEE Trans. Auto. Cont.* **14** (1969): 270–276.

[44] W. A. Porter, Decoupling of and inverses for time-varying linear systems, *IEEE Trans. Auto. Cont.* **14** (1969): 378–380.

[45] W. A. Porter, Diagonalization and inverses for nonlinear systems, *Int. J. Cont.* **11** (1970): 67–76.

[46] R. M. Hirschorn, Invertibility of multivariable nonlinear control systems, *IEEE Trans. Auto. Cont.* **24** (1979): 855–865.

[47] H.-W. Wohltmann, Function space dependent criteria for target path controllability of dynamical systems, *Int. J. Cont.* **41** (1985): 709–715.

[48] U. Kotta, Right inverse of a discrete time nonlinear system, *Int. J. Cont.* **51** (1990): 1–9.

[49] V. D. Tourassis and C. P. Neuman, Inverse dynamics applications of discrete robot models, *IEEE Trans. Sys. Man and Cybern.* **15** (1985): 798–803.

[50] H. Seraji, Design of feedforward controllers for multivariable plants, *Int. J. Cont.* **46** (1987): 1633–1651.

[51] H. Seraji, Minimal-inversion feedforward-and-feedback control system, *NASA Tech Brief*, vol. 14, no. 7, Item 1, JPL Invention Report NPO-17701/7205, July 1990.

[52] G. Zames, Feedback and optimal sensitivity: model reference transformations, multiplicative seminorms, and approximate inverses, *IEEE Trans. Auto. Cont.* **26** (1981): 301–320.

[53] M. Araki and K. Yamamoto, Decoupling of sampled-data systems: two input two output case, *Int. J. Cont.* **35** (1982): 417–425.

[54] K. L. Doty and H. Frank, Invertibility, equivalence, and the decomposition property of abstract systems, *IEEE Trans. Circuit Theory* **16** (1969): 162–167.

[55] S. Even, On information lossless automata of finite order, *IEEE Trans. Comput.* **14** (1965): 561–569.

[56] R. P. Paul, *Robot Manipulators: Mathematics, Programming, and Control*, MIT Press, Cambridge, 1981, pp. 157–164.

[57] J. Albus and A. Meystel, Foundations of cognitive architecture for intelligent machines, NIST/Drexel University Technical Report, 1991.

[58] J. Albus, Outline for a theory of intelligence, *IEEE Trans. Sys. Man Cybern.* **21** (1991).

[59] A. Meystel, Theoretical foundations of planning and navigation for autonomous robots, *Int. J. Intell. Sys.*, no. 2, 1987.

[60] P. Kokotovic and R. Yackel, Singular perturbations of linear systems: basic theorems, *IEEE Trans. Auto. Cont.* **17** (1972): 29–37.

[61] P. Kokotovic and J. Chow, Eigenvalue placement in two time scales systems, in *Proc. IFAC Symp. Large-Scale Systems*, 1976, pp. 321–326.

[62] P. Kokotovic and J. Chow, A decomposition of near-optimum regulators for systems with fast and slow modes, *IEEE Trans. Auto. Cont.* **21** (1976).

[63] A. Chammas and C. Leondes, Pole assignment by piecewise constant output feedback, *Int. J. Cont.* **29** (1979).

[64] T. Hagiwara, T. Fujimura, and M. Araki, Generalized multirate output controllers, *Int. J. Cont.* **52** (1990).

[65] H. Seraji, Minimal inversion, command matching and disturbance decoupling in multivariable systems, Technical Report, Jet Propulsion Laboratory, CALTECH.

[66] Y. Maximov and A. Meystel, Optimum design of multiresolutional hierarchical control systems, *Proc. Int. Conf. Intell. Cont.*, Glasgow, Scotland, 1992.

[67] S. G. Mallat, A theory for multiresolution signal decomposition: the wavelet representation, *IEEE Trans. Pattern Anal. Mach. Intell.*, July 1989.

PROBLEMS

9.1. Demonstrate that representation of a system by the automata formalism and by differential equations is equivalent.

9.2. A mass should be moved by an applied force from one point to another in minimum time. The values of maximum speed, maximum acceleration, and maximum jerk are limited. Draw the time diagrams for path, speed, acceleration, and jerk for the complete cycle of motion.

9.3. In Problem 9.2, initial conditions are different from "0." Given that solution for Problem 9.2 has already been obtained, solve Problem 9.3 without engaging in any additional computations.

9.4. Given a dynamic system of the second order with one input and one output variable, the required output trajectory (in time) is a triangle. Find the input required (the magnitude of input should not exceed some particular constraint). Since δ-functions will be present in the solution, one would be interested in an approximation that would minimize the output error. Find this approximation.

9.5. For the results of solving Problem 9.4, find a PD-feedback controller that will compensate for the error.

9.6. Given a dynamic system of the second order with one input and one variable, the required output trajectory (in time) is a triangle. Find a PD-feedback controller that will provide for tracking the required output trajectory.

9.7. Compare gains in solutions of Problem 9.5 and 9.6. Interpret the results of comparison.

9.8. Construct a hierarchical architecture for solving Problem 9.4.

CHAPTER 10

LEARNING

10.1 INTELLIGENT SYSTEMS AND LEARNING

This chapter will discuss the ideas governing the research in the area of intelligent systems with learning control systems. The number of ideas and concepts of learning is very large, but not one of them pretends to encompass the whole problem; indeed, many of them contradict each other. If one collects them all into one document, they do not constitute a consistent body and are confusing for engineer and researcher working in the field. Areas of adaptive control systems, pattern recognition, production systems, learning controllers, and neural networks state different goals of learning and promulgate different approaches to the analysis and design of learning systems. Engineers and psychologists, cognitive scientists and logicians-philosophers, computer analysts, and mathematicians, interpret the term "learning" in slightly different ways depending on the discipline.

This situation is not surprising, since there exist plenty of different frameworks within the different disciplines of science entailed by the different avenues of natural science development. This situation does not seem to be productive for the area of intelligent systems. It can be considered detrimental for both the results of scientific research and for the emerging technologies exploring the fruitful area of intelligent control, intelligent robotics, intelligent communications, and others where the idea of making learning systems (in a clear layman understanding of learning) is very promising.

We discussed already learning in earlier chapters. In Chapter 2 the primary learning procedures of clustering and classification were introduced, learning semiosis was presented, and the triplet of GFACS was described. In Chapters 3 through 8, the elements of learning were demonstrated to be ingrained in the algorithms of knowledge organization, sensory processing, world modeling, and behavior generation. This chapter attempts to develop a unified framework for putting together the pieces of the various theories of learning scattered in different parts of the intelligent system and in various disciplines of science. Many of the theories have been developed in areas that never exchanged their ideas and techniques earlier. To be sure, they could have been mutually beneficial had they been available to each other. We should note that engineering descriptions of learning control systems are based on user specifications, while psychological treatment of learning processes is based on subtle and elusive human mental activities. Our interest is to make these diverse approaches and results consistent with each other. Adding to that philosophical thought which has collected numerous insights concerned with learning processes, we want to make a treasury of concepts available for the designer of computer architecture, or for the engineer contemplating an intelligent controller with learning capabilities.

We will use terminology that is potentially acceptable for all of the areas involved. Learning is demonstrated as a result of consecutive applying of the triplet of operations: focusing of attention, grouping, and combinatorial search (GFACS; see Chapter 2) to the incoming information. All existing concepts of learning are covered by this general view. The algorithmic structures described in this chapter are intended to be instrumental in unifying the diversified results from different areas investigating learning. The algorithms presented here strengthen the concept of multiresolutional architecture of the intelligent system, and they fit within the concept of multiresolutional knowledge representation, hierarchical system of goals, and hierarchical behavior generation.

10.2 DEFINITIONS OF LEARNING

It is an enormously difficult problem to formulate anything related to the area of learning at the present time. The term "learning" has many definitions; those definitions have even a larger number of interpretations. Nevertheless, anyone using this term has a good chance of being inadequately understood by different scientific communities. Let us review several definitions of learning existing in the literature (including the one that we propose to satisfy at least the community of researchers working in the area of intelligent control, robotics, machine intelligence, theory of control, communication systems, and autonomous robotics.

DEFINITION 1 Webster's Dictionary: "Learning is knowledge or skill acquired by study in any field; *the process of obtaining*[1] *skill*[2] *or knowledge*" [1].

This definition is an important one for understanding the situation: "I learned something from him," which is commonly understood as follows: "He communicated to me some knowledge that I (presumably) understood[3], and/or memorized. So, I learned it!" To "know something" is presumed to be always a result of "to learn something." Since we can get some knowledge by looking in a dictionary, this way of getting the knowledge is also considered "learning." In fact there is some gigantic "underwater" part of this type of learning process. When he told me *this*, it became *knowledge* only if I trust him, for example, as a result of many times that verified my experience of seeing him talking truth. Otherwise, I would suspect that what he told me might not be a truth and it won't qualify to be considered *knowledge*; that is, I won't be able to claim that I *learned* it. So *learning presumes getting the information validated by an acceptable level of belief*, which is information that underwent a process of belief validation. (It is tempting to state that in the process of belief validation, the initial information, or an initial hypothesis, or a tentative concept, is being transformed into knowledge.)

[1]Presumably, and storing so memory is involved.

[2]Understood as useful operational knowledge.

[3]The issue of *understanding* is supposed to be one of the key indications of *intelligence*.

DEFINITION 2 Encyclopedia Britannica: "Learning (animal)—*the alteration of an individual behavior as a result of experience*"[4] [2].

This definition alludes to development of reflexes, emergence of evolutionary changes related to behavior, and so on. Interestingly, there is no mention of knowledge acquisition. This definition reflects the frequent conviction among psychologists that animals can "know" only in a form of behavior; an animal's knowledge not reflected in behavior is not considered to be knowledge. Many psychologists are still trying to make a drastic distinction between cognitive processes in a human and in an animal.

DEFINITION 3 G. A. Kimble: "Learning is *a relatively permanent change in a behavioral potentiality*[5] that occurs as a result of reinforced practice"[6] [3].

Here a new factor is mentioned: *behavioral potentiality*. This invokes a need for a place where this *potentiality* is stored, and the process of learning leads to changes in this storage. In the meantime the source of this potentiality is explicated: *reinforced practice*.

DEFINITION 4 A. Newell, J. C. Shaw, and H. A. Simon: "A learning situation[7] requires another program, called the learning program, that operates on the performance program as *its object*[8] *to produce a performance program better adapted to its task*" [4].

Apparently, in speaking of a *learning situation*, this definition refers to a process of *learning* whose storage changes from the previous definition. In order to drive this process, the existence of *an algorithm of learning* is virtually postulated.

DEFINITION 5 H. A. Simon: "Learning denotes changes in the system that are adaptive in the sense that they *enable the system to do the same task or tasks drawn from the same population more efficiently and more effectively the next time*" [5].

One cannot talk about learning exhaustively without mentioning what it is for: for improving the creature that learns and correct its prior behavior, or improving the cases of behavior in the future. We start getting used to the situation that learning is something that depends on learning process, that both presume a creature that demonstrates some behavior, that this behavior can be more or less efficient and the process of learning serves as a tool of increasing its efficiency.

DEFINITION 6 M. Minsky: "*Learning is making useful changes in the working of our minds*" [6].

No rigid definition can be expected from this source, but it does divert attention from behavior as an object of learning and focuses on the carrier of the acquired knowledge. The mind is mentioned as a metaphor: The acquired knowledge resides not necessarily only in the brain, or even in the CNS; each subsystem, each module of the system, each parameter of the body is a carrier of the acquired knowledge, and so on.

[4]It is tacitly understood that the knowledge of an experience is being stored.

[5]Behavioral potentiality looks like a code word for memory of behaviors that can be exercised. A provisional definition of *behavior*; a program of available (purposeful) operations.

[6]Again, something is supposed to be memorized.

[7]This is not a direct definition of *learning*: The learning process seems to be more important than the concept of learning.

[8]An important loop: the *learning* program is supposed to exist before the *operation* starts in order to enable this current *operation* to improve something in the future *operations*.

DEFINITION 7 R. Michalski: "*Learning is constructing or modifying representations of what is being experienced. ...* A fundamental problem in any research on machine learning concerns the *form and method used to represent and modify the knowledge or the skills being acquired*"[9] [7].

This definition focuses only on the representation. Neither the purpose of learning nor the mechanisms are addressed. The important detail is our experiences: This is what is supposed to be reflected in the representation.

DEFINITION 8 J. P. Guilford: "*Learning is essentially discovery.*[10]...The achieved cognition may be an acquaintance with a new unit, the formation of a new class, the formation of a new relationship or a new system, the awareness of a transformation, or the extension to new implications" [8].

This 1961 paper demonstrates that many particular learning processes can be interpreted within this definition of learning: memorizing, skill learning, reinforcement learning, learning while problem solving, and so on. The special novelty of the acquired knowledge is underlined in this definition. The results of learning are expected to be not something that can be deduced by our minds but something that emerges unexpectedly, cannot be routinely foreseen, so the phenomenon of "emergence" is required. The concept of "emergence" is not well-defined. However, in many cases it is known to be associated with the phenomenon of generalization (see Chapter 2).

DEFINITION 9 M. I. Shlesinger: "Learning of image recognition is a *process of changing the algorithm*[11] of image recognition *in such a way as to improve, or maximize a definite preassigned criterion characterizing the quality* of recognition process" [9].

After all previously discussed definitions, this focusing on a novel algorithm seems to be a narrowing of the class of phenomena defined as learning: Indeed, a discovery of an unexpected entity within an image may be considered an "emergence" and thus qualify as learning too.

DEFINITION 10 Y. Tsypkin: "Under the term learning in a system, we shall consider a process of forcing the system to have a particular response to a specific input signal (action) by repeating the input signals[12] and then correcting the system externally"[13] [10].

This definition, given by a prominent specialist in control theory, contains remarkable conceptual bridges to biological learning, such as reinforcement learning and the development of reflexes. The resemblance of the definition to the "Hebbian (or Pavlo-

[9]Again, memory is presumed. However, the location and organization of the storage are yet to be explicated.

[10]This does not look like a definition of learning. The author is concerned with bringing about a message of the importance of the phenomenon of innovation which is associated with learning. Actually he wants to say that just a simple memorization is not a final result of learning. The product of learning is expected to be a discovery as a result of the intermediate process of memorization.

[11]This is a remarkable statement: Learning presumes a change in the algorithm of operation!

[12]It seems the testing is presumed, with subsequent storing of the results of the testing

[13]The ability of intelligent systems to produce goals at a level of resolution should be bounded by the set of subgoals for the goal which was prescribed from the level of lower resolution.

vian) learning" principle is remarkable. Sometime after Pavlov discovered the process of unconditional reflex learning in 1934, in 1949, D. Hebb formulated a similar condition for neurons: "When an axon of cell A is near enough to excite a cell B and repeatedly or persistently takes place in firing it, some growth process or metabolic change takes place in one or both cells such that A's efficiency, as one of the cells firing B, is increased" [11].

The many existing definitions, their various angles and scopes in approaching the problem of learning, create a feeling of dissatisfaction, and some authors admit the need to reconsider the accepted definitions. P. S. Churchland has expressed this dissatisfaction as follows:

> The general category of learning has already fragmented into a variety of kinds of processes, and indeed the term "learning" is now often replaced by the broader and less theoretically burdened expression "plasticity" [which includes] habituation, sensitization, classical conditioning, operant conditioning, imprinting, habit formation, post-tetanic potentiation, imitation, song learning (in birds), one-shot learning to avoid nausea producing foods, and cognitive mapping, in addition to which are the apparently high-level phenomena distinguished in terms of what is learned, such as learning language, learning who is a conspecific, learning to read, learning social skills, learning mathematical skills, learning to learn more efficiently, learning to lower blood pressure, and heaven knows what else ([12] [pp. 151–152]).

Clearly, Churchland chooses to characterize the situation with learning as a pretheoretical situation. Some of the definitions imply that the learning process should incorporate information about prior operations of the system, possibly should generalize and make classifications on the information collected. All refer to potential changes in operation. Some talk of modifying the results; one expects discovery from learning. All are based on an idea that a process of collecting information about experiences can be organized in such a manner that some similarity of interest will be discovered in the information. Then the results of our observations can be unified by this similarity. The results of this grouping will later allow us to change an algorithm of operation to bring an improvement.

Let us recapitulate the main points of all above definitions. Learning should have the following features:

1. Be a process of obtaining skill (the term "skill" presumes the existence of a program that produces a useful behavior; this program as well as the corresponding knowledge can be represented somewhere).
2. Produce alteration of an individual behavior (the term "alteration" implies that the program to produce the behavior can be changed).
3. Produce change in a behavioral potentiality (the term "potentiality" alludes to a storage of the programs).
4. Develop an inner program better adapted to its task (the term "better" implies that the measurement of "goodness" should be built into the mechanisms of learning).
5. Enable a task to be performed more efficiently (the term "efficiency" implies that the goodness of performance should be measured).
6. Change the quality of the output behavior (the term "quality" implies that the "goodness" of the output behavior can be evaluated).
7. Make useful changes in [mind] (the term "useful" refers to "goodness"; the term "mind" alludes to the ability to store and modify the programs).

8. Construct or modify representations (the term "construct" implies not only an ability to store but also an ability to create new programs).
9. Be essentially an act of discovery (the need to create new programs is emphasized).
10. Form new classes and generalized categories (the formation of new programs should be achieved via generalization).
11. Change the algorithm (the term "algorithm" should be understood in the same way as the term "program" was used in the observations above).
12. Force the system to have a particular response to a specific input signal (action) by repeating the input signals (two concepts are introduced: "response" and a "repetition" of the experience that requires this response).

A definition that addresses most of these features was given by G. Edelman.

DEFINITION 11 G. M. Edelman: "In an environment containing unforeseen juxtapositions of events that may affect survival, it is learning, not just perceptual categorization, that ensures successful adaptation. . . . Perceptual categorization and memory are therefore considered to be necessary for learning but obviously are not sufficient for it." [pp. 56–57] "Learning arises as a specific linkage between category and value in terms of adaptive responses that lead to changes in behavior." ([13, p. 152]).

The concept of "survival" can be interpreted within our vocabulary as the ultimate (and very special from the point of view of tools applied) way of evaluating the "goodness."[14]

We will synthesize a definition that blends these ideas of knowledge, and skill acquisition and/or modification, change of behavior via change in the program, efficiency, and quality of operation as a goal of learning, and achieving these changes in a program as a result of discovery. This definition is formulated as to be applied for a coupled control system–controlled system. This couple is adequate to a multiplicity of technological problems, and allows for powerful interpretation in a number of biological, psychological, and sociological cases.

DEFINITION 12 *Learning is a process based on the experience of intelligent system functioning*[15] *(their sensory perception, world representation, behavior generation, value judgment, communication, etc.) which provides higher efficiency which is considered to be a subset of the (externally given) assignment*[16] *for the intelligent system.*

Development consists of *modification* and *restructuring*. Development can be a part of design prior to system operation, and it can be a part of normal system operation. Thus it can be said that learning is a perpetual redesigning of a system. Modification can be understood as a parametric adjustment of the algorithms with no change to their structure (e.g., making corrections in the range of the rules without changing the rules). The cost-function is assigned by the upper level of resolution (external user, population

[14]We replace the Darwinian term "survival" by the term "goodness," since we want to avoid (or substantially reduce) references to Darwinism. A discussion of Darwinism and learning awaits the writing of a separate book.

[15]In using the term "functioning," we aim to address all processes pertinent to intelligent systems including those of design and development (modifying and/or restructuring), planning, and the like.

[16]Later we will call this a "goal set."

of intelligent systems, etc.) which is assumed to be a source of the specifications for functioning.

The more complicated case is when the user assigns only general guidelines, and the lower-level cost-functions are formulated and then modified and restructured by the system itself. We won't talk about this type of system in this chapter; our immediate goal is to discuss learning control systems that can work under a preassigned cost-function.

Restructuring presumes not only modification of existing rules but also the creation of the new rules,[17] or the new meta-rules. This is more sophisticated learning. The highest level of learning takes place when the goals of the functioning can be reconsidered based on the results of the operation of the subsystem of learning. The following analogy can be of importance: the system with learning is permanently undergoing the online process of re-design. A system with no learning is a system that was designed in the past once and forever does not show (nor afford) any improvement.

Both *learning via modifying*, and *learning via restructuring* are based on two types of knowledge: knowledge obtained from an external source of knowledge ("learning by being told"), and knowledge created within the learning system ("learning by discovery"). External source of knowledge is evaluated by the degree of belief, and therefore it just substitutes for the internal subsystem that creates the knowledge within the learning system (and whose results of operation are also evaluated by the degree of belief). Thus the systems with external sources of learning can be considered a subset of systems that learn by discovery, in which the process of discovery is externalized.

Thus formulated, this definition of learning can be helpful in defining learning systems, or learning control systems. We will try to show this step by step, and we will begin by defining the control systems we are going to deal with.

The ultimate form of learning will lead to change in the design of the creature (this apparently is the process of evolution). We will understand learning as this process in evolving a model of the world (including the models of sensory processing and behavior generation) for the benefit of the system under consideration.

Decision making as a process of selecting among the alternatives is not a process of learning (learning ends as soon as the decision algorithm is in place). However, the development of the alternatives is connected with learning. The same can be said about planning. To construct a plan of actions is not learning if this plan is obtained from the existing lookup table. However, planning is learning when the process of designing new rules is a part of the construction plan. This always happens if search is involved, and thus the result is not a combination of the previously stored components but a discovery of new models of behavior that have not been explored previously.

Subsequently the following processes employ learning: concept generation, data classification, image (object, entity) recognition. Logical schemes of all corresponding algorithms are similar.

10.3 IMPLICIT AND EXPLICIT LOGICAL AND PSYCHOLOGICAL SCHEMES OF LEARNING

10.3.1 Need in "Bootstrap" Knowledge: Axioms and "Self-evident" Principles

Knowledge acquisition cannot be done without existence of some initial "background" knowledge, or "bootstrap" knowledge. This "axiomatic" knowledge is a matter of faith: It cannot be learned; it should exist in the system prior to the learning process, and the system should just to believe that "this is so." It is instructive therefore to determine

[17] See the definition of "discovery" in definition 8 from Section 10.2.

the axiomatic structure within the existing learning schemes and to build the advanced learning theory based on a set of explicitly stated axioms. The following general properties are expected from the general truths, a priori laws, axioms [14]:

1. Unrestricted generality ("always true and true of everything").
2. Independence of sense experience ("cannot be proved or disproved by an examination of sensory experience").
3. Self-evidence ("cannot possibly be doubted").

Less evident principles cannot be considered axioms, especially when they are instilled by the context. We will list several such semi-axioms (SA).

SA 1 Every observation is evaluated for the degree of truthfulness, degree of belief, probability, likelihood, goodness, possibility, and so on. It is presumed that the results of such an evaluation are necessary (1) to decide on memorization and subsequent retention of an observation, (2) to decide on the validity of using an observation for subsequent generalization, or to assign the value of belief to the results of the generalization, and (3) to decide on the action to be undertaken after an observation. In all three cases some *process of decision making is presumed* after the observation took place.

SA 2 Every observation is presumed to be associated with prior observations of the same object, as well as with observations concerned with the associated objects.

"Bootstrap" knowledge is partially embodied in various incarnations of the GFACS-triplet: algorithms of clustering [15], classification [16], and approximation [17]. Together with the axioms and semi-axioms the procedures of grouping, focusing of attention, and combinatorial search can serve a mathematical explanation for the various *gestalt principles*.

10.3.2 Gestalt Principles: Entity Discovering Insights

Gestalt theory has helped explicate at least some of the self-evident principles in a pre-algorithmic form. Its intention is to look for wholeness [18]. Gestalt principles are considered to be important in all psychological treatments of learning processes [19, 20]. We can easily recognize within each of these principles a structure of GFS (see Chapter 2), each based on procedures of grouping, focusing of attention, and combinatorial search.

Gestalt Principle 1: The Law of Figure-Ground Relationships

The existence of an entity is determined by the existence of the rest of the world. This means that the entity is discernible from the rest (extracted from the "chaos") in the same way as the *figure* is discernible from its *background*. This is possible due to the existence of an operation of distinguishing some unspecified (not initially discovered) properties of the entity as opposed to the properties of the background. This means that the following important facts are implicitly accepted:

1. Possibly there exists an entity (as opposed to the rest of the world which is considered background, an amorphous environment).
2. Possibly there exists a property or a set of properties that can be perceived for each particular space (or location) at each moment of time.

3. There exists a procedure of identifying properties.
4. Some properties are stable (not changing) for some locations.
5. An entity can be characterized by a particular set of properties typical for this particular entity.
6. There exists a procedure of grouping together "spaces" ("locations") with similar stable properties.
7. There exists a procedure of characterizing the group of these locations by some meta-feature (a basis for assigning a name to this group).
8. The set of procedures above can be stored as a *procedure of identifying* the entity with its particular property (or properties).

This set of procedures apparently consists of the operations of attention focusing (on the particular locations), grouping (associating together in a single information structure), and searching for the locations that fit better for the subsequent grouping.

The existence of a property (or a set of properties) means that prior to learning the entity we should learn the property. If no property is known, no entity can be distinguished from the chaos. It is important also to stress the fact that the operation of distinguishing the entity from the background without prior knowledge of the entity existence, its properties, its relation to other entities, and/or similarity with other entities can be qualified as *discovery*.[18]

Gestalt Principle 2: The Law of Similarity

We already exercise this law when the locations with similar properties are grouped together. Entities are perceived as a group too if they are similar to each other in some way. This property of similarity is sometimes referred to as a gestalt property. "When the memory-images of successive notes are present as a simultaneous complex in consciousness, then an idea of belonging to a new category, can arise in consciousness, a unitary idea, which is connected in a manner peculiar to itself with the ideas of the complex of notes involved. The idea of this whole belongs to a new category" [20–22]. This new category is not a simple sum of the initial set of the elementary properties it is formed of: It is a quality that transcends any particular set of such elements [23, 24].

In this view the Pavlovian conditional reflexes using ideas of association, generalization, and rule memorization [25] can be easily interpreted as creation of a new category based on similarity between experiences observed and memorized. (The issue of repetitiveness of the similar experience will be discussed later in the sections dedicated to induction and abduction.)

A property of being similar (the *similarity property*) is postulated here. An *operation of determining the property of similarity* as a relationship between a pair of entities is presumed. A procedure is available that *groups together* (*clusters*) all coupled entities by the relation of similarity. This cluster is claimed to form a pattern,[19] that is, to be perceived as another entity.[20] In other words, *entities gathered together based on their similarity form other entities*: The entities are nested. (This is actually Aristotle's principle of similarity in establishing associations).

[18]This may be considered a definition of *discovery*.

[19]*Pattern* can be defined as an entity for which a characteristic property has been declared or discovered.

[20]Actually, we are again talking about *entity discovery*.

Gestalt Principle 3: The Law of Nesting

Specialists in the gestalt theory of learning refer to a hierarchy of gestalt properties, and to entities that are constructed as a result of nesting as "formations" of higher order [22]. Both the law of similarity and the law of nesting describe a converging process of learning in that the nested hierarchy of entities emerge from chaos. Nesting is a result of recursiveness in applying the GFS triplet.

Gestalt Principle 4: The Law of Proximity

Entities are grouped in patterns (e.g., other entities) if they are proximal to each other in space and/or time. (This is Aristotle's law of contiguity.) Certainly the idea of proximity implies the presence of *distance* in the space under consideration, and so on. In order to apply the law of proximity, the idea of distance must be learned beforehand as well as the skill of evaluating it (likely the notion and the skill are simultaneously acquired). Proximity therefore is a property-generating idea [23–25]. It implies that the similarity of properties is a basis for grouping entities together. Proximity in space is also a property that is taken into account in the process of cluster generation. Proximity in time is a basis for unifying changes related to each other into a single *discrete event.*

Gestalt Principle 5: The Law of Good Figure

Among all possible patterns, *the best* one is selected as a pattern is being formed from a set of entities. An ability to evaluate some criterion of *goodness* (e.g., of fit) is expected. So this idea of goodness must be learned in advance. When the problem of grouping by similarity is ill-posed, and a multiplicity of solutions is expected, another (goodness) criterion is applied for the selection. Implicit is the notion of goodness being outside the circle of logical arguments: It may be something more like beauty of solution, symmetry, or simplicity [20, 24–28]. The notion of simplicity brings to mind efficiency. The notion of efficiency brings to mind beauty. Similarity to prior positive experiences can also be perceived as a thing of "beauty."

Gestalt Principle 6: The Minimum Principle

By the minimum principle, the simplest or most homogeneous organization is perceived to fit a given stimulus pattern [20]. The presumed procedure is as follows: Alternative entities are generated. Their homogeneity (uniformity of associated properties) is evaluated. The reduced set of alternatives contains only highly homogeneous candidates. This set is compared using a criterion of simplicity. The entity selected is what we must learn. An ability to recognize ideas of homogeneity and simplicity is implied as well as the skill to measure, evaluate, and compare them. If the efficiency is achieved, it is perceived as a thing of beauty too ("A thing of beauty is a joy forever.")

Gestalt Principle 7: The Law of Closure

If the pattern formed seems to be incomplete, it must be completed. The concept that "things are meant to attain completeness" is the result of spatial and temporal predictions that are ingrained in all procedures of learning. We expect closure because we see many closures in situations around us. It is one more confirmation that satisfies our feel for beauty. In the procedure, a set of patterns known from prior experiences is presumed to exist for possible comparison. The entity formed is meant to be compared with the entities from a *set of good patterns*. This means that sets of good patterns are learned in advance [26–28].

Gestalt Principle 8: The Law of Field

Any system can be considered a point in a multiplicity of fields where each field generates its own forces. The behavior of a system can be understood only if all the forces are considered as a whole [24]. An ability to consider multiple factors simultaneously is presumed.

10.3.3 Repetitiveness: Induction, Abduction, and Deduction

Prediction (in time and space) is an important component of decision making as well as an important component of any gestalt experience. Since repetitiveness is ingrained in the characterization of objects and processes we are interested in, our desire is to evaluate possibility, probability, and likelihood [29]. Thus repetitiveness generates anticipation which stimulates prediction. All existing algorithms of prediction employ the phenomenon of statistical consistency (hidden within the concept of "repetitiveness"). This is why *induction* underlies all our behavior-generating processes (see Chapter 2). If the experience of repetition is insufficient, the risk of failed prediction increases. Nevertheless, *abduction* (an induction with insufficient basis for expecting repetitiveness) is both frequent and necessary part of learning processes.

Induction as a source of concept generation, and or parameters adjustment, was connected by researchers to the theory of probability. For example, "the theory of probability shows how far we go beyond our data in assuming that new specimens will resemble the old ones, or that the future may be regarded as proceeding uniformly with the past," says W. S. Jevons [30]. Another component of induction is a judgment of what is cause and what is the effect. Bertrand Russell substituted two principles: "(a) The greater the number of cases in which a thing of the sort A has been found associated with a thing of the sort B, the more probable it is (if no cases of failure of association are known) that A is always associated with B; (b) under the same circumstances, a sufficient number of cases of the association of A with B will make it nearly certain that A is always associated with B, and will make this general law approach certainty without limit" [14, 31].

Deduction is required because reasoning should be applied to arrive at conclusions from repetitive phenomena. However, predicate calculus of the first order might not be sufficient. Some researchers envision the growing need in the predicate calculus of the second order, which is the calculus in which we do not infer statements by the virtue of predicates; rather, predicates become variables and subject to reasoning about themselves [32].

10.3.4 Recursion and Iteration

When a law L is discovered, it can be applied using the laws of recursion (L_r) and/or iteration (L_i). Recursion is defined as a statement where the law L_r is applied to a number of variables among which there are results of prior application of this law to a reduced number of variables. Consecutively applying L_r to a reduced number of variables, we eventually approach the "root" situation in which the immediate results of applying L_r can be obtained directly and do not require any additional computation and/or action ("converging recursion"). Most of the learning schemes appeal to recursion, since the computation of L_r at each step requires prior computation of L_r related either to another resolution level, or prior computational result. (The number of levels involved as well as the number of prior computations required is assumed to be finite.)

On the contrary, the computational schemes employing L_i do not require any consecutive nested process described above. The law of iteration is formulated in such a way as to appeal only to previous computation that has already been definitely performed. In comparison with recursion, the iteration requires less memory, although the symbolic representation might be more cumbersome.

10.3.5 Typology of Learning

The following different types of learning processes and schemes are visualized by specialists in psychology [26]. We reformulated and rearranged the list of types so that they fit within the subject of our book.

Type 1. *Signal learning*. This involves finding an association by similarity or making a discrimination by the lack of resemblance.

Type 2. *Learning of stimulus–response* (S–R) couples[21] as a result of generalization of experiences. Learning a chain of S–R couples, or events, is called *chaining*. During chaining, the system is learning a series of responses in a definite order.

Type 3. *Learning associations between S–R units* or stimulus–response couples, and labeling the associations. *Conditioning* is an important class of the type 3. It contains two subclasses:

  a. *Classical conditioning* (discovered by I. Pavlov). This was illustrated by the following example: A neutral stimulus, such as a bell, was *repetitively* presented just before the delivery of some effective reward such as food placed in the mouth of a dog. The response (salivation), usually evoked only by the effective stimulus, eventually appeared after the neutral stimulus was presented.

  b. *Instrumental conditioning*. This is a variation on classical conditioning and is oriented toward obtaining a reward or avoiding a punishment. Animals can be taught to press levers for food; they also can learn how to avoid electric shock.

  Skill learning can be interpreted as belonging to one or both of these conditioning classes.

Type 4. *Learning S–R classes*, or clustering the stimulus–response couples. This involves associating these labels (words) and the formation of their classes. Psychologists frequently call it "principle learning." A subject may be shown sets of figures belonging to different classes (some rectangles and some circles). After repetitive rewards, the subject learns to distinguish any "oddity." The formation of S–R classes is closely related to types 5 and 7 below.

Type 5. *Discriminating objects* with the same labels (later it gives an opportunity to discover new concepts). In discrimination learning the subject is reinforced

[21] S–R is understood as follows: A measurable sensation or stimulus; usually precedes some action or response.

to respond only to selected sensory characteristics of stimuli. Pigeons can learn to discriminate differences in colors that are indistinguishable to humans without special optical devices.

Type 6. *Rule learning*. The S–R couples can be inverted so that causal links are determined (what "action" should be produced in order to achieve an "effect").

Type 7. *Concept learning*[22]. The stimulus–response couples can be analyzed, and concepts can be obtained as a result. The concepts are discovered as "objects-actors" within the S–R couples. This type is closely related to type 4. Actually the learning processes belonging to type 4 become concept learning after the oddity is recognized and given a label (a concept is constructed and defined).

Type 8. *Problem solving*. The process of learning should include how to solve a problem. We will consider this type of learning part of planning.

Numerous new types of learning have emerged recently as a result of new research wave within control systems, artificial intelligence, machine learning, and the like. Below is a brief list of the most important of them.

Type 9. *PAC concept learning*. PAC means "probably approximately correct" [33]. This type of learning was introduced within computational learning. PAC is an abstract paradigm of concept learning formulated as follows: One discovers a "target concept" by taking samples from a set of concepts (with some probability measure) and receiving from an "oracle" a judgment whether the sample contains the target concept. The concept is considered PAC learnable if the accuracy of hypotheses is growing as the number of samples is growing, and the absolutely accurate hypothesis is achieved when the number of samples approaches infinity.

Type 10. *Model-free PAC learning*. This involves analyzing the source of information about a concept rather than deciding what to do next [34].

Type 11. *Distribution-free PAC learning*. Hypotheses are developed where no probability measures are available [34].

Type 12. *Learning concepts from ABI* (attribute-based instances). Information of properties of single objects (concepts) are processed by a deductive analysis of samples (instances) which are determined within the version space [35], inductive analysis [36], and/or clustering methods [37].

Type 13. *Learning concepts from structured instances*. Information on associations among the objects (concepts) is explored for possible clustering [38].

Type 14. *Learning by complementary discrimination*. This research explores the notion that generalization by clustering similar samples is equivalent to generalization by discriminating among complementary samples. This type of learning is introduced in [39], although it was admitted that exercising this approach is a common practice in almost all papers on learning when generalization is employed.

Type 15. *Learning by testing the environment and generalizing* (TEG learning). Common practice is to properly store and organize the results of expe-

[22]"Concept learning" from the psychological or practical viewpoint should not be confused with "PAC concept learning" (see type 9).

riences. This is known for cases of early learning [40, 41] and therefore is explored for the domain of automata theory [39], and simulated for a variety of applications [42–44].

Type 16. *Q-learning*. This is learning from delayed rewards [45]. This technique has dynamic programming embedded in it. In other words, it relies on local tests of the environment, attempts to envision the proper continuation of behavior, and computes the probability of rewards by taking into account the function of discount: The value of future rewards is discounted in comparison with the value of immediate rewards.[23] Cases combining Q-learning with generalization are surveyed in [46].

All types of learning can be applied in supervised and unsupervised settings. We are particularly interested in unsupervised learning because it is more instructive in discussing intelligent systems. In unsupervised learning, all types—classes and subclasses of learning—are based on the concept of reward (positive or negative) and thus all are reinforcement learning.

We can easily see that types 9 through 16 do not add much to types 1 through 8. They rather explore further decompositions of these eight classes into subclasses based on the certain algorithms under certain circumstances. This decomposition is actually unbounded: Many circumstances can be taken into account in different combinations. In addition the stage of learning is important: It matters whether the process of learning develops at an early stage (early learning) when there is no previously collected information, or whether the base of rules and experiences is substantial. The typology is illustrated in Figure 10.1.

There are yet many terms related to learning that are left out. The area of learning is still in the making. Researchers working in the area of machine learning include everything linked with information processing. In one recent survey, the following topics are considered in a list of methods of learning: dynamic programming, game theory, optimal control, methods of regression, Markovian and non-Markovian decision problems, problems of estimation and identification, theories of logical reasoning and other techniques. In [32], search methods, including genetic algorithms are considered to be methods of learning (not the tools utilized in various methods of learning). In the same source, a machine learning algorithm is considered that includes "examples" and "background knowledge" in the input, and the result is a "concept description" as output.

We will conclude our typology with some other additional terms used in the literature on intelligent systems.

Type 17. *Repetition learning*. Repetition can occur without any feedback from action results. Repetition as a basis of learning was discovered by I. Pavlov in his research on conditional reflexes. For neurons, learning by repetition was first hypothesized by Hebb, and for this reason it is sometimes called Hebbian learning. Repetitiveness is an aspect of most learning schemes when statistics (induction) is applied.

Type 18. *Reinforcement learning*. Reinforcement incorporates feedback from the action results. Reward reinforcement learning in the BG system is a form of positive feedback. The more rewarding the task, the greater the probability is that it will be selected again. Reinforced punishment, or error correcting learning, occurs when $g(t)$ is negative. Every time a situation occurs and

[23] This is one of those enigmatic maxims that is never questioned. Why must future rewards be discounted?

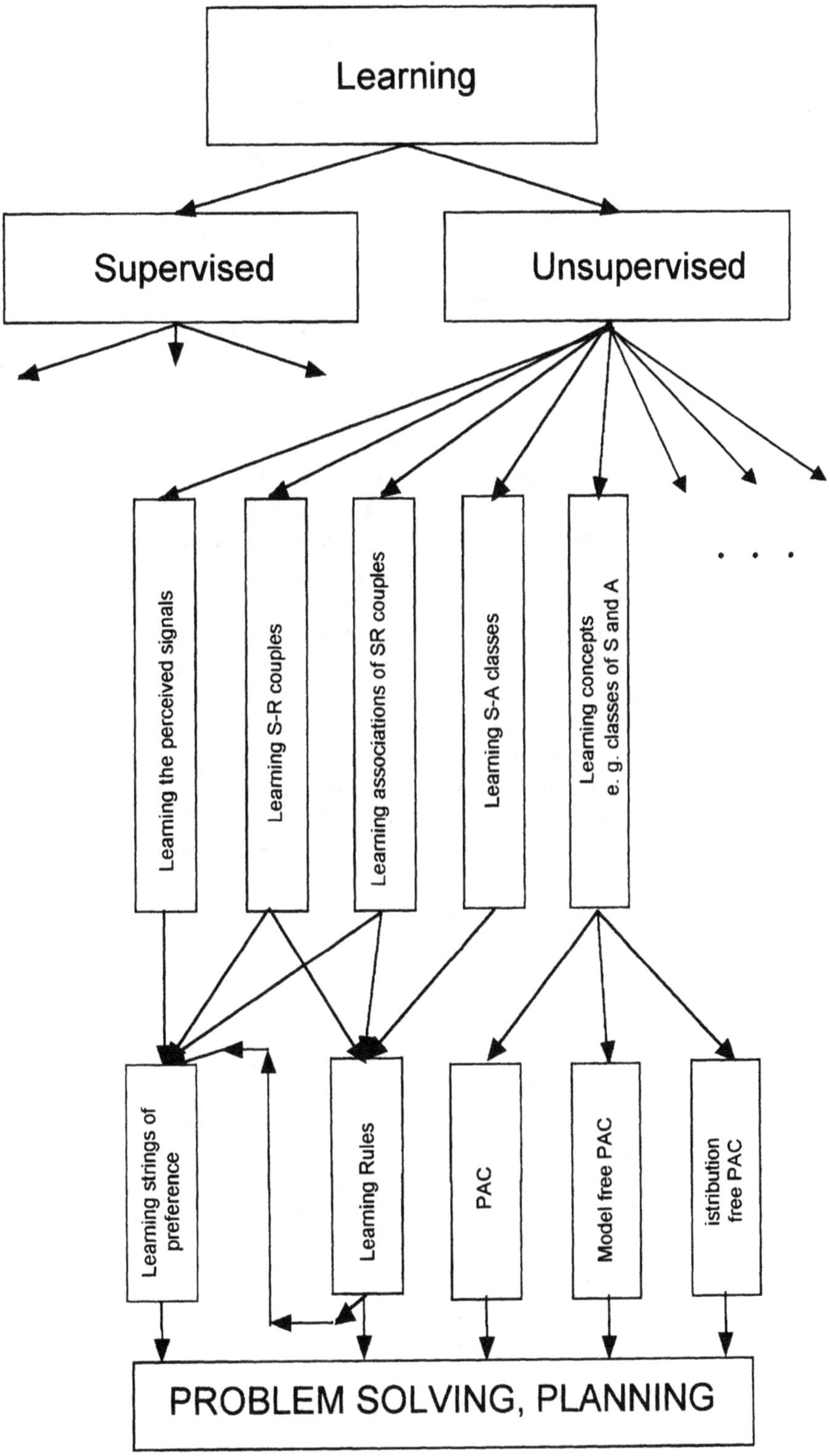

Figure 10.1. Classification of learning.

the evaluation gives punishment, the same synapses are weakened and the output (or its occurrence probability) is reduced.

Type 19. *Error correction learning*. This is a subclass of reinforcement learning and is a form of negative feedback. With every training experience, an amount of error is reduced, and hence the amount of punishment. Error correction is therefore self-limiting, and it tends to converge toward a stable result. It produces no tendencies toward addiction.

Type 20. *Specific error correction learning*. Reinforcement learning has enabled this subclass of the error-correcting learning to evolve. In specific error correction, sometimes called *teacher learning*, the correct or desired response of each output neuron is provided by a teacher. Teacher learning tends to converge rapidly to stable results because it has knowledge of the desired firing rate for each neuron. Teacher learning is always error correcting. The teacher provides the correct response, and anything different is an error.

The confusion in terminology and classification is linked with the multidisciplinary character of the issue of learning. It can be illustrated by considering AI penetration into this area. AI became (officially) involved in the issues of learning in 1956 at the famous Dartmouth Conference. At that moment the problem of learning control and computer learning already existed, and a broad and diversified body of literature on learning control systems and learning automata had begun to be generated at multiple research facilities all over the world. The body of literature dedicated to simulation of neurocircuits and networks has been growing. In our opinion, the diversity of interests and approaches in the area of learning has remained as rich, but it would be unfair not to mention the importance of "machine learning" in the AI community. Due to the combined efforts of AI reseachers, many important advancements have been made, although the area itself is not directly related to learning control systems.

We will incorporate machine learning results using the labels for key issues[24] from [47, 48] that describe the operation components of learning control systems. We provide some comments on these labels below as they apply to similar problems in learning control systems.

Perception in Learning

We assume that perception in the process of acquiring knowledge satisfies some definite properties. One is that information must be redundant (and even highly redundant). This redundancy is the source of pure learning procedures that acquire *variety* (in W. R. Ashby's words [49]), and this enables the application of combinatorial algorithms of learning. The following topics do not reflect the actual problem of information acquisition but only some particular facets of this process:

- Acquisition of proof skills.
- Learning from observation.
- Learning by experimentation.

The phenomena of generalization and nesting are not fully appreciated in AI work on perception.

[24] It is a matter of further discussion whether the particular source of the key issues is relevant for the problem under consideration.

Representation in Learning

Researchers in the entity-relational organizations of data have developed a powerful body of implementable results (see [50]). The mathematical background is available in a number of papers (e.g., [51]). This is supplemented by results in semantic networks (e.g., [52]) and conceptual modeling (e.g., [53]) which, when combined, can serve as a good theoretical background for representation in learning. Some other issues are painfully underdeveloped (e.g., issues of similarity among linguistic and quantitative hierarchies, or descriptive and numerical space tessellations). In the literature on learning (see subsequent topical labels) these results are often ignored:

- Learning as organizing information (e.g., about physical domain).
- Learning to classify.
- Goal-oriented classifications of structured objects.
- Chunking the goal (goal hierarchy).
- Acquiring knowledge from information management.

Nevertheless, the many existing results of classification theory are not fully utilized because they are published in the domain of automatic control theory. (Significant contributions on the general theory of classification are by M. Aizerman, E. Braverman, and L. Rosonoer [54–56].)

Storing for Learning

From learning experiences, as well as from other components of learning systems, we can infer that continual information is stored for learning. However, the problem of storing has not been explored in sufficient detail. Indeed, we do not know of any suggested organization of this storage, how long information is stored, nor when one forgets what one has remembered.

The following statements which are related to the human memory (or any anthropomorphic memory) can be fruitfully interpreted for other kinds of information storages:

1. Perception presumes a storing of the images and other representations (symbolic structures). Some images are recognized prior to storing; others are recognized afterward. In either case, the process of selecting images for subsequent storing is unclear.[25]
2. Perception is not supposed to give a verifiable view of the environment. We are not interested in the question whether the stored images represent the *truth* about the external world.
3. Perception does not include recognition as a necessary part; recognition may be considered an independent subsystem.
4. It is argued that the brain categorizes stimuli in accordance with past experiences and present needs. This categorization constitutes the basis of perception and recognition. In the constantly changing world, we prefer not to store images and other world representations but rather *procedures* that will help us to reconstruct, analyze, understand, and manipulate the world. We rely not on fixed images but *reconstructions* of the past remolded in the context of the present [57].

[25]No doubt, in certain instances intentionally and in others randomly.

5. Long-term (lower-temporal resolution, coarse) and short-term (higher-temporal resolution, fine) memories must be separated: "There are obvious difficulties involved in supposing that one and the same system can accurately retain modifications of its elements and yet remain perpetually open to the reception of fresh occasions for modifications" [58]. One can assume that generalization, and the associated subsequent lowering of resolution of the information stored, is an intrinsic property of the retention of the multiplicity of the information about modifications.
6. The phenomenon of stored (retained) information is that of *stratification* or of retaining information in nested hierarchies: "our psychical mechanism has come into being by a process of stratification: the material present in the form of memory-traces[26] being subjected from time to time to a *re-arrangement* in accordance with fresh circumstances—to a *re-transcription*. ... memory is present not once but several times over, ... it is laid down in various species of indications" [59]. One can expect that the higher the resolution of representation, the lower would be retention provided by the system ("ephemeric memory at the lowest levels" [60]).
7. Nested hierarchical representation can be based on more than one principle of organization of information, such as different modalities of sensing. In this sense it is prudent to expect a redundant system of (nested hierarchical) representation.[27]
8. A mechanism of categorizing (mechanism of clustering, grouping, classification mechanism, taxonomic technique) based on finding similarities (or dissimilarities) can be an important tool of preparation for subsequent information storing.
9. The mechanism of categorizing with subsequent use in the system of representation has a resemblance to the Darwinian theory of natural selection. It is also based on a *principle of determining and retaining* properties (information in an appropriate form) or *principle of natural selection* of these properties (the same), which works better for survival. Existence of this mechanism on the molecular genetic levels as well as on the level of neuronal grouping has been proved experimentally [61–64].
10. The Pavlovian conditional reflex experiments demonstrate that only generalizations upon series of similar experiments are retained, although the ephemeral memories of each particular instantiation might also exist.

Comparing in Learning

Comparison is a key procedure in learning. More precisely, the basic operation is a search for similarity in order either to unify entities in a particular class or to determine the value and/or the character of changes. The notions of learning by analogy and learning by example refer to "analogy" and "examples" as standards for comparison. The substance of search for similarity is not propagated in an explicit manner. These notions can be aligned within the structure given above for the learning control systems.

[26] Or time sequences.

[27] These representations are known (in psychology of the brain) as maps. Brain maps of sensory–motor stimuli were discovered in 1870 by G. T. Fritsch and E. Hitzig in Germany. The complexity and variation of sensory maps in monkeys was analyzed by M. Merzenich and his collaborators at the University of California at San Francisco in 1983.

Evaluation in Learning

The numerical evaluation of objects, relations, and events of the world is an adequate form of representation and is a subject of learning at a corresponding level of resolution. (For more details, see section 10.3.6.) Comparison hypotheses require the introduction of relevant metrics which depend on the nature of the particular method of hypotheses generation.

Response Generation (Decision Making) in Learning

Decision making is perhaps the most interesting topic in the theory of learning systems. The General Problem Solver (GPS), and a broad variety of decision recommending production systems (including SOAR), are becoming commonplace in AI. GPS architecture is associated with the following topics:

- Transformation of advice into heuristic search.
- Learning as knowledge compilation.
- Acquiring heuristics.
- Constructing a production system.
- Learning by augmenting rules.
- Learning from discovery.
- Discovery and heuristics.

In fact the major ideas have not changed since the first GPS in 1959 [4]. Learning in this architecture is performed by augmenting rules [65]; although P. Winston's research paradigm is far from the idea of automated learning. The initial scheme of automated theory formation was created by S. Amarel (1961 [66]) who introduced the "discovery" that led to a combinatorial synthesis of solutions for subsequent selection. However, the use of heuristics (though reflected in a massive literature stream) has not been substantiated by a theoretical analysis and has not been attempted in automated learning.

The AI approach to learning processes is to treat them in terms of different logical structures for the inference processes: deduction, induction, abduction, and the like. In this treatment the key feature of learning—generalization over the redundant information sets (collected prior to the system operation)—becomes blurred (as we show below).

Actuation in Learning

Actuation is not only a tool for producing a diverse number of behaviors: it is also a tool for diversifying the research of behavior within a particular environment. That is to say, the more flexible the ability to produce a behavior, the broader is the range of responses and the more adequate is the model obtained as a result. The resolution and dimensionality of the vocabulary of actuation must correspond to the actual level of resolution for which the model is supposed to be obtained. The correspondence between the values of resolution of actuation and of the model to be produced is a matter for further research.

Generalization in Learning

Generalization is performed within the subset of information determined by the scope of attention. Within this scope, a search for similar entities is performed that can form a *unifiable* group of similar entities. A variety of combinations must be explored before the best grouping is achieved. The groups are hypotheses for the future evaluation as alternatives of S–R couples, rules, or particular concepts. The spatial and/or temporal similarity cluster is a basis for declaring (or discovering) unnoticed earlier regularity (or discovering a new concept). A variety of logical methods is used for arriving at the conclusion that a new cluster is a new concept. Clustering by focusing attention, grouping, and combinatorial search supported by the available techniques (inductive, deductive, abductive, etc.) is called generalization. The list of associated topics from the literature follows; these are represented in the variety of papers dedicated to particular research instantiations:

- Mechanisms of generalization.
- Conceptual clustering.
- Inductive learning.
- Learning to predict sequences.
- Effect of noise on concept learning.
- Learning concepts by asking questions.
- Using derivational histories.
- Search for regularity.
- Generalizing plans from past experience.
- Generalizations upon rich memory.
- Concept formation by incremental analogical reasoning.
- Theory formation.

The literature in this area cannot satisfy the fundamental questions that naturally emerge: What is the common element in different methods of generalization? Can we learn something from a nongeneralizable information? Are "generalization" and "concept creation" comparable? Can a multiresolutional system of world representation be constructed by consecutively applying algorithms of generalization?

The problem starts at the stage of attention focusing. We really do not know how to select the body of information that will be used for subsequent generalization. Preliminary "filtering" of the available information has to be done as a trade-off between the danger of losing a correct result by cutting off "good" information and the danger of losing efficiency by processing unnecessarily excessive information.

We believe that all generalization processes are based on a set of primitives for recognizing similarities/dissimilarities and that they are built into every known system of learning. In our view, this determines the construction of the world model which then boils down to a generalization upon available information. It helps in understanding computer simulation of learning processes, since the processes are viewed as independent of any logical concepts entertained in a computational system.

Abstraction is commonplace in the literature on artificial intelligence. In fact generalization is always considered whenever abstraction levels are mentioned. The time has come to bring some order to the situation: to determine the degree of equivalence between generalization and concept formation, to determine the degree of equivalence between abstraction and concept formation, to determine induction and combinatorial search as the only mechanisms of generalization, to accept that recursive application of the mechanisms of generalization leads to the generation of multiresolutional world representations.

10.3.6 On the Difference between Adaptive and Learning Systems

We have seen that new numerical data can be acquired as a result of learning and that systems will adjust to a changing environment. This is *quantitative*, or *parametrical, learning*. It is usually associated with the domain of *adaptive systems*. This leads us to *parametric learning controllers*, or *parametric (adaptive) controllers*. The alternate situation would be generalizations that reject further tuning of the system and call for a restructuring of (1) the model of the system or (2) the model of the controller. We saw this process consistently effected in a model where numerical nested hierarchical processes of generalization were supplemented by *cognitive learning* or by linguistic nested hierarchical processes (which can be understood as an extension of numerical generalization processes at low resolution of representation). On this basis the controllers are renamed here *cognitive learning controllers*.

In using the term *learning* rather than the term *structural adaptive controllers* or *cognitive adaptive controllers*, we believe that we allow for further systems development: learning control is beneficial in any engineering domain. We are interested in a cross-fertilization of control and AI areas. Machine learning, as a branch of AI exists and has demonstrated a vast number of interesting and important results that cannot be underestimated or neglected. Specialists in control should admit the existence of *machine learning* and incorporate its results in a form that is productive and harmonious. Translation processes are never painless; however, we have plenty of evidence that machine learning is already scientifically fruitful.

10.4 INFORMATION ACQUISITION VIA LEARNING: DOMAINS OF APPLICATION

10.4.1 Estimation and Recognition as Components of Learning

In essence, learning is exemplified by processes of recognition and estimation at all its stages. We now will take a closer look at the numerous estimation methods based on specific *logical schemes*. They can be visualized as follows:

- Information is acquired repeatedly about the same zone in the state space (e.g., the same object of the world); the repetitiveness is characterized by a particular time period of sampling that pertains to a particular level of resolution.
- The information is stored (memorized) for the subsequent dealing with the whole set.
- A law of clustering is assumed for the stored information about the same object of the world.
- If the information is not known to belong to the same zone in the time–state-space (object of the world), then existence of the law of clustering is assumed to be evidence of belonging to the same object of the world.
- All known estimation procedures use recursive processes of computation with subsequent increasing of resolution of information used and produced.
- The results of prior estimation steps are assumed to be a good reason for expecting similar results in information acquisition. These results are the "predictors" employed in a number of estimation methods.
- The groups found as a result of clustering are considered single objects and the multiple information related to the cluster is generalized (when numerical data are

involved, averages or probabilistic expectations are taken, etc.); this concludes the process of estimation.

- The generalization for the cluster is compared with similar cases in the past for which an interpretation was constructed (and confirmed). This concludes the process of recognition.

Information acquisition involves receiving *samples* of data from sensors and transforming the data into a form that is understandable for the subsystems that use the information. Sensory data samples $x_i(a_i(O), v)$ that carry measured or assessed values v (a number of units of a particular scale) of an attribute a (name of characteristic feature of an object O_j, whose quality determines the physical units of the scale) may be represented as propositions of the form: the ith attribute of the jth object is value, or (attr-obj); value), for example, [temperature (cylinder, coordinate), 86°C].

Here the *attribute* is determined by a modality of sensing (e.g., temperature), namely by the feature being measured by this sensor. The object is the entity of the external reality to which this feature is known to be attributed (cylinder, coordinate). The modality of sensing (i.e., the physical essence of the particular sensor based on the nature of energy transformation process performed by the sensor) is known, since the preliminary design of the machine to be controlled is known.[28] The object is often known (in the generally applied automated machines), but in the intelligent machines, the object remains to be recognized—this is a part of the overall learning process. The value is a scalar quantity of an n-tuple quantified over some numerical or linguistic scale that pertains to the resolution level under consideration.

There are two types of sets for x_i: space-related sets $x_i(x_k)$, $k = 1, 2, \ldots, m$, $k > i$, m is the total number of attributes reflected in the sensory data, and time-related sequences $x_i(t)$, and/or $x_i(x_k, t)$. These sets and sequences contain a number of patterns $\{P_\mu\}$, ($\mu = 1, \ldots, m$, a number of patterns that can be recognized), which are inclusive of a set of objects $\{O_j\}$ existing in the world $\{P_\mu\} \supset \{O_j\}$. Recognition mapping $C_1 : \{x_i(a_i(O), v)\} \to \{P_\mu\}$ is known to usually precede the interpretation mapping $C_2 : \{P_\mu\} \to \{O_j\}$. Together, these two mappings (recognition and interpretation) constitute an operator of cognition $C = C_1^* C_2$. When the set of objects (entities) of the world is found (C_2) and the relationships upon this set are determined, the problem of control can be attempted.

We do not consider here the mechanism of perception that is based on the recognition and interpretation mappings. However, we would like to mention two general alternative rules characteristic for the overall C-operator: (1) the vocabulary of the output of C can contain only those words (and relationships among them) that are understandable for the control system; (2) otherwise, it must include a request for establishing new words with the subsequent incorporation of these new words into the operation of the control system. Either rule does not change anything in the content of the input vocabulary $\{x_i(a_i(O), v)\}$. However since the set of $\{O\}$ is subject to change, we will denote the input $\{x_i(a_i(H), v)\}$ where the set of H is understood as a set of all hypotheses generated about the objects of the world.

Consider observed input $[\{x_{is}(a_i(H), v_{iks})\}, \{H\} = 0] \to x_i(x_k), v_{iks}$, where s is the number of samples $s = 1, \ldots, N$. It is assumed that there exists a *true* value of the input variable v_{iks}^*. The deviation of each observation from the true value will be called

[28] An interesting (but highly complicated) problem can be formulated of learning the modalities of sensors. We do not consider this problem here.

error of observation

$$\varepsilon_{iks} = v_{iks} - v^*_{iks},$$

which gives the following models for judging (estimating) the value of the input variable:

1. Linear model (LM) estimation [67]:

$$\text{est}(v_{iks}) = av_{iks} + \varepsilon_{iks}$$

   where

$$a = \frac{\sum^N [\text{est}(v_{iks})] \cdot v_{iks}}{\sum^N (v_{iks})^2}$$

2. Maximum a posteriori probability (MAP) estimation:

$$a \quad \text{provides for} \quad \max\left\{\frac{p[\{\text{est}(v_{iks})\}|a]p(a)}{p\{\text{est}(v_{iks})\}}\right\},$$

   where is the probability density function (pdf) of the input samples.
3. Maximum likelihood (ML) estimation: If the a priori pdf is unknown, then it can be assumed, for example, to take the form of a likelihood function, and the rule of ML estimation is formulated as

$$a \quad \text{provides} \quad \max\left\{\frac{N}{2}\ln 2\pi + N\ln\sigma_\varepsilon + \frac{1}{2}\sigma_\varepsilon^2\sum^N[\text{est}(v_{iks}) - av_{iks}]^2\right\}.$$

Each of these estimation postulates presumes that a number of measures are executed. Their results are memorized, and based on the set of those memorized values, a judgment about the true value is made, which is called an *estimate*.

10.4.2 Domains of Application for the Theory

The discovery of control systems had the practical importance that (design) decisions could be considered a reflection of the relationships existing within representations. Intelligent systems emerged as a response to the virtual stochastic character of the world in which nonintelligent techniques limit the opportunity to receive competitive levels of efficiency. In other words, in organizing our *knowledge representation*, we can discover relationships within the representation that can be expected to exist within reality. Early stages of the control theory were aiming toward "regulation" and stable operation of machines, toward providing stable and accurate operation of dynamical systems with feedback, and eventually toward providing required functioning of any preassigned object of control, or *plant*, including machines in a variety of environments, man-machine systems, and even human teams, economical systems, for example. There was no doubt that a model of the plant could contribute to control system design. So the *knowledge* of the system (the *plant*) was presumed.

Another thing presumed at the early stages was that the control system should be designed in advance in full. At least the structure was expected to be known, that is, obtained with knowledge of the control systems process (with the major parameters of this structure) before the controller is built and functioning of the control system started. Some of the parameters were allowed to be determined imprecisely and were subject to constantly updating and correcting within the so-called *adaptive* control systems. Gradually control specialists discovered that there is a substantial difference between the two groups of solutions: online control solutions, and off-line control solutions. They can both be combined, but the component *online control* should always remain.

Conventional control was dealing with a wide variety of problems in a range of devices, from simple speed regulators in the early steam engines to devices for stabilizing a goal-oriented spacecraft. However, the following problems are still unequivocally considered to be difficult in conventional control theory with no intelligence and no learning:

1. The optimum control of nonlinear systems does not have a well-organized solution because the established paradigms were created for linear systems only. Nonlinear control theory is perpetually in its embryonic state, and optimum control solutions cannot typically be found for important nonlinear systems.
2. The optimum control of stochastic systems can be addressed only within a simulation paradigm because the models of stochastic systems presume knowledge of probabilistic parameters and characteristics of systems that cannot be provided in advance in the existing practice of design.
3. The control of multilink manipulators (either 6-DOF, or the redundant[29] ones) is performed only in an approximate manner because the models of *plants* turn out to be so huge that even the off-line solutions can become problematic to a control engineer, not to mention the online control which in fact is required.
4. The control of redundant systems leads to so-called ill-posed problems; in order to solve them, one has to introduce a regularizing function that is based on desirable assumptions rather than on information known about the system.
5. The control of autonomous robots does not allow for using any conventional techniques because most of the information is not known in the beginning. Thus they cannot be supported by off-line solutions but require online interpretation.
6. The control of systems with multi-sensor feedback information cannot be done just by using standard multi-variable approaches because the multi-sensor information must initially be conceptually integrated. This means that some generalization activities are presumed.
7. Online control of systems with incomplete initial knowledge of the model, and/or of the environment, is not supported by any conventional techniques because we do not know how to incorporate new knowledge of the world into the model of the plant, and/or the model of the controller.

In all of these problems, neither the knowledge assumed at the stage of design can be considered complete and satisfactory nor the process of design can be completed unless new knowledge is additionally supplied. There has appeared an opportunity to view all these systems in a different, unconventional way as systems with a never-ending design stage as a result of a broad range of computer applications, and in particular, industrial computer systems equipped with a number of transducers-sensors.

[29]The word "redundant" means allowing for infinite number of the optimum trajectories, containing more degrees of freedom than is necessary for performing the assignment

Indeed, all of the systems mentioned in the list above can be presented as a loop that contains all the means required for dealing with the complicated problems: *perception* for organizing a diversified information set coming from the numerous sensors, *cognition* for enabling the system not only to interpret the results of perception but also to put them into perspective in determining strategies, and policies of future operation as well as to submit all necessary information for planning/control processes, *planner/controller* to generate proper control sequences, and so on (Fig. 10.1a).[30] This system is a result of a finite design process shown as a mapping of the controlled system into the control system.

10.5 AXIOMATIC THEORY OF LEARNING CONTROL SYSTEMS

In order to discuss axiomatic learning in the learning control system the following definitions are necessary.

10.5.1 Axioms

AXIOM OF REPRESENTATION

(Representation Set) Any reality of the world can be represented[31] (or modeled) by a representation set. The latter is defined as a list of characterizations for the entities (both with their qualitative and quantitative characterizations) composing this reality, and a list of relationships among these entities.[32] (This definition is a recursive one, and it can be applied to the word *entity* within the definition; the words *world*, *reality*, and *entity* are not defined, common thesaural definitions can be applied.) So the label for the entity "position" does not belong to the representation set, but the statement "position is six" does belong to it. Therefore the lists consist of couples "entity → characterization," and "entity → entity." Any desire to represent the world requires memory (storage of the representation). Each entity of the world is a carrier of its own information and could be considered a "memory of itself." This is not the kind of memory that is embodied within the representation. Representation presumes that the memory of an object is separated from this object.[33]

Transformations of the representation sets (of entities, and/or of their relationships) are presumed. The transformed sets can be a list in the form of, for example, differences, and then differential equations (transformations are expected to be based on assumptions that are part of the initial lists). Any consistent result of the transformation of the initial lists is still considered to be a representation set. All lists are considered to be open: They allow for endless growth, so when new labels appear, they are added in corresponding lists of entities and relationships, and they generate their own new lists of decomposition. Since consecutive decomposition is presumed, this type of relational

[30]A question could be raised about how this system operates. Each subsystem is considered an independent entity with its own knowledge base. Of course the knowledge bases of all subsystems can be organized in a network and allow for communication, negotiation of mutual consistency of operation, and so on. Using feedback loops can be expected from cognition to perception, from planning/control to cognition, and so on (see [57]).

[31]Or modeled; we will use the words *representation* and *model* interchangeably.

[32]Unfortunately, we cannot talk about the world referring to it in another way rather than use words such as *objects*, or *entities*, and *relationships* among them. Sure all these entities might turn out to be just our imagination, a result of our desire to superimpose some compulsory organization up on the free and chaotic world. However, this is the only way that we can talk about the world. What I name here as *reality* is just a meta-representation of it.

[33]As opposed to, say, memory distributed within the object.

structure is identical to a tree structure. Representation set is considered to be applicable only to a particular moment in time.

The recursive character of the definition for the representation set implies decomposition of each entity in its parts with the set of relationships among them, and so forth. This decomposition focuses on parts of the initial entity which require higher resolution for their consideration than can be provided within the initial representation set. Thus any representation set is a multiresolutional hierarchy by definition (for details on the representation set, see [68, secs. IV.E–IV.J on the D-structure]. The time representation at each level of a representation set also has a different resolution. Each representation set related to a particular moment in time is a *state* of the representation set[34] (so a state is a snapshot of the representation set at a definite resolution level). The consecutive string of states is called a *process*.

AXIOM 1 OF CONTROL

(Control System, or Controller) The control system is a set of interconnected devices operating in such a way as to provide behavior that succeeds in satisfaction of a *goal set*, or a *set of output specifications* formulated for a controlled system, and judged by a user.[35] The control system can be represented as a set of logical statements (of existence, and of belonging to a class) that can be made about the implications that hold among units of the *representation set of the control system*. For example, if the representation set contains subsets of the parameters and variables of a dynamic system, then a set of logical statements can be represented as a system of differential equations for this dynamic system, or a system of difference equations when the variables are tessellated, or a list of implications if these difference equations are written in linguistic form)[36]. The so-called control problem (see Axiom of System) is about how to build the control system.

AXIOM OF PLANT

(Controlled System, or Plant) The system of interest for a user is expected to be managed by a control system in such a way as to provide behavior that ends with the satisfaction of a *goal set*, or to provide a *set of output specifications* formulated and judged by a user. Controlled system can be represented as a set of logical statements about the implications that hold among the components of the *representation set for the controlled system*. (The example from the definition of the control system applies.)

AXIOM OF GOAL

(Goal Set and Output Specifications) A subset of the representation set is called the *goal set*; a set of output specifications is a list of logical statements about the implications that hold among the subsets of the qualitative and/or quantitative *set of observables* (e.g., the *input subset* and the *output subset* as well as the set of *inner states* that allows for measuring) of the system, which are accepted by a user to be characteristic for the controlled system and which must be satisfied as a result of using the control system. (Like the rest of the representation sets of a system, goals are also nested hierarchies.)

AXIOM OF OBSERVABLES

(Observables) A subset of the representation set is called the *set of observables*; it is a set of all characteristics (variables) that can be observed, measured (with a definite accuracy), and *recorded*. The set of observables contains (not necessarily consists of) three major subsets: the input subset, the output subset, and the *set of inner states* that allows for measuring. The input subset includes those observables that can be assigned and executed by the user's external means, the output subset includes those observables that cannot be executed by external means but can be executed only by the controlled

[34]The event?

[35]A goal set presumes a user. No technological problem can be understood when there is no user. So all controllers have a goal set. The problems not considered here appear outside the technological domain.

[36]That is, statements of existence, statements of belonging to a class, and statements of implication; mathematical statements are the same, just written in a different notation.

system, while the set of inner states includes those variables that are the components of the existing model of the plant.

CONTROL PROBLEM

The standard problem of control is defined as follows: For each event from the set of initial events (t_0, x_o) a control should be determined $u(\cdot)$ that transforms this initial event into a goal set, and minimizes the cost-function simultaneously [69]. In fact this definition is a tautological one: The solution for the control problem is a control system as was defined earlier (see Axiom 1 of Control).

Another definition of a control problem has a clear reference to *conventional control theory* [70]. The control problem is recommended to be divided into the following steps: (1) establishing of a performance set, (2) writing down the performance specifications, (3) formulating a model of the system as a set of differential equations, (4) *using conventional control theory* to find the performance of the original system and, if it does not satisfy the list of requirements, then *adding cascade or feedback compensation to improve the response*, (5) using modern control theory to assign the entire eigenstructure, or to minimize the structure's specified performance index (which is understood as a quadratic performance index). In general, one can easily find that there is a surprising lack of uniformity in the existing views on the *control problem*.

The following definition of control law can be considered consistent with the practice of control (G. Saridis, [71]): *Control of a process implies directing a process effectively toward a specified goal.* One can see that the notion of a goal is included in this definition. Cost-functions and cost-functionals are considered a part of the goal set. In a multiresolutional setting we talk of a hierarchy of goals and a hierarchy of cost-functionals.

Thus, the recommended steps for solving a control problem can be reformulated as follows: (1) establishing of a set of cost-functionals for performance evaluation, (2) writing down the performance specifications, (3) formulating a model of the system, (4) finding the performance of the original system and, if it does not satisfy the list of requirements, then determining the input which is required to achieve the performance specifications with minimizing and/or maximizing the cost-functionals in a sequence that make sense for the user (programmed control), (5) evaluating how the unexpected factors and quantitative uncertainties can affect the operation of the open loop system, and introducing the feedforward and the feedback controls as required.

Our definition of a system will differ from the classical definition [72] by our introduction of a controller as a part of the system. The system is understood as a couple $\{\Sigma, \Delta\}$ consisting of the descriptive set characterized by the vector of evaluations Σ and the vector Δ of corresponding values of the distinguishability zones (of the values of accuracy). In all subsequent presentations any variable is presumed to be assigned as a couple of its evaluation (e.g., average value), and its accuracy can be considered a source for determining a corresponding distinguishability zone (or the size of the corresponding tessellatum[37]). For simplicity we will omit the couple notation when possible; however, the concept of accuracy for each component of Σ will be invoked as the need emerges. The following notations will be used: $\mathbf{T}$ for the time set, $\mathbf{X}$ for the input set, $\mathbf{U}$ for the set of instantaneous input values, $\boldsymbol{\Omega}$ for the set of acceptable inputs, $\mathbf{Y}$ for the set of outputs, $\boldsymbol{\Gamma}$ for the set of instantaneous output values, $\boldsymbol{\varphi}$ for the state transition function, $\boldsymbol{\eta}$ for the output function, and $\mathbf{K}$ for the control law exercised in the system.

[37]This means that the multiple sampling took place prior to the discussion (or should be performed) within some scope (size of window of attention). This multiplicity of samples has been generalized by applying procedures of attention focusing, combinatorial search, and grouping. As a result both the evaluation and the accuracy should have been determined.

AXIOM OF SYSTEM (System) A system$\{\Sigma, \Delta\}$ is a representation structure formed over nine-tuple representation sets[38] $(T, X, U, \Omega, Y, \Theta, \varphi, \eta, K)$ defined by the following concepts:

1. *Existence*. There exists a given *time set* **T**, a *state set* **X**, a *set of instantaneous input values* **U**, a nonempty set of acceptable *input functions*, (or *command sequence*, or *control string*) generated by the control law **K**:

$$\Omega = \{\omega : K \to U\},$$

a set of *instantaneous output values* **Y**:

$$\boldsymbol{\Theta} = \{y : \mathbf{T} \to \mathbf{Y}\}$$

and an *output function*

$$\boldsymbol{\eta} = \mathbf{T} \times \mathbf{X} \to \mathbf{Y}$$

2. *Direction of time*. T is an ordered subset of the reals.
3. *Organization*. There exists a *state-transition function* (or trajectory of motion, or solution curve)

$$\varphi : \mathbf{T} \times \mathbf{T} \times \mathbf{X} \times \Omega \to \mathbf{K} \to \mathbf{X}$$

whose value is the state $\mathbf{x(t)} \in \mathbf{X}$ resulting at time $\mathbf{t} \in \mathbf{T}$ from the initial state (or event) $\mathbf{x}_0 = \mathbf{x(t)}$ at the initial time $\tau \in \mathbf{T}$ under the action of the input $\omega \in \Omega$.

Usually, the control law **K** is predetermined: It is either a time-dependent program (open-loop control, feedforward control), a mapping from the set of variables (closed-loop control, feedback control), or both (OLC/CLC-control, combined feedforward–feedback control).

In the reality of learning control, we are interested in defining the state-transition function *before* the definite moment in time ($t \geq \tau$): The problems of prediction are solved by analyzing the behavior of the system at the interval T_m. The concepts 1 through 3 from Axiom of System imply that the time scale of the system (at a level of resolution) should be preselected.

This set of definitions and axioms is usually supplemented by the demand for stationarity, linearity, and smoothness if the representation is expected to be used in the form of differential equations. The requirement of smoothness determines the acceptable granularity (the law of tessellation). Properties of the state-transition function lead to the following theorem proven in [72]. Every system Δ with a state-transition function is defined as shown above, and with the norm $||\omega|| = \sup||u(t)||$ has a transition function at a level of resolution in a form

$$\frac{dx}{dt} = f(t, x, \pi_t \omega),$$

where the operator π_t is a mapping $\Omega \to U$ derived from $\omega \to u(t) = \omega(t)$. The suggestion to select π_t in a form $\pi_t : \omega \to \left(u(t), u'(t), \ldots, u^{(n)}(t)\right)$ (by [73]) is rejected. This further narrows the domain of the system under consideration. In the smooth, lin-

[38] Which are nested multiresolutional hierarchies as are all other representation sets in this section.

ear, finite dimension case, the transition function obeys the simplified relations. The simplification is determined by selecting norms of the corresponding spaces with no derivatives of the time functions for controls. Only now we are coming to realize that the expectation of [73] might be right; we are dealing with systems that need a norm based on the entire set of $u(t), u'(t), \ldots, u^{(n)}(t)$. The complete representation of the system can be written as follows

$$\frac{dx}{dt} = A(t)x + B(t)u(t),$$
$$y = C(t)x(t) + D(t)u(t),$$

where $A(t)$ and $B(t)$ are parts of the expression $f(t, x, u(t)) = A(t)x + B(t)u(t)$, and $C(t)$ is a mapping which is obtained from the equation

$$y(t) = \eta(t, x(t)) = C(t)x(t).$$

Here $T = R_1$ and X, U are normed spaces, $A(t)$ is a mapping $A : T \to \{n \times n \text{ matrices}\}$, $B(t)$ is a mapping $B : T \to \{n \times m\text{matrices}\}$, n is a dimensionality of states $x \in R^n$, m is a dimensionality of controls $u \in R^m$, p is a dimensionality of outputs $y \in R^p$. Proper adjustments and modifications are made for a variety of cases: discrete systems, systems with nonvarying parameters, and so on.

We can see that the familiar forms for the system representation are not affected by using the paradigm determined by definitions 1 through 7, which means that the theory of learning control systems can be constructed using these familiar form however having in mind their renewed meaning.

Now we can refine the Axiom of Control.

AXIOM 2 OF CONTROL

(Revised: Control System or Controller) The control system (see components of Σ) is a subsystem $\mathbf{K}(\Theta, \Gamma)$ of the system under consideration that generates the input according to a definite (required) control law.

The *control law* in conventional control is defined as a permanent mapping $\mathbf{T} \times (\mathbf{X} \times \mathbf{Y} \times \mathbf{Y}^*) \to \mathbf{U}$ that puts in correspondence $\mathbf{x(t)}$, $\mathbf{y(t)}$, and $\mathbf{u(t)}$ for each moment of time, and thus determines (1) the operator $\mathbf{G}$ of the open-loop control subsystem (which provides $\mathbf{U} \to \mathbf{Y}$, $\mathbf{Y} - \mathbf{Y}^* = 0$, i.e., the operation of the system under the assumption that no deviations of conditions and parameters, no uncertainties, can exist; see the matrix $\mathbf{G(t)}$ below), (2) the operator of the feedback ($\mathbf{F}$) and feedforward ($\mathbf{D}$) control subsystems (which are supposed to compensate the uncertainties entailed by applying the open-loop control law). Control command sequences can be computed online (primarily matrices $\mathbf{D}$ and $\mathbf{F}$), and/or off-line (primarily the matrix $\mathbf{G}$) in order to satisfy a specified list of control requirements. Control is computed according to the equation

$$u(t) = F(t)y(t) + G(t)v(t),$$

where $v(t)$ is a program of operation that provides $U \to Y$, $Y - Y^* = 0$ if no deviations from the expected conditions of operation are observed. The control system is shown in Figure 10.2.

DEFINITION OF LEARNING CONTROL SYSTEM

(Learning Control System) The learning control system $\mathbf{K}[\Theta, \Gamma_L(\mathbf{t})]$ is a control system that generates the input according to a control law that is constantly redefined[39]

[39] This means that the goal set is reconsidered.

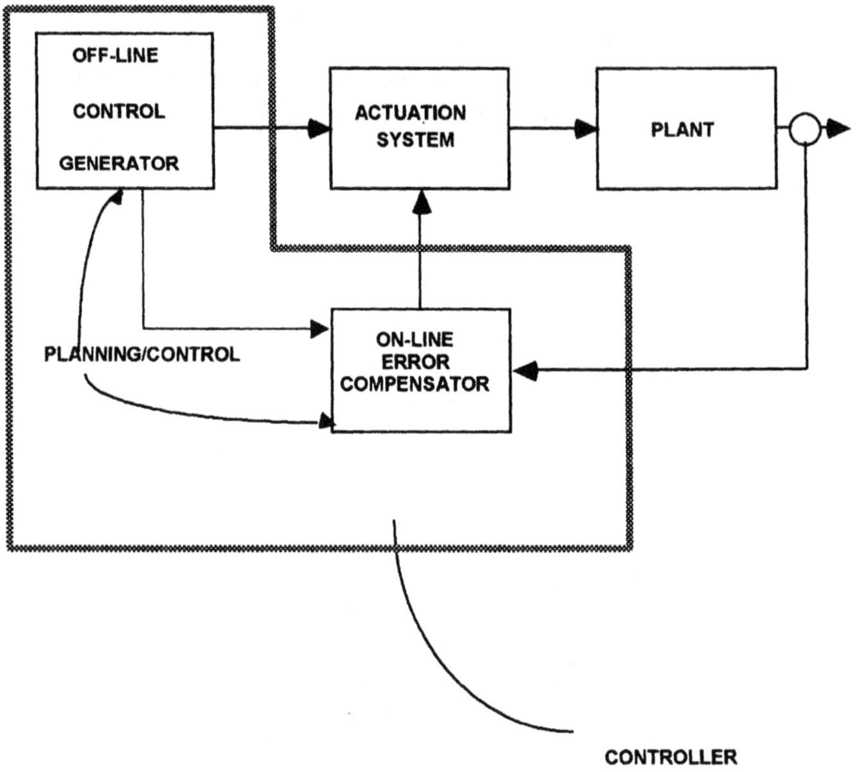

Figure 10.2. Control system.

and recomputed.[40] Therefore the goals $\Gamma_L(\mathbf{t})$ are evolved in time and the requirements concerning the cost-functionals (which are a part of the goal set) are better satisfied. (We could say that the control system is constantly redesigned online, but in this case we have to define what is design; this is a very complicated matter because of multiple meanings this word can have in a variety of areas.)

The *learning control law* is defined as a variable mapping $T \times (X \times Y \times Y^*) \rightarrow U$ that puts in correspondence $x(t)$, $y(t)$, and $u(t)$ for each moment of time, and thus determines (1) the operator G of the open-loop control subsystem (which provides $U \rightarrow Y$, $Y - Y^* = 0$, i.e., the operation of the system under assumption that no deviations of conditions and parameters, no uncertainties, can exist), (2) operators of the feedback (**F**) and feedforward (**D**) control subsystems (which are supposed to compensate the uncertainties entailed by applying the open-loop control law). Control command sequences should be computed online (primarily, for **F** and **D**), and/or off-line (primarily, for **G**) as to satisfy the specified list of control requirements. Control is computed according to the equation

$$u(t) = F(t)y(t) + G(t)v(t),$$

where $v(t)$ is a program of operation that provides $U \rightarrow Y$, $Y - Y^* = 0$ if no deviations from the expected conditions of operation are observed. All operators are reconsidered

[40]This means that under the accepted goal set, the required mappings are determined.

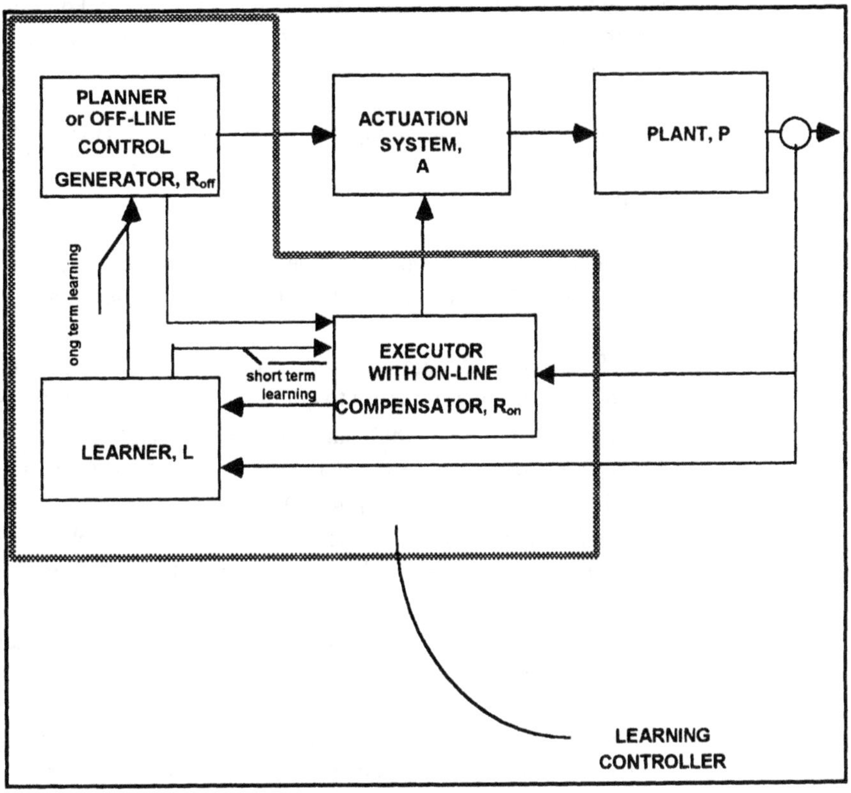

Figure 10.3. Learning control system

on a regular basis, and algorithms that change them are introduced. These algorithms are called *algorithms of learning*, and they are considered to be part of the overall learning control system (see Fig. 10.3).

10.5.2 Learning Control Systems

The axiomatic learning approach was introduced so that principles of learning could be received and evolve in control systems. Still our knowledge of the system cannot be considered exhaustive, or presented in the adequate form. Our knowledge of the EXO system (see Section 1.5) is rudimentary in the vast number of cases. Since the goal-set is subjected to modification, to evolution, and to change in general, a subsystem is required that takes care of all these changes and keeps the operation of the system consistent with the goal-set.

In Figure 10.2 a feedback compensator is shown that adjusts the control commands to the unpredictably changing circumstances of the operational environment. This compensator demonstrates one of the greatest commonsensical principles ever utilized in technology: the feedback principle. Feedback compensation functions as a rudimentary system of learning: Feedback exercises the *reflex principle* which characterizes creature actions occurring as direct and immediate response to particular stimuli uniquely correlated with them. Two types of reflexes are known: unconditional and conditional. Unconditional reflexes are responses that are stored in the memory of system prior to the operation (stored in a look-up table of the transition and output mappings). Conditional reflexes are responses that are learned as a result of experience. The latter is understood as a chain

$$\mathbf{R} : \mathbf{S}_0 \rightarrow (\mathbf{S} - \mathbf{S}_0) \rightarrow \mathbf{E} \rightarrow \mathbf{DM} \rightarrow \mathbf{A},$$

where

$\mathbf{S}_0$ = what has happened, the observed state (an ability to *perceive*, *represent*, and *store* the representation temporarily, and *focus* on the subset of interest is presumed),

$(\mathbf{S} - \mathbf{S}_0)$ = change observed, the difference between the previous state and the observed state (an ability to *compare* the representations is presumed),

$\mathbf{E}$ = evaluation of the change observed (an ability to evaluate representations is presumed, e.g., to assign a measure of "goodness," which is required to extract a decision from the lookup table),

DM = generation of the response, or decision making (an ability is presumed to *generate* the response that means that either a storage of responses is used or the response can be inferred or otherwise synthesized),

A = response (an ability to act corresponding to a particular G, i.e., actuation is presumed).

The simple act of feedback compensation then is an operation of the response controller which consists of elementary (computational) operations of (1) perceiving, (2) representing, (3) storing, (4) comparing, (5) evaluating, (6) generating the response, and (7) acting correspondingly. It is also implied that the difference $(\mathbf{S} - \mathbf{S}_0)$ is computed over a time interval $\Delta\mathbf{t}$ within which the response is not yet being generated. It is implied too that there exists a mapping (a production system, or pre-solved differential equations) that evokes the transforms

$$(\mathbf{S} - \mathbf{S}_0) \rightarrow \mathbf{E} \rightarrow \mathbf{G}.$$

However, a question emerges: Why should every small action be determined by observation of a small change in a state? Should not there be a *sequence of states*, a *string of states* rather than a single state? The answer is simple: We do consider a sequence of states, since a single state at the ith level of resolution can be expressed as a sequence of states at the higher $(i + 1)$th resolution level in discrete time $\Delta\mathbf{t}_i < \Delta\mathbf{t}_{i+1}$. The chain $\mathbf{R}$ is applicable to each of the states at the lower level, which in turn is a string of states of the lower level, and so on. So at any level of resolution the reflexive chain is valid in the above form.

Clearly, then, the chain $\mathbf{R}$ presumes one more operation (the 8th one): generalizing time strings at the lower level into single states at the upper level. This consecutive refinement is continued top-down and creates the *nested hierarchical controller* [68]. It is possible to show that the familiar procedure of recursive estimation is based on the reflexive chain $\mathbf{R}$ too.

Interestingly, we cannot include this operation in the set of elementary operations listed above, since it is a special operation: It is part of all the other seven operations in the list. Indeed, the first operation (perception) is based on grouping the units of sensory information into entities, these entities into entities of even lower resolution, and so on. This recursive grouping of units with their subsequent formation of entities is also a generalization (not in temporal domain but in the spatial one). Operations of representing and storing are utilizing the operation of generalization in absolutely the same way as it is shown for perception.[41] The operations of evaluation and of comparison do not incorporate generalization per se but rather are generalization dependent.[42] (Generalization is the subject of [44]).

The reflexive chain is described as if it is applied online. One can easily imagine that a similar chain can be activated off-line, which brings us to preplanned program of operation. This program is generated by a subsystem of planning (the off-line control generator in Fig. 10.2). This subsystem reproduces the *reflexive plan-chain* before the actual process begins:

$$\mathbf{R_p} : \mathbf{S_e} \rightarrow (\mathbf{S} - \mathbf{S_e}) \rightarrow \mathbf{E}_n \rightarrow \mathbf{DM}_n \rightarrow \mathbf{P},$$

where

$\mathbf{S_e}$ = what is expected to happen, the expected state,

$(\mathbf{S} - \mathbf{S_e})$ = the change expected, the difference between the previous state and the expected state,

$\mathbf{E}_n$ = evaluation of the expected change observed, planning evaluation,

$\mathbf{DM}_n$ = generation of the response which is performed off-line (planning decision making),

$\mathbf{P}$ = planned response (a plan of action corresponding to a particular G when the expected change takes place).

One can see that $\mathbf{E_p}$, $\mathbf{DM_p}$, and $\mathbf{P}$ do not require any changes for the rest of preplanned process of motion. In fact other operators can be used, since the operation is being done off-line with no computational time limitations; therefore more precise and thorough computational procedures can be applied. It is also easy to understand that planning reflects the so-called open-loop control.

Here we can see that planning presumes off-line learning in the sense that information about expected circumstances is stored and analyzed prior to operation, a variety of expected situations are analyzed, and the recommended solutions for these situations are *learned* in advance (off-line). Statistically speaking, the plan represents a (time-dependent) statistical expectation of the motion to be executed; in other words, the plan is an expectation of the nonstationary stochastic (vector) process.[43] Any deviations of the reality from the preplanned process are taken care of by the online reflexive feedback of the so-called closed-loop control.

The learning control system is a development of the control system that includes one more subsystem: the LEARNER whose functions are as follows:

[41] Obviously a more graphic treatment of these processes (perception, representation, storing) could illustrate the universal character of the generalization process.

[42] These operations can be also described in more detail.

[43] It is well known that there is no theory for finding expectations of nonstationary stochastic processes. Yet there are many applied methodologies for dealing with this issue.

1. To enhance the vocabulary[44] of *perception*, that is, to enable the system to perceive more entities than were available for perception at the beginning of its operation.
2. To enhance the vocabulary of *representation*, that is, to enable the system to represent more entities than were available for representation at the beginning of its operation.
3. To enhance the vocabulary of storage, that is, to enable the system to *store* more entities than were available for storing at the beginning of its operation.
4. To enhance the vocabulary and the arsenal of rules for *comparison*, that is, to enable the system to compare more entities than were available at the beginning of the operation.
5. To enhance the vocabulary, the arsenal of rules, and the precision of *evaluation*, that is, to enable the system to evaluate more entities than were available, and to take into account more factors than it had in the beginning of the operation.
6. To enhance the vocabulary, the arsenal of rules, and the set of meta-rules of the *response-generating* production system, that is, to enable the system to generate more adequate responses than were available in the beginning of the operation.
7. To enhance the vocabulary of responses available for *actuation*.
8. To enhance the number and the adequacy of the techniques for generalization.

Learning control systems are introduced axiomatically, and they are consistent with the following definition by G. Saridis [71, p. 22]: "The system is called learning if the information pertaining to the unknown features of the process or its environment is learned, and the obtained experience is used for future estimation, classification, decision, or control such that the performance of the system will be improved." (The term "learned" is interpreted as in definition 11 in Section 10.1.)

It is useful also to consider the following definition containing a condensed representation of the processes characteristic for learning control systems.

DEFINITION OF PROCESS OF LEARNING

(Functioning of Learning Control System) The functioning of learning control systems consists in nested hierarchical generalizations performed over redundant stored information about current and/or prior experiences of operation. This system of generalizations changes the world representation as well as the algorithms of control available for selection during the problem solving.

Figure 10.3 has been redrawn in simpler form in Figure 10.4. The three blocks of the elementary learning controller (R_{off}, L, R_{on}) are replaced by a single R_{on} controller for the upper level of resolution due to the fact that the time sequence of states at the ith level correspond to a single generalized state at the $(i-1)$th level of resolution. However, a new learner and a new R_{off} subsystem is required. This process can be continued bottom-up until the required resolution of planning is achieved.

10.5.3 Emerging Nonconventional Issues

Given both definitions of the classical approach to control problems, and of the learning control system, the following issues can be raised:

[44]We understand now that vocabulary enhancement can be considered a recursive procedure, and when initiated at one of the resolution levels, it can trigger an avalanche of corresponding changes in the rest of the levels.

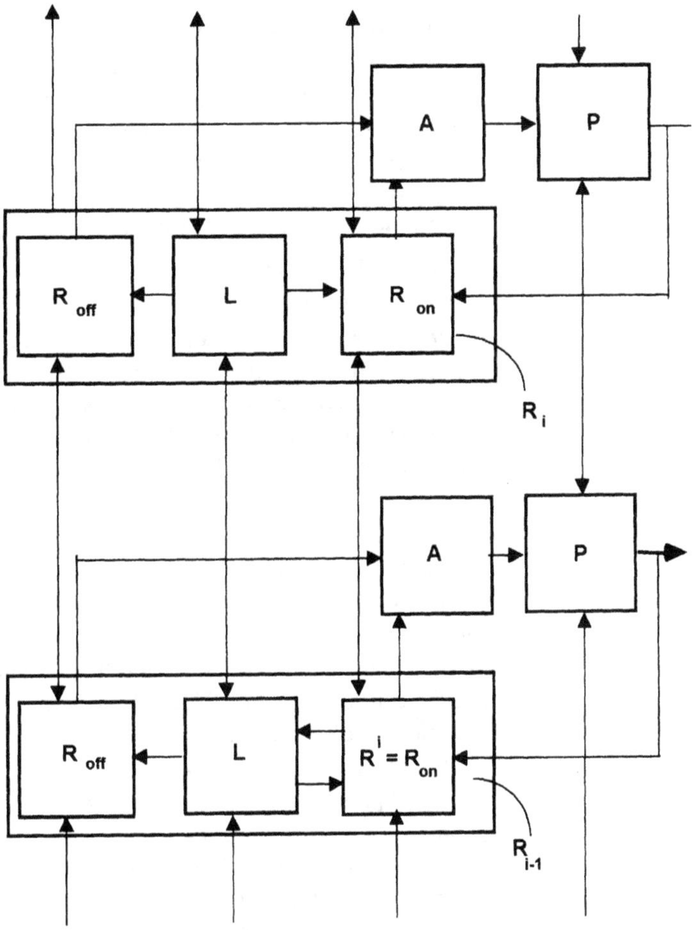

Figure 10.4. Nested hierarchy of the learning control system.

1. *At the present time a model is implicitly considered to be the source of the theories for problem solving, and not vice versa.* Conventional control practice requires having a model of a system as a starting point for the process of design, and then methods of dealing with this formal body for a particular problem within the boundaries of the selected model. The user is more interested in a theory of control that will not depend on the model of system. It is a fact that the *required mapping is capable of inducing the process of interest*, and *we must drive this process toward the goal*. As a result the deficiencies of the assumed off-line model will affect the structure of the learning system and the learning processes. The models formulated from general premises do not satisfy the conditions of the online learning process: they were created for the off-line use.
2. *The problem of constructing the required mapping capable of inducing the process under consideration* is becoming a central problem of the research in the area of learning control systems. Indeed, if the model is not supposed to induce the methods of dealing with the system, another way of constructing can be formulated as follows:

- The goal-set is formulated by the user. This induces the initial descriptive structure for the subsequent solution.
- The implementation set is selected with the participation of the user. These sets of information (the goal-set and the implementation set) supplement the initial descriptive structure. The new descriptive structure can be called the *pre-learning knowledge structure*.
- It would be prudent to use this pre-learning knowledge structure as the starting point for the process of finding the initial control mappings.
- The initial control mappings should be exercised unless they contradict some external knowledge of the designer.[45]
- The first results of the operation will contain information that should be considered the source for a (hopefully) converging process that enhances the descriptive structure and thus *improves the control mappings*.[46] We will assume that improvement of control mapping is a primary issue of the problem, as opposed to the issue of "adequate descriptive structure."

It is important to realize that the strategy proposed is not a strategy *to learn how to learn*. It is just an evolution of the existing methods of consecutive approximation, and recursive estimation to the less uniform domain.

3. *Usually the formulation of control problem is not reconsidered during the control operation*. Thus the goal subsets, and their negotiation, are not considered to be part of the control problem. "Natural" cost-functions are considered to be outside the functioning of a control system, so cost-functions are usually not negotiated within the functioning of the controller.
4. *Problems of dealing with information (knowledge?) not considered part of the functioning of existing control systems*. A substantial innovation came when K.-S. Fu started openly talking about the control systems with *recognition in the loop* [74–78]. Recognition[47] processes can affect control processes remarkably, since the vocabulary of the controller depends on the recognition results. Learning controllers are expected to work with a constantly changing vocabulary which is updated as a matter of normal operation. Every change in the vocabulary results in an avalanche of generalizations which can completely alter the structure of information and the result of control computations.
5. *The initial problem formulation and assignment of the goal-subset is also part of the control problem for learning control systems*. As a rule, and not as an exception to the rule, initially the task is posed imprecisely. The description of many functions is imprecise or incomplete, and they are left to be provided by the computer control system. The expected situations of the operation are given to the control engineer by the user in also an approximate way. Frequently the assignment is determined by the very fact of the impossibility to transfer the knowledge of a user to the designer. A problem is assigned because of its intractability, as

[45]The question to be raised is whether to submit this knowledge base of the future machine by supplementing the initial knowledge set before making any trial. A way around this is for the designer to make a mental experiment, and if the result shows that this knowledge is required, then the initial knowledge base must be enhanced.

[46]The question is: What is the general method of improving them?

[47]It is not always clear what is expected to be "recognized in the loop." The simplest answer is "the errors of control." However, the clusters of errors might testify for more interesting and more important phenomena-subjects for recognition, too.

it seems, for the user. The control engineer therefore is faced with designing a substantial part of the future system, from task formulation to the operation of actuators. But often the designer is not familiar with the context of operation. Quite a while back, it was understood that the design of the control process trades off the parameters of a system. Later it became clear that it trades off the model as well as the actual structure of the system. It is becoming clear now that the tradeoff includes the processes of task formulation. Moreover the negotiation of the task-set is part of the design. We can expect that in future systems a continuous process of task negotiation will be part of the computer control system.

10.5.4 Exosystem and Multi-actuation

Information about an exosystem (EXO) often cannot be given in the beginning. This information becomes evident within the reality of EXO, and after recognition, it can be handled properly. EXO information contains facts of different relevance, and these facts are processed in different ways. Some have attention focused on them; others do not and can be considered as a general entity unifying a number of facts. Thus the idea in studying the reality of EXO under different resolutions depends on our interest in the details. Consistency must be provided for the overall system to use the different parts of EXO under different resolutions.

The easiest way of providing this consistency is to consider the EXO as a hierarchy of decomposition in parts or as a multiresolutional hierarchy, which is the same thing (this concept coincides with the *frame* concept in AI). This is how the perceptual subsystem delivers the EXO to the controller. The hierarchy of EXO appears in a way which is quite similar to the way in which the hierarchy of PL appears: The bulk of the information (knowledge?) about the world, or the *world description* should be decomposed (tessellated) into a hierarchical multiresolutional set of parts. The mechanism of the new knowledge acquisition is provided by a system of multiple sensors, and their signals must be integrated in order to enrich our knowledge of EXO—in other words, EXO should be constantly *perceived*. Perception of EXO is considered normal system operation; thus learning is a natural procedure that requires recognition of familiar features and objects as well as discovery of unfamiliar entities and concepts.

On the other hand, since most of the systems are *multi-actuator* systems [79], their actuators must be assigned individual controllers, and each controller must be dealt with as an individual control system within its *scope of attention*. At the same time, all this multi-actuator system can be considered as an object of control: a single system is supposed to coordinate the activities of the multiple controllers subsystems. A separate controller is supposed to submit to the *coordinator* a control assignment that is obtained at a higher level of the control hierarchy (lower level of *resolution*), and it deals with the objects, parameters, variables, and controls formulated at a higher level of *generality*.

Thus we come to the hierarchically intelligent controller which has organization, coordination, and hardware control levels as described in by G. Saridis [77, 78, 80]. The *organization level* accepts and interprets the input commands and related feedback from the system, defines the tasks to be executed, and segments it into subtasks in their appropriate order of execution [81]. At the appropriate organization levels appropriate *translation and decision schemata* linguistically implement the desirable functions [81–83].The coordination level receives instructions from the organizer and feedback information for each subtask to be executed and coordinates execution at the lowest level. The lowest-level control process usually involves the execution of a certain motion and requires, besides the knowledge of the mathematical model of the process, the assignment of end conditions and a performance criterion (cost-function) defined by the coordinator [84]. This effectively allows for automatic generation of control strategies [85].

10.6 LEARNING AND BEHAVIOR GENERATION: CONSTRUCTING AND USING MR REPRESENTATION AND GOAL HIERARCHIES

We now introduce an approach for the analysis of learning systems as behavior-generating processes. This approach can be applied to both living and artificial intelligent systems. We will introduce an automaton with joint learning and behavior generation that comprises a class of learning algorithms for unsupervised learning. The learning algorithms employ recursive processes of grouping, focusing of attention, and combinatorial search that lead to the top-down and bottom-up computations with growing multiresolutional hierarchy of knowledge representation. In generating behavior, learning searches for an appropriate set of rules (control law K) and thus for a preferable motion trajectory. The planning/control processes are based on and include further development of multiresolutional knowledge representation and lead to a temporal evolution of the system. Analysis of this automaton demonstrates the benefits of a joint analysis of spatial and temporal processes of behavior and learning.

10.6.1 Learning from Multiple Experiences

In 1985 a "baby-robot" was introduced in [40]. More theoretical discussion followed in 1986 [86]. The idea of a baby-robot helped demonstrate the processes of unsupervised learning which is associated with the joint evolution of behavior and knowledge incorporated within a system. It is likely that learning from multiple experiences cannot be separated from the architecture of acquired knowledge that shapes desirable behavior. However, we will not focus on an analysis of the evolution of living creatures, nor on problems of knowledge acquisition and the growth of the intelligence that result in this evolution. All survival-oriented matters are beyond the discussion in this book. Rather, we want to introduce a formal system—a learning automaton—that can evolve as a result of its own experiences. It has the faculties required to obtain and use knowledge: sensors, subsystems for storing and organizing information, and subsystems for generating commands. It has a goal that is prescribed externally. It also has faculties to produce its own behavior: actuators for changing the world. However, initially it has neither a world model nor any rules to achieve its goal. This knowledge has to be learned, and we are interested in understanding how it happens.

The notion that experiences influence behavior is connected with the idea of "goodness." The automaton has to be capable of evaluating the desirability of its own behavior and aspects of this behavior (its components which are elementary experiences) as discrete events. In this way we can begin to judge how beneficial the "thinking" processes of an automaton are, since the output behavior is shaped by actions that emerge as a result of plans: strings of decisions on how to achieve goal-states, on the required actions leading to the goal-states, and on how they should be organized into strings. This process of finding the string of desirable states and actions leading to them was introduced in Chapter 8 in our discussion of *planning*. Planning would be impossible without the ability to judge the degree of "goodness": the desirability of states and actions are acquired in advance. *Execution* as the process of generating and applying proper commands to execute the planned trajectory and compensate for errors (see Chapter 9) is the generator of "experiences." Both planning and execution are a part of control system functioning. The process of acquiring the relevant information and processing it so that the appropriate behavior can be generated is called learning. Planning/control and learning are complementary computational procedures, merging into a joint process of intelligent control.

The process of planning starts with focusing attention on a subset of representation: the automaton must select a "subset of interest": the initial representation of the world that will be explored or the subspace with its boundaries. The latter is fragmented ("discretized") into tessellata that determine the resolution of maps, or their granularity (scale). Recall from Chapter 2 that combinatorial search is performed as a procedure of constructing all possible strings of consecutive strings of states and choosing one of them (the minimum cost string). Focusing of attention is instrumental in limiting the multiplicity of all possible strings formed out of the space tessellata at a particular level of resolution.

GFS AS A TOOL OF LEARNING

Grouping the tessellata in a variety of feasible strings and the subsequent selection of the "best" string allows for focusing attention on the vicinity of the chosen string. This initiates the process of "zooming in" and leads to the evolution of the behavior by consecutive refinement of the planning process top-down. The latter is performed by limiting the scope via constructing an envelope for planning at higher resolution around the minimum cost string. This envelope is submitted to the next level of resolution where the next cycle of computation starts. Focusing of attention presumes proper distribution of nodes in the state-space so that no unnecessary search be performed. This is how combinatorial search forms the alternatives at all levels of resolution. Grouping, focusing attention and combinatorial search together can be considered *generalization* which in the case of our learning automaton is a process of generating plans at a particular level of resolution and generation maps for the subsequent planning at the higher level of resolution. The intelligent properties of the learning automaton are produced by joint functioning of grouping, focusing of attention and combinatorial search (GFACS). These operations allow the automaton to "behave" as a result of its planning decisions and actions of the actuators, create experiences, evaluate them and learn from them. The overall automaton[48] can be represented using the concept of elementary loop of functioning as shown in Chapter 2 [87–89].

A novel mechanism of unsupervised learning, with combinatorial enhancement and generalization, is explored in this section. It is demonstrated that learning leads to multiresolutional representation, a hierarchy of subgoals, and a hierarchy of control commands. The learning system provides an explanation of evolutionary processes both in technical objects and in living creatures. We will imply that the evolution of living creatures is a special case of the evolution of knowledge representation and their behavior generation.

The evolution process of the learning automaton can be characterized by a gradual sophistication of its knowledge and behavior. This gradualness is not monotonic in all its derivatives. The evolution of knowledge is rather "punctuated"[49] by steps that emerge when a collection of hypotheses gets transformed into a rule. The evolution of behavior

[48]"Learning automaton" is based on concepts of ELF and GFACS and subsumes the concept of a "learning agant" known in the domain of AI and consisting of a "problem generator," a "performance element," a "critic," and a "learning element."

[49]S. J. Gould and N. Eldredge, Punctuated equilibria: the tempo and the mode of evolution reconsidered. *Paleobiology*, v3, 1977; pp 115–151.

has similar steps due to the emergence of new skills, which happens also in different moments in time. We intend to use our model for processes of early as well as mature learning.

10.6.2 Algorithms of Unsupervised Learning

Learning Automaton

An automaton is a state machine that generates outputs by using its transition function and state-output function tabulated in advance for all possible situations. Automata are capable of demonstrating "reactive behaviors" according to the prescriptions stored in their transition and output functions. A class of learning automata is introduced which are state machines whose transition and state-output functions are updated and modified based on prior experiences stored in the memory and transformed into sets of rules. Thus a meta-control structure is presumed that modifies the transition and output functions. The transition and output functions of the learning automata have open lists of rules. New rules can be added to these lists. The aim of a learning automaton is to acquire rules that will increase the total reward in achieving the goal, or to increase the value of an objective criterion of optimality associated with the goal achievement. This concept is similar to the one described in [90]. We will equip the learning automaton with a new capability to synthesize its output by combining together previously stored rules in search of the most appropriate behavior. Thus, in addition to reactive responses, our learning automata will demonstrate the skill of deliberation, or planning.

This new type of automaton will be called *learning and planning automaton* (LPA). LPA is presumed to have a learning system LS that allows for enriching both the transition and the output functions and a planning system which is a part of the mechanism of behavior generation (BG). We will demonstrate that as the system of rules develops, it becomes hierarchical and creates corresponding developments in the input and output vocabularies of LPA. This is equivalent to formation of the hierarchy of automata as a result of the evolution of a single learning automaton. Operators of LS are equipped by minimum initial set of tools, or "bootstrap knowledge." This includes the ability to form strings, to construct hypothetical implications, and to infer tautologies.

The learning system (LS) can be defined as a system of acquisition of experiential information, transformation of the information of experiences into rules of action, derivation of new concepts, and organization of these concepts into knowledge and decision-making structures suitable for achieving the goal.

In order to apply the results of learning, each LPA is equipped with a set of actuators that implements control commands and creates the output of learning automaton—or changes in the world. The changes are measured by the sensors. A set of sensors is the only source of information for LPA about the state of the world (situation.) The automaton is also equipped by subsystems of sensory processing and world model that allow for interpreting the input from sensors (see ELF in Chapter 2). Each of these modules has been discussed in the previous chapters. We will treat all modules of LPA as discrete event systems.

Many questions may arise at this moment: Is this learning automaton based on a Mealy or a Moore automaton? Is this a Markov controller? How do we "equip" it with actuators and sensors? Is the output of the automaton submitted to the actuators, or *are* the actuators the output of the automaton? Is the output of sensors submitted at the input of LPA, or *is* it the input of LPA? What does it mean to equip the automaton with actuators and sensors? Does it mean that the world model emerges within the automaton? How does its transfer function look, and what are its states?

The learning system can judge the truthfulness of the sensor's (input) information only by the action results (behavior) which are undertaken to achieve the goal. Therefore LPA should be able to evaluate the results of its behavior. The subsystem of behavior generation (BG) contains transition function and output function [87, 88, 91] (see Chapter 7). This section is a further development of the approach presented in [93, 94].

Experiences

SITUATION

We use the term *situation* to describe the state of the world represented as a set of n sensors outputs which arrive at a moment of time i:

$$S_i = \{s_{ji}\}, \qquad j = 1, \dots, n, \ i = 1, \dots, T, \tag{10.1}$$

where s_{ji} is a numerical (or logical) value for the j sensor at the time i.

Different levels of resolution will have different sensors, and the representation of the situation can be different depending on the type of representation used (intervals, NNs, splines, decision trees, etc.). An example using an interval representation could be demonstrated as follows:

$$\{s_{11}1 = [10, 10.1], s_{21} = [5.6, 6.0], s_{31} = [-7.0, -7.0], \\ s_{41} = [15, 15.1], s_{51} = [\text{ON}, \text{ON}]\},$$

where

$$\begin{aligned} s_{11} &= \text{the distance to the target at time step 1}, \\ s_{21} &= \text{the heading at time step 1}, \\ s_{31} &= \text{the angle to target at step 1}, \\ s_{41} &= \text{a laser range distance to the obstacle at step 1}, \\ s_{51} &= \text{a touch sensor.} \end{aligned}$$

In the expression above the interval represents the value of the sensor within time step 1. If, on the other hand, we are representing with NNs, the situation is written as

$$\begin{matrix} S_{11} - - - - | \\ S_{21} - - - - | \\ S_{31} - - - - | \\ S_{41} - - - - | \\ S_{51} - - - - | \end{matrix} \quad S_1 - - \text{true/false},$$

where S_1 is an element of a NN that gives a true or false response by thresholding its feedforward output when the sensors values are connected to the input.

The action is a set of m action outputs generated by the system at moment of time i:

$$A_{i,i+1} = \{a_{ji}\}, \qquad j = 1, \dots, m, \ i = 1, \dots, T, \tag{10.2}$$

where a_{ji} is a numerical value that has been observed as an output of actuator j. There are two possible actions that can be represented in the learning automaton:

1. A set of output values sent to the actuators, for example,

$$\{A_{11} = 10, A_{21} = 15\}$$

where

$$A_{11} = \text{output to the "go forward" actuator at time step 1},$$
$$A_{21} = \text{output to the "rotate" actuator at time step 1}.$$

2. Values of a subgoal situation to be achieved, for example,

$$\{\text{achieve } S_3 = [10, 11]\}.$$

In executing this action, a set of actions of the determined type is found and submitted to the actuators.

Situations can be represented as sets (lists) or as vectors within a particular system of coordinates. Actions are understood as causes of changes that are sensed after the actions are applied. This allows for correlations between these causes and the sensed changes to be recorded, and it introduces the concept of a rule.

GOODNESS OF EXPERIENCES

We use the term *experience* instead of *cause–effect relationship* because we want to underscore the fact that no prior cause–effect knowledge is available. Sometimes the functioning of learning automata is described in terms of the quadruplets: states-before-action, states-after-action, actions, and rewards. For example, a "single move" experience can be described as

$$E_{i,i+1} = \{S_i, A_{i,i+1}, S_{i+1}, J_{i,i+1}(G)\}, \qquad i = 1, 2, \ldots, T, \tag{10.3}$$

where S_i is a vector of situation at the time i, $A_{i,i+1}$ is a vector of action applied to S_i, S_{i+1} is a vector of the next situation in which $A_{i,i+1}$ has been applied. In dynamic systems we are interested in longer strings of moves that we regard as experiences. Any catenation of single moves generates a multiple-move experience of the type "situation-action-situation-· · ·-situation." This serves as a basis for the subsequent learning.

$J_{i,i+1}(G)$ is the evaluation of goodness of the action under the goal G. It is the final situation to be achieved as a result of the behavior. The overall goal of functioning is presumed to be given to a system externally. From [87–91] we know that goals can emerge also as a result of planning. From [88–89] we know that the process of learning generates subgoals.

It seems reasonable to evaluate experiences not for just one goal but for as many goals as possible. Tentative experiences during the learning process are obtained as a result of random actions; thus they carry equal amount of rule-creating power for each possible goal. By including the goal in the experience set, we abandon the universal part of the information obtained.

An example of an experience is

$$\{\{S_{11} = [10, 10.1], S_{21} - [5.6, 6.0], S_{31} = [-7.0, -7.0],$$
$$S_{41} = [15, 15.1], S_{51} = [\text{ON}, \text{ON}]\},$$
$$\{A_{11} = 10, A_{21} = 15\}\},$$
$$\{S_{21} = [10.1, 10.7], S_{22} = [5.6, 6.0], S_{23} = [-7.0, -7.3],$$

$$S_{24} = [15.1, 16], S_{25} = [\text{ON}, \text{ON}]\}\},$$

$$J_{21,11}(G) = 677.5.$$

The descriptions of "situation-before-action," "action," and "situation-after-action" should be supplemented by evaluation of the "goodness" of the result achieved. Evaluation of goodness is a complicated feat: Even when the simple distance to the goal is the measure of our accomplishments, we want to evaluate the goodness of the move by taking in account the following:

- How much closer is LPA to the goal after the move than before?
- Is there any particular reward associated with the move we performed?
- How expensive was this move (the "negative reward," or "punishment")?
- How much effort will it take to complete the whole assignment after this particular move?
- What will be the sum of all future rewards collected during the motion?

As rewards and punishments are evaluated, LPA applies a general comparative attitude toward present and future rewards: Are the future rewards discounted when taken in consideration "now," and what is the value of discount, and so on?

GOODNESS DETERMINES REWARD

$J(G)$ is a "goodness" (or reward) attained under the goal G as a result of a single move. It is presumed that any valued experience is associated with a certain measure of "goodness" $J_{i,i+j}(G)$. It can be interpreted as a reward for the pursuit of the goal G. In a multiple-move experience the value of reward is determined not only by the initial and final situations but by the collection of moves that brought LPA from the initial to the final situation. Both single-move and multiple-move experiences are regarded as a unit of experiences. A goal is imposed upon the experiences, and the goodness is calculated for each one of them. The value of the reward will be used for the subsequent process of hypotheses evaluation, generation, and selection. Ultimately this determines the rules necessary for "survival."

Rules

A rule can be obtained as a result of transforming the cause–effect relations discovered within repetitive experiences. From an experience

$$E_{i,i+1} = \{S_i, A_{i,i+1}, S_{i+1}, J_{i,i+1}(G)\}, \qquad i = 1, 2, \ldots, T,$$

two types of rule hypotheses can be formally deduced with no additional information required:

Type CR (Control Rules):

IF present situation S_i and goal G, THEN the action required is $A_{i,i+1}$.

Type ER (Event Rules):

IF present situation S_i and action $A_{i,i+1}$, THEN the new situation is S_2.

In the literature, one can find two extreme cases related to these two rules:

1. *Systems that contain only CRs*. Some examples of the behavior that these systems create can be appreciated in multiagent behavior. The behavior is almost always generated by a fixed set of CRs. Since there are no ERs, which would allow many possible states to be contemplated, no planning can be conducted.)
2. *Systems that contain only ERs*. This is the case of a system employing only path-planning algorithms. The algorithms find the appropriate behavior by testing with random points the model dictated by the ERs. Any error in the model will cause the system to replan, since there are no reactive rules that give a recommendation of what to do if the required rule is not within the rule base.

We will discuss an LPA where both CRs and ERs are being learned. The ERs are used for finding the next state in the planning algorithm, and the CRs are used to guide the search algorithm toward actions that were more successful in the past.

There are some other important differences between ERs and CRs. ERs represent knowledge related to the properties of an environment; they are not related to any particular goal. However, they can be used to create behavior for different goals by testing the model and searching for the appropriate trajectory leading toward the required goal. On the other hand, CRs represent knowledge about the ability of LPA to move within a particular environment toward the particular goal. CRs are valid for a specific goal, and they can be easily used for creating behavior, without testing the environment, only by reproducing what was successful in the past.

Some example of the two rules are given for illustration:

CONTROL RULES

CR Type of Rules. CR rules can be constructed as follows:

1. *Using intervals*. If $S_2 = [4.5, 4.8]$ and $S_3 = [4.3, 4.7]$ and goal $G = \{S_4 = [1.1, 1.2]\}$, then $A_{23}[10]$, which can be interpreted as if the heading and the angle to the target were "equal to α and β correspondingly," and then DO[A_{23}], or "turn until 'heading' is equal to 'angle-to-target.' "
2. *Using an NN*. The same rule is learned and stored for all $S_2 = S_3$ necessary in the interval representation. The CR rule can be represented as

$$\text{IF } NNSFF_1(S_1, S_2, S_3, \ldots) = 1$$

$$\text{AND GOAL} = \{NNSFF_2(S_1, S_2, S_3, \ldots) \rightarrow \text{THEN } A_{23}[10],$$

where $NNSFF_1$ is a neural network state in the feedforward mode. Another way to do this would be to train a separate NN to learn the mapping between the situation and the recommended action:

$$\text{IF } NNSFF_1(S_1, S_2, S_3, \ldots) = 1$$

$$\text{AND GOAL} = \{NNSFF_2(S_1, S_2, S_3, \ldots)$$

$$\rightarrow \text{THEN } NNRFF_1(S_1, S_2, S_3, \ldots),$$

where $NNRFF_2$ is the NN rule mapped in the feedforward mode which maps the situation recognized in $NNSFF_1$ to a set of actuator commands.

EVENT RULES

ER Type of Rules. ER rules can be constructed as follows:

1. *Using intervals.* If $S_{2i-1} = [4.5, 4.8]$ and $S_{3i} = [4.3, 4.7]$ and $A_{1i} = 1.1$, then $S_{2i-1} = [5.1, 5.2]$, where i is a number of each time step.
2. *Using NNs.* The same ER can be represented:

$$NNSFF_1(S_{1i-1}, S_{2i-1}, S_{3i-1}, \ldots, S_{1i}, S_{2i}, S_{3i}, A_{1i}, A_{2i}, \ldots) = S_{i+1}.$$

In cases where it is not possible to train just one NN to represent this mapping, the space can be broken up into smaller parts so that a group of NNs can accurately represent the required mapping.

To find new rules, an operation of inductive generalization is applied as follows: Clusters of similar experiences are created. Then the hypotheses of control rules and event rules are postulated by transforming (inverting) the generalized experiences. These rule hypotheses are used by the automaton as provisional rules for behavior generation. A frequency of valuable rewards estimates the validity of the newly created rules and confirms their correctness if the frequency is high.

After the rules are confirmed, some of their parts that emerged as a result of generalization may not belong to the initial list of input and output vocabularies (VI and VO.) These parts are considered to be new concepts. A concept is understood as a label for an entity that is part of a rule (obtained recursively as a cluster or group of higher-resolution inverted experiences). All new concepts are obtained from two available classes of rules (see Fig. 10.5). Situations and actions are concepts, and so are clusters of situations, actions, and their components. They are organized into an "ontology" base that stores them together with their relationships.

The main reason for creating concepts is the need to remember concept-clusters of higher resolution (in this case, the clusters of sensor readings) and to use them in generating future behavior efficiently. Clusters are created only because the system can behave more efficiently by drawing on constructed clusters. It is prudent to assume that intelligent creatures (e.g., humans) also create clusters and assign labels to them because this helps them to change their behavior in an efficient manner.

Taking the idea of "labeled clusters" as a behavior base, it becomes easier to understand that concepts are created as a result of interaction between the learning creature and the environment. All interactions lead to new concepts, but obviously different environments and interactions with different goals create different sets of concepts (the Sapir-Wharf hypothesis). Some experiences can be collected from random actions by actuators (no particular goal is stated). Let us assume that experiences achieve different

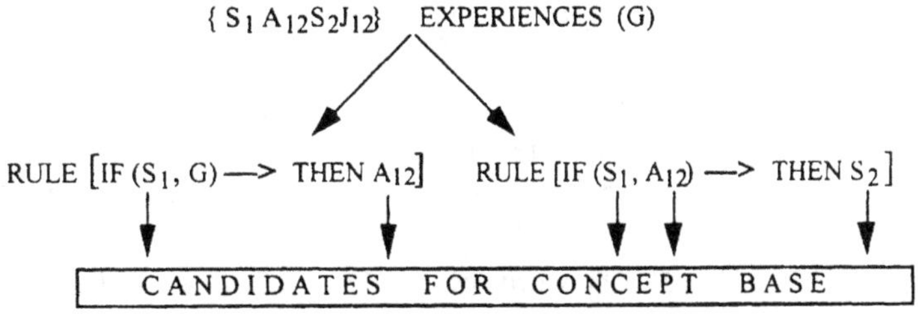

Figure 10.5. Experiences → rules → concepts.

goodness measures based on a goal at $S_1 = 0$. The best experiences are selected, and the sets of experiences are stored as follows:

S_1	S_2	S_3	S_4	S_5	A_1	A_2
11	35.6	*35.3*	**120**	**off**	0	**10.0**
98	**35.3**	**35.9**	13	**off**	**0.1**	**10.0**
58	**35.8**	35.8	12	**off**	**−0.2**	9.8
235	35.5	35.5	**138**	**off**	0.3	9.9

The values with a tendency to cluster are printed in bold letters. The following hypothesis can be extracted from the table:

$$\text{IF } S_2 = [35.3, 35.8] \text{ and } S_3 = [35.3, 35.9] \text{ and } S_5 = \text{OFF} \rightarrow$$

$$\text{THEN } A_1 = [-0.2, 0.1] \text{ and } A_2 = [9.8.10.0].$$

Note that in S_1 there are no clusters to be found, and in S_3 the cluster has shifted. The reason for this is that the values in this example cover most of the range. The situation would be different if the clustering threshold were different. Here the triplet $\{S_2 = [35.3, 35.8], S_3 = [35.3, 35.9], S_5 = \text{OFF}\}$ and the couple $\{A_1 = [-0.2, 0.1], A_2 = [9.8.10.0]\}$ are the candidates for new concepts.

The new concepts, obtained from the generalized experiences as a result of their transformation into rules, are new object-labels and action-labels that were not present in the previously defined input and output vocabularies. The new labels are thus words of new vocabularies at a lower, more generalized level of resolution. They describe clustered experiences: The phenomena are related to groups of initial units of experiences; these groups are regarded now as new entities belonging to a lower level of resolution. In this way a single sensor value becomes a group of sensor values which are unified by the fact that a particular action or string of actions can lead to highly rewarded results.

Goodness, or reward (J), is computed in a different way depending on the particular application. For example, it can be computed as a difference between increments of cost $C_{i,i+1}$ accumulated for the interval from state i to state $(i+1)$, and $C_{i-1,i}$ accumulated for the interval from state $(i-1)$ to state i. The following sequence of activities can be explicated from the definition of learning:

PROCESS OF LEARNING

Step 1. Experiences $\{E_i\}$ are collected and stored in memory.

Step 2. Experiences are compared, the values of similarity (resemblance) are determined, and clusters are formed by the virtue of resemblance.

Step 3. Clusters of experiences that already have proved their cause–effect relationships are transformed into rules, and control rules and events rules are separated.

Step 4. The clusters of situations and actions that are part of the newly created rules are stored as concepts of lower resolution; then the growth of the concept base begins. New concepts emerge as a result of clustering. These concepts are labeled and receive the status of a new word.

Step 5. The words of the clusters form a vocabulary. This vocabulary is regarded as a lower-resolution vocabulary. Thus, in addition to the previous vocabularies VI and VO, two new VI and VO vocabularies are obtained.

Step 6. The new experiences are stored in parallel in two new vocabularies: the original VI and VO, and another formed by the newly created concepts VI and VO .

Step 7. The process of consecutive operations {collection of experience → hypothesis formation → generation of rules → concepts emergence} is repeated each time as a new experience arrives.

Step 8. As vocabularies VI and VO grow, they allow the system to represent and control its functioning by using the new words. Thus, new generalized experiences are recorded for generalized situations and actions. The sequence of steps 1 through 6 is repeated in building up a new lower level of representation.

This algorithm describes a recursive process that leads to the development of a multiresolutional system of world representation and a multiresolutional system of action rules, both of which are acquired as a result of learning. The described learning process applies to the experiences focusing of attention, search, and grouping using *existing* features of resemblance. However, when this is not satisfactory for successful learning, the third component is simplified: No new combinations are formed. The formation of new combinations is actually a form of *combinatorial enhancement*.

Learning with generalization allows for experiences to be used in a very efficient way. Multiple clustered experiences are treated as group phenomena. Rules generated by them are group rules. When the number of group rules becomes large, the algorithm of generalization can be applied again, so other group rules emerge that have even lower resolution.

Nevertheless consecutive generalization is just a tool of reducing complexity. In order to receive innovations, *new combinations* must be introduced for generating words beyond the existing experiences. Similar mechanism have proved to be effective for design purposes [95].

Combinatorial Enhancement: Searching for Hidden Implications

Each experience can testify only on behalf of some part of an overall situation. This is the idea behind the combinatorial enhancement of rules and concepts introduced in the theory of learning to explain enhanced actions and situations.

Alone, the available sensor information is not necessarily a good basis for generating a hypothesis. For example, take the autonomous mobile robot which must learn how to act in a particular environment. For any given movement, a single actuator command cannot elicit the proper response. A combined command (steering + propulsion) must be assigned. The sensor's reading of an angle ["heading," $-a$] or [value of "angle to the goal," $-s$] cannot alone be an antecedent in a rule on what to do. However, their difference $(-a-(-)s)$ is a perfect variable of control.

Without these two persuasive experiences of learning, the evolution of living creatures and the evolution of knowledge would be impaired, since no combinatorial enhancement would be possible. The factors considered critical in stimulating the evolution of living creatures are reproduction, mutation, competition, and selection [94]. These factors can be applied to knowledge in a similar way. There is the reproduction of statements formed from experiences. The generalization of the best results with the highest values of rewards involve competition and selection (see Section 2.3). The process of search enables a combinatorial enhancement that is the direct analogue of mutations in species. Likewise there is crossover and the adaptation of mechanisms in forming new entities along with some element of randomization.

We will be able to explore only the most primitive combinations, to which we will apply arithmetic rules for all possible couples. We will call on formations of combinations of available data for consecutive searching and grouping combinatorial enhancement. For example, if measurable values v_1 and v_2 are known at the beginning of an experience, neither v_1 nor v_2 might be indicative of the need to use a particular action, while the value $v_2 - v_1$ or $v_2 + v_1$, or some other combination, might be important in interpreting the results of measurements and in constructing meaningful clusters.

The enhancement takes as input a set of values and gives as output a combination among these values:

$$\text{Comb}\{v_1, v_2, v_3, \ldots, v_n\} = \{v_i, (v_i \pm v_j)\}, \qquad i = 1, \ldots, n, \ j = 1, \ldots, n, \ i \neq j. \tag{10.4}$$

This operation creates a new set containing the initial set and all possible combinations of the elements of the initial set. In (10.4) only combinations of additions and subtractions are illustrated. The creation of classes of situation and action introduces two challenges:

1. A suitable language must be found that can accurately represent the classes that the intelligent system is trying to learn. This representation must be compatible with the behavior generation process and must save space (since we cannot store all experiences) and save time in the computation (the generator of behavior must use these classes for finding the control).
2. A suitable algorithm must be found that can learn efficiently these classes and store them in the defined form. The discovery of classes entails the creation of boundaries that separate the entities in the class from the rest of the set (from other classes). Different approaches have been found for creating these boundaries, from the introduction of simple intervals by thresholding or constructing hypercubes to using neural networks.

The following heuristic approaches are used to perform the tasks of creating boundaries:

FINDING BOUNDARIES OF CLASSES

1. *Use of intervals (by tessellating the space).* This is probably the simplest way of discovering and storing the classes. There are many methods for finding these intervals. The simplest is to look at the situation dimension by dimension and locate places without good experiences; then use these spaces to separate classes. An extension of this process is to combine sensor values to "enhance" the representation of the experience. For example, by looking at the addition or subtraction of two dimensions, we are allowing the system to be more flexible in finding the shapes of boundaries between adjacent classes. Another good method of finding intervals is described by Quinlan in [96]. But a disadvantage of his method is that if the class does not fit the hypercube shape imposed by his algorithms, the class gets divided into smaller classes occupying more space and causing the behavior generator to spend more time in finding the control.
2. *Closest neighbor algorithm.* This is a common method of clustering. The algorithm represents its classes in hyperspheres. The algorithm has several drawbacks: An initial (desired) number of clusters must be specified, and classes that do not fit

the spherical shape that the algorithm imposes wind up divided into many classes. Again, we end up occupying more space than necessary.

3. *Neural networks (NNs).* Neural networks can be trained (supervised training) for a complete set of experiences, and they can be trained so that the output distinguishes the inside from the outside boundary. The advantage of NNs is that given a sufficient number of neurons, there can be learned a variety of decision boundaries. The disadvantages are that it may take a long time to train sequential computers, and the results of the learned classification may be hard to interpret (see Section 10.6).

An enhanced representation is constructed by considering enhanced vectors of the situations and actions applied. For example, if the vector of a situation is built upon readings of three sensors $S = \{s_1, s_2, s_3\}$, the enhanced vector of the situation will include nine coordinates $\{s_1, s_2, s_3, (s_1+s_2), (s_1+s_3), (s_2+s_3), (s_1-s_2), (s_1-s_3), (s_2-s_3)\}$. Similar change will occur to the vector of actions. Instead of $A = \{a_1, a_2, a_3\}$, a new vector will emerge: $\{a_1, a_2, a_3, (a_1+a_2), (a_1+a_3), (a_2+a_3), (a_1-a_2), (a_1-a_3), (a_2-a_3)\}$. It would be prudent to consider also a new hybrid vector $\{(a_1+s_1), (a_1+s_2), (a_1+s_3), (a_2+s_1), (a_2+s_2), \ldots, (a_2-s_3), (a_3-s_1), (a_3-s_2), (a_3-s_3)\}$. However, in the simple examples that we have used from robotics, it was hard to come up with physical interpretation of every hybrid combination.

Thus the enhanced situation is a set formed by a unique situation, taking into account all possible combinations of components and all possible combinations between components of the action A and the situation S. Each experience presented by (10.3), or a string of single-move experiences (10.3), must be stored together with the synthesized enhanced situation S_{Enh} and enhanced action A_{Enh}. "Enhancement" is intrinsically linked with the "augmenting rules" (see [65]).

Algorithm of Inductive Generalization

In addition to forming clusters typical for any generalization, inductive generalization uses information about dynamic properties of the statistics of cluster formation. The maxim of inductive generalization claims that multiple occurrences of similar experiences testify for the existence of a rule if most of these occurrences have the same (or similar) explanation of the causes [98]. If the number of occurrences is not statistically persuasive, then we can talk about the case of hypothesis generation by means of abductive generalization. In both cases it is important to account for the list of attributes/variables of a situation and also for the relations among them [99].

The process of unsupervised learning employs the algorithm of generalization which presumes a multiple iteration of the triplet (1) focusing of attention on the subset of experiences, (2) searching among them, and (3) their combining the experiences until a grouping can be made and this cluster of similar units can be described by a single generalized hypothesis [95].

This hypothesis is then stored in the database of hypotheses. If the subsequent experiences confirm the hypothesis, it becomes stronger. This is not a trivial operation. Automata with operation of generalization have not been previously discussed in the literature on learning. The operation of generalization includes steps 1 through 4 of the algorithm of learning.

The stored experiences and the goal are both inputs to the process of generalization only in the beginning. Then a third input is made to the process of generalization; this is information from the database of hypotheses for behavior generation, which is not

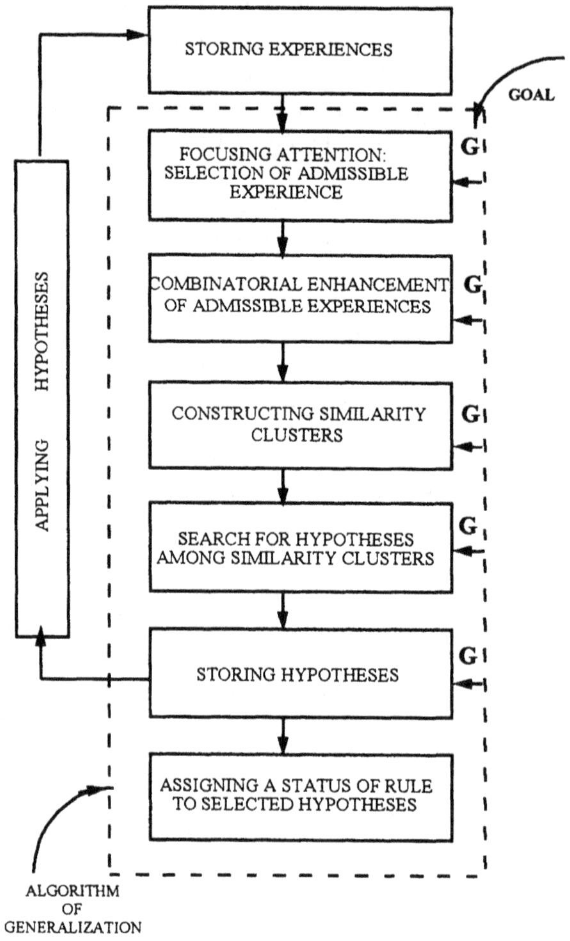

Figure 10.6. The generalization algorithm of learning.

included into Figure 10.6. This database is useful for the algorithm of generalization, and it has two major functions:

1. It allows for information lost in erased experiences to be restored. The hypotheses included in the database are generalizations (or compressions) of lost units of information.
2. It does not allow for creation of hypotheses already created, or of hypotheses that were created earlier but demonstrated to be not valid. Since the algorithm of learning is applied recursively to its own results, each two adjacent levels of resolution use the generalization applied to the initial information that can be considered a set of experiences and to generalized information that can be called hypotheses. Thus we can distinguish two kinds of generalizations pertaining to the level at which they are applied: generalization from experiences before any hypothesis is available and generalization from hypotheses.

The following factors are characteristic for the process of generalization and should be taken into account in any computer algorithm. They are applicable both for generalizing experiences and hypotheses. We can treat hypotheses as experiences themselves, since hypotheses are validated as a result of their application. So they become subjects for the subsequent generalization which is characterized by the following factors:

1. The most straightforward method of clustering is based on extraction from the database of two important subsets: admissible and nonadmissible. Admissible experiences are "good," while nonadmissible are "bad" with respect to a given goal which is determined by values of goodness/reward the experiences deliver. The degree of goodness which determines the threshold of clustering can vary. It will determine the productivity of learning and eventually the success of functioning.
2. The admissible set is used to generate hypotheses of rules that prescribe "what to do," while the nonadmissible set generates hypotheses with a warning content: "this should not be done" in this situation.
3. The representation of each experience is enhanced by combining components that can be validated as new concepts. In effect, enhancement acts as "imagination" in the overall process of generalization. Its role is similar to that of behavior generation which creates strings of anticipated trajectories in the subsystem (see Chapter 7).
4. Searching is performed to determine groups of similarity among the enhanced experiences, and the clusters are created.
5. Each of the clusters of similar enhanced experiences is considered to be a candidate for becoming a rule hypothesis.
6. Experiences are tested, and the results of testing enter the base of experiences.
7. Hypotheses are validated by statistics of their use and then enter the base of hypotheses.

All manipulations with sensor/actuator values are performed after the values are normalized. The reason for normalization comes from the fact that different sensors and actuators work over different ranges. Thus normalization is a function that maps a value from a particular interval into the normalized value interval [0–1]. De-normalization is the inverse function of normalization. Re-normalization is applied when a new value arrives from outside the existing interval. It is necessary to re-normalize all the values previously normalized with the old interval.

10.6.3 Evolution of Knowledge

The algorithm of generalization creates new hypotheses, rules, and concepts that become new words in vocabularies. The same algorithm is applied to its own results. Now the previously generalized experiences are generalized again, and hypotheses and rules of "lower resolution" are obtained. This evolution of acquired knowledge is illustrated in Figure 10.7.

As the figure shows, the joint system of knowledge representation and behavior generation converges to a multiresolutional one. This convergence is completed only after many new concepts are obtained and the consecutive generalization processes run a sufficient number of times. Then levels are formed with the clustering of multiple outcomes of these consecutive generalizations.

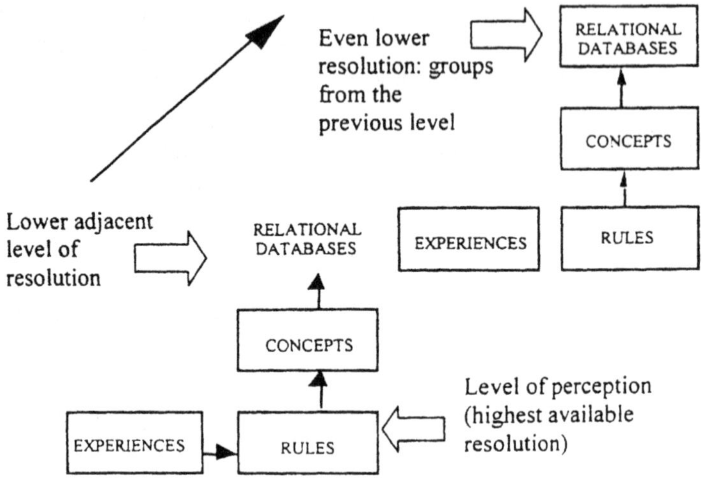

Figure 10.7. Evolution of acquired knowledge.

10.6.4 Focusing of Attention

This procedure is applied to the complete database of experiences. It outputs the set of admissible experiences. The procedure of attention focusing narrows down the bulk of available data by using the most general similarity feature, their "goodness" or "badness."

There are different options in selecting experiences suitable for generalization:

1. Select good/bad experiences that are close to each other in their value of goodness:
   (a) Experiences with the value of goodness higher/lower than a certain threshold.
   (b) Experiences that are part of a particular top/bottom fraction of best/worst experiences (percentage threshold).
   (c) Experiences that form a group of n best/worst experiences (quantity threshold).
2. Select good/bad experiences that are close to the current situation:
   (a) Experiences that are closer than a certain threshold to the current situation.
   (b) Experiences that are the part of some top/bottom fraction of closer experiences.
   (c) Experiences that form a group of n closest experiences.
3. Intermediate strategies are possible. For example, if option 1 is applied and doesn't find any rules with recommendations about the current situation, then option 2 can be applied. Option 2 can be used with different thresholds. If a large value of a threshold of closeness is chosen and no rules are found, then for the current situation a smaller threshold should be used. If no rule has been found after a few iterations with option 2, then the learning algorithm should collect more experiences for the current situation.

When the database of experiences is large, the percentage threshold could choose too many experiences. A reasonable approach is to select the n best experiences. In this way setting a minimum value of goodness can be avoided, while retaining the ability to set the maximum amount of computational burden for the learning algorithm.

At the stage of combinatorial enhancement, attention focusing is applied by using some selected experiences to create enhanced situations and others to create enhanced actions. In this book only addition and subtraction were mentioned for constructing enhanced situations and actions. Other operations can be used, and they will generate larger sets of rules.

10.6.5 Formation of Similarity Clusters

The process of grouping previously chosen good/bad experiences into clusters of similarity takes as an input the sets of experiences and enhanced experiences created in the previous steps, and it finds inner clusters among them. As shown in Figure 10.8, the process inputs a set of groups of selected experiences and outputs clusters.

Clustering allows for the reduction of computational complexity for the following reasons:

- The system cannot store all particular experiences; the creation of groups allows for the compression of data.
- Many experiences are ambivalent; the system needs only those experiences that can imply a set of actions or restrictions.

The following requirements must be satisfied for the set of experiences used by clustering:

1. There are no repeated experiences in any of the output clusters.
2. Every experience at the inputs is included to one of the output clusters; that is, the cluster-forming algorithm does not eliminate any experience.
3. No new experiences are generated by the clustering algorithm.

A comprehensive arsenal of theoretical tools for generating clusters can be found in [100]. We explore two simple approaches for formation of classes: the "jump" approach and the closest neighbor approach. More sophisticated approaches are known [101, 102]. We can expect that in choosing the proper algorithm of clustering, all the many factors present in the description of the environment within which the learning system is functioning are taken into account.

The "jump" approach is similar to coordinate-by-coordinate approach introduced in [98]. Instead of "sparseness," we prefer to use the term "density of good (or bad) experiences." Separating clusters by using the jump threshold leads to a large percentage of meaningless recommendations if clustering is performed on all experiences. However,

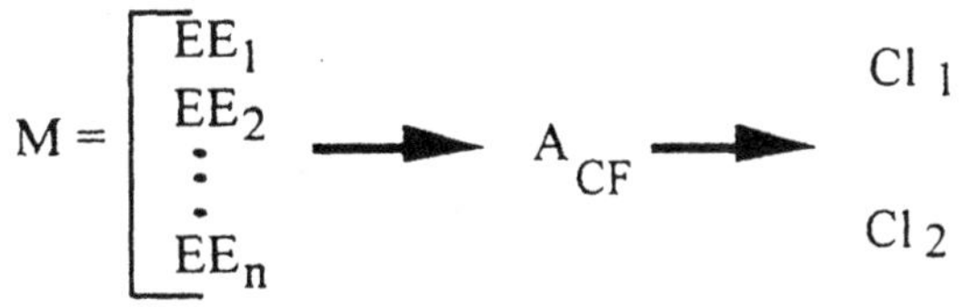

Figure 10.8. Class-forming algorithm.

in the initial set of experiences, only the good (or bad) ones are represented. So we find jumps because of inner cause–effect correlations. Only certain actions can be applied from a repertoire of possible actions. These actions are applied in certain clusters of situations to give acceptable (or unacceptable) values of goodness.

In this way the algorithm will find rules of action only for unequivocally advantageous (or totally unacceptable) outcomes that generate positive and/or negative rules. If no applicable rules are found in a situation, then a new goal must be declared. This assigns a situation for which a good rule is found as a subgoal. Under this subgoal a rule is found or again a new subgoal is declared.

The jump threshold is a measure of density of good experiences in the domain of a particular variable because it separates the classes by finding domains of space that separate places where the good examples are situated densely. Most of the clusters separated by a jump become parts of meaningful rule hypotheses.

Another approach to clustering is merging based on a minimum distance or the closest neighbor method. Its strategy is to merge the most similar pairs of experiences at each step. The criterion used to select the closest experiences is similar to the idea applied in the stepwise optimal hierarchical methods. This makes the smallest possible stepwise increase in the sum of squared errors which is also a very common algorithm for finding clusters.

10.6.6 Searching for Valid Hypotheses among Clusters

It is our belief that each cluster as it is being created is a candidate for becoming a rule of different behavior. We assume that there exists a class of enhanced representation actions A_{Enh} that has already been applied to a class of enhanced representation situations S_{Enh} and that produces values of goodness in the interval from $J_{\min}$ to $J_{\max}$. If we have a situation belonging to a class S_{Enh}, then an action can be determined within class A_{Enh} that provides goodness $J_{\max}$. So each class of the enhanced experiences becomes a hypothesis.

The hypotheses are stored in the database of hypotheses as a tree where each hypothesis is related to its "parent" by a goal. If no hypothesis is found for a certain situation, then the situation in the hypothesis with the closest situation to it becomes a subgoal. This is one more source for the emerging hierarchies of acquired knowledge.

Figure 10.9 shows at the top a parent hypothesis and at the bottom a child hypothesis that takes its goal from the parent's situation. Since the child hypothesis has a new goal (which is the parent's situation), it has a new measure of goodness.

It is not possible to find rules for all situations if only good experiences are assigned to the classification procedure. Other parts of the space will use as a goal the situations with an existing "direct" rule of action toward the original goal. Two types of hypothesis formation are to be considered: based on situations (case A) and based on actions (case B).

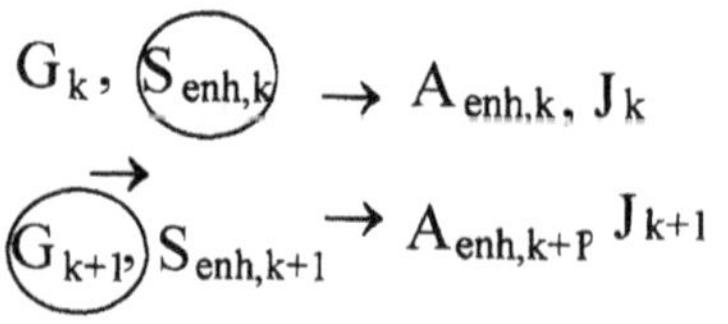

Figure 10.9. Parent and child hypothesis.

Case A

Suppose that two or more hypotheses in the database of hypotheses have the following characteristics:

1. They have the same goal.
2. They have the same enhanced action set.
3. They have different enhanced situation sets.

They have a number of options. If their situations do not intersect, there are the following options:

1. Leave the situations as separate hypotheses.
2. Create a joint hypothesis $\{S_{\mathrm{Enh1}}, S_{\mathrm{Enh2}}\} \rightarrow \{A_{\mathrm{Enh}}\}$.

If they intersect, the two options are:

1. Create a new joint hypothesis with situation $\{S_{\mathrm{Enh1}}, S_{\mathrm{Enh2}}\}$.
2. Create a new joint hypothesis with situation $\{S_{\mathrm{Enh.new}}\}$.

Case B

Suppose that we have two or more hypotheses in the database of hypotheses that have the following characteristics:

1. They have the same goal.
2. They have the same enhanced situation.
3. They have the different enhanced action.

Here it is necessary to merge their actions, which produces two cases:

1. If their actions do not intersect, then they will be left as separate hypotheses.
2. If they do intersect, a new joint hypothesis with action $\{A_{\mathrm{Enh.new}}\}$ is created.

10.6.7 Learning in Behavior Generation

The second process characteristic for the learning automaton is behavior generation (BG): synthesizing the preferable behavior (see Chapter 7 and references [89, 103, 104]). This process is understood as a sequence of top-down planning activities that converges while receiving a set of control commands. The purpose of *learning* is to enable the subsequent process of *planning* to the BG process. We therefore will consider the process of deliberative planning based on exploration of "imaginary alternatives" of motion. Conventional automata are only capable of reactive decisions; they are not capable of "look-ahead" decision-making processes that are typical for deliberative planning. Unlike learning, which develops bottom-up and works via generalization (with subsequent coarsening), the process of planning develops top-down and works via instantiation (with subsequent refinement). The process of planning means choosing desirable behavior by anticipating admissible alternatives among possible behaviors and selecting the best of them by comparing tentative trajectories in the state-space (finding the planned trajectory, PT).

Earlier in this chapter we came to a conclusion that all acquired experiences and generated hypotheses contain knowledge of some reactive rules. For example, "if it is necessary to get to S_2 from S_1, apply A_{12}." By following this rule, LPA reacts by evoking the action A_{12} in response to the need of getting into S_2 from S_1 at the ith level of resolution. Since this rule is applied top-down, it turns out that the space between $S_{1,i-1}$ and $S_{2,i-1}$ contains more nodes $s_{j,i-1}$, where $j = 1, 2, 3, \ldots$, are nodes of the $(i-1)$th level of resolution. The process of selecting any string from them requires deliberation. Thus multiple-step search procedures were recommended in [98].

PT is a trajectory which satisfies the specifications (the latter are determined by the goal). The trajectory consists in a string of adjacent admissible elementary subdomains (or tessellata, tiles of the discretized space) denoted as $\{W_{ij}\}$, where i is the level of resolution and j is the number of the tessellatum in the string $\{W_{ij}\}$. If necessary, the PT can also be represented as a sequence of the subdomain indexes $T_{1\{j(m)\}}$ where m is a number of elementary subdomains in a string $\{m = 1, 2, \ldots, z\}$. We will address only a well-posed problem (wpp) of planning. Any wpp of planning should (1) start by assigning the initial point (SP_{i-1}) and the final point (FP_{i-1}) of the trajectory to be determined at the higher resolution $(i-1)$ within the tessellatum of the resolution under consideration (i) and (2) determined first the feasible trajectory at the lower level of resolution (i).

The PT is called a feasible trajectory if it has an initial point $SP_{i-1} \in \Omega_{i,1}$, a final point $FP_{i-1} \in \Omega_{i,n}$, and all tiles of the string contained within an envelope formed by the feasible trajectory at the lower level of resolution (see Chapter 8). Thus a feasible trajectory for the level i is always represented as a string of tiles for the $i+1$ level of resolution and is a subspace of the i level of resolution. We determine the subspace in which the wpp of planning should be resolved and the optimum trajectory should be found. Indeed, the feasible trajectory determined at the $i+1$ level of resolution becomes an "envelope," a bound domain of space at the i level of resolution.

The recursive definition of the feasible trajectory does not lead to any infinite incomputable computational procedure. As we extend our search for the feasible trajectory to the consecutively lower levels of resolution, we eventually arrive at a level where both the initial point SP_i and the final point FP_i belong to the same tessellatum of the lower level of resolution: $FP_{i-1}, SP_i, FP_i \in \Omega_{i+1}$. This tessellatum is the initial space for a feasible trajectory at level i.

The sequence of feasible trajectories can be considered a nested system of spaces for multiresolutional search. To increase the reliability of searching for optimal trajectory, we increase the envelope of search by surrounding the feasible trajectory by one or more adjacent tiles along its boundary that will determine the "width" of the envelope. This results in a system of enhanced volumes $V_1 \supset V_2 \supset \ldots \supset V_k \supset \ldots \supset V_m$ which we will use later for searching. The optimum trajectory at the i level of resolution is a minimum-cost path in the graph; it is built upon all center-points of the tessellata at the i level of resolution. These tessellata are constructed within the envelope formed around the feasible trajectory found for the $i+1$ level of resolution.

The definitions for feasible and optimum trajectories imply the off-line method that has been introduced to find the best trajectory of motion to be followed by the control system. Search in the state space (S^3 search; see Chapter 8 and [93, 98–107]) is done by synthesizing the feasible trajectories for the i level of resolution and then building the alternatives of possible motion trajectories for the $i-1$ level within the envelope cost space.

A particular volume of the state-space is designated for the subsequent search for a solution. The contraction operation puts constraints on this volume and should be properly justified. To find the optimum path trajectory, we need to reduce the probability (the risk) that contraction eliminates some or all of the opportunities. The following heuristic strategy of contraction is chosen. After the search at the lowest-resolution level is performed, the optimum trajectory is encompassed by an envelope. It is a convex hull that has a width w determined by the context of the problem. Then at the next level of resolution the random points are generated only within this envelope of search. This strategy has often been demonstrated to be effective in practice. However, the problem of consistency of representation under the contraction heuristic has to be addressed in the future.

LPA merges deliberative planning with learning. They are both procedurally kindred and match from the software engineering point of view. They produce and use the same multiresolutional system of representation. Both use the same Ω state-space in which the start and final points, SP and FP, are given. The path from SP to FP is to be found with the final accuracy ρ. The operator of learning $L(\Omega, \rho)$ constructs the system of representation via the creation of generalized maps $M(\Omega, \rho)$ at each level of resolution and obtained by generalization of the higher-resolution map (ρ corresponds to the level of resolution of this map determined by the density of the search-graph).

Even if the learning automaton were not given any concept of "self," this concept would be developed automatically by the algorithm of learning. At the lower-resolution level the deliberation of strings is expected, combined from cells of the higher-resolution level. Concatenation of these strings is possible only via the concept of the "current state." It is possible to demonstrate that this concept will evolve into the concept of "self." The complexity of computations is drastically reduced when both the surrounding objects and the LPA are represented in the same map in some global system of coordinates. Thus the learning automaton is able to put itself in a map.

10.6.8 Evolution of Multiresolutional Learning Automata

Figure 10.10 shows a detailed, enhanced elementary loop of functioning (see Chapter 2 and [87, 88]). This is a symbolic representation of the information processing during various learning activities typical for a system equipped by modules of unsupervised learning (UL) and behavior generation (BG). This configuration remains the same in all goal-oriented cases of automata equipped with the joint ULBG modules. Perception allows for a set of recent experiences to be recorded in symbolic form. In the grouping of the experiences, classes of similarity are discovered. This processs induces hypotheses that explain the similarities, or it instigates new experiences belonging to the same class of similarity. Within the semiotic paradigm, the loop of ULBG is called "a loop of semiosis" [87].

The system is presumed to function under an externally assigned goal. The initial set of experiences (which might be obtained by random actions) is generalized into hypotheses. The hypotheses enter the subsystem of behavior generation as a substitute for the rules. The decision for an action is made, the action is performed, changes in the world occur, and the transducers (sensors) transform them into a form that can be used by perception. The long and complicated process of moving from signs to meaning starts again. Now the enhanced set of experiences brings about another set of hypotheses that can confirm or refute the tested hypotheses. This is when the symbol grounding occurs.

After multiple tests, if the hypotheses cross the threshold of "trustworthiness," a new rule is created. A rule (or a set of rules) within a context is considered to be "a theory." At each development step the unit under consideration undergoes a comparison

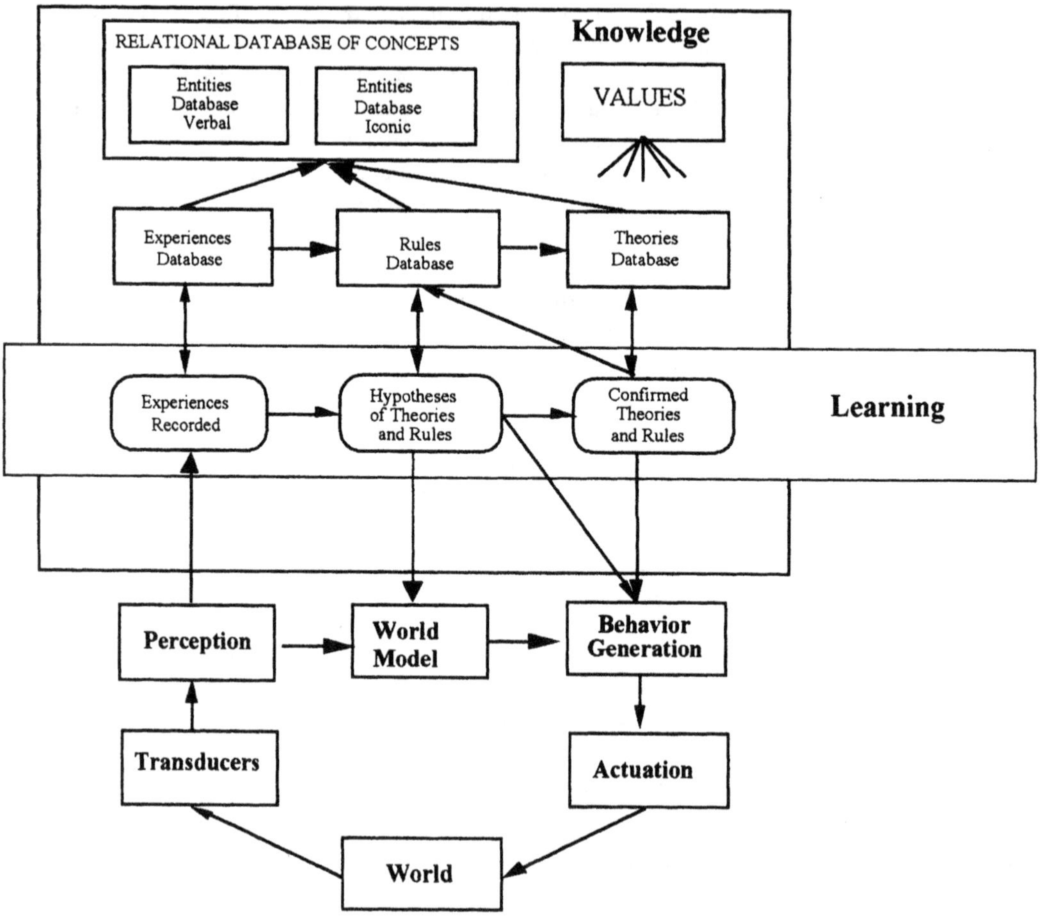

Figure 10.10. Functioning of learning automaton: ELF with learning.

with other kindred units confined in corresponding databases (of experiences, of rules, and of theories.) Then the symbols tentatively assigned to some "unities," "entities," or "concepts" enter their place within the database of concepts (which is a relational network of symbols).

Rules (or the hypotheses that will become rules) are formed when experiences cluster together unified by their similarity. For the prior state S_1 the applied action A_{12} leads to the emergence of the new state S_2; the value of reward J_{12} is the result. After gathering a sufficient number of the experiences and generalizing them properly, the rules of the following form can be constructed: IF the value J is desired upon achievement of the goal-state SG from the present state S_1, THEN the action A_1G should be applied.

An interesting and unique feature of generating rules is that every component of a rule is a generalized component of experience. This means that to obtain a component of a rule, several similar components of experiences must be grouped together into a class, a cluster. This requires applying a set of procedures using the triplet of grouping, focusing of attention, and combinatorial search. The label attached to this cluster signifies the process and the result of generalization. The premises behind the process of generalization can be different. But the result is always the same: the creation of a new object for the lower level of resolution.

For example, let G_i symbolize the phenomenon of generalization upon i similar experiences ($i = 1, 2, \ldots, n$), then

$$[\text{zone of states } S \text{ with } J_1 < J < J_2] \rightarrow G_i\{S_i, J_1 < J_i < J_2\}. \quad (10.5)$$

$$[\text{action } A \text{ to achieve a desired zone}] \rightarrow G_i\{S_i, A_{i,i+1}\}. \quad (10.6)$$

Only the desired state is not subject to generalization. It is always individual, pertaining to a concrete system and problem.

10.6.9 Further Research in LPA

The theoretical analysis presented in this chapter is based on the experimental results in [86, 92, 93]. Baby-robot was able to learn how to reach an arbitrarily situated goal only after implementing the algorithm of generalization with combinatorial enhancement (see Section 10.7).

Before this algorithm was implemented, baby-robot was able to learn how to reach a particularly situated goal. If the location of the goal was changed, the successful learning process for the previous goal could not help it find a new one. Generalization n with combinatorial enhancement enabled the robot to make the discovery, and initiate a process of hierarchical learning. The following salient issues are important:

1. An algorithm of multiresolutional unsupervised learning with inductive generalization and a search for hidden implications was introduced and tested. This algorithm was applied recursively to its own results at the output. Therefore it built up a multiresolutional structure of results. This type of unsupervised learning (UL) is demonstrated to be a mechanism of development of an evolving multiresolutional system of representation. It enables the system of behavior generation (BG) also to evolve. Together, ULBG provide for an evolution of the automata equipped with such systems. An automaton equipped by ULBG becomes a multiresolutional automaton, and its levels of resolution can change as the evolution of knowledge and behavior proceeds. This evolution is illustrated in Figure 10.11.
2. From Figure 10.11, we can see that as behavior evolves, it produces different plans and motion trajectories. These are shown as a horizontally developing tree. At the same time, knowledge evolves as shown in the vertical hierarchical structures. This important process has not yet been analyzed in the evolution of living creatures. We believe that the automaton equipped with ULBG allows for an analysis of its evolution processes (automata with ULBG) as species. It is possible to equip the automaton with a system of reproduction. It should therefore be possible to analyze how the process of knowledge evolution is affected by different mechanisms of reproduction.
3. This line of research takes advantage of the uniqueness of automata with ULBG among other known systems of automata with learning. The mechanism of unsupervised learning ultimately allows for freedom in the way the learning process organizes acquired knowledge. It is possible that as the knowledge base evolves, the knowledge becomes utterly diversified. Rules concerning the external world will emerge, and rules concerning processes of inner knowledge organization and procedures of processing will follow. A simple analogy with mental processes suggests that a system that starts by perceiving the world as a set of values $\{v_{ij}\}$, that is, having an "ego-focused" representation, will learn how to develop maps

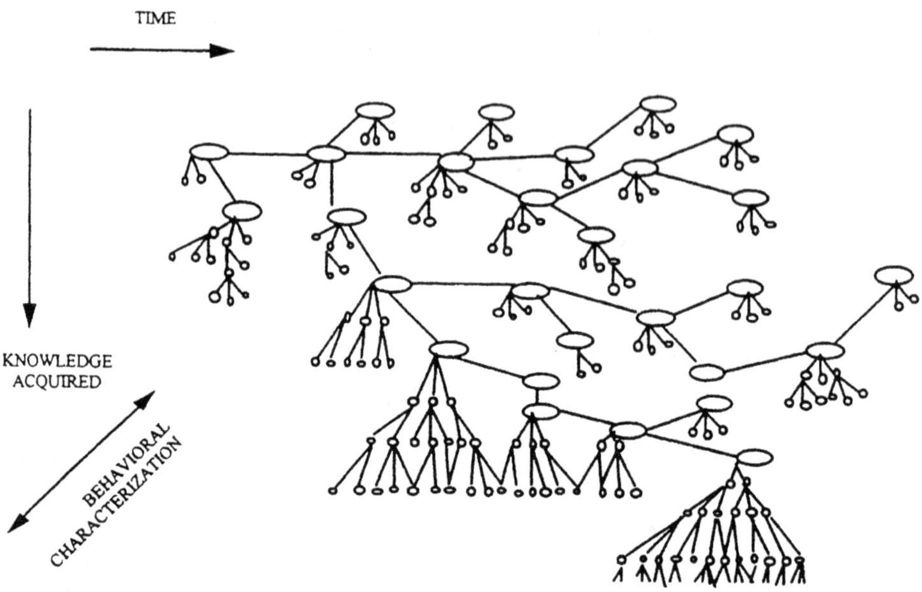

Figure 10.11. Evolution of knowledge and behavior of the automata with ULBG.

in externally-fixed coordinates that will allow for the map to be put on the system itself (this might be interpreted as the emergence of "consciousness" in some applications).

4. The automaton with ULBG can be used to analyze all stages of learning including the "early learning" stage. Certainly some initial knowledge ("bootstrap knowledge") is presumed. The early organization of knowledge can strongly affect subsequent knowledge evolution. Nevertheless, the learning system is presumed to be free in building its subsequent knowledge organization. How does it organize the knowledge acquired and why? This is still a research issue in the area of automata with ULBG.
5. The processes of knowledge acquisition are affected by stored knowledge. These processes start by creating some bias in subsequent knowledge acquisition, since the results of automaton functioning are induced by previously stored knowledge. If the results of functioning were "good" or led to a "better" behavior, the system might assume that its goodness was due to the knowledge used. It might happen that the experiences the system acquires are limited by its predisposition. This is another research topic of interest in the ULBG area.
6. Since the system has been developed to demonstrate a particular behavior, the evaluation of this behavior is the ultimate measure for the process and results of knowledge organization as well as the processes and results of the ways knowledge is acquired from the external world and knowledge is used for behavior generation. This evalution thus interconnects the further developing hierarchy of functions for evaluation of goodness as it precipitates at different levels of resolution. The interconnection between learning, behavior, and development of the "system of values" in automata with ULBG is an important research issue.
7. In concert the processes of acquisition and organization affect the overall system's functioning. Therefore other systems of learning can modify and alter the

system of ULBG. At the present time the following mechanisms of knowledge acquisition are known from the literature (other than UL):

Learning by transfer. All knowledge that is subsequently required for behavior generation is transferred from another source where it was stored and organized in advance based on existing design decisions and experiences of functioning. This method of knowledge acquisition presumes that the structure of the system of interest and its functioning in required circumstances are previously known, and the knowledge is organized so that this structure is properly supported.

Learning by example. We presume the existence of a "teacher" who has substantial knowledge about possible cases of functioning, stores knowledge of previous experiences, and spells out a set of possible scenarios in which the functioning of the system is expected. Undoubtedly, a set of tests can be developed in which the behavior of a system is entertained, and after each case of behavior, the system receives the teacher's evaluation whether it was good and how good it was. These tests provide exercises for known solutions. The interaction of the ULBG system with other systems of learning is, however, a separate research issue.

The system is taught responses to the test by demonstrating a particular behavior. At the end of the test, it is informed by the teacher whether this behavior was right or wrong. In more complex schemes, it can be informed of how good it was and why. Unlike the the situation of mechanism 1, mechanism 2 does not interfere with how the learning system organizes the required knowledge. Nevertheless, although the teacher's interference into the process of knowledge acquisition can be deep, mechanism 2 does not say anything about the way knowledge should be organized within the learning system.

8. Learning and behavior generation produce structural and behavioral hierarchies. It is possible to state that using ULBG reduces computational complexity by increasing the structural complexity.

10.7 BABY-ROBOT: ANALYSIS OF EARLY COGNITIVE DEVELOPMENT

The section is dedicated to the processes of early cognitive development (ECD) in intelligent systems. Similarities and differences exist at the ECD stage in humans and robots for the two types of learning: quantitative (QL) and conceptual (CL). The baby-robot was devised as a tool for investigating ECD processes. The baby-robot has been instrumental in the analysis of mechanisms of autonomous knowledge bases generation. The structure of the baby-robot provided a foundation for constructing self-organizing cognitive processes in intelligent systems. The theoretical basis of this section evolves from the analysis presented in Section 10.6. However, both evolve from the premises of the axiomatic theory of probability [78].

10.7.1 ECD: Its Significance for Learning in Intelligent Systems

Interest in early cognitive development (ECD) intensified as the new areas of research evolved in intelligent systems (IS), including robotics and automation, systems for intelligent control, autonomous mobile systems, unmanned industrial devices, and unmanned systems of integrated manufacturing. All these areas of application of IS are characterized by the following distinctive properties:

1. The IS is given a "goal" that must be achieved in the IS functioning. This goal is given as a description of the required changes in the world, as a cost-functional to be maximized (or minimized) in the process of operation, and as a set of constraints to be satisfied during the operation. The "operation" is understood as a combination of actions to be undertaken in order to achieve the goal. In this section we will consider only one class of IS: the one that achieves the goal by developing a specific motion trajectory.
2. The notion of success in achieving the goal is incorporated in the IS vocabulary, but the reality of successful achievement can be beyond the IS "understanding" at the moment of task formulation. Realistically, the IS operates beyond the reach of any specification. Thus the operation is performed under condition that the initial information is incomplete, the model cannot be captured in all detail, the sensor information is not sufficient (contains errors), and the means of actuation are limited.

The distinctive properties mentioned above are typical for most situations in which the application of unmanned autonomous intelligent robots is expected and desired. Clearly, the existing concepts of control based on hierarchical systems of intelligence [79, 109] and using knowledge-based controllers to generate execution commands [110] are supposed to be utilized for intelligent robots in general, and for intelligent mobile autonomous system (IMAS) in particular. However, the system of knowledge required for actual operation of these devices has to be enormous.

Even in a case of limited operation (e.g., computer simulation of a 2D world with multiple spatial relationships in the system IMAS Set of Obstacles), it is impossible to imagine and put into the knowledge base all situations that will generate meaningful rules [110]. The IMAS must be able to recognize some unexpected spatial situations. In order to recognize a "trap" situation and a corresponding rule of action for a particular motion trajectory, the IMAS must remember the trajectory of its motion during at least some past limited time period. Thus learning from experience is a vital part of dealing with the reality even for a simulated IS. In constructing the rule bases, a method of "abductive" inference was used (from insufficiently frequent particular to general).

The recognition of familiar situations stored in the memory is therefore not sufficient. In the easiest "trap" or "difficult" situation, the IS starts moving in circles and repeats itself. But it would be desirable for it to *learn from an experience with no prearranged patterns given.* Since the number of situations that the IS will encounter is expected to be immense, all the patterns cannot be provided to the IS "at birth." In this section we will explore a different approach: Making the IS capable of learning different situations whose solutions can be found depending on the present situation of a given goal. We will consider a class of robots at the initial stage of their existence when they do not have a teacher nor any prior experience. We address first the problem of designing the structure of an IS with this early learning capability.

10.7.2 Resemblances and Differences between ECD in Human and ISs

Resemblances

A limited repertoire of unlearned behaviors exists "at birth" in both humans and ISs. These behaviors consist of a limited list of action-primitives (suck, grasp, blink, accelerate, turn, etc.). Piaget had labeled such primitive behavior as schema. "Schema" is understood to be an elementary unit of a cognitive structure; in many cases it is understood as an activity A with its structural connotation.

Another basis of resemblance is related to the so-called sensorimotor activities that connect sensation and physical movement; the sensorimotor schemata presumably do not include complex forms of cognitive activities such as imagination and problem solving. Thus these schemata can be written in a form of clauses. Each of these clauses is understood as a conception of behavior (C_B). The structural connotation of clauses is not explicitly recognized by an actor and does not participate in the control of the sensorimotor activities. This representation does not contradict the existing views of the scholars in the area of cognitive development [111–113], since ECD can be described adequately enough within the framework of automata theory. The structural framework of sensorimotor clauses corresponds to the closed-loop theories of motor learning [112]. Thus a set of senses is presumed in the set of the above-mentioned primitive actions. From these two sets of schemata, a set of rules can be formed where each rule is represented by a clause.

Psychologists talk about the logical weakness of a child's reasoning, which is said to be syncretic and transductive. Syncretic reasoning reflects processes of classification when the criterion of classification is changing in time. Transductive reasoning (from particular to particular) is a supplement to more common types of reasoning: deductive (from general to particular) and inductive (from particular to general). At the initial stages of ECD, a human child is characterized by egocentric mentality, whereby perception is the dominating way of thinking. An IS at the ECD stage (baby-robot) is expected to define its state in relation to surrounding objects, and vice versa. As the system evolves, baby-robot develops the need for a capability to represent the world in some global coordinates. The reasoning from insufficiently frequent particular to general (called abductive) is learned too. This ability evolves as it is being developed in the human child. The capability to choose the most appropriate reference frame is given at birth as one of the important "seeds" of the development. It seems that for the functioning GFACS, this capability is a by-product.

It is expected that for the baby-robot, most of the situation (S) and goal (G) descriptions allow for a translation into a set of rules $[R]$ for which the following holds: $\{[R] \wedge S \wedge G\} \rightarrow A$. These implications are developed during the "teaching tests" which are analogous to sensorimotor play in a human child. It is possible also that for a number of problems, an analogue for so-called imaginative play will be found.

The distinction between exafference and reafference in the baby-robot is the same as in a human child.[50] Since it was shown that learning based on an individual's own movements is important for the development of accurate visually guided behavior [114], the strategy of programmed procedures of reafference are especially important for the baby-robot.

Differences

As was mentioned above, a human child is egocentric initially, but a baby-robot is not necessarily egocentric. Many different frames of reference are easily given to the baby-

[50]The motion control of animals is considered a feedback loop. It works as follows: The *afferent* nerves carry signals toward the central nervous system while *efferent* ones carry signals from the central nervous system to the motor areas. Afferences are divided into receptor excitations of two types: *reafference* caused by internal changes in the musculature and *exafference* produced passively by external stimulation.

robot at birth. Many of the general postulates of ECD that are discoveries for a human child are given to it in a "burnt-in" form (e.g., rules of conservation, reversibility, identity, and ability to create combinations). Hence, unlike a human child, the baby-robot might not have a period of "proportional thought," and the ability to perform concrete and even formal operations is supposed to be given to the baby-robot at birth. The same is related to a number of reflexes and responses that might be conditioned at birth if the manufacturer chooses so. The same is related to the infant states including alert situations and focused activities.

Most of the differences are related to the processes of development linked to communication. Since we limit out analysis by consideration of a single IS, the social play is not expected to be reproduced. (The problem of robotic team is not discussed here.) However, the most important difference is linked with the drive to activity. An IS is typically built for use in a definite spectrum of goal-oriented activities. In other words, the hierarchy of tasks can always be formulated and the world description can always be structured with respect to the hierarchy of tasks, and the spectrum of available means of performing these tasks. Certainly we can consider human behavior within a task-focused paradigm ("task-driven behavior"). However, it is impossible to do so when ECD is analyzed. Indeed, baby-robot is given its task vocabulary at birth, whereas a human child cannot condition his or her behavior by a clearly cognized task in most of the ECD situations.

One of the major differences that at the present time seems to be out of reach is the innate ability of humans to organize input information according to the context and meaning. Gestalt rules might be programmed, but this cannot ensure their proper application within the structures of learning and problem solving unless we know the mechanism of their development (possibly in the form of a GFACS triplet).

10.7.3 Quantitative Learning Domain

Frequently the "knowledge" of an IS is understood as something that must be provided at the beginning [115]. Nevertheless, the need in learning for a mechanism of online knowledge acquisition was recognized, and in the 1960s an active effort was undertaken to develop this mechanism. One of the first accounts of learning systems is given in [116] where the results are collected of earlier work in the area of learning. The learning network quickly became a dominating concept [117]. The learning system was simplified into four component networks in the form of sensors that provide data for learning networks (LNs). An LN responds to the input (stimulus) by developing an output (response) that generates the commands for the actuators. Their action is evaluated by the goal network that generates either reward or punishment. In [118] this concept is disclosed in more detail. The elements of the theory of learning systems are explained in [119].

Adaptive neural network concepts such as "perception" [120, 121] or "adaline" [122] are involved supervised learning pattern classification. They form discriminative rules if the "teacher" can supply each adaptive element with its individual desired response. It was not immediately recognized that the learning process within the adaptive network (associative network, perception, etc.) does not create any new concepts; only a quantitative adjustment takes place. We will name this process *quantitative learning* (QL).

On the other hand, after the learning process is completed, each elementary network can be considered a particular "concept." Using the neural network (NN) for concept formation is proposed in [123]. It was next recognized that after the learning process in the NN converges, the result of learning must be symbolized by a new label corresponding to the new concept [124]. We will name the process of learning a *conceptual learning* (CL) if a system of concepts is obtained as a result of learning.

In the meantime the processes of QL have been developed substantially and gained many important properties and a high level of sophistication. In [125] NN is treated as a form of perception, and it is shown that the learning process converges for a large class of distributions. An attempt to provide a learning process without a teacher has yielded some positive results [126]. Use of the NN for control of autonomous mobile systems is described in [127, 128].

References [129–133] propose an associative search element (ASE) and an associative search network (ASN) that learn from experience, without the presence of a teacher. Each generates actions and, by a random component of the generation process, introduces the variety that is necessary to serve as the basis for subsequent selection by evaluative feedback.

Contemporary models of multilayered perceptrons exemplify some of the most advanced concepts of hierarchical world representation in visual domain [134]. These models might be a source of a concept-generating structures. However, this problem of cognition is beyond the focus of this section. A number of applications of the NN are known in the area of manipulator control [135, 136]. In all of the known works, QL is based on teaching tests. A good example is given in [137] where after repeated path trials compensation is achieved for effects such as friction and torque due to velocity and gravity effects. QL has been applied to a number of intelligent systems. Usually it involves changing values of the coefficients in the pattern classifier as a result of statistical learning under an accepted reinforcement law and an accepted technique of memory adjustment [138].

It is important to observe the following fact: QL demonstrates two general concepts found in the experience of living nature. One of them is based on using statistics to induce a tendency to change [139]. Another is the idea of choice, since the discriminators based on various decision rules are utilized in the system of QL. Assuming a probabilistic description of the uncertainties, a classical decision rule is to choose the strategy that minimizes the expected cost. It was noticed in [140] that this way did not reduce the probability of getting a bad performance in the realization, and a decision rule was proposed that consists of minimizing a combination of the expectation and variance of the cost [141].

Based on these two ideas, a method of random search has been created (one of the first instrumental descriptions applied to learning processes is given in [142].) In practice, the multivariate probability distribution is used for a "thoughtful" guided search [143]. Most of the genetic algorithms and systems with evolutionary programming fall under this rubric.

10.7.4 Conceptual Learning Domain

After the process of concept formation is completed by the algorithms of classification and clustering (see Sections 10.3–10.5), or after the particular NN representation for this concept can be "frozen" and transformed into a single neuron of a "higher order," the process of conceptual learning can be considered completed. In fact NNs are capable of learning boolean functions [144]. These concepts represented in the form of boolean functions describing a hierarchy of action-primitives are to be combined in response to a task [145].

QL systems are also based on a principle of inductive learning in building programs, which operates in two phases [146]:

1. *Training phase*. The system decides online by interacting with the sensors whose motions are to be executed. By performing several instances of the same task, it generates multiple traces of execution.
2. *Induction phase*. The system applies transformation rules to the experiences generated by the training phase, and builds a manipulator level program including symbolic variables, conditional statements, and loops.

Induction, as the process of inferring the description of a class from the description of some individual of this class, was an object of attention in AI for quite a while [147, 148]. In [147] learning is understood as filling in the prearranged boxes of possible relationships among the entities of the vocabularies that are not subjected to change. However, the actual application to motion control was first presented in [146]. A typical motion statement in LM language is "move F by T until C," where F is an object to be moved, T is a transformation by which F is to be moved, and C is a sensory based condition on which the motion is to be stopped even if T is not completely achieved. All of the planning rules are of the form <conditions> → <plan> and all of the execution monitor rules are of the form <conditions> → <relation>.

The purpose of the "induction phase" is to construct an LM program from a set of execution traces that has been produced by applying the same strategies, that is, "by selecting the same planning rules in the same situations." The rules of induction are intended to unify in a cluster of different motions that result in the same state of different states that can be generated by the same motion.

Clearly, the state description is critical for the situation understanding (interpretation) and the plan generation. The most advanced results on the world description are given a priori in English language. The schema-based approach is presumed which is also generally accepted in most of the substantial results from [149–153].

Although, the idea of schemata-based analysis of the image (situation) has been proved to be beneficial, working with the complete image did not seem to be efficient and the idea appeared of the "focus of attention" [154]. Using the state description for the plan generation was shown in [155]. In a number of papers the "schema-based description of robotic skills is used to generate plans [156–157].

Only recently, works have appeared that put the CL problem in the science of intelligent systems on a level of substantially new capabilities [158–159], especially when the world representation is incomplete [160]. The following conclusions can be made about the state-of-the-art in the area of IS motion planning and learning:

1. Training is understood to be a generation of multiple traces of execution.
2. The transformations of motion execution are given in advance.
3. Sensory-based conditions of motion stopping conditions are predetermined.
4. Expert knowledge used in the process of trace generation is taken from a knowledge base that is set up by the user. When some knowledge is lacking, the system requests for an additional rule.
5. All of the rules employed by the system are given in the form of implications, and these implications are never questioned; they are considered as ultimate laws.
6. Strategies of selecting the same rules in the same situations are based on the assumption that "sameness" cannot be evaluated quantitatively. In the meantime it is not clear how to select a rule even for the identical situation if more than one rule might be recommended.

7. The logic of dealing with information (the inference engine) is based primarily on the rules of deductive logic implanted in the mechanisms of programming languages.

This list of properties may vary from IS to IS, and from problem to problem; however, the core remains the same and is never challenged.

10.7.5 Baby-Robot: Simulation Tool for Analysis of ECD Processes

Hidden within the state-of-the-art are some interesting and important problems for the IS programming decision making, and planning and learning techniques:

1. Training must be treated as knowledge acquisition based on intentionally organized activities or as knowledge acquisition through redesigned experiences.
2. A full list of concepts [113] does not need to be given in advance. A minimum number of rules can be expected to be given to an IS at birth; all other rules are learned (and discovered) within the framework of IS activities. In other words, there is a minimum list of rules-primitives to be determined, which depends on the minimum list of action primitives.
3. Sensory-based conditions of stopping (or other changes in the motion mode) are not predetermined. We are not sure that the user is always capable of giving the best solution of how to respond by selecting a particular action when the information is changing. So

$$< i\text{th image } j\text{th change}> \rightarrow < A_k >$$

   is the best rule for all of the situations the robot can face in reality. There can be expected to be a minimum set of rules to be recommended after the change in perception is recorded.
4. The expert knowledge to be applied for the trace generation is generated within the IS and not provided by a user. The necessity to address a user is considered a system failure. We expect that such failure will appear for any of these reasons:
   (a) Insufficient list of rules.
   (b) Insufficient logical capabilities.
   (c) Insufficient mutual understanding of perception and planning subsystems.
   (d) Insufficient list of initial rules in the expert knowledge.
   (e) Natural limits for the concrete class of IS.
5. Although a form of "implication" is a major tool in the inference machines of systems for knowledge-based control, it is unclear that this mechanism should dominate in intelligent autonomous robots. We can assume that "implication" is not necessarily required, it can emerge within an IS in a natural way (the need for "implication" must otherwise be proved.)
6. "Sameness" or "degree of similarity" between two states and/or two situations cannot be assigned. The complete equivalence between two structural descriptions should be considered as an expectation rather than as a typical situation. The question here is: What is the value of structural (and quantitative) similarity that predisposes the system to a definite rule?"
7. The logic of dealing with information cannot be assigned by user. The user has created his logic (whatever it is) in different problems. The question to be asked

is: What is the structure of the logic to be employed by an IS? Is the predicate calculus of the first order a sufficient tool?

All these problems relate to the idea of baby-robot. It should be given capabilities of perception, conception, planning, and control actuation. However, the vocabulary of perception, conception, and motion planning and control must be acquired via operational experience. Everything has to be questioned before submitting it to the baby-robot as "inherited information." Indeed, it is a question of whether the ability to generalize (to unify descriptions in a cluster by some token) should it be "inherited" or whether it should be created in the process of operation (or maybe both?). This ability to generalize will affect the ability to form "descriptive structures" and to decompose them. Further we need to know if the process of alternatives selection (during decision making) is determined by the stored implications or if it is based on ambivalent similarities. Likewise we need to know if the relative frequency of cause–effect coupling makes the couple valid for assigning it a rule status.

10.7.6 Baby-Robot: A Mental Experiment

The baby-robot behavior in a bug-world[51] can be illustrated as follows [40, 86, 161, 162]. In Figure 10.12*a* is shown a world containing three objects: a robot *R* (triangle), a goal *G* (star), an obstacle *B* (rectangle). The dotted line that connects *R* and *G* indicates the drive that achieves the goal as soon as possible.

In the beginning our baby-robot *R* does not know how to determine and execute ("plan and control") its own motion toward *G*. *R* decides to explore "what would happen if" and starts making tentative moves and evaluating their consequences. It initiates the first move in a random direction and estimates the distance to *G*, which happens to become long (position 2). This is "bad" concludes our robot; let me try something else.

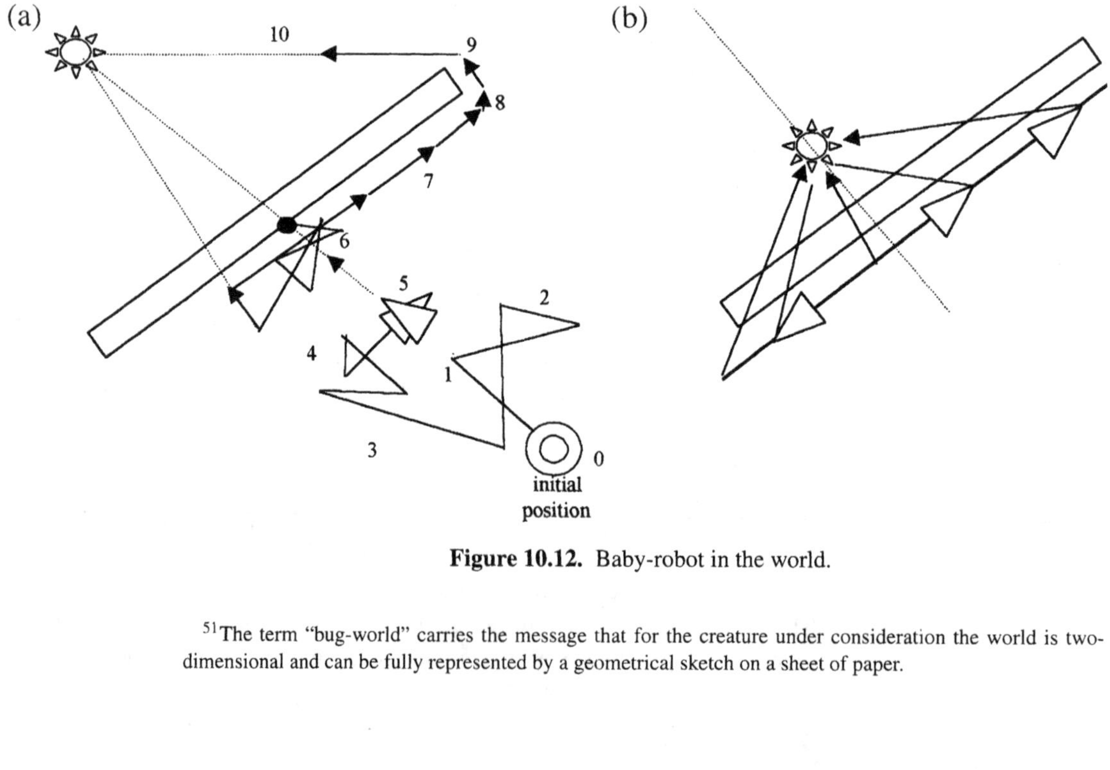

Figure 10.12. Baby-robot in the world.

[51]The term "bug-world" carries the message that for the creature under consideration the world is two-dimensional and can be fully represented by a geometrical sketch on a sheet of paper.

R takes another move in a random direction which seems to be a "good" one (position 3), but the move also turns out "bad." After a number of random moves baby-robot can deduce what should be done and assigns a "proper direction" to the system of actuation. It is our goal to find the bootstrap-knowledge that helped it deduce the right direction from a variety of random moves. It would be even more desirable to determine the general rule of finding direction in any situation, not only in the present state. This might require more bootstrap-knowledge. This is exactly what we want to explore next.

After the process of learning has generated a concept of direction, the TME of motion toward the goal based on a conception of motion direction starts from the position 5 toward the goal. The arc of the "range of vision" $O_1 O_2$ intersects the growing image of the obstacle and is recorded in memory until the impact in the position 6. An "impact" can be manifested by the appearance of a second modality of perception (from the tactile or force sensors), and this modality can play a role of "pain." The impact can be with the growing image of an obstacle, and it receives the label "bad."

On the other hand, it is not necessary to multiply entities when there is no need for them (Okham's *parsimony* principle). After impact with the obstacle, the robot cannot merely continue its motion toward the goal; other possibilities must be explored as in the initial position. However, now R has more knowledge and skills. It already has a concept of direction, and it knows that an obstacle is "bad." The difference between positions 7 and 8 (Fig. 10.12*b*) is not understood yet. After the trial movement toward 7 and 8, R may learn that the end 7 of the "bad" becomes closer while the "end" 8 remains at the same distance. The robot should link it with the limit of its range of vision, thus acquiring the first evidence of this phenomenon (before the statistical conclusion of the viability of some rule is obtained). After position 7 is achieved, the completion of the "task" is easily performed.

A problem to be overcome is hidden in the fact that although subgoal 7 supposedly reduces the "badness" of the situation of facing an obstacle, moving toward the position 7 increases the "badness" of being remote from G. This contradiction can be better illustrated by Figure 10.12*c* where moving toward 8 seems to be more preferable than moving toward 7. Indeed, although the end 7 becomes closer, the "badness" of distance to the goal is growing. At the same time, although the border 8 does not become closer, the "badness" of distance to the goal is reducing. (To be sure, moving toward 8 will give a better result if the end is in 8 and not in $8'$; otherwise, a return back to 7 is imminent.)

The adequacy between symbols for visual representation and representation that is necessary for decision making is obvious. No sophistication of the alternatives is required; however, descriptive and quantitative (statistical) generalizations are expected. Gradually increasing the complexity of the world, we expect to arrive at the self-generation of the major logical axioms, rules of inference, and even to such sophisticated concepts as strategies of search (and even A^* algorithm).

10.7.7 A Possible Algorithm of Learning

One of the prerequisites for learning is the capability of a controlled system to sense the environment, and to arrive at conclusions using the results of sensing. This sensing input, and the output with conclusions on the result of learning, are the input and output of an information processing system that is the object of our consideration.

Let us try to restore the stages of this information processing by analysis of these stages backward. The output of the learning system can be represented as a clause and ISA statement. A clause (IF–THEN statement) is a form in which any rule might be represented. All other statements of belonging to a known class, and the statement of forming a new class, can be unified as ISA statements. On the other hand, an ISA statement is a particular case of the clause IF ANYTHING, THEN A IS A B.

We will constrain our research by considering only clause statements at output. The following prerequisites for clause generation must be satisfied:

1. Words for antecedent as well as for consequence should already exist in the vocabulary of the system.
2. The implication link (IF–THEN) must be taught or learned from an experience. For the stage of early cognitive development we do not assume any teaching. Thus the following questions must be answered:
   (a) How do the words appear in the vocabulary?
   (b) How is an implication link learned from experience?

Labels can be assigned to any value of the sensor output, to any pixel of the CCD matrix, and so on. Some measure must be taken for not labeling a distinguishable signal (at a given level of resolution). The only mechanism that seems to be applicable is the possible repetitiveness of the signal values at the sensor output. Since the results with any type of similarity are clustered on the basis of their *proximity* in space, time, magnitude, sign, and the like, the ability to make the comparison is presumed. The comparison determines differences among members of the similarity set. The following definitions can be introduced:

DEFINITION OF SET OF SIMILARITY

The *set of similarity* is a set in which a feature can be declared that is similar for all members of the set.

The set of similarity will be named a "class" after more than one set of similarity appears in the database. In other words, the similarity set is an initial array of nonclassified signals. The classification procedure is applied to determine the similarity feature and the similarity set. The same procedure serves for distinguishing classes within a similarity set.

Let us assume a set of objects with A property as a feature of similarity ($A_1, A_2, \ldots, A_n$) which is a similarity set by definition (other features are not known yet). The quantitative value v of A property can be put in correspondence generating set of couples $\{(A_1 v_1), (A_2 v_2), \ldots, (A_n v_n)\}$. In fact we are dealing with a sequence of numbers given upon numerical axis that might be distributed randomly, uniformly have clots, and so on.

A statement of existence can be formulated:

"(there exists) a cluster of A objects with a center at v_x."

The relation between the clusters can be found:

"(there exists) a number of clusters with centers forming clusters..."

The behavior of these clusters may be analyzed in time. An attempt to find implication can be made on the basis of experiences such as

if $t > 5$ am, then cluster at V_1.
if $t > 11$ am, then cluster at V_2, etc.

In order to do this, a mechanism of finding correlation should be devised that enables the system to notice that clot i happens to coincide with some period of time (comparison of moments of time is presumed.) Clearly, in order to make any implication, more than one modality of sensing should be considered (in this case sensing feature A and sensing time).

10.7.8 Conditions of Clause Generation

Clauses can be formed when more than one modality of sensing is considered. The following procedures must precede the appearance of a clause if the condition of clause generation is satisfied:

1. Labeling the set by all existing sensing modalities.
2. Putting into correspondence statements of the ith modality's existence with statements of the jth modality's existence.
3. Determining the frequency of these associations and the relative frequencies.
4. Linking the associated high frequencies ("dots") as a clause.
5. Putting the clause into long-term memory.

Thus the existence of a short-term memory, and a long-term memory is implied. The short-term memory deals with auxiliary information at the stage preceding clause formulation; the long-term memory stores the clauses. The question is whether the short-term memory should be cleared after the clause is formulated, or whether the material must be remembered for the feature associations. We will not discuss this question here, since our interest in this section is solely the problem of early cognitive development.

The selection of a class with a double feature is equivalent to spotting the entity, or the object (e.g., finding the edges on the basis of intensity and chain adjacency, or finding the blobs on the basis of intensity and spatial plane adjacency).

The algorithm of early cognition can therefore be presented as follows:

EARLY COGNITIVE ACTIVITIES

Step 1. Recognize the features of the object from the sensing results.
Step 2. Label the features.
Step 3. Compare objects by their labels.
Step 4. Group objects with the same label.
Step 5. Validate the groups (e.g., by their frequency of appearance).
Step 6. If more than one validation criterion is present, then assign the status of "class" to this group.
Step 7. Label each class by a class feature.
Step 8. Compare the classes by the class features.
Step 9. Compare the objects within classes by the labels.
Step 10. Group objects with the same label.
Step 11. Validate the groups.
Step 12. If more than one validation criterion is present, then assign each to a subclass.
Step 13. Count the frequencies of the classes.
Step 14. Validate the appearance of frequencies.

Step 15. If a frequency is greater than a frequency threshold, and if this phenomenon appears in more than one class, then put into correspondence the discovered frequent statements as a clause.

Step 16. Count the frequencies of the subclasses... and return to step 14.

10.7.9 Learning When the Goal Is Given

The process of cognition described above is related to "unmotivated" cognition which is oriented toward the efficient organization of information received from the sensors. The efficiency of the organization effort is determined by the required memory space, and the time of storage and retrieval. The question of adequate representation is not relevant, since presumed in this process is a response to the question "adequacy for what?" So the process implies the existence of a goal.

In robots the goal is strictly linked with mechanical motion. Let us assume, without loss of generality, that the goal is given in the form: Position the end-effector (or the mobile-system) in a definite point in a space and do this by satisfying some goodness criterion (e.g., minimum time or minimum path). This means that the features of class formation are already explicitly defined:

- Relevance to the goal (measure of adjacency, closeness, etc.).
- Relevance to the goodness criterion (e.g., minimum time or distance).

The labels that characterize the goal of the goodness criterion are the initial features of class formation, and the classes (and the clauses) that are formed first are not built upon an independent frequency validation but rather use the goal and goodness labels as their initial "centers of crystallization." The consequent generation of subclasses can follow the procedure that has been described above.

The algorithm of early cognition for the goal-oriented case can therefore be written as follows:

Step 1. Assign features to the object by sensing and by goal words.

Step 2. Label features.

Step 3. Validate objects by the goodness criterion.

Step 4. Compare the objects by their labels.

Step 5. Compare the objects by their goodness.

Step 6. Assign classes by label status (A_{LS}).

Step 7. Assign subclasses by goodness (A_{G}).

Step 8. State the clause:

$$A_{\text{LS}} \rightarrow A_{\text{G}} \text{ or } A_{\text{G}} \rightarrow A_{\text{LS}}.$$

Step 9. Count the frequencies of the classes.

Step 10. Validate the appearance of the frequencies. ...

10.7.10 Simulation and Physical Experiments with Baby-Robot

A simulation experiment was developed to test baby-robot using the algorithm of generalization. The sensors return values of the distances to the target D, the angle of heading H, and the angle to the target A. The actuators are "go forward" (G) and rotate (R). The

Table 10.1. Enhanced Representation Sensors and Actuators in Baby-Robot

1. Goodness	10. $D - R$	19. $D + G$
2. Distance (D)	11. $H - A$	20. $D + R$
3. Heading (H)	12. $H - G$	21. $H + A$
4. Angle to Target (A)	13. $H - R$	22. $H + G$
5. Go forward (G)	14. $A - G$	23. $H + R$
6. Rotate(R)	15. $A - R$	24. $A + G$
7. $D - H$	16. $G - R$	25. $A + R$
8. $D - A$	17. $D + H$	26. $G + R$
9. $D - G$	18. $D + A$	

assigned goal is "Make $D = 0$." This statement will lead the learning subsystem to develop control rules that will allow baby-robot to achieve the goal.

Following the algorithm of generalization, the algorithm searches the list of hypotheses. Since it does not find any rule for performing the assignment, it applies a random sequence to the actuators collecting experiences. It next turns to the generalization algorithm. The first step is to separate the "good" experiences from the bad. In the case where the goal is "Make $D = 0$," the goodness of an experience is defined "delta D." The next step is to enhance the representation of these "good" experiences. The enhanced representation is ordered as shown in Table 10.1. This notation will be used for all the data [170–172].

Step-by-Step Recursive Generalization

The first step that the algorithm of generalization does is to check the database of hypotheses (Fig. 10.13). The only schema present in the database says that it should "Make $D = 0$." Since there is nothing in the database about what to do in order to "Make $D = 0$," it assigns "$D = 0$" as the goal, and collects a random sequence of experiences. The next step in the algorithm of generalization is the creation of the classes of experiences. The clustering algorithm discovers two classes given the following rules of unification of classes:

> If there are two consequent classes that have arbitrarily close maximums and the minimum separating them is arbitrarily close to the maximums, then the two classes are unified.

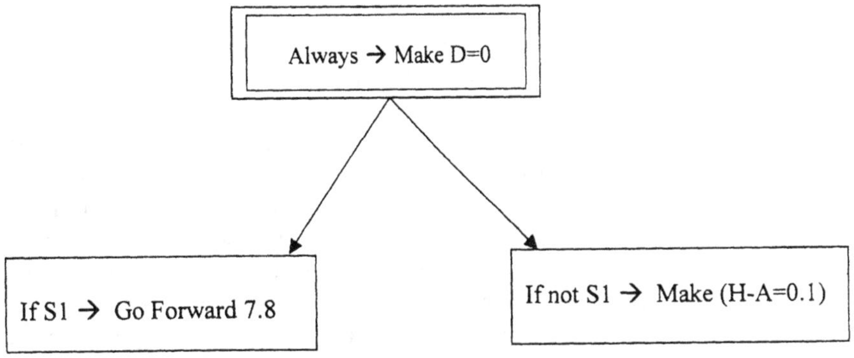

Figure 10.13. Rules for "Make $H - A = 0.1$."

If a class occupies the complete range in that coordinate (i.e., there is only one class that includes all experiences in that coordinate), then the coordinate is discarded as a rule candidate.

They correspond to the actuator "Go forward" and the enhanced representation sensor "Heading–angle to target." These two hypotheses have an important difference. The first one gives a recommendation about what actuator to use in this situation. The second hypothesis says that if we are not in this situation, we should get into this situation.

The problem with the second hypothesis is that there is nothing in the database of hypotheses that gives instructions about how to "Make ($H - A = 0.1$)." Thus the generalization algorithm starts again:

Step 1. Re-rank all the experiences using the new goodness measure for the new goal.

Step 2. Select the "best" experiences using one of the previously selected methods.

Step 3. Send again these experiences to the classification algorithm; in other words, create new eventgrams for the new goal so that the classification algorithm can create new clusters that will form rules to follow the new goal.

Step 4. Take a new goodness measure by checking whether the experience brought it closer or further to the new goal.

Step 5. Set the new goal to "Make $H - A = 0.1$."

In Figure 10.14, we show the eventgrams for the same experiences and the new goal "Make $H - A = 0.1$."

The experiences are now sent to the averaging algorithm, which finds two clusters in coordinate 6 and corresponds to the actuator "Rotate." These two clusters become two hypotheses that get incorporated into the database, as shown in Figure 10.15. In this figure "Rotate 0.9" and "Rotate 9.8" correspond to rotate right and rotate left.

Baby-Robot for 2D Operation

The robot for 2D operation is called the land baby-robot (LBR). The start of the simulation is shown in Figure 10.16. On the left we can see the mobile platform, and on the right it is possible to see the target. Figure 10.17 shows the random path taken initially by LBR to collect experiences and create hypotheses, and eventually the schemata.

We can see that the trajectory of motion is sophisticated. It includes many possible single moves and their strings. Thus, not only the commands are verified, but also the ability to perform these commands. In other words, the strings of moves demonstrate the ability of our dynamic system to handle the input commands. The clauses inferred from the strings contain the dynamic model of the robot implicitly. When the number of random moves is large, it gives the richness of experiences which is satisfactory for inferring many possible rules. We have tested the quality of rules that can be inferred from the set of experiences of different volume. The results demonstrate that from the low total number of experiences, the robot can infer the rule that has a large error. Increasing the total number of moves makes the rule more accurate. If the system is capable of inferring from the changes in the rule that evolve when the number of test moves grows, then the system can notice the asymptotic character of these changes, and the expected "correct" rule can be inferred before the total number of test moves approaches the infinity.

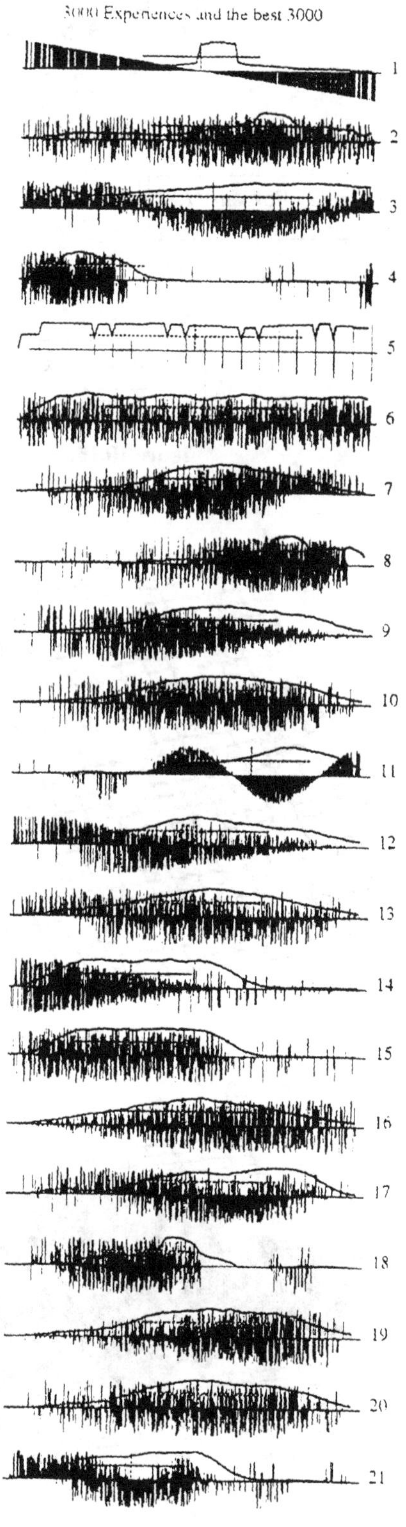

Figure 10.14. Eventgrams collected statistically: 3000 best experiences (the numbers of eventgrams correspond to Table 10.1).

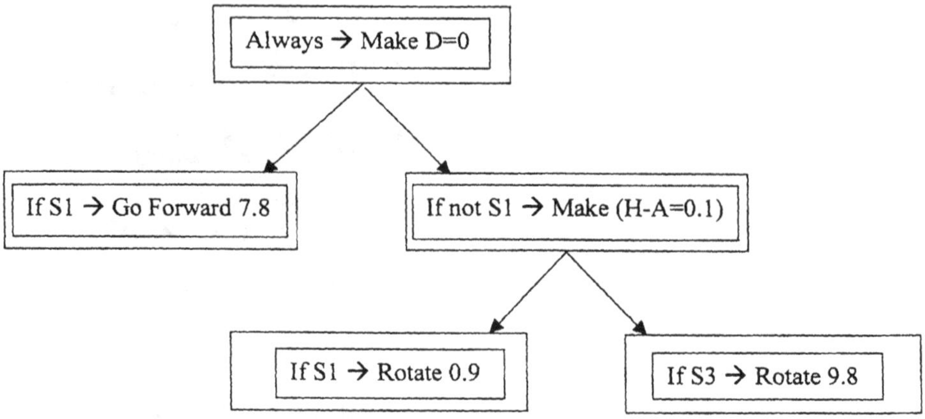

Figure 10.15. Rules obtained for two clusters.

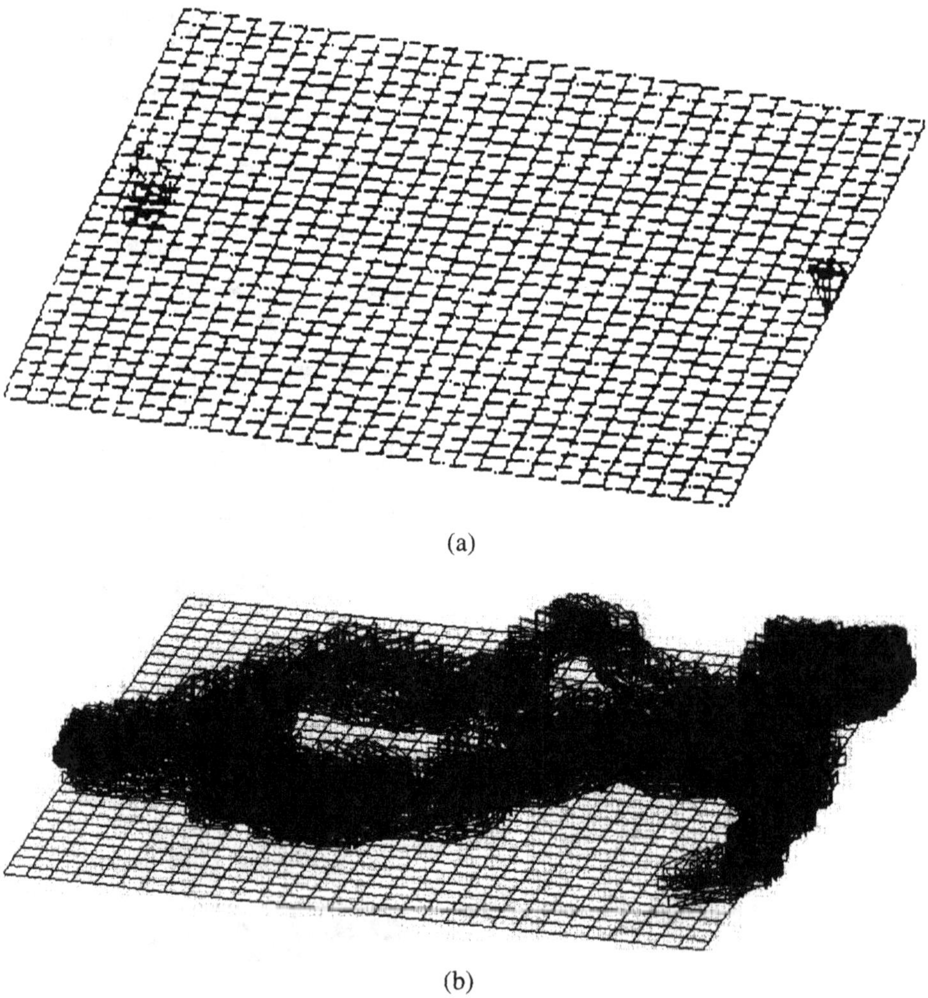

Figure 10.16. Trajectory of motion formed by multiple random moves. (*a*) At the beginning; (*b*) at the end.

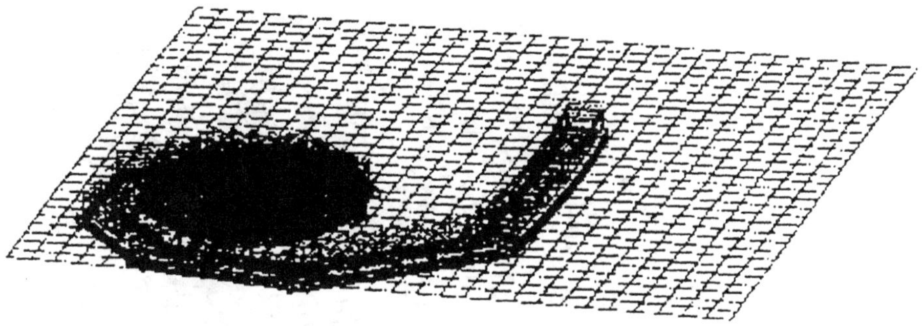

Figure 10.17. New trajectory as the first set of hypotheses is applied.

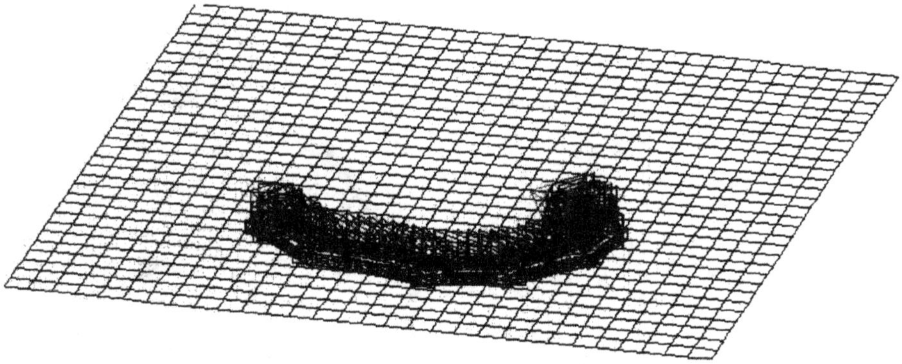

Figure 10.18. Motion after the further improvements.

The experiences are processed, and the first set of hypotheses allows for the robot to apply and verify them. Figure 10.17 shows the behavior of the LBR after the first set of hypotheses are collected. Clearly, the direction of the required motion is not computed with sufficient accuracy. In the meantime the new experiences of motion increase the database, and the new hypotheses arrive. The trajectory of motion becomes more focused. Figure 10.18 shows the numerical improvement on the original hypothesis that yields better paths. Figure 10.19 shows a quasi-optimal path to which the LBR converges. It is possible to see that with the growth of the experience base, the error of the rule gets smaller.

Baby Robot for 3D Operation

The platform for 3D operation is called the underwater baby-robot (UBR). The two sensors given to the UBR are: (1) distance to the target and (2) the three Euler angles to the target. The actuators given to UBR are: (1) propulsion and (2) the angles of the two rudders.

Figure 10.20 provides an example of a learned hypothesis. At the beginning there are some random movements. Then a hypothesis is generalized, showing a spiral movement of an incorrect hypothesis of spiraling. After more experiences are collected, the robot realizes the performance can be improved, and a new hypothesis is generalized; this one may be thrown away.

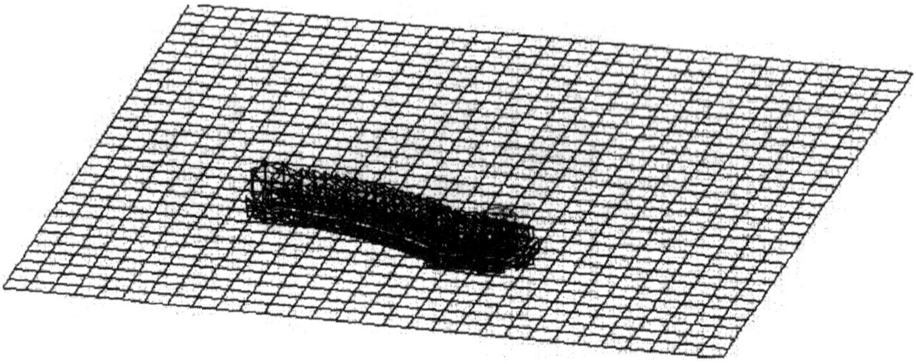

Figure 10.19. Quasi-optimum trajectory of motion.

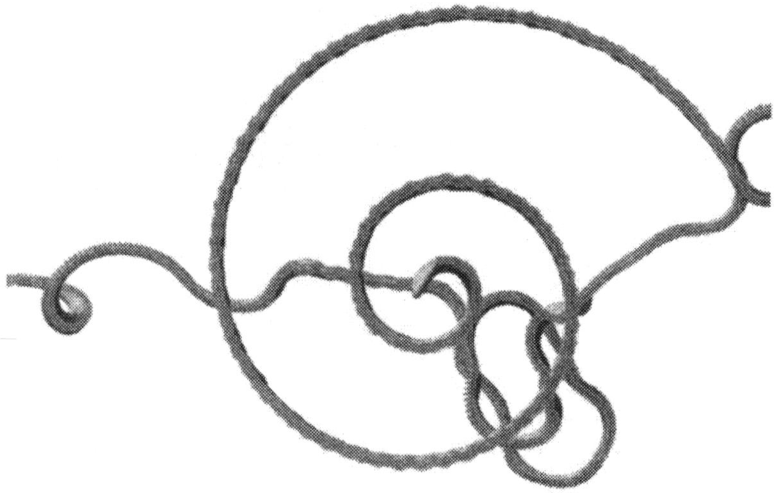

Figure 10.20. Visual discovery of hypotheses.

Figure 10.21 shows the first trial to go to the goal. The traces are left on to show how complicated and superfluous is the path, including random movements and wrong hypotheses of spiraling control. More details on Baby-Sub and AstroBaby can be found in [163]. At the beginning of this trial the learning algorithm does not have any knowledge about the environment. It is possible to see four or five different hypotheses that create different motion.

Figure 10.22 shows some random movements and then the bang-bang control. Note that we did not teach our robot the concept of bang-bang control; it discovered it during the learning process. After the first set of random movements, we can see a sharp change in behavior. The first set of schemata is learned, and as Figure 10.22 shows, the submarine applies a bang-bang strategy of control. Interestingly the bang-bang strategy is attributed to a theoretically inspired method of optimization. The experience with baby-robot demonstrates that bang-bang can be discovered by the robot at a very early stage of its learning.

Figure 10.21. First trial motion trajectory.

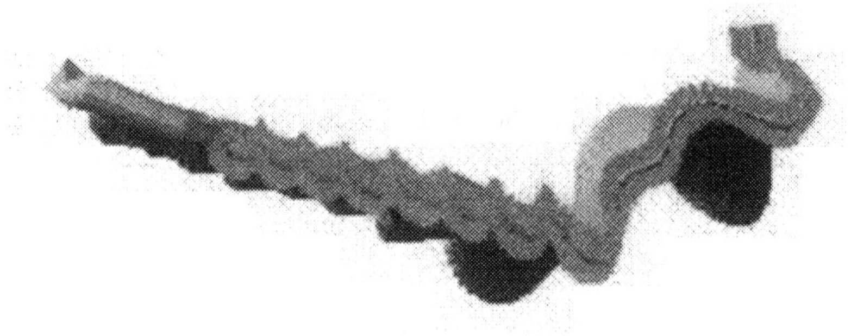

Figure 10.22. Bang-bang control.

10.7.11 Implications of Baby-Robot Research

1. A comparison of ECD processes in human and intelligent systems has provided a key to determining the structures of learning processes. The existence of bootstrap knowledge is determined to be a component whose contents can be delineated in the learning process.
2. QL relates to the numerical values of the modules used for structures that are known prior to the initiation of learning, while CL relates to new concepts constructed as a result of generalization of previously existing concepts. The proposed structure of CL is for use in intelligent systems.
3. The concept of baby-robot is introduced as a tool for analysis of ECD processes in CL. Its semiotic mechanism is an elementary loop of functioning (ELF) that contains sensors and actuators as well as mechanisms of sensory processing and behavior generation connected to the storage of information. ELF equipped not only with an ability to store the experiences but also to organize and generalize them.

4. An algorithm that serves as a subsystem for unsupervised learning is proposed for the intelligent controller's structure. It is based on the algorithm of generalization and therefore contains operations of grouping, focusing of attention, and combinatorial search.
5. The algorithm of generalization works recursively. It is applied to the result of its own functioning and employs the method of nested clustering.
6. The system simulates and executes decision making under a constantly growing set of rules. Thus, although the final goals of the system are known, the precise actions that the system chooses cannot always be predicted by the external observer, but the decisions can always be justified within the system of reasoning and set of values submitted to the IS at the stage of design.
7. The system organizes input information into a multiresolutional structure of world representation. The structure of its knowledge base is prepared for the subsequent growth into the hierarchy of representation.
8. Simulation experiments have confirmed the theoretical premises. The Baby-Robot Senior Design Project at Drexel University has been manufactured as a hardware unit.

10.8 APPLYING NEURAL NETWORKS FOR LEARNING

So far we have established a vision of learning as a generalization upon multiple information. This multiple information can be collected within a temporal window around an instant of time, and it can be collected within a spatial window (volume) around a point in space and/or moment of time. The process of generalization is understood as recovering entities of lower resolution from information presented in a higher resolution. It is done usually by some probabilistic averaging (determining the moments), and the computation can include special architectural modules known as neural networks, which are specialized multiprocessors with all the advantages of parallel processing versus single-processor computation.

10.8.1 Neural Networks: Architectures for Generalization of Multiple Information

Neural networks are conceptual architectures for dealing with multiple information. Each architecture is constructed from *modules* which are designed to solve the bigger problem of generalization. This allows one to substitute a representation using many samples of high resolution by another representation of lower resolution. Since the processing can be performed on a single-processor computer as well as with a multiprocessor module, the conceptual importance of neural networks is as a mechanism of generalization. Thus neural nets can be considered as both a metaphorical and hardware way of dealing with multiple information.

The standard solutions developed in multiple information processing modules (NN-modules) are for associative memory, pattern recognition, and category learning. Various ideas are explored in developing these solutions. The following milestones in NN development are considered:

- McCulloch-Pitts neuron
- Perceptrons
- Adaline and madaline

- Back propagation
- Learning matrix
- Linear associative memory
- Embedding fields
- Instars and outstars
- Competitive learning
- Adaptive resonance theory
- Cognitron and neocognitron

McCulloch-Pitts Neuron

The McCulloch-Pitts model [164] describes a neuron whose output is the sum of inputs that arrive via weighted pathways. The input from a particular pathway is an incoming signal multiplied by the weight of that pathway. These weighted independent inputs are summed. By assigning weights, one can obtain the weighted average for any probabilistic law. The outgoing signal can be a nonlinear function-binary, sigmoid, threshold, or linearly growing input signal in that cell. The McCulloch-Pitts neuron can also have a bias term, which is formally equivalent to the negative of a threshold of the outgoing signal function.

The McCulloch-Pitts neuron can handle grouping (by weighted summation) and focusing of attention (by fanning) but not combinatorial search. A convenient notation is known for describing the McCulloch-Pitts neuron, called the *adaptive* filter [165]. The elementary adaptive filter has the following components:

- A level that registers an input pattern vector; it performs a function of focusing attention by "fanning" the signals that pass through weighted pathways (i.e., focusing of attention is performed).
- A level that summarizes these inputs (i.e., grouping is performed.)

This principle of operation has proved useful because the adaptive filter computes a pattern match by producing an output function that can be a binary step (a decision), a linear growth, or a sigmoid signal. In 1943 McCulloch and Pitts analyzed this device *without adaptation* (i.e., without adjustment of the weights [174]). This adjustment (adaptation) can be considered a search for *proper* grouping, and thus it is elementary learning.

Perceptron: Learning with Feedback Compensation

The next generation of researchers started exploring the McCulloch-Pitts neuron for repetitive learning and adaptation. The perceptron was developed by Rosenblatt in 1958 [166]. The idea behind the perceptron is the development of rudimentary learning by assigning weights. In the perceptron matrix the results of collective learning by multiple neurons are grouped as the results allow, and searched again for matching submodules within them. The submodules of the perceptron were called the sensory unit, the association unit, and the response unit.

One of the many perceptrons that Rosenblatt studied, was the *back-coupled perceptron* [167]. This model was equipped with a feedforward adaptive filter and had a binary output signal. The weights fanning in to the particular node were adjusted in proportion to the error at that node. The actual output vector was subtracted from the target output vector. Their difference was defined as the error, and that difference is then fed back to adjust the weights, according to some probabilistic law (back-coupled error correction). This model of two-level perceptron could sort linearly separable inputs into two classes.

Minimizing the Group Value of Error

A new set of perceptron models was created by B. Widrow and his colleagues, which were called *adaline* and *madaline* perceptrons. The adaline model had just one neuron at the intermediate level, while the madaline, or many-adaline, model had any number of neurons in that level. An adaline/madaline model compares the *analog* output x with the target output b. This comparison provides a more subtle judgment of error than a law that compares the *binary* output with the target output [168–169]. The error is fed back to adjust weights using a Rosenblatt back-coupled error correction rule. Their system is sometimes referred to as *least mean squared* error correction rule, or LMS.

Multilevel Perceptrons and Backpropagation

Very soon multiple-level perceptrons emerged. First, they were introduced in 1962 by Rosenblatt. He also introduced the term "backpropagation" [167] in the section called "Back-propagating error correction procedures." His backpropagation algorithm used a probabilistic learning law. Present versions of backpropagation combine the McCulloch-Pitts linear filter both with a sigmoid output signal function and with Rosenblatt backcoupled error correction. The modern version was created by P. Werbos in 1974 [170] (and independently developed by D. Parker in 1982 [171]). Frequently the back-propagation scheme performs associative learning. During training one vector pattern is associated with another. After completion of training, the second pattern is recalled in response to input of the first [172]. The backpropagation system is trained under *slow learning* conditions, whereby each pattern is presented repeatedly (at this stage a combinatorial search is performed). In a multilevel perceptron, the input signal vector converges on a hidden unit level after passing through the first set of weighted pathways. Fanning between the levels plays the role of attention focusing. Output signals fan out to the level, and this generates the actual output of this feedforward system (the results of grouping). A backcoupled error correction system compares the actual output with a target output and feeds back their difference to all the weights.

Learning Matrix

Many models followed the perceptron. The "Hebbian rule of learning" made a correlation between presynaptic and postsynaptic signals. The Hebbian rule provides for a qualitative description of increases in path strength that occur when one cell helps to fire another. In the adaptive filter formalism, this hypothesis is often interpreted as a weight change that occurs when a presynaptic signal is correlated with a postsynaptic activity [173]. One of the earliest NN modules based on Hebbial rule is the learning matrix developed by K. Steinbuch [174]. The function of the learning matrix is to sort, or partition, a set of vector patterns into categories (classification). This model is the precursor of a fundamental module widely used in present-day neural network modeling and called *competitive learning*. Steinbuch's learning rule can be translated into the Hebbian formalism, with weight adjustment during learning a joint function of a presynaptic and postsynaptic signals. A comparative analysis of the learning matrix, the madaline models, and their electronic implementations can be found in [175]. Many developments incorporate a linear matrix and Hebbian rule of learning. This is related to the linear associative memories developed and described in [176–180].

Real-time Modeling

The desire to avoid the external control of NN-modules led to the theory of *embedding fields* [181] which has allowed *fast* nodal activation to be combined with *slow* weight adaptation. Two key architectural components of embedding field systems are the *instar* and *outstar* units. Instars appear in systems designed to carry out adaptive coding, or content-addressable memory (CAM). The outstar, which is the counterpart to the instar, carries out spatial pattern learning. Powerful computational properties arise when neural network architectures are constructed from a combination of instars and outstars. The outstar and the instar have been studied in great detail and with various combinations of activation, or short-term, or long-term memory and equations. A review of neural models that are versions of the additive equation can be found in [182]. A series of theorems encompassing neural network pattern learning by systems employing a large class of these and other activation and learning laws (including *outstar learning theorems*) was proved in [183–184]. In general, the order of activation of the outstars, as well as the spatial patterns themselves, need to be learned. This can be accomplished by using autoassociative networks, as in the theory of serial learning [185–186].

Competitive Learning

A module for *competitive learning* brings the properties of the learning matrix into the real-time setting. The basic competitive learning module combines the instar pattern-coding system with a competitive network that contrasts and enhances its filtered input. The basic competitive learning architecture consists of an instar filter and a competitive neural network. The competitive learning module can operate with or without an external teaching signal, and the learned changes in the adaptive filter can proceed indefinitely or cease after a finite time interval. If there is no teaching signal at a given time, then the net input vector is the sum of signals arriving via the adaptive filter. Then, if the category representation network is designed to make a choice, and if its weight vector matches the signal vector the best, the node will automatically become active. If there is a teaching signal, the category representation will still depend on past learning, but this will be balanced against the external signal, which may or may not overrule the past in the competition. In either case, an instar learning law allows a chosen category to encode a pattern in its learned representation. Systems were designed to learn computational maps, producing an output vector in response to an input vector. In the core of many of these computational map models there is an instar-outstar combination [187–194]. The self-organizing feature map [179] and the counter-propagation network [195] are also examples of instar–outstar competitive learning models.

Adaptive Resonance Theory

The analyses of instability cases for feedforward instar–outstar systems led to S. Grossberg's introduction of adaptive resonance theory (ART) [189] and to the development of neural network systems ART 1 and ART 2 [188, 189]. ART networks are designed, in particular, to resolve the stability-plasticity dilemma: They are stable enough to preserve significant past learning but remain adaptable enough to incorporate new information when it appears.

An ART module contains all properties required to qualify it as a GMAC unit: it has focusing attention and performs grouping by the virtue of combinatorial search (see Chapter 2). The minimal ART module includes a bottom-up competitive learning system combined with a top-down outstar pattern-learning system. When an input is presented to an ART network, the system dynamics initially follows the course of competitive learning with bottom-up activation leading to a category representation with enhanced contrast. In the absence of other inputs, the active category is determined by the past learning as encoded in the adaptive weights in the bottom-up filter. But now, in contrast with feedforward systems, signals are sent via a top-down adaptive filter. This feedback process allows the ART module to overcome all of the sources of instability. A minimal ART module is a category learning system that self-organizes a sequence of input patterns into various recognition categories. It is not an associative memory system. However, like the competitive learning module in the 1970s, a minimal ART module can be embedded in a larger system for associative memory. ART systems can also be used to pair sequences of the *categories* self-organized by the input sequences. The symmetry of the architecture implies that pattern recall can occur in either direction during the performance. This scheme brings to the associative memory paradigm the code compression capabilities of the ART system, as well as its stability properties [195–197].

Cognitron and Neocognitron

Two classes of models are variations on the themes previously described. The first class consists of the cognitron and the second, the larger-scale neocognitron [198–200]. This class of neural models is distinguished by its capacity to carry out translation-invariant and size-invariant pattern recognition. This is accomplished by redundantly coding elementary features in various positions at one level, and then cascading groups of features to the next level, then groups of these groups, and so on. Learning can proceed with or without a teacher. Locally the computations are a type of competitive learning that uses combinations of additive and shunting dynamics.

10.8.2 CMAC: An Associative Neural Network Alternative to Backpropagation

Most of the developments described above contributed to the cerebellar model arithmetic controller (CMAC).[52] CMAC is based on a theory developed independently by D. Marr [237] and J. Albus. CMAC has properties of local generalization, rapid algorithmic computation based on LMS training, incremental training, functional representation, output superposition, and fast practical hardware realization. The next section is dedicated to an explanation of how CMAC works, and to descriptions of CMAC applications in robot control, pattern recognition, and signal processing.

[52]CMAC was originally interpreted as a cerebellar model articulation controller. It was invented for practical application in control systems of multilink manipulators where the problem of motion articulation has the highest priority. Later it became clear that CMAC has broader application, and it was renamed in order to emphasize its general computational capabilities.

Learning with CMAC

CMAC is a neural network with a function very similar to the one symbolized by GFACS.[53] The cerebellar model arithmetic controller was designed for real-time control of an industrial robot and in other applications, typically employing neural networks. The CMAC can represent and use hundreds of thousands of adjustable weights that can be trained to approximate nonlinearities that are not explicitly written out nor even known. The CMAC can learn relationships from a very broad category of functions. Furthermore the learning algorithm generally converges in a small number of iterations.

The CMAC neural network is an alternative to the backpropagation-trained analog multilayer neural network. An alternative is useful, since backpropagation has a number of disadvantages. It requires many iterations to converge and therefore is inappropriate for online real-time learning. It produces a large number of computations per iteration so that the algorithm runs slowly unless implemented in expensive custom hardware. It has an error surface with possible local minima hazardous to backpropagation training that is based on gradient search techniques. It does not allow for productive incremental learning: All inputs must be seen before any weight change can take place if convergence should be quickly achieved.

In the 1970s J. Albus reported work on the CMAC [201, 202]. D. Marr published a similar work independently [237]. In the literature the model is frequently called Marr-Albus model. The CMAC was applied for rote learning in movements of an artificial arm [207, 208]. After all these years the practicality of the CMAC has many practical confirmations. Many additional advantages have been discovered. For example, the CMAC can be used to learn general state-space-dependent control responses [209]. The Robotics Laboratory at the University of New Hampshire has been investigating and using CMAC with considerable success [209–220]. Other groups working with CMAC include Ersu and colleagues [222–226], and Moody at Yale [227]. Several industrial research groups use the CMAC. The CMAC has become a key neural module for many works in machine learning and neurocontrol.

The CMAC is an associative neural network. Only a small subset of the network influences any instantaneous output, and that subset is determined by the input to the network. The associative mapping built into the CMAC assures local generalization: Similar inputs produce similar outputs, while distant inputs produce nearly independent outputs. As a result of the built-in associative properties, the number of training passes required for network convergence on real problems is orders of magnitude smaller with the CMAC than with backpropagation.

Architecture of the CMAC

We will treat the CMAC as an automaton with input space, outputs, inner functions, and a mechanism of generalization. Figure 10.23 gives an overview of the CMAC [228]. Its input, state-space, and the conceptual memory A are N-dimensional. The actual memory A' has as many dimensions as there are output components. An input vector is the collection of N appropriate sensors of the real world and/or measures of the desired goal. The input space consists of the set of all possible input vectors. The number N of input vector components and the number of outputs is arbitrary within some practical

[53]CMAC is an abbreviation for cerebellar model articulation controller. After some period of application its name changed: instead of "articulation" it became an "arithmetic" unit. The term "cerebellar" was given because of some resemblance between CMAC and the cerebellum in the brain.

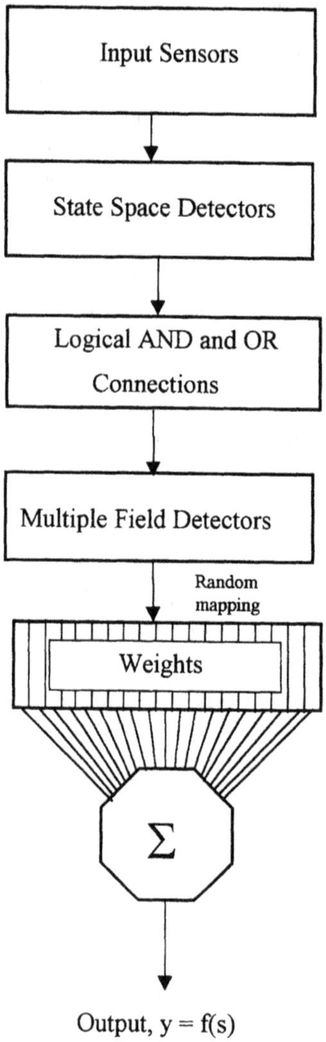

Figure 10.23. Overview of the CMAC.

limits. The vector **S** consists of an input command from higher motor centers combined with feedback from sensors in the joints, muscles, and skin of a limb. Each CMAC separately computes a $\mathbf{S} \rightarrow \mathbf{A}^*$ mapping. $\mathbf{A}^*$ is the set of locations in memory selected by the mapping. The selected locations in $\mathbf{A}^*$ contain weights corresponding to synaptic strength between parallel fibers and dendrites on a cerebellar Purkinje cell. The computed output is the sum of the weights sorted in the $\mathbf{A}^*$ address set.

The CMAC algorithm maps any inputs it receives into a set of C locations in a large "conceptual" memory (A in Fig. 10.23) in such a way that two inputs $s_1 > s_2$ that are adjacent in input space will have locations in C_1 overlap in the A memory with locations from C_2, with more overlap for closer inputs. If two inputs are far apart in the input space, there will be no overlap in their C element sets in the A memory, and therefore no generalization will be possible.

For practical systems the input space is extremely large. For example, a system with 10 inputs, each of which can take on 20 different values, would have 10^{13} points in its input space, requiring a correspondingly large number of locations in the memory A. Since most learning problems do not involve all of the input space, the memory requirement is reduced by mapping the A memory onto a much smaller physical memory A'.

As a result any input presented to the CMAC will generate only C memory locations, whose contents will be added in order to obtain an output. Notice that the nonlinearity, expected from all neural networks, is in the associative input mapping, not in the sigmoid/threshold function normally at the output of each neuron.

Figure 10.24 gives a diagram of CMAC as implemented in the laboratory at the University of New Hampshire. Each variable in the input state vector **s** is fed to a series of input sensors with overlapping receptive fields. Each input sensor produces a binary output which is ON if the input falls within its receptive field and is OFF otherwise. The width of the receptive field of each sensor produces input generalization, while the offset of the adjacent fields produces input quantization. Each input variable excites exactly C input sensors, where C is the ratio of generalization width to quantization width ($C = 4$ in Fig. 10.24; $C = 32$ to 256 in typical implementations).

The binary outputs of the input sensors are combined in a series of threshold logic units (called state-space detectors) with thresholds adjusted to produce logical AND functions (the output is ON only if all inputs are ON). Each of these units receives one input from the group of sensors for each input variable, and thus its input receptive field is the interior of a hypercube in the input hyperspace (the interior of a square in the two-dimensional input space of Fig. 10.24). The state-space detectors of Figure 10.24 correspond logically to the individual memory locations of the A memory in Figure 10.23.

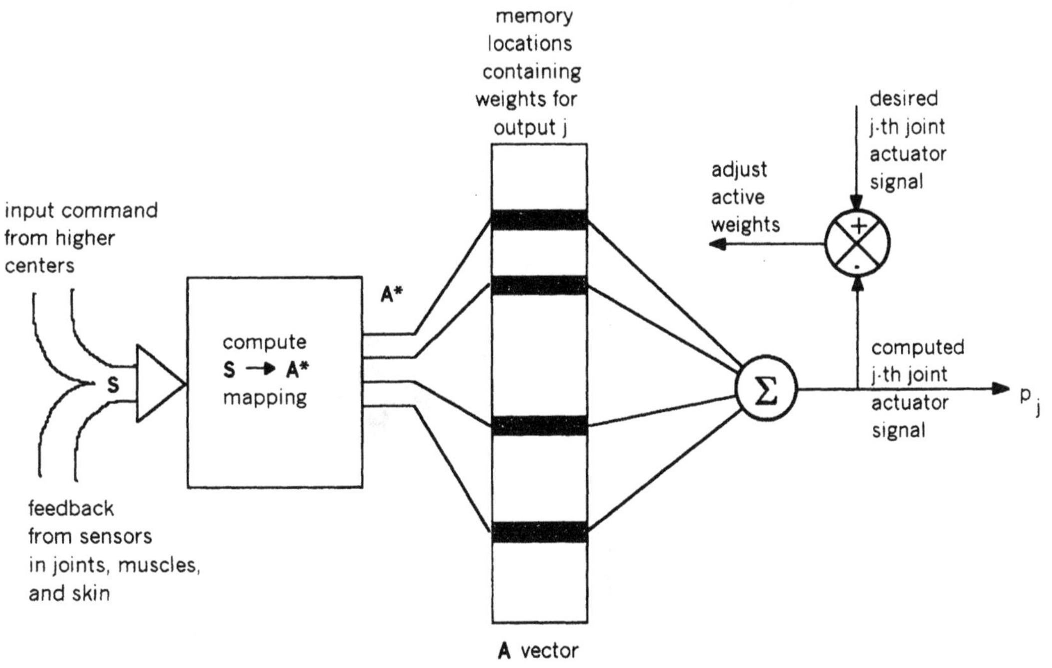

Figure 10.24. Simple example of a CMAC neural network with two inputs and one output. The generalization parameter C has the value 4. Only a partial set of the state-space detectors is shown.

If the input sensors were fully interconnected, there would be a very large number of state-space detectors, and a large subset of these detectors would be excited for each possible input. However, the input sensors are interconnected in a sparse and regular fashion in such a way that each input vector excites exactly C state-space detectors. The total collection of state-space detectors is divided into C subsets. The receptive fields of the units in each of the subsets are organized so as to span the input space without overlap. Each input vector excites one state-space detector from each subset, for a total of C excited detectors for any input. There are many ways to organize the receptive fields of the individual subsets that produce similar results. In our implementation, each subset of the state-space detectors is identical in organization but is offset relative to the others along hyperdiagonal s in the input hyperspace (adjacent subsets are offset by the quantization level of each input variable).

Grouping in the Compressed Tessellated State-Space

Ideally the associative mapping within the CMAC network assures that nearby points in the input space generalize while distant points do not. The effect of the converging connections between the state-space detectors and the multiple field detectors, however, is to create randomly distributed low-magnitude generalization with distant points in the space. Figure 10.25 illustrates this effect for a simple two-input CMAC. The size of the input space in this example is 128 points along the horizontal axis and 96 points along the vertical axis, for a total of 12,288 points in the input space. The figure shows the degree to which a central point in the space generalizes with all other points in the space, for a CMAC with 1000 adjustable weights and $C = 16$. Note that even when using this small memory (relative to the size of the input space), the characteristic of local generalization in the space is largely preserved. Note also that the generalization region is not symmetrical about the horizontal or vertical axes. The generalization of a

Figure 10.25. Generalization of a single point within 128 × 96 input space.

single point can be seen within a 128 × 96 input space, using a CMAC with $C = 16$ and 1000 memory locations. Greater generalization is indicated by darker shading.

Network training is typically based on observed training data pairs of the input and the desired output (supervised learning), using the least mean square (LMS) training rule [229].

Properties of the CMAC

Let us summarize the main properties of the CMAC:

1. *CMAC accepts real inputs and gives real outputs.* The input components are quantized, but the number of levels can be as large as desired, so high accuracy is achievable.
2. *CMAC has a built-in local generalization.* This means that input vectors that are "close" in the input (state-) space will give outputs that are close, even if the input has not be trained, as long as there has been training in that region of the state-space. The measure of "closeness" is the Hamming distance (the sum, over all components, of the absolute value of the differences of each component). Locally generalizing networks have less learning interference than globally generalizing networks such as the multilayer perceptron.
3. *Large CMAC networks can be used and trained in practical time.* This is true even with the software version of the system. The reason is that there is a small number of calculations per output even though there is a large number of weights. In CMAC realizations at UNH, on average, tens and hundreds of thousands of weights and 10 to 128 additions per output were employed. For both hardware and software this is a small amount of computation in comparison to that required in an equivalent multilayer perceptron. For example, in the pattern recognition problem CMAC required about 50 iterations while the multilayer network required about 12,000 iterations. Similar results were obtained in [227].
4. *CMAC uses the LMS adaptation rule of Widrow-Hoff* [203]. This least squares algorithm is equivalent to a gradient search of a surface that is quadratic and therefore has a unique minimum [229].
5. *CMAC can learn a wide variety of functions.* It is easy to show that a one-input CMAC can learn any discrete one-dimensional single-valued function, given a few mild conditions on the parameters of the CMAC.
6. *CMAC obeys superposition in the output space.* This means that a multidimensional discrete Fourier series can be used to show that a broad class of functions is learnable.

Problems That Require Functioning of the CMAC

The CMAC is an adaptive system that allows for a broad class of control functions to be computed by referring to a lookup table rather than by mathematical solution of simultaneous equations (for many degrees of freedom). The CMAC can combine the variables into an input vector which is used to address a memory where the appropriate output variables are stored. Each address consists of a set of physical memory locations, whose arithmetic sum of the contents is the value of the stored variable. The memory-addressing algorithm takes advantage of the deterministic nature of the control function: It allows for finding a unique correspondence of this type. The CMAC does it in a way that promises to make it possible to store the necessary data in a physical memory of practical size.

In order to carry out any control problem, it is necessary to drive the system through a sequence of states as a function of time. The control input to each subsystem actuator is, in general, a function not only of time but of other state variables as well. The states depend on higher-level input variables that identify the particular task to be performed as well as on many parameters and variables originating in the external environment. In order to deal with a problem of this complexity without sidestepping the computational difficulties as in direct human control systems [230–231], either one should ignore most of the relevant variables as in point-to-point industrial robot control [232], or one should partition the control problem into manageable subproblems [233], or both (see Chapter 7).

For controlling intelligent systems, the computations required to coordinate individual subsystems so as to produce particular motion trajectories of subsystems are usually solved by computations based on concrete relationships between architectural subsystems of the IS (e.g., see [234, 235]). In the resolved motion rate control system, end-effector motion is expressed as a function of all the individual joint motions. The computations performed for this are typically based on incomplete and erroneous mathematical approximations (the difficulties in representing real factors are described in [233, 236]. Since many factors are introduced, the formalisms of systems of this type become less and less tractable. It is simply not possible to deal with many degrees, of flexing and twisting or a very broad range of force, touch, and acceleration inputs, by systems of simultaneous equations that can be solved by computer programs of practical speed and size.

When one *examines* the type of manipulation tasks routinely performed by biological organisms such as squirrels jumping from tree to tree, birds flying through the woods, and humans playing tennis or football, one is left with the distinct impression that the solution of mathematical equations is a totally inadequate method for producing truly sophisticated motor behavior. To be sure, the present mathematical formalisms for manipulator control are in deep trouble when addressing the type of mechanical control problems that are obviously trivial for the brain of the tiniest bird or rodent.

Cerebellum as the Inspiration for the Neural Network Learner

Early attempts to model the structural properties of the biological brain were notoriously unsuccessful in producing any significant results. The subsequent disillusionment strongly prejudiced the intellectual community against seeking any guidance from the numerous existence theorems provided by nature. We believe, however, that it may be possible to duplicate the *functional* properties of the brains without modeling the structural characteristics of the neuronal substrate.

One part of the brain that seems to be intimately involved in generating motion trajectories is the cerebellum. Recent anatomical and neurophysiological data have led to a detailed theory concerning the functional operations carried out by the cerebellum [237, 238]. Input to the cerebellum arrives in the form of sensory and proprioceptive feedback from the muscles, joints, and skin together with commands from higher-level motor centers concerning what motion trajectories are to be performed. According to the theory, this input constitutes an address, whose contents are the appropriate muscle actuator signals required to carry out the desired movement. At each point in time the input addresses an output that drives the muscle control circuits. The resulting motion produces a new input, and the process is repeated. The result is a trajectory of the limb through space. At each point on the trajectory, the state of the limb is sent to the cerebellum as input, and the cerebellar memory responds with actuator signals that drive the limb to the next point on the trajectory.

A neurophysiological theory of how the cerebellum accomplishes these tasks has been published in [232, 239]. We will describe the mathematical concepts of how the cerebellum structures input data, how it computes the addresses of control signals, how the memory is organized, and how the output control signals are generated. Certain features of the neurophysiological and anatomical structure of the cerebellum has led to the theory that the cerebellum is analogous in many respects to a perceptron [238].

CMAC Mapping Algorithm

The CMAC algorithm functions by breaking the $S \rightarrow A$ mapping into two sequential mappings: $S \rightarrow M$ and then $M \rightarrow A$. Each R-ary variable s_i in the input vector $\mathbf{S}_i = (s_1, s_2, \ldots, s_N)$ is first converted into a binary variable m_i according to the following rule:

Rule of Conversion

1. Each digit of the binary variable m_i must have a value 1 over one and only one interval within the range of $\mathbf{S}_i$ and must be 0 elsewhere.
2. There are always m^* equal to 1 in the binary variable m_i for every value of the variable $\mathbf{S}_i$.

Multidimensional Mappings

The complete mapping $S \rightarrow M$ consists of N individual mappings $S_i \rightarrow m_i^*$ for all the variables in the input vector $\mathbf{S}_i = (s_1, s_2, \ldots, s_N)$.

$$
\begin{aligned}
S_1 &\rightarrow m_1^* \\
S_2 &\rightarrow m_2^* \\
S &\rightarrow M \\
&\vdots \\
S_N &\rightarrow m_N^*
\end{aligned}
$$

Mapping into a Memory of Practical Size

Earlier we described a means of performing the mapping $f : S \rightarrow A$ in a manner that is well suited to producing generalization where generalization is desired, and dichotomization where that is desired. We have not, however, explained how this transformation can be accomplished with a reasonable number of association cells. The concatenation of subscript names m_i^* produces a potentially enormous number of association cell names. If each variable in has R distinguishable values, then there are R distinguishable points in input-space. After the mappings, the concatenation to obtain A^* yields a potential number of association cell names on the same order of magnitude as R. For any practical manipulator control problem, the number of input variables N is likely to exceed 10, and the number of distinguishable values R of each variable will probably be 30 or more. The number 30^{10} is clearly an impossibly large number of association cells for any practical control device. If, however, it is not required for input vectors outside of the same neighborhood to have zero overlap, but merely a small probability of significant overlap, then it is no longer necessary to have RN association cells. Assume that an additional mapping $A \rightarrow A'$ is performed such that the RN association cells in the very large set A are mapped onto a much smaller, physically realizable set A. One way in which this can be done is by hash-coding [213].

Hash-coding is a commonly used computer technique for reducing the amount of memory required to store sparse matrices and other data sets where a relatively small amount of data is scattered over a large number of memory locations. Hash-coding operates by taking the address of where a piece of datum is to be stored in the larger memory and using it as an argument in a routine that computes an address in the smaller memory. For example, any address in the larger memory might be used as an argument in a pseudorandom number generator whose output is restricted to the range of integers represented by the addresses in the small memory. The result is a many-to-few mapping of locations in the larger memory onto locations in the smaller. Any association cell name (address) in A can be used as the argument in a hash-coding routine to find its counterpart in A'. The number of association cells in A' can be chosen arbitrarily equal to the size of the physically available memory. In practice, A' may be orders of magnitude smaller than A. Thus the $A \rightarrow A'$ mapping is a many-into-few mapping.

The many-into-few property of the hash-coding procedure leads to collision problems when the mapping routine computes the same address in the smaller memory for two different pieces of data from the larger memory. Collisions can be minimized if the mapping routine is pseudorandom in nature so that the computed addresses are as widely scattered as possible. Nevertheless, collisions are bound to occur eventually, and a great deal of hash-coding theory is dedicated to the optimization schemes that deal with them. The CMAC, however, can simply ignore the problem of hashing collisions because the effect is essentially identical to the already existing problem of cross-talk, or learning interference, which is handled by iterative data storage.

Inputs from Higher Levels

Commands from higher centers are treated by the CMAC in exactly the same way as input variables from any other source. The higher-level command signals appear as one or more variables in the input vector **S** reference variable affecting the selection of A^*. The result is that input signals from higher levels, like all other input variables, affect the output and thus can be used to control the transfer function. If, for example, a higher-level command signal changes its value, then the sample set will change. If the change is large enough, then the concatenation process will converge to a different result.

Thus, by changing the signal, the higher-level control signal can effectively change the CMAC transfer function. This control can be either a discrete or a continuous variable (i.e., it can vary smoothly over its entire range). Examples of the types of discrete commands that can be conveyed to the CMAC by higher-level input variables are "reach," "pull back," "twist," and "scan along a particular surface"(see [209–218]). An example of the types of continuously variable commands that might be conveyed to the CMAC are velocity vectors describing the motion components desired of the manipulator end-effector. Three higher-level input variables might represent the commanded velocity components of a manipulator end-effector in a coordinate system defined by some work space.

Conclusions

The CMAC form of neural network has advantages and disadvantages in comparison to other forms of neural networks. The CMAC has the advantage that it is fast in software and can be realized in high-speed hardware, thereby allowing the practical use of more weights and larger systems in solving problems. Furthermore the CMAC is able to learn a large variety of nonlinear functions, reducing the need to use slow backpropagation-based networks. The CMAC provides for a local generalization which gives it a certain edge over global generalization: There is little or no learning interference due to recent learning in remote parts of the input space, and local generalization usually requires a smaller number of additions, therefore giving fast computation speeds. Of course, CMAC/LMS training takes fewer iterations than backpropagation. The speed and network size advantage of CMAC means that real systems and real problems can be undertaken with results beyond the quality of other standard methods.

Collisions due to the hash-coding necessary to reduce the memory size to something realizable can cause a noise or interference if care is not taken in the design to prevent that from happening. Finally some care in its design must be exercised to ensure that a low-error solution will be learned in a specific application.

We can conclude that the CMAC neural network is a reliable alternative to the multilayer backpropagated network for learning situations with analog inputs and outputs, especially where speed of convergence and speed of computation are important, and where large numbers of weights are needed. The CMAC has proved able to learn unknown nonlinear functions quickly and to generalize on inputs it has never seen.

The CMAC computes control functions by referring to a look-up table rather than by solving analytic equations or using any other conventional techniques. Functional values are stored in a distributed fashion such that the value of the function at any point in input-space is derived by grouping together the contents over a number of memory locations.

The unique feature of the CMAC is the mapping algorithm that converts distance between input vectors into the degree of overlap between sets of addresses where the functional values are stored. The CMAC is thus a memory management generalizing technique that causes similar inputs to group into new units so as to produce similar outputs; yet dissimilar inputs result in outputs that are independent.

REFERENCES

[1] *New Webster's Dictionary of the English Language*, Delair Publ., 1985.

[2] *Encyclopedia Britannica. Micropedia*. vol. 6, 1978.

[3] G. A. Kimble, *Hilgard and Marquis' Conditioning and Learning*, 1961.

[4] A. Newell, J. C. Shaw, and H. A. Simon, A variety of intelligent learning in a general problem solver, in M. C. Yovits, and S. Cameron, eds., *Self-organizing Systems, Proc. Interdisciplinary Conf.*, May 1959, vol. 2, Pergamon Press, New York, 1960.

[5] H. A. Simon, Why should machines learn, in R. S. Michalski, J. G. Carbonell, and T. M. Mitchell, eds., *Machine Learning: An Artificial Intelligence Approach*, vol. 1, Tioga, Palo Alto, CA, 1983.

[6] M. Minsky, *The Society of Mind*, Simon and Schuster, New York, 1985.

[7] R. S. Michalski, Understanding the Nature of learning: issues and research directions, in R. S. Michalski, J. G. Carbonell, and T. M. Mitchell, eds., *Machine Learning: An Artificial Intelligence Approach*, vol. 2, Morgan Kaufmann, Los Altos, CA, 1986.

[8] J. P. Guilford, Factorial angles to psychology, *Psych. Rev. 69* (1961): 1–20.

[9] M. I. Shlesinger, Learning of image recognition, *Encyclopedia of Cybernetics*, vol. 2, Kiev, 1975.

[10] Y. Tsypkin, *Adaptation and Learning in Automatic Control*, Academic Press, New York, 1971.

[11] D. 0. Hebb, *The Organization of Behavior*, Wiley, New York, 1949.

[12] P. S. Churchland, *Neurophilosophy: Toward a Unified Science of the Mind/Brain*, MIT Press, Cambridge, 1986.

[13] G. M. Edelman, *The Remembered Past: A Biological Theory of Consciousness*, Basic Books, New York, 1989.

[14] B. Russell, *The Problems of Philosophy*,

[15] B. Mirkin, *Mathematical Classification and Clustering*, Kluwer Academic, 1996.

[16] S. Mitra and S. Pal, Self-organizing neural network as a fuzzy classifier, *IEEE Trans. SMC 24* (1994): pp. 385–398.

[17] J. C. Spall, Multivariate stochastic approximation using a simultaneous perturbation gradient approximation, *IEEE Trans. Auto. Cont. 17* (1992): pp. 332–341.

[18] D. F. Bowers, *Atomism, Empiricism, Scepticism*, Princeton University: Princeton, 1941.

[19] K. Koffka, *Principles of Gestalt Psychology*, Harcourt Brace, New York, 1935.

[20] A. Neel, *Theories of Psychology: A Handbook*, Schenkman, Cambridge, MA, 1977, pp. 319–332.

[21] C. Ehrenfels, On Gestalt-qualities, *Psych. Rev. 44* (1937): pp. 521–524.

[22] W. S. Sahakian, *History and Systems of Psychology*, Wiley, New York, 1975, pp. 189–193.

[23] R. Lowry, *The Evolution of Psychological Theory*, Aldine Atherton, Chicago, 1971, pp. 205–208.

[24] D. McCarthy, Gestalt as learning theory, G. I. Brown, ed., *The Live Classroom: Innovation Through Confluent Education and Gestalt*, Viking Press, New York, 1975, pp. 46–51.

[25] G. H. Bower and E. R. Hilgard, *Theories of Learning*, Prentice Hall, Englewood Cliffs, NJ, 1981, pp. 49–57.

[26] R. M. Cagne, *The Conditions of Learning*, Holt, Rinehart and Winston, New York, 1965.

[27] J. L. Brown, *The Evolution of Behavior*, W. W. Norton and Company, New York, 1975.

[28] C. L. Hull, *Principles of Behavior*, 1943; survey of the results is given in [21].

[29] H. L. Van Trees, *Detection, Estimation, and Modulation Theory*, Wiley, New York, 19.

[30] W. S. Jevons, *The Principles of Science: A Treatise on Logic and Scientific Method (1873)*. Reprint Dover, New York, 1958, pp. 218–269.

[31] K. R. Popper, *The Logic of Scientific Discovery (1959)*. Reprint Harper Torchbooks, New York, 1968.

[32] M. Kubat, I. Bratko, and R. Michalski, A review of machine learning methods, in R. Michalski, I. Bratko, and M. Kubat, eds., *Machine Learning and Data Mining: Methods and Applications*, Wiley, New York, 1996, pp. 3–71.

[33] L. Valiant, A theory of the learnable, *Comm. ACM 27* (1984): pp. 1134–1142.

[34] M. Vidyasagar, *A Theory of Learning and Generalization*, Berlin, Springer-Verlag, 1997.

[35] T. Mitchell, *Version Space*, Ph.D. Dissertation, Stanford University, 1978

[36] R. Michalski et al., The Multi-purpose incremental learning system AQ15 and its testing application to three medical domains, in Proc. Nat. Conf. on AI, MIT Press, Cambridge, 1986.

[37] R. Quinlan, Learning Efficient Classification procedures and their application to chess endgames, in R. Michalski, J. Carbonell, and T. Mitchell, eds., *Machine Learning*, Morgan Kaufmann, Los Altos, CA, 1983.

[38] S. Vere, Multilevel counterfactuals for generalizations of relational concepts and productions, *Artificial Intelligence 14*: 139–164.

[39] A. S. Poznyak, K. Najim, E. Gomez-Ramirez, *Self-learning Control of Finite Markov Chains*, Marcel Dekker, 1999.

[40] A. Meystel, J. Eilbert, L. Venetsky, and S. Zietz, Baby-robot: learning to control goal oriented behavior, *Proc. IEEE Workshop on Intelligent Control*, 1985, pp. 182–186.

[41] D. Angluin, Learning regular sets from queries and counterexamples, *Information and Computation*, vol. 75, 1987.

[42] A. Lacaze, M. Meystel, A. Meystel, Multiresolutional schemata for unsupervised learning of autonomous robots for 3D space application, *Proc. Goddard Conf. on Space Applications of AI*, Greenbelt, MD, 1994, pp. 103–112.

[43] J. Albus, A. Lacaze, and A. Meystel, Algorithm of nested clustering for unsupervised learning, *Proc. 10th IEEE Int. Symp. on Intelligent Control*, Monterey, CA, 1995.

[44] J. Albus, A. Lacaze, and A. Meystel, Multiresolutional intelligent controller for baby-robot, *Proc. 10th IEEE Int. Symp. on Intelligent Control*, Monterey, CA, 1995.

[45] C. Watkins, *Learning from Delayed Rewards*, Ph.D. Dissertation, Cambridge University, 1989.

[46] W.-M. Shen, *Autonomous Learning from the Environment*, Computer Science Press, New York, 1994.

[47] R. S. Michalski, J. G. Carbonell, and T. M. Mitchell, eds., *Machine Learning: An Artificial Intelligence Approach*, vol. 1, Tioga, Palo Alto, CA, 1983.

[48] R. S. Michalski, J. G. Carbonell, and T. M. Mitchell, eds., *Machine Learning: An Artificial Intelligence Approach*, vol. 2, Morgan Kaufmann, Los Altos, CA, 1986.

[49] W. R. Ashby, *Introduction to Cybernetics*, Chapman and Hall, London, 1956.

[50] C. G. Davis, S. Jajodia, P. A. Ng, and R. T. Yeh, eds., *Entity-Relationship Approach to Software Engineering*, North-Holland, Amsterdam, 1983.

[51] H. Gallaire, J. Minker, and J. M. Nicolas, eds., *Advances in Data Base Theory*, vol. 1, Plenum Press, New York, 1981.

[52] N. V. Findler, *Associative Networks: Representation and Use of Knowledge by Computer*, Academic Press, New York, 1979.

[53] M. L. Brodie, J. Mylopoulos, and J. W. Schmidt, eds., *On Conceptual Modelling*, Springer-Verlag, New York, 1984.

[54] M. Aizerman, E. Braverman, and L. Rosonoer, Theory of the method of potential functions for the problem of automatic classification of input functions, *Auto. Remote Cont., 25* (1964).

[55] M. Aizerman, E. Braverman, and L. Rosonoer, Probabilistic problem of the automata learning to recognize classes and the potential functions method, *Auto. Remote Cont. 25* (1964).

[56] M. Aizerman, E. Braverman, and L. Rosonoer, Method of potential functions applied for the problem of restoration of the functional characteristics of a transducer using random observations, *Auto. Remote Cont. 25* (1964).

[57] I. Rosenfeld, *The Invention of Memory: A New View of the Brain*, Basic Books, New York, 1988.

[58] S. Freud, The interpretation of dreams, in *The Standard Edition of the Complete Psychological Works of Sigmund Freud*, vol. 5, Hogart Press, London, 1959, p. 538.

[59] S. Freud, Letter 52, in *The Standard Edition of the Complete Psychological Works of Sigmund Freud*, vol. 5, Hogart Press, London, 1959, pp. 233–235.

[60] A. Meystel, Theoretical foundations of planning and navigation, *Int. J. Intell. Sys.* (1986).

[61] G. M. Edelman, Group selection as the basis for higher brain function, in F. O. Schmitt et al., eds., *The Organization of the Cerebral Cortex*, MIT Press, Cambridge, 1981.

[62] G. M. Edelman and L. H. Finkel, Neuronal group selection in the cerebral cortex, in G. M. Edelman, W. E. Gall, and W. M. Cowan, eds., *Dynamic Aspects of Neocortical Function*, Wiley, New York, 1984, pp. 653–695.

[63] G. M. Edelman and G. N. Reeke Jr., Selective networks and recognition automata, *Ann. NY Acad. Sci. 426*, (1984): pp. 181–201.

[64] G. M. Edelman, *Neural Darwinism: The Theory of Neuronal Group Selection*, Basic Books, New York, 1987.

[65] P. H. Winston, Learning by augmenting rules and accumulating censors, in R. S. Michalski, J. G. Carbonell, and T. M. Mitchell, eds., *Machine Learning: An Artificial Intelligence Approach*, vol. 2, Morgan Kaufmann, Los Altos, CA, 1986.

[66] A. Amarel, An approach to automated theory formation, in H. Von Foerster and G. W. Zopf Jr., eds., *Principles of Self-organization*, Pergamon Press, Oxford, 1962.

[67] A. A. Giordano and F. M. Hsu, *Least Square Estimation with Applications to Digital Signal Processing*, Wiley, New York, 1985.

[68] A. Meystel, Intelligent control in robotics, *J. Robot. Sys. 5* (1988).

[69] P. L. Falb, *Direct Methods in Optimal Control*, McGraw-Hill, New York, 1969.

[70] J. D'Azzo and C. Houpis, *Linear Control Systems: Analysis and Design*, McGraw Hill, New York, 1981.

[71] G. N. Saridis, *Self-Organizing Control of Stochastic Systems*, Marcel Dekker, New York, 1977.

[72] R. E. Kalman, P. L. Falb, and M. A. Arbib, *Topics on Mathematical Systems Theory*, McGraw Hill, New York, 1969.

[73] L. A. Zadeh and C. A. Desoer, *Linear Systems Theory*, McGraw-Hill, New York, 1963.

[74] K. S. Fu, Learning control systems, in J.T. Tou, ed., *Advances in Information System Science*, Plenum Press, New York, 1969.

[75] R. S. Fu, Learning control systems—review and putlook, *IEEE Trans. Auto. Cont. 15* (1970).

[76] K. S. Fu, Learning control systems and intelligent control systems: an intersection of artificial intelligence and automatic control, *IEEE Trans. Auto. Cont. 16* (1971).

[77] G. N. Saridis, Intelligent robotic control, *IEEE Trans. Auto. Cont. 28* (1983).

[78] G. N. Saridis, Toward the realization of intelligent controls, *Proc. IEEE 67* (1979).

[79] A. Meystel, Intelligent control of a multiactuator systems, in D. E. Hardt, ed., *IFAC Information Control Problems in Manufacturing Technology 1982*, Pergamon Press, Oxford and New York, 1983.

[80] G. N. Saridis, Intelligent control for robotics and advanced automation, in *Advances in Automation and Robotics*, vol. 1, JAI Press, Greenwich, CT, 1985.

[81] J. H. Graham and G. N. Saridis, Linguistic design structures for hierarchical systems, *IEEE Trans. Sys. Man Cybern., 12* (1982).

[82] G. N. Saridis, J. Graham, and G. Lee, An integrated syntactic approach, and suboptimal control for manipulators and prostheses, *Proc. 18th CDC*, Ft. Lauderdale, FL, 1979.

[83] G. N. Saridis and C-S. G. Lee, An approximation theory of optimal control for trainable manipulators, *IEEE Trans. Sys. Man Cybern. 9* (1979).

[84] G. N. Saridis, An integrated theory of intelligent machines by expressing the control performance as entropy, *Control-Theory and Advanced Technology*, vol. 1, MITA-Press, Tokyo, Japan, 1985, pp. 125–138.

[85] A. Meystel, Nested hierarchical intelligent module for automatic generation of control strategies, in U. Rembold, and K. Hormann, eds., *Languages for Sensor-Based Control in Robotics*, NATO ASI Series, vol. 29, Berlin: Springer-Verlag, 1987.

[86] A. Meystel, Baby-robot: on the analysis of cognitive controllers for robotics, *Proc. IEEE Int. Conf. Man Cybern.*, Tuscon, AZ, Nov. 11–15, 1986, pp. 327–222.

[87] A. Meystel, Intelligent systems: a semiotic perspective, *Int. J. Intell. Cont. Sys., 1* (1996): pp. 31–57.

[88] J. Albus and A. Meystel, A reference model architecture for design and implementation of semiotic control in large and complex systems, *Proc. IEEE ISIC Workshop, Architectures for Semiotic Modeling and Analysis in Large Complex Systems*, Monterey, CA, 1995, pp. 33–45.

[89] A. Meystel, Nested hierarchical control, in P. Antsaklis and K. Passino, eds., *An Introduction to Intelligent and Autonomous Control*, Kluwer Academic, Boston, MA, 1992.

[90] K. Rajaraman and P. S. Sastry, Finite time analysis of the pursuit algorithm for learning automata, *IEEE Trans. Sys. Man Cybern. Part B: Cybernetics 26* (1996): pp. 590–598.

[91] J. Albus, Outline for a theory of intelligence, *IEEE Trans. Sys. Man Cybern. 21* (1991): pp. 473–509.

[92] J. Albus, A. Lacaze, and A. Meystel, Autonomous learning via nested clustering, *Proc. 34th IEEE Conf. Decision and Control*, New Orleans, LA, 1995, pp. 3034–3039.

[93] J. Albus, A. Lacaze, and A. Meystel, Theory and experimental analysis of cognitive processes in early learning, *Proc. IEEE Int. Conf. SMC*, vol. 4, Vancouver, BC, Canada, 1995, pp. 4404–4409.

[94] D. B. Fogel, *Evolutionary Computation: Toward a New Philosophy of Machine Intelligence*, IEEE Press, New York, 1995.

[95] A. Meystel and M. Thomas, Computer aided conceptual design in robotics, *Proc. IEEE Int. Conf. on Robotics*, Atlanta, GA, March 13–15, 1984, pp. 220–229.

[96] J. R. Quinlan, Combining instance-based and model-based learning, in *Proc. 10th Int. Conf. on Machine Learning*, Amherst, MA, 1993, pp. 236–243.

[97] J. H. Holland, K. J. Holyoak, R. E. Nisbet, and P. R. Thagard, *Induction: Processes of Inference, Learning, and Discovery*, MIT Press, Cambridge,1986.

[98] R. S. Michalski, A theory and methodology of inductive learning, *Artificial Intell. 20* Elsevier, Amsterdam.

[99] L. Goldfarb, J. Abela, V. C. Bhavsar, and V. H. Kamat, Can a vector space based learning model discover inductive class generalization in a symbol environment, *Pattern Recognition Letters 16* (1995): pp. 719–726.

[100] B. Mirkin, *Mathematical Classification and Clustering*, Kluwer Academic, Dordrecht, 1996.

[101] B. Mirkin, The method of principal clusters, *Auto. Remote Control* (1988): pp. 1379–1388.

[102] B. Mirkin, Method of fuzzy additive types for analysis of multidimensional data, *Auto. Remote Control* (1990): pp. 683–821.

[103] A. Meystel, Planning in a hierarchical nested controller for autonomous robots, *Proc. IEEE 25th Conf. on Decision and Control*, Athens, Greece, 1986.

[104] G. Grevera and A. Meystel, Searching for a path through Pasadena, *Proc. IEEE Symp. on Intelligent Control*, Arlington, VA, 1989.

[105] A. Meystel, Planning in a hierarchical nested control system, in W. Wolfe and N. Marquina, eds., *Mobile Robots, Proc. of SPIE*, vol. 727, Cambridge, MA, 1986, pp. 42–76.

[106] G. Saridis, An integrated theory of intelligent machines by expressing the control performance as entropy, in *Control: Theory and Advanced Technology*, vol. 1, MITA-Press, Tokyo, Japan, 1985, pp. 125–138.

[107] A. Meystel, S. Uzzaman, G. Landa, S. Wahi, R. Navathe, and B. Cleveland, State space search for an optimal trajectory, *Proc. IEEE Symp. on Intelligent Control*, vol. 2, Philadelphia, 1990.

[108] A. N. Kolmogorov, *Foundations of the Theory of Probability (1933)*. Reprint Chelsea, New York, 1950.

[109] G. N. Saridis, Intelligent robotic control, *IEEE Trans. Auto. Cont. 28* (1983).

[110] C. Isik and A. Meystel, Knowledge-based pilot for an intelligent mobile autonomous system, *Proc. IEEE 1st Conf. of AI Applications*, Denver, CO, 1984.

[111] G. R. Lefrancois, *Psychology*, Wadsworth, Belmont, CA, 1983.

[112] J. A. Adams, *Learning and Memory: An Introduction*, Dorsey Press, Homewood, IL, 1976.

[113] C. P. Ward, *Sensation and Perception*, Academic Press, Orlando, FL, 1984.

[114] A. Hein, The development of visually guided behavior, in C. Harris, ed., *Visual Coding and Adaptability*, Laurence Elbaum Associates, Hillsdale, NJ, 1980.

[115] A. M. Farley, Issues in knowledge-based problems solving, *IEEE Trans. Sys. Man Cybern., 10* (1980).

[116] E. B. Carne, *Artificial Intelligence Techniques*, Spartan Books, Washington DC, 1965.

[117] R. L. Beurle, Storage and manipulation of information in the brain, J. IEEE, *5* (1959).

[118] R. L. Beurle, Storage and manipulation of information in random networks, in C. A. Muses, ed., *Aspects of the Theory of Artificial Intelligence*, Plenum Press, New York, 1962.

[119] M. D. Mesarovich, On self-organizational systems, in M. C. Yovitts, G. T. Jacobi, and G. D. Goldstein, eds., *Self-organizing Systems*, Washington DC, 1962.

[120] F. Rosenblatt, *Principle of Neurodynamics*, Spartan Books, New York, 1962.

[121] M. Minsky and S. Papert, *Perceptions: An Introduction to Computational Geometry*, MIT Press, Cambridge, 1969.

[122] B. Wizdow and M.E. Hoff, Adaptive switching circuits, WESCON Conv. Record, Part IV, 1960.

[123] S.-I. Amari, Neural theory of association and concept formation, *Biol. Cybern. 26* (1977).

[124] T. J. Nelson, A neural network model for cognitive activity, *Biol. Cybern. 49* (1981).

[125] L. Bobrowsky, Learning processes in multilayer threshold nets, *Biol. Cybern. 31* (1978).

[126] L. Bobrowsky and E. R. Caiariello, Comparison of two unsupervised learning algorithms, *Biol. Cybern. 37* (1980).

[127] A. D. Golt'sevy and E. M. Kussul, Controlling the motion of a moving robot using a NN, *Auto. Remote Cont. 5* (1982).

[128] D. F. Cuhn and S. R. Phillips, ROBNAV: A range-based robot navigation and obstacle avoidance algorithm, *IEEE Trans. Sys. Man Cybern. 5* (1975).

[129] A. G. Barto, R. S. Sutton, and C. W. Anderson, Neuron-like adaptive elements that can solve difficult learning control problems, *IEEE Trans. Auto. Cont. 13*, No. 5, 1983.

[130] A. G. Barto and P. Anandan, Pattern-recognizing stochastic learning automata, *IEEE Trans. Sys. Man Cybern., 15* (1985).

[131] A. G. Barto, C. W. Anderson, and R. S. Sutton, Synthesis of nonlinear control surfaces by a layered associative search network, *Biol. Cybern. 43* (1982).

[132] A. G. Barto and R. S. Sutton, Landmark learning: an illustration of associative search, *Biol. Cybern. 42* (1981).

[133] A. G. Barto, R. S. Sutton, and P. S. Browser, Associative search network: a reinforcement learning associative memory, *Biol. Cybern., 40* (1981).

[134] K. Fukushima, S. Miyake, and T. Ito, New cognition: a neural network model for a mechanism of visual pattern recognition, *IEEE Trans. Sys. Man Cybern., 13* (1983).

[135] R. R. Gawzonski and B. Macukow, The adaptive neuron like layer net as a control learning system research, Vol. 3, Adv. Publ. Ltd. London, 1977.

[136] R. R. Gawzonski, Two stage learning algorithm for optimal control and a manipulator, *Proc. Conf. on Robotic Research: The Next Five Years and Beyond*, Bethlehem, PA, 1984.

[137] J. J. Craig, Adaptive control of manipulators through repeated trials, *Proc. ACC*, vol. 3, San Diego, CA, 1984.

[138] J. Simons et al., A self-learning automation with variable resolution for high precision assembly by industrial robots, *IEEE Trans. Auto. Cont. 27* (1982).

[139] D. Stove, Hume, probability, and induction, *Philosophical Rev.* (April 1965).

[140] G. Cohen and P. Bernhard, On the rationality of some decision rules in a stochastic environment, *IEEE Trans. Auto. Cont., 24* (1979).

[141] S. R. Liberty and R.C. Hartwig, Design-performance-measure statistics for stochastics linear control systems, *IEEE Trans. Auto. Cont., 23* (1978).

[142] S. M. Brooks, A comparison of maximum seeking methods, *Operations Res. 7* (1959).

[143] A. N. Mucciardi, Self-organizing probability state variable parameter search algorithms for systems that must avoid high-penalty operation regions, *IEEE Trans. Sys. Man Cybern. 4* (1974).

[144] S. Hampson and D. Kibler, A Boolean complete neural model of adaptive behavior, *Biol. Cybern. 49* (1983).

[145] L. Siklossy and J. Dreussi, An efficient robot planner which generates its own procedures, in *Robotic Research, Proc. 3-d IJCAI*, Stanford, CA, 1973.

[146] B. Dufay and J. C. Lalombe, An approach to automated robot programming based on inductive learning, in M. Brady and R. Paul, eds., *Robotic Research, 1st Intell. Symp.*, MIT Press, Cambridge, 1984.

[147] R. E. Fikes, P. E. Hart, and N. J. Nilsson, Learning and executing generalized robot plans, *Artificial Intell. 3* (1972).

[148] P. H. Winston, Learning structural description from examples, in *The Psychology of Computer Vision*, McGraw-Hill, New York, 1975.

[149] P. H. Winston, T. O. Binford, B. Katz, and M. Lowry, Learning physical descriptions, examples and precedents, in M. Brady and R. Paul, eds., *Robotic Research, 1st Intell. Symp.*, MIT Press, Cambridge, 1984.

[150] W. S. Havens, Recognition mechanisms for schema-based knowledge representations, *Comp. Maths. Appls. 9* (1983).

[151] J. M. Tenenbaum and H. G. Barrow, IGS: A paradigm for integrating image segmentation and artificial intelligence, ed. by C. C. Chen, Academic Press, New York, 1976.

[152] M. D. Levene, A knowledge-based computer vision system, in A. R. Hanson and E. M. Riseman, eds., *Computer Vision Systems*, Academic Press, New York, 1978.

[153] A. R. Hanson and E. M. Riseman, A design of semantically directed vision processor, *Comp. and Info. Sci. Dept. Tech, Rep. 75C-1 University of Mass.*

[154] R. Bajcsy, Visual and conceptual focus of attention, in *Structured Computer Vision Systems*, Academic Press, New York, 1980.

[155] T. Lozano-Perez, A language for automatic mechanical assembly, in P. H. Winston and R. H. Brown, eds., *Artificial Intelligence, an MIT Perspective*, vol. 2, MIT Press, Cambridge, 1979.

[156] H. Farreny and H. Prade, On the problem of identifying an object in a robotic scene from a verbal imprecise description, in A. Danthine and M. Geraldin, eds., *Advanced Software in Robotics*, North-Holland, Amsterdam, 1984.

[157] H. Jappinen, A model for representing and invoking robot bodily skills, in A. Danthine and M. Geraldin, eds., *Advanced Software in Robotics*, Elsevier, North-Holland, Amsterdam, 1984.

[158] J. H. Graham and G. N. Saridis, Linguistic decision structures for hierarchical systems, *IEEE Trans. Sys. Man. Cybern.*, *12* (1982).

[159] G. N. Saridis and J. H. Graham, Linguistic decision schemata for intelligent robots, *Automatica 20* (1984).

[160] T.-S. Chen, Image processing by incomplete information, *IEEE Workshop on Languages of Automation*, Mallorca, Spain, 1985.

[161] A. Meystel, Multiresolutional architecture for autonomous systems with incomplete and inadequate knowledge representation, in H. B. Tzafestas and H. B. Verbruggen, eds., *AI in Industrial Decision Making, Control and Automation*, 1995.

[162] A. Lacaze, M. Meystel, and A. Meystel, Schemata for unsupervised learning for autonomous robots for 3d space operation, *NASA Conf. on Space Applications of AI*, Greenbelt, MD, 1993, pp. 103–112.

[163] A. Lacaze, Unsupervised early conceptual learning in mobile robots. MS thesis, Drexel University, 1994.

[164] G. A. Carpenter, Neural network models for pattern recognition and associative memory, *Neural Net. 2* (1989).

[165] W. S. McCulloch and W. Pitts, A logical calculus of the ideas immanent in nervous activity, *Bull. Math. Biophys. 9* (1943): pp. 127–147.

[166] F. Rosenblatt, The perceptron: a probabilistic model for information storage and organization in the brain, *Psych. Rev. 65* (1958): pp. 386–408.

[167] R. Rosenblatt, *Principles of Neurodynamics*, Spartan Books, Washington DC, 1962.

[168] B. Widrow and M. E. Hoff, Adaptive switching circuits, *1960 IRE WESCON Convention Record*, part 4 (1960): pp. 96–104.

[169] B. Widrow and R. Winter, Neural mets for adaptive filtering and adaptive pattern recognition, *Computer 21* (1988): pp. 25–39.

[170] P. J. Werbos, Beyond regression: new tools for prediction and analysis in the behavioral sciences, Ph.D. Thesis. Harvard University, Cambridge, 1974.

[171] P. J. Werbos, Generalization of backpropagation with application to a recurrent gas market model, *Neural Net. 1* (1988): pp. 339–356.

[172] D. E. Rumelhart, G. E. Hinton, and R. J. Williams, Learning internal representations by error propagation, in D. E. Rumelhart and J. L. McClelland, eds., *Parallel Distributed Processing: Explorations in the Microstructures of Cognitions*, vol. 1, MIT Press, Cambridge, 1986, pp. 318–362.

[173] S. Grossberg, Some nonlinear networks capable of learning a spatial pattern of arbitrary complexity, *Proc. Nat. Acad. Sci., USA 59* (1968): pp. 368–372.

[174] K. Steinbuch, Die lernmatrix. *Kybernetik 1* (1961): pp. 36–45.

[175] K. Steinbuch and B. Widrow, A critical comparison of two kinds of adaptive classification networks, *IEEE Trans. Electr. Comput.*, *14* (1965): pp. 737–740.

[176] T. Kohonen, Correlation matrix memories, *IEEE Trans. Comput. 21* (1972): pp. 353–359.

[177] J. A. Anderson, A simple neural network generating an interactive memory, *Math. Biosci. 14* (1972): pp. 197–220.

[178] T. Kohonen, *Content-Addressable Memories*, Springer-Verlag, Berlin, 1980.

[179] T. Kohonen, *Self-organization and Associative Memory*, Springer-Verlag, Berlin, 1984.

[180] N. Nakano, Associatron: a model of associative memory. *IEEE Trans. Sys. Man Cybern. 2* (1972): pp. 381–388.

[181] S. Grossberg, *The Theory of Embedding Fields with Applications to Psychology and Neurophysiology*, Rockefeller Institute for Medical Research, New York, 1964.

[182] S. Grossberg, Nonlinear neural networks: principles, mechanisms, and architectures, *Neural Net. 1* (1988): pp. 17–61.

[183] S. Grossberg, Pattern learning by functional-differential neural networks with arbitrary path weights, in K. Schmitt, ed., *Delay and Functional Differential Equations and Their Applications*, Academic Press, New York, 1972, pp. 121–160.

[184] S. Grossberg, Neural expectation: cerebellar and retinal analogs of cells fired by learnable or unlearned pattern classes, *Kybernetik 10* (1972): pp. 49–57.

[185] S. Grossberg and J. Pepe, Schizophrenia: possible dependence of associational span, bowing, and primacy vs. recency on spiking threshold, *Behavioral Sci. 15* (1970): pp. 359–362.

[186] R. Hecht-Nielsen, Neural analog information processing, *Proc. Society of Photo-Optical Instrumentation Engineers, 298* (1981): pp. 138–141.

[187] S. Grossberg, On the development of feature detectors in the visual cortex with applications to learning and reaction diffusion systems, *Biol. Cybern. 21* (1976): pp. 145–159.

[188] S. Grossberg, Adaptive pattern classification and universal recoding, I: Parallel development and coding of neural feature detectors, *Biol. Cybern. 23* (1976): pp. 121–134.

[189] S. Grossberg, Adaptive pattern classification and universal recoding, II: Feedback, expectation, olfaction, and illusions, *Biol. Cybern. 23* (1976): pp. 87–202.

[190] S.-I. Amari, Neural theory of association and concept formation, *Biol. Cybern. 26* (1977): pp. 175–185.

[191] S.-I. Amari and A. Takeuchi, Mathematical theory on formation of category detecting nerve cells, *Biol. Cybern. 29* (1978): pp. 127–138.

[192] E. Bienenstock, L. N. Cooper, and P. W. Munro, A theory for the development of neuron selectivity: orientation specificity and binocular interaction in the visual cortex, *J. Neurosci. 2* (1982): pp. 32–48.

[193] C. von der Malsburg, Self-organization of orientation sensitive cells in the striate cortex, *Kybernetik 14* (1973): pp. 85–100.

[194] D. E. Rumelhart and D. Zipser, Feature discovery by competitive learning, *Cog. Sci. 9* (1985): pp. 75–112.

[195] R. Hecht-Nielsen, Counterpropagation networks, *Appl. Optics 26* (1987): pp. 4979–4984.

[196] G. A. Carpenter and S. Grossberg, A massively parallel architecture for a self-organizing neural pattern recognition machine, *Comput. Vision, Graphics, Image Process. 37* (1987): pp. 54–115.

[197] G. A. Carpenter and S. Grossberg, ART 2: self-organization of stable category recognition codes for analog input patterns, *Appl. Optics 26* (1987): pp. 4919–4930.

[198] K. Fukushima, Cognitron: A self-organizing multilayered neural network, *Biol. Cybern. 20* (1975): pp. 121–136.

[199] K. Fukushima, Neocognitron: A self-organizing neural network model for a mechanism of pattern recognition unaffected by shift in position, *Biol. Cybern. 36* (1980): pp. 193–202.

[200] K. Fukushima, Neocognitron: a hierarchical neural network capable of visual pattern recognition, *Neural Net. 1* (1988): pp. 119–130.

[201] J. S. Albus, A theory of cerebellar functions, *Math. Biosci. 10* (1971): pp. 25–61.

[202] J. S. Albus, Theoretical and experimental aspects of a cerebellar model, Ph.D. Dissertation, University of Maryland, 1972.

[203] B. Widrow and M. E. Hoff, Adaptive switching circuits, *Proc. IRE Western Electronic Show and Conv. 4* (1960): pp. 96–104.

[204] B. Widrow, Generalization and information storage in networks of adaline neurons, in M. C. Yovits, ed., *Self-organizing Systems*, Spartan Books, Washington DC, 1962, pp. 435–439.

[205] J. J. Hopfield, Neurons with graded response have collective computational properties like those of two-state neurons, *Proc. Nat. Acad. Sci. USA 81* (1984): pp. 3088–3092.

[206] J. J. Hopfield, Neural networks and physical systems with emergent collective computational abilities, *Proc. Nat. Acad. Sci. USA 79* (1982): pp. 2554–2558.

[207] J. S. Albus, Data storage in the cerebellar model articulation controller, 1. *ASME Trans. Dynamic Sys. Meas. Cont.* (Sept. 1975): pp. 228–233.

[208] J. S. Albus, A new approach to manipulator control: the cerebellar model articulation controller (CMAC), (Sept. 1975): pp. 220–227.

[209] W. T. Miller, A nonlinear learning controller for robotic manipulators, *Proc. SPIE, Intelligent Robots and Computer Vision*, vol. 726, Oct. 1986, pp. 416–423.

[210] W. T. Miller, A learning controller for nonrepetitive robotic operations, *Proc. Workshop on Space Telerobotics*, Publication 87-13, vol. 11, Pasadena, CA, Jan. 10–12, 1987, pp. 273–281.

[211] W. T. Miller, Sensor based control of robotic manipulators using a general learning algorithm, *IEEE Trans. Robot. Auto. 3* (1987): pp. 157–165.

[212] W. T. Miller, F. H. Glanz, and L. G. Kraft, Application of a general learning algorithm to the control of robotic manipulators, *Internat. J. Robot. Res. 6* (1987): pp. 84–98.

[213] W. T. Miller and R. P. Hewes, Real time experiments in neural network based learning during high speed nonrepetitive robotic operations, *Proc. 3rd IEEE Int. Symp. on Intelligent Control*, Aug. 24–26, 1988, pp. 513–518.

[214] W. T. Miller, Real time learned sensor processing and motor control for a robot with vision, *1st Ann. Conf. of the Int. Neural Network Society*, Sept. 1988, p. 347.

[215] W. T. Miller, Real time application of neural networks for sensor based control of robots with vision, *IEEE SMC 19*, July–Aug. 1989, pp. 825–831.

[216] F. H. Glanz and W. T. Miller, Shape recognition using a CMAC based learning system, *Proc. SPIE Conf. on Robotics and Intelligent Systems*, vol. 848, Nov. 1987, pp. 294–298.

[217] F. H. Glanz and W. T. Miller, Deconvolution using a CMAC neural network, *Proc. 1st Ann. Conf. of the Int. Neural Network Society*, Sept. 1988, p. 440.

[218] F. H. Glanz and W. T. Miller, Deconvolution and nonlinear inverse filtering using a neural network, *Int. Conf. on Acoustics and Signal Processing*, vol. 4, May 23–29, 1989, pp. 2349–2352.

[219] L. G. Kraft and D. P. Campagna, A comparison of CMAC neural network and traditional adaptive control systems, *Proc. 1989 American Control Conf.*, vol. 1, June 21–23, 1989, pp. 884–889.

[220] D. Herold, W. T. Miller, L. G. Kraft, and F. H. Glanz, Pattern recognition using a CMAC based learning system, *Proc SPIE, Automated Inspection and High Speed Vision Architectures 11*, vol. 1004, Nov. 10–11, 1989, pp. 84–90.

[221] W. T. Miller, R. P. Hewes, F. H. Glanz, and L. G. Kraft, Realtime dynamic control of an industrial manipulator using a neural-network-based learning controller, *IEEE Trans. Robot. Auto. 6* (1990): pp. 1–9.

[222] E. Ersu and H. Tolle, Hierarchical learning control—an approach with neuron-like associative memories, *Proc. IEEE Conf. on Neural Information Processing Systems*, Nov. 1988.

[223] E. Ersu and H. Tolle, A new concept for learning control inspired by brain theory, *Proc. FAC 9th World Congress*, July 2–6, 1984.

[224] E. Ersu and J-Militzer, Real-time implementation of an associative memory-based learning control scheme for nonlinear multivariable processes, *Proc. 1st Measurements and Control Symp. on Applications of Multivariable Systems Techniques*, 1984, pp. 109–119.

[225] E. Ersu and X. Mao, Control of pH using a self-organizing control concept with associative memories, *Proc. Int. IASTED Conf. on Applied Control and Identification*, 1983.

[226] J. Moody, Fast learning in multi-resolution hierarchies, in D. Touretzsky, ed., *Advances in Neural Information Processing*, Morgan Kaufmann, Los Altos, CA, 1989.

[227] W. T. Miller, B. A. Box, and E. C. Whitney, Design and implementation of a high speed CMAC neural network using programmable CMOS logic cell arrays, University of New Hampshire, Report no. ECE.IS.90.01, Feb. 6, 1990.

[228] B. Widrow and S. D. Stearns, *Adaptive Signal Processing*, Prentice-Hall, Englewood Cliffs, NJ, 1985.

[229] E. G. Johnsen and W. R. Corliss, *Teleoperators and Human Augmentation*, NASA SP-5047, 1967.

[230] W. R. Corliss and E. G. Johnsen, *Teleoperator Controls*, NASA SP-5070, 1968.

[231] J. R. Ashley and A. Pugh, Logical design of control systems for sequential mechanisms, *Int. J. Prod. Res. (IPE) 6* (1968).

[232] R. Paul, Modeling, Trajectory calculation and servoing of a computer controlled arm. Ph.D. Thesis. Stanford University, 1972.

[233] D. E. Whitney, Resolved motion rate of manipulators and human prostheses, *IEEE Trans. Man-Mach. Sys. 10* (1969): pp. 47–53.

[234] First Annual Report for the Development of MultiModed Remote Manipulator Systems, Charles Stark Draper Laboratory (Division of Massachusetts Institute of Technology). Report C-3790.

[235] M. E. Kahn and B. Roth, The near minimum-time control of open loop articulated kinematic chains, *J. Dynamic Sys. Meas. Cont., Trans. ASME*, series G, *93* (Sept. 1971).

[236] S. P. Grossman, The motor system and mechanics of basic sensory-motor integration, *Textbook of Physiological Psychology*, Wiley, New York, 1967 Ch. 4.

[237] D. Marr, A theory of cerebellar cortex, *J. Physiol.*, **202** (1969): pp. 437–470.

[238] J. S. Albus, A robot conditioned reflex system modeled after the cerebellum. *Proc. Fall Joint Computer Conf. 61*, 1972, pp. 1095–1104.

[239] W. T. Miller, III, F. H. Glanz, and L. G. Craft, CMAC: An associative neural network alternative to backpropagation, *Proc. IEEE 78* (1990): pp. 1561–1567.

PROBLEMS

10.1. Give your analysis of the existing definitions of learning.

10.2. Determine which of the Gestalt principles could be procedurally satisfied by using the set of procedures GFS (GFACS).

10.3. Compare induction, abduction, and deduction. Give examples.

10.4. Give a list of existing types of learning. Which of them are different in principle?

10.5. Explain functioning of an algorithm of inductive generalization.

10.6. Demonstrate that when unsupervised conceptual learning is used, it will develop a hierarchy of concepts, or a hierarchy of rules.

10.7. Is it possible that baby robot (employing the algorithm of inductive generalization) would arrive at a conclusion (i.e., would learn) that baby robot itself should be represented in the map with all other phenomena of the world?

10.8. Compare the learning properties of (a) NN with back propagation and (b) CMAC.

CHAPTER 11

APPLICATIONS OF MULTIRESOLUTIONAL ARCHITECTURES FOR INTELLIGENT SYSTEMS

11.1 AN INTELLIGENT SYSTEMS ARCHITECTURE FOR MANUFACTURING

The intelligent systems architecture for manufacturing (ISAM) is a reference model architecture for intelligent manufacturing systems. It is intended to provide a theoretical framework for the development of standards and performance measures for intelligent manufacturing systems. It also intended to provide engineering guidelines for the design and implementation of intelligent control systems for a wide variety of manufacturing applications. ISAM consists of a hierarchically layered set of intelligent processing nodes organized as a nested series of control loops. In each node, tasks are decomposed, plans are generated, world models are maintained, feedback signals from sensors are processed, and control loops are closed. In each layer, nodes have a characteristic span of control, with a characteristic planning horizon, and corresponding level of detail in space and time. Nodes at the higher levels deal with corporate and production management, while nodes at lower levels deal with machine coordination and process control. ISAM integrates and distributes deliberative planning and reactive control functions throughout the entire hierarchical architecture, at all levels, with all spatial and temporal scales.

11.1.1 ISAM Standard

The ISAM model addresses the manufacturing enterprise at a number of levels of abstraction:

1. At the highest level of abstraction, ISAM provides a conceptual framework for viewing the entire manufacturing enterprise as an intelligent system consisting of machines, processes, tools, facilities, computers, software, and human beings operating over time and on materials to produce products.
2. At a lower level of abstraction, ISAM provides a reference model architecture to support the development of performance measures and the design of manufacturing systems and software.

3. At a still lower level of abstraction, ISAM is intended to provide engineering guidelines to implement specific instances of manufacturing systems such as machining and inspection systems.

11.1.2 ISAM as a Conceptual Framework

The ISAM conceptual framework spans the entire range of manufacturing operations, from those that take place over time periods of microseconds and distances of microns to those that take place over time periods of years and distances of many kilometers. The ISAM model is intended to allow for the representation of activities that range from detailed dynamic analysis of a single actuator in a single machine to the combined activity of thousands of machines and human beings in hundreds of plants comprising the operations of a multinational corporation.

To span this wide range of activities, ISAM adopts a hierarchical layering with different range and resolution in time and space at each level. This permits the definition of functional entities at each level within the enterprise such that each entity can view its particular responsibilities and priorities at a level of spatial and temporal resolution that is understandable and manageable to itself. At any level within the hierarchy, functional entities receive goals and priorities from above and observe situations in the environment below. In each functional entity at each level, there are decisions to be made, plans to be formulated, and actions to be taken that affect peers and subordinates at levels below. Information must be processed, situations analyzed, and status reported to peers and supervisors above. Each functional entity needs access to a model of the world that enables intelligent decision making, planning, analysis, and reporting activity to be carried out despite the uncertainties and unwanted signals that exist in the real world. At each level, there are values (often not explicitly stated) that set priorities and guide decision making.

Typically a manufacturing enterprise is organized into management units such that each management unit consists of a group of intelligent agents (humans or machines). Each of these possesses a particular combination of knowledge, skills, and abilities. Each has a job description that defines duties and responsibilities. Each management unit accepts tasks from higher-level management units and issues subtasks to subordinate management units. Within each management unit, agents are given job assignments and allocated resources with which to carry out their assignments. These intelligent agents schedule their activities to achieve the goals of the jobs assigned to them. Each agent is expected to make local executive decisions to keep things on schedule by solving problems and compensating for minor unexpected events.

Typically each unit of management has a model of the world environment in which it must function. This world model is a representation of the state of the environment and of the entities that exist in the environment, including their attributes and relationships, and the events that take place in the environment. The world model also typically includes a set of rules that describes how the environment will behave under various conditions. Each unit of management also has access to sources of information that keep its world model current and accurate. Finally each management unit has a set of values, or cost functions, that it uses to evaluate that state of the world and by which its performance is evaluated.

11.1.3 ISAM versus Current Practice

For the most part, current industry practice assumes that manufacturing consists of largely predictable processes that can safely proceed without the benefit of, or need for, on-line measurement and real-time feedback control. Most adjustments in manufacturing processes are made by operators that often use intuition and experience to tune parameters. In control parlance, most production processes operate "open loop." There is little or no consideration given to the need for real-time planning or replanning, automatic error recovery, on-line optimization, or adaptability to changing conditions. On-line schedule changes, and process modifications are handled mostly by manual ad-hoc methods.

Current interface standards are mostly limited to data exchange standards for static data such as IGES and STEP [1, 2]. Most communication protocol standards such as MMS and TCP/IP [3, 4] do not deal with the semantics or pragmatics of the processes being controlled. A few steps toward standards for dynamic interfaces that convey meaning are appearing, such as APIs and CORBA [5, 6]. However, open architecture interface standards based on standard functional decomposition, and that can enable software from a variety of vendors to work together dynamically, are still well in the future. ISAM embodies the concepts described in [7, 13–17]. Their theoretical underpinning was outlined in [12]. Their manufacturing implementation has a variety of forms. The specification of interfaces is presented in [8].

Future manufacturing will be characterized by the need to adapt to the demands of agile manufacturing, including rapid response to changing customer requirements, concurrent design and engineering, lower cost of small volume production, out-sourcing of supply, distributed manufacturing, just-in-time delivery, real-time planning and scheduling, increased demands for precision and quality, reduced tolerance for error, in-process measurement and feedback control. These demands generate requirements for adaptability and online decision making that cannot be met with current practice. Future manufacturing systems will require intelligent control concepts that for the most part are being developed outside of the field of industrial engineering. The ISAM conceptual framework attempts to apply intelligent control concepts [7–9] to the domain of manufacturing so as to enable the full range of agile manufacturing concepts.

11.1.4 ISAM as a Reference Model Architecture

ISAM defines the functional elements, subsystems, interfaces, entities, relationships, and information units involved in intelligent manufacturing systems.

DEFINITION *Functional elements* are the fundamental computational units of a system.

AXIOM 1 The functional elements of an intelligent system are behavior generation, sensory perception, world modeling, and value judgment.

AXIOM 2 World modeling maintains and uses a distributed dynamic store of knowledge that collectively forms a world model and includes both a model of the manufacturing environment and a model of the system itself.

AXIOM 3 The functional elements and knowledge database of an intelligent system can be represented in an architectural node by a set of modules interconnected by a communication system that transfers information between them.

AXIOM 4 The complexity inherent in intelligent systems can be managed through multiresolutional hierarchical layering.

The BG modules form a command tree. The information in the KD is shared between the WM modules in the nodes within the same subtree. The right side shows examples of the functional characteristics of the BG modules at each level. The left side shows examples of the types of entities recognized by the SP modules and stored by the WM in the KD knowledge database at each level. The sensory data paths flowing up the hierarchy typically form a graph, and not a tree. An operator interface provides input to, and output from, the modules in every node.

A typical machining center has four subsystems: tool motion, parts handling, communication, and inspection. Each of the four subsystems has one or more mechanisms. Each of these has one or more actuators and sensors. For example, the manipulation subsystem may consist of a spindle and tool path controller with several axes of continuous-motion path control, plus a tool changer consisting of a tool carousel and a manipulator. Each of these has two or more actuators and sensors. The parts handling subsystem might be a pallet feeding mechanism, which consists of a conveyor and buffer, a pallet shuttle, and a pair of turn tables. The communication subsystem might consist of a message encoding subsystem, a protocol syntax generator, and communications bus interface.

The inspection subsystem might consist of mechanisms that use cameras and touch probes to detect and track objects, surfaces, edges and points, and compute trajectories for probe drive motors, and pan, tilt, and focus actuators. All of these functions need to be coordinated in order to successfully achieve behavioral goals.

The operator interface provides the ability for the operator to start or stop the system at any time, to single step, to override the feed rate of the motion controller, to actuate or monitor any of the tool change or pallet-feeding mechanisms. The operator interface can send or display information from the communications subsystem, or display any of the state variables in the world model in real time. Using the operator interface, a human operator is able to run diagnostic programs in the case of failures, display control programs (plans) while they are being executed, generate graphic images of tool paths, display volumes to be removed from parts by numerical control (NQ) programs, and portray shaded images or wire frame models of parts with dimensions and tolerances.

The functionality of each level in the ISAM reference model hierarchy is defined by the characteristic timing, bandwidth, and algorithms chosen for decomposing tasks and goals at each level. Typically these are design choices that depend on the dynamics of the processes being controlled. The numbers shown in the figure are representative of those appropriate for a machining center. For other types of systems, different numbers would be derived from design parameters.

11.1.5 An ISAM Example

To illustrate the types of issues that can be addressed by the ISAM reference model architecture, an example of a seven-level ISAM hierarchy for a machine shop is given below.

Level 7: Shop

The shop level receives input task commands of the form <manufacture_orders>. It decomposes those commands into output subtask commands to the cell level of the form <machine_orders>, <assemble_orders>, <ship_orders>, and <maintain_inventory>.

The shop level plans activities and allocates resources for one or more manufacturing cells for a period on the order of one eight-hour shift. At the shop level, orders are sorted into batches, and a production schedule is generated for the cells to process the batches. The world model maintains a knowledge database containing names, contents, and attributes of batches and the inventory of tools and materials required to manufacture them. Maps may describe the location of, and routing between, manufacturing cells. Sensory perception processes compute information about the flow of parts, the level of inventory, and the operational status of all the cells in the shop. Value judgment computes the cost and benefit of various batching and routing options and calculates statistical quality control data. An operator interface allows human operators to visualize the status of orders and inventory, the flow of work, and the overall situation within the entire shop facility. Operators can intervene to change priorities and redirect the flow of materials and tools. Executors keep track of how well plans are being followed, and modify parameters as necessary to keep on plan. The output from the shop level provides work flow assignments for the cells.

Level 6: Cell

The cell level receives input task commands of the form <machine_orders>, and <assemble_orders>. It decomposes those commands into output subtask commands to the workstation level of the form <mill_batch>, <drill_batch>, <grind_batch>, <weld_batch>, <assemble_batch>, <inspect_batch>, and <transport_batch>.

The cell level plans activities and allocates resources for one or more workstations for a period of about an hour into the future. Batches of parts and tools are scheduled into particular workstations. The world model symbolic database contains names and attributes of batches of parts and the tools and materials necessary to manufacture them. Maps describe the location of, and routing between, workstations. Sensory perception determines the location and status of trays of parts and tools. Value judgment evaluates routing options for moving batches of parts and tools. An operator interface allows human operators to visualize the status of batches and the flow of work through and within the cell. Operators can intervene to change priorities and reorder the plan of operations. Executors keep track of how well plans are being followed, and modify parameters as necessary to keep on plan. The output from the cell level are commands issued to particular workstations to perform machining, inspection, or material handling operations on particular batches or trays of parts.

Level 5: Workstation

The workstation level receives input task commands of the form <mill_batch>, <drill_batch>, <grind_ batch>, <weld_batch>, <assemble_batch>, <inspect_batch>, and <transport_batch>. It decomposes those commands into output subtask commands to the workstation level of the form <load_part>, <fixture_part>, <mill_part>, and <unload_part>.

The workstation level schedules tasks and controls the activities within each workstation with about a five-minute planning horizon. A workstation may consist of a group of machines, such as one or more closely coupled machine tools, robots, inspection machines, materials transport devices, and part and tool buffers. Plans are developed and commands are issued to equipment to operate on material, tools, and fixtures in order to produce parts. The world model symbolic database contains names and attributes of parts, tools, and buffer trays in the workstation. Maps describe the location of parts, tools, and buffer trays. Sensory perception determines the position of parts and tools in trays and buffers. Value judgment evaluates plans for sequencing machining and parts handling operations within the workstation. An operator interface allows human operators to visualize the status of parts and tools within the workstation, or to intervene to change priorities and reorder the sequence of operations—within the workstation. Executors keep track of how well plans are being followed and modify parameters as necessary to keep on plan. Output commands are issued to particular machine tools, robots, and tray buffers to perform tasks on individual parts, tools, and fixtures.

Level 4: Equipment Task

The equipment level receives input task commands of the form <load_part>, <fixture_part>, <mill_part>, and <unload_part>. It decomposes those commands into output subtask commands to the E-move level of the form <locate_part>, <grasp_part>, <maneuver_part>, <set_clamps>, <mill_face>, <mill_pocket>, <mill_slot>, and <mill_champfer>.

The equipment level schedules tasks and controls the activities of each machine within a workstation with about a 30-second planning horizon. (Tasks that take much longer may be broken into several 30-second segments at the workstation level.) Level 4 decomposes each equipment task into elemental moves for the subsystems.

Plans are developed that sequence elemental movements of tools and grippers, tool changers, and pallet shuttle systems. Commands are formulated to move tools and grippers so as to approach, grasp, move, fixture, cut, drill, mill, or measure parts. The world model symbolic contains names and attributes of parts, such as their size and shape (dimensions and tolerances) and material characteristics (mass, color, and hardness). Maps consist of drawings that illustrate part shape and the relative positions of part features. Sensory perception measures part dimensions and tolerances. Value judgment evaluates part quality and supports planning for part handling and fixturing sequences.

An operator interface allows human operators to visualize the status of operations of the machine, to intervene to change priorities, or to interrupt the sequence of operators. Executors keep track of how well plans are being followed and modify parameters as necessary to keep on plan. Output command are issued to level 3 for machining, manipulating, and inspecting part features.

Level 3: Elemental Move (E-move)

The E-move level receives input task commands of the form <grasp_part>, <maneuver_part>, <set_clamps>, <mill_pocket>, and <mill_slot> It decomposes those commands into output subtask commands to the primitive level of the form <move_gripper_along_path>, <move_part_along_path>, and <move_tool_along_path>. The E-move level schedules and controls simple machine motions requiring a few seconds. (Motions that require significantly more time may be broken up at the task level into several elemental moves.) Plans are developed and commands are issued that define safe pathway points for tools, manipulators, and inspection probes so as to avoid collisions and singularities and ensure part quality and process safety. The world model symbolic database contains names and attributes of part features such as surfaces, holes, pockets, grooves, threads, chamfers, and burrs.

Maps consist of drawings that illustrate feature shape and the relative positions of feature boundaries. Sensory perception measures dimensions of individual features, and computes surface properties. Value judgment supports planning of machine motions and evaluates feature quality. An operator interface allows a human operator to visualize the state of the machine or to intervene to change mode or interrupt the sequence of operations. Executors keep track of how well plans are being followed, and modify parameters as necessary to keep on plan. Output consists of commands to move along trajectory segments between way points.

Level 2: Primitive

The primitive level receives input task commands of the form <move-[part | tool | gripper |-along-path>. It decomposes those commands into output subtask commands to the servo level of the form

<go_to_(x, y, z, roll, pitch, yaw) at feedrate_v>.

The primitive level plans paths for tools, manipulators, and inspection probes so as to minimize time and optimize performance. It computes tool or gripper acceleration and deceleration profiles taking into consideration dynamical interaction between mass, stiffness, force, and time. Planning horizons are on the order of a few hundred milliseconds. The world model symbolic database contains names and attributes of linear features such as lines, trajectory segments, and vertices. Maps (when they exist) consist of perspective projections of linear features such as edges, lines, and tool or end-effector trajectories. Sensory perception computes observed motions of tools and grippers. Value judgment supports trajectory optimization. An operator interface allows a human operator to visualize the state of the machine, and to intervene to change mode or override the feed rate. Executors keep track of how well plans are being followed and modify parameters as necessary to keep within tolerance. Output consists of commands to move tool or grippers at desired velocities and accelerations, to exert desired forces, or to maintain desired stiffness parameters.

Level 1: Servo Level

The servo level receives input task commands of the form <go_to_(x, y, z, roll, pitch, yaw) at feedrate_v>. It decomposes those commands into output subtask commands to the actuators in the form of desired force, torque, or power.

The servo level transforms commands from tool path to joint actuator coordinates. Planners interpolate between primitive trajectory points for each actuator with a planning horizon of a few tens of milliseconds. The world model symbolic database contains values of state variables such as joint positions, velocities, and forces; proximity sensor readings; position of discrete switches; state of touch probes; as well as image attributes associated with camera pixels. Maps consist of camera images and displays of sensor readings. Sensory perception scales and filters data from sensors that measure actuator positions, velocities, forces, torques, and touch.

An operator interface allows a human operator to visualize the state of the machine or to intervene to change mode, set switches, or jog individual axes. Executors servo individual actuators and motors to follow interpolated trajectories. Position, velocity, or force servoing may be implemented, and in various combinations. Output commands to power amplifiers specify desired actuator torque or power. Outputs are typically produced every few milliseconds (or whatever rate is dictated by the machine dynamics and servo performance requirements). The servo level also commands switch closures that control discrete actuators such as relays and solenoids.

At the servo and primitive levels, the output command rate is typically clock driven on a regular cycle. At the E-move level and above, the command output rate becomes irregular because it is event driven.

At each hierarchical level, world modeling, sensory perception, and value judgment modules provide to the behavior generation modules the information needed for decision making and control. At each hierarchical level, behavior generation modules decompose tasks into subtasks for subordinate behavior generation modules. At each level world model knowledge is shared between knowledge databases at the same level, and relational pointers are established between knowledge data structures at both higher and lower levels.

At each level, commands and status information are transmitted up and down the control hierarchy and between behavior generation and operator interface modules. Operator interfaces are able to display information from any of the functional modules and the knowledge database.

At each level, sensory perception modules accept sensory observations from sensory perception modules at lower levels, and output processed and clustered sensory observations to sensory perception modules at higher-level nodes.

At all levels, SP modules compare and correlate predictions from the WM with observations from lower level SP modules. Differences are used to update the estimated state of the world stored in the world model knowledge database. Correlations are used to recognize correspondence between what is stored in the internal world model knowledge database and what is observed in the external real world.

At each level, knowledge is represented in a form and with a spatial and temporal resolution that meets the processing requirements of the node. At each level, there is a characteristic loop bandwidth, a characteristic planning horizon, a characteristic type of task decomposition, a characteristic range of temporal integration of sensory and a characteristic window of spatial integration. At each level, information is extracted from the sensory data stream to keep the world model knowledge database accurate and up to date.

At each level, sensory data are processed, entities are recognized, world model representations are maintained, and tasks are deliberatively decomposed into parallel and sequential subtasks to be performed by cooperating sets of agents within the BG modules. At each level, feedback from sensors reactively closes a control loop allowing each agent to respond and react to unexpected events. At each level, tasks are decomposed into subtasks and subgoals, and agent behavior is planned and controlled. The result is a system that combines and distributes deliberative and reactive control information throughout the entire hierarchical architecture with both planned and reactive capabilities tightly integrated at all levels of space and time resolution.

The specific numbers and functions given in this example are illustrative only. They are meant only to illustrate how the generic structure and function of the ISAM reference model architecture might be instantiated at the lower levels of a manufacturing enterprise. The point of this example is to illustrate how the ISAM multilevel hierarchical architecture integrates real-time planning and execution of behavior with dynamic world modeling, knowledge representation, and sensory perception. At each level, behavior generation is guided by value judgments that optimize plans and evaluate results. The system architecture organizes the planning of behavior, the control of action, and the focusing of computational resources. The overall result is an intelligent real-time control system that is both driven by high-level goals and reactive to sensory feedback.

A methodology for implementing the ISAM architecture for this and other types of specific manufacturing applications is currently being addressed at NIST as part of an ongoing project. As part of that effort, the internal structures of the BG, WM, SP, VJ, and KD modules are being developed.

ISAM as an Implementation Guide

A first step in designing an ISAM architecture for a large-scale intelligent manufacturing system is to develop a generic processing node that can be duplicated many times at many different levels. The generic node can then be populated with the knowledge, skills, and abilities required by its particular duties and responsibilities, and interconnected with other generic nodes so as to produce the desired functionality.

The relationships and interactions among the BG, WM, KD, SP, and VJ modules in a generic node of the ISAM architecture were discussed earlier in Chapters 1, 2, and 7. The behavior-generating (BG) modules contain planner (PL) and executor (EX) submodules. The planner has sub-submodules of job assignment (JA), scheduling (SC), and plan selector (PS). The world modeling (WM) module contains the knowledge database (KD), with both long-term and short-term symbolic representations and short-term iconic images. In addition WM contains a simulator where the alternatives generated by JA and SC are tested for comparison in PS. The sensory processing (SP) module contains filtering, detecting, and estimating algorithms, plus mechanisms for comparing predictions generated by the WM module with observations from sensors. It has algorithms for recognizing entities and clustering entities into higher level entities. The value judgment (VJ) module evaluates plans and computes confidence factors based on the variance between observed and predicted sensory input.

Each node of the ISAM hierarchy closes a control loop. Input from sensors is processed through sensory processing (SP) modules and used by the world modeling (WM) modules to update the knowledge database (KD). This provides a current best estimate of the state of the world x. Depending on the concept of estimation used, a particular control law is applied to this feedback signal arriving at the EX submodule. The EX submodule computes the compensation required to minimize the eventual difference between the planned reference trajectory and the trajectory that emerges as a result of the control process. The "best estimate" is also used by the JA and SC sub-submodules and by the WM plan simulator to perform their respective planning computations. It is further used in generating a short-term iconic image that forms the basis for comparison, recognition, and recursive estimation in the image domain in the sensory processing SP module.

Current work at NIST is directed toward the development of the generic node shown in Figure 7.22 (Chapter 7). A control hierarchy for a machine shop, including inspection machines, machine tools, and a material handling system is being constructed from such nodes interconnected through application program interfaces (APIs) which are intended to serve as a prototype for future standards [10]. An enhanced machine controller has been developed using APIs to serve as a testbed for open architecture controllers for machine tools, coordinate measuring machines, and robots [11]. Future efforts will be directed toward the development of engineering methods for implementation of intelligent manufacturing systems.

11.2 PLANNING IN THE HIERARCHY OF NIST-RCS FOR MANUFACTURING

Planning in the manufacturing environment can be approached consistently as a part of the general behavior generation process generated by the BG module. We compare here different planning strategies and describe a planner for the hierarchical reference architecture NIST-RCS. The suggested planner is simulated and the results are compared with other algorithms.

11.2.1 Planner within a Single RCS Module

The function of a planner is to find "the best" possible plan:

1. At the assigned level of resolution.
2. With respect to an assigned cost function.
3. Following all the assigned constraints.
4. In the allocated computational time.

In general, planning consists of the following steps:

Step 1. Create (using combinatorics or other methods) sets of future decisions of space (resource-task) assignments.
Step 2. Calculate the next state after each decision was made.
Step 3. Assign time tags to the decisions.
Step 4. Predict the cost of the strings of future decisions. Simulate the strings to check for possible inconsistencies (or conflicts).
Step 5. Select the time tagged string of decisions (plan) that follows the constraints and better fits the assigned cost function.

To perform the assigned tasks, the NIST-RCS planner (PL) is composed of three modules: JA, SC, and PS. It receives a task from the upper level (lower resolution) and transforms this task into a time-tagged set of subtasks (plan) which is passed to EX for execution. These three modules were described in the previous sections.

The same modules (JA, SC, PS) can be identified in all planning algorithms. By modules we are implying that there are conceptually different functions that the planner must perform. This does not mean that they must be separated pieces of software, and we are not implying any interchangeability properties. Although there is a wide variety of planning algorithms, most planners mainly differ on the interaction between the JA and the SC.

Tightly Coupled JA -SC

In these cases the spatial and temporal search is performed jointly by the JA-SC couple. JA gives one possible assignment (one decision) at a time to the scheduler. Based on the scheduler's response, JA makes a new possible assignment. Dynamic programming fits this category.

Loosely Coupled JA-SC

There are some cases where JA and SC can be loosely coupled. Most of them are just special cases of the tightly coupled case:

- There are no "possible" alternative job assignments. The JA makes an actual instead of possible assignment, and gives it to SC. Because no alternatives exist, the complete assignment can be given to SC. There is no need for a JA-SC loop. All dispatching rules fit under this category.
- The cost cannot be evaluated on a decision-by-decision basis. Complete strings of job assignments are created by JA and then given to SC. In this case there is a JA-SC loop. It is done in plan-by-plan steps. These cases are rare and make

the problem of finding "the best" plan more cumbersome because it is necessary to calculate the cost of the complete string. In most cases it is possible to find decision-by-decision cost approximations that allow us to use more efficient tightly coupled JA-SC algorithms.

In some cases, task decomposition precedes SC. But in most cases, only after scheduling is completed, the task can be considered decomposed.

11.2.2 Planner in the RCS Hierarchy

There is a common misconception that says that the structure of planners at each manufacturing level has to be radically different. We have found that at all the levels where we have a fairly well-defined WM representation, the planning problems can be transformed into a multidimensional state-space search that can be conceptually solved with the same algorithms. The difference between one level and the next is that the type of information represented changes, but not the representational structure. So the planner can search through a decision graph in a similar way, although the information is conceptually different. We believe that the same planning algorithm can be applied at the resource planning level, shop scheduling level, work cell planning level, machine control level, and actuator control level. It can also be applied to the product and production planning level in cases where the world model is thoroughly defined. This is rarely the case in production scheduling.

The planning problems are typically represented as follows:

1. There are t tasks to be performed.
2. There are r resources that can perform those tasks.
3. There is a cost-function to be minimized.
4. There is a set of ordering constraints that must be followed.

Under these assignments, the objective is to find the distribution of tasks among the resources (JA) and times of performing these tasks (SC) so that the cost function is minimized and the constraints followed.

Assume that we have two resources (r_1 and r_2), and three tasks (t_1, t_2, and t_3) that must be processed. Let us say that t_2 and t_3 can only be performed if t_1 is finished. For simplicity, we will assume that in our optimization problem we will minimize time. PL should plan jobs in such a way that their distribution will allow for evaluation of the overall time of performing all tasks that are assigned so that the cost function is minimized.

At $t = 0$ there are only two possible (fully allocated) assignments: t_1 assigned to r_1 (alternative 1) or t_1 assigned to r_2 (alternative 2). If PL explores the alternative 1, then there are two different decisions that can be made after the r_1 is done: either t_2 to r_1 and t_3 to r_2 (alternative 1–1) or t_3 to t_1 and t_2 to r_2 (alternative 1–2). Two other alternatives are possible if we explore alternative 2. All these alternatives can be stored in a tree form (the decision tree). The string of decisions between the root of the decision tree and a leaf is a plan. The decision tree is useful in comparing planning algorithms to see what decisions are explored. This model may not be correct for production scheduling where there are penalties for not processing tasks, but the "optimal" plan may not process all of them. In this example, we assume that all tasks must be processed.

11.2.3 Algorithm of Planning

Our strategy has the following characteristics (this list allows for comparison with other existing techniques of schedule generation, e.g., with utilization of GA algorithms):

1. We assume that a decision-by-decision cost-function is available. The cost of the complete course of action will be the sum of all individual costs for all decisions made along that decision path. Decision-by-decision cost-functions (or "incremental cost-functions") are not always available; for example, the cost of not delivering work within a given window for production planning might not be known.
2. The JA-SC couple of algorithms constructs the decision tree as it generates and explores the alternatives instead of first creating the complete decision tree and then searching through it.
3. Only in the worst computational cases the complete decision tree will be evaluated, usually the decision is achieved before the whole tree is constructed.
4. No decision path is studied more than once. It keeps track of previously explored paths, unlike most planners utilizing GA techniques ("genetic algorithms" of searching).
5. Theoretical premises and the fundamentals of computational complexity have been thoroughly tested in different cases of motion path planning. Although motion path planning is not considered as a classical scheduling problem, the algorithm that we present can be used for both cases.
6. Even if the available time does not allow for completion of the algorithm operation, and the tree of decisions is not constructed in full, the first decisions that have already been obtained (the ones that need to be executed sooner) as a part of the best *incomplete* schedule have a low probability of being changed if the process of computations would not be interrupted.

The algorithm that we apply frequently in various NIST-RCS engineering systems is a mathematical programming technique rooted in Dijkstra or A^* techniques of search. It has the following steps:

Step 1. If the decision tree is empty (or "selected plan" $pl_{\text{sel}} = \emptyset$), then go to step 3.

Step 2. If the decision tree is not empty, then find the cheapest leaf in the decision tree. The cost of the leaf is the cost of the decision path until the leaf (by Dijkstra algorithm). If A^* was applied as a search algorithm, then the cost is the result of summation of two costs: the cost of the path from the starting node to the least expensive leaf and the estimated cost of the least expensive possible path to the end of the schedule (A^*). The second choice requires an estimate of the least possible losses of time to the end of the schedule (it needs to be always smaller or equal to the actual path to warranty optimality). Label the plan created by the decisions made from the root (or roots) of the decision tree to the selected leaf as pl_{sel}.

Step 3. Identify the number of tasks for each task type ($nt_1, nt_2, \ldots$) that has all its constraints satisfied at t_{sel} given that all the decisions in pl_{sel} were taken, where t_{sel} is the next time at which a resource becomes free or a task is finished (or both) along pl_{sel}.

Step 4. Identify the number of resources available for each task type ($nr_1, nr_2, \ldots$) at t_{sel} given that exhaust all decisions in pl_{sel}.

Step 5. For each task type (i):

a. Create all combinations of the nr_i resources and the nt_i tasks as well as combinations of the nt_i tasks and the nr_i resources if $n, r_i > nt_i$.

b. Create all permutations of each combination and store them in a list (L_i).

Step 6. Create a new leaf in the decision tree for each possible assignment $l_{n,1}, l_{m,2}, l_{o,3}, \ldots, l_{z,i}$ where $l_{n,1}$ is an element of L_1, $l_{m,2}$ is an element from L_2, and so on. These leaves are parts of the tree pl_{sel}.

Step 7. Find which is the cost before the next event in each of these new leaves of pl_{sel} by determining places where an event is a resource that becomes free or a task being finished (or both). Store this cost and the time of the event in the leaf (to be used within step 2).

Step 8. (Optional) Prune the new leaves if they achieve a state that was achieved by another path in a cheaper manner. This step is optional because searching for these states may be more computationally expensive than not pruning.

Step 9. If all assigned tasks are finished, select the completed plan with the best cost by backtracking through the decision tree; exit successfully; otherwise, go to step 2.

Steps 2 through 7 demonstrate that JA-SC computational processes are tightly coupled. Also it is possible to interpret step 9 as a sketch of the PS module. Step 7 implies that functions from VJ (value judgment) are invoked. This algorithm will find the optimal cost because all paths that are shorter than the optimal cost have already been explored in step 3 both in the A^* version and in the Dijkstra version. No path longer than the optimal one will be explored to the end.

H. Simon states in [18]:

> ...while OR [operations research] has been able to provide a theoretical analysis of scheduling problems for very simple situations—a shop with a very small number of machines and processes—automatic scheduling of large shops with many machines and processes has been beyond the scope of exact optimization methods. (p. 13)

This is true if we attack the problem of planning as a one-level problem. But NIST-RCS provides an architecture where each level has a reduced focus of attention (search envelope), thus allowing the use of optimal algorithms like the one presented above. Because PL creates plans only for one level, it should always be a limited amount of combinations to study.

It is not always clear that the planning problems can be decomposed into subproblems with constrained (preferably nil) interactions. An assumed decomposition does not necessarily warrant the best solution achieved by this decomposition. If the decomposition is inappropriate, then the solution found will be an expensive one. On the other hand, we know that complex planning problems are always subdivided into different levels from the very beginning (on an intuitive level). We have seen examples of these decompositions within the control layers in many practical examples. In the case of manufacturing, we can go from the actuator control level, where the tasks could be the application of a voltage to a motor, to a machine control level, where the tasks can be the milling of a part and the resources are the different machines that can do that milling. In the case of the work cell planning level, the tasks could be small batches of parts and the resources are the different work cells within the shop floor. In most cases it is possible to identify the tasks and resources at each level of the manufacturing control hierarchy.

If the algorithm takes more time than allocated to find the optimal plan, the following alternatives are always available:

1. Introduce a new intermediate NIST-RCS level to the control hierarchy that makes the search space (envelope) smaller and the search time reduced.
2. Allow the algorithm to search as deep into the future as the time allocated, and if it is not completed at the moment of time when the result is immediately required, then the best path found at that moment should be extended with a dispatching rule. Of course, it will not be able to warrant optimality.

The addition of the dispatching rule will create one more "just-in-time" planner, which is a heuristic. Still the algorithm warrants finding a plan that is better than other available alternatives of pursuit (or equal to them, in the worst case) and better than the dispatching rule by itself.

11.2.4 RCS Approach within the System of Contemporary Views on Scheduling in Manufacturing

Production scheduling is concerned with the effective allocation of the manufacturing resources over time for the manufacture of goods. These resources include agents, materials, and time. The agents are a combination of machines and the labor force. Most organizations, both large and small, must solve scheduling problems on a recurrent basis. This creates considerable demand for good scheduling techniques. Starting in the 1950s, operations researchers began advocating formal optimization approaches to scheduling [19].

Project scheduling allocates resources at specific times to tasks that together complete a project or accomplish a mission. Although production scheduling involves recurrent operational decisions, project scheduling normally involves a unique project. The project may be the construction of a house or a deep space expedition by an unmanned probe. The principles involved in project scheduling carry over from one project to another. But we never schedule exactly the same objects for the same reasons as in production scheduling [19].

Two earlier methods of project scheduling became universally accepted: the critical path method (CPM) and the project evaluation and review technique (PERT). The CPM considers activity times as a function of cost and seeks to trade off cost with completion time. The PERT models activity times as random variables and is used to analyze project completion time [37, 38, 39] explain both models.

Production Scheduling from the AI Point of View

Knowledge-based systems (KBS) provide the major AI approach to scheduling, in general, and production scheduling, in particular. These systems employ domain-specific information to derive schedules and in the process do not aspire nor guarantee optimality [19].

KBS can be divided between those that use dispatching rules (most) and those that don't. Among those that use dispatching rules we can find [20, 21, 22]. GENREG [23] can modify a set of dispatching rules to better suit the problem. A different approach to scheduling within the Al paradigm is a constraint-based reasoning. Within this category we can find ISIS [24].

ISIS creates a "beam" search where a few alternatives are created based on some arbitrarily selected operators. ISIS is hierarchical with the following levels predefined: order selection, resource selection, and reservation selection. Note that what is meant by hierarchy is very different from the NIST-RCS hierarchy. If none of the best-first itineraries satisfies all the constraints, then a "repair" algorithm is called to try to reschedule the "level" with a more relaxed set of constraints. ISIS evolved into OPIS [25]. Its methodology is almost identical to that of the ISIS ("beam"). It is a reactive scheduling algorithm in the sense that it concentrates upon how to modify the schedules already found to satisfy new constraints and new rules. OPIS is composed of the following parts: order scheduler, resource scheduler, right shifter, left shifter, and demand swapper. The current versions of OPIS are not suited for real time operations because of the time that it takes to schedule [19].

A hybrid system KBSS is presented in [26] that combines KBS (reactive) systems with search. It is claimed that KBSS finds schedules of good quality, usually within a couple of percentage points from the optimum.

Computational Intelligence (CI) Approach to Production Scheduling

CI methods for production scheduling are mainly based on a genetic algorithm (GA) [27], simulated annealing, and neural networks. The genetic algorithm approach is basically a search technique that mimics the general idea of genetic evolution. GAs are capable of quickly identifying possible solution areas, but they may not converge rapidly to the optimal solution [28]. GAs require different operators to create the new generation of solutions. These operators are a set of heuristics that cross the current generation of solutions or schedules. Two different kinds of heuristic crossover operators are shown in [29, 30].

Musser in [31] uses a simulated annealing technique that is currently used at the Texas Instruments PC manufacturing facility. It produces a weekly schedule for a process containing 50 operations. The average time to generate the schedule is 15 minutes, which precludes it from operating real time [19]. Simulated annealing is another search technique that has its roots in the formation of crystals and is related to mechanical engineering. Laarhoven in [32] presents an argument where he concludes that simulated annealing is not as efficient as some other tailored heuristics, since they show large running times.

Vaithyanthan and Satake in [33, 34] present two Hopfield network schedulers. They have different methods to deal with the problem of early conversion of their networks. But both claim they have no way to ensure that the final results are optimal or reasonably close to the optimum. In [35, 36] Willems and Aoki use neural networks in conjunction with integer programming to schedule.

Project Scheduling from the AI Point of View

In [40] project scheduling is done by way of an expert system where previous solutions to the problem are stored and the schedule is adapted by a set of heuristics that rearrange a set of operations that apply to the schedule. Barber in [41] describes a prototype for an intelligent tool for project control and scheduling known as XPERT. It considers the problem of expediting selected activities if a project is to break some timing constraint. Again this is a set of heuristics used for "fixing" a previous schedule.

Computational Intelligence (CI) Approach to Project Scheduling

A number of classical algorithms (Simplex, network flow, Dijkstra) are used under the CPM approach in [38]. McKim in [42] presents three case studies in neural networks and compares them to classical methods.

Comparison between Searching for the "Best" Schedule and Using Dispatching Rules

The following question should be addressed while creating an optimal planning algorithm: Is the cost saved by the optimal planning algorithm significant enough to make the effort to look for more alternatives than a simple dispatching rule? The following experiment was conducted to answer this question: A number of planning problems were created with t tasks and r resources.

- The length of the tasks and the efficiency (speed) of the resources were assigned randomly.
- No precedence constraints were imposed.
- Bidding (best first) was used to plan all the created planning problems. Approximately 20 planning problems were created for each combination of tasks and resources, $t = 2, 4, 6$ and $r = 2, 4, 6$.
- The Dijkstra algorithm was used to plan the same problem.
- The results were compared.

From the experience of these applications of ISAM, the following conclusions can be made:

1. Planning is realizable on-line as a joint search procedure of JA, SC, and PS for a wide variety of numerical examples.
2. The NIST-RCS allows us to use optimal algorithms by reducing the space of search.
3. Search in the state space (S_3-search based on Dijkstra or A^* algorithms) is a good choice among the optimal algorithms
4. The optimal algorithms give significant cost advantages over dispatching rules.
5. The proposed strategy and algorithm of planning allows for unfinished results to be used ("just-in-time planning").

11.3 INSPECTION WORKSTATION-BASED TESTBED APPLICATION FOR THE INTELLIGENT SYSTEMS ARCHITECTURE FOR MANUFACTURING

A multiresolutional architecture is being developed to support the development and implementation of intelligent control systems for manufacturing. The diversity of manufacturing architectures entertained in industry is notoriously rich, and this diminishes advantages of possible, but not yet implemented in full, application of standards. We provide here some background on the approach being used and describe the initial implementation testbed.

11.3.1 HPCC Program: A Paradigm for ISAM Implementation

The High-Performance Computing and Communications (HPCC) program was formally established following passage of the High-Performance Computing Act in 1991. The Systems Integration for Manufacturing Applications (SIMA) program activity is one activity under the Information Infrastructure Technology and Applications (IITA) component of the HPCC program. Since 1994 the SIMA program has partially funded work on the development of a standard reference model architecture for the control of complex intelligent systems. This effort builds on earlier related work in several projects at the National Institute of Standards and Technology (NIST) dating back to the early 1980s and leverages current efforts in other programs.

In particular, the NIST efforts have included developing the real-time control system (RCS) architecture [7], manufacturing systems integration (MSI) architecture [44], and the quality in automation (QIA) architecture [45]. Each of these was developed with specific attention given to different areas of application in the manufacturing domain. For example, while RCS primarily focused on real-time control issues in machine tools, robots, autonomous and tele-operated vehicles, underwater vehicles, mining machines, cranes, and others, MSI focused on manufacturing activities above the shop floor. At the same time, application of QIA concepts focused on problems of in-process and process-intermittent inspection on machine tools.

The reference model control architecture used in this project includes important concepts from the reference model known as Intelligent Systems Architecture for Manufacturing (ISAM) and presented in Section 11.1. ISAM is a reference model architecture for intelligent manufacturing systems. It is intended to provide a theoretical framework for developing standards and performance measures for intelligent manufacturing systems, as well as engineering guidelines for the design and implementation of intelligent control systems for a wide variety of manufacturing applications. ISAM is intended to anticipate the future needs of industry.

The implementation of a testbed to evaluate the applicability of the reference architecture is currently under way. The testbed development leverages the work of several industry/government consortia. This ensures attention to current industry issues. These relationships are described briefly in our discussion of the approach below. The primary goals of this project are to develop a detailed design of the NIST reference model architecture for intelligent control of manufacturing processes and then to demonstrate, validate, and evaluate the NIST reference model architecture through analysis and performance measurements of a simulated and prototype implementation.

These two objectives address the NIST Manufacturing Engineering Laboratory's (MEL) intention of providing U.S. industry with state-of-the-art manufacturing architectures and models, fostering the development and implementation of advanced manufacturing systems, and anticipating and addressing the needs of U.S. manufacturing industry for the next generation of advanced manufacturing systems and standards.

11.3.2 The Approach

The approach is to proceed with the reference model design, while simultaneously choosing and developing background and theory, and then the reference model itself. At this point, enough of the architecture has been defined to warrant evaluation in a testbed implementation. The initial implementation was chosen to be a coordinate measuring machine– (CMM–) based inspection workstation (IWS) at NIST. The choice of this facility permits this project to take advantage of capabilities in advanced inspection techniques currently being developed under the Next Generation Inspection System (NGIS) [46] program. That use of vision and scanning probe technologies provides a sensory rich environment for testing aspects of the architecture.

This workstation is also a test node in the computer integrated manufacturing (CIM) framework project [47]. It operates in the National Advanced Manufacturing Testbed (NAMT) that has been developed at NIST. Ultimately the reference model architecture will be tested with this workstation and other simulated and implemented factory components. Another key element of the approach is to leverage the work of the DOE Technologies Enabling Agile Manufacturing (TEAM) programs [48] in message API definitions and enhanced machine controllers (EMC) [11]. The development approach includes parallel activity areas. Several important ones are described next. This effort is to create a software structure, interface definitions, and task frame specifications that permit the core control node model to be standardized to the extent possible. As such, the software organization and development support for the behavior generation, world modeling, value judgment, and sensor processing components of a control node and their interfaces are being examined and developed.

The initial implementation employs a set of standard C++ software templates that simplify node development. These were derived from earlier versions of templates used in an EMC open architecture control system for machine tool control developed by NIST and installed for evaluation at the GM Powertrain.

11.3.3 Application Design for Prototype System Implementations

The inspection workstation was chosen as the initial implementation, in part because of the complexity of the control issues and richness of its sensor suite, including vision. Through the joint efforts of RCS, MSI, and QIA and NGIS experts, a draft scenario has been written for the activities of the IWS that can exercise the architecture in important ways. First, a list of capabilities that will sufficiently exercise the architecture was generated. Activities that required these capabilities were then included in the scenario.

These capabilities include online task planning, processing, and handling; explicit quality loops; advanced machine–human interaction; interface with legacy software and hardware; simulation design, plan, and production data interface; administrative functions; learning; multiple simultaneous tasks per node; and interaction between CAD and sensed features.

The scenario identifies visible activities and data exchanges that reflect characteristics of typical inspection activities determined through various factory visits. A subset of the activities was used for the initial implementation in September 1996. The scenario served as an indispensable starting point in the implementation design. From this point there were developed task descriptions and task trees, control node hierarchies, and state representations for all tasks for each intended testbed mission. Part of the effort was the continued interaction and collaboration with software tool vendors that produce products that may be adapted to assist control system developers in this phase of implementation.

In addition to providing the baseline control module templates, the EMC program has also provided additions to the template structure to support neutral manufacturing language (NML) communications interfaces, an API specification, a development environment with version control and source check-in and check-out mechanisms, operator interfaces, and advanced control system diagnostic capabilities.

The intelligent closed-loop processing (ICLP) activity of TEAM can define a set of API messages for manufacturing. As a member of the reference model architecture project team, it actively participates in that effort and applies and extends applicable results to the APIs used in the IWS testbed implementation. These specifications support communication between control nodes.

11.3.4 Modules

Planning

Planning at an ISAM node is conducted in the behavior generation module. An extensive study of planning/scheduling algorithms has been performed, and performance experiments have been conducted. Further work to develop standard interfaces to planners, plan simulators and evaluators, mid schedulers is needed. These interfaces are to be tested in the IWS testbed. Issues in discrete event planning are also being examined in a number of research efforts.

Feature-Based Control

Initial developments in feature-based machining are being extended to the inspection domain. The method will employ solid models in step form and use step tools where possible.

World Modeling/CAD

This effort seeks to formalize world modeling definition within the generic template and to design possible API access for world model information. Experiments are planned for integration of product. This work will also include developing knowledge base entities of ISAM among which are task frames, command frames, plans, algorithms, and resources.

Operator Interfaces

The operator interface work has included integration of EMC operator interface technologies and integration of a contractor supplied diagnostic tool. This work also includes the analysis of operator interface specification requirements with respect to the ISAM generic control module and world model and MSI guardian. Longer-term developments are expected to include remote access capabilities via developments, under the SIMA Operator Interface project.

Prototype Development and Implementation

The development sequence for the initial prototype employs the existing RCS-based control systems for the control of the CMM motion and vision systems. These were developed under the NGIS program. New higher levels of control are introduced at the task and workstation levels. These first test control nodes are being extended from initial EMC-based versions. This configuration will provide a working testbed for initial evaluation. Over time the remaining nodes will be made consistent with ISAM concepts.

More implementation details are provided in the following section. As mentioned earlier, the IWS testbed is a node in the NAMT. A portion of the NAMT is being used to study integration issues within the IWS testbed as a part of that effort. Figure 11.1 shows the Framework project system configuration for FY96, including the IWS. Among other

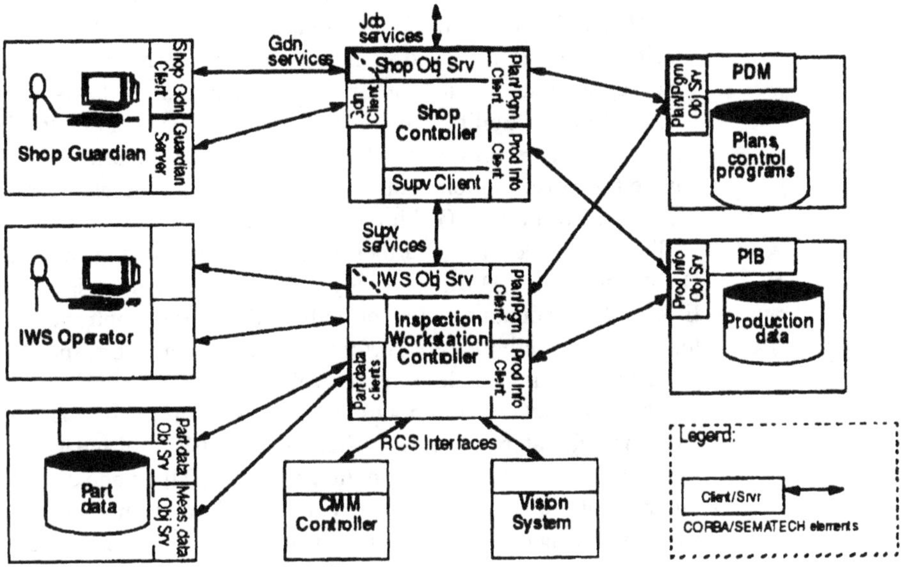

Figure 11.1. NAMT framework FY96.

things, the Framework project is studying issues which are derived from experiments with the SEMATECH CIM Framework and implementation issues related to use of CORBA.

The NAMT plan for FY97 builds on this initial testbed and adds additional workstations and services. This provides a more complex testbed that can be used in refining the ISAM reference model. Figure 11.2 depicts its configuration.

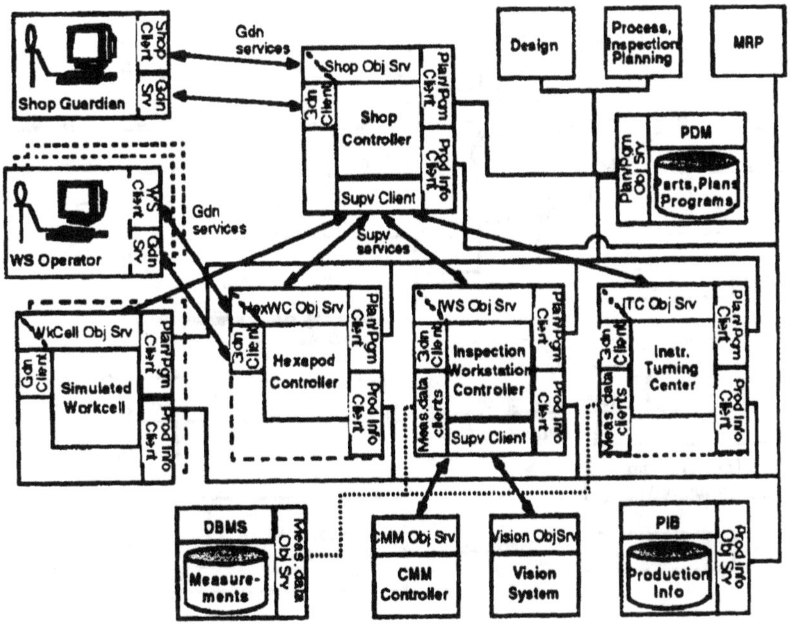

Figure 11.2. NAMT framework FY97.

IWS Implementation Proceeding from the overview of the configuration of the NAMT and the role of the IWS provided in Figures 11.1 and 11.2, we now look more specifically at the IWS and its near-term configuration. The hierarchy of control that supports integrated vision and inspection activities nodes is also operational. Integration of these capabilities into the testbed will occur after the initial implementation described here. From Figure 11.3 we can see that the primitive (Prim) and servo levels handle low-level motion control. These levels and the elemental move (E-move) level are adapted from the NGIS implementation.

The E-move level was further developed to handle the command set provided by the DMIS interpreter. DMIS is a language for specifying inspection plans. Note that the E-move level contains nodes for both CMM motion control and for tool changing. The tool-changing activities are performed by a human who responds to the commands and indicates the status to close the control loop.

The task level contains control nodes to support the measurement subsystem and the fixturing subsystem of the workstation. The task level measurement node coordinates the CMM tool changing and motion activities. This node also is responsible for interpretation of the DMIS inspection plan. The fixturing node handles loading, unloading, and re-fixturing parts. In this implementation, these are executed by a human. The work-

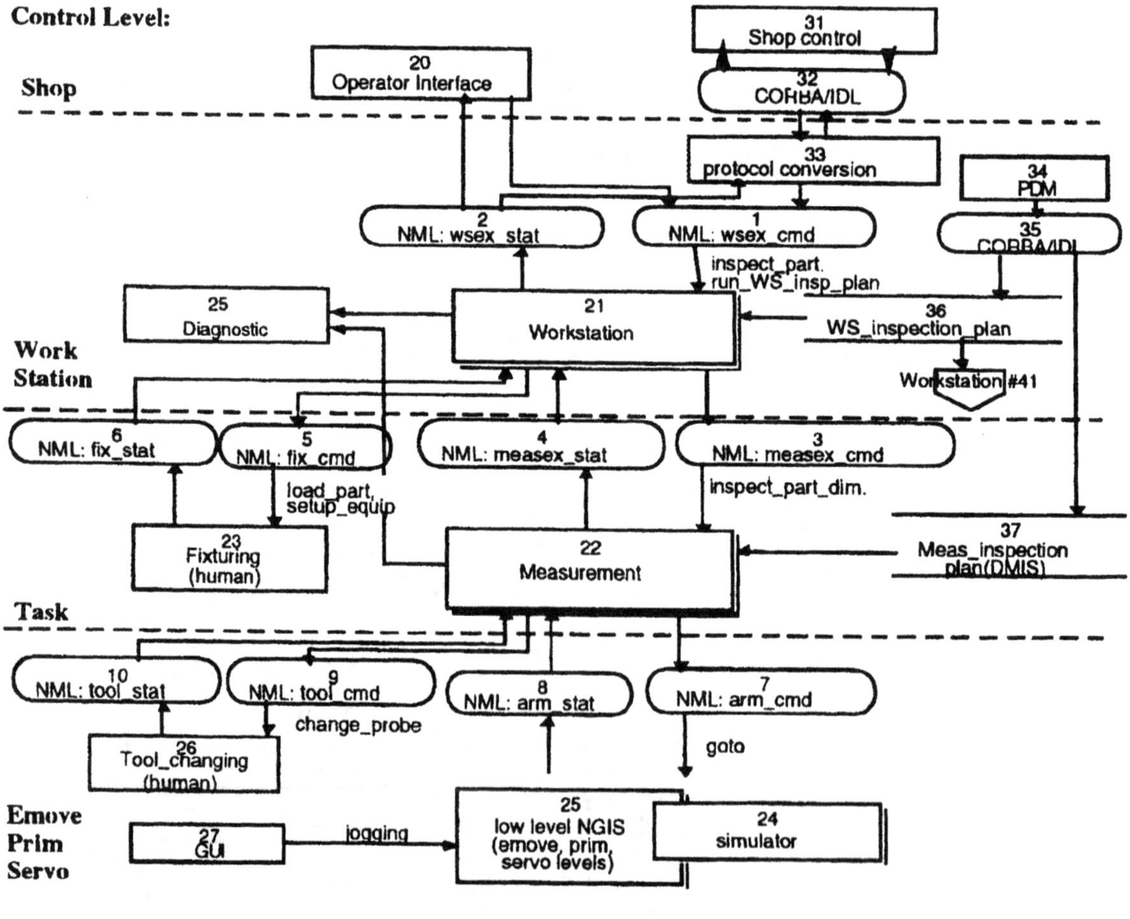

Figure 11.3. IWS control node hierarchy.

station level control node accepts jobs from the shop level and coordinates the activities of the measurement and fixturing nodes to carry out the job on a particular part lot via a workstation level plan. The shop level obtains orders and develops the jobs for the workstation based on these orders.

The workstation node, task measurement node, task fixturing node and E-move–tool change node employ the generic control module template extended from EMC. Control node communication is supported by NML at the workstation level and below. A diagnostic tool that permits a display of the hierarchy and other information in real-time for control nodes employing the generic control module template has been developed by Advanced Technologies Research Corporation (ATR).

As part of the NAMT Framework project, communication between the shop and workstation and between the workstation and task measurement nodes, and the production database (PDM) are being implemented using CORBA.

Shop Level BG: The schedule for the making of products specified in orders is constructed primarily off-line. Orders received during a day, week, or or other time periods (as the business practice of the shop specifies) are accumulated.

- How are new batch schedules installed?
- What happens to existing queues, work in process?

For each product to be made, the set of types and entities is gathered of workpieces (some of which may be assemblies). The total quantity of workpieces of each type is totaled up.

The workpieces are then grouped in production order units, taking into consideration the time at which the workpieces must be ready for assembly into the final products. This process is called *batching*. For each product type, a plan is created for making a production order unit of this product type relative to the workcells in the shop. Appropriate material handling operations to transport production order units from one workcell to another and assembly operations to combine workpieces into the product described in the order must be included in the plan. At a single level of control, plan decomposition may occur. This allows the workpieces to be re-batched any number of times at one control level. For example re-batching may be desired for material handling purposes. If during the production day/week/period, an urgent order arrives, the plans for the shop will be immediately modified to include the making of parts for that new order.

Plan

For each product type, a plan is created for making (shop-level) production order units of the constituent workpieces and assemblies. Each node in this plan calls for the production of a specific number of workpieces of a single product type, for assembling a mix of workpieces into a final product, or for the routing of a container. A node may refer to an expanded plan at the same level of control thus allowing re-batching any number of times.

Commands

Execute task commands are issued to expand plans for making workpieces and assemblies at various workcells.

SP

No sensory processing occurs at this level; rather, human operators collect and record the information (virtual sensing).

WM

The following classes of information form the world model.

Product type descriptions and orders for products.

Relationships of product type to a number of assemblies and workpieces required to manufacture that product type. Relationships of workpieces to the various scheduling units.

Plans for making production order units of a single product type at workcells.

The routing of workpieces and complex orders may be tracked through the various processing order units.

Workcell Level

Plan

The plans at this level schedule the inspection of workpieces (or assemblies) in (workcell) production order units. This includes the coordination of parts removed from a tray and delivered by the material handling system onto the table and coordination of the task-level controller which in turn coordinates the CMM arm, probe, and camera. The production order units at this workcell level do not have a simple relationship to those in higher control levels. The workpieces may be regrouped into totally different production order units by the workcell scheduler. Part shuffling is accomplished by appropriate robot operations.

BG

Rebatching of workpieces occurs for convenient material handling and processing within the inspection workcell. Execute task commands are given with associate plans to the robot and task one lower-level controllers.

Commands

Execute task commands can be issued to a subordinate with a plan reference given as a parameter.

SP

No sensory processing occurs at this level.

WM

A plan is created for the integrated effort of the robot and CMM controllers.

Task Level, Task 1 (Controller 1)

Plan

An inspection plan is created for a single part to be inspected. This is essentially a process plan which is scheduled dynamically. The plan is rough, containing such items as the order in which features of the part can be inspected.

BG

The BG function dispatches the *execute task* commands with associated plans.

Commands

An execute task command can be issued to a subordinate with a plan identifier given as a parameter.

WM

As inspection proceeds, a model of the part *as inspected* in a coordinate system independent of the orientation and placement of the part in external space is constructed from information shared with the task-level world models. Models of these features are formulated by the next lower level of control.

SP

No sensory processing occurs at this level.

Task Level, Task 1 (Controller 2)

Plan

Plans are created for moving parts off the tray and onto the table in proper orientation. This is a process plan, which is scheduled dynamically.

BG

The BG function dispatches the *execute task* commands with associated plans.

Commands

An execute task command can be issued to a subordinate with a plan identifier given as a parameter.

WM

As inspection proceeds, a model of the part *as inspected* in a coordinate system independent of the orientation and placement of the part in external space is constructed from information shared with the task level world model. Models of these features are formulated by the next lower level of control.

SP

No sensory processing occurs at this level.

Task Level, Task 2 (Controller 1)

Plan

An inspection plan is created for single feature sequences in the making of inspection points for this single feature.

BG

The BG function dispatches *execute task* commands with associated plan references.

Commands

The *execute task* command has an associated plan and one parameter, the approach point. The approach point is located in space at the point where the probe is expected to start. The approach point is typically calculated off-line by a planner, relative to the part's geometry.

WM

The world model at this level contains a model of the inspected feature in a coordinate system independent of the orientation and placement of the part in external space

SP

None occurs at this level.

Task Level, Task 3 (Controller 1)

Plan

The plan coordinates the activities of the arm, probe, and camera for the inspection of a single line from the given approach point. The plan is the same each time, with adjustments made in the execution parameters. Since the approach point is passed down as a plan parameter, the approach vector is calculated by the vision system, and it can be passed down to the arm in a command. The measurement then becomes part of the model of the inspected feature. The resolution is a measure of how closely the feature is to be inspected.

BG

The inspection plan is executed and commands are sent according to which E-move level should function; this plan regards the gross activities of camera, probe, and arm.

Commands

Commands locate a part, select a coordinate system for the arm, and move the arm a distance delta in the direction of the approach vector.

Vision

The workstation initializes, extracts edges, finds the probe's start point, finds probe's endpoint, tracks the probe, and finds the edge point.

WM

The WM contains the approach point and associated approach vector pairs. The approach vectors are created using the vision system and the approach point by the E-move level activity. The approach vectors give the direction and velocity of the probe's approach from the approach point. The probe gives readings of the inspection points for the path. The world modeling function is responsible for translating these points into a representation that is independent of the placement and orientation of the part in external space.

SP

None occurs at this level.

E-move

Comments

The position of the probe with regard to the edge of the part is calculated by the vision system and the results of the difference between the probe position and the part's edge are shared laterally through a shared information mechanism, such as common or shared memory.

Controller 1

Plan

The plan enables dynamic calculation from the approach point to the vector.

BG

The part is located in table coordinates. This activity may involve operator assistance in calculating the probe speed when visual servoing is performed.

Commands

The objective is to go to point x with the vector, with constraints on the path, and with average velocity. The vector is calcuated to the destination point (in table coordinates). After leaving the initial point, it must keep going until the location of the destination point. Commands are issued to go to the end of the feature and gather data along the way, and then to return done when it is done.

WM

The position is from probe servo x, y, and z (in coordinates relative to the arm). The location of the part and feature are given in table coordinates. The approach point and vector are translated into table coordinates from the vision sensory processing system.

SP

None occurs at this level.

Controller 2

Plan

None is given.

BG

The BG function determines the edges of the part and the approach vector for tracking the probe.

Commands

No commands are issued.

WM

No probe position is given. The world model is empty.

SP

Two-dimensional features are used from the prim level, with 3D feature extraction.

Prim Level: Controller 1

Plan

The path is calculated as needed from the approach point and vectors in table coordinates.

BG

The actual path is calculated in arm coordinates with time components (typically points on the path are separated by n milliseconds), using sensory input for the processed position from the probe servo wm and arm servo wm.

Commands

Commands are issued to the arm to move delta in x, y, or z direction with specific kinematics. Commands to the probe are select x-kind-of probe (laser, 3d, midas).

WM

Robot geometry is involved; with robot performance characteristics.

SP

None occurs at this level.

11.4 RCS-BASED ROBOCRANE INTEGRATION

The Intelligent Systems Division (ISD) of the National Institute of Standards and Technology (NIST) has been researching new concepts in robotic cranes for several years. These concepts use the basic idea of the Stewart platform parallel link manipulator. The unique feature of the NIST approach is to use cables as the parallel links and to use winches as the actuators. Based on this idea, a revolutionary new type of robot crane has been developed and aptly named the RoboCrane. The RoboCrane provides six degrees of freedom load stabilization and maneuverability.

This is accomplished through the following control modes: master/slave, joystick input, operator panel input, preprogrammed trajectory following (teach programming, graphical off-line programming, or part programming), and sensor-based motion compensation. The current control system includes both the controller and user interface within the same control level, which makes controller enhancements and modifications difficult and error prone. A real-time control system (RCS) [50] is currently being developed to incorporate the control modes in an open system architecture. This development will support hierarchical control modules, along with a separate user interface. A remote tele-presence system is also being implemented to provide foveal/peripheral stereo images and other necessary data to enable a remote operator to perform a variety of tasks with the RoboCrane.

Our objective is to look at the past and future efforts toward integration of an RCS open system architecture controller within the RoboCrane Integration Testbed (RIT). The RCS-based RoboCrane controller will allow continuing research into parallel-link manipulator controllers and application-oriented controller capabilities. We begin by describing the RoboCrane concept and the current control system, and then consider the envisioned RCS-based RoboCrane control system and the intended integration procedure.

11.4.1 The Prototypes

The RoboCrane prototype (Figure 11.4) was developed by NIST in the late 1980s [51]. A NIST program on robot crane technology, sponsored by the Defense Advanced Research Projects Agency (DARPA), developed and tested several potential robot crane designs to determine the desired performance characteristics. Initial testing of these prototypes showed that a six-cable design results in a remarkably stable platform capable of performing accurately six degrees-of-freedom (DOF) manipulations. This stabilized platform can be used to improve typical crane operations or as a maneuverable robot/tool base.

The RoboCrane is based on the Stewart platform parallel-link manipulator [51], but it uses cables as the parallel links and winches as the actuators. By attaching the cables to a suspended work platform and maintaining tension in all six cables, the load is kinematically constrained. Moreover the suspended work platform resists perturbing forces and moments with a mechanical stiffness determined by the angle of the cables, the suspended weight, and the elasticity of the cables. Based on these concepts, the RoboCrane is a revolutionary robot crane that can control the position, velocity, and force of tools and heavy machinery in all six degrees of freedom (x, y, z, roll, pitch, and yaw).

NIST's research into Stewart platforms also produced an innovative structure from which to suspend the RoboCrane work platform. The structure is of an octahedral tubular construction, containing the three upper support points necessary to suspend the work platform, and providing exceptional structural stiffness in a lightweight frame. The structure's legs are connected in an octahedron configuration, so the forces and torques incurred by the work platform are translated into pure compression and tension in the legs.

With only slight bending moments in each of the structure's legs (due to self-weight), the RoboCrane's octahedron structure can be made extremely lightweight compared to conventional gantry structures. This fact, along with the RoboCrane's ability to lift

Figure 11.4. Robocrane development (six-meter NIST RoboCrane prototype).

very heavy loads, produces a much higher lift-to-weight ratio than conventional serial link manipulators. Also the RoboCrane stable structure is well-suited for mobility. By affixing independent wheeled vehicles under each of the structure's three feet, the RoboCrane can be made to traverse rough terrain. This was demonstrated using a two-meter, radio controlled prototype [53].

The six-meter RoboCrane prototype (Figure 11.4) and a compact facility-mounted prototype called TETRA-Robocrane have been subjected to a variety of performance measurements and computer simulations. Experimental tests were conducted to verify their functional work volume, static loading capability, and load positioning accuracy. These experimental results compared favorably to associated computer analysis [51]. The RoboCrane work platform has been equipped with tools such as a gripper, grinder, welder, saw, and inspection equipment (stereo vision and laser scanner). These tools have been used to demonstrate a variety of tasks. Each new application has contributed to the overall functionality of the RoboCrane controller and to the design of the human/computer interface.

The current controller implementation provides for intuitive and robust control of the RoboCrane through the following control modes: master/slave, joystick input, operator panel input, preprogrammed trajectory following (teach programming, graphical off-line programming, or part programming), and sensor-based motion compensation. Individual winch control is also possible. Potential application areas for the RoboCrane technology can be found in the construction industry [54], nuclear/toxic waste cleanup, the subsea arena [55], and in planetary exploration [53].

NIST has also developed an adaptation of the RoboCrane technology in order to investigate its effectiveness for six-phase quadrature signal inputs from winch motor encoders (500 pulses per revolution), six-analog 10-turn potentiometers (pot) on each winch motor through speed reducers, six-analog pots making up a Stewart platform (SP) joystick, and six-analog output tension sensors into digital information. The sensor interface outputs can be digital or analog. They are used to turn on/off linear actuators, tools, and lights for a variety of applications such as gripping, grinding, and welding. The Macintosh outputs are analog signals from a 12-bit digital/analog converter board that provides input to the six power amplifiers.

Pulse width modulated (20 kHz) power amplifiers can be set in velocity or torque control modes. These are typically configured for velocity mode. For the operator interface, the RoboCrane control panel provides interactive control over RoboCrane functions, settings, and status. Because the operators of these controllers are not expected to be computer literate, and they might wear protective gloves, a simple and intuitive graphic interface was developed using LabView software. Control modes and motion types can be activated or deactivated by computer mouse actions and/or touch screen actions. For direct manipulation of the RoboCrane platform, a 6 DOF force sensing joystick (Spaceball) is input serially through the computer modem port; it communicates at 19.2 kbaud.

11.4.2 RoboCrane Subsystems

Remote Operator/Observer Interface

A remote operator/observer interface was developed to allow networked communications with the RoboCrane controller. This graphical interface consists of a touchscreen panel that looks like the actual RoboCrane controller panel. It communicates with the RoboCrane controller via ethernet and serves as a remote operator control station anywhere the network can reach. In addition this remote interface can act simply as an observer of RoboCrane operations, without any control functionality. The interface's front panel buttons allow a remote operator, with the proper permissions, to take control of RoboCrane functions or to simply act as a passive observer. The executable code for this remote operator/observer interface can be run on a computer located anywhere on the network.

Off-Line Graphical Programmer

The off-line graphical programmer allows for safe and easy generation of platform trajectories (move commands) along with the timely actuation of tools. The graphical programmer runs on a Silicon Graphics (SGI) computer and controls the operation of a TGRIP (teleoperative graphical robot instruction program) simulation of the RoboCrane workspace. The RoboCrane's platform and tools are intuitively represented as three-dimensional solid models. Graphical simulations of RoboCrane motions, tool status, and other information are stored in a standard text file and then made available to the RoboCrane through the network.

System Complexity

The RoboCrane controller consists of a complex assortment of over 900 electrical wires that can be difficult, tedious, and time-consuming to troubleshoot. The current configuration requires strong electrical signals to pass through long cables connecting the controller to motors, sensors, and tools. For example, the cables connecting the power amplifiers to the motors measure more than 25 meters in length. This type of electrical system complexity will be remedied in the RCS-based RoboCrane control system.

In addition to maintaining the RoboCrane's current control modes and capabilities, the proposed RCS-based RoboCrane control system contains several targeted improvements: for example, a hierarchical open system architecture with standard control module interfaces that separate user interface into each level of the controller and a more adaptable electronic design that scales well for larger systems. These and other enhancements will form a modular, reconfigurable control system that will allow the system to be easily optimized for particular applications.

Computer Platforms

The proposed computer platforms for the RCS-based RoboCrane control system differ from the current system. The controller hardware will be based on PC-compatible machines running the Windows NT operating system with real-time extensions. This combination will support the RoboCrane's computing requirements, while maintaining consistency with a de facto industry standard platform. Similarly the graphical operator interface will be an implementation running on a PC-compatible/Windows NT machine. The off-line graphical programmer will remain on the SGI (running UNIX) due to heavy graphics and rendering requirements. All of these computer subsystems will be connected via ethernet.

Component Changes

One major change to the physical components of the system will be the power amplifiers for the winches. In the proposed system the same six winches will be driven by device network amplifier linking the power amplifiers. These device network amplifiers will be located at each of the winches. The device network amplifiers will communicate with the host computer through a device network interface card residing within the controller computer. The data transfer rate for the device network system is up to 1 Mbit/s. for networks up to 40 meters long and can contain 0 to 8 bytes of data without segmentation. These device network amplifiers will remove the need for RoboCrane's existing sensor interface, power amplifiers, power supplies, and isolation transformers. In addition an emergency stop system will be independently connected to the six device network amplifiers, thereby eliminating the need for a separate electronics rack.

Most other major components, such as the winches, joystick, and sensors (tension, encoder, potentiometer), will be incorporated into the proposed design. The addition of "jog pots" will be necessary for occasional direct axis control of each winch, since the amplifiers will be colocated with the winches. Jog pots will be used during noncomputer-controlled tasks such as calibration and cable replacement.

Hierarchical Controller Modules

The core concept behind the RCS-based RoboCrane controller architecture centers around a hierarchical decomposition of tasks required to perform a particular application (Figure 11.5) [58]. The basic decomposition of commands can be loosely thought of in terms of time needed to perform the action. That is, at the bottom most servo level, time (t) can be considered instantaneous. As one traverses up the hierarchy, each command level requires roughly an order of magnitude of greater time ($10\times$) to perform its command. Therefore, the *primitive (Prim) level* would require roughly $10t$, the

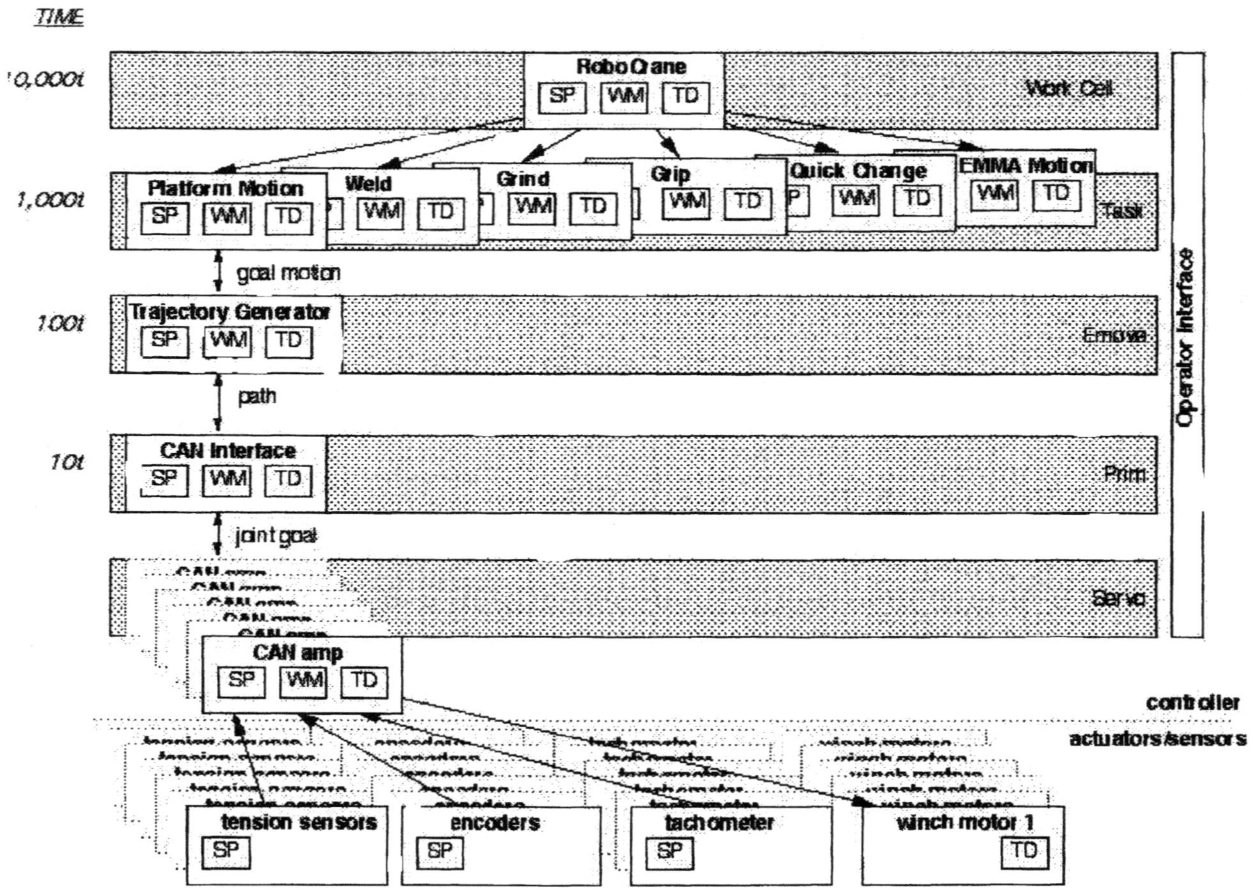

Figure 11.5. Task decomposition for RoboCrane.

elemental-move (E-Move) level would require $100t$, the *task level* would require $1000t$, and so on.

Similarly the RCS model supports upward passing of sensor data as necessary for a particular module to perform its function. Since each decomposed command is (or is not) successfully performed, a status is sent back up the hierarchy to the appropriate controller module at any given level. At the higher levels of the hierarchy, such as the *task level*, commands take longer to perform and status indications return less frequently than at the lower levels. While at the *servo level*, commands, status messages and sensor readings are carried out almost continually.

In addition the concept of a world model database is maintained so that any controller module may access a particular piece of information if and when it is necessary. For instance, the *Workcell Level: RoboCrane* module might need to check the status of a gripper (functional or broken) before agreeing to perform some task that involves parts manipulation.

Integral to an open system architecture is the definition and standardization of controller module interfaces. In this way developers of particular modules with enhanced capabilities or experimental algorithms are able to "plug" their module into the overall control system and test it seamlessly. In the case of the RCS-based RoboCrane controller, the neutral manufacturing language (NML) [59] will be used to perform all communications between control modules.

Task Narrative

Consider the task decomposition of a RoboCrane application such as welding (Figure 11.5). For the RoboCrane to perform a welding application, the RoboCrane's *Workcell Level: RoboCrane* control module must receive a command to "weld a part." This high-level command could be generated by a *shop level* control module (not shown in Figure 11.5), just as a part was placed in the RoboCrane work volume. Alternatively, that same command could be issued via the operator interface for that level, as will be the case for the RoboCrane.

The "weld a part" command is then decomposed by an intelligent *Workcell Level: RoboCrane* module that understands what tools are available and knows how to decompose the given command into its constituent elements. As a result it passes down two simultaneous task level commands to its subordinates, *Task Level: Platform Motion* and *Task Level: Weld*. Since the *Task Level: Platform Motion* module receives its command, it decomposes it into a series of goal motions and passes them down to the *E-move Level: Trajectory Generator* module. This module understands the RoboCrane platform kinematics and can turn goal motions into an appropriate path. If the RoboCrane's kinematics are changed or if a completely different robot is used, this is the only controller module that needs to be updated.

The *E-move Level: Trajectory Generator* module passes down path information that the *Prim level: device network Interface* module knows how to turn into joint goals for each *Servo Level: device network amplifier* module. Then each *Servo Level: device network amplifier* receives its goal information from above and interfaces directly with its specific hardware (a winch in this case). Each *Servo Level: device network amplifier* sends incremental move commands to its winch and listens for feedback from associated sensors (encoder or tachometer). This way, the *Servo Level: device network amplifier* closes the servo loop right at the winch, providing a rapid response. All these servo level actuators are synchronized by the levels above so that the resulting platform motion traces the intended robot trajectory.

Meanwhile the *Task Level: Weld* module has been waiting for the status of the platform motion to show that it is near the intended weld seam. The *Task Level: Weld* module may check the world model database to see if a particular point has been passed. Once that status is true, the *Task Level: Weld* module will begin to deploy the weld tip and, at the appropriate time, power the welder on and off to create a synchronized welding path.

Integration Procedure

The RCS-based RoboCrane controller will support the current RoboCrane control modes, motion types, and targeted improvements listed above. However, it is being developed in a number of novel ROBOCRANE platforms prior to integration into the RoboCrane testbed for several tactical reasons. First, it is preferable to maintain RoboCrane demonstrations of tools and applications while the RCS-based controller is being developed. Second, since the TETRA-ROBOCRANE and RoboCrane platforms differ slightly, adapting the controller to the RoboCrane will provide a good first test of the controller's modularity and portability. Third, TETRA-ROBOCRANE's compact design provides easy access to all hardware and will simplify development and testing of the new controller.

TETRA-ROBOCRANE's current controller configuration consists of six conventional amplifiers driving six winches. Both TETRA-ROBOCRANE's onboard controller and off-board operator interface computers will use the Windows NT operating system, supplemented with real-time extensions that provide a 1 microsecond cycle time. The onboard controller can incorporate RCS templates, NML communication protocols, shared memory, and other concepts consistent with an RCS controller [57]. The graphical user interface will be developed in C++.

NIST has contracted with an industry partner, Advanced Technology and Research Corp. (ATR), to work closely in the development of the RCS-based RoboCrane controller aimed at producing a commercially available system. ATR will provide control module development and consulting, similar to their efforts on NIST's enhanced machine controller [57], a four-axis machine tool controller currently in use in a General Motors manufacturing facility.

11.5 MOBILE ROBOT GOES MULTIRESOLUTIONAL

One of the first experiences of running an autonomous mobile vehicle with a multiresolutional control system was the Drexel Dune-Buggy developed in 1983 to 1986 and tested in 1986 to 1987. A dune-buggy was transformed into an intelligent mobile autonomous system (IMAS). It was equipped with a multiresolutional RCS-type architecture with a three-level behavior generation module: planner–navigator–pilot. The U. S. Army Dune-Buggy (see Fig. 11.6) was used as a hardware upon which the multiresolutional intelligence was tested.

Figure 11.7 shows the process of assembly. Computers are placed on a special mobile "table" in the seat area, and a computer vision CCD camera was attached to the metal frame. Ultrasonic sensors for the rough "pilot" vision can be seen in the area of front bumper. Thus the results of computer vision with this camera location (Fig. 11.8*a*) were arriving to the SP subsystem in a very skewed form (Fig. 11.8*b*). Nevertheless, this image was good enough for interpretation, and it was able to unify multiple images into a world model.

Figure 11.6. Dune-Buggy as IMAS.

SP captured objects by constructing boxes around the hypotheses about objects. Thus the image delivered by the CCD camera appeared as shown in Figure 11.9.

In the simple setting of IMAS, many important concepts have been tested, First, there was tested the concept of multiresolutional control architecture (planner–navigator–pilot) in which the decision making was gradually refined. The BG hierarchy was supported by a hierarchy of representation in which we had a predetermined map at the planner level, a subarea in the map with visible objects inserted at the navigator level, and some objects with ultrasonic clarifications at the pilot level. The structure of this system is described in [63–72].

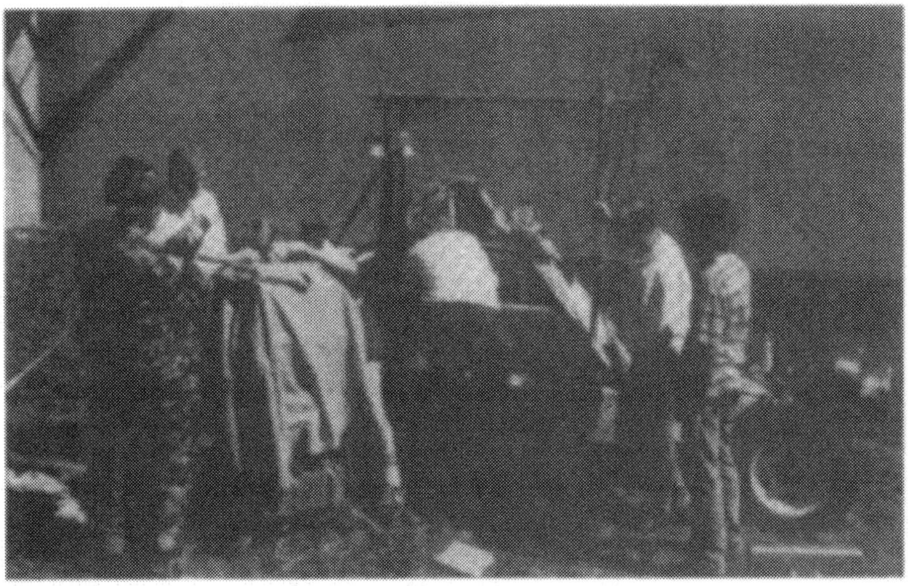

Figure 11.7. IMAS: The process of assembly.

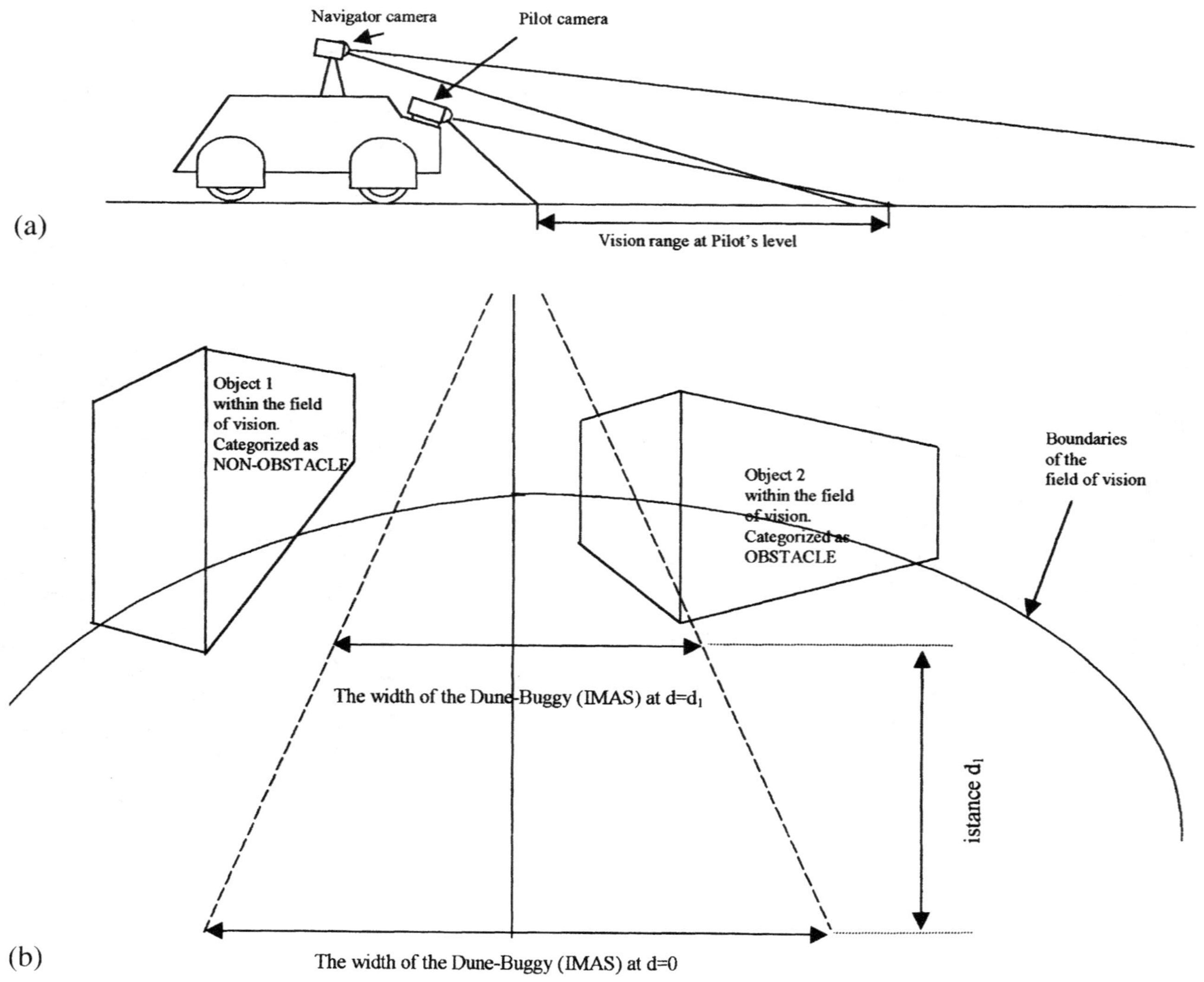

Figure 11.8. Image distortion.

(a)

(b)

Figure 11.9. Capturing the objects: (*a*) Incoming image; (*b*) constructed objects.

This section describes a system of guidance for an intelligent mobile autonomous system (autonomous robot) based upon an algorithm of pilot decision making that incorporates various strategies of operation. Depending on the set of circumstances, including the level of "informedness," initial data, concrete environment, and so on, the "personality" of the pilot is selected from two alternatives: (1) a diligent strategist that tends to explore all available trajectories off-line and is prepared to follow one of them precisely, and (2) a hasty decision maker inclined to make a choice of solution in a rather reckless manner based on short-term alternatives not regarding long-term consequences. The simulation shows that these two personalities support each other in a beneficial way. Let us consider first the concept of a nested hierarchical controller [60, 63]. The intelligent mobile autonomous system (IMAS) is considered to be controlled by a planner–navigator–pilot type of intelligent controller [62]. Only the pilot's level is analyzed. The pilot level was thoroughly simulated with behavioral duality in the Laboratory of Applied Machine Intelligence and Robotics (LAMIR) at Drexel University and applied in the IMAS (Drexel-Dune-Buggy). The concept of behavioral duality is a development of our prior work exploring the pilot level of a multiresolutional controller [65, 72].

11.5.1 Types of Decision Making

Even with the few existing different approaches to motion control, each algorithm has its strengths and weaknesses. It does not appear that a method can be devised that is entirely appropriate for all combinations of the environment, situation, and physical system [61]. An obvious remedy then is to design many special purpose algorithms, each with clearly defined operating limits, and to use meta-control or a dispatcher to choose among the available algorithms depending on the circumstances. Each algorithm results in a definite behavioral portrait of the intelligent mobile autonomous system (IMAS).

The use of algorithms with a range of operational strategies and thus, with different behavioral portraits, should result in behavioral plurality. We take this approach with two algorithms, one which is best for known or slowly changing environments, and the other which is more appropriate for unknown or highly dynamic worlds. Our motivation is to analyze a particular case of behavioral duality.

For our vocabulary of situations we use a simplified predicate calculus which can classify a given world description so that a familiar set can be recognized and the appropriate control found. This language must support "fuzzy" descriptions, since we would quickly be swamped, computationally, if we try to describe every detail of a situation uniquely. These fuzzy descriptions should differentiate between aspects of the situation that *most affect control decisions* and obscure differences that do not.

11.5.2 Rule Bases

In the case of a priori known environments, the problem of motion control reduces to that of planning a complete trajectory, and then tracking it, with possible compensation for error. Since the world is given, the complete trajectory can be found *before motion begins* (ultimately, off-line) and the more time-consuming methods are initiated.

In most applications, the situation is a partially known world. Although the robot may have a complete map of its surroundings (e.g., the factory), it will inevitably be required to deal with unforeseen (moving or temporary) obstacles. Thus compromises are required, and a trajectory must be planned *that is achievable*. This reduces the computation required for each trajectory and ensures that the path found will not soon become invalid. The Precise Preplanning algorithm (see Section 11.5.3 below) does this job.

The search is considered as the most general approach to the trajectory planning problem in known worlds. Often various heuristics are used to reduce computational complexity. Our approach has been to use generalized results of a search done *off-line* (with no constraints on the time of search) to create a generalized rule base online. The generalized search results then can be seen as a pruning *after the fact*, where each pruned region is transformed into a description of (class of) the situation common to the pruned set, and a trajectory representative of the pruned set.

The other extreme is a completely unknown world. Undoubtedly, any robot architecture is based on certain assumptions about its intended environment, and in that sense the world is never *completely* unknown. However, we are interested in the case where no information is given to the robot other than some built-in assumptions. In the unknown world situation, planning complete trajectories, at any stage of the motion, is not an option, since the trajectories can quickly become invalid.

There are far fewer existing solutions to the problem of motion planning in unknown worlds. A big reason for this is that there is no clear method for evaluating the performance of any algorithm. Whatever parameter the decision mechanism is attempting to "optimize," it must be evaluated in terms of the information the system has at a given moment, since the robot's knowledge is constantly changing.

11.5.3 Precise Preplanning (PP) Algorithm

General Structure

In a completely known world, the problem of planning a dynamically feasible (minimum time) trajectory in the presence of obstacles can be formulated as an optimal control problem, with the obstacles acting as constraints on position. One way to approximate a solution to this problem is by using search in a state-space consisting of the state variables of the vehicle. Positions of obstacles can then be superimposed on this space, making those positions "illegal" states. Using a model of the vehicle, a successor generating function can be defined; it will generate the next possible states, given any one state. The time of motion from one state to the next can be used as the cost of each arc in a graph generated recursively by applying the successor generator. A search algorithm can then be applied to find a minimum time trajectory. Details of an implementation of this trajectory search procedure can be found in [71].

Although the method mentioned above does indeed yield a minimum-time solution to the control problem, it is too computationally expensive to be used online. The following observations led us to a solution to transform results of search performed *off-line* into a rule form usable *online*. First, in the similar situations, optimum trajectories are similar. Second, only a limited number of characteristics of the situation determine the resulting trajectory. Thus a language of situations puts into correspondence a rule base of trajectory classes, and the base of associated situation descriptions. Then the rule base can be defined as a set of rules in a form:

$$\text{(situation description)} \rightarrow \text{(trajectory)},$$

where the trajectory is an actual trajectory instance meant to be representative of the entire set of trajectories within the bounds of acceptable error.

More formally we can state this relationship as follows: Let the planning universe (U) be a discrete space represented at some resolution Δ. Then any trajectory planned on this space will have accuracy, or resolution (Δ), of at most Δ. We will use D_i to denote the ith situation description, from the set of all possible situation descriptions given the vocabulary V. A situation description is intended to be an n-tuple, with each component having a range of discrete values that it can assume. The set of all real-world situations (at resolution Δ) described by a given D_i will be expressed as $S(D_i)$.

Notice that this implies that there is a many-to-one mapping between *situations* and *situation descriptions*. This is because we generally assume that our situation descriptions are at a lower resolution than the representation from which they are derived. In some cases this representation may be the world itself. In this case our resolution of the world means the resolution at which we can measure or sense the world. Actually the resolution at which one represents situation descriptions will depend on the accuracy of motion that is ultimately required. This relationship between the various discretizations and required accuracy of motion will be discussed in Section 11.6.

Using the notation given above, we can describe a PP motion control rule as follows: $\mathbf{T}_i$ = a representative trajectory. $\mathbf{T}_i \in \mathbf{T}(\mathbf{D}_i)|\text{Error}(T_i, T_k) < e$ for all $k|T_k \in \mathbf{T}(\mathbf{D}_i)$ where $\text{Error}(T_i, T_k)$ is some measure of difference between T_i and T_k in terms of the cost-function being used (e.g., time of motion). Notice that $e < \text{Error}(T_j, T_k)$, where $T_j, T_k \in T(D_i)$ and T_j is highest-cost trajectory, and T_k is lowest-cost trajectory.

Then the rule can be represented as

$$\mathbf{D}_i \rightarrow \mathbf{T}_i$$

which means that if the situation can be described by a set D_i, then use trajectory T_i.

Rule Language

The rule language has several important features:

1. It must be able to distinguish between situations that require solution trajectories differing by more than the boundaries of allowable error.
2. It must be concise and easy to compute. In other words, there must exist simple, fast procedures for instantiating the "words" or "sentences" of the language from some more rule-primitives, based on situtation-primitives and action-primitives.
3. The "sentences" must be of uniform structure so that the language is easy to use. (This is analogous to model-based rules.)
4. The language must allow different "description accuracies." There must be some fuzzy, but quantitative, aspect nature to the descriptions in order to make it easy to refine the descriptions and thus impart more accurate control.

Keeping these considerations in mind, added to our knowledge of the behavior of our system gained from experience, we can develop a language with which to implement the control rules. We will begin with a look at the main components of this language. In the next subsection we will give the details of their implementation.

Earlier we explored one of the ways to derive the "word classes" directly from the variables that appear in a differential equations formulation of the problem [63]. We have yet to determine the range of discrete values that each of these words (variables) may assume. A hidden problem is the characterization of obstacle positions. Even in the analytical formulation of the problem, there is more than one way to represent obstacle positions.

Thus in the "translation" to a linguistic form, we have some flexibility in how we will represent obstacle positions. In creating the part of the rule language that describes obstacle positions, we can use language criterion 2 above to characterize the obstacle configurations, as well as knowledge of the kinematics of the type of vehicle we are considering. The following descriptions of the word classes are used to form the PP rules:

- *Goal angle* (goal-ang). The *goal angle* describes the direction to the goal relative to the current position of the robot. All measurements such as goal angle are done in a polar or rectangular coordinate frame, with the origin being the current robot position. Notice that these values may not be distributed uniformly over the full 360° range. This is because we found that goal angles within certain adjacent regions resulted in the same control recommendations, regardless of the values of the other precondition variables. Thus those adjacent regions (angles) could be consolidated for the sake of criterion 2.
- *Goal range* (goal-range). The goal range describes the distance to the *subgoal* from the current robot position. Recall that the subgoal is determined by selecting a position in an immediately surrounding passageway that allows achievement of the NAVIGATOR's currently planned subgoal at minimum cost.
- *Phantom points* (phantoms). This can be thought of as a discrete artificial potential field *surrounding the robot.* Phantoms may also be seen as virtual proximity sensors. They are used to help classify obstacle configurations in constrained situations. Since the underlying representation of obstacles is a local grid surrounding the robot, this structure is fast and easy to use. For each phantom point we only need to reference the obstacle array to determine if it is stimulated. Alternatively, the phantom points could be implemented as *actual* proximity sensors, thus bypassing the need for an underlying representation of obstacle positions. Figure 11.10 gives the configuration of phantom points we have used.
- *Feasibility of trajectory* (traj-feasible). This "word" is a binary variable which is true if the region swept by the trajectory of the right side of the rule is obstacle-free. The swept region is determined by the width of the vehicle plus some safety margin. In all cases this must be true for the given rule to fire.
- *Current state* (curr-state). The state of the vehicle (system) itself is represented via pertinent state variables, such as steering position, velocity, acceleration. In the PP algorithm this is likely to determine the first few commands of a given trajectory. In some cases it can be a dominating variable in that, for a given situation, the trajectories found will be quite different based on the value of curr-state.

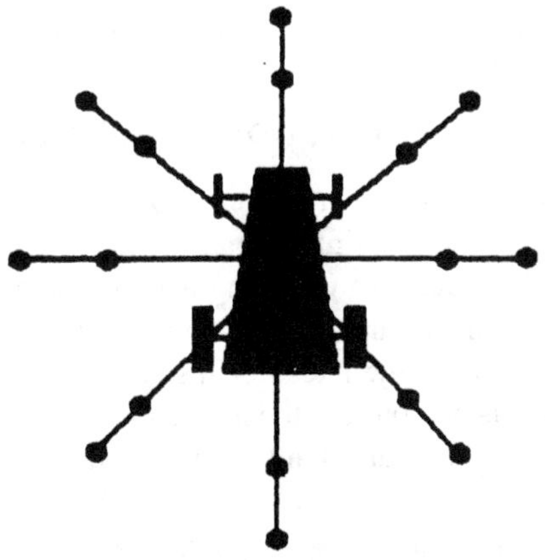

Figure 11.10. Configuration of the phantom points.

With these "words" defined, we can now give the actual form of a PP rule.

$$\text{IF } [(\text{goal-ang} = a) \,\&\, (\text{goal-range} = r) \,\&\, (\text{phantoms} = p) \,\&\, (\text{traj-feasible } (\mathbf{T}_i))],$$
$$\text{THEN } (\text{trajectory} \leftarrow \mathbf{T}_i).$$

11.5.4 Implementation

The PP algorithm operates with some modifications. The first step is to transform the NAVIGATOR's currently planned subgoal to a locally reachable passageway. This is done using two simple criteria of minimum deviation from the path and safe width of the passageway. Next the precondition variables are instantiated. Rule matching is facilitated by having the rules organized in a loose tree form. Common parts of the preconditions are extracted and are incorporated into a dummy parent rule. The dummy rule determines which subordinate set of rules should be tested. This organization enables fast matching, but at the expense of the addition of more difficult rules. Clearly, an automatic classification tree could be formed from a full set of rules.

Most of the rule vocabulary components described above depend on an underlying representation. We have found a simple binary grid to work satisfactorily. Obstacle boundaries are projected onto the grid, and thus each cell is marked as either "free-space" or "obstacle." This grid is the basis of the pilot's "context map." Although the sensors provide less than a 360° view, the pilot context map is 360°, so certain trap situations can be avoided. The mapping of obstacle boundaries onto the grid can be facilitated by using built-in graphics functions for drawing lines, which many systems have.

Goal angle and *goal range* are computed using integer division and a simple lookup table. A minimum division is defined, and then a table is defined that maps each division to a linguistic variable. An alternative would be to use the integer division only. However, it would then be necessary to have substantial redundancy in the rule base.

Phantom points are easily implemented with the grid representation of obstacles. A ray of the phantom point is activated (or said to be true) if any point along the ray corresponds to a cell tagged as an obstacle in the underlying grid trajectory $\mathbf{T}_i$. The robot position is "conceptually" moved to each position of the trajectory. For each position the phantom points are checked against the obstacles grid. A trajectory is collision-free if there are no phantom points activated during such an iteration.

Current state is normalized into a linguistic value by simple integer division of the full range of values that are used to measure the state.

11.5.5 Instantaneous Decision Maker (ID) and Path Monitor

After PP produces a sequence of commands, the execution controller issues actuation positions. In our simulation system we call this mechanism the *path monitor*. Its functions include, first, checking the consistency of the next command with the current position estimate; then it will make small adjustments if necessary. Second, the path monitor compares the latest local sensor information with the currently planned trajectory. If there is no conflict, the next command is issued. The following criteria cause the path monitor to exit and send a "replan" message to the pilot:

- A new plan is given by the NAVIGATOR.
- The trajectory is blocked by the latest sensor information.

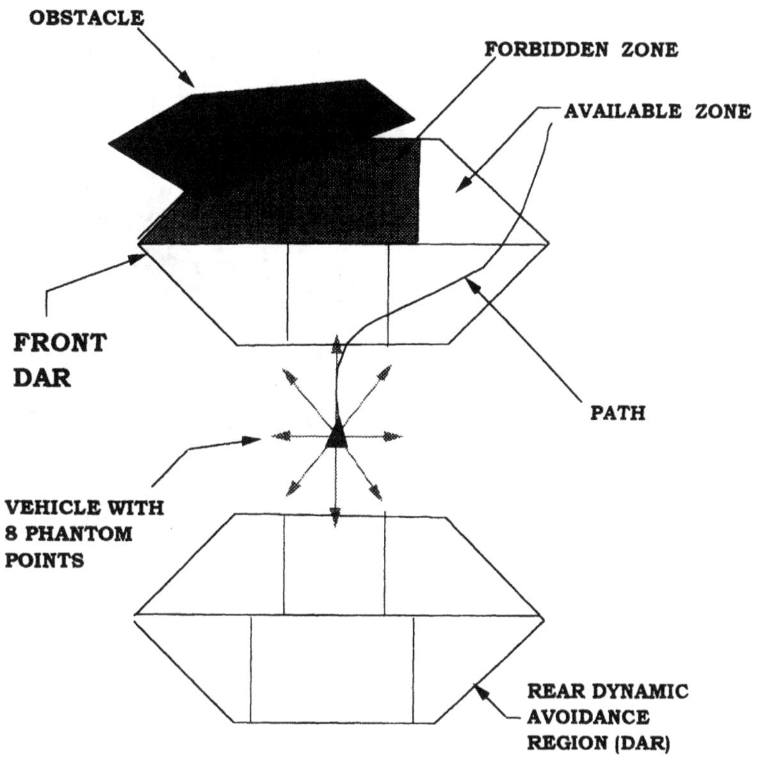

Figure 11.11. Representation of the front and rear DARs.

- The currently planned trajectory has been completely executed.
- The current vehicle position is significantly off course from the current trajectory.

This system uses linguistic variables in two forms. ID uses a construct called *dynamic avoidance regions* (DARs). This is the tool that detects and characterizes the position of obstacles. Depending on the fact in which DAR an obstacle is detected, the rules of motion are applied so that the appropriate changes could be achieved (Fig. 11.11).

Another generator of linguistic variables is the construct shown in Figure 11.12. Each direction is associated with the required steering command.

ID uses a lookup table (Table 11.1) to generate motion commands (speed of propulsion and direction of motion). ID uses all available linguistic variables and generates the most beneficial command. To provide correctness of suggested changes, the table should be carefully trained (by exercising many meaningful scenarios of situations and using human judgment to figure out how to deal with each situation).

11.6 MISSION STRUCTURE FOR AN UNMANNED VEHICLE

The 4-D/real-time control systems architecture (4-D/RCS) defines a hierarchical decomposition for intelligent systems, with corresponding command and control pathways. In designing the vehicle's behavior generation functions, we developed a set of task commands, goals, and actions that are organized hierarchically to form the basic

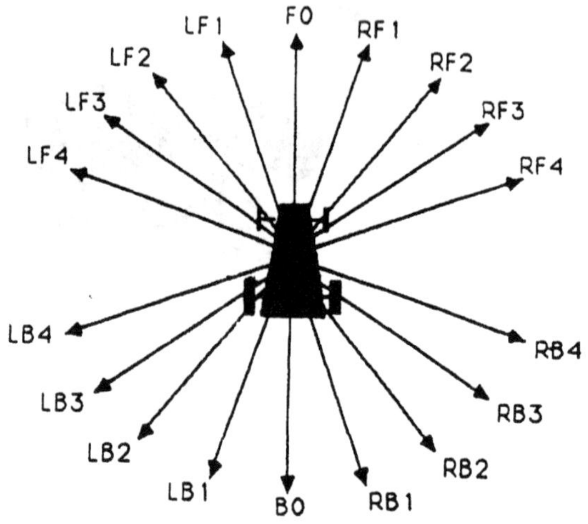

Figure 11.12. Linguistic (fuzzy) directions.

vocabulary for the planning and control of the vehicle. This section documents the basic tasks and vocabulary for a scout platoon implemented according to 4-D/RCS.

11.6.1 Levels of the System

Let us begin with the set of tasks that a scout platoon and its components are expected to perform during the course of a mission [60] and define a library of task commands that the scout platoon must execute in producing the necessary actions. The tasks, their goals, as well as the actions are organized in corresponding hierarchies, and together they form a vocabulary of the planning/control language used for behavior generation. The scout platoon is presumed to consist of the autonomous vehicles (equipped by the 4-D/RCS system). The functioning and architecture of the 4-D/RCS system is described in Section 11.7 and [63]. Here we will briefly address the general information relevant to the mission structure.

Level 7: Mission

The mission level is understood as the assignment to the scout platoon formulated at the battalion. The mission is a set of sequential, or parallel, tasks that the scout platoon should accomplish during a particular period of time within a particular designated space. This assignment is decomposed in the hierarchy of tasks and subtasks for the other six levels of the architecture. Scout platoon missions are typically defined by the battalion headquarters (HQ) and conveyed to a platoon through a platoon leader via a set of task commands that refer to knowledge residing in maps and reports that describe the state of the battle-space. This set of task commands assumes that the scout platoon has certain skills and its components have certain capabilities and knowledge of how to perform tasks corresponding to the commands.

Commands and knowledge input from the battalion HQ to a scout platoon are combined with knowledge resident in the scout platoon itself. In 4-D/RCS, task knowledge is formally represented in task frames. The battalion-level planning process may consider the exposure of each unit's movements to enemy observation, and the traversabil-

Table 11.1. Lookup table: input → direction.

Vars.	B0	LB1	LB2	LB3	LB4	LF4	LF3	LF2	LF1	F0	RF1	RF2	RF3	RF4	RB4	RB3	RB2	RB1	V0	V1	V2	V3
F0	0	0	2	4	6	8	12	15	40	70	40	15	12	8	6	4	2	0	20	20	20	20
R1	0	0	0	2	4	6	8	12	15	40	70	40	15	12	8	6	4	2	20	20	20	20
R2	2	0	0	0	2	4	6	8	12	15	40	70	40	15	12	8	6	4	20	20	20	20
R3	4	2	0	0	0	2	4	6	8	12	15	40	70	40	15	12	8	6	20	20	20	20
R4	6	4	2	0	0	0	2	4	6	8	12	15	40	70	40	15	12	8	20	20	20	20
R5	15	12	8	6	4	2	0	2	4	6	8	12	15	40	70	70	70	40	20	20	20	20
D0	70	70	40	15	12	8	6	4	2	0	2	4	6	8	12	15	40	70	20	20	20	20
L5	15	40	70	70	70	40	15	12	8	6	4	2	0	2	4	6	8	12	20	20	20	20
L4	6	8	12	15	40	70	40	15	12	8	6	4	2	0	0	0	2	4	20	20	20	20
L3	4	6	8	12	15	40	70	40	15	12	8	6	4	2	0	0	0	0	20	20	20	20
L2	2	4	6	8	12	15	40	70	40	15	12	8	6	4	2	0	0	0	20	20	20	20
L1	0	2	4	6	8	12	15	40	70	40	15	12	8	6	4	2	0	0	20	20	20	20
NEAR	20	20	20	20	20	20	20	20	20	20	20	20	20	20	20	20	20	20	5	20	15	10
MEDIUM	20	20	20	20	20	20	20	20	20	20	20	20	20	20	20	20	20	20	5	10	20	15
FAR	20	20	20	20	20	20	20	20	20	20	20	20	20	20	20	20	20	20	5	10	15	20
F-L1-CLOSE	0	0	0	0	0	0	0	−40	−70	−40	0	0	0	0	0	0	0	0	5	10	15	20
F-L2-CLOSE	0	0	0	0	0	−20	−70	−70	−20	0	0	0	0	0	0	0	0	0	5	10	20	15
F-L1-FAR	0	0	0	0	0	0	0	0	−10	0	0	0	0	0	0	0	0	0	5	10	15	20
F-L2-FAR	0	0	0	0	0	0	−10	−10	0	0	0	0	0	0	0	0	0	0	5	10	15	20
F-C-CLOSE	0	0	0	0	0	0	0	0	−40	−70	−40	0	0	0	0	0	0	0	5	10	15	20
F-C-FAR	0	0	0	0	0	0	0	0	0	−10	0	0	0	0	0	0	0	0	5	10	15	20
F-R2-CLOSE	0	0	0	0	0	0	0	0	0	0	−20	−70	−70	−20	0	0	0	0	5	10	15	20
F-R1-FAR	0	0	0	0	0	0	0	0	0	0	−10	0	0	0	0	0	0	0	5	10	15	20
F-R2-FAR	0	0	0	0	0	0	0	0	0	0	0	−10	−10	0	0	0	0	0	5	10	15	20
NONE	0	0	0	0	0	0	0	0	0	0	0	0	0	0	0	0	0	0	20	20	20	20
B-L1-CLOSE	−20	−70	−70	−20	0	0	0	0	0	0	0	0	0	0	0	0	0	0	5	20	10	5
B-L2-CLOSE	0	0	−20	−70	−70	−70	−20	0	0	0	0	0	0	0	0	0	0	0	5	20	10	5
F-R1-CLOSE	0	0	0	0	0	0	0	0	0	−40	−70	−40	0	0	0	0	0	0	5	10	15	20

(continued)

Table 11.1. (Continued)

Vars.	B0	LB1	LB2	LB3	LB4	LF4	LF3	LF2	LF1	F0	RF1	RF2	RF3	RF4	RB4	RB3	RB2	RB1	V0	V1	V2	V3
B-L1-FAR	−5	−10	−10	−5	0	0	0	0	0	0	0	0	0	0	0	0	0	0	5	20	10	5
B-L2-FAR	0	0	−5	−10	−10	−5	0	0	0	0	0	0	0	0	0	0	0	0	5	20	10	5
B-C-CLOSE	−70	−20	0	0	0	0	0	0	0	0	0	0	0	0	0	0	0	−20	5	20	10	5
B-C-FAR	−10	−5	0	0	0	0	0	0	0	0	0	0	0	0	0	0	0	−5	5	20	10	5
B-R1-CLOSE	−20	0	0	0	0	0	0	0	0	0	0	0	0	0	0	−20	−70	−70	5	20	10	5
B-R2-CLOSE	0	0	0	0	0	0	0	0	0	0	0	0	0	−20	−70	−70	−70	−20	5	20	10	5
B-R1-FAR	−5	0	0	0	0	0	0	0	0	0	0	0	0	0	0	−5	−10	−10	5	20	10	5
B-R2-FAR	0	0	0	0	0	0	0	0	0	0	0	0	0	−5	−10	−10	−5	0	5	20	10	5
NONE	0	0	0	0	0	0	0	0	0	0	0	0	0	0	0	0	0	0	20	20	20	20
F	0	0	0	0	0	−80	−80	−80	−80	−80	−80	−80	−80	−80	0	0	0	0	5	20	10	5
FR	0	0	0	0	0	0	0	0	0	−80	−80	−80	−80	−80	−40	−10	0	0	5	20	10	5
R	0	0	0	0	0	0	0	0	0	0	0	0	−10	−80	−80	−40	−10	0	5	20	10	5
BR	−80	−80	−80	−80	0	0	0	0	0	0	0	0	0	−10	−40	−80	−80	−80	5	20	10	5
B	−80	−80	−80	−80	−80	0	0	0	0	0	0	0	0	0	−80	−80	−80	−80	5	20	10	5
BL	−80	−80	−80	−80	−80	−10	0	0	0	0	0	0	0	0	−80	−80	−80	−80	5	20	10	5
L	0	0	−10	−40	−80	−80	10	0	0	0	0	0	0	0	0	0	0	0	5	20	10	5
FL	0	0	0	−10	−80	−80	−80	−80	−80	−80	−10	0	0	0	0	0	0	0	5	20	10	5
F-FL	0	0	0	−10	−80	−80	−80	−80	−80	−80	−80	−80	−80	−80	0	0	0	0	5	20	10	5
F-FR	0	0	0	0	0	−80	−80	−80	−80	−80	−80	−80	−80	−80	−80	−10	0	0	5	20	10	5
R-BR	−80	−80	−80	−80	0	0	0	0	0	0	0	−10	−80	−80	−80	−80	−80	−80	5	20	10	5
BR-B	−80	−80	−80	−80	−80	0	0	0	0	0	0	0	0	−10	−80	−80	−80	−80	5	20	10	5
Bl-B	−80	−80	−80	−80	−80	−20	−10	0	0	0	0	0	0	0	−80	−80	−80	−80	5	20	10	5
L-BL	−80	−80	−80	−80	−80	−80	−40	−10	0	0	0	0	0	0	−80	−80	−80	−80	5	20	10	5
FL-L	0	0	−10	−40	−40	−80	−80	−80	−80	−80	−80	−10	0	0	0	0	0	0	5	20	10	5
FR-R	0	0	0	0	0	0	0	−10	−80	−80	−80	−80	−80	−80	−80	−40	−10	0	5	20	10	5
NONE	0	0	0	0	0	0	0	0	0	0	0	0	0	0	0	0	0	0	5	10	15	20
PERMANENT	20	20	20	20	20	20	20	20	20	20	20	20	20	20	20	20	20	20	5	10	15	20
TEMPORARY	20	20	20	20	20	20	20	20	20	20	20	20	20	20	20	20	20	20	5	20	10	15

ity of roads and cross-country routes. At the battalion level, the 4-D/RCS world model maintains a knowledge database containing names, contents, and attributes of friendly and enemy forces and of the force levels required engaging them. Maps have a range of 1000 km.

Level 6: Platoon

A scout platoon is a unit that consists of 10 vehicles organized into one or more sections. The platoon commander and section leaders plan activities and allocate resources for the sections in the platoon. At the platoon level, plans are computed for a period of about 2 hours into the future, and replanning is done about every 10 minutes, or more often if necessary. Section waypoints are computed about 10 minutes apart.

Level 5: Section

A scout section is a unit that consists of a group of individual scout vehicles. A 4-D/RCS node at the section level corresponds to a squad leader plus its vehicle commanders (humans or intelligent software). This command team assigns duties to vehicles and schedules the activities of each vehicle within a section. Orders are decomposed into assignments for each vehicle, and a schedule is developed for vehicles to maneuver in formation relative to enemy forces and large obstacles. Plans are developed to conduct coordinated maneuvers and to perform reconnaissance, surveillance, or target acquisition functions. At the section level, plans are computed for about 10 minutes into the future, and replanning is done about every minute, or more often if necessary. Vehicle waypoints about one minute apart are computed.

Level 4: Individual Vehicle

The vehicle is a unit that consists of a group of subsystems, such as locomotion, attention, communication, and mission package. A manned scout vehicle may have a driver, a vehicle commander, and a lookout. The vehicle commander assigns a job to the subsystems (possibly in collaboration with the subsystem controllers) and schedules the activities of all the subsystems within the vehicle. A string of waypoints is developed for the locomotion subsystem to be traversed with a particular schedule while avoiding obstacles, maintaining position relative to nearby vehicles, and achieving the desired vehicle heading and speed along the desired path on roads or cross-country. A schedule of tracking activities is generated for the attention subsystem to track obstacles, other vehicles, and targets. Waypoints about five seconds apart out to a planning horizon of one minute are replanned every five seconds or more often if necessary.

Level 3: Subsystem Level

The subsystem is a unit consisting of a group of related primitive systems, such as steering and braking, engine and transmission, sensor stabilization and pointing, message encoding and decoding, and weapons loading and aiming. A schedule of steering and braking commands is developed to avoid obstacles. A schedule of pointing commands is generated for aiming cameras and sensors. A schedule of messages is generated for communications, and a schedule of actions is developed for loading and aiming weapons. For each primitive system, a plan consisting of a trajectory of waypoints about

500 milliseconds apart is generated out to a planning horizon of about 5 seconds in the future. A new plan is generated about every 500 milliseconds.

Level 2: Primitive Level

The primitive level is a unit consisting of a group of controllers that plans and executes velocities and accelerations to optimize dynamic performance of components such as steering, braking, acceleration, gear shift, camera pointing, and weapon pointing, taking into consideration dynamical interaction between mass, stiffness, force, and time. Velocity and acceleration setpoints are planned every 50 milliseconds out to a planning horizon of 500 milliseconds.

Level 1: Servo Level

Each node at the servo level is a unit consisting of a group of controllers that plan and execute actuator motions and forces, and generate discrete outputs. Communication message-bit streams are produced. The servo level transforms commands from component to actuator coordinates and computes motion or torque commands for each actuator. Desired forces, velocities, and discrete outputs are planned for 5 millisecond intervals out to a planning horizon of 50 milliseconds.

The existing arrangement for the concept of "commander" will be analyzed and might be reconsidered for the organizational units consisting of autonomous vehicles. Indeed, when the group of autonomous vehicles travels together, there is no mandatory need to decide which one is the leader vehicle, since their knowledge is the same and the plans they compute are equivalent. As the mission progresses, their knowledge changes, yet because they share and exchange it, their decision-making capabilities remain equivalent, and the need to consider one of them a leader might not emerge. If they are not able to share their knowledge, and if their position is not equally convenient for the mission continuation, the battalion command might choose one of them to fit the role of a leader. These issues are to be discussed in the future, when some experience of testing will allow us to better understand the capabilities of the autonomous vehicles.

The scout platoon serves as the eyes and ears of the battalion as part of the intelligence collection process. The scout platoon's mission is to confirm or deny the battalion commander's intelligence preparation of the battlefield (IPB) and to provide information as assigned in the reconnaissance and surveillance plan. In unmanned ground vehicles, platoon command and control is done via the leader vehicle.

11.6.2 Classification of Missions

A mission is a set of sequential and/or parallel tasks that the scout platoon should accomplish during a particular period of time within a particular designated space. This section describes the types of mission tasks that can be assigned to the scout platoon (consisting of both manned and/or unmanned vehicles). Each unit receives a set of commands that define the task and the goal of this task. A mission is decomposed into tasks and subtasks. Correspondingly, the mission goal is decomposed into the hierarchy of goals for the lower levels of the hierarchy of tasks and subtasks. Each goal and subgoal is the designed final state of the accomplished task and/or subtask (e.g., a destination point). The overall mission results are the outcome of the overall goals/subgoals hierarchy (e.g., a detection of the enemy from the observation post). Mission plans are composed of sequential and/or parallel tasks that progress jointly, or follow each other upon the satisfaction of a "transition condition," which allows completion of the preceding task and progression to the subsequent one.

Each task is decomposed into the set of subtasks that form a sequential/parallel set. The sequence of tasks presumes completion of the previous one before initiation of the next (e.g., "crossing the bridge after approaching the river"). The parallel subtasks require having independent actuators for doing this (e.g., "watching" while "moving"). This information is taken into consideration during the decision-making process within the behavioral generation module (see Chapter 7 and [17]).

The task assignments are defined and organized into hierarchical levels according to the seven levels of the 4-D/RCS architecture. The following discussion includes a list of tasks from the level of the unit (platoon) through the level of single vehicle to the level of the main subsystems in a vehicle.

Mission Tasks Assigned to a Platoon

Travel Task Navigate from one point on the terrain to another. The goal of this task is to make sure that the unit (platoon) will reach the destination point or area. The path-planning algorithm determines the trajectory of motion and thus has the responsibility for the accomplishments of the unit mission. The commander may add check points to the task if he wants to correct and/or guide the path chosen and to be sure that the unit is on the right way to the destination point. As necessary, the platoon adjusts the route (planned trajectory), formation, and/or the movement techniques chosen in response to emerging obstacles, minefields, changing probability of the contact with enemy, and/or changes in the tactical situation. The platoon unit maintains movement security at all times taking into account a variety of issues like safe travel or stealthiness, if needed. The designed result is that the platoon reaches the given destination.

Establishing an Observation Post Task This task is assigned to a platoon when reporting of information is required that will have a potential intelligence value. The OP (observation post) is positioned to allow the scout platoon to observe the assigned sector, likely enemy avenues of approach, and/or named area(s) of interest (NAI) with required depth of observation through the sector. Based on the commander's guidance, a number of needed OPs are chosen in order to cover the area of operation. The task may include a list of reactions for events like "hide" if you are detected or "transit to track mode" if a certain type of vehicle is detected. The designed result is that the OP observer provides early warning of enemy activity.

Task activities (subtasks)

1. Terrain analysis to determine OP's location and to plan the trajectories from the current location.
2. Zone reconnaissance in case the area is not clear.
3. Observation Post occupation according to the plan.
4. Recognition of a threat and battlefield organizations.
5. Reporting.

Task of advance/withdraw to a position

1. This is a war task when the scout platoon is a part of the larger fighting unit (parent unit). The scout platoon is relocated when the "line of the battle" is advanced or withdrawn in a way to assist the present assignment. The vehicles can be at

positions that are hidden from the enemy, or they can relocate to *observation position or fighting position.*

2. A fighting position is a location in which the vehicles are hidden but, if necessary, could observe and fire upon enemy position. The designed result: cooperation with parent unit and increased safety and efficiency.

Task activities (subtasks)

1. Go to fighting position/support by fire position.
2. Go to hiding position.
3. Go to observing position.

Screen Operation Task This task is for local security, or for self-protection. In this case, the goal of the scout platoon is to provide the security assistance to another unit. The scout force is located between a parent force unit and the enemy. The unit tasked with performing a screen operation must provide an early warning of the enemy activity to the rest of the platoon unit and to the parent unit. The scout force can be located in a hidden observation post around the stationary force unit or perform an overwatch movement leading a heavy force in an assault operation. The designed result is no enemy passage through the assigned area.

Convoy Escort Operation Task The scout platoon goal is to ensure the security of a moving convoy. The escort may be tasked to provide front, rear, or flank security. Another option is to integrate it within the convoy. The platoon should have enough combat power to handle the situation and it can request assistance from engineers, military police, dismount infantry, or a heavy reaction force.

Task activities (subtasks)

1. Tactical road march.
2. Tactical cross-country movement.
3. Route reconnaissance.

Surveillance Task This is a security operation that is called by the army "reconnaissance by fire." The scout platoon goal is to make sure that an area is clear from enemy forces. The scout vehicles conduct area reconnaissance, which means covering an area while under movement. In case of contact, the units react according to the rules of engagement. This may be self-fire or indirect fire.

Reconnaissance (Reconnoitering) Task This task is to report if route/area is traversable for manned vehicles. The unit/single vehicle will navigate through a series of a preplanned waypoints along the assigned route or will cover an assigned area. As necessary, the platoon adjusts the route in the same way that is employed when the traveling task is performed. In this mission the vehicle mobility sensors are dedicated, for mapping and performing the traversability assessment for the vehicle. Additional sensors may be operated on-board for environment probing such as NBC (nuclear, biological, or chemical). The plan will ensure a complete coverage of the area in an efficient way by the vehicles. Online planning would be needed for coordinating the vehicle's movements, since the movements are required in real-time in order to accomplish the mission on time. The designed result is that the platoon reports about the complete area/zone.

Task activities (subtasks)

1. Route reconnaissance
2. Area/zone reconnaissance
3. NBC (nuclear, biological, chemical) reconnaissance (with radiation monitoring).

Search for the Enemy (Reconnaissance) Task The task goal is to detect and locate the enemy. The scout platoon is assigned to search for the enemy in a specified area. The vehicle's starting location may be at some distance from the search area, so the vehicles may have to perform a travel task to the search area. Sometimes the task is extended to include surveillance, so the scouts track the enemy while updating its location. The task requires stealth, and the enemy location and movements are not known initially. The designed result is that enemy units are detected and located in assigned area or their absence is confirmed.

Linkup with an Adjacent (nearby) Moving or Stationary Ground Unit Task The scout platoon unit acts always with a parent unit. There are situations where the scout unit or even a single vehicle is ordered to support another parent unit. The process of traveling to another area of operation and exchanging commanders must be done with extra considerations. Autonomous vehicles can be linked up to another unit through communication without any visual contact, if necessary. The designed result is a linkup with another parent unit.

Coordinated Passage of Lines Forward/Rearward Task This task is needed for the cases when a scout unit has to be active in the enemy territory behind the front line. The scout unit must coordinate with its front force on the way in and out of the border. This is a complicated task for an autonomous unit. It requires definitions of two-way communication and information transfer. As in the linkup task, crossing lines of passage demands a definition of the identification process between both sides of the line. The designed result is that the unit successfully passes the line.

Re-supply Task This task is concerned with the refueling and other maintenance needs for the unmanned scout vehicles. It is performed always while the process of preparation is in progress for the next task. Platoon vehicles that are on duty for several days must be able to travel to an assembly area for refueling, or for the equipment replacement. This subtask must be planned in advance or executed by field decision. The designed result is that vehicles have the fuel and equipment to carry on their operations.

Halt Task The task to stop may be preplanned or can be a part of the reactions to events. A halt task requires additional instructions about the position and the behaviors of the vehicles. Among these instructions for the unit is type of formation to use. Halt task can be a reaction to air attack, or can be due to the required plan changes. Since the vehicles exchange plans, the halt tasks help to bring the vehicles to another set of way points to be traversed while keeping the prior formation. The designed result is that the unit stops movement and protects itself.

1.6.3 Activities of the Platoon Leader

The platoon leader performs seven steps in troop-leading procedure:

1. Receive and analyze the set of mission tasks.
2. Issue a warning order—pass the information to the subordinate leaders.
3. Make a tentative plan.
4. Initiate a movement—to the new assembly area or closer to the operational area.
5. Conduct the reconnaissance—proceeding with the motion requires the routes of the plan to be verified by a preliminary watch.
6. Complete the plan—refinement of the plan and informing the parent unit commander and other platoon leaders (this is required when an alternative of the plan is close to completion after the reconnaissance).
7. Issue the order—after walking through the rehearsal with all subordinate leaders, the platoon leader should ensure the complete understanding of the upcoming operation.

Prepare

The platoon leader prepares for a mission by converting the platoon assignment to a plan/program that includes tasks for each of the sections. Each task consists of a string of subtasks and subgoals to be achieved. The plans are updated according to the newly arrived information. The platoon leader translates the task to a sequence of subtasks. These subtasks are translated from the army tactical language (platoon tasks) to more simple, technical terms with the particular meaning refinement.

Command, Control, and Communication

The platoon leader receives assignments from the battalion supervisor and transfers tasks to each section. Command and control are supplemented by communication of tasks, commands, and reports of the accomplishments.

React and Re-plan

The platoon leader decides how to react to an event or to re-plan in cases where the necessary deviation from a plan must exceed a certain threshold.

Travel

Platoon leader is supposed to lead the travel subtask of the platoon assignment. This rule might undergo some corrections and/or changes for unmanned vehicles. It will be demonstrated that in some situations the role of a "leader" will become a provisional one quickly "moved" from one vehicle to another. The leader determines the suitable type of formation and adjusts his movement to allow the platoon to keep in formation.

Observe

The platoon leader coordinates the subtask of "observing" when it is a part of the platoon assignment. Typically this is the second basic activity of a platoon.

Supervise and Monitor

The platoon leader has to supervise and monitor the activities of the platoon vehicles and the status of operation. This will require issuing a set of assessment algorithms associated with deviation from a plan, task accomplishments, and the mission results. The platoon leader uses the assessment result as an input to his decision-making procedure.

Learn

The platoon leader has the ability to learn from experience during the whole process of mission accomplishment. Learning consists of collecting experiences, categorizing them, clustering them as needed by the goal and situation, and deriving the appropriate map updates and rules of action.

11.6.4 Tasks Assigned to a Section

Section tasks are the result of plan decomposition at the platoon level into subtasks necessary to accomplish the platoon plan. An assignment given to a platoon typically breaks down into a set of sequential/parallel actions (move to first location, move to second location, move and report, activate mission package, and so on). Subsequently it breaks down according to tasks submitted to the subordinates of the platoon leader.

Prepare

Section leader prepares for a mission by converting the task assignment to the best alternative of plan that includes tasks for each of the vehicles. The task descriptions include goals to be achieved. Plan is created by the module of behavior generation. Consider the examples: Go to assembly area, refuel, load necessary databases, perform diagnostics, and so on. In each of these cases the task decomposition with formulation of the subtasks for the vehicles is required.

Command, Control, and Communication

Components of the platoons are part of the command and control chain. They receive assignments from their supervisor (the platoon leader) and the assignments are transferred to each vehicle. Command and control are supplemented by the corresponding communication part.

Travel

A section conducts the travel task in concert with the rest of the platoon. Most of the assignments include travel as their component. Section travel is done in formation. Examples of travel include overwatch movement, traveling overwatch, and follow the leader.

Observe

A section observes the environment looking for terrain features, enemy vehicles, and other objects that are required, or can be helpful for accomplishing the task assignment.

Learn

Sections have the ability to learn from experience during mission activation and observation. Collecting the experiences, categorizing them, and performing the operation of clustering and deriving the map updates and new rules does it.

11.6.5 Tasks Assigned to a Single Vehicle

Lead the Platoon/Section

The meaning of the assignment to be a leader is that vehicle has to plan for the upcoming periods of time and command to reaction for the real-time events. The vehicle assigned to be a leader plans for the entire section.

Move

Each vehicle performs travel in formation, which requires constantly computing and re-computing the motion trajectory. Reactive deviations from the planned motion trajectory requires online computing of the required short-term motion trajectories, such as in obstacle avoidance where obstacles are not demonstrated in the map.

Sense and Scan

The vehicle senses the environment, as scheduled and assigned in the mission program (type of object to observe, azimuth, elevation, range, rate of scanning, etc.). The vehicle constantly collects data that contain terrain features, obstacles, other vehicles, and more.

Detect and Locate

The vehicle has to detect other vehicles, or other objects not shown in the map. It has to estimate their location based on data from the sensors (including the inertial navigation system).

Communicate

The vehicle will communicate with other vehicles or communication stations according to the plan and its reactive behaviors.

11.6.6 Tasks Assigned to a Vehicle's Subsystems

Plan

The vehicle level generates plans for all the subsystems in the vehicle. Commands to the vehicle level include lead, travel, sense and scan, and others. The behavior generation module performs planning (see Chapter 7 and [62]).

Drive

Driving plans for the next 10 minutes are generated by the vehicle-level driving planner. The driving subsystem executor is activated according to the plan. Its desired motion trajectory (including the path with associated values of speed and acceleration) is provided by the vehicle-level drive planner.

Sense

The vehicle sensors are activated as planned by the vehicle-level attention planner. Each of the sensors is dedicated to an appropriate task such as: searching for obstacles, searching for other vehicles, or probing for NBC (nuclear, biological, and chemical).

Search

The vehicle-level plans how to point or scan the activated sensors toward the environment. The vehicle-level attention executor carries out the plan.

Track

The vehicle-level attention executor locks the sensor on a detected object, keeping it in the field of view. This provides for tracking motion of the sensor pointing subsystem.

Detect and Locate

The vehicle-level planner applies an algorithm to locate a detected object in the sensed data using data from the navigation system. This subtask is directly associated with the subtask of search.

Communicate

The communication subsystem reports status to the supervisor and receives commands from it. Received commands are routed to the appropriate level of the control hierarchy.

Data Collection and Map Update

The vehicle saves data in the database as a by-product of some or all of its assigned tasks. The data are used for registering with the maps, initiating the map-update procedure, correcting the world model, or supporting other requirements. Data are provided from the database as input to the other subsystems, such as terrain maps for the planning of motion trajectories, tactical knowledge for the platoon-level planning, and the model for the detection procedure.

11.6.7 Invariant Elements of Mission Programs

So far we have not determined the mission vocabulary and its hierarchical organization. From the previous sections, the taxonomy of the objects and actions has become clear. The intention is to provide the battalion level (level 7) with a tool for communicating the mission to the platoon. The mission order is expected to be in a form suitable to the mission program. The mission order includes the mission assignments/tasks and updated information relevant to the mission as described later in this section. Most of the mission program is the output of the planner. The information provided by planner depends on the level of hierarchy.

Missions programs are divided into the following classes of components:

- Movement requirement: Task knowledge
- Geographical area relations: World model
- Relation to other agents (our vehicles): World model
- Relation to the enemy: World model
- Communication requirements: Task knowledge
- Sensing activities: Task knowledge
- Reaction and re-planning requirements: Task knowledge
- Transition condition to the next mission task: Task knowledge

Movement Requirement

This part of the mission program contains the information about movement requirements for a scout platoon during the mission: Movement task types (road march, cross-country, follow a vehicle (leader), overwatch movement, traveling overwatch), movement parameter (assigned road route, assigned off-road route, assigned combination of road route and cross-country route, assigned destination as point, assigned destination as area waypoints), definition type (non-mandatory/mandatory waypoint, check point/not check point, with/without action, with/without time flag), with/without heading direction point, tolerances (coordinate, heading, time).

Formation Definitions

The plan includes a chosen set of definitions and parameters for the way of keeping formation from a larger set that a scout unit may be able to perform: chosen formation chosen leader vehicle to be followed, follow the leader offset (degrees), spacing (open, close, user define).

Relation to Other Vehicles and to Parent Unit

The scout platoon is organized, equipped, and trained to conduct activities assigned by the parent unit. The program includes a chosen status of activity of the parent unit to be related by the scout's behavior: screen operation, fighting activity, non war operation, damage report, and ammunition).

Object Detection and Allocation

In a situation where an enemy unit is within the scope of the scouts, each scout vehicle has its own activities, like tracking, related to one, selected, enemy vehicle. This part of the plan includes a chosen methods of target detection for the single scout vehicle. In addition the plan includes instructions about the allocation of vehicles to objects. For example, a section of two vehicles has detected 10 enemy vehicles; according to their orders, they have to track the enemy vehicles. The allocation instruction will include the priority of the object to be tracked (tank, armored vehicle, tanker, or houses with people). In cases where the objects are similar (e.g., all are tanks), the allocation can be based on geometry or leadership (e.g., lead vehicle, center vehicle). This part of the mission program is delivered from the mission order.

Geographical Area Relations

This part of the program includes the information related to maps of the operation area. All the data of this part can be shown as a layer in a geographical (or a topographical) map.

Area Observation Mode

The chosen definitions and parameters are for the observation tasks: cover from static observation post, cover while moving (to protect the parent unit), generate movement in order to cover area, and cover while enemy movement.

Geographical Information Database

This part is concerned with the geographical database of the vehicle. It includes general geographical maps and specific information on entities like coordinates, height, and velocity.

Sensing and Focusing Attention (SFA) Activities

Onboard sensors that allow traveling, mapping, and observing to perform SFA activities. Algorithms for the detection and recognition of the external world (including enemy vehicles) do it. This part of the program includes instructions and definitions for the subsystems of sensing and sensory processing.

11.6.8 Detection and Recognition of Sensed Objects

This part of the program includes tasks for the image processing subsystem to detect and track a specific target or to allocate targets between our vehicles. This part of the program is the planned sensing activity including special instructions, scheduling, and operational condition.

Sensing/Observation Acquisition

This part of the program includes tasks to acquire data from the sensors, including the navigation system, for purposes such as localizing objects. This task includes the state estimation with the subsequent apprehension and evaluation of static and moving objects or general information.

We have listed the results of task commands, goals, and actions from a set of behaviors for a military scout platoon in Table 11.2. The resulting set of tasks and vocabulary for the scout platoon can be used in implementing a 4-D/RCS control hierarchy for autonomous vehicles that exhibit these required behaviors.

11.7 4-D/RCS AS AN IMPLEMENTATION GUIDE FOR DEMO III

11.7.1 Subset of 4-D/RCS for Demo III

A subset of the 4-D/RCS that was implemented on the NIST HMMWV in December 1999 includes ELFs of the highest level of resolution for actuating four different motions: steering, braking, throttle, and gear shift. At the lower left is a suite of cameras including a LADAR, a stereo pair of FLIRs, a stereo pair of CCD cameras, and one or more color CCD cameras with different focal lengths and fields of view. A sensory processing system produces a set of range images overlaid with entity class labels such as obstacles and targets. These images are transformed into three maps at three different scales (i.e., 256 pixels = 5 meters at the primitive level, 256 pixels = 50 meters at the subsystem level, and 256 pixels = 500 meters at the vehicle level) with labeled regions. On these three maps, map-based path planning can be performed at three levels of resolution in space and time.

An attention subsystem with a pan/tilt platform points the cameras at the parts of the egosphere that contain regions of interest and schedules saccades between points of attention. The attention subsystem also coordinates tracking of entities of attention. The image processing hierarchy works in the following manner.

Table 11.2. The hierarchy of tasks and their results

Task Assignment	Task Goal	Task Result
Travel	Navigate from one point to another while adjusting route, formation, or movement technique	Reach destination area
Establish an observation post	Reach the destination points from where the assigned area can be seen	Report from observation post on enemy activity within the assigned area
Advance/withdraw to position	Adjust position according to battle location to allow hiding/fighting	Cooperate with parent unit during battle to increase safety and efficiency
Screen operation	Provide security assistance to another unit by overwatch and surveillance action	Allow early warning about the enemy activities
reconnaissance	Travel through an assigned area or route while avoiding obstacles and detection	Report if the route/area is traversable for other vehicles
Search for the enemy (reconnaissance)	Cover an assigned area by traveling and observing	Detect and locate the enemy in an assigned area
Linkup to nearby unit	Travel to nearby area of operation with a coordination to change the parent unit	Linkup to another parent unit accomplished
Coordinate passage of lines	Travel to pass line while coordinating with border guards	Unit has passed the line successfully
Re-supply	Vehicle travels to assembly area to maintenance and refuel	Vehicles have fuel and equipment to carry on the operation
Halt	Vehicles stop current task and get together in an assigned static formation	Units stop traveling and gather in a static formation to allow self-protection

At level 1, image processing algorithms compute spatial and temporal gradients of intensity and range, and x- and y-flow values at each pixel. The output of level 1 is a labeled level 1 entity image. Pixels are labeled as range-edge pixels, brightness-edge pixels, color-edge pixels, surface patch pixels, or unlabeled pixels. For each pixel, its set of attributes constitutes a pixel attribute vector.

At level 2, labeled pixels are grouped into level 2 regions, or list entities. For each level 2 region, attributes are computed and stored in level 2 entity frames. Attributes of level 2 regions may be compared with list entity classes and a labeled level 2 entity image generated. Pointers define the correspondence between the labeled level 2 entity frames and labeled level 2 regions in the image.

At level 3, labeled level 2 entities are grouped into level 3 regions, or surface entities. Level 3 region attributes are computed and stored in level 3 entity frames. Level 3 entity attributes are entities. The size of each window is determined by the confidence factor computed by the level 2 recursive filtering process. If the estimated entity confidence value is high, each window will be narrowed to only slightly larger than the set of pixels in the corresponding list entity region. If the estimated entity confidence value is low, the windows will be significantly larger than the level 2 list entity regions. Until there exists a set of recognized list entities, level 2 windows are set wide open.

At level 2, grouping is a process by which neighboring pixels of the same class with similar attributes can be grouped to form higher level entities. Information about each pixel's class is derived from the recognized pixel entity image. Information about each pixel's orientation and magnitude is derived from the predicted pixel attribute images. Grouping is perfomed by a heuristic algorithm based on gestalt principles such as continguity, similarity, proximity, pattern continuity, or symmetry. The choice of which gestalt heuristic to use for grouping may depend on an attention function that is determined by the goal of the current task, and on the confidence factor developed by the recursive estimation process at level 2. Grouping causes an image to be segmented, or partitioned, into regions (or sets of pixels) that correspond to higher order entities.

In 4-D/RCS implementations beyond version 1.0, these image processing algorithms may involve recursive estimation for noise rejection, motion tracking, and confirmation of grouping hypotheses. However, the version 1.0 implementation of 4-D/RCS will use more conventional image processing algorithms that are well-known, readily available, and easy to implement for real-time operations.

Version 1.0 sensory processing algorithms will be performed once per second on LADAR images that are compensated for vehicle motion, and possibly with stereo images computed 10 times per second. Beyond version 1.0, sensory processing operations may be performed 30 times per second on images from both wide and narrow FOV cameras that are registered and fused with images from the LADAR camera and stereo systems.

Version 1.0 LADAR images will contain only 8,000 pixels. Hence, the initial version 1.0 system will require only about 800 regions at level 2, 80 regions at level 3, and 8 regions at level 4. Beyond version 1.0, CCD images will consist of 64,000 pixels. Level 2 regions may consist of up to 16 pixels. Level 3 regions may consist of up to 16 level 2 regions. Level 4 regions may consist of up to 16 level 3 regions. Thus the 4D/RCS architecture should eventually be capable of classifying and tracking up to 16 objects at level 4, up to 256 surfaces at level 3, 4,000 list entities at level 2, and 64,000 pixels at level 1.

The systen uses two types of map representations: world maps and plan maps. Because each pixel in the LADAR image has a range value, labeled entity images can be easily transformed into world map coordinates. Self-vehicle state (position, velocity, and heading) derived from GPS and INS sensors enable the self-vehicle entity to be represented on this world map. There are several possible points at which the transformation from image to world coordinates can be performed. It has not been decided yet where the best point (or points) is in the SPWM hierarchy for transforming between image and world map coordinates.

The three world maps provide the information needed for path and route planning at levels 2, 3, and 4. At level 2, the world map is derived entirely from data generated by the cameras. The range is too small at level 2, and the scale is too high resolution for any useful information to be derived from a digital terrain map database.

At level 3, the world map is derived primarily from entity images generated by the cameras but can be registered with and overlaid on maps obtained from a digital terrain database. The digital terrain database contains information from sources other than the cameras on the vehicle. There can be a number of world map overlays that represent obstacles, terrain elevation, ground cover, traversability, risk, and visibility from a given observation point. Other overlays may prove useful.

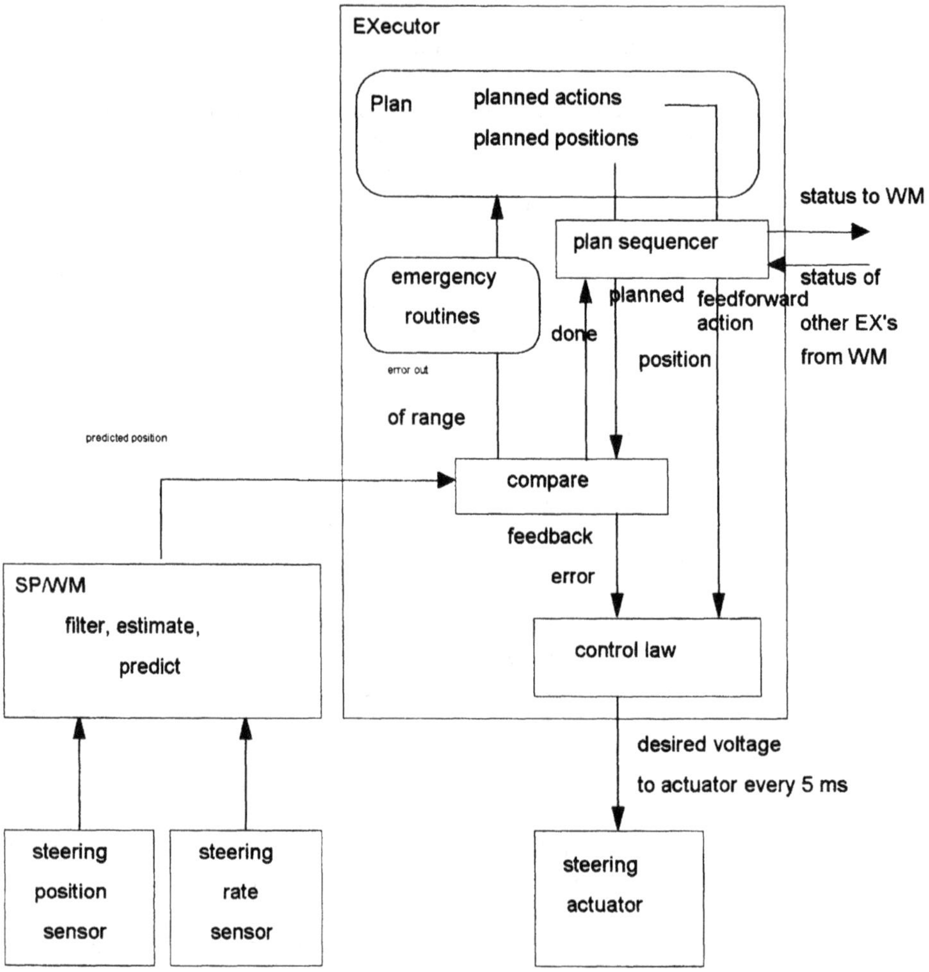

Figure 11.13. BG hierarchy for steering.

The hierarchy of behavior generation for steering is shown in Figure 11.13. Similar pictures can be built for braking, throttle, and gear shift. At the vehicle, subsystem, and prim levels, the planners hypothesize planned actions and the world model simulators generate planned waypoints. (Alternatively, the planners hypothesize waypoints and the inverse world model simulators generate planned actions.) At the vehicle level, the world map has a range of 500 meters and the WM simulator uses a point/velocity model (or inverse model) of the vehicle. At the subsystem level, the world map has a range of about 50 meters and the WM simulator uses a 3D (x, y, heading) kinematic model (or inverse model) of the vehicle. At the prim level, the world map has a range of about 5 meters and the WM simulator uses a 6-DOF dynamic model of the vehicle. At the servo level, there is no map representation, and the WM simulator uses 1-DOF dynamic models of actuators and loads.

The task command frame includes the following:

1. Task name (from the vocabulary of tasks the receiving BG process can perform).
2. Task identifier (unique for each commanded task).

3. Task goal (a desired state to be achieved or maintained by the task).
4. Task object(s) (on which the task is to be performed).
5. Task parameters (e.g., speed, force, priority, constraints, coordination requirements).
6. Plan (in which the commanded task is the first step).

At the vehicle level, task goals are desired states to be achieved approximately 50 seconds in the future. The plan (item 6 in the task frame) enables the vehicle level planner to operate in the context of the squad level plan that extends about 10 minutes into the future.

Within the vehicle-level KD, there exists a library of tasks that the vehicle-level BG process is capable of performing. For each task in this library there exists task knowledge that describes how to do the task, or how to select or generate a plan to do the task. For each task in the library, there exists a task frame that specifies the materials, tools, and procedures for accomplishing a task. For example, a vehicle level task frame may include the following:

1. Task name (from the library of tasks the system knows how to perform). The task name is a pointer or an address in a database where the task frame can be found.
2. Task identifier (unique ID for each task commanded). The task identifier provides a method for keeping track of tasks in a queue.
3. Task goal (a desired state to be achieved or maintained by the task). The task goal is the desired result to be achieved from executing the task.
4. Task objects (on which the task is to be performed). Examples of task objects include map waypoints and states to be achieved, objects to be observed, sectors to be reconnoitered, and targets to be designated.
5. Task parameters (that specify, or modulate, how the task should be performed). Examples of task parameters are priority, speed, time of arrival, degree of stealth, level of aggressiveness, amount of cost, and risk that can be accepted.
6. Agents (that are responsible for executing the task). Agents are the subsystems (locomotion, attention, communication, and mission package) that carry out the task.
7. Task requirements (tools, resources, conditions, state information). Tools may include sensors and weapons. Resources may include fuel and ammunition. Conditions may include weather, visibility, soil conditions, daylight, or darkness. State information may include the position and readiness of friendly forces and other vehicles, position and activity of enemy forces, time of day, or stage in the battle plan.
8. Task constraints (upon the performance of the task). Task constraints may include visibility of sectors of the battlefield, location of sector lines, timing of movements, and requirements for covering fire.
9. Task procedures (plans for accomplishing the task, or procedures for generating plans). At all levels, the KD contains a library of plans for various routine contingencies as well as procedures that specify what to do under various kinds of unexpected circumstances or emergency conditions. This enables the executors to react quickly to sensed conditions.

10. Control laws and error correction procedures (defining what action should be taken for various combinations of commands and feedback conditions). These may be developed during system design and possibly modified through learning from experience.

The vehicle-level planner generates a plan in the form of a state graph (or state table) that consists of a series of planned actions interspersed with a series of planned states (e.g., waypoints in world map coordinates). About every 10 seconds, the vehicle-level planner generates a new plan for the next 50 seconds. The vehicle-level executor executes the current plan by generating a series of task commands and sending them to the set of BG processes (for locomotion, attention, communication, and mission package) at the subsystem level. Task commands to the subsystem-level BG processes are issued about every 5 seconds. Goals in subsystem-level task commands are defined as desired states approximately 5 seconds into the future.

The subsystem-level planners generate a new plan about every second. The subsystem-level executors issue task commands to the primitive level BG processes about every 500 milliseconds. The primitive-level planners generate a new plan about every 150 milliseconds. The primitive-level executor issues task commands to the servo-level BG processes every 50 milliseconds. Servo-level planners generate a new plan about every 15 milliseconds. The servo-level executors issue commands to the actuator drive electronics every 5 milliseconds.

The sensory processing system updates the maps used for planning at a rate that is primarily determined by the scanning rate of the cameras and the computational speed of the SP processes. For the current Dornier LADAR, the scanning rate is limited to one full image per second. The current stereo processing speed of the JPL system is approximately 10 images per second. The utility of using the 5 meter map for path planning depends on vehicle speed. If the vehicle is traveling at speeds of greater than 10 meters per second, the utility of map-based planning on the 5 meter map is small. However, at speeds of less than 1 meter per second, there may be significant benefits in path planning on the 5 meter map.

A draft version of the 4-D/RCS generic shell for behavior generation is shown in Figure 11.13 at the servo level for the velocity subsystem. The commands to the servo level indicate the desired acceleration and yaw rate to be achieved 50 milliseconds into the future. The job assignor breaks this task into a commanded position for the steering wheel actuator, a commanded brake pedal force, a commanded throttle position, and a commanded gear shift setting—all to be achieved 50 milliseconds into the future. For each actuator, a scheduler generates a schedule of planned actions and planned states over the 50 millisecond interval. Selected job plans are placed in a plan data structure in the executor where they are accessed by a plan sequencer. On each cycle of the controller, an executor selects a planned action and a desired resulting state from its respective plan (stored as a state table). The planned action is sent to the control law as a feedforward action. The desired resulting state is simultaneously compared with the feedback predicted state from the world model. The difference between the feedback predicted state and the planned state is a feedback error that is also sent to the control law where compensation is computed and added to the feedforward planned action.

A generic shell for behavior generation makes it relatively easy to program at this level of complexity. The generic shell provides the timing and communication functions and allows the system designer to think in terms of planning and control algorithms rather than C++ code constructs. At least two tools look promising for constructing generic shell modules. One is the control shell tool developed by Real Time Innovations.

John Horst is the resident NIST expert on the control shell. The second is the Java tool developed by Will Shackleford.

Figure 11.14 shows the organizational structure for behavior generation generic shells at the prim and subsystem levels integrated with those at the servo level. The driving system at the subsystem level is decomposed into the camera subsystem and

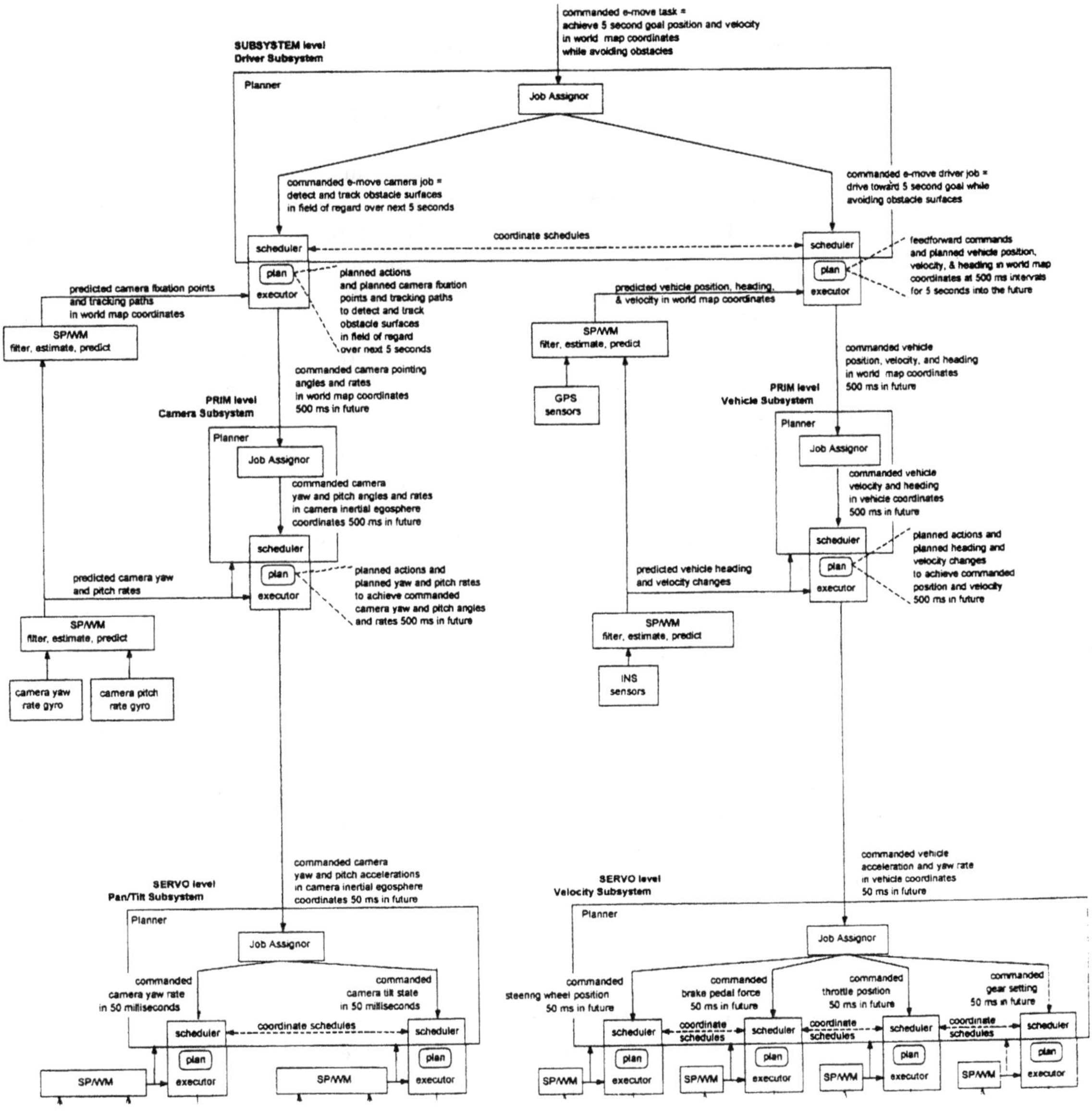

Figure 11.14. Generic shell BG modules at the subsystem, primitive, and servo levels. Any planner is no better than the simulator, the models, and the maps it uses. In an uncertain domain such as cross-country driving on the battlefield, models and maps can be very dynamic, and the cost-function may change from time to time. Therefore the replanning function must be fast and efficient so that replanning can be performed many times faster than the real-time execution of the plan. In extremely dynamic situations, a new plan should be generated by the time the first step or two in the old plan is completed.

the vehicle subsystem at the prim level, and into the pan/tilt subsystem and the velocity subsystem at the servo level. The servo planner looks ahead 50 milliseconds into the plan generated by the prim level. The prim planner looks ahead 500 milliseconds into the plan generated by the subsystem level. The subsystem level planner looks ahead 5 seconds in the plan generated by the vehicle level (not shown in Figure 11.14). At each level the planner always looks ahead at least to its own planning horizon in the higher-level plan.

The WM simulator permits the planners to hypothesize actions and anticipate expected results. Tentative plans can be subjected to a cost function so that a variety of possible behaviors can be analyzed out to the planning horizon at each level. Planners at all levels have cost functions that they attempt to minimize. The choice of cost functions determines the choice of plans. Thus a cost-function that places high value on achieving mission goals and low value on minimizing risk will produce aggressive behavior. A different cost-function that places high value on stealth and assigns high cost to risk will produce cautious behavior. The planners thus can use cost-functions to choose between competing simulated behaviors in advance of action. The ability to assign different cost-functions will be included in the September 1999 demonstration on the NIST HMMWV.

The complexity of planning can vary over a wide range. Planners can be very simple (with stored plans) or very complex (with the ability to heuristically search the space of possible futures). At either extreme or in the middle, planners anticipate the future by simulating what might be expected to happen as a result of the various possible behavioral choices. The planner replans as fast as possible, or as often as necessary, to make sure that the best achievable plan resides in the plan buffer where it can be executed by the executor. Replanning at all levels enables the entire system to respond quickly and effectively to dynamically changing conditions on the battlefield.

Demo IIIA Camera Suite

The 4-D/RCS architecture allows a variety of camera systems and image processing software and hardware to be integrated into the Demo III vehicles. The baseline NIST HMMWV vision system will consist of a Dornier LADAR and two CCD cameras mounted on a pan/tilt platform. The Dornier LADAR produces a range image once per second consisting of 128×64 pixels in an approximately 64×32 deg field of regard. This yields an image with about 0.5 deg per pixel as shown in Figure 11.15.

For the laser range imager, at least six attribute images will be maintained:

1. r = range.
2. dr/dx = range x gradient, or surface slope along x direction.
3. dr/dy = range y gradient, or surface slope along y direction.
4. dr/dt = range rate, or surface motion along the pixel line of sight.
5. r − edge = $|dr/dx|$ or $|dr/dy| >$ threshold.
6. r − edge-angle = orientation of r-edge (from 0 through 360° counterclockwise around the pixel line of sight).

The LADAR image will be analyzed to extract range, range rate, x and y range gradients, and range discontinuities at each pixel.

Pixels in the LADAR image will be grouped or clustered into image regions that can be segmented and recognized based on the following attributes:

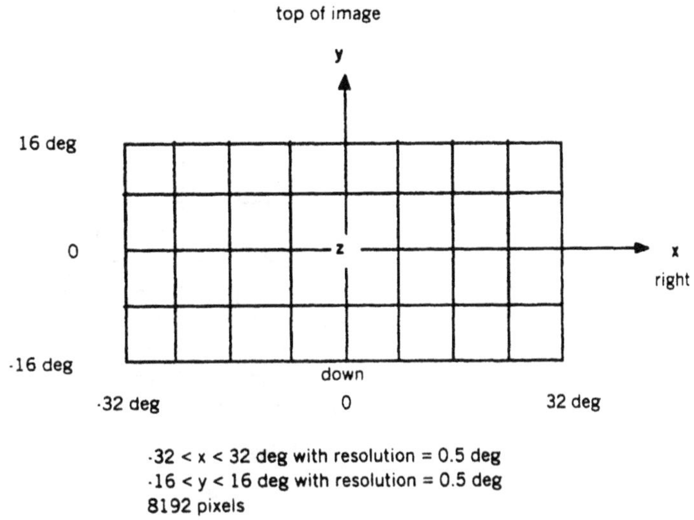

Figure 11.15. Laser range imager field of view.

1. Smooth surfaces where range values vary relatively smoothly from pixel to pixel. Smooth surfaces may consist of the smooth ground, sides of buildings, walls, sand, snow, short grass, rocks, trunks of trees, barrels, boxes, vehicles, and so on. For these types of surfaces it is possible to compute surface range and surface slope at each LADAR pixel.
2. Surface discontinuities where range values change abruptly from one smooth surface to another. Surface discontinuities may correspond to the edges and roof lines of buildings, sides of tree trunks, or the sides and tops of rocks, barrels, boxes, vehicles, and so on. Surface discontinuities also exist at the crest of hills, or the near edges of cliffs or gullies.
3. Surface joins where surface slopes change abruptly from one smooth surface to another. Surface joins may correspond to the bottom edge of walls, curbs, ditches, or where rocks and tree trunks emerge from the ground.
4. Rough surfaces where range values vary abruptly from one pixel to the next. Rough surfaces may correspond to tree foliage, branches, bushes, tall grass, piles of rubble, and so on.
5. Long thin surfaces where near range values only occasionally are returned. Thin surfaces may correspond to ropes, wires, and single thin tree branches.
6. Specular surfaces where range values are absent altogether. Specular surfaces may correspond to water, glass, or polished metal.

The wide-angle CCD camera produces a red, green, blue, and intensity image of 256 × 256 pixels with a 32 × 32 degree field of view. This yields an image with about 0.125 degree/pixel as shown in Figure 11.16.

The foveal CCD camera produces a red, blue, green, and intensity image of 256 × 256 pixels with a 8 × 8 degree field of view. This yields an image with about 0.031 degree/pixel as shown in Figure 11.17. This is slightly lower resolution than normal human foveal vision. Additional cameras can be added or substituted for these. For

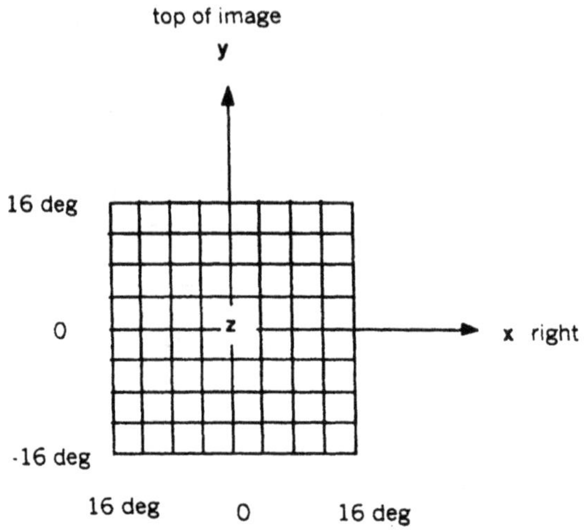

-16 < x < 16 deg with resolution = 0.125 deg
-16 < y < 16 deg with resolution = 0.125 deg
65536 pixels

Figure 11.16. Wide-angle CCD camera field of view.

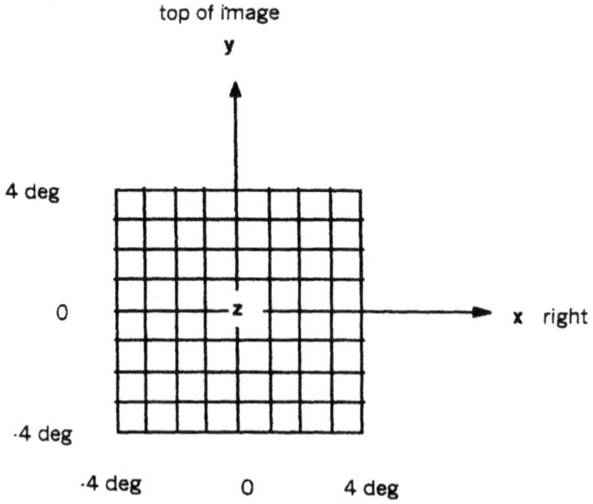

-4 < x < 4 deg with resolution = 0.031 deg
-4 < y < 4 deg with resolution = 0.031 deg
65536 pixels

Figure 11.17. Foveal CCD camera field of view.

example, stereo cameras and stereo FLIRs from JPL will be integrated for testing as will the UBM/Sarnoff stereo system.

Combining LADAR and CCD Images

The fields of view (FOV) of the laser range imager, and the wide-angle, foveal CCD cameras can be registered and projected onto a Mercator projection of the retinal egosphere as shown in Figure 11.18. The result of combining LADAR and CCD images will add color and brightness attributes to the LADAR image, and add range and surface attributes to the CCD image. There are $4 \times 4 = 16$ CCD pixels per range pixel in the wide-angle CCD, and $16 \times 16 = 256$ CCD pixels per range pixel in the foveal camera, as shown in Figure 11.19. The correspondence between pixels of these two sensors is shown in Figure 11.20.

Assigning Range and Color

Since there are many CCD pixels within each LADAR pixel, it is easy to assign color and brightness attributes to the LADAR image. Color and brightness information can simply be averaged over all the CCD pixels within the spatial region occupied by the LADAR pixel. On the other hand, adding range attributes to the brightness pixels is more complicated. For smooth surfaces the problem is simplest. A first approximation of range for each pixel can be computed from the LADAR pixel attributes of range plus x and y range gradients. For most purposes this is sufficient. If more detail is desired, shape-from-shading algorithms might be used to refine range estimates for each CCD pixel.

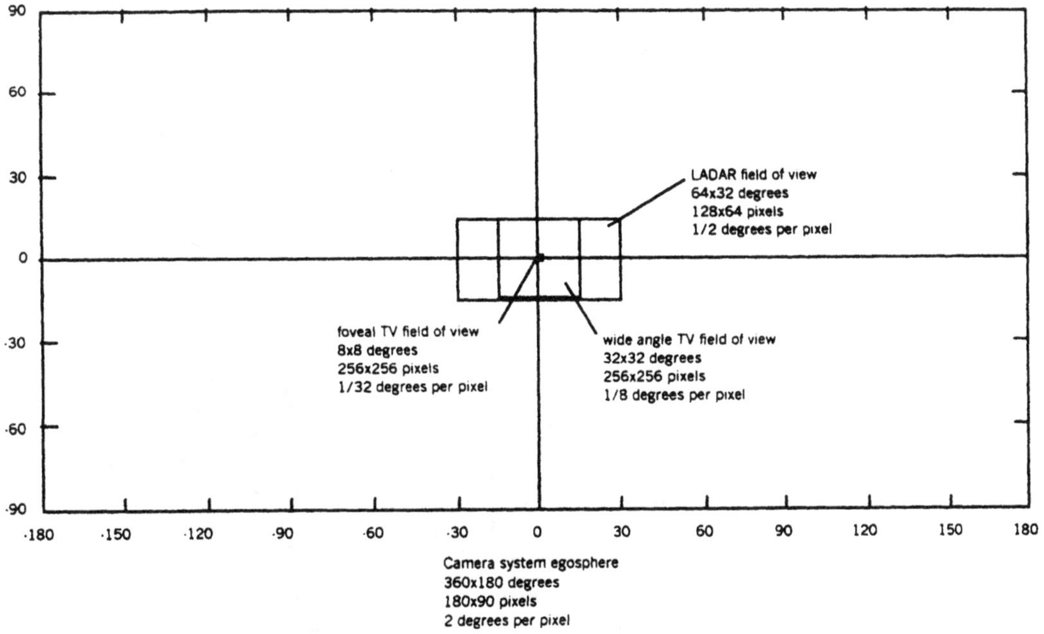

Figure 11.18. Retinal egosphere in Mercator projection coordinates.

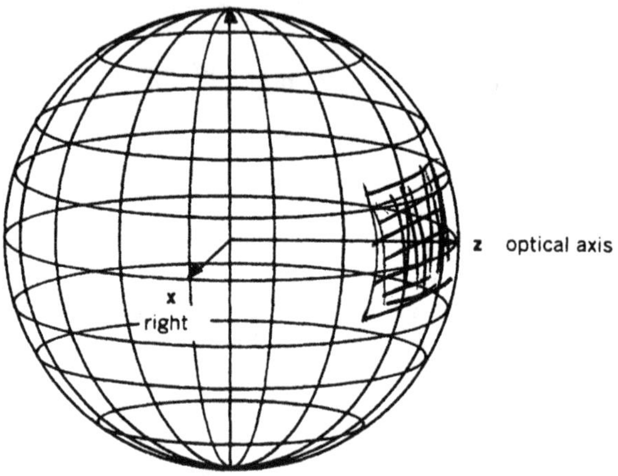

Figure 11.19. Retinal egosphere in spherical egosphere coordinates.

At surface boundaries, two adjacent LADAR pixels have significantly different range values. This implies that somewhere between the centroids of the two LADAR pixels there exists a surface boundary. It is quite likely that there will also be a sharp brightness discontinuity at the same surface boundary. Therefore brightness edges between the two range edge pixels that have the same orientation as the range edge can be assumed to correspond to the surface edge. CCD pixels on the surface side of the brightness edge can be assigned range values corresponding to the surface range value. CCD pixels on the background side of the brightness edge can be assigned range values corresponding to the background range value.

For rough surfaces, such as tree or bush leaves and branches, CCD pixels with color or brightness attributes corresponding to the branches or leaves can be assigned the nearest range values. CCD pixels with color or brightness attributes corresponding to the background can be assigned the more distant range values. Differential CCD image flow rates may also be used in making range assignments to CCD pixels for rough surfaces.

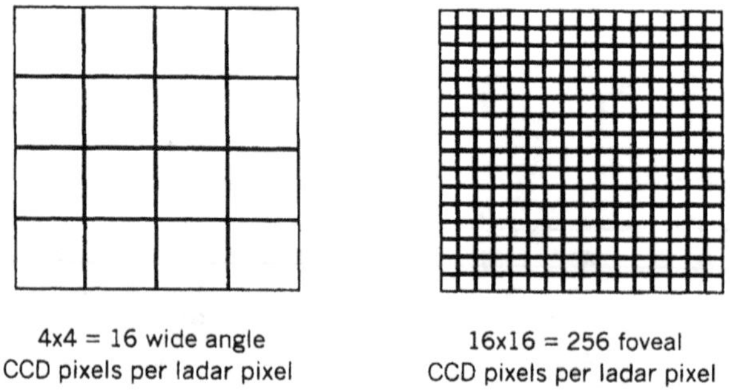

Figure 11.20. Ratio of CCD pixels to one laser range pixel.

For thin surfaces, CCD pixels with color or brightness attributes corresponding to a rope or wire or narrow branch can be assigned the nearest range values. CCD pixels corresponding to the background will be assigned the more distant range values.

For specular surfaces, CCD pixels at the boundaries of the specular surface can be assigned range values the same as the nearest valid range pixel. The range to interior CCD pixels on specular surfaces can be assumed to vary smoothly over the region between the specular surface boundaries.

Relative Timing

The laser range imager produces a range image every second with 0.5 degree resolution. The two CCD cameras produce 30 images per second with 0.125 and 0.031 degree resolution, respectively. This means that each CCD pixel is updated 30 times for each corresponding laser range imager update. This is illustrated in Figure 11.21. It thus becomes necessary for the CCD camera pixels to flywheel their estimate of range at each pixel through 30 frames of CCD images between each range estimate update from the laser range imager. This can be done using knowledge of camera motion and measured image flow.

Combined Attribute Images

Once the attributes from the LADAR images have been combined with the two CCD camera images, a set of attribute images can be computed for the fields of view of the two CCD images. In each attribute image the estimated value of each attribute can be initially set to some a priori value, and subsequently updated by a recursive estimation procedure of the form

$$\hat{a}(x, y, t|t) = a^*(x, y, t|t-1) + K(a(x, y, t) - a^*(x, y, t|t-1)),$$

where

$\hat{a}(x, y, t|t)$ = the estimated value of the attribute a at pixel (x, y) at time t after the measurement at time t,

$a^*(x, y, t|t-1)$ = predicted value of a at pixel (x, y) at time t after the measurement at $t-1$,

$a(x, y, t)$ = the observed value of a at pixel (x, y) at time t,

$K(t)$ = the confidence in the observed value at time t, with $K(t) = 0 \rightarrow$ no confidence in the observed value, and $K(t) = 1 \rightarrow$ complete reliance on the observed value, for $0 < K(t) < 1$.

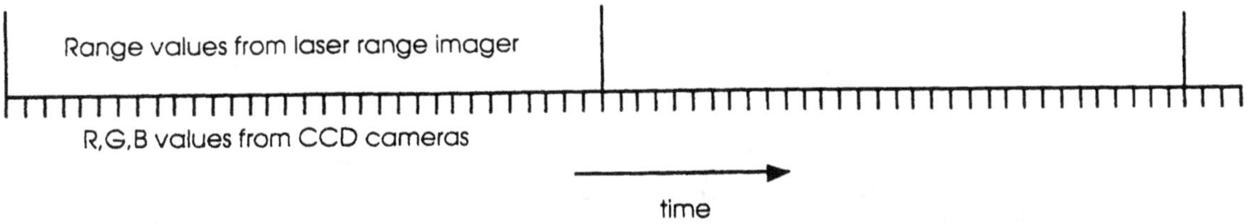

Figure 11.21. Timing of range images and CCD image compared. For each pixel, a CCD camera pixel update occurs 30 times for every LADAR pixel update.

Image Segmentation and Entity Recognition

Once a full set of range, brightness, and edge attributes are assigned to each pixel at each resolution, it becomes possible to perform grouping of pixels and segmentation of the image into regions based on similarity and contiguity between pixel attributes. Edge pixels can be grouped into edge and vertex entities. Surface pixels can be grouped into surface entities that are delimited by boundary and vertex entities. Surface, boundary, and vertex entities can be further grouped into object entities. Entity attributes such as size, shape, orientation, position, velocity, and rotation can be computed over the image regions occupied by the entities. Pixels that belong to entities define entity images. Entity images can be used to define masks and windows that facilitate efficient tracking and rapid processing of entity images.

Attention

The estimated value of each attribute for every pixel need not be computed every frame. Attention mechanisms driven by prior knowledge and the task command can control when and where various entity images will be processed, and they can select which processing algorithm will be applied. Windows may be used to mask out portions of the image, based on expectations and task requirements. There will be a list of entities of attention. Entities will be added to this list on the basis of several factors:

1. Entities in an active task frame are entities of attention. The attention subsystem will search for these entities, and once they are discovered, they will be tracked until they no longer appear in an active task frame.
2. Regions of the image that do not conform to expectations may be labeled unexplained and entered into the list of entities of attention. These entities may become the object of an inspection task, depending on competing task priorities.
3. Regions on the egosphere corresponding to areas on the world map where the path planner has made a tentative plan to go but that have not been recently observed. These regions should be inspected for obstacles before a plan that calls for traversing these areas is selected.

When there are items on the list of entities of attention, windows can be placed around the entities of attention, and detailed processing of all pixels will take place only within the windows. This will speed up processing of entities of attention. When there are no items on the list of entities of attention, all pixels in the image will be processed completely. This may require a lengthy processing period, but (by definition) there is nothing else that needs doing.

11.7.2 Examples of 4-D/RCS Entities

In 4-D/RCS the collection of entities stored in long-term memory are the set of entities that are relevant to the tasks that the system may be required to perform. Some object entities that are relevant to Demo III tasks are [ground], [sky], [hill crest], [ditch], [rock], [tree], [bush], [road], [water], [vehicle], [human], and [animal]. Each of these objects has attributes and subentities such as surfaces, and edges that can be observed in, and computed from, the image.

For example, the ground consists of one or more surfaces, including the surface directly under the vehicle and extending outward from beneath the vehicle. Each of these surfaces has attributes of slope, roughness, and ground cover. Each surface has boundaries, or edges that may be fuzzy or sharp. The slope of the ground directly under the vehicle can be measured by a tilt meter in the vehicle. Directly in front of the vehicle, the slope of the ground can be measured by the LADAR camera, by vergence and stereo disparity, or by image flow calculations. Ground surfaces may blend smoothly into each other, or be separated by walls, embankments, curbs, fences, ditches, hill crests, or other discontinuities.

A generic entity frame for the ground might have the form:

name = ground

Attributes

slope
roughness
ground cover

Subentities

surface 1
surface 2
surface 3
etc.

Criteria for recognition

surface 1 is directly beneath the vehicle
all surface subentities are roughly horizontal

Ground subentities have the form:

name = surface 1

Attributes

slope
roughness
ground cover
shape
size
map position (in egosphere coordinates)
map position (in world coordinates)
map orientation (in egosphere coordinates)
map orientation (in world coordinates)

Subentities

boundaries
surface patches

Criteria for recognition

directly beneath the vehicle

name = surface 2

Attributes
- slope
- roughness
- ground cover
- shape
- size
- map position (in egosphere coordinates)
- map position (in world coordinates)
- map orientation (in egosphere coordinates)
- map orientation (in world coordinates)

Subentities
- boundaries
- surface patches

name = surface 3

etc.

An entity frame for the sky might have the form:

name = sky

Attributes
- color
- cloud types
- position
- shape

Pointer to entity image
- location of image with segmented sky entity

Criterion for recognition
- elevation = above the horizon
- range = infinite
- color = blue or gray (by day), black (at night)
- may contain clouds of various shapes

An entity frame for a hill crest or ditch might have the form:

name = hill crest

Attributes
- dr/dy = discontinuity
- elevation of discontinuity

Criteria for recognition
- magnitude of dr/dy decreases as vehicle approaches
- range of discontinuity increases as vehicle approaches

Both a [hill crest] entity and a [ditch] entity have an attribute of a large dr/dy disc tinuity. The criteria for recognition enable the SP processes to distinguish a [hill cr from a [ditch].

name = ditch

Attributes

dr/dy = discontinuity
elevation of discontinuity
dr/dt of discontinuity
depth
slope of banks
type of bottom (water | rocks | dirt)
map position (in egosphere coordinates)
map position (in world coordinates)
map orientation of near edge (with respect to vehicle motion)
map orientation of near edge (in world coordinates)

Subentities

bank surfaces
bottom surface
near edge
far edge

Criteria for recognition

magnitude of dr/dy = constant as the vehicle approaches
map position of dr/dy remains fixed as the vehicle approaches

Entity frames for other entities relevant to an unmanned vehicle might have the form:

name = rock

Attributes

height
width
map position (in egosphere coordinates)
map position (in world coordinates)
shape
color

Subentities

surfaces
edges

Criteria for recognition

range texture
color
temperature relative to ground (depending on solar illumination)

name = tree

Attributes

shape
size

diameter of trunk
clearance under lowest branch
map position (in egosphere coordinates)
map position (in world coordinates)

Subentities

trunk
branches
foliage

Criteria for recognition

size
shape

name = bush

Attributes

width
height
density
map position (in egosphere coordinates)
map position (in world coordinates)
color

Subentities

foliage

Criteria for recognition

size
shape

name = road

Attributes

surface roughness
color
width
shape (curved or straight, banked, etc.)
map position (in egosphere coordinates)
map position (in world coordinates)
map orientation (with respect to vehicle motion)
map orientation (in world coordinates)

Subentities

lanes
lane markings

Criteria for recognition

size
shape
position

name = water

Attributes

surface roughness
width
length
slope of banks
map position of nearest edge (in egosphere coordinates)
map position of nearest edge (in world coordinates)
map orientation of nearest edge (with respect to vehicle motion)
map orientation of nearest edge (in world coordinates)
etc.

Subentities

bank surfaces
water surface
near edge
far edge

Criteria for recognition

position
reflectivity
surface normal to gravity

name = vehicle

Attributes

height
length
width
orientation
color
map position (in egosphere coordinates)
map position (in world coordinates)
velocity (in egosphere coordinates)
velocity (in world coordinates)

Subentities

surfaces

Criteria for recognition

size
shape
motion
temperature

name = human

Attributes

height
width
orientation
color
map position (in egosphere coordinates)
map position (in world coordinates)
velocity (in egosphere coordinates)
velocity (in world coordinates)

Subentities

surfaces

Criteria for recognition

size
shape
motion
color
temperature

name = animal

Attributes

height
width
orientation
color
map position (in egosphere coordinates)
map position (in world coordinates)
velocity (in egosphere coordinates)
velocity (in world coordinates)

Subentities

surfaces

Criteria for recognition

size
shape
motion
color
temperature

In most cases each of the attributes defined for the entities above can be estimated from the image. Our list above is only an illustrative example. A complete knowledge database of entities and their attributes will be compiled as the Demo III project progresses.

11.7.3 Integration of the System

An integrated system for intelligent autonomous driving was demonstrated and evaluated as a collaborative enterprise of UBM, Sarnoff, and NIST. The system included Sarnoff's stereo-based real-time image-processing module, UBM's 4D control structure module, and an integration of these modules into the baseline LADAR system of 4D/RCS HMMWV (see Figure 11.22).

The integrated HMMWV system will drive autonomously on cross-country roads with the following capabilities:

1. Speeds up to 10 mph will be autonomously controlled as appropriate for the terrain and vehicle dynamics.
2. Obstacles and rough terrain conditions will be detected in time to enable decelerating to a safe speed or to steer around the problem condition.
3. A control station will provide an interface for limited mission planning and data collection. It will perform system checks and command simple missions.

Sensors

The sensors used in the system include a Dornier 1 Hz LADAR range imaging camera (EBK), two CCD cameras, a GPS sensor, an INS sensor, and two pan/tilt units. All

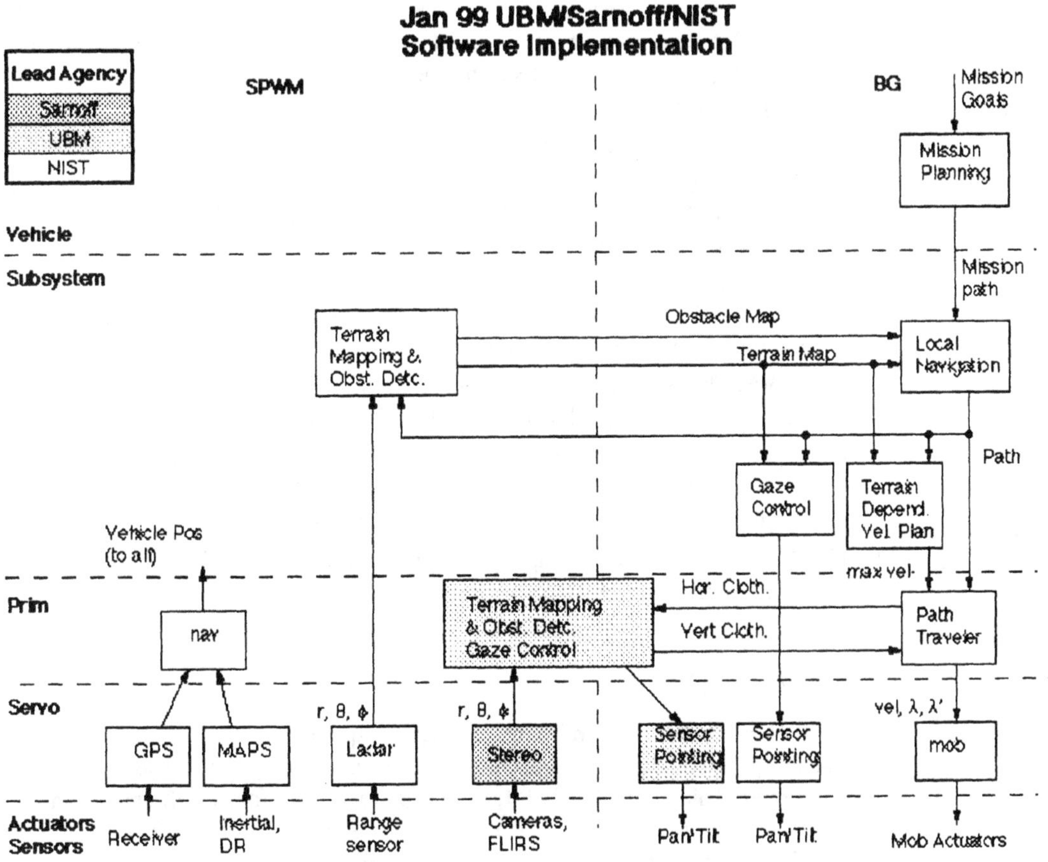

Figure 11.22. Software implementation.

sensors are mounted on the HMMWV. For additional information about the GPS, INS, and LADAR sensors see [62].

Software Modules

Figure 11.22 illustrates the software configuration of the integrated UBM-Sarnoff-NIST system. We present an overview of the functionality of each of these modules below:

1. The GPS module receives signals from satellite stations and computes the global position of the vehicle.
2. A MAPS module reads inertial sensors and computes vehicle position and orientation based on dead reckoning.
3. A NAV module computes the vehicle position and orientation based on the integration of the outputs of the GPS and the MAPS modules.
4. A NIST terrain mapping and obstacle detection module computes and maintains a local terrain map and an obstacle map at a resolution of 50 by 50 meters. The maps are recursively updated as each scan line of the LADAR is processed (every 1/128 second.)
5. A local navigation module uses the mission path, and the local obstacle and terrain maps to compute a clear path at 0.4 meter resolution.
6. A path traveler decomposes the output of the navigation module to produce a smooth trajectory represented as horizontal clothoids. They are sent to the terrain mapping and obstacle detection module (11) which returns the perceived terrain roughness in the form of vertical clothoids.
7. A module controls vehicle actuators and reads vehicle sensors on the HMMWV.
8. A NIST gaze control and sensor-pointing module sends pan/tilt control commands to the active LADAR carrier to redirect the LADAR sensor and reads the current pan/tilt shaft encoder.
9. A UBM gaze control and sensor pointing module sends pan/tilt control commands to the active camera carrier in order to redirect gaze and to obtain the current pan/tilt shaft encoder position.
10. A VFE 200 stereo module extracts and localizes image features in stereo images and estimates image disparities of feature points.
11. A UBM terrain mapping module computes a local terrain model for 10 meters in front of the vehicle, and the vehicle ego-state. The terrain model and the ego-state are recursively updated every 1/25 second.
12. A mission planning module generates a mission path and displays all major states and functions of the HMMWV.

Hardware Configuration

The NIST HMMWV shown in Figure 11.23 is equipped with electric actuators added to the steering, brake, transmission, transfer case, and parking brake. Dashboard-type feedback provides the controller RPM, speed, temperature, fuel level, and so on. Multiple navigation sensors are used. A Kalman filter derives vehicle position and orientation using data from an inertial dead reckoning system and a carrier phase differential GPS unit, the Ashtech Z12. The inertial sensor is the U.S. Army's Modular Azimuth and

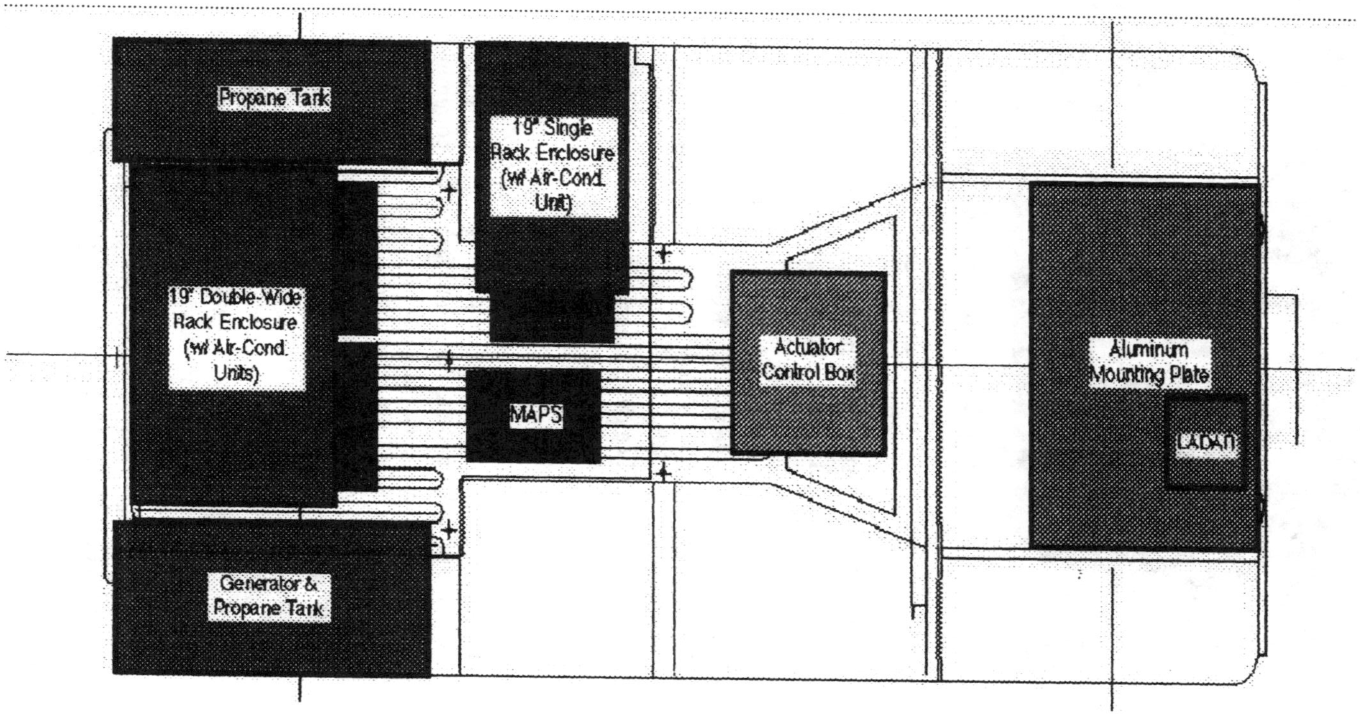

Figure 11.23. Hardware mounted on HMMWV (top view).

Figure 11.24. Van containing the mission planner hardware.

Positioning System which contains three ring laser gyros, three accelerometers, and a rear axle odometer.

Figure 11.23 illustrates the hardware configuration used on the NIST HMMWV1. A 19-inch single-rack enclosure with an air-conditioning unit holds a vxWorks system, a transputer, and inertial sensors for low-level actuator control. Software modules 1, 2, 3, 6, and 7 run in this chasis. A 19-inch double-width rack enclosure (with an air-conditioning unit) contains a SPARC 10 with 3 processors and an SGI-02 with R5000 chips. Software modules 4 and 5 run on this system. The van shown in Figure 11.24 contains an ULTRA SPARC used by software module 12. The van and the HMMWV are linked together via a radio-ethernet connection.

Communication

The neutral manufacturing language (NML) (http://isd.cme.nist.gov/proj/rcs_lib/NML cpp.html) is used for communication between NIST-RCS processes and UBM modules. NML provides a higher level interface to the communication management system (CMS). It provides a mechanism for handling multiple types of messages in the same buffer as well as simplifying the interface for encoding and decoding buffers in neutral format and the configuration mechanism.

CMS provides access to a fixed-size buffer of general data to multiple reader or writer processes on the same processor, across a backplane, or over a network. Regardless of the communication method required, the interface to CMS is uniform. Methods are provided to encode all of the basic C data types in a machine independent or neutral format, and to return them to the native format. CMS uses a configuration file so that users can change communications protocols or parameters without recompiling or relinking the applications. It has been compiled and tested on several platforms including MS-DOS, MS Windows, Windows NT, vxWorks, Lynx, IRIX, and several UNIX Workstations.

REFERENCES

[1] B. Smith and J. Wellington, IGES: Initial graphics exchange specification, Version 3.0, NBSIR 86-3359, April, 1986.

[2] STEP—Standard for Exchange of Product Model Data, http://www.steptools.com/welcome.html.

[3] MMS—Manufacturing Messaging Specifications, http://www.sisconet.com/mmsease.html.

[4] TCP/IP—Transmission Control Protocol/Internet Protocol, http://www.rsasecurity.com/news/pr/960917.html.

[5] API—Application Programming Interfaces, http://java.sun.com/products/jdk/1.2/docs/api/overview-summary.html.

[6] CORBA—Common Object Request Broker Architecture in http://www.acl.lanl.gov/CORBA/.

[7] J. S. Albus and A. M. Meystel, A reference architecture for design and implementation of large complex systems,' *Int. J. Intell. Cont. Sys.* **1** (1996): 15–30.

[8] S. Wallace, M. K. Senehi, E. Barkmeyer, M. Luce, S. Ray, and E. Wallace, Control entity interface specification, NISTIR 5272, 1993.

[9] M. A. Domnez, Ed., Progress report of the quality in automation project for FY90. NISTIR 4536, 1991.

[10] Technologies Enabling Agile Manufacturing (MAM) API Standards http:Hisd.cme.gov/info/team.

[11] F. M. Proctor and J. Michaloski, Enhanced machine controller architecture overview. NISTIR 5331, 1993.

[12] J. Albus, Outline for a theory of intelligence, *IEEE Trans. Sys. Man Cybern.* **21** (1991): 473–509.

[13] J. Albus, RCS: A reference model architecture for intellignet control, *COMPUTER, Special Issue: Computer Architecture for Intelligent Machines*, (May 1992): 56–59.

[14] J. Albus, RCS: A reference model architecture for intellignet machine systems, in *Proc. of the Workshop on Intelligent Autonomous Control Systems*, Haifa, Israel, 1992.

[15] J. Albus and A. Meystel, A reference model architecture for design and implementation of semiotic control in large and complex systems, in *Semiotic Workshop of 10th IEEE Int. Symp. on Intelligent Control*, Monterrey, CA, 1995, pp. 33–45.

[16] J. Albus, An intelligent system architecture for manufacturing (ISAM), Technical report, NIST, 1996.

[17] J. Albus and A. Meystel, Behavior generation. NISTIR-NIST, Gaithersburg, MD, 1995.

[18] H. Simon, Two heads are better than one: The collaboration between AI and OR, *Interfaces* **17** (1987):8–15.

[19] D. Brown, J. Maxin, and W. Scherer, A survey of intelligent scheduling systems, *Intelligent Scheduling Systems*, 1995.

[20] R. Conway, Priority dispatching and job lateness in a job shop, *J. Indust. Eng.* **16** (1965): 123.

[21] D. Elvers, Job shop dispatching rules using various delivery dates, *Prod. Invent. Management* **14** (1973): 61.

[22] J. Blackstone, D. Phillips, and G. Hogg, A state of the art survey of dispatching rules for manufacturing job shop operations, *Int. J. Prod. Res.* **20** (1982): 27–45.

[23] H. Pierreval and H. Ralambondrainy, A simulation and learning technique for generating knowledge about manufacturing systems behavior, *J. Operation Res.* **41** (1990): 461–473.

[24] M. Fox, ISIS: A retrospective, *Intelligent Scheduling*, 1994.

[25] S. Smith, OPIS: A methodology and architecture for reactive scheduling, *Intelligent Scheduling*, 1994.

[26] A. Kusiak. A knowledge optimization based approach to scheduling in automated manufacturing systems, in D. Brown and C. White, eds., *Operations Research and Artificial Intelligence*, pp. 453–479, Kluger Academic, Dordrecht,1990.

[27] J. Holland, *Adaptation in Natural and Artificial Systems*, University of Michigan Press, Ann Arbor, 1975.

[28] S. Uckun, S. Bagchi, K. Kawamura, and Y. Miyabe, Managing genetic search in job shop scheduling, *IEEE Expert*, Oct. 1993, pp. 15–24.

[29] D. Whitley, T. Starkweather, and D. Shaner, The travelling salesman and sequence scheduling, in L. Davis, ed., *Handbook of Genetic Algorithms*, Van Nostrand Reinhold, New York, 1991.

[30] G. Syswerda, Schedule optimization using genetic algorithms, in L. Davis, ed., *Handbook of Genetic Algorithms*, Van Nostrand Reinhold, New York, 1991.

[31] K. Musser, J. Dhingra, and G. Blakenship, Optimization based job scheduling, *IEEE Trans. Auto. Cont.* **38** (1993).

[32] V. Laarhoven and E. H. L. Aarts, *Simulated Annealing: Theory and Applications*, Reidel Publ., Dordrecht, The Netherlands, 1989.

[33] S. Vaithyanthan and P. Ignizio, A stochastic network for resource constrained scheduling, *Computers Ops. Res.* **19** (1992): pp. 241–254.

[34] Y. Satake, K. Morikawa, and N. Nakamura, Neural-network approach for minimizing the makespan of the general job shop, *Int. J. Prod. Econ.* **33** (1994): pp. 67–74.

[35] T. Willems and J. Rooda, Neural networks for job-shop scheduling, *Control Eng. Practice* **2** (1994): pp. 31–39.

[36] K. Aoki, M. Kanesashi, M. Itoh, and H. Matsuura, Power generation scheduling by neural networks, *Int. J. Systems Sci.* **23** (1992):1977–1989.

[37] S. Elmaghraby, *Activity Networks: Project Planning and Control by Network Models*, Wiley, New York, 1977.

[38] J. Moder, C. Phillips, and E. Davis. *Project Management with CPM, PERT and Precedence Diagramming, 3rd ed.*, Van Nostrand Reinhold, New York, 1983.

[39] S. Gupta and L. Taube, A state of the art survey of research on project management, in *Project Management: Method and Studies*, Elsevier Science, Amsterdam, 1985.

[40] Y. Zong, T. Yang, and J. Ignizio, An expert system using an exchange heuristic for the resource constrained scheduling problem, *Expert Sys. Appli.* **8** (1993): pp. 327–340.

[41] T. Baxber and J. Boardman, Knowledge based project control employing heuristic optimization, *IEEE Proc.* **135** (1988): pp. 529–538.

[42] R. McKim, Neural network applications for project scheduling: Three case studies, *Proj. Mgmt. J.* **24** (1993): .

[43] S. Wallace, M. K. Senehi, E. Barkmeyer, M. Luce, S. Ray, and E. Wallace, Control entity interface specification, NISTIR 5272, Sept. 1993.

[44] Progress report of the quality in automation project for FY90, M. A. Domnez, ed., NISTIR 4536, Mar. 1991.

[45] M. Nashman, T. Hong, W. G. Rippey, and M. Herman, An integrated vision touch-probe system for inspection tasks, *Machine Vision Applications*, Cleveland, OH, June 1996.

[46] M. Bloom and N. Christopher, A framework for distributed and virtual discrete part manufacturing, *Proc. CALS EXPO 96*, Oct. 1996, Long Beach, CA.

[47] Technologies Enabling Agile Manufacturing (TEAM) Application Programming Interfaces, web location http://isd.cme.nist.gov/info/te-,im.

[48] F. M. Proctor and J. Michaloski, Enhanced machine controller architecture overview, NISTIR 5331, Dec. 1993.

[49] J. Albus, H. McCain, and R. Lumia, NASA/NB S standard reference model for telerobot control system architecture (NASREM), NIST Technical Note 1235. National Institute of Standards and Technology, July, 1987.

[50] J. S. Albus, R. V. Bostelman, and N. G. Dagalakis, The NIST RoboCrane, a robot crane, *J. Robot. Sys.*, July 1992.

[51] D. Stewart, A platform with six degrees of freedom, part I *Proc. Inst. Mech. Eng.* **180** (1965–1966): pp. 371–386.

[52] R. V. Bostelman, J. S. Albus, et al., A Stewart platform lunar rover, *Engineering Construction and tions in Space IV Proc.*, Albuquerque, NM, Feb. 26–Mar. 3, 1994.

[53] R. Bostelman, J. Albus, N. Dagalakis, and A. Jacoff, RoboCrane project: An advanced concept for large scale manufacturing, *Association for Unmanned Vehicles Systems Int. Proc.*, Orlando, FL, July 15–19, 1996.

[54] R. V. Bostelman. and J. S. Albus, Stability of an underwater work platform suspended from an unstable reference, *Engineering in Harmony with the Ocean Proc.*, Oct. 1993.

[55] N. G. Dagalakis, J. S. Albus, R. X. Bostelman, and J. Fiala, Development of the NIST robot crane teleoperation controller, *Robotics and Remote Handling Proc. 5th*, Knoxville, TN, April 1993.

[56] R. X. Bostelman, J. S. Albus, and N. G. Dagalakis, A robotic crane system utilizing the Stewart platform configuration, *Int. Symp. on Robotics and Manufacturing Proceedings*, Santa Fe, NM, Nov. 10–12,1992.

[57] R. Lumia, The enhanced machine controller architecture, *Int. Symp. on Robotics and Manufacturing Proc.*, Maui, HI, Aug. 14–18, 1994.

[58] W. Shackleford, The NML programmer's guide, *http//:Hisd.cme.nist.-2ov/t)roi/rcs.lib/nmi.html*, 1995.

[59] Mission training plan for the scout platoon, ARTEP 17-57-10-MTP, Headquarters, Department of the Army, July, 1996.

[60] J. Albus, *4-D/RCS: A Reference Model Architecture for Demo III*, NIST, Gaithersburg, MD, 1998.

[61] A. Meystel, Y. Moskovitz, E. Messina, Mission structure for an unmanned vehicle, *Proc. of the 1998 IEEE ISIC/CIRA/ISAS Joint Conference*, Gaithersburg, MD, September 14–17, 1998, pp. 36–43.

[62] K. Murphy and S. Legowik, GPS aided retrotraverse for unmanned ground vehicles, *Proc. SPIE 10th Annual AeroSense Symp., Conference 2738, Navigation and Control*, Technologies for Unmanned Systems, Orlando, FL, April 1996.

[63] S. Waldon, D. Gaw, and A. Meystel, Updating and organizing world knowledge for an autonomous control system, *Proc IEEE Intl Symp. on Intelligent Control*, Philadelphia, PA, 1987, pp. 423–430.

[64] A. Meystel and S. Uzun, Computer system for structuring images with no edge detection, *Proc. IEEE Intl Symp. on Intelligent Control*, Philadelphia, PA, 1987, pp. 266–270.

[65] A. Meystel, R. Bhatt, L. Venetsky, D. Gaw, and D. Lowing, A real time pilot for autonomous robot, *Proc. IEEE Intl Symp. on Intelligent Control*, Philadelphia, PA, pp. 135–139.

[66] P. Graglia and A. Meystel, Planning minimum time trajectory in the traversability space of a robot. *Proc. IEEE Intl Symp. on Intelligent Control*, Philadelphia, PA, 1987, pp. 82–87.

[67] R. Bhatt, D. Gaw, and A. Meystel, A real-time guidance system for an autonomous vehicle, *Proc. IEEE Intl Conf. on Robotics and Automation*, Raleigh, NC, March 31–April 3, 1987, pp. 1785–1791.

[68] A. Meystel, Planning/control architectures for master dependent autonomous systems with nonhomogeneous knowledge representation, *Proc. IEEE Intl Symp. on Intelligent Control*, Philadelphia, PA, 1987, pp. 31–41.

[69] A. Meystel, M. Montgomery, and D. Gaw, Navigation algorithm for a nested hierarchical system of robot path planning among polyhedral obstacles, *Proc. IEEE Intl Conf. on Robotics and Automation*, Raleigh, NC, March 31–April 3, 1987, pp. 1616–1622.

[70] C. Isik and A. Meystel, Pilot level of a hierarchical controller for an unmanned mobile robot, *IEEE J. of Robotics & Automation*, **4**, No. 3, 1989, pp. 244–255.

[71] A Meystel, Knowledge-based nested hierarchical control, in *Knowledge-based Systems for Intelligent Automation*, Vol. 2, Ed. by G. Saridis, JAI Press, Greenwich, CT 1990, pp. 63–151.

[72] A. Meystel, *Autonomous Mobile Robots: Vehicles With Cognitive Control*, World Scientific, Singapore, 1991.

CHAPTER 12

INTELLIGENT SYSTEMS: PRECURSOR OF THE NEW PARADIGM IN SCIENCE AND ENGINEERING

An interesting phenomenon can be observed at the present time. On one hand, it is unlikely that an autonomous creature with a powerful intelligence would be needed (and funded) in the observable future. On the other hand, willingly or unwillingly, most of the effort in the prospective research in the areas of computer science and systems engineering is devoted to the building blocks for the Frankenstein of the twenty-first century. Only one example will suffice to illustrate this statement: The most challenging problems in the area of intelligent systems is creation of a theory, or a device capable of generalizing upon texts in natural language. We even do not understand yet what exactly it means to generalize the text in natural language. One would be surprised to discover that hundreds of companies are fervently working on devices that summarize the contents of papers, categorize collections of documents, try to extract their hidden meaning, and the like. A theory of intelligent systems plays an important role in solving this problem.

In this sense the creation of a Frankenstein in the twenty-first century is unavoidable. There is a strong need to support the limited intellectual capabilities of humans. The "terra incognita" of the still concealed mechanisms of life and intelligence poses an exciting problem, and we are tempted by the opportunity to make a computer testbed for everything we are still unable to understand. However, an unusual obstacle emerges that occludes this tempting perspective. This obstacle is the demand for *multidisciplinarity*.

12.1 MULTIDISCIPLINARITY: THE MOST PROMISING BUT A BUMPY ROAD

We will outline some thoughts concerning the new emerging scientific area of intelligent systems. The area is active; not too many areas can boast such a level of interest, the frequency of discussions, the density of innovative ideas in the papers, books, e-mail discussions, and the like. Of course, there is a room for criticism too. Many of emerging concepts and results are repetitive, and the density of innovative ideas in them sometimes exceeds the level of their scientific persuasiveness. The focus of the information flow related to intelligent systems sometimes gets too blurred and contradictory.

The study of intelligent systems is an active field of research. Schools of thought are growing like mushrooms after a long rain. Students, engineers, and laymen frequently are confused about existing and unresolved contradictions. The situation in the multidisciplinary aspect of intelligent systems can be improved substantially by delineating the topics for attack.

Let us remind you that at the foundation of intelligent systems, especially within the area of design and subdomain of their architectures, is an effort to create explanatory and predictive models of the computational mechanisms associated with intelligence. Intelligent systems offer a novel flourishing method of scientific exploration: If the research confronts the dangers of introspection, the mathematical and engineering tools can be applied in dealing with this introspection. Indeed, the hallmark of a good science is a good symbolic representation and introspection via mathematics. And if the symbolic/mathematical model is difficult to encompass and compute, let us explore the engineering model of the system. The focus shifts toward the theory and practice of designing the intelligent systems. We have already mentioned that our approach boils down to using primarily a set of interrelated engineering and mathematical tools. It is very desirable to understand how to approach this design, so many people work on this research issue. On the other hand, the architecture of intelligence is a part of us, human beings, and possibly, a part of other living creatures. This understanding seems to have broader scientific and mathematical implications. Both scientific and mathematical tasks dwell upon a common platform of reasoning and constructing a system of symbolic representation. This is where the idea of *semiotics* enters the picture.

A BODY OF AICS KNOWLEDGE

It turns out that different approaches compete in the area of architectures for intelligent systems. Some researchers are fully satisfied to have a robot thoroughly automated and to call it "intelligent," while others aspire to create a system similar to human intelligence and look forward to simulation of robotic emotions. On the other hand, the intelligence of living creatures should be considered a part of our problem, too. Some want to have a unified model of analysis that can encompass both living creatures and robots. Others believe in a nonformalizable nature of natural intelligence and talk about it in a more poetical and spiritual manner. Third group tries to synthesize the previous two platforms. And so on it goes.

The area of AICS (architectures of intelligent control systems) from the very beginning has carried the germ of curiosity: Can we design a control system with the following characteristics?

1. A rich representation.
2. An ability to make decisions creatively.
3. An ability to learn with or without a teacher.
4. An analogue of "self."

Until recently control theory has not allocated much attention to the concept of autonomy. We believe that the concept of autonomy is an inseparable part of the area of intelligent systems. We are especially interested in determining which way of modeling is friendly to our novel research target of the autonomous control system. We want to find out whether the same tools of modeling that are suggested for automatic control systems can be applicable for modeling the intelligence of living creatures. All of us (no matter whether we are scientists, engineers, or mathematicians) are full of curiosity: Can we develop an autonomous intelligent control system and, if so, then how?

The field of intelligent systems has many rich and exciting scientific, engineering, and mathematical research topics. For example:

Topic 1. What are the links between the degree of intelligence and the value of complexity?[1]

Topic 2. How to evaluate the degree of the system's intelligence.

Topic 3. How does the degree of intelligence affect the productivity of intelligent systems, the survivability of the military autonomous vehicle, the reliability of functioning of the autonomous intelligent sentry guard-robot, and the efficiency of human activities?

Topic 4. How should we design intelligent systems with a specified degree of intelligence?

Topic 5. Can "intelligence" be acquired via learning, or is learning a structural (architectural) property?

Topic 6. What are the principles of designing the system of knowledge representation for an autonomous intelligent robot?

Topic 7. How many levels of hierarchy should be provided to achieve a trade-off between the performance and complexity of the system?

Topic 8. How does the degree of intelligent affect the specifications of servo-controllers (e.g., should they be more accurate, or less)?

We can continue this list, and it seems that all topics mentioned above are related to science, engineering, and undoubtedly, mathematics. But we have to decide how we can discuss all these topics and remain equally comprehensive for all disciplines involved. We have to negotiate a general theoretical paradigm that will allow us equal access to the scientific treasures collected by all area participants.

We do not think that just because of such implicit terms as "intelligence" and "self" we should relegate intelligent systems to pure psychology for these are legitimate topics of cognitive science and computational semiotics. It would likewise not be particularly farsighted to relegate the problem of representation to AI or data base researchers, for most of them have no deep understanding of control theory, especially if the dynamic issues are involved. Certainly the set of topics we listed has a component of multidisciplinarity. So a community of scientists, engineers, and mathematicians is best qualified to study these problems, outline research programs, and implement the results.

The issue should be addressed of existing habits of using terminology related to intelligence, intelligent control, and intelligent systems. Many researchers identify control based on neural networks, fuzzy logic, and the application of artificial intelligence as intelligent control. Others believe that this interpretation is unnecessarily broad and suggest that a control implementation is considered intelligent control if and only if it includes a significant set of the logic equivalent of "if... then... else" statements. The idea is that logical decisions are the essence of "intelligence," and hybrid control systems consisting of the subsystem of logic plus continuous dynamic subsystems seem to be intelligent ones (see Fig. 12.1). They pose many problems for sophisticated theoretical and experimental research.

However, there is much more to intelligent systems and intelligent control than just blending logic with continuous models. The early view of intelligent control blended

[1]We would like to remind the reader that computational complexity is a measure of the amount of computation required for solving a problem. On the other hand, the architectural complexity (or "sophisticatedness") is determined by the number of hierarchical levels, density of associative connections at a level, and so on. "Sophisticatedness" can be manifested in the richness of architectural vocabulary and grammar leading to the more meaningful statements in some language. Increases in "sophisticatedness" can actually reduce the computational complexity.

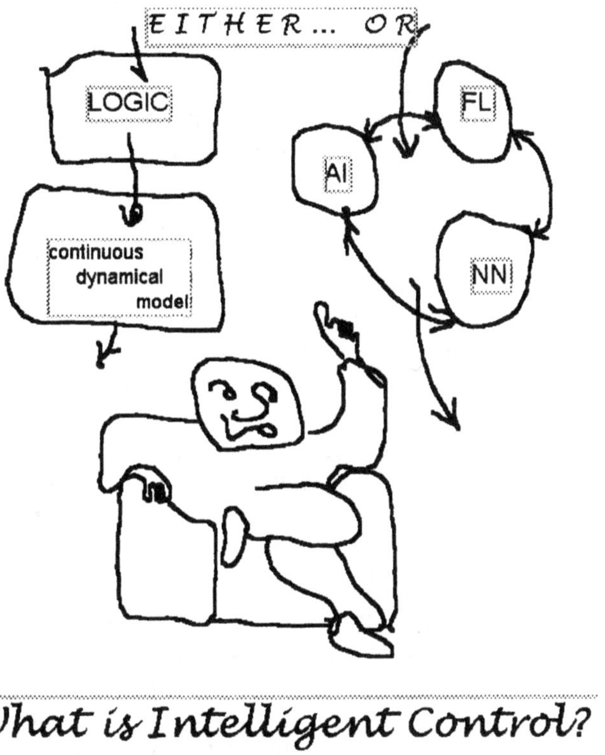

Figure 12.1.

control theory, AI, and operation research methods. In 1993 IEEE organized a task force to address the question of intelligent control. Despite many heated discussions, the task force could not arrive at any unanimity [1]. The main feature, the multidisciplinarity, of the issue was the dividing factor. The backgrounds of the participants from various disciplines were based on premises that were too diverse and thus precluded consensus. However, the unanimous feeling is that intelligent control is somewhere within the circular interaction between artificial intelligence, fuzzy logic, and the connectionist schemes of the neural network area (see Fig. 12.1).

It appears that the existing "conventional" algorithms of control do not allow for features to be demonstrated that are indicative of *intelligent control* either. Yet we have to determine the properties that allow us to distinguish intelligence and intelligent control from other phenomena. It is not that conventional control is out of the loop: it is simply insufficient for addressing the richness of the world of intelligent systems. Although, it is a mandatory prerequisite. In this book we have tried to demonstrate that algorithms of intelligent control and other computational constructs applied in intelligent machines have the following easily recognizable properties:

KNOWLEDGE OF INTELLIGENT SYSTEMS

1. Intelligent control systems have always several levels of resolution (granulation, scale). The need to minimize the computational complexity of searching and generating combinations in large databases is commonplace. Thus the architectures of intelligent systems tend to become multiresolutional. This is why connectionist (NN) models can be discovered underneath the lumped logic of general ELF structures. High levels of resolution operate with a narrow scope and multiple

details. Low resolution reduces the level of detail and allows for broadening the scope. The engineering conception of multiple resolution levels can vary from system to system. However, in principle, there is no difference between the multilevel architecture of a flexible manufacturing system and any multilayer neural network. Both are particular cases of multiresolutional systems.

2. Generalized information is passed bottom-up, while instantiated information proceeds top-down. This bidirectional flow is running in all three major hierarchies of an intelligent system: sensory processing, world model, and behavior generation. Information (knowledge) of a lower-resolution level is always obtained as a result of generalization of the information of the higher-resolution level. Otherwise, it is the same information about the same domains of the state-space, just at a different scale. The core tool of generalization is *grouping*. Grouping is ubiquitous. Most systems we explore, design, and use are permeated with procedures of clustering, averaging, group estimation, and so on. Once the operation of grouping is performed, the results sublimate bottom-up to become a single unit of information at the level of lower resolution. Usually the inverse movement of information top-down is applied in a variety of instantiation procedures, e.g., via decomposition procedures.

3. Intelligent systems are characterized by generating hypotheses while forming alternatives in the search for a decision at each level of resolution. Hypothesizing is a part of the next core procedure of the generalization process: *combinatorial search*. The search among heaps of old information is performed primarily to synthesize units (e.g., strings) of new information. All algorithms of search are based on "successor generation," which is a rudimentary form of "hypothesis generation." This simulation of processes expected to be performed amounts to "imagination," and it is a part of all decision-making situations.

4. Intelligent systems have a mechanism of selecting proper subsets of information (knowledge) from a wide range of available information; the operations of selectors are coordinated at all levels. This leads us to the third fundamental component of generalization: *focusing of attention*.

   An important detail: hypotheses generation is required for both bottom-up and top-down processing. Bottom-up hypotheses are the alternatives of grouping, i.e. generalizations. Top-down hypotheses are the alternatives of "ungrouping," i.e. alternatives of instantiation, e.g., by using operators of decomposition. Neither of them is covered by the schemes of conventional control.

   The ability to make a choice cannot be taken for granted; the richness of the set of alternatives is the main issue. Even air conditioners choose whether to wait or whether to start cooling. Their choices are limited, their decisions are predetermined. If air conditioners were given more choice and more freedom of choice, we would upscale them to the domain of intelligent systems.

5. Intelligent control is a multilevel control system with a parallel operation of loops that are similar in structure. The loops tend to involve similar operators of control at each level. These operators choose among hypotheses generated by the control system. These hypotheses are produced using the triplet of generalization, focusing of attention, and combinatorial search (see Fig. 12.2).

Many questions remain: What if a system already has a preferable way of functioning and its controller has been programmed based on this choice of functioning? Can this

Figure 12.2.

controller be considered *intelligent*? If in the preparation of the program, a lot of human intelligence was used, how should we categorize the controller in this case?

The situation is clear: on one hand we have a storage filled with intelligent solutions and clues how to recognize when each could be selected. On the other hand we have a system which is capable of finding these solutions if the situation requires. We don't think that an intelligent reader would have any difficulty in choosing which system is intelligent and which one is a fake. Certainly, the intelligent solutions will probably employ GFS and use operators of grouping, focusing attention, and combinatorial search.

This is a problem similar to J. Searle's "Chinese room" problem [2]. Indeed, the program incorporates a lot of intelligence, but the controller is not intelligent. It mimics functioning of the "would-be-intelligent" controller, but it is dumb. If the situation deviates from the "blueprint," it won't know what to do.

12.2 WHAT IS THE CORE ACTIVITY OF INTELLIGENT SYSTEMS?

As dictionaries attest, among the family of words related to or derived from the term "intelligence," there cannot be found a verb. So how do we name an action that results in intelligence? How do we label the process of producing the results respected as "intelligent" choices? Let us create a provisional verb that will allow us to consider intelligent systems to be natural or (artificially) constructed *formations* that can *intelligize*. This is an odd term, and we do not suggest its use in conversation, but it does seem to satisfy our present need. We want to define intelligent systems based on their processes rather than based upon their properties. Thus, to *intelligize* means to process information by nested use of GFS procedures, create and explore a multiresolutional system of knowledge representation, and strive for success.

Any real-world formation is understood as a sign (or an interrelated set of signs) that reflects a high level of value judgment that contains random habitual opinions. In order to provide for and ensure IS survival, intelligizing should be computationally efficient.

The process of intelligizing should end up with the following:

1. A plausible and efficient interpretation of the external world.
2. A construction of the meaningful and rich while efficient system of world representation.
3. An efficient procedure for constructing and choosing the beneficial courses of action.
4. Means for satisfactory communication.
5. The ability to learn and improve behavior.

We anticipate a flurry of comments and question-answering crossfire, something like this:

Q1. So, is the representation required for *intelligizing*?

A1. How can it be otherwise? One cannot name any meaningful (not a scholastic) existing creature without a representation! Representation is always there: just in a different disguise...

Q2. Even a virus has a representation? In this case, maybe the whole epidemic should be considered a "creature," not just a single "virus?"

A2. A virus? Even a molecule of water! Think about this wealth of knowledge represented in it by the virtue of having two atoms of hydrogen cooperate with one atom of oxygen in bringing to us the joy of water! But the population of viruses must contain a representation of what is good and what is bad for it.

Q3. Wait, why then do all the complex creatures emerge? Is it not cheaper to have billions of sturdy surviving viruses with their pointed representation that quickly learn how to beat our antibiotics, rather than "per aspera ad astra" to develop a fragile *Homo Loquent* with its unstable and murky, though very multiresolutional representation?

A3. Look at less puzzling analogies: the complexity of chemical elements grows steadily and Nature found it to be rational, a kind of "energy based." But the buck stops at 92d element. Maybe it stops at the human beings too (at least at their conditions we know of). Somehow, there is a merit in a limited growth of complexity, but the limit seems to always exist. Well, maybe this is not a growth of "complexity" but rather a growth in sophisticatedness.

Q4. So, at least, the limited growth of complexity seems to be a nice thing, isn't it?

A4. Simple life forms survive by producing millions of offsprings. The probability of survival for the individual is minimal. Intelligence moves the chances of survival from multiple agent coalitions to a single individual. Want to survive?

Intelligizing starts with a given goal,[2] i.e., with a desired state and the process of getting the representation of the present state of the world. It ends with the command

[2]One could argue that we focus on the goal-oriented intelligence almost indiscriminately. But let us try to find examples of "non-goal-oriented intelligence." Looking into multiple areas where different manifestations of intelligence reside, one may be surprised to notice that all cases of intelligence ascend to some particular goal (see the definition of intelligence). No matter how unrelated to a particular goal an intelligent activity might look, it is not so difficult to discover a string of causal relationships that explain the concrete goal (or even a set of goals) in all cases.

for the action to be done. It employs all skills of dealing with symbols endowed in a particular intelligent system. It uses the power of system's semiosis to develop the representation (see Chapter 2). In a simple case of an air conditioner, "intelligizing" is done with the help of a lookup table. In more complex cases, it requires simulating various alternatives of solutions, comparing them and choosing one of them by the virtue of processes that we call *thinking*.

Intelligizing dwells upon the semiosis. It is possible to demonstrate that the semiosis of intelligizing is equivalent to *multiresolutional recursive sign processing with learning* (MRSPL-processing). The MRSPL occurrences can be described as follows:

- For a single run of a system (creature).
- For a statistics of runs in a single creature.
- For a statistics in a population of creatures.
- For a statistics upon the creatures.
- For a statistics upon the set of population.

All items in this list *must* be considered at different scales introduced for time and space. Thus the whole totality of multiscale processes of intelligizing requires constructing a multiresolutional system of representation. It is possible to demonstrate a single entity, or a population of single entities, or a part of the entity to which the elementary loop of functions can be formulated. Intelligizing is understood as getting and transporting the external reality represented in a symbolic form so as to generate actions for survival. Survival encompasses a class of outcomes for the multiplicity of actions that can be considered successful.

12.3 MORE QUESTIONS THAN WE CAN HANDLE

Scientifically, however, we have only begun to appreciate what "intellligence" means. Thus, we are not prepared for dealing with processes such as *intelligizing*. Indeed, most of the well-established, customary concepts turn out to be insufficiently developed to satisfy the needs of the area of "intelligent systems." We need to construct a theory that will determine these systems, without constantly alluding to comparisons with humans.

12.3.1 Logic and Automata: Reasoning about Truth and Equivalence

Equivalences in math and logic are not the same. The math equivalence is based on two major ideas: equivalence relation and equivalence of sets. The equivalence relation is a triplet of transitivity, symmetry, and reflexivity (and generally presumes "equality"). The equivalence of sets presume one-to-one correspondence. All usages of the term "equivalence" in the mathematical sources ascend to these two major ideas and somehow blend them.

P. Suppes introduces equivalence as a part of biconditional sentences [3]. We use the words "if and only if" to obtain from two sentences a biconditional sentence. A biconditional sentence is also called equivalence. However, even before we start talking about equivalence, we introduce the term "truth" which should be understood with no reference to the subsequent symbol grounding (see Chapter 2; Fig. 12.3). The statement are proclaimed to be "true" or "false," and no checking is required by putting in correspondence the statements and the reality. The following three statements are equivalent:

Figure 12.3.

P if and only if Q.
P if Q and P only if Q.
If P, then Q, and if Q, then P.

In all three cases we write $P \leftrightarrow Q$ and interpret it as Q is a necessary and sufficient condition for P. (Notice that the implication $P \rightarrow Q$ is understood as

P only if Q,
Q if P ,
Q provided that P.

That is, P is a sufficient condition for Q, and Q is a necessary condition for P). From this equivalence relation we might conclude that "all true formulas of predicate calculus are equivalent" [4]. There are two schools of thought in this connection. On the one hand, it is known that two predicates $P(x_1, \ldots)$ and $Q(x_1, \ldots)$ are equivalent if in a formula obtained from the statement $P \leftrightarrow Q$, by substitution of free variables by numbers, they can be derived in bounded arithmetic (where "bounded arithmetic" is obtained from axiomatic arithmetic by eliminating from it the axiom of complete induction).

On the other hand, "equivalence" turns out to be a very complex concept depending on the context and therefore hard to explain in closed form. Sometimes equivalence is interpreted as "homomorphism of constructive models" [5]. Even more applicable for our case is defining equivalence for automata via the idea of "covering" to us. G. Birkhoff and T. Barett demonstrate that two automata M_1 and M_2 are equivalent if M_1 covers M_2 and M_2 covers M_1. The relation of covering is reflexive and transitive, and if covering were not bidirectional, we would not receive the symmetry required for equivalence [6].

This allows for approaching the definition of equivalence for multiscale systems. Each of the levels of resolution is an automaton. Typically an automaton covers only in one direction. However, in the realistic systems there are different relations of inclusion and therefore different coverings. One might consider two types of covering: "inclusion by grouping" and "inclusion by attention" [7]. They provide for covering in opposite directions, but this is covering of different modality; it therefore does not provide for equivalence.

We have arrived at a conclusion that the phenomenon of understanding typical for intelligent systems needs to evolve the automata theory from a logic with a simple "undergraduate" equivalence into a logic of models where the simple equivalence might not hold. Yet it is easy to demonstrate that any meaning requires a knowledge of the context or at least a statement of a "goal" to fully exercise the existing capabilities of symbol grounding required for discussing the fitness of the proposed interpretation.

For this reason our concept of the automaton includes the notion of a "goal." The presence of a goal allows for the introduction of "costs" and a measure of "goodness." These automata will allow to construct multiresolutional hierarchies of models demonstrating a system of nested goals.

12.3.2 Models That Incorporate the Concept of Goal

Consider an autonomous vehicle (AV). No model for AV can be proposed that ignores the phenomenon of its goal. Frequently more than one goal can be contemplated; quite likely a system of nested goals will surface after a short analysis. It is not easy to think simultaneously about several issues belonging to different levels of resolution and to keep their nestedness. This is one of the reasons that the concept of nested goals is difficult even for large complex systems (LCS) to grasp.

The temptation may be to perceive phenomena as if they just happen. Nevertheless, nothing just happens. As was stated above, the behavior of living creatures is determined by a system of nested goals. All actions are driven by particular goals, and each of these goals is always nested within a more general goal. Let us consider the discovery of robotic "flocking" by some MIT researchers. This phenomenon was discovered accidentally in the behavior of little robots and it looked like something that emerged spontaneously. It turned out, however, that this phenomenon was determined by a particular design of the little robots that made them capable of "flocking." In putting together a particular set of rules, there were represented hidden preferences (e.g., minimizing some cost-function) that predetermined the emergence of the unexpected phenomenon of "flocking." In general, rules and preferences are reflected in the input/output vocabularies and transition/output functions the designers have put in the automata. The modified automata will encode a goal and keep track of its performance. "Flocking" is one of the various incarnations of "swarming" behavior. The purpose of the swarm is a potentiality represented within the bee's automaton.[3]

Certainly, the bee's automaton contains the world model implicitly. Indeed, what are the transition and output functions of an automaton, if not its world model supplied by designers to enable the automaton functioning in a particular context?

For the living creature, the goal of the lower adjacent level of resolution might be formulated in terms of its family, tribe, or the whole population. Undoubtedly, each of these entities forms its own ELF[4] with its goal, resolution, time scale, and behavior generation.

The concept of a goal does not presume a group of commanders sitting in their armchairs and capriciously creating "goals" to the lower levels. When we talk about the nestedness of an *objectively existing* goal system, it is a system determined by conditions of viability for the ELF formulated for a particular level of granularity (resolution).

This system of concepts allows us to think simultaneously about automata and processes situated at different levels of resolution and to maintain a mental picture of their nestedness and their goals which propagate top-down within our system. We could consider it as a "swarm of swarms of swarms" [8]. The phenomenon of autonomy analyzed at any level of resolution is linked with the goal-orientedness of the larger picture which is discovered via additional research of the multiresolutional system of nested goals.

12.3.3 An Ability to Do Unprescribed Actions

At each level of resolution we can talk about the autonomy within a particular goal-oriented activity. The orientation is coming from the larger picture where the conditions of viability are delineated. Sometimes these conditions take the form of constraints. Some autonomy is required to meet the high-resolution eventualities that at a lower-resolution level are not defined and have aspects of uncertainty.

12.4 MULTIRESOLUTIONAL RECURSIVE SIGN PROCESSING WITH LEARNING (MRSPL PROCESSING)

12.4.1 What Are the Tools?

We spoke a lot about semiotics. However, semiotics does not offer specific tools of information processing in any specific domain or discipline. Rather, it is a meta-theory that allows for metaphor generation and analysis within a specific domain and allows multiple domains to talk with each other. Although all domains of knowledge speak seemingly different languages, they use the same laws in their symbolic languages to represent the meaning. Recent research results in data mining, text and image processing, and automated planning demonstrate that most of the problems boil down to meaning extraction, which is impossible to accomplish without extracting these laws. Semiotics studies and interprets these laws.

So the goal of this meta-theory is the pursuit of meaning, which is hidden in symbolic system but can be confirmed by symbol grounding within the reality. This requires knowledge-intensive processes with focus on learning in intelligent systems. These pro-

[3]We addressed the issue of modeling each system as an automaton as a matter of proper organization of information.

[4]See Chapter 2 for more on the elementary loop of functioning.

cesses allow for discovering the architectures of intelligence. They are similar to evolution of biological systems, development in economics, and design in engineering. These three areas are semiotically related. Evolution of living creatures is easily interpretable as learning. Part of the evolution is the design of the "nodes-successors," and this is where the similarity with engineering design emerges. Engineering has already accepted as a tool for planning/control and design creation of the nodes successors with their consecutive interpretation. Economics can be advanced only by using the models similar to the models of biological evolution and engineering design combined with studies of optimal behavior.

The art and science of constructing models for interpretation—and thus, for meaning extraction—is what we refer to as semiotics in this book. Meaning is typically elusive and tends to disappear in the process of multiple language-to-language translation. Semiotic laws of generalization lead to a poetic metaphor, successful move in the market, and to a scientific law in a similar way.

The core of this meta-theory is the process of constructing multiresolutional representation via learning from experience (in this book, we call this process semiosis). As a result, we conduct a multigranular (multiresolutional, multiscale) modeling of the world no matter which concrete language is applied at a particular level of granularity. When a researcher realizes that the meaning is multigranular, the researcher becomes a semiotician.

Semiotics alludes to the multidisciplinary approach. It proclaims that it is not enough to have the symbolic representation; understanding is required that would extract the meaning. Understanding is achieved via the process of interpretation.

Over the last 30 years, semiotics, linguistics, and cybernetics have moved toward a possible merger in the future. This merger can be expected to occur in the very near future, and the tendency toward this merger has been definite and steady. When AI was officially born in 1956, semiotics had already existed for centuries. It was hard to say what was a subset of what: Was semiotics a subset of AI, or vice versa? AI papers referred to the same logical, philosophical, and mathematical roots as semiotic papers.

Mathematics is a laboratory of our mind, a laboratory of science, a laboratory of all human endeavors. Mathematics does not have a tendency to ground itself to reality at each step—otherwise it would be transformed into semiotics. Mathematics is where semiotics makes its theoretical search of novel constructs, mutants of thought, and catastrophes of unexpected alternatives.

12.4.2 Semiotics and Object-Oriented Engineering: Kindred Strangers

Semiotics is *the* language of engineering. Engineers have exercised object-oriented thinking and practices for centuries. Consciously they are practicing during the last several decades. This is why theory of automata and languages turned out to be so appropriate as formal computer engineering methods. When the entity–relational approach to system representation emerged in the 70s, it was understood by engineers as something natural which was overdue. Thus, when programmers "discovered" object-oriented method, it immediately became a working tool of engineering design.

The central procedure of semiotics is the exchange of information between the object, the system of signs and the user (similar to multiresolutional ELF). This is a cycle of symbol (or matter) processing and fits within our concepts of object-oriented engineering techniques. Semiosis described in Chapter 2 gave a rough idea of what is the cycle of semiosis and what are its components. In Chapter 10, semiosis was interpreted as the algorithm of learning via generalization. But there is more to this artifact of semiotics that can be seen by an eye.

It has now become clear that semiosis is not "flat" as the Figure 2.13 in Chapter 2 shows. It is a multiscale hierarchy to which further concepts can be added. This is why we introduced in Section 12.2 an engineering model of semiosis called *multiresolutional recursive sign processing with learning* (MRSPL-processing). More important, engineers only partially depend on formal methods. Much engineering knowledge concerns understanding cases, clustering cases, and interpreting and reasoning about cases. In this respect, object-oriented engineering has many parallels and counterpoints with semiotics.

12.4.3 Unit of Intelligence

Let us consider the most convenient way to build self-organizing structures with minimum computational complexity. Recall the unit of intelligence shown in Figure 2.6*b* for a particular case of three repetitions of a set of procedures, or for three resolution levels (in fact it can have as many levels as it is required). This set of cognitive procedures can be considered the molecule of intellect.

It is shown in this diagram that at each of the three levels of the system under consideration, the same cycle operates of three consecutive GFACS procedures. The cycle can start in any point and proceed circularly also to an arbitrary stage. In Figure 2.6, it starts with focusing attention, evolves into combinatorial search, and ends after the last grouping is performed. This sequence can be frequently found in cognitive systems if we interpreted them as a generalization process. The components of generalization can be arranged in different sequence depending on the situation; the rearrangement does not change the functional meaning of GFACS. It will be clear later that the intelligent systems use GFACS in all of their components. Thus GFACS can be considered a unit of intelligence. This unit performs an elementary MRSPL operation.

Throughout the book, we have addressed in detail the procedures and components. The unit of intelligence contains the algorithms of searching for resemblance and focusing attention. It is engaged in combinatorial search. The latter includes generation and comparison of combinations synthesized out of components at hand. Randomness is a component of normal functioning of this unit.

12.4.4 Unit of Life?

All we know about "life," a strange phenomenon observed in the nature, is that it fights for itself, learns, self-reproduces, and tends to increase its complexity; nobody knows why. Let us also add that life is an anti-entropic phenomenon. In analyzing this phenomenon, we can arrive at the expected conclusion that life demonstrates itself algorithmically as grouping (G), focusing of attention (FA), and combinatorial search (CS), taken as a joint unit (GFACS). This performs MRSPL operation.

Some of the computational combinations of the operation of GFACS operation, have a striking similarity with the phenomenon of life. There might be a distinct resemblance between GFACS and the protein that emerged within the prebiotic soup, the ancient ocean in which protein molecules appeared as an initial stage of life. We will focus not upon this chemistry of biology, and not upon different forms of lives that emerged. We will rather concentrate upon the algorithmic structure of the mechanism of life. MRSPL operation (e.g. multiresolutional semiosis) allows for analysis and interpretation of the algorithmic structures of life and intelligence.

Biologists are close to discover life's hidden algorithmic structure, which is expected to be close to the MRSPL. The semiotic nature of the elementary life units (cells) is foreseen in [9–10] since cells demonstrate the following behavior:

- Cells look for resemblances among themselves.
- Cells try different alternatives of possible combinations.
- Cells stick together with other cells when such resemblance has been found.
- Cells organize generalized units.

These observations testify to the fact that the cells do it all: they focus attention, generalize, and perform combinatorial search! After the larger groups of cells are formed they also tend to do the same GFACS-like set of procedures. Each group of cells does it as a unit. Not only cells but all kinds of living organisms demonstrate the same set of GFACS procedures. The process of evolutionary development of living forms has plenty of proofs for this statement.

12.4.5 How to Use It

The unit of intellect is meant to be applied to any type of knowledge representation, including knowledge representation in the form of concrete biological realities (we can imagine a sign system based on biological realities as symbols). The "unit" is a "software package" based on the GFACS phenomenon and performing the set of MRSPL processing. Sometimes this package operates implicitly, in a hidden or camouflaged form that becomes clear only after interpretation of the biological or technological results (e.g., living creatures and intelligent machines).

We can interpret in terms of GFACS the very ability of simple molecules to move. The motion can be interpreted in the following way: A randomized source of mechanical motion produces tentative alternatives, attention is focused on a subset of preferable results, and generalization leads to learned rules of behavior. This sounds like a description of a programmed device, but all this is known for regular activities of the living creatures.

Indeed, many bacteria demonstrate clear random move generating behavior, a tendency to group the best of these moves (or to group the bacteria that exercised the best moves), and decision-making behavior based on the results of grouping: We know this for the *E. coli* structure and behavior. Certainly, for a biologist, the GFACS interpretation of the behavior of the bacteria sounds somewhat artificial. However, some biologists are accepting this model. The same behavioral structure is demonstrated in many other cases too.

This allows us to conclude that all living creatures have a subsystem of knowledge representation. We know that GFACS should be applied to knowledge representation—this is where it is supposed to work. How does the system of knowledge representation look? What does it represent? Can we imagine *the world according to E. coli*?[5]

[5]Construction of worlds according to each particular creature (*umwelts*) is one of the problems belonging to the domain of intelligent systems. J. Hoffmeyer raised this issue in his book *Signs of Intelligence in the Universe*. See also A. Meystel, "Multiresolutional *Umwelt*: Toward a Semiotics of Neurocontrol." *Semiotica*, vol. 120, No. 3–4, 1998, pp. 343–380.

12.4.6 Invariance

Invariance should be addressed in more detail. One may ask: Does invariance really exist, or did we invent it, and headlong infused it into bits and pieces of knowledge? Here we approach the problem of finding and measuring *resemblance*—the true core of semiotic research, since we will deal with the symbolic structure of world representation, which constitutes knowledge processing, including the issue of intelligence.

Indeed, generalization involves items of knowledge that are in some way grouped (after the fact, we think they are "properly" grouped). The grouping process cannot be done unless resemblance is found among the items of knowledge. Actually we cannot think about knowledge representation unless we are able to search and build a knowledge organization consisting of clusters based on a variety of resemblances. We would be extinct as a species long ago if we did not have an ability to find resemblances. The search for resemblance (for the purpose of subsequent class-generation and combinatorial search) is a key property of intelligence and life. We thus search for the *invariance*, which is *the* substance of resemblance.

When we build clusters or classes we presume that the invariance exists. The issue of invariance has been appreciated in different ways at the different stages of human cultural history. While the need and broad use of the skill of finding similarity and forming classes has always and everywhere existed, the ability to appreciate the significance of finding a resemblance requires appropriate skill or even talent, since recognition of resemblance requires some qualifications. Often we can detect some skill; in complex cases it takes talent to be able to determine resemblance, to find the invariances. If a particular talent has been repetitively demonstrated to be beneficial for the species, it is assumed by the population (hopefully), incorporated into the tissue of the culture, and subsequently transformed into a special habit.

The habit of generalizing based on resemblances within a sphere of attention has protected living creatures and probably created human cultures. Likely, the search for invariance may underlie the phenomenon of culture, and it is a result of this search that a system of symbols was created. For example, the eternal search for rhyme and rhythm has led to the system of symbols which we call "poetry" and "music." The search for scientific explanations and predictions, which require looking for resemblance among phenomena, has led to numerous systems of symbols which we call "theories," "laws," and "scientific disciplines." Everything from medical diagnostics to weather forecasting looks for, makes use of, and tries to arrive at some resemblances to build a system of symbols. This is what semiotics is all about.

Great discovery comes when we arrive at the understanding that the resemblances within a particular process is just a class of simple resemblances. All processes in all domains have a resemblance (this is why such things as the theory of discrete events is possible, to take just one example). When we start looking for more and more resemblances among different domains, we arrive at a conclusion that this is the same ubiquitous invariances that were striking for all researchers of all time.

We find resemblances everywhere. In 1821, the poet Percy Bysshe Shelley wrote a paper wherein he compared the activities of the scholars with the activities of poets: both categories of thinkers try to discover resemblances within the surrounding world. In some cases, these resemblances were coming from the resemblances of the nature, not our mind. "These similitudes of relations are finely said by Lord Bacon[6] to be 'the

[6]Sir Francis Bacon (1561–1626), lord chancellor of England (1618–1621). A lawyer, statesman, philosopher, and master of the English tongue, he was the author of multiple great philosophical concepts that proved to be ahead of the time.

same footsteps of nature impresses upon the various subjects of the world'; and he considers the faculty which perceives them as the storehouse of axioms common to all knowledge" [11].

12.5 MULTIRESOLUTIONAL INTELLIGENCE

Intelligence is the ability of a system to act appropriately in an uncertain environment, where appropriate action is that which increases the probability of success, and success is the achievement of behavioral subgoals that support the system's ultimate goal.

This definition has several powerful premises:

- Uncertainty of situations. Intelligence evokes the need for a special tool in dealing with uncertainty. Otherwise, either no intelligence would be necessary in dealing with deterministic situations or some external intelligence could provide the system with prescriptions in advance.
- Repetitive efforts. The notion of "probability" is evoked in the validation of potentially successful repetitive efforts in dealing with uncertainty. Thus a probabilistic evaluation might be connected with the uncertain environment.
- Intelligence is a matter of degree. With the different probabilities come different degrees of intelligence which can be distinguished quantitatively.
- Hierarchy of the system and its goals. The hierarchy of goals–subgoals leaves open the question where these goals are coming from. It contains reference to the system's "ultimate goal." It also declares that "intelligence" is for a system with a goal, that systems with no goals probably need no intelligence.

Further definition of intelligence focuses on the ability to make right choices, the main reason why the faculty of intelligence is needed at all. No wonder, intelligence can have many interpretations, and this is a potential problem for many practical applications. These applications also generate a number of questions:

- How do we qualify cases where knowledge is lacking? Say that the situation is deterministic, but we have no knowledge that will allow us to properly construct solution alternatives and make appropriate choices. Since there is missing knowledge, the situation can be considered uncertain. Thus most situations are uncertain, since no device or technique can serve as an omniscient knowledge base.
- What to do with unique situations? It would be presumptuous to expect that all situations demanding intelligence have comparable prehistory, or can be expected to ever occur again. Most situations are unique, and some are so unique that they will never repeat to allow us to correct a wrong solution. So the solution depends on finding a resemblance with other known cases and taking a prudent analogical approach.
- Do different degrees of intelligence employ the same mechanisms? Our definition does not say anything about the mechanisms of intelligence. It considers intelligence to be a *class label* for an action with a particular effect and not for a generic mechanism.
- Is an existing goal a prerequisite for applying intelligence, or is the goal a result of intelligence? Very often the goal is assigned externally. However, subgoals are

produced by the system alone. It seems that intelligence is a factor in producing subgoals.

The definition of intelligence we have discussed above is clearly a functional definition. It opens an opportunity for introducing an instrumental definition. Let us consider the definition of *computational intelligence*:

DEFINITION OF COMPUTATIONAL INTELLIGENCE

Computational intelligence is a technique of processing information via joint functioning of several operators; the phenomenon of *intelligence* emerges as a result of applying these operators jointly. These operators generate and transform symbolic representations of the external world. They include grouping, focusing attention, searching, formation of combinations, and selecting a working mode of "bottom-up" (for generalization) and "top-down" (for instantiation). When the operators are applied to the system of symbolic representation of knowledge, the multiresolutional systems of knowledge develop, and nested loops of knowledge processing emerge. Given a goal, the system demonstrates properties that we call *intelligence*.

The phenomenon of computational intelligence has its own interesting directions of interpretation:

- The instrumental definition of intelligence centers on symbols generation and processing. It is less important what information is received and more important how to process it.
- Most of engineering and AI decisions and devices, like neural networks, neuro-fuzzy systems, genetic algorithms of search, or evolutionary intelligence programming, fall into the rubric of computational intelligence because they all perform the procedures mentioned in the definition.
- The assertion that the joint functioning of operators (or grouping, focusing attention, searching, and formation of combinations) leads to the nesting of loops of knowledge is an important one. The existence of an initial loop (the elementary loop of functioning) is a prerequisite for computational intelligence.

Further exploration of this definition will no doubt lead to multiple interpretations and perhaps generate useful practical applications.

As MRSPL processing (multiresolutional semiosis) driven by GFACS constructs the world's symbolic representations, it definitely enters the domains of design and control. The concepts of MRSPL processing and the architecture of intelligence have become educational tools for specifying and coordinating the engineering problems in the large complex systems for the subsequent analysis and solution. This includes the problem of general assignment formulation and the subsequent task decomposition. Both issues became important only in the 1980s. However, there is a big gap between the ability to recognize and visualize a goal and the ability to formulate and specify the assignment. How to actually achieve this goal is a fundamental problem, both from the point of view of corporate engineering/business practices and a university education. Until now, task decomposition was considered an issue lying beyond the normal circle of engineering concerns.

We have demonstrated that a voluntary, unsystematic approach to the design problem entails activities that reduce the efficiency of the system, although each separate activity may be an excellent one and commendable within some narrow scope of consideration. We have collected many examples of this phenomenon related to the design, development, and functioning of manufacturing systems, power systems, and other large complex systems.

A methodology of the multiresolutional modeling and simulation for complex systems, which is implied by the concepts of MRSPL processing and the architecture of intelligence, is presented in this book. This methodology is shown to support the individualized methods of optimum design and control using the preliminary analysis of texts and creation of syntactic and semantic structures (D-structure described in Chapter 3), methods of off-line and online identification including semiotic modeling, methods of optimization including use of recursive multiresolutional search on a semiotic model, methods of planning and programming based on semiotic methodology, and so on. The area of multiresolutional modeling is just emerging. Most professional engineers recognize that problems of large and complex systems cannot be solved without decomposing them into a set of subproblems related to the level of resolution. However, the inertia of the past is very strong, and the premises of MRSPL processing should be repeated again and again. This is one of our aims, to make this knowledge available and understandable to the engineering community.

MRSPL processing has turned out to be effective from the engineering point of view and educationally beneficial. We have developed a novel concept that enhances the existing object-oriented analysis by incorporating into it self-organization and learning (SOL). This concept allows us to recommend a novel architecture of a multiresolutional knowledge base (MRK) for large complex systems (MRK-SOL). When information is stored in MRK-SOL, it clusters into entities of the lower-resolution levels, thus building up a multiresolutional model for the system. The subsequent processes of task decomposition turn out to be fully supported by the emerging hierarchy of the system reflected in its knowledge base structure.

12.6 LEARNING HOW TO KNOW: SEMIOTICS AND MULTISCALE CYBERNETICS

Intelligent systems acquire knowledge through various mechanisms. Among these mechanisms are preprogrammed tools of information acquisition and more sophisticated tools of learning (see Chapter 10). In every case the system deals with information represented in signs (labels, elementary codes) and symbols (generalizations of signs) which invokes not only an encoded object but a class of objects. The synapses of our nervous system generate signs and symbols, and they are signs and symbols too. The formation and use of signs and symbols is analyzed within the discipline of *semiotics* (see Chapter 2). In semiotics, the central process under consideration is called *semiosis*, which is interpreted within the engineering paradigm as MRSPL processing. The latter takes the form of a multiresolutional signal-processing loop which represents both *functioning* and *learning* (see Chapter 10). MRSPL processing (multiresolutional semiosis) consists of the following subprocesses:

1. Encoding of sensations (generating sets of signals from transducers interpretable as states of the reality) by using elementary signs.
2. Associating encoded sensation with codes of actions (sets of phenomena that entail the commands generated by the system).

3. Constructing strings of codes for consecutive "states-actions-..."

$$(\ldots \rightarrow S_i \rightarrow A_{i,i+1} \rightarrow S_{i+1} \rightarrow A_{(i+1)(i+2)} \rightarrow \ldots)$$

stored in the memory.

4. Assigning to these strings of signs values of goodness (cost-function) J interpretable under specific goals G, which allow interpretation of "experiences"

$$E{:}[G, (\ldots \rightarrow S_i \rightarrow A_{i,i+1} \rightarrow S_{i+1} \rightarrow A_{(i+1)(i+2)} \rightarrow \ldots)] \rightarrow J$$

5. Discovering classes of experiences $\{E\}$ under a particular goal G and a cost-function J, assigning labels (signs, symbols) to them, and assigning signs to their components (generation of the concepts of objects and actions).
6. Forming hypotheses of new behavioral rules using previously stored results of prior realizations of MRSPL processing cycles.

The loop of MRSPL processing (or the loop of multiresolutional semiosis) has a striking similarity to cybernetic views of the system. The difference is in demonstrating not only what is called "feedback" but also what is called "feedforward" control with its imagination and planning.

The sequence above describes six computational activities at a single level of granularity in a system for knowledge acquisition and utilization (e.g., "action"), that is, in a *semiotic system*. It is typical for the analysis of semiotic system to be focused on the *meaning* of the system and the process: After answering the *What?* question, the next question an engineer attempts to answer is *Why?*

FUNDAMENTAL ROLE OF CLOSURE

In cybernetics, the same elementary signal-processing loop is a source of different insights. *How?* and *How to control?* are the typical questions for the cyberneticist. As a result, semiotic systems usually are equipped with means to generate feedforward commands that contain the goals and tasks, and this turns the semiotic system practically into a conventional *cybernetic system*.

In other words, the cybernetic loop of system functioning is almost equivalent to the semiotic loop of semiosis with the following difference: the first describes control processes in physical terms, while the second analyzes the processes in the sign–symbol domain. A more serious difference is in the emphasis: Since semiotics is pursuing the meaning of the system, the goal is always questioned, and another loop should be discussed at which the goal is just a variable. This new loop (a coarser granularity loop) has a different goal, and the goal of the first loop is just a variable for this new loop. The pursuit of meaning leads us to discovery of loops of different granularity. Actually, the emergence of the objects and processes of coarser granularity was ingrained in the description of the process of learning (see Chapter 9).

As m levels of granularity (resolution, scale) emerge during learning in the semiotic system, similar levels emerge in the corresponding cybernetic system. A *cybernetic system of the mth order*, or *multiscale cybernetics*, emerges. The multiscale cybernetic systems can be characterized by a multigranular (m loops) MRSPL-processing loops that invoke a variety of interesting and important phenomena including *reflection* ("reflexia"), self-organization, and emergence (usually attributed to the situation's "complexity"[7]) and others.

[7]Some research schools use the term *complexity* to characterize the need in multiple interactions within the body of the system to enable the phenomena of "emergence." This use of the term should not be confused with computational complexity as we discuss it throughout the book.

Can we talk about virtual equivalence of semiotic and cybernetic systems just by considering a single-level loop? We can, but the powerful concepts of the semiotic–cybernetic scientific paradigm will be wasted. We do not need the tools and techniques of either in a single-loop case. On the other hand, if learning is involved, if planning is of interest, if processes of self-organization and self-description are of significance, no single-loop (single-level) discussion can be productive. Any such simplification will have hidden mistakes. The joint semiotic–cybernetic paradigm allows us to address all these issues and to explore the implied research questions, and the applications of the expected theoretical results.

People were always respectful of and comfortable with activities related to knowledge acquisition and organization. This appears to have started with broadminded Aristotle. Some two dozen centuries later the problems related to knowledge have narrowed to the domain of intelligent systems. We have realized that within a single problem we must address the modeling of the system at several levels of resolution (granularity, scale). We have also started analyzing processes of self-reflection, since, in addition to the world produced by sensations, we encounter a world produced by our own representations which included representation of ourselves.

A long time ago people discovered that intelligent systems and learning processes depend on architectures of sign–symbol processing. These architectures determine how intelligent systems perceive the world, recognize objects in it, and interpret the results of recognition for the subsequent actions. The multigranular architectures of our intelligence and the multigranular representations generated by our intelligence are affected by the processes of single and multi-level self-reflection.

LEARNING HOW TO KNOW

Architectures of sign processing affect the ways intelligent systems behave in the world. They depend on the way these systems *think they behave*. Careful analysis demonstrates that all our knowledge depends on our self-reflection. Our own brain-architecture is reflected in all the knowledge we acquire, when we analyze an electron, a spray casting, or a bacterium. All our representations reflect the architecture of our brain. Architectures of intelligent systems incorporate the recursion of self-reflection which becomes crucially important for describing interactions among multiple intelligent systems.

Cybernetics was a step in learning *how to know*. We have discovered the phenomenon of self-organization via planning and error compensation during functioning. Self-organization was a step toward self-reflection. With realization of the phenomenon of self-reflection we started paying more attention to the techniques of interpretation. The results of our interpretation are strongly affected by our "self" hidden in the techniques of our intelligence. Then the cybernetics of the second order emerged (this term was introduced by H. von Foerster to describe the level produced by self-reflection).

In intelligent systems *knowing* always presumes *changing resolution* or *granularity* or *scale* (which is the same). Therefore learning brings on changing the scale of representation and subsequent changes in processes of decision making, planning, and error compensation within this scale of representation. Each next step of learning leads to the generation of a next order of cybernetic analysis. Once a higher level (level of coarser granularity) is achieved, all previously known (finer resolution) subsystems are subsumed by (aggregated within the model of) the level of coarser granularity.

Various disciplines, including knowledge engineering, artificial intelligence, intelligent control, and semiotics, are stages in the exciting search within and traveling toward multiresolutional cybernetics, or cybernetics of the mth order. Such systems produce representations in multiple scales, supplement control with self-organization, and throw us into a totally new world, a multigranular world, in which we always live, even though we do not pay too much attention to its multiresolutional nature.

Various artifacts of life and technology should be discussed from the point of view of multiscale cybernetics. In engineering problems and poetry, scientific research and visual arts, everywhere we will continue to discover the multiresolutional world of semiotic–cybernetic models.

12.7 ENGINEERING OF MIND

One of the future projects linked with the science of intelligent systems is engineering of mind. This project can be conducted as a very large project of national, or even international, magnitude, and therefore requires national and international coordination. Working on this project has already started; however, it will take some time to bring all research endeavors to a degree of coordination that such a project deserves.

While the mind itself remains a set of mysterious and inaccessible phenomenon, many of its components, such as perception, behavior generation, knowledge representation, value judgment, reason, intention, emotion, memory imagination, recognition, learning, and intelligence, are becoming well-defined and amenable to analysis. Not only such projects like baby robot (Chapter 10) and autonomous mobile robots (see Chapter 11) are considered as introduction to mind engineering: dealing with the Internet is too. Bio-engineering projects, such as modeling of vision, blood circulation, and central nervous system, are sub-projects of mind engineering. Projects in text processing, including text compression, translation, summarization, and categorization, are also mind engineering, as well as projects related to data mining. The efforts and results related to mind engineering are plenty. Altogether they illustrate a body of multidisciplinary research.

Multidisciplinarity is a blessing when we talk about the factor of creativity, but it is a curse as far as coordination of multiple research efforts is concerned. Even the short list of ongoing projects we have mentioned demonstrates that the curse of multidisciplinarity is here. The results in each domain are typically presented as being created within a particular discipline, and claim to be fully self-contained as a part of this particular domain of science.

Moreover, each of these projects demands the use of lingo of its scientific domain. It is not that the results of other domains are negated, rejected, or considered to be unacceptable. They simply do not exist because each domain tests its results against itself only. Whatever is done in other domains is neglected. It is not easy to talk about coordination under these circumstances.

What is lacking besides coordination (and is determined precisely because of the lack of coordination) is a general theoretical model that ties all these separate areas into a unified framework that includes engineering, social, biological, psychological, and other embodiments of the components of the mind. We expect that this will be a simulation model. In this book, we have partially illustrated how some of the semiotic concepts developed for intelligent machines apply to biological intelligence, and have suggested how engineering principles might be developed for the design and analysis of practical intelligent systems. Semiotics is a powerful unifying paradigm for this model because it allows for different disciplines to talk the same language, understand each other, and use each other's results.

Our goal is to facilitate a merger between traditional scientific and semiotic activities, unified by the goal to understand intelligent systems. Semiotics strives to close the triangle of reasoning by practice and its interpretation as an inseparable part of reasoning ("symbol grounding"). This is why semiotics became so widespread initially in the humanities (discipline). The contention was always emerging because of the tendency in all domains to be "like science," where "science" was understood in a straightforward single-resolutional way.

It turned out that intelligent systems demand using semiotics to be analyzed. Now we know it is because of the fundamental significance of this "molecule of intellect" that determines the core of all processes characteristic for intelligent systems. It determines also all processes characteristic for semiotics.

Scientists are becoming open to the semiotic vision of the reality and semiotic tools. Similarly, semioticians are becoming incorporated within the body of a broad scientific process. This bilateral process seems to be natural. Semioticians *are* scientists in the best sense of this word, while successful scientists and engineers *are* always semioticians in the depth of their hearts. Frequently, they are semioticians–amateurs. It is our goal to develop scientists into professionals–semioticians, and to let semioticians be professionals–scientists, which they are.

Engineering of mind considers the mind to be a practical intelligent system, although it is very large and complex. Large complex systems (LCS) can be characterized by the following semiotic features (see Chapter 2):

COMMON SEMIOTIC FEATURES

1. LCS have a very large number of components. Often these components are too complex and too large for all of them to be considered in modeling, analyzing, or controlling the system. Thus any conventional symbolic system for encoding and manipulating the knowledge of LCS is insufficient by itself. There is a need in studying *self-organizing and self-referential symbolic systems.*
2. The large number of symbols can be lumped into a smaller number of generalized entities. This generalization is related to objects and situations, and actions and processes. This is not a simple "clusters" or "classes" formation. To start generalizing, one needs to discover new ("coarser") entities that reduce the initial number of symbols (not the number of components in the original LCS). These entities might not exist within our present vision of things. We are talking rather about *metaphors generation* that would both satisfactorily describe and is compatible with our present body of knowledge.
3. The generalization is especially successful if it is applied repetitively and recursively, so that the final representation of the system has several levels of resolution. This is the reason why the symbolic systems related to LCS are primarily *multiresolutional* ones. Consistent representation would be conducive to generating elementary functioning loops (ELFs) compatible with observed and/or predicted reality.
4. This process of bottom-up generalization requires sacrificing some of the properties during the required computations; we focus our attention on the subsets of importance (or relevance) and constantly conduct top-down symbol grounding by instantiating within ELFs the generalizations that have been obtained. A new logic must be developed that would allow for verifiable analysis of systems and processes, both at a level of resolution (scale) and within a hierarchy of resolutions.
5. These joint processes of bottom-up generalization and top-down instantiation result in different sets of variables at different levels of resolution. We do not know yet how to compose or decompose variables. We are still at the stage of intu-

itive generation of our visionary descriptions for each level of resolution. There is nothing wrong about being visionary. However, at this stage we need to be able to formalize the process of generation of our visions and develop software packages for generating hierarchies of variables.

6. The entities selection is not based on mathematical rules of matrix decomposition; it is phenomenological, and sometimes it is determined by the nonlinearities that emerge in a particular system. Already we feel dissatisfied with such euphemistic adjectives as "nonlinear" and are ready to constructively use an object-oriented (multiscale) techniques in this case, too.
7. If the new representation does not fit within some particular mathematical model, a linguistic method is used to determine the same results with the same accuracy. The need to shift our research paradigms toward active use of descriptive tools is felt strongly. The need in *computing with words* will probably determine the most active research techniques in the area of intelligent systems.

The resemblance between LCS and intelligent systems is a well-known fact. Intelligent systems have been in the focus of scientific studies for a long time. LCS came to specific attention only recently as a result of the intelligent activities of humans. Surprisingly, natural LCS demonstrate similar properties. Although the concept of an *intelligent system* is not fully understood, it affects the interpretation of the existing research results as well as the choice of new research directions in the LCS area.

It could be discovered that LCS are symbol-processing systems where information is carried by physical carriers that are inseparable from signs. If these signs (or physical carriers) are processed by the joint functioning of three GFACS (GFS) operators (grouping, focusing attention, and combinatorial search), the same conceptual structure would explain the processes of both LCS and intelligent systems. The body of knowledge on the theoretical tools of semiotics is scattered in the diverse literature, including conference proceedings [12]–[16], unpublished manuscripts such as [17], several excellent books on intelligent control and soft computing [18]–[22], conceptual papers [23]–[27], and numerous sources in cognitive science and artificial intelligence, and intelligent control.

At present we in fact have an abundance of diverse answers to questions that have not yet been formulated in an organized fashion. In all instances the analyses ascend to a symbolic system that is supposed to be a carrier of intelligence no matter which particular model is selected by the researcher. This is why semiotics can be recommended as a natural framework, an invariance paradigm for the intelligent systems including LCS.

The fundamental role of signs and other symbolic systems for analysis of intelligent systems is unquestioned. Sign systems enter our activities under different disguises—in the form of automata languages, computer languages, systems of indexing related to data and knowledge bases, systems of secure communication, technological notations for the CAD/CAM systems with the "discovery" feature, and sign systems in human/human, human/machine, and machine/machine communications in which the communicating parties *understand* each other. Undoubtedly, sign/symbol studies will be in the core of the projects on engineering the mind.

MAXWELL DEMON?

The following phenomena are to be addressed and interpreted in this project from the point of view of fundamental semiotic concepts:

- Conscious decision-making processes that are observed and registered by the "mind-observer" which monitors mind activities.

- The balance between rational reasoning and the emotional foundations of decision-making processes as an indicator of mind functioning.
- Decision processes in multiagent systems; cases where each agent has a faculty of rational thinking, and cases where it does not.
- Subconscious decision-making processes that are conducted in parallel with those conscious ones but are not registered by "mind-observer."
- Both conscious and subconscious voluntary and/or involuntary processes of combinatorial exploration of the memories of the past in order to determine the alternatives that have not been created and therefore have not been analyzed.
- Subconscious and involuntary processes of combinatorial exploration of the memories of the past feelings and activities either related to the particular registered goals, or related to the set of possible unexplored goals.

The multiscale model of semiotic analysis allows not only various conjectures to be advanced, but actually apply them in solving practical (engineering) problems in the area of intelligent systems. Semiotics will enable us to provide answers concerning other exploratory questions, including the following:

What is our "awareness" and how is it related to "consciousness?"

What are the mechanisms that give rise to "imagination?"

What is perception and how is it related to the object perceived?

What are emotions, why do we have them, and can we control them?

What is individual will and how do we choose what we intend to do?

Is there such a thing as a collective will?

How do we convert intention into action?

How do we plan and how do we know what to expect from the future?

Each of these questions epitomizes a new technology not available today and contains tremendous potential for the contemporary economy. Our social and technological capabilities are seriously restricted by our inability to develop self-referential and self-organizing systems with understanding, our computer vision systems do not understand the scenes they see, our software packages for information processing do not understand the articles they read, and our automated planning systems do not understand the environment they plan for. The answers to all these questions are looming, and all are related to the problem of engineering the mind.

Answering these questions seems to be no less important than leaving our footprints on the moon, colonizing Mars, and solving the mystery of black holes.

REFERENCES

[1] *IEEE Control System Magazine*, June 1944.

[2] J. Searle, Minds, brains, and programs, *Behavioral and Brain Sciences*, No. 3, 1980, pp. 417–424.

[3] P. Suppes, *Introduction to Logic*, Van Nostrand, Princeton, 1957.

[4] N. Kondakov, *Logical Dictionary*, *Nauka*, 1971 (in Russian).

[5] Yu Ershov, Algorithmic problems of the theory of fields in *Handbook of Mathematical Logic* (J. Barwise, ed.), *Nauka* in 1982.

[6] G. Birkhoff and T. Bartee, *Modern Applied Algebra*, McGraw-Hill, New York, 1970.

[7] A. Meystel, *Autonomous Mobile Robots: Vehicles With Cognitive Control*, World Scientific, 1991.

[8] J. Hoffmeyer, *Signs of the Meaning in the Universe*, Indiana University Press, Bloomington, IN, 1996.

[9] G. M. Tomkins, The metabolic code, *Science*, Vol. 189, September 5, 1975, pp. 760–763.

[10] K. L. Bellman and L. V. Goldberg, *Common Origin of Linguistic and Movement Abilities*, Published by the Americal Physiological Society, 0363–6119/84, 1984, pp. R915–R921.

[11] P. B. Shelley, "In the Defense of Poetry," 1821.

[12] J. Albus, A. Meystel, D. Pospelov, and T. Reader (eds.), *Architectures for Semiotic Modeling and Situation Analysis in Large Complex Systems*, Proc. of the 1995 ISIC Workshop, 10th IEEE International Symposium on Intelligent Control, Monterey, CA, August 1995, 441 pp.

[13] *Intelligent Systems: A Semiotic Perspective*, Proc. of the 1996 Int'l Multidisciplinary Conference, Vol. 1: Theoretical Semiotics (298 p), Volume 2: Applied Semiotics, Gaithersburg, MD, October 1996.

[14] A. Meystel, *Proceedings of the 1997 International Conference on Intelligent Systems and Semiotics: A Learning Perspective*, Gaithersburg, MD, September 22–25, 1997.

[15] *Proc. of the 1998 IEEE Int'l Symp. on Intelligent Control, A Joint Conference on the Science and Technology in Intelligent Systems*, Gaithersburg, MD, September 1998, 899 pp.

[16] *Proc. of the 1999 IEEE International Symposium on Intelligent Control, Intelligent Systems and Semiotics*, Cambridge, MA, September 1999, 464 pp.

[17] D. A. Pospelov, *Applied Semiotics*, unpublished manuscript distributed by Battelle Institute, 1995.

[18] P. J. Antsaklis and K. M. Passino (eds.), *An Introduction to Intelligent and Autonomous Control*, Kluwer Academic, Boston, 1993, p. 427.

[19] S. Tzafestas and H. Verbruggen (eds.), *Artificial Intelligence in Industrial Decision Making, Control, and Automation*, Kluwer Academic, Boston, 1995.

[20] M. M. Gupta and N. K. Singh (eds.), *Intelligent Control Systems: Theory and Applications*, IEEE Press, New York, 1995, p. 820.

[21] D. A. White and D. A. Sofge (eds.), *Handbook of Intelligent Control: Neural, Fuzzy, and Adaptive Approaches*, Van Nostrand Reinhold, New York, 1992, p. 568.

[22] L. J. Fogel, *Intelligence Through Simulated Evolution*, Wiley, New York, 1999.

[23] S. K. Pal and S. Mitra, *Neuro-Fuzzy Pattern Recognition: Methods in Soft Computing*, Wiley, New York, 1999.

[24] P. Antsaklis, Intelligent control, *IEEE Control Systems Magazine*, June 1994.

[25] J. S. Albus, Outline for a theory of intelligence, *IEEE Transactions on Systems, Man, and Cybernetics*, Vol. 21, No. 3, May/June 1991, pp. 473–509.

[26] A. Meystel, Multiscale systems and controllers, *Proceedings of the IEEE/IFAC Joint Symposium on Computer-Aided Control System Design*, Tucson, AZ, 1994, pp. 13–26.

[27] J. Albus and A. Meystel, A reference model architecture for design and implementation of semiotic control in large and complex systems, in *Architectures for Semiotic Modeling and Situation Analysis in Large Complex Systems*, Proc. of the 1995 ISIC Workshop, Monterey, CA, 1995, pp. 33–45.

[28] A. Meystel, Multiresolutional umwelt: toward a semiotics of neurocontrol, *Semiotica*, Vol. 120, No. 3–4, 1998, pp. 343–380.

ABBREVIATIONS

A

A*	algorithm of heuristic search that evaluates not only the cost of the state chosen during the search but evaluates the remaining cost of achieving the goal
AAAI	American Association of Artificial Intelligence
ABI	attribute-based instances
ABSTRIPS	same as STRIPS (see) but applied in abstraction spaces
ACM	Association of Computing Machinery
AI	artificial intelligence
AICS	architectures of intelligence control systems
AIS	adaptive intelligent systems
AMRF	automated manufacturing research facility
ART	adaptive resonance theory
ASE	associative search element
ASN	associative search network
ATLANTIS	an exploratory architecture in behavior-based mobile robotics that is intended to combine a reactive behavior with planning
AV	autonomous vehicle

B

BG	behavior generation [esp. in RCS]

C

CAM	content-addressable memory
C-E	cost-effect
CCD	charge-coupled device
CL	conceptual learning
CLC	closed-loop controller
CMAC	cerebellar model arithmetic controller
CMM	coordinate measuring machine
CMS	communication management system

CNS	central nervous system
CORBA	Common object request broker architecture
CPM	critical path method
CR	control rules

D

D	D-structure descriptive structure, an entity relational network (ERN) consisting of descriptive units and obtained from a descriptive source, e.g., natural language text
DAR	dynamic avoidance regions

E

ECD	early conceptual development
ELF	elementary loop of functioning
EMC	enhanced machine controller
ER	event rules
ERN	entity-relational network, a graph with entities as its nodes and with relations as its edges

F

FBC	feedback control
FFC	feedforward control
FLIR	forward looking infrared
FLR	fuzzy linguistic representation
FOV	field of view
FST	fuzzy sets theory

G

GA	genetic algorithm
GDF	goal-directed functioning
GENREG	generating regular systems of objects, e.g., graphs
GFACS	operator of Generalization, consisting of Grouping (G), Focusing Attention (FA), and Combinatorial Search (CS), (see GFS)
GFS	operator of Generalization, consisting of Grouping (G), Filtration or Focusing Attention (F) and Search (S), often Combinatorial Search (see GFACS)
GII	global information infrastructure
GLA	generalizing learning agent
GLAC	generalizing learning agent with combiner
GLAS	generalizing learning agent with selector
GPS	general problem solver [AI domain]
GPS	global positioning satellite [mobile robots domain]
GS	goal state
GSM	generalized subsistence machine

H

HLR	higher level of resolution
HMMWV	high mobility multipurpose wheeled vehicle, see URL: http://www.hqmc.usmc.mil/factfile.nsf/7e931335d515626a8525628100676e0c/deef19f7262589df8525627a0052894a?OpenDocument
HPPC	high performance computing and communication
HRA	higher resolution action
HRB	higher resolution behavior
HRG	higher resolution goal
HRN	horizontal relational network
HRPL	higher resolution plan
HRWM	higher resolution world model

I

ICARUS	an exploratory architecture for mobile robots using agent based negotiations during the process of reasoning for planning
ISIS	information storage and interconnect systems
IGES	initial graphics exchange specifications
IMAS	intelligent mobile autonomous system
IO	intelligent observer
IS	initial state [planning/control domain]
IS	intelligent system [general science domain]
ISAM	intelligent systems architecture for manufacturing
IWS	inspection work station

J

JA	job assignor [esp. in RCS]

K

KBS	knowledge based systems
KBSS	knowledge based systems with search
KD	knowledge database [esp. in RCS]
KS	knowledge of states

L

LADAR	laser radar
LAR	logic of associative relations
LBR	land baby-robot
LCS	large complex systems
LLR	lower level of resolution
LPA	learning and planning automaton
LN	learning network
LRA	lower resolution action

LRB	lower resolution behavior
LRG	lower resolution goal
LRPL	lower resolution plan
LRWM	lower resolution world model
LS	learning system
LTI-system	linear time invariant system

M

MAUV	multiple autonomous undersea vehicle
MCA	multiscale coalition of agents
MIMO	multi-input-multi-output
MMS	manufacturing messaging specification
MOLGEN	a knowledge driven program (1980) that was supposed to assist molecular geneticists in planning their experiments, it is one of the efforts to introduce hierarchical planning
MPA	multiscale planning agent
MPAC	multiscale planning agent with combiner
MPAS	multiscale planning agent with selector
MPPA	multiscale predicting and planning agent
MPPAC	multiscale predicting and planning agent with combiner
MPPAS	multiscale predicting and planning agent with selector
MPSE	multidisciplinary problem solving environment
MR	multiresolutional, having multiple resolution levels
MRSPL	multiresolutional recursive sign processing with learning
MSI	manufacturing systems integration

N

N-gram	a group of n symbols (e.g., words)
NAMT	national advanced manufacturing testbed
NASREM	NASA/NBS standard reference model
NBS	National Bureau of Standards
NGIS	next generation inspection system
NH	nested hierarchical
NIST	National Institute of Standards and Technology
NML	neutral manufacturing language
NN	neural network
NOAH	nets of action hierarchies, a consultant system (1975) that was supposed to advise a human amateur in a repair task
NSC	network scientific computing

O

OAR	operation research
OLC	open-loop controller

OOP	object oriented programming
OPIS	opportunistic intelligent scheduler; a methodology and architecture for reactive scheduling

P

PAC	probably approximately correct
PANDORA	a planning system using common sense and reactive reasoning knowledge (1982); also, a mission support designed for the purpose of military planning (National Aerospace Lab, The Netherlands)
PDE	partial differential equations
PERT	project evaluation and review technique Ph phenomena
PID control	proportional integral derivative control
PL	planner [esp. in RCS]
PRS	procedural reasoning system
PS	present state [planning/control domain]
PS	plan selector [esp. in RCS]
PT	planned trajectory

Q

QIA	quality in automation architecture
QL	quantitative learning

R

RALPH-MEA	real-time decision-theoretic agent architecture with multiple execution capabilities; the execution architecture that uses dynamic influence diagrams for reactive reasoning in planning
RCS	real-time control system. This system was created at NIST, and is frequently referred to as NIST-RCS
4D-RCS	a modification of RCS developed for autonomous vehicle program DEMO III
RE	recursive estimation
RHR	recursive hierarchy of representation
RPC	remote procedure call
RSF	regular subsistence functioning

S

S3	state space search
S-R	stimulus-response
SA	simple agent
SAC	simple agent with combiner
SAS	simple agent with selector
SC	scheduler [esp. in RCS]
SCHEMA(ta)	a technique that structures the representation of activities in a well organized fashion similar to the one exercised in the theory of automata

	(both AUTOMATA and schemata domains were vastly enriched by M. Arbib).
SCRIPT(s)	"Scripts" and "schemata" are techniques of describing generic events, e.g., for the purposes of planning and natural language understanding SigArt Special Interest Group in Artificial Intelligence
SIMA	systems integration for manufacturing applications
SIPE	system for interactive planning and execution monitoring
SISO	single input single output
SLA	simple learning agent
SLAC	simple learning agent with combiner
SLAS	simple learning agent with selector
SOAR	state, operator and result is a paradigm that presumes searching in the problem space and applying the operator to the present state until this process ends up with a result. Works in SOAR area are permeated by A. Newell's unified theory of cognition.
SP	sensory processing [esp. in RCS]
SPh	sub-phenomena
SRI	Stanford Research Institute
STA	statistical text analysis
STD	statistical text decomposition
STEP	Standard for Exchange of Product Model Data; a standard for the computer-interpretable representation and exchange of product data, providing a neutral mechanism capable of describing product data throughout the life cycle of a product
STRIPS	Stanford Research Institute Planning System introduced in the 70s by R. Fikes and N. Nilsson and based on finding the rules that reduce the difference between the goal and the present state
SUF	sensory uncertainty field

T

TCP-IP	transmission control protocol/Internet protocol
TEAM	technologies enabling agile manufacturing
TEG	testing the environment and generalizing
TMPM	trajectory of motion required for performing a particular maneuver

U

UBR	underwater baby-robot
UL	unsupervised learning
UM	uniform media
UOD	universe of discourse

V

VJ	value judgment [esp. in RCS]

W

WBK	words based knowledge
WBKR	words based knowledge representation
WM	world model [esp. in RCS]

NAME INDEX

INDEX

www.ingramcontent.com/pod-product-compliance
Lightning Source LLC
LaVergne TN
LVHW081130290926
848685LV00003B/9
* 9 7 8 0 4 7 1 1 9 3 7 4 6 *